Handbook of Cognitive Behavioral Therapy by Disorder

Handbook of Cognitive Behavioral Therapy by Disorder
Case Studies and Application for Adults

Edited by

Colin R. Martin
Professor of Clinical Psychobiology and Applied Psychoneuroimmunology & Clinical Director, Institute for Health and Wellbeing, University of Suffolk, Ipswich, United Kingdom

Vinood B. Patel
Reader in Clinical Biochemistry, University of Westminster, London, United Kingdom

Victor R. Preedy
Emeritus Professor of Nutritional Biochemistry, King's College London, London, United Kingdom
Professor of Clinical Biochemistry and Pathology (Hon), King's College Hospital, London, United Kingdom

Academic Press is an imprint of Elsevier
125 London Wall, London EC2Y 5AS, United Kingdom
525 B Street, Suite 1650, San Diego, CA 92101, United States
50 Hampshire Street, 5th Floor, Cambridge, MA 02139, United States
The Boulevard, Langford Lane, Kidlington, Oxford OX5 1GB, United Kingdom

Copyright © 2023 Elsevier Inc. All rights reserved.

No part of this publication may be reproduced or transmitted in any form or by any means, electronic or mechanical, including photocopying, recording, or any information storage and retrieval system, without permission in writing from the publisher. Details on how to seek permission, further information about the Publisher's permissions policies and our arrangements with organizations such as the Copyright Clearance Center and the Copyright Licensing Agency, can be found at our website: www.elsevier.com/permissions.

This book and the individual contributions contained in it are protected under copyright by the Publisher (other than as may be noted herein).

Notices

Knowledge and best practice in this field are constantly changing. As new research and experience broaden our understanding, changes in research methods, professional practices, or medical treatment may become necessary.

Practitioners and researchers must always rely on their own experience and knowledge in evaluating and using any information, methods, compounds, or experiments described herein. In using such information or methods they should be mindful of their own safety and the safety of others, including parties for whom they have a professional responsibility.

To the fullest extent of the law, neither the Publisher nor the authors, contributors, or editors, assume any liability for any injury and/or damage to persons or property as a matter of products liability, negligence or otherwise, or from any use or operation of any methods, products, instructions, or ideas contained in the material herein.

ISBN 978-0-323-85726-0

SET ISBN 978-0-443-18556-4

For information on all Academic Press publications
visit our website at https://www.elsevier.com/books-and-journals

Publisher: Nikki P. Levy
Acquisitions Editor: Natalie Farra
Editorial Project Manager: Timothy Bennett
Production Project Manager: Swapna Srinivasan
Cover Designer: Miles Hitchen

Typeset by STRAIVE, India

Dedication

I dedicate this book to my wonderful daughter,
Dr. Caragh Brien, of whom I am so incredibly proud.
Colin R. Martin

Contents

Contributors   xxi
Foreword   xxv
Preface   xxvii

1. The context of mental health disorders in comparison to other diseases: Interlinking cognitive behavioral therapy

Rajkumar Rajendram, Vinood B. Patel, and Victor R. Preedy

Introduction	1
Global burden of disease attributable to disorders of mental health	1
Placing mental disorders in context	1
Cognitive behavioral therapy as a treatment for disorders of mental health	4
Trends in publications on cognitive behavioral therapy	5
Application to other areas	6
Key facts of mental health disorders	6
Summary points	7
References	7

Section A
Setting the scene and introductory chapters

2. Mental health concerns in primary care: Applications of cognitive behavioral therapies

Danielle L. Terry and Michelle A. Nanda

Mental health concerns in primary care	11
Common conditions in primary care	12
Barriers to MH in primary care	12
Behavioral health treatment in primary care	13
Coordinated care	13
Co-located care	13
Integrated care	15
Models of care	15
Barriers to CBT in primary care	15
Adaptation of CBT for the primary care setting	16
The setting	16
The CBT provider's role	16
Billing and CBT in primary care	16
Informed consent	17
Documentation	17
Brief CBT: A general approach in primary care	17
Therapist skills	17
Patient characteristics	18
Specific considerations in therapy	18
Insomnia	18
Depression and anxiety	18
Summary	18
Mini-dictionary of terms	18
Key facts	19
Applications to other areas	19
Summary points	19
References	19

3. Mechanisms of change in cognitive-behavioral therapy for weight loss

Loana Comșa and Oana David

Introduction	21
Treatments in weight loss	21
CBT for weight loss	22
Characteristics of CBT for weight loss interventions	22
Theories behind CBT for weight loss	23
The research on CBT for weight loss	23
Association between proposed mechanisms of change and outcomes effect sizes	23
Mechanisms of change	23
Protocols in CBTWL	24
Limitations and conclusions	24
Practice and procedures: Outcome measures	24
Mini-dictionary of terms	24

Key facts 25
   Key facts of obesity 25
   Key facts in CBTWL 27
Applications to other areas 27
Summary points 28
References 28

4. Ethno-cognitive behavioral therapy and ethnopsychotherapy: A new narrative

Farooq Naeem

The ethnopsychotherapy and ethno-CBT 31
Culturally adapted psychotherapies-current evidence 31
   Types and processes of cultural adaption 32
   Effect of cultural factors on therapy outcomes 32
Need for ethnopsychotherapy 33
Culture 34
The ethno-CBT: Cultural adaptation of CBT 34
   Southampton adaptation framework for CBT 34
The fundamental areas of cultural competence in CBT 36
Future directions 38
Applications to other areas 38
Key facts 39
Summary points 39
References 39

5. Cognitive-behavioral therapy and cancer survival

Špela Miroševič and Zalika-Klemenc Ketiš

Introduction 43
   Linking psychosocial morbidity to cancer progression 43
   Psychosocial interventions for cancer patients 44
   How does CBT in cancer survival works? 45
Conclusion 46
Practice and procedures 47
Mini-dictionary of terms 47
Key facts of CBT in cancer survival 47
Application to other areas 47
Summary points 48
References 48

Section B
Cognitive behavioral therapy in adults

6. Acrophobia and consumer-based automated virtual reality cognitive behavior therapy

Tara Donker and Markus Heinrichs

Introduction 53
   Cognitive behavioral therapy for specific phobia 53
Virtual reality 54
Virtual reality cognitive behavioral therapy for specific phobias 54
   The problem of scalability and accessibility of virtual reality CBT 55
Practice and procedures 55
   Selection and inclusion of studies 55
   Characteristics of included studies 55
   Effects and quality of automated VR studies for specific phobia 57
Conclusions 60
   Limitations and recommendations for further research 61
Mini-dictionary of terms 61
Key facts 62
   Key facts of consumer-based automated VR-CBT for specific phobia 62
Applications to other areas 62
Summary points 62
References 63

7. Cognitive behavioral therapy and adjustment disorder

Soledad Quero, Sara Fernández-Buendía, Rosa M. Baños, and Cristina Botella

Introduction 65
Classification and diagnostic criteria of AjD 66
Prevalence 66
Course and impact 66
Assessment 67
Psychological treatment 67
   Cognitive behavioral therapy 68
ICTs for the treatment of AjD 68
   Virtual reality: EMMA's world 68
   Treatment components 69
   Efficacy data 69
Internet-based treatments for AjD 70
   Internet-based treatment for AjD: TAO 70
   Treatment components 71

Efficacy and acceptability data	72
Future directions	**72**
Blended treatments	72
Practical case	**72**
Chapter 1: Acceptance	73
Chapter 2: Coping	73
Chapter 3: Change of meaning	73
Mini-dictionary of terms	**73**
Applications to other areas	**74**
Key facts of adjustment disorder	**74**
Key facts of ICTs for the treatment of adjustment disorder	**74**
Summary points	**74**
References	**74**

8. Anxiety disorders: Mindfulness-based cognitive behavioral therapy

Jennifer Apolinário-Hagen, Marie Drüge, Roy Danino, and Siegfried Tasseit

Introduction	**79**
Mindfulness-based interventions	80
Mindfulness-based cognitive therapy	80
MBCT group format, scope and extensions	80
MBCT: Application to anxiety disorders	**81**
MBCT for anxiety: Current evidence	83
MBCT for generalized anxiety disorder: Rationale and case illustration	84
Mindfulness-based interventions: Adverse events and contraindications	85
Newer developments: Mindfulness apps and Internet-delivered MBCT	86
Conclusions	**86**
Practice and procedures	**87**
Mini-dictionary of terms	**87**
Key facts of MBCT	**87**
Applications to other areas	**87**
Summary points	**88**
References	**88**

9. Avoidant/restrictive food intake disorder: Features and use of cognitive-behavioral therapy

P. Evelyna Kambanis, Christopher J. Mancuso, and Angeline R. Bottera

Introduction	**91**
The three ARFID profiles	91
Cognitive-behavioral conceptualization of ARFID	**92**
Sensory sensitivity	93
Fear of aversive consequences	93
Lack of interest in food or eating	93
Cognitive-behavioral therapy for avoidant/restrictive food intake disorder	**95**
Overview of treatment	95
Assessment and treatment preparation	96
Treatment protocol	97
Sensory sensitivity	98
Fear of aversive consequences	98
Lack of interest in food or eating	98
CBT-AR case example	**99**
Presenting complaint	99
Assessment and treatment preparation	99
Outcome and conclusion	100
Practice and procedures	**101**
Mini-dictionary of terms	**101**
Key facts of ARFID and CBT-AR	**101**
Applications to other areas	**102**
Summary points	**102**
References	**102**

10. Diabetes-related distress and HbA1c: The use of cognitive behavioral therapy

Peerasak Lerttrakarnnon, G. Lamar Robert, Puriwat Fakfum, and Kongprai Tunsuchart

Introduction	**105**
DRD and associated factors	**106**
Definition and prevalence of DRD	106
Factors associated with DRD	107
CBT on DRD or other outcomes in DM patients	**109**
Practices and procedures	**113**
DRD assessment tools	113
CBT intervention programs in DM with DRD	114
Mini-dictionary of terms	**115**
Key facts about diabetes mellitus epidemiology	**115**
Key facts about psychosocial issues to diabetes mellitus	**116**
Applications to other areas	**116**
Summary points	**116**
References	**116**

11. Dizziness: Features and the use of cognitive behavioral therapy

Masaki Kondo

Introduction	**121**
Transition of the concept of the functional vestibular disorder	**121**

Persistent postural-perceptual dizziness 122
   Prevalence and clinical course 122
   Diagnosis and primary symptoms 122
   Exacerbation factors 123
   Precipitant factors (onset factors) 124
   Exclusion of other disorders, and comorbidity 124
   Hypothesis of pathophysiology 124
   Psychological factors involved in pathophysiology 125
   Treatment: Antidepressant 125
   Treatment: Vestibular rehabilitation 125
Cognitive behavioral therapy for persistent postural-perceptual dizziness 126
   Effectiveness studies 126
   Predictors of effectiveness 127
Practice and procedures 127
   Measurement instruments 127
   Cognitive behavioral model (formulation) 127
   Psychoeducation and sharing the cognitive behavioral model 128
   Relaxation techniques 129
   Cognitive techniques 129
   Behavioral techniques 130
Acceptance and commitment therapy for persistent postural-perceptual dizziness 130
Mini-dictionary of terms 131
Key facts of persistent postural-perceptual dizziness 131
Applications to other areas 131
Summary points 132
References 132

12. Epilepsy, sexual function, and mindfulness-based cognitive therapy

Zainab Alimoradi, Mark D. Griffiths, and Amir H. Pakpour

Introduction 135
Effects of epilepsy on various aspects of life 136
Effects of epilepsy on sexual function among individuals with epilepsy 136
Prevalence of sexual dysfunction among patients with epilepsy 137
Mechanisms of sexual dysfunction in epilepsy 137
Strategies for the treatment of sexual disorders in patients with epilepsy 138
Mindfulness-based cognitive therapy 138

Practice and procedures of mindfulness-based cognitive therapy 139
Application of mindfulness-based cognitive therapy to promote sexual function 140
Evidence-based experiences of mindfulness-based cognitive therapy interventions to enhance sexual function 140
Evidence of the effectiveness of mindfulness-based cognitive therapy intervention to improve sexual function among epileptic patients 142
Practice and procedures for MBCT-S 142
Conclusion 143
Mini-dictionary of terms 143
Key facts 143
Summary points 144
References 144

13. Female sexual dysfunction: Applications of cognitive-behavioral therapy

Françoise Adam and Elise Grimm

Introduction 147
The female sexual response 148
Sexual dysfunction model 148
Practice and procedures 149
   Diagnosing female sexual dysfunctions 149
   Assessing female sexual dysfunctions 152
   Treating female sexual dysfunctions 156
Mini-dictionary of terms 158
Key facts of female sexual dysfunctions 158
Applications to other areas 158
Summary points 159
References 159

14. Cognitive behavior therapy for insomnia in adults

Susmita Halder and Akash Kumar Mahato

Introduction 163
Why targeting insomnia is important 164
Etiological factors of insomnia and its nonpharmacological management 164
Whether evidence base of CBT-I is universal? 165
Practice and procedures 165
Points to consider 166
   Case 1 168
   Case 2 169
   Case 3 169
Psychotherapeutic intervention 170

Mini-dictionary of terms	171
Key points	172
Applications to other areas	172
Summary points	172
References	172
Further reading	174

15. Internet-based cognitive behavioral therapy for loneliness

Anton Käll and Gerhard Andersson

Loneliness: Introduction and prevalence	175
Connection to physical and mental health	175
Reducing loneliness: Previous research and current directions	176
Loneliness: A CBT framework	176
CBT via the internet: ICBT	177
Practice and procedure	178
Part 1: Assessment and activation	178
Part 2: Dealing with obstacles and improving chances of success	179
Part 3: Maintaining gains and modular considerations	180
Mini dictionary of terms	180
Key facts about loneliness	180
Applications to other areas	181
Summary points	181
References	181

16. Mild traumatic brain injury, cognitive behavioral therapy, and psychological interventions

Karen A. Sullivan

Introduction	185
Models of mTBI recovery	186
Psychological approaches for mTBI recovery	187
CBT for mTBI recovery	188
CBT effectiveness and efficacy for mTBI and PPCS	189
Client considerations	189
Recommendations for future CBT studies	190
Conclusion	190
Practice and procedures	190
Mini-dictionary of terms	190
Key facts of mTBI	191
Applications to other areas	191
Summary points	191
References	191

17. Multiple sclerosis fatigue and the use of cognitive behavioral therapy: A new narrative

Moussa A. Chalah and Samar S. Ayache

Introduction	195
Cognitive behavioral therapy	195
Multiple sclerosis fatigue: Definition and impact	196
Multiple sclerosis fatigue: Underlying mechanisms and management	197
Studies applying cognitive behavioral therapy in multiple sclerosis fatigue	198
Face-to-face individual versus group interventions	198
Face-to-face, telephone-delivered, and web-based interventions	198
Response to cognitive behavioral interventions	199
Discussion	200
Practice and procedures	201
Mini-dictionary of terms	201
Key facts	201
Applications to other areas	201
Summary points	202
References	202

18. In-patient/residential treatment for obsessive-compulsive disorder

Madhuri H. Nanjundaswamy, Lavanya P. Sharma, and Shyam Sundar Arumugham

Introduction	205
Behavioral therapy	205
Cognitive therapy for OCD	206
Components of CBT	207
Family interventions in OCD	208
Treatment setting	208
In-patient/residential treatment for OCD	208
Treating team	209
Components of in-patient/residential treatment	209
Cognitive-behavior therapy	209
Group therapy	210
Medication management	210
Family education and support	210
Discharge planning	210
Other interventions	210
Evidence for in-patient/residential treatment	211
Long-term outcome	211

Predictors of response	211
Indications for in-patient residential treatment	211
B4DT: Revolutionizing ERP for OCD	212
Practice and procedures	213
Mini-dictionary of terms	213
Key facts	214
Application to other areas	214
Summary points	214
References	214

19. Postpartum depression and the role and position of cognitive behavioral therapy

Rachel Buhagiar and Elena Mamo

Introduction: The perinatal period & perinatal mental health	219
Maternal postpartum depression	220
Definition & prevalence	220
Clinical presentation	220
Diagnosis & diagnostic systems	221
Screening	221
Adverse effects of maternal PPD	221
Severe PPD	221
Paternal postpartum depression	224
Definition & prevalence	224
Paternal PPD and the family system	224
Evidence-based treatments for PPD	224
Psychological and pharmacological treatments	224
Preferred treatment in maternal PPD	225
Barriers to care for mothers	225
Barriers to care for fathers	225
The role of CBT in the treatment of PPD	226
The effectiveness of CBT in PPD	226
Adapting CBT for PPD	226
Early phase of treatment and case conceptualization for PPD	226
Middle phase of treatment: Modifying CBT interventions for PPD	227
Late phase of treatment and relapse prevention	229
Beyond traditional individual CBT treatment	229
Internet-based and application-based CBT	229
Group-based CBT	229
Conclusion	230
Practice and procedures	230
Mini-dictionary of terms	230
Key facts of postpartum depression (PPD)	230
Applications to other areas	231
Summary points	231
References	231

20. Applications of cognitive-behavioral therapy to posttraumatic stress disorder: A focus on sleep disorders

Morohunfolu Akinnusi and Ali A. El-Solh

Definition and pathophysiology of posttraumatic stress disorder	235
Current treatment modalities for PTSD	236
Trauma-focused cognitive-behavioral therapy	236
Pharmacologic treatment of PTSD	236
PTSD and sleep disorders	237
CBT for insomnia (CBT-I) in PTSD	237
CBT for nightmares in PTSD	237
CBT for obstructive sleep apnea	238
Barriers to effective treatment of PTSD with CBT	239
Improving outcomes of CBT application to PTSD treatment	240
Practice and procedures	240
Mini-dictionary of terms	241
Summary points	241
References	241

21. Psychosocial interventions for occupational stress and psychological disorders in humanitarian aid and disaster responders: A critical review

Cheryl Yunn Shee Foo, Helen Verdeli, and Alvin Kuowei Tay

Introduction	245
Occupational mental health of humanitarian aid and disaster responders	246
Existing MHPSS guidelines and their limitations	247
Comprehensive and systematic MHPSS framework for psychological wellbeing of humanitarian aid personnel	247
Organizational policy and standards of practice	251

Reducing workplace psychosocial stressors	251
Predeployment	253
Peri-deployment mental health monitoring and support	254
Perideployment crisis intervention and psychological treatment	255
Postdeployment resilience-building and posttraumatic growth	256
Limitations of research and considerations for implementation	258
Reducing barriers to help-seeking and stigma	259
Applications and adaptations to vulnerable groups and cultural contexts	259
Cost-effectiveness and funding	259
Conclusions	259
Mini-dictionary of terms	260
Key facts of burnout in humanitarian personnel	261
Summary points	261
References	261
Further reading	263

22. Social anxiety: Linking cognitive-behavioral therapy and strategies of third-generation therapies

Isabel C. Salazar, Stefan G. Hofmann, and Vicente E. Caballo

Introduction	265
Social anxiety dimensions	266
Age of onset, associated factors and course	267
Prevalence and comorbidity	267
Treatment	267
Practice and procedures	268
Key elements of the MISA program application	268
Instruments for assessing program outcome	269
Therapeutic strategies composing the MISA program	269
Training in the dimensions of the MISA program	272
Working with the "Criticism and embarrassment" dimension (1st part)	274
Key facts	276
Mini-dictionary of terms	276
Applications to other areas	276
MISA program applied to BPD	276
Example of a therapeutic session	277
Summary points	277
References	277

23. Implementing mindfulness-based cognitive therapy on dynamics of suicidal behavior: Understanding the efficacy and challenges

Debasruti Ghosh, Saurabh Raj, Tushar Singh, Sunil K. Verma, and Yogesh K. Arya

The diverse aspects of suicidal behavior	281
Underpinnings of suicidal behavior from various perspectives	282
Why MBCT is relevant in the context of suicidal behavior?	285
The process and variations of MBCT	286
Evidence of effectiveness of MBCT on suicidal ideation and other psychological dynamics related to suicide	288
MBCT on suicidal ideation	288
MBCT on cognitive mechanisms underlying suicide	288
MBCT on associated psychological mechanisms related to suicide	288
Implementing MBCT with suicidal clients: Challenges and way ahead	289
Mini-dictionary of terms	289
Key fact on suicide	290
Application to other areas	290
Summary points	290
References	290

24. Cognitive-behavioral therapy for tobacco use disorder in smokers with depression: A critical review

Alba González-Roz, Sara Weidberg, and James MacKillop

Introduction	293
Explanatory models for the link between tobacco use and depression	293
Clinical assessment: Diagnostic tools and screening measures for smoking and depression	294
Empirical evidence on CBT for smokers with depression	294
Evidence of other behavioral combination treatments in smokers with depression	295
Core components and strategies in CBT for smoking and depression	295
General considerations	295
Treatment rationale	296
Nicotine fading	297
Self-monitoring of cigarette consumption	297

Feedback of cigarette smoking	297
Cognitive restructuring	298
Stimulus control	298
Behavioral activation	298
Social skills assertiveness	299
Anger management	299
Weight concerns and exercise planning	299
Social support	299
Relapse prevention	299
Conclusions	300
Practice and procedures	300
Mini-dictionary of terms	300
Key facts	301
Key facts of cognitive-behavioral therapy for smokers and depression	301
Applications to other areas	302
Summary points	302
References	302

Section C
International aspects

25. Psychopathophysiology and compassion-based cognitive-behavior group therapy for patients with coronary artery disease

Chia-Ying Weng, Tin-Kwang Lin, and Bo-Cheng Hsu

Introduction	307
Psychological factors and coronary artery disease	307
Hostility and anger	308
Anxiety	308
Depression	308
Stress	308
Psychopathological mechanisms of coronary artery disease	309
Autonomic dysfunctions	309
Hypothalamic-pituitary-adrenal (HPA) axis	310
Hypertension and high blood pressure	310
Endothelial dysfunctions	311
Altered unhealthy behaviors and lifestyle	311
Psychotherapy and coronary artery disease	311
Practice and procedures	311
Mini-dictionary of terms	317
Key facts	317
Applications to other areas	317
Summary points	318
References	318

26. Application of mindfulness-based cognitive therapy and health qigong-based cognitive therapy among Chinese people with mood disorders

Sunny Ho-Wan Chan and Charlie Lau

Introduction	321
Prevalence of depression and anxiety	321
Limitation of traditional cognitive behavioral therapy	321
Alternative forms of therapy: Mindfulness-based cognitive therapy	321
Alternative forms of therapy: Health qigong cognitive therapy	322
Study objective	322
Candidates for the interventions	322
Reduced mood symptoms	322
Improved health status	323
HQCT & physical health	323
HQCT led to more benefits than MBCT	323
Somatization tendency in Chinese culture	323
Conclusion	323
Practice and procedures	323
MBCT	323
HQCT	325
Mini-dictionary of terms	326
Key facts	326
Applications to other areas	327
Summary points	327
References	327

27. Bipolar disorder in Japan and cognitive-behavioral therapy

Yasuhiro Kimura, Sayo Hamatani, and Kazuki Matsumoto

Introduction	331
Effectiveness of CBT for bipolar disorder	332
Epidemiology of bipolar disorder in Japan	332
Social status of bipolar disorder in Japan	332
Medical care for patients with bipolar disorder in Japan	333
Psychosocial treatment for bipolar disorder in Japan	334

Psychoeducation program	334
Rework program	334
Family support	335
CBT for bipolar disorder in Japan	335
Relationship between CBT and universal health insurance in Japan	336
Eclectic therapies in Japan	336
Conclusion	336
Practice and procedures	336
Assessment	337
Psychoeducation	337
Case formulation	338
Therapeutic goal setting	338
Self-monitoring	338
Establishment of daily routine and sleep habits	338
Behavioral experiment	338
Development of coping skills at prodrome	339
Summarize his/her CBT treatment	339
Mini-dictionary of terms	340
Key facts	340
Key facts of suicide in Japan	340
Applications to other areas	340
Summary points	340
References	340

28. Cognitive-behavioral therapy for anxiety disorders in Italian mental health services

Laura Giusti, Silvia Mammarella, Anna Salza, and Rita Roncone

Introduction	343
The Italian scenario and CBT treatments	344
CBT anxiety in mental health service in Italy: The first steps	344
A vital training	344
The first effectiveness studies	347
The Tuscany replication studies	347
The Italian "second generation" of CBT intervention for ADs	348
Postdisaster distress transdiagnostic CBT	349
Online and digitized CBT intervention for ADs: What experiences in Italy?	350
The impulse given by COVID-19 pandemic to digitalized CBT for ADs	350
Conclusions	352
Practice and procedures	352
Mini-dictionary of terms	352
Key facts	352
Applications to other areas	352
Summary points	353
References	353

29. Mood and anxiety disorders in Japan and cognitive-behavioral therapy

Naoki Yoshinaga and Hiroki Tanoue

Introduction	355
Mood and anxiety disorders in Japan	355
Development of academic societies for CBT in Japan	355
Research on CBT for mood and anxiety disorders in Japan	356
Randomized clinical trials for mood disorders ($n=4$)	356
Randomized clinical trials for anxiety disorders ($n=2$)	358
Clinical practice under the health insurance scheme in Japan	358
CBT training in Japan	360
Future direction	360
Practice and procedures	361
Mini-dictionary of terms	361
Key facts	361
Key facts of mood and anxiety disorders in Japan and CBT	361
Applications to other areas	361
Summary points	363
References	363

30. Cognitive behavioral therapy for posttraumatic stress disorder in Pakistan

Anwar Khan

Introduction	365
Posttraumatic stress disorder: A brief overview and history	365
Cognitive behavioral therapy: As a first line treatment choice	366
Pakistan aspects of posttraumatic stress disorder	367
Cognitive behavioral therapy in Pakistan	368
Cognitive behavioral therapy for posttraumatic stress disorder in Pakistan	370
Conclusion	371
Practice and procedure	371
Mini-dictionary of terms	371
Key facts	371
Applications to other areas	372
Summary points	372
Funding	372
References	372

31. Schizophrenia in Japan and cognitive behavioral therapy

Hiroki Tanoue and Naoki Yoshinaga

Introduction	377
Schizophrenia and its current status in Japan	377
The challenge of long-term hospitalization for schizophrenia in Japan	378
Where CBT is provided to people with schizophrenia	379
Spread of CBT for schizophrenia in Japan	379
Research on CBT for schizophrenia in Japan	380
Research on group CBT for schizophrenia ($n=3$)	380
Research on individual CBTp for at-risk status of schizophrenia ($n=1$)	381
Clinical practice in Japan	381
Future directions	382
Practice and procedure	382
Mini-dictionary of terms	382
Key facts	382
Key facts of CBT for schizophrenia in Japan	382
Applications to other areas	383
Summary points	383
References	383

32. Tinnitus and psychological and cognitive behavioral therapies in Japan

Sho Kanzaki, Mami Tazoe, Chinatsu Kataoka, and Tomomi Kimizuka

Introduction	387
CBT and tinnitus treatment	387
Evidence level and recommended strength	387
CBT for tinnitus in Japan	388
Practice and procedures	389
Mini-dictionary of terms	392
Examples of mini-dictionary of terms	392
Key facts	392
Purpose of treatment for tinnitus is to reduce attention to tinnitus	392
Applications to other areas	393
Summary points	393
References	393

33. Cognitive-behavioral interventions for mental health conditions among women in sub-Saharan Africa

Huynh-Nhu Le, Kantoniony M. Rabemananjara, and Deepika Goyal

Introduction	395
HIV and comorbid mental health conditions	396
HIV comorbidity and CBT	396
Trauma and gender-based violence against women	398
Trauma and GBV against women and CBT	400
Perinatal depression	401
Perinatal depression and CBT	401
Summary of CBT interventions across mental health conditions	401
Future directions	403
Practice and procedures	403
Mini-dictionary of terms	403
Key facts	403
Applications to other areas	403
Summary points	404
References	404

Section D
Case studies

34. Application of online cognitive-behavioral therapy for insomnia among individuals with epilepsy

Zainab Alimoradi, Mark D. Griffiths, and Amir H. Pakpour

Introduction	409
Cognitive-behavioral therapy	410
Cognitive-behavioral therapy for insomnia	410
Use of technology in the treatment of insomnia	411
Procedure	411
CBT-I program content	411
Application to other factors	411
Key facts	412
Summary points	412
References	414

35. CASE STUDY: Borderline personality disorder and cognitive behavioral therapy in an adult

Jaiganesh Selvapandiyan

Introduction	417
Psychotherapy and borderline personality disorder	417
Application of cognitive behavior therapy in borderline personality disorder	417
Examples	418
Examples	418
Treatment setting	419
Conclusion	419
Summary points	419
References	419

36. CASE STUDY: Cognitive behavioral therapy for an adult smoker receiving substance use treatment

Alba González-Roz, Gema Aonso-Diego, and Roberto Secades-Villa

Introduction: Prevalence rates of cigarette smoking in persons with substance use disorder and its associated consequences	421
Case study	422
Case formulation	422
Clinical assessment	422
Cognitive behavioral therapy: Treatment implementation	422
Applications to other areas	427
Key facts	427
Summary points	428
References	428

37. CASE STUDY: Cultural diversity and cognitive-behavioral therapy

Esteban V. Cardemil, Sarah J. Hartman, and José R. Rosario

Introduction	431
Definitions of disparities	432
Brief review of cultural adaptation literature	432
Brief review of implementation science literature	433
Cultural adaptations and implementation science: Working together to reduce disparities	434
Attending to culture and context through adaptations in intervention content	435
Attending to disparities through adaptations to the delivery of the intervention	436
Attending to disparities with culturally sensitive interventionists	436
Concluding thoughts	437
Summary points	437
References	438

38. CASE STUDY: Cognitive behavior therapy for body dysmorphic disorder in an adult

Marie Drüge and Birgit Watzke

Introduction	441
Treatment options for BDD	441
CBT for BDD	442
Introduction to the case study—The initial phase	442
Case history	443
The diagnostic process and treatment options	443
The beginning of therapy	443
BDD-specific psychoeducation	444
Exposure and response prevention (ERP)	444
The reconstruction of thoughts	444
Value-focused interventions	444
The end of therapy	445
Applications to other areas	445
Key facts	445
Summary points	445
References	445

39. Case study: The role of cognitive behavioral therapy in the treatment of postpartum depression

Elena Mamo and Rachel Buhagiar

Introduction	447
The identification of perinatal mental health issues	447
Perinatal mental health services—Initial intake and screening	448
Screening procedure and outcomes	448
Previous history	448
Clinical presentation—From pregnancy to postpartum	448
Multidisciplinary support	449
Psychotherapeutic support—Lisa's CBT journey	449

Initial phase—Goal setting, case conceptualization and treatment plan	449
Middle phase—Delving into CBT interventions	450
Late phase—Focus on maintenance and relapse prevention	452
Perinatal psychiatric treatment	452
Conclusion	452
Summary points	453
References	453

40. CASE STUDY: Compassion-based cognitive-behavior group therapy for patients with coronary artery disease

Tin-Kwang Lin, Chin-Lon Lin, Shu-Shu Wong, and Chia-Ying Weng

Introduction	455
Methods	456
Study participants	456
Experimental protocol	456
Compassion-based cognitive-behavior group therapy	456
Measurement of outcome variables	457
Data analysis	457
Results	457
The validation of the experimental manipulation of anger on autonomic nervous system activities	457
Therapeutic effects of CBGT program on hostility levels and psychophysiological reactions	457
Effects of CBGT program on autonomic nervous system activities and hostility	458
Conclusion	459
Summary points	460
References	460

41. Application of mindfulness-based cognitive therapy and health qigong–based cognitive therapy among Chinese people with mood disorders: A case study

Sunny Ho-Wan Chan and Charlie Lau

Introduction	463
Prevalence of depression and anxiety	463
Limitation of traditional cognitive behavioral therapy	463
Alternative forms of therapy: Mindfulness-based cognitive therapy	463
Alternative forms of therapy: Health qigong cognitive therapy	464
Case study: Personal background and history	464
Case study: Description of recent circumstances	464
Case study: Description of thought, feelings, and behaviors	464
Case study: Initial assessment	465
Case study: Interventions	465
Summary points	465
References	465

42. Case study: Mechanisms of change in cognitive-behavioral therapy for weight loss

Loana Comşa and Oana David

Introduction	467
Intervention and treatment overview	467
Case example	471
Susan's results	471
Intervention—Sessions 1–8	472
Discussion	472
Summary points	472
References	472

43. CASE STUDY: Cognitive-behavioral therapy for Japanese Bipolar II disorder patients

Yasuhiro Kimura

Introduction	475
Case 1	475
Patient information and visit history	475
Growth history and clinical history	475
Case formulation	476
Intervention	476
Case 2	477
Patient information and visit history	477
Growth history and clinical history	477
Case formulation	477
Intervention	478
Discussion	479
Summary points	479
References	480

44. Treating social anxiety with the MISA program: A case study

Isabel C. Salazar and Vicente E. Caballo

Case presentation	481
Anamnesis	482
Previous treatments	482
Assessment and diagnosis	482
Clinical diagnosis	484
Functional analysis	485
Treatment goals	486
At the cognitive level	486
At the emotional and physiological level	486
At the behavioral level	487
Intervention	487
Posttreatment assessment	487
One-year follow-up	489
Discussion	489
Key facts	489
Applications to other areas	489
Summary points	489
References	490

45. Application of mindfulness-based cognitive therapy on suicidal behavior: A case study

Debasruti Ghosh, Saswati Bhattacharya, Saurabh Raj, Tushar Singh, Sunil K. Verma, and Yogesh K. Arya

Introduction	491
Case summary of the client	491
Interview and assessment	492
Case conceptualization	492
Planning of techniques	494
Session summary	494
Treatment outcome and critical evaluation	495
Concluding remarks	496
Application to other areas	496
Key fact	496
Summary points	496
References	497

46. Recommended resources for cognitive-behavioral therapy in different disorders

Vinood B. Patel, Rajkumar Rajendram, and Victor R. Preedy

Introduction	499
Resources	499
Other resources	500
Summary points	500
Mini-dictionary of terms	504
Acknowledgments	508
References	508

Index	511

Contributors

Numbers in parentheses indicate the pages on which the authors' contributions begin.

Françoise Adam (147), Psychological Sciences Research Institute, Faculty of Psychological and Educational Sciences, UCLouvain, Ottignies-Louvain-la-Neuve, Belgium

Morohunfolu Akinnusi (235), VA Western New York Healthcare System; Division of Pulmonary, Critical Care, and Sleep Medicine, Department of Medicine, Jacobs School of Medicine and Biomedical Sciences; Department of Epidemiology and Environmental Health, School of Public Health and Health Professions; University at Buffalo, Buffalo, NY, United States

Zainab Alimoradi (135,409), Social Determinants of Health Research Center, Research Institute for Prevention of Non-Communicable Diseases, Qazvin University of Medical Sciences, Qazvin, Iran

Gerhard Andersson (175), Department of Behavioral Sciences and Learning, Linköping University, Campus Valla, Linköping, Sweden

Gema Aonso-Diego (421), Addictive Behaviors Research Group, Department of Psychology, Faculty of Psychology, University of Oviedo, Oviedo, Principality of Asturias, Spain

Jennifer Apolinário-Hagen (79), Institute of Occupational, Social and Environmental Medicine, Faculty of Medicine, Centre for Health and Society, Heinrich Heine University Düsseldorf, Düsseldorf, Germany

Shyam Sundar Arumugham (205), Department of Psychiatry, National Institute of Mental Health and Neurosciences (NIMHANS), Bangalore, India

Yogesh K. Arya (281), Department of Psychology, Faculty of Social Sciences, Banaras Hindu University, Varanasi, UP, India

Samar S. Ayache (195), EA 4391, Excitabilité Nerveuse et Thérapeutique, Université Paris-Est Créteil, Créteil, France; Service de Physiologie – Explorations Fonctionnelles, Hôpital Henri Mondor, Créteil, France

Rosa M. Baños (65), Department of Personality, Evaluation and Psychological Treatment, Universitat de València, Valencia; CIBER Fisiopatología de la Obesidad y Nutrición (CIBERON), Instituto Salud Carlos III, Madrid, Spain

Saswati Bhattacharya (491), Tara Neuropsychiatry Clinic and Counseling Center, Ghaziabad, Uttar Pradesh, India

Cristina Botella (65), Department of Basic, Clinical Psychology, and Psychobiology, Universitat Jaume I, Castellón; CIBER Fisiopatología de la Obesidad y Nutrición (CIBERON), Instituto Salud Carlos III, Madrid, Spain

Angeline R. Bottera (91), Department of Psychology, University of Wyoming, Laramie, WY, United States

Rachel Buhagiar (219,447), Department of Psychiatry, Mount Carmel Hospital, Attard, Malta

Vicente E. Caballo (265,481), Department of Personality, Assessment and Psychological Treatment, Faculty of Psychology, University of Granada, Granada, Spain

Esteban V. Cardemil (431), Frances L. Hiatt School of Psychology, Clark University, Worcester, MA, United States

Moussa A. Chalah (195), EA 4391, Excitabilité Nerveuse et Thérapeutique, Université Paris-Est Créteil, Créteil, France; Service de Physiologie – Explorations Fonctionnelles, Hôpital Henri Mondor, Créteil, France

Sunny Ho-Wan Chan (321,463), School of Health and Social Wellbeing, University of the West of England, Bristol, United Kingdom

Loana Comşa (21,467), Department of Clinical Psychology and Psychotherapy, Babeș-Bolyai University, Cluj-Napoca, Romania

Roy Danino (79), Institute of Psychology, Heinrich Heine University Düsseldorf, Düsseldorf, Germany

Oana David (21,467), Department of Clinical Psychology and Psychotherapy; Data Lab: Digital Affective Technologies in Therapy and Assessment, Babes-Bolyai University, Cluj-Napoca, Romania

Tara Donker (53), Department of Psychology, Laboratory of Biological and Personality Psychology, Albert-Ludwigs University of Freiburg, Freiburg im Breisgau, Germany

Marie Drüge (79,441), Department of Psychology, University of Zurich, Zurich, Switzerland

Ali A. El-Solh (235), VA Western New York Healthcare System; Division of Pulmonary, Critical Care, and Sleep Medicine, Department of Medicine, Jacobs School of Medicine and Biomedical Sciences; Department of Epidemiology and Environmental Health, School of Public Health and Health Professions; University at Buffalo, Buffalo, NY, United States

P. Evelyna Kambanis (91), Department of Psychology, University of Wyoming, Laramie, WY, United States

Puriwat Fakfum (105), Aging and Aging Palliative Care Research Cluster, Faculty of Medicine, Chiang Mai University, Chiang Mai, Thailand

Sara Fernández-Buendía (65), Department of Basic, Clinical Psychology, and Psychobiology, Universitat Jaume I, Castellón, Spain

Cheryl Yunn Shee Foo (245), Department of Counseling and Clinical Psychology, Teachers College, Columbia University of New York, New York, NY, United States

Debasruti Ghosh (281,491), Department of Psychology, MDDM College (Babasaheb Bhimrao Ambedkar Bihar University), Muzaffarpur, Bihar, India

Laura Giusti (343), Department of Life, Health and Environmental Sciences, University of L'Aquila, L'Aquila, Italy

Alba González-Roz (293,421), Department of Psychology, University of Oviedo, Oviedo; Addictive Behaviors Research Group, Department of Psychology, Faculty of Psychology, University of Oviedo, Oviedo, Principality of Asturias, Spain

Deepika Goyal (395), The Valley Foundation School of Nursing, San José State University, San Jose, CA, United States

Mark D. Griffiths (135,409), International Gaming Research Unit, Psychology Department, Nottingham Trent University, Nottingham, United Kingdom

Elise Grimm (147), Psychological Sciences Research Institute, Faculty of Psychological and Educational Sciences, UCLouvain, Ottignies-Louvain-la-Neuve, Belgium

Susmita Halder (163), Department of Psychology, St. Xavier's University, Kolkata, India

Sayo Hamatani (331), Research Center for Child Mental Development, Chiba University, Chiba; Research Center for Child Mental Development, University of Fukui, Fukui, Japan

Sarah J. Hartman (431), Frances L. Hiatt School of Psychology, Clark University, Worcester, MA, United States

Markus Heinrichs (53), Department of Psychology, Laboratory of Biological and Personality Psychology, Albert-Ludwigs University of Freiburg, Freiburg im Breisgau, Germany

Stefan G. Hofmann (265), Department of Clinical Psychology, Philipps-University Marburg, Germany, Marburg/Lahn, Germany

Bo-Cheng Hsu (307), Department of Psychology, National Chung Cheng University, Chiayi County, Taiwan

Anton Käll (175), Department of Behavioral Sciences and Learning, Linköping University, Campus Valla, Linköping, Sweden

Sho Kanzaki (387), Department of Otolaryngology Head and Neck Surgery, Keio University School of Medicine, Tokyo, Japan

Chinatsu Kataoka (387), Department of Otolaryngology Head and Neck Surgery, Keio University School of Medicine, Tokyo, Japan

Zalika-Klemenc Ketiš (43), Department of Family Medicine, Faculty of Medicine Ljubljana, University of Ljubljana, Ljubljana; Department of Family Medicine, Faculty of Medicine Maribor, University of Maribor, Maribor; Primary Healthcare Research and Development Institute, Community Health Centre, Ljubljana, Slovenia

Anwar Khan (365), Department of Psychology and Management Sciences, Khushal Khan Khattak University, Karak, Khyber Pakhtunkhwa, Pakistan

Tomomi Kimizuka (387), Department of Otolaryngology Head and Neck Surgery, Keio University School of Medicine, Tokyo, Japan

Yasuhiro Kimura (331,475), Department of Welfare Psychology, Faculty of Welfare, Fukushima College, Fukushima, Japan

Masaki Kondo (121), Department of Psychiatry and Cognitive-Behavioral Medicine, Nagoya City University, Nagoya, Japan

Charlie Lau (321,463), Department of Rehabilitation Sciences, The Hong Kong Polytechnic University, Hung Hom, Hong Kong

Huynh-Nhu Le (395), Department of Psychological and Brain Sciences, George Washington University, Washington, DC, United States

Peerasak Lerttrakarnnon (105), Aging and Aging Palliative Care Research Cluster, Department of Family Medicine, Faculty of Medicine, Chiang Mai University, Chiang Mai, Thailand

Chin-Lon Lin (455), Department of Internal Medicine, School of Medicine, Tzu Chi University, Hualien; Department of Internal Medicine, Dalin Tzu Chi Hospital, Buddhist Tzu Chi Medical Foundation, Chiayi, Taiwan

Tin-Kwang Lin (307,455), Department of Internal Medicine, School of Medicine, Tzu Chi University, Hualien; Department of Internal Medicine, Dalin Tzu Chi Hospital, Buddhist Tzu Chi Medical Foundation, Chiayi, Taiwan

James MacKillop (293), Department of Psychiatry and Behavioral Neurosciences, McMaster University & St. Joseph's Healthcare Hamilton, Hamilton, Canada

Akash Kumar Mahato (163), Amity Institute of Behavioural Health and Allied Sciences, Amity University Kolkata, Kolkata, West Bengal, India

Silvia Mammarella (343), Department of Life, Health and Environmental Sciences, University of L'Aquila, L'Aquila, Italy

Elena Mamo (219,447), Psychology Department, Mater Dei Hospital, Msida, Malta

Christopher J. Mancuso (91), Department of Psychology, University of Wyoming, Laramie, WY, United States

Kazuki Matsumoto (331), Research Center for Child Mental Development, Chiba University, Chiba; Division of Clinical Psychology, Kagoshima University Hospital, Kagoshima, Japan

Špela Miroševič (43), Department of Family Medicine, Faculty of Medicine Ljubljana, University of Ljubljana, Ljubljana, Slovenia

Farooq Naeem (31), Department of Psychiatry, University of Toronto, Toronto, ON, Canada

Michelle A. Nanda (11), Sayre Family Medicine Residency at Guthrie Robert Packer Hospital, Sayre, PA, United States

Madhuri H. Nanjundaswamy (205), Department of Psychiatry, National Institute of Mental Health and Neurosciences (NIMHANS), Bangalore, India

Amir H. Pakpour (135,409), Department of Nursing, School of Health and Welfare, Jönköping University, Jönköping, Sweden

Vinood B. Patel (1,499), School of Life Sciences, University of Westminster, London, United Kingdom

Victor R. Preedy (1,499), Faculty of Life Science and Medicine, King's College London, London, United Kingdom

Soledad Quero (65), Department of Basic, Clinical Psychology, and Psychobiology, Universitat Jaume I, Castellón; CIBER Fisiopatología de la Obesidad y Nutrición (CIBERON), Instituto Salud Carlos III, Madrid, Spain

Kantoniony M. Rabemananjara (395), Department of Psychological and Brain Sciences, George Washington University, Washington, DC, United States

Saurabh Raj (281,491), Department of Psychology, Ramdayalu Singh College (Babasaheb Bhimrao Ambedkar Bihar University), Muzaffarpur, Bihar, India

Rajkumar Rajendram (1,499), College of Medicine, King Saud bin Abdulaziz University for Health Sciences; Department of Medicine, King Abdulaziz Medical City, King Abdullah International Medical Research Center, Ministry of National Guard Health Affairs, Riyadh, Saudi Arabia

G. Lamar Robert (105), Aging and Aging Palliative Care Research Cluster, Faculty of Medicine, Chiang Mai University, Chiang Mai, Thailand

Rita Roncone (343), Department of Life, Health and Environmental Sciences, University of L'Aquila, L'Aquila, Italy

José R. Rosario (431), Frances L. Hiatt School of Psychology, Clark University, Worcester, MA, United States

Isabel C. Salazar (265,481), Department of Personality, Assessment and Psychological Treatment, Faculty of Psychology, University of Granada, Granada, Spain

Anna Salza (343), Department of Life, Health and Environmental Sciences, University of L'Aquila, L'Aquila, Italy

Roberto Secades-Villa (421), Addictive Behaviors Research Group, Department of Psychology, Faculty of Psychology, University of Oviedo, Oviedo, Principality of Asturias, Spain

Jaiganesh Selvapandiyan (417), Department of Psychiatry, All India Institute of Medical Sciences, Vijayawada, Andhra Pradesh, India

Lavanya P. Sharma (205), Department of Psychiatry, National Institute of Mental Health and Neurosciences (NIMHANS), Bangalore, India

Tushar Singh (281,491), Department of Psychology, Faculty of Social Sciences, Banaras Hindu University, Varanasi, UP, India

Karen A. Sullivan (185), School of Psychology and Counselling, Queensland University of Technology, Brisbane, QLD, Australia

Hiroki Tanoue (355,377), School of Nursing, Faculty of Medicine, University of Miyazaki, Miyazaki, Japan

Siegfried Tasseit (79), Psychotherapeutic Private Practice, Bad Gandersheim, Germany

Alvin Kuowei Tay (245), Faculty of Medicine, School of Psychiatry, University of New South Wales, Sydney, NSW, Australia

Mami Tazoe (387), Department of Otolaryngology Head and Neck Surgery, Keio University School of Medicine, Tokyo, Japan

Danielle L. Terry (11), Sayre Family Medicine Residency at Guthrie Robert Packer Hospital, Sayre, PA, United States

Kongprai Tunsuchart (105), District Public Health Assistance, Sarapee District Public Health Office, Chiang Mai, Thailand

Helen Verdeli (245), Department of Counseling and Clinical Psychology, Teachers College, Columbia University of New York, New York, NY, United States

Sunil K. Verma (281,491), Department of Applied Psychology, Vivekananda College, University of Delhi, New Delhi, India

Birgit Watzke (441), Department of Psychology, University of Zurich, Zurich, Switzerland

Sara Weidberg (293), Department of Psychology, University of Oviedo, Oviedo, Spain

Chia-Ying Weng (307,455), Department of Psychology, National Chung Cheng University, Chiayi, Taiwan

Shu-Shu Wong (455), Department of Social Work, Chaoyang University of Technology, Taichung, Taiwan

Naoki Yoshinaga (355,377), School of Nursing, Faculty of Medicine, University of Miyazaki, Miyazaki, Japan

Foreword

Cognitive behavioral therapy (CBT) has contributed significantly to both the understanding and treatment of a wide range of mental health problems and psychological issues. CBT not only derives from a cogent theoretical account of human functioning but is also underpinned by concepts and principles that have enabled the development of a range of effective treatment protocols and techniques, often within the context of time-limited intervention plans. Consequently, CBT is now a critical component of the services offered to those experiencing mental health problems and may be used in combination with other treatment approaches such as pharmacotherapy. The efficacy of CBT from an evidenced-based perspective has been consistently illustrated in the academic and clinical research literatures over the past 50+ years. Moreover, the spectrum of clinical presentations for which CBT has been found to be effective has expanded exponentially in recent years, including in its application to physical illness that may be accompanied by significant psychological distress.

Although the fundamental principles of CBT are well established and readily understood, the breadth of clinical presentations for which CBT interventions is relevant may be less familiar to many practitioners and scholars. Given the accruing evidence base for CBT across a broader range of presentations, a comprehensive and evidence-based resource that focuses on these innovative applications of CBT is both timely and highly desirable for practitioners and researchers alike. Professor Martin, Dr. Patel, and Professor Preedy have, in my view, done a terrific job of addressing this need with this exciting new book titled *Handbook of Cognitive Behavioral Therapy by Disorder,* which complements the second book in the series titled *Handbook of Lifespan Cognitive Behavioral Therapy.*

Leading experts from around the world have contributed chapters that integrate the best evidence-based practice in both traditional and novel areas where CBT may be applied with confidence and therapeutic optimism. Thus, chapters within this innovative new volume include the application of CBT in primary care, weight loss, oncology, diabetes, adjustment disorder, CBT and mindfulness, food intake disorder, dizziness, sexual dysfunction, epilepsy, and brain injury, among others. The delivery of CBT using contemporary technology such as the Internet is also explored in dedicated chapters. The latest evidence and application in the more traditional areas of application are also represented, for example, smoking cessation, social anxiety, and insomnia. As a CBT practitioner and academic myself, I believe that the editors are to be congratulated on attending to this gap in the literature with an evidence-based, easily accessible, and most importantly clinically applied comprehensive book on CBT stratified by disorder, both traditional and more contemporary presentations. I have no doubt that this new book will be of keen interest to practitioners and researchers alike.

Sarah Corrie
Professor Sarah Corrie works at the University of Suffolk, where she is the Lead for the development and delivery of training in CBT and counseling. She is Consultant Clinical Psychologist and Fellow of the British Association for Behavioural and Cognitive Psychotherapies, where she is also currently Chair of the Course Accreditation Committee. Among her other publications, Sarah is the coauthor of *First Steps in Cognitive Behaviour Therapy* (Sage), *Treating Relationship Distress and Psychopathology in Couples: A Cognitive-Behavioural Approach* (Routledge), *Assessment and Case Formulation in Cognitive Behavioural Therapy* (Sage), and *CBT Supervision* (Sage).

Preface

Overall, the disease burden of mental illness is considerable. Recent estimates suggest that the global burden of mental illness is just over 10% of disability adjusted life years (DALYs) and about 30% of years lived with a disability (YLDs). There are a variety of conditions, disorders, and phobias associated with these burdens and include those within early life stages (e.g., pregnancy), childhood, adolescence, adulthood, and aging. Mental illness impacts not only the individual but also the family unit, the community, and society at large.

Hitherto, some conditions were treated by pharmacological agents or conventional psychotherapy. More recently, however, cognitive behavioral therapy has seen extensive usage. The material on cognitive behavioral therapy disseminated in the past 5 years outstrips the material produced in the previous 30 years.

Cognitive behavioral therapy has been shown to be more cost- and outcome-effective than conventional or drug-based therapies. However, there are many ways in which cognitive behavioral therapy can be carried out. Particular practitioners have their own techniques. The response rates and outcome indices can also be variable depending on the condition being treated. Furthermore, there are unexplored translational aspects from different studies. For example, evidence-based information, limitations, and practitioner-based experiences in one particular situation may be applicable to other conditions of mental ill-health. However, the material on cognitive behavioral therapy is published across vastly different scientific domains. Furthermore, some texts are devoted exclusively for the expert or those from a particular discipline. As a consequence, important information from one particular scenario may not be readily available to other practitioners. Thus, different cognitive behavioral therapy approaches provide the foundation for future studies.

To address the aforementioned issues, the editors have compiled the **Handbook of Cognitive Behavioral Therapy**. The **Handbook of Cognitive Behavioral Therapy** is divided into two books

Handbook of Cognitive Behavioral Therapy by Disorder: Case Studies and Application for Adults
Handbook of Lifespan Cognitive Behavioral Therapy: Childhood, Adolescence, Pregnancy, Adulthood, and Aging

This book, *Handbook of Cognitive Behavioral Therapy by Disorders*, comprises four sections. Section I: Setting the Scene and Introductory Chapters covers topics such as cognitive behavioral therapy in mental health, cancer, and weight loss. Section II: Cognitive Behavioral Therapy in Adults covers topics such as cognitive behavioral therapy in acrophobia, anxiety disorders, insomnia, loneliness, female sexual dysfunction, postpartum depression, posttraumatic stress disorder, suicidal behavior, epilepsy, and mild traumatic brain injury. Section III: International Aspects covers topics such as cognitive behavioral therapy in coronary artery heart disease, schizophrenia, bipolar disorder, mood disorders, and anxiety disorders. Section IV: Case Studies covers topics such as cognitive behavioral therapy in borderline personality disorder, cultural diversity, body dysmorphic disorder, weight loss, and social anxiety.

This book not only bridges the intellectual and disciplinary divides but also describes translational aspects. It has a far thinking and evidence-based ethos. Each chapter has the following six sections:

- Abstract (published online)
- Practice and Procedures
- Applications to Other Areas
- Key Facts
- Mini-dictionary of Terms
- Summary Points

The section Key Facts covers focused areas of knowledge written for the novice. The Mini-dictionary of Terms section explains terms that are frequently used in the chapter. The section Applications to Other Areas describes how cognitive behavioral therapy is applied to other fields that the chapter may be relevant to. The Practice and Procedures section covers a cognitive behavioral therapy methodology or technique that is applied in the chapter. The section Summary Points encapsulates each chapter in a succinct way.

Contributors to the *Handbook of Cognitive Behavioral Therapy by Disorders* are innovators, authors of international and national standing, and leaders in the field. Emerging fields of science and important discoveries relating to cognitive behavioral therapy are incorporated in this book. This book represents indispensable reading for practitioners of cognitive behavioral therapy, physicians, psychologists, psychiatrists, behavioral scientists, councilors, and health scientists. It is also suitable for novice students and experts, including lecturers, professors, and institutional leaders. The book will be a valuable resource for academic libraries, medical departments, and colleges.

Editors
Colin R. Martin
Vinood B. Patel
Victor R. Preedy

Chapter 1

The context of mental health disorders in comparison to other diseases: Interlinking cognitive behavioral therapy

Rajkumar Rajendram[a,b], Vinood B. Patel[c], and Victor R. Preedy[d]
[a]College of Medicine, King Saud bin Abdulaziz University for Health Sciences, Riyadh, Saudi Arabia, [b]Department of Medicine, King Abdulaziz Medical City, King Abdullah International Medical Research Center, Ministry of National Guard Health Affairs, Riyadh, Saudi Arabia, [c]School of Life Sciences, University of Westminster, London, United Kingdom, [d]Faculty of Life Science and Medicine, King's College London, London, United Kingdom

Abbreviations

DSM Diagnostic and Statistical Manual of Mental Disorders (DSM)
ICD International Classification of Diseases

Introduction

In the 21st century, psychiatry probably still has greater links to the past than any other medical specialty (Hilton, 2021). Though described in ancient texts (e.g., Galen, circa 129–216 current era) and Shakespeare's plays (Hilton, 2021), disorders of mental health remain stigmatizing (Hilton, 2021). They induce fear in the afflicted and those around them (Hilton, 2021).

To define and diagnose these diseases and conditions, clinicians and researchers use the Diagnostic and Statistical Manual of Mental Disorders (DSM) (American Psychiatric Association, 2013). The International Classification of Diseases (ICD) can also be used to classify mental health disorders. These disorders, which include anxiety, depression, schizophrenia, developmental disorders, and dementia, cause significant morbidity and mortality. The Global Burden of Disease study has attempted to quantify these conditions.

Global burden of disease attributable to disorders of mental health

Fig. 1 illustrates the 2019 data for the total global disease burden, measured in Disability-Adjusted Life Years (DALYs) (Roser & Ritchie, 2021). The data are stratified by subcategory of disease (communicable or noncommunicable) or injury. One DALY equals one lost year of healthy life. It can be seen that disorders of mental health were responsible for approximately 125 million DALYs in 2019 (Fig. 1; Roser & Ritchie, 2021). This was the 7th highest cause of DALYs worldwide (Roser & Ritchie, 2021).

Placing mental disorders in context

It is interesting to place DALYS associated with mental disorders in context and compare the data with other conditions. Cardiovascular diseases, cancers, neonatal disorders, other noncommunicable disease, respiratory infections, and TB and musculoskeletal disorders outrank mental disorders in terms of the total DALYS (Fig. 1). On the other hand, DALYs for mental disorders are higher than diabetes and kidney diseases, unintentional injuries, respiratory diseases, and neurological disorders (Fig. 1).

Figs. 2–4 show the 2019 data for the total disease burden, measured in DALYs in England, the United States of America (USA) and Qatar, respectively (Roser & Ritchie, 2021). Similar, but not identical, positions for DALYS associated with mental disorders are also obtained for England, and the USA. In England, mental disorders were ranked fourth, above neurological disorders, respiratory disease, other NCDs, digestive disease and diabetes (Fig. 2). In the USA, mental disorders

2 Handbook of cognitive behavioral therapy by disorder

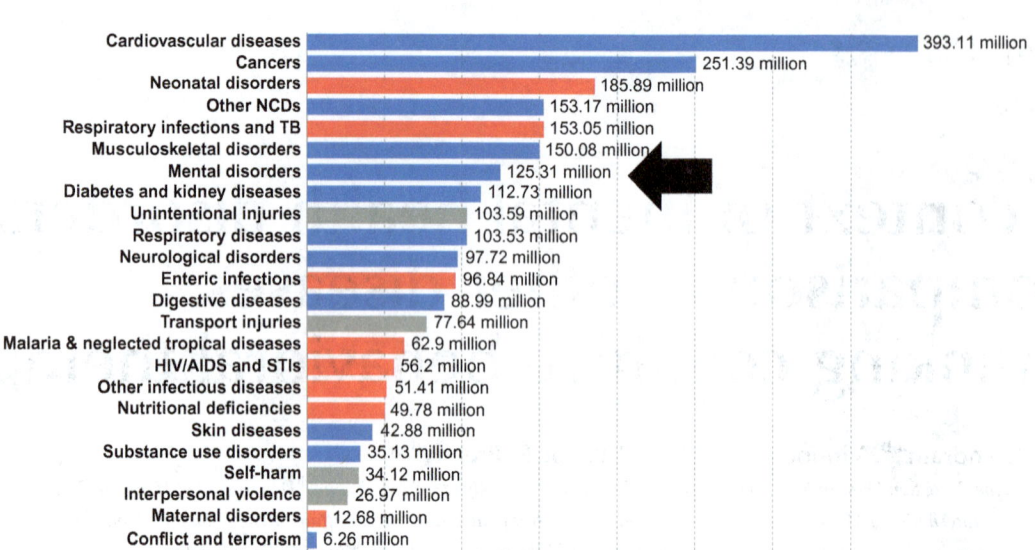

FIG. 1 Burden of disease: Global data. Global 2019 data for the total disease burden, measured in DALYs by subcategory of disease or injury. DALYs measure the total burden of disease—both from years of life lost due to premature death and years lived with a disability. One DALY equals one lost year of healthy life. *Blue bars*: noncommunicable diseases. *Red bars*: communicable, maternal, neonatal and nutritional diseases. *Green bars*: injuries. The black arrow identifies the position of disorders of mental health. Figures produced by Our World in Data are completely open access under the Creative Commons BY licence. *(The text for the legend was taken directly from the source of figures. Roser, M., & Ritchie, H. (2021). Burden of disease. Published online at: OurWorldInData.org. Retrieved from: https://ourworldindata.org/burden-of-disease (online resource).)*

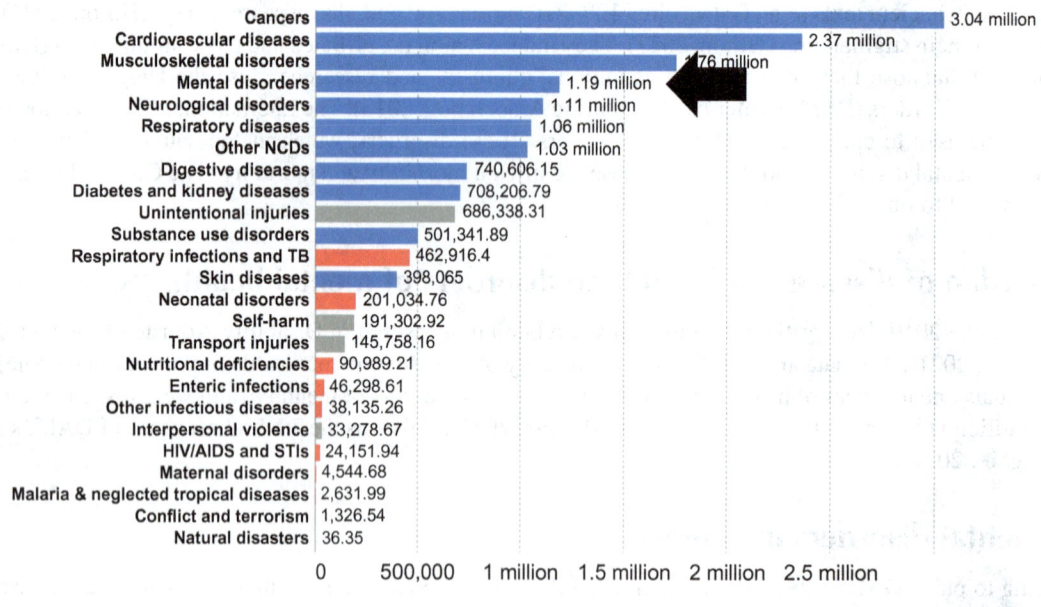

FIG. 2 Burden of disease: England. England 2019 data for total disease burden, measured in Disability-Adjusted Life Years (DALYs) by subcategory of disease or injury. The black arrow identifies the position of disorders of mental health. *(From Roser, M., & Ritchie, H. (2021). Burden of disease. Published online at: OurWorldInData.org. Retrieved from: https://ourworldindata.org/burden-of-disease (online resource). For other details see legend to Fig. 1.)*

ranked 5th (Fig. 3). In absolute terms, disorders of mental health caused approximately 1.2 million DALYs in the UK (Fig. 2; Roser & Ritchie, 2021) and 7.3 million DALYs in the USA (Fig. 3; Roser & Ritchie, 2021). While less than 62,000 DALYs were attributable; disorders of mental health caused the most DALYs in Qatar in 2019 (Fig. 4; Roser & Ritchie, 2021), i.e., in Qatar, mental disorders rank higher than cardiovascular disease and cancers. This illustrates a fundamental point about DALYs. The rank and thus relative contribution of different conditions show substantial

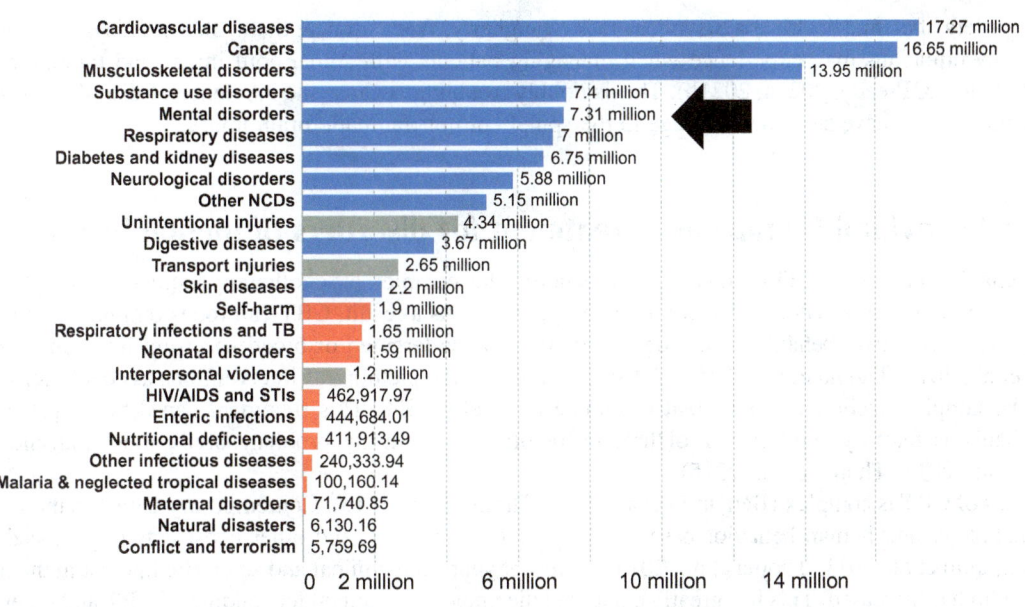

FIG. 3 Burden of disease: USA. USA 2019 data for total disease burden, measured in Disability-Adjusted Life Years (DALYs) by subcategory of disease or injury. The black arrow identifies the position of disorders of mental health. *(From Roser, M., & Ritchie, H. (2021). Burden of disease. Published online at: OurWorldInData.org. Retrieved from: https://ourworldindata.org/burden-of-disease (online resource). For other details see legend to Fig. 1.)*

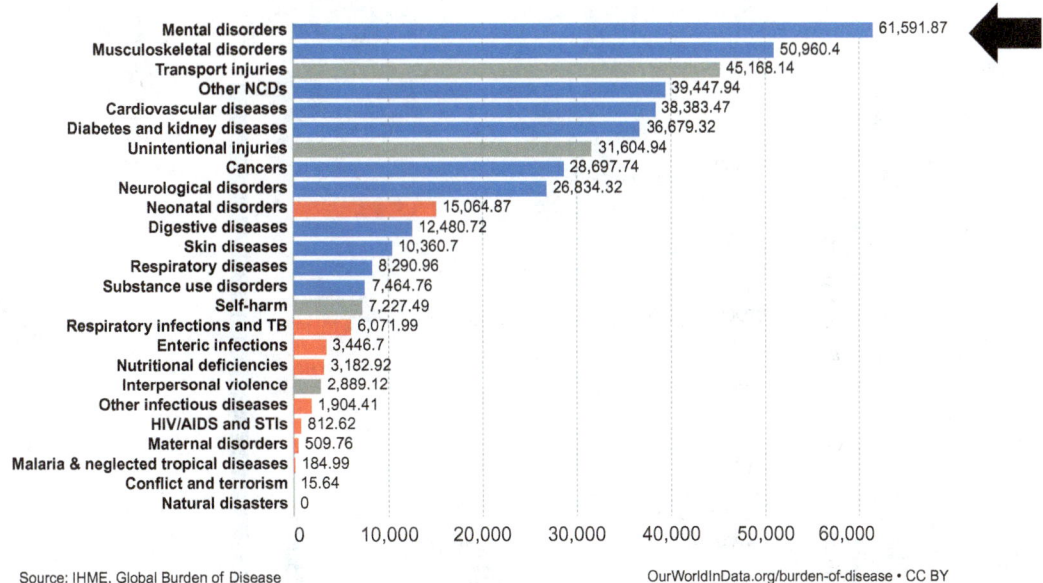

FIG. 4 Burden of disease: Qatar. Qatar 2019 data for total disease burden, measured in Disability-Adjusted Life Years (DALYs) by subcategory of disease or injury. The black arrow identifies the position of disorders of mental health. *(From Roser, M., & Ritchie, H. (2021). Burden of disease. Published online at: OurWorldInData.org. Retrieved from: https://ourworldindata.org/burden-of-disease (online resource). For other details see legend to Fig. 1.)*

variability between countries. Furthermore, it is also important to point out that the rank of disease burdens is not reflective of their ranking in terms of research expenditure (Rajendram, Lewison, & Preedy, 2006).

More detailed analysis of disease burdens attributable to mental disorders can be found elsewhere for global data (GBD 2019 Mental Disorders Collaborators, 2022) and country specific information (Alva-Diaz et al., 2020; Amendola, 2022).

Thus, disorders of mental health clearly cause significant morbidity and mortality. Yet, regrettably, in the 21st century, besides those few specialists in psychiatry and psychotherapy, many clinicians still feel a sense of therapeutic nihilism approach when faced with a patient with a disorder of mental health. Indeed, some authors have described the treatment of severe disorders of mental health as a Sisyphean task (Daugherty, Warburton, & Stahl, 2020). This is somewhat similar to the melancholy clinicians may experience when managing patients with spinal cord injury and traumatic brain injury (Rajendram, Patel, & Preedy, 2021a, 2021b). This is doubly troubling because the incidence of disorders of mental health is high in patients who have sustained damage to the spinal cord or traumatic brain injury.

Cognitive behavioral therapy as a treatment for disorders of mental health

Cognitive behavioral therapy (CBT) can be of great benefit. This complex psychotherapy includes several treatment strategies (Benjamin et al., 2011; Thoma, Pilecki, & McKay, 2015). As a result, CBT can focus on many points of potential vulnerability (i.e., affective, behavioral, or cognitive) and provide targeted or broad-spectrum interventions as required (Benjamin et al., 2011; Thoma et al., 2015). Not surprisingly, of the plethora of interventions available for mental health disorders; the nonpharmacological treatment of choice is usually CBT. This treatment can develop patients' personal coping mechanisms, mastery, and self-control through cognitive restructuring, problem solving modeling, and conditioning (Benjamin et al., 2011; Thoma et al., 2015).

The history of CBT is complex (Benjamin et al., 2011; Thoma et al., 2015). Contemporary CBT has integrated theories of psychopathology and human behavior based on cognitive, behavioral, and other constructs (e.g., social or developmental) (Benjamin et al., 2011; Thoma et al., 2015). In recent years, the clinical and scientific interest in the improvement of mental health has increased. This has greatly expanded the knowledge and understanding of CBT and other psychotherapies (Figs. 5–7).

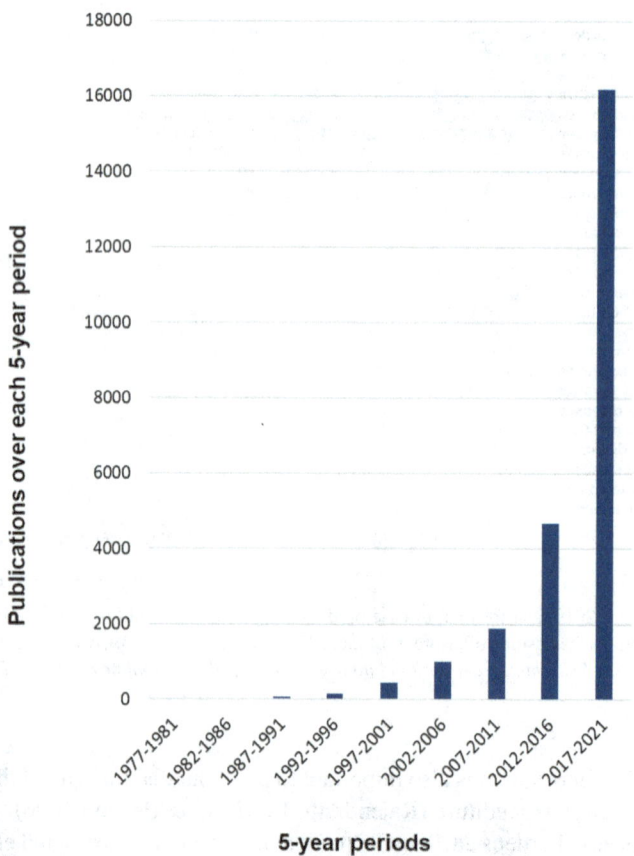

FIG. 5 Time related changes in publications pertaining to cognitive behavioral therapy. The figure shows the number of publications relating to cognitive behavioral therapy over time. The data was generated using the Excerpta Medica dataBASE (EMBASE). A cautionary approach should be taken in interpreting the data which is not designed to show absolute numbers, but rather relative changes. Figures generated by the authors.

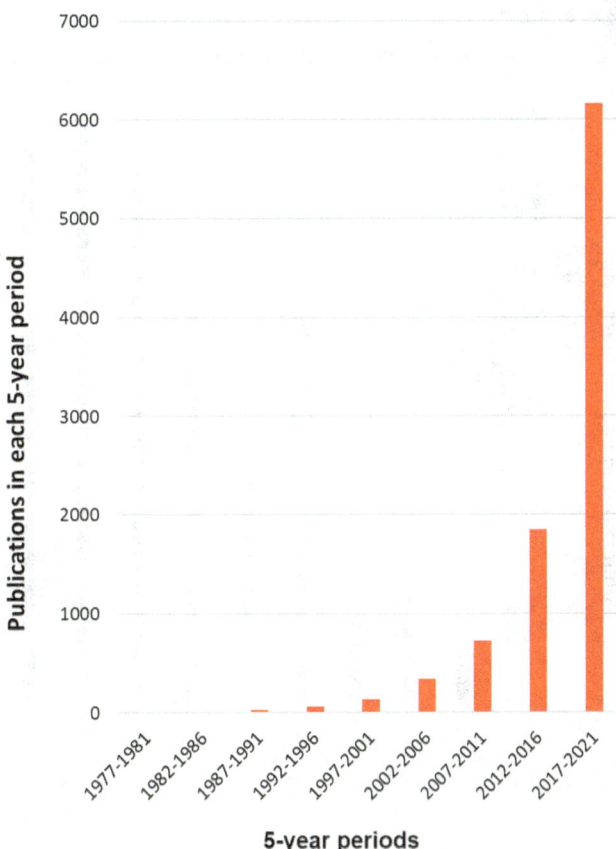

FIG. 6 Trends in publications pertaining to cognitive behavioral therapy and depression over time. The figure shows the number of publications relating to cognitive behavioral therapy and depression. The data was generated using Excerpta Medica dataBASE commonly known as EMBASE. For other details see the legend to Fig. 5. Figures generated by the authors.

Trends in publications on cognitive behavioral therapy

Fig. 5 shows the trends in publications pertaining to CBT over approximately 40 years. The data was generated using the Excerpta Medica dataBASE (commonly known as EMBASE). It is not exhaustive in that different databases will generate different data, albeit with similar sequential trends. Nevertheless, Fig. 5 clearly shows the remarkable increase in CBT-related publications over the past decade. Similar trends were also recorded in CBT and depression (Fig. 6). Another way to look at this data is in relation to studies on the use of antidepressants (Fig. 7). Fig. 7 compares the number of publications pertaining to antidepressants versus CBT and depression. The proportion of publications on CBT and depression has increased relative to the proportion of publications on antidepressants.

However, it must be emphasized that such analysis is rather crude and can be criticized. For example, a proportion of the published literature on antidepressants will include preclinical animal data (such as studies conducted on mice or rats) or research using in vitro modeling systems. Also, the quality of the publications and the relative amounts of material identified using the search terms (i.e., CBT, depression, and antidepressant) are not considered. Another criticism relates to the database searched. While EMBASE was used, other databases include PubMed, Medline, Scopus, Google Scholar, and so on. Nevertheless, the figures provide information on changing trends.

It is important to point out that publication trends, derived from data bases, are very rudimentary measures of research interest. They do not reflect ongoing research, usage, applicability, translation to clinical practice or efficacy.

In the United Kingdom, the National Health Service compiles data on the use and efficacy of different therapies. These data are presented every year in the Improving Access to Psychological Therapies (IAPT) Annual report (IAPT Team and NHS Digital, 2021). The most recent report (2020–21) was published in 2021 (IAPT Team and NHS Digital, 2021).

Table 1 shows the relative treatments for NICE-approved therapies for treating people with anxiety or depression in England. The table clearly shows the usage of different types of therapies and the prominent role of CBT. Material relating to the use of CBT with or without medication has been reviewed elsewhere (for examples see Chen et al., 2021;

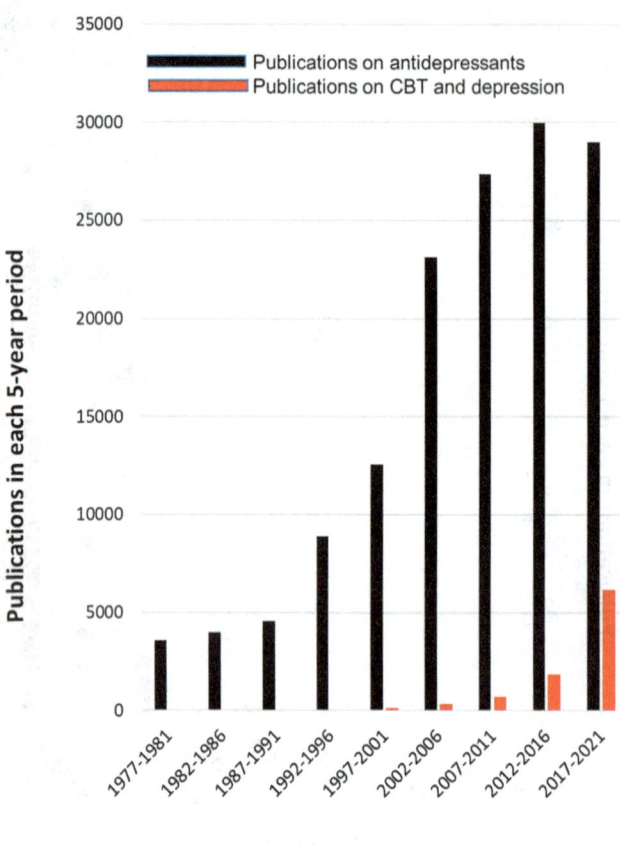

FIG. 7 Trends in publications pertaining to cognitive behavioral therapy and depression versus publications pertaining to antidepressants over time. The figure shows the number of publications relating to cognitive behavioral therapy and depression compared to those relating to antidepressants. The data was generated using the Excerpta Medica dataBASE which is commonly known as EMBASE. For other details see legend to Fig. 5. Figures generated by the authors.

Hoppen et al., 2021; Liu, Li, Wang, Wang, & Yang, 2021; Matsumoto, Hamatani, & Shimizu, 2021; Nakao, Shirotsuki, & Sugaya, 2021; Xu, Cardell, Broadley, & Sun, 2021; Zamiri-Miandoab, Hassanzadeh, Kamalifard, & Mirghafourvand, 2021).

Overall, therapeutic nihilism toward disorders of mental health is unwarranted. Regardless, much more work is required to elucidate the moderators and mediators that can increase the efficacy of CBT in specific settings and accurately predict treatment outcomes.

Application to other areas

The principles needed to understand mental disease are broadly similar to the behavioral and psychosocial constructs that underpin CBT. Thus, the data on CBT presented in this chapter are also relevant to disorders of mental health in general and other psychotherapies.

Key facts of mental health disorders

The Diagnostic and Statistical Manual published by the American Psychiatric Association is used to define and diagnose mental disorders worldwide.
  The International Classification of Diseases (ICD) can also be used to classify mental health disorders.
  Disorders of mental health cause significant morbidity and mortality worldwide.
  Cognitive behavioral therapy is one form of psychotherapy for mental health disorders.

TABLE 1 Relative proportions of therapies for treating people with anxiety or depression (percent) in England 2020–21.

NICE-approved therapies for treating people with anxiety or depression (percent)	
Other low intensity treatments	28.2
Guided self-help book	23.8
Cognitive behavioral therapy (CBT)	17.6
Nonguided self-help book	8.1
Other high intensity	8.0
Counseling for depression	3.9
Not known/not stated/invalid	3.4
Guided self-help computer	2.7
Collaborative care	2.3
Psychoeducational peer support	0.8
Non guided self-help computer	0.4
Employment support	0.2
Couple therapy for depression	0.1
Eye movement desensitization and reprocessing therapy	0.1
Interpersonal psychotherapy (IPT)	0.1
Mindfulness-based therapy	0.1
Short-term psychodynamic therapy	0.1
Applied relaxation	<0.1
Structured physical activity	<0.1

Data from Psychological Therapies, Annual report on the use of Improving Access to Psychological Therapies (IAPT) services, 2020–21 contains public sector information licensed under the Open Government Licence v3.0. Open Government Licence (nationalarchives.gov.uk) IAPT Team, NHS Digital, NHS Digital, Government Statistical Service.

Summary points

Disorders of mental health can cause life-changing disabilities that result in significant morbidity and mortality.

On a global basis, DALYs for cardiovascular diseases, cancers, and neonatal disorders are higher than mental disorders.

DALYs for mental disorders are higher than diabetes and kidney diseases, unintentional injuries, and respiratory diseases.

Cognitive behavioral therapy can be used to treat of disorders of mental health so therapeutic nihilism is unwarranted.

The understanding of cognitive behavioral therapy has increased as a result of the vast amount of research conducted on this topic in recent years.

Much remains unknown about disorders of mental health, cognitive behavioral therapy, and other psychotherapies.

References

Alva-Diaz, C., Huerta-Rosario, A., Molina, R. A., Pacheco-Barrios, K., Aguirre-Quispe, W., Navarro-Flores, A., et al. (2020). Mental and substance use disorders in Peru: A systematic analysis of the global burden of disease study. *Journal of Public Health, 30*, 629–638.

Amendola, S. (2022). Burden of mental health and substance use disorders among Italian young people aged 10–24 years: Results from the Global Burden of Disease 2019 Study. *Social Psychiatry and Psychiatric Epidemiology, 57*, 683–694.

American Psychiatric Association. (2013). *Diagnostic and statistical manual of mental disorders* (5th ed.). Arlington, VA: American Psychiatric Association.

Benjamin, C. L., Puleo, C. M., Settipani, C. A., Brodman, D. M., Edmunds, J. M., Cummings, C. M., & Kendall, P. C. (2011). History of cognitive-behavioral therapy in youth. *Child and Adolescent Psychiatric Clinics of North America, 20*(2), 179–189.

Chen, J., Chen, X., Sun, Y., Xie, Y., Wang, X., Li, R., & Hesketh, T. (2021). The physiological and psychological effects of cognitive behavior therapy on patients with inflammatory bowel disease before COVID-19: A systematic review. *BMC Gastroenterology, 21*(1), 469.

Daugherty, B., Warburton, K., & Stahl, S. M. (2020). A social history of serious mental illness. *CNS Spectrums, 25*(5), 584–592.

GBD 2019 Mental Disorders Collaborators. (2022). Global, regional, and national burden of 12 mental disorders in 204 countries and territories, 1990–2019: A systematic analysis for the global burden of disease study 2019. *The Lancet Psychiatry, 9*(2), 137–150.

Hilton, C. (2021). A history of psychiatry in 1500 words. *BMJ, 374*, 1443.

Hoppen, L. M., Kuck, N., Burkner, P.-C., Karin, E., Wootton, B. M., & Buhlmann, U. (2021). Low intensity technology-delivered cognitive behavioral therapy for obsessive-compulsive disorder: A meta-analysis. *BMC Psychiatry, 21*(1), 322.

Liu, W., Li, G., Wang, C., Wang, X., & Yang, L. (2021). Efficacy of sertraline combined with cognitive behavioral therapy for adolescent depression: A systematic review and meta-analysis. *Computational and Mathematical Methods in Medicine*, 5309588.

Matsumoto, K., Hamatani, S., & Shimizu, E. (2021). Effectiveness of videoconference-delivered cognitive behavioral therapy for adults with psychiatric disorders: Systematic and meta-analytic review. *Journal of Medical Internet Research, 23*(12), e31293.

Nakao, N., Shirotsuki, N., & Sugaya, N. (2021). Cognitive–behavioral therapy for management of mental health and stress related disorders: Recent advances in techniques and technologies. *BioPsychoSocial Medicine, 15*, 16.

National Health Service (NHS) Digital. (2021). *Improving access to psychological therapies (IAPT) team: Psychological therapies, annual report on the use of IAPT services 2020–21*. UK: NHS Digital. Available at: https://digital.nhs.uk/data-and-information/publications/statistical/psychological-therapies-annual-reports-on-the-use-of-iapt-services/annual-report-2020-21.

Rajendram, R., Lewison, G., & Preedy, V. R. (2006). Worldwide alcohol-related research and the disease burden. *Alcohol and Alcoholism, 41*, 99–106.

Rajendram, R., Patel, V. B., & Preedy, V. R. (2021a). Recommended resources and sites for the neuroscience of traumatic brain injury. In R. Rajendram, V. B. Patel, & V. R. Preedy (Eds.), *Neuroscience of traumatic brain injury*. USA: Elsevier.

Rajendram, R., Patel, V. B., & Preedy, V. R. (2021b). Recommended resources and sites for the neuroscience of spina cord injury. In R. Rajendram, V. B. Patel, & V. R. Preedy (Eds.), *Neuroscience of spinal cord injury*. USA: Elsevier.

Roser, M., & Ritchie, H. (2021). *Burden of disease*. Published online at: *OurWorldInData.org*. Retrieved from: https://ourworldindata.org/burden-of-disease (online resource).

Thoma, N., Pilecki, B., & McKay, D. (2015). Contemporary cognitive behavior therapy: A review of theory, history, and evidence. *Psychodynamic Psychiatry, 43*(3), 423–461.

Xu, D., Cardell, E., Broadley, S. A., & Sun, J. (2021). Efficacy of face-to-face delivered cognitive behavioral therapy in improving health status of patients with insomnia: A meta-analysis. *Frontiers in Psychiatry, 12*, 798453.

Zamiri-Miandoab, N., Hassanzadeh, R., Kamalifard, M., & Mirghafourvand, M. (2021). The effect of cognitive behavior therapy on body image and self-esteem in female adolescents: A systematic review and meta-analysis. *International Journal of Adolescent Medicine and Health, 33*(6), 323–332.

Section A

Setting the scene and introductory chapters

Chapter 2

Mental health concerns in primary care: Applications of cognitive behavioral therapies☆

Danielle L. Terry and Michelle A. Nanda
Sayre Family Medicine Residency at Guthrie Robert Packer Hospital, Sayre, PA, United States

Abbreviations

ADHD   attention deficit hyperactivity disorder
CBT   cognitive behavior therapy
CCM   collaborative care model
PCBH primary care behavioral health
SBIRT screening, brief intervention, and referral to treatment

Mental health concerns in primary care

Adults with mental illnesses are more likely to receive behavioral health care through primary care providers rather than mental health specialists. The primary care setting provides a fertile ground to access and treat those that are struggling with mental health conditions. Given the most common conditions seen by health care providers and the availability of shorter-term, evidence-based mental health treatments, it seems only logical that both medical and mental health personnel might look to the primary care setting as a viable place to access patients and provide care. The purpose of this chapter is to highlight the potential ways in which primary care settings might access and serve the needs of individuals presenting with mental health concerns, briefly review existing models that might allow for application of evidence-based treatment in primary care settings, and identify specific ways in which cognitive behavioral treatments can be, and have been, adapted in these settings.

Among the 45 million American adults with mental illness, primary care physicians are the main source of behavioral health care (Centers for Disease Control and Prevention, 2018). Continuity is a hallmark of primary care, which helps to foster trust between patients and providers over time. A strong patient-provider relationship can provide an environment in which patients feel comfortable discussing symptoms and behaviors. Recognition and treatment for mental health conditions in primary care has improved since the introduction of selective serotonin reuptake inhibitors in the 1980s. As SSRIs offer improved safety, efficacy, and ease of dosing as compared to the predecessors, primary care physicians have been more likely to diagnose and treat mental health disorders since their advent.

Primary care physicians manage some type of psychiatric disorder for roughly one-third of their entire patient panels (American Academy of Family Physicians, 2018), with other data suggesting that 20% of patients visits are mental-health related (Centers for Disease Control and Prevention, 2011). According to the National Ambulatory Medical Care Survey data, 34.9% of mental health related physician visits in 2012 were with primary care physicians (Centers for Disease Control and Prevention, 2012). Within outpatient care practice, antidepressants are commonly used, with some data suggesting that the three most prescribed medications are antidepressants (Rui, Hing, & Okeyode, 2014).

Though many adults with mental health conditions see a primary care provider, many may not receive mental diagnoses or treatment. There are several potential causes for the disconnect between mental health diagnoses and treatment in the primary care setting. Patients may be reluctant to disclose psychiatric symptoms to their primary providers. Furthermore,

---

☆ All authors have approved the submission of this manuscript and adhere to the ethics relating to copying and plagiarism.

time constraints within the primary care outpatient setting may lead to the prioritization of the most pressing medical needs with little time to investigate psychiatric symptoms. Primary care providers may have inadequate training to recognize and diagnose specific mental health disorders. In the case that a behavioral health diagnosis is made in the primary care setting, providers may subsequently have insufficient resources and/or training to provide proper treatment.

Despite often being the first point of contact for those that are seeking treatment specific for mental health, evidence has clearly identified the negative impact of mental health conditions on medical outcomes. Individuals with chronic illnesses such as heart disease, diabetes or cancer are more likely to suffer from depression (Centers for Disease Control and Prevention, 2012). Coexisting psychological conditions can also worsen health outcomes in common chronic diseases such as heart disease, cancer and chronic respiratory disease (American Psychological Association, 2014). In order to properly treat chronic disease, and enhance medical outcomes, primary care physicians must achieve competency in recognition and treatment of mental health conditions. Furthermore, patients with mental health disorders are more likely to receive inferior health care which may perpetuate worsened health outcomes (Daré et al., 2019). Thus, there are multiple reasons to identify and care for the mental health needs of patients in primary care.

Common conditions in primary care

Primary care physicians may encounter a variety of mental health illnesses in everyday practice. According to a University of Michigan study, primary care physicians most commonly diagnose depression, anxiety and substance use disorders among their patient populations, with prevalence rates of 36%, 31%, and 25%, respectively, among visits surveyed (Beck, Page, Buche, Schoebel, & Wayment, 2019). These physicians also diagnosed bipolar disorder, attention deficit hyperactivity disorder (ADHD) and serious mental health conditions such as schizophrenia, though the conditions were less common, with prevalence rates of 17%, 16%, and 17%, respectively. In this study, primary care physicians expressed confidence in management of depression, anxiety, and ADHD. However, those surveyed expressed a lack of confidence in treating substance use disorders, bipolar disorder, and severe mental health conditions (Beck et al., 2019).

In the past decade, national recommendations have encouraged depression screening in primary care settings among adults, to enhance the accuracy of diagnosis. These recommendations also encouraged "accurate diagnosis, effective treatment, and appropriate follow-up" (Siu and US Preventive Services Task Force, 2016). Close to half of those that die by suicide have seen their primary care providers the month leading up to their deaths (Ahmedani et al., 2014; Luoma, Martin, & Pearson, 2002). However, The USPSTF recommendations specify that depression screening in adults should be performed in settings with adequate diagnostic and treatment capabilities (US Preventive Services Task Force, 2016).

Barriers to MH in primary care

Generally, primary care physicians refer a large portion of patients for behavioral health services, most often to psychiatrists, community health centers, substance use treatment, or psychiatric hospitals (Beck et al., 2019). Many patients who are referred for mental health services do not actual receive them. Causes for this disconnect may relate to the discussion provided by the referring primary care physician, patient attitudes of psychiatric services or the availability of the recommended service. Primary care physicians may not thoroughly discuss recommended psychiatric referrals, which may result in confusion or misunderstanding from patients. Mental health disorders also remain a source of social stigma. As a result, patients receiving behavioral health referrals may find the services unnecessary or question the validity of a mental health diagnosis (Singh, Subodh, Mehra, & Mehdi, 2017).

A major barrier to mental health management with behavioral health specialists is geographic availability. Over 75% of all U.S. counties are considered mental health shortage areas. Half of all counties have no mental health professionals (Substance Abuse and Mental Health Services Administration, 2016). The density of mental health services is lower in rural areas as compared to urban and metropolitan areas. Outpatient substance abuse treatment is available in just 12.1% of rural hospitals compared to 43.7% in urban hospitals (Substance Abuse and Mental Health Services Administration, 2016). The overall percentage of mental health visits to psychiatrists is lower in rural areas (Centers for Disease Control and Prevention, 2018). Rural patients may have to travel long distances to behavioral health services or may lack access to transportation.

Another deterrent to specialty behavioral health care is cost. Those with Medicaid are less likely to find participating providers in proximity to their homes. Patients with mental illness who have private insurance are more likely to receive care by a psychiatrist (56%) compared to patients with Medicaid (34%; Centers for Disease Control and Prevention, 2018). Low Medicaid reimbursement rates disincentivize behavioral health professionals from providing care to patients of low socioeconomic status. Only 60% of U.S. counties have substance abuse services which accept Medicaid (Substance Abuse and Mental Health Services Administration, 2016). Cost of travel to behavioral health appointments and the expense of copays may further deter patients from seeking mental health services.

Mental illness continues to carry social stigma through the U.S. Such stigma can prevent individuals in need of mental health treatment from seeking needed care. In small communities, patients may not have adequate privacy with regards to the services they receive. They may avoid behavioral health care to avoid social stigma in their communities (Substance Abuse and Mental Health Services Administration, 2016). The level of social stigma associated with mental illness may vary based on factors such as geography, socioeconomic status or ethnicity. Negative attitudes toward mental illness have been particularly challenging among Black Americans. Up to 63% of Black Americans believe that mental illness is a sign of personal weakness (National Alliance on Mental Illness, 2021). Social stigma is also a challenge to mental health treatment in rural communities. Some rural communities may normalize mental illness and substance abuse which may provide social acceptance of untreated mental illness. Rural populations may also be more likely to adopt the ideology of "self-reliance" which may be rooted in social standards within agricultural populations. This notion may perpetuate avoidance of mental health treatment (Substance Abuse and Mental Health Services Administration, 2016).

Many Americans with mental illness are not receiving needed treatment and may have inadequate access to behavioral health specialists. Though expansion of mental health services and improving accessibility of such services is necessary, primary care providers may help to bridge this care gap. Primary care physicians and/or behavioral health providers working in primary care provide a unique opportunity to provide mental health treatment for patients who may not otherwise have access to such care. The continuity of provider-patient relationships in primary care can foster trust and may allow patients to feel comfortable disclosing sensitive mental health symptoms. A strong provider-patient relationship may also encourage otherwise hesitant patients to seek out mental health services if recommended by a trusted provider.

Behavioral health treatment in primary care

Integrated behavioral health care has emerged as one way to improve access to care in a setting where access to patients in need is plentiful. It has been defined an approach to care that allows providers to address patient needs collaboratively in a primary care setting, including physical and mental needs, as well as substance use disorders (Peek, 2013). In 1995, Doherty and colleagues first identified formal classifications of primary-care and behavioral health collaboration. These classifications targeted the extent of collaboration, but not the how and why of specific interactions. The levels of interaction ranged from 1 (minimal collaboration) to 5 (fully integrated). As time has progressed, the five levels of integration served as a foundational framework for a variety of adaptations. Generally, it is understood that integration exists along a continuum (Collins, Hewson, Munger, & Wade, 2010; Miller, Teevan, Phillips Jr., Petterson, & Bazemore, 2011; Peek, 2009; Reynolds, Chesney, & Capobianco, 2006; Seaburn et al., 1996; Strosahl, 1998).

In 2013, the Substance Abuse and Mental Health Services Administration published *A Standard Framework for Levels of Integrated Healthcare*. See Fig. 1. The updated integration framework identifies six levels of collaboration or integration between the medical care provider and behavioral health provider. There are three primary categories that the levels fall into, coordinated care, co-located care, and integrated care. As the levels advance, there is an increasing frequency of communication with stronger relationships and a greater mutual understanding of the roles that other providers play in patient care.

Coordinated care

Coordinated care is divided into two different levels. Level one, *minimal collaboration*, the primary care team and behavioral health teamwork in separate systems and facilities, with limited communication. Level two, Basic *Collaboration at a Distance*, providers are still in separate locations and systems, however, communication may be more frequent, and prompted by specific issues (e.g., request for a psychiatric report).

Co-located care

Care that is co-located can be divided into *Basic Collaboration Onsite and Close Collaboration with Some System Integration*. Both levels include medical and behavioral health providers who are in the same facility. In basic collaboration, both teams are using separate systems, but given the proximity of their location, the referral process increases the likelihood of movement between the medical and behavioral healthcare providers, however, decision-making is still conducted independently.

Close Collaboration with Some System Integration includes a co-located behavioral health practice that has at least some shared systems, such as reception, medical records, or scheduling. There is increased mutual understanding of the roles of medical and behavioral health providers, and greater use of consultation with complex medical issues.

COORDINATED KEY ELEMENT: COMMUNICATION		CO-LOCATED KEY ELEMENT: PHYSICAL PROXIMITY		INTEGRATED KEY ELEMENT: PRACTICE CHANGE	
LEVEL 1 Minimal Collaboration	LEVEL 2 Basic Collaboration at a Distance	LEVEL 3 Basic Collaboration Onsite	LEVEL 4 Close Collaboration Onsite with Some System Integration	LEVEL 5 Close Collaboration Approaching an Integrated Practice	LEVEL 6 Full Collaboration in a Transformed/ Merged Integrated Practice
Behavioral health, primary care and other healthcare providers work:					
In separate facilities, where they:	In separate facilities, where they:	In same facility not necessarily same offices, where they:	In same space within the same facility, where they:	In same space within the same facility (some shared space), where they:	In same space within the same facility, sharing all practice space, where they:
» Have separate systems » Communicate about cases only rarely and under compelling circumstances » Communicate, driven by provider need » May never meet in person » Have limited understanding of each other's roles	» Have separate systems » Communicate periodically about shared patients » Communicate, driven by specific patient issues » May meet as part of larger community » Appreciate each other's roles as resources	» Have separate systems » Communicate regularly about shared patients, by phone or e-mail » Collaborate, driven by need for each other's services and more reliable referral » Meet occasionally to discuss cases due to close proximity » Feel part of a larger yet non-formal team	» Share some systems, like scheduling or medical records » Communicate in person as needed » Collaborate, driven by need for consultation and coordinated plans for difficult patients » Have regular face-to-face interactions about some patients » Have a basic understanding of roles and culture	» Actively seek system solutions together or develop work-a-rounds » Communicate frequently in person » Collaborate, driven by desire to be a member of the care team » Have regular team meetings to discuss overall patient care and specific patient issues » Have an in-depth understanding of roles and culture	» Have resolved most or all system issues, functioning as one integrated system » Communicate consistently at the system, team and individual levels » Collaborate, driven by shared concept of team care » Have formal and informal meetings to support integrated model of care » Have roles and cultures that blur or blend

FIG. 1 Six levels of collaboration/integration (core descriptions).

Integrated care

Finally, levels five and six highlight specific practice changes that share a culture, without one dominant discipline. Level five, *Close Collaboration Approaching an Integrated Practice*, involves frequent communication as a team. Members of the team recognize the value of the other professions; however, there may be specific barriers to integrating care that have not yet been resolved. *Full collaboration in a Transformed/Merged Practice,* is the final level of integration. This level of integration allows for provision of services that focuses on whole-person care. Outcomes and performance indicators minimize the importance of service data and individual staff productivity in favor of population-based outcomes. See Heath, Wise Romero, and Reynolds (2013) for further reading and discussion of the standard classifications.

Models of care

While there are many different models and approaches to integrated care, the three most prominent models of integrated care are the collaborative care model (CCM), screening, brief intervention, and referral to treatment (SBIRT) and the primary care behavioral health (PCBH) model.

The CMM is arguably the most studied team-based approach to integrated care. It is comprised of three primary aspects, collaboration between providers, step-care approaches, and includes outcomes-driven improvements. This model has been shown to be clinically effective (Goodie & Hunter, 2014; Hunter, Goodie, Oordt, & Dobmeyer, 2009; Pigeon, Funderburk, Cross, Bishop, & Crean, 2019) improving clinical outcomes for at least 24 months, and is cost-effective (Archer et al., 2012; Camacho et al., 2016, 2018; Green et al., 2014). Generally, the CCM requires more than co-location. Primary care providers and mental health providers collaborate to identified shared patient goals, adjust treatment plans, all based on symptom-related outcomes. Studies of Project IMPACT (Improving Mood: Providing Access to Collaborative Treatment for Late-Life Depression) included a multidisciplinary team of clinicians who collaborate to treat depression within a patient. This included the primary care provider, the depression specialist, and a team psychiatrist. These models have achieved improved outcomes in depression across a variety of settings (Thota et al., 2012).

The SBIRT model was originally developed as a model of treatment based in public health models for alcohol and other substance use disorders. Generally, evidence-based screening tools are used to identify at-risk individuals. Its public health approach aims to reduce the burden of nonmedical use of substance. Most often, those using the SBIRT model will have a screening process and procedure that identifies individuals who are at-risk for substance use. Those who are identified as low-to-moderate risk would be candidates for brief intervention. Individuals who are considered to be higher risk, or who are currently showing signs of a more severe substance use disorder are referred to treatment.

The PCBH model integrates the behavioral health providers into the primary care team. Like the SBIRT model, it is a population-based approach to health that involves co-location and collaboration within an integrated team. In doing so, this allows the primary care team to provide a broader range of services. An essential part of this model involves the co-location of medical record. The PCBH model faces numerous barriers related to reimbursement and the need for an integrated infrastructure, like an electronic health record (Freeman, Manson, Howard, & Hornberger, 2018). Typical billing codes can be used in this model, and many organizations might attempt to bill for services. However, organizations may also choose to forgo billing as part of the organizational philosophy and/or access to grant funding. For example, the Veterans Administration funds PCBH programming within the budget, rather than within billing.

There are numerous other evolving models of integrated care, which target a variety of populations. Again, the purpose of highlighting the potential models is to note that should primary care more extensively implement evidence-based treatments like cognitive-behavioral therapy, there are effective and streamlined models that could be used to integrate cognitive behavior therapy (CBT), although each face their own barriers.

Barriers to CBT in primary care

Despite a conceptually sound framework, there are many barriers to the integration of services. In addition, use of collaborative or integrated care does not necessarily involve the use of psychotherapy or cognitive behavioral therapy. Among the greatest barriers to implementation and use of CBT in primary care is the availability of trained specialists who can provide treatment. This can be costly, as well as unavailable. This chapter does not aim to provide and exhaustive discussion of collaborative and integrated models of care, nor the barriers to implementation. Rather, it is recognized that there are conceptually-sound frameworks for integration of care, in which mental health services could be integrated into primary care. Given the dearth of guidance on how to use CBT in primary care settings, this chapter provides a brief review of some of the practical considerations of this work.

Adaptation of CBT for the primary care setting

Given its wide application in the treatment of various mental health conditions it only seems logical that CBT would, or could, be applied in primary care settings. Indeed, there is evidence that suggests that CBT can be effectively used for the treatment of a variety of conditions often seen in primary care, including low back pain, sleep difficulties, and irritable bowel syndrome. In addition, CBT can be effective for more complex patients, and can be useful for both high intensity and low intensity treatment.

The setting

The setting of the practice may impact how CBT can be carried out, and interventions may have to be adjusted accordingly. Dependent upon the level of integration, providers may work within an office in a primary setting, this may include an office with all the usual amenities, or it may be a simple exam room commonly used in primary care practice. The length of the visit may be dictated by need for space, and the CBT agenda may have to be adjusted due to time. Given the challenges of shared office space, knowing the length of patient visits may impact the focus of the visit and/or the number of agenda items or tasks that can be comfortably completed. For example, in a busy primary care clinic, there may be time or space limitations, particularly when typical primary care visits may be as short as 15 min. In addition, to aid in efficiency and streamlining care, providers might consider efficient ways in which they will access common written measures they might use for outcome assessment. For example, to measure anxiety over time, does the (potentially shared) office space have access to an anxiety inventory, clipboard and writing utensil? Is there nursing staff that can assist with administration of commonly used screening tools, like the Patient Health Questionnaire 9-item version?

The CBT provider's role

The CBT provider may serve a variety of functions within a primary care setting. First, they may provide care to patients during individually scheduled visits, which is most consistent with traditional approaches to CBT. For those with the luxury of time and office space, the implementation of the protocol may strongly resemble traditional, manualized CBT. More often, individual sessions may be provided in an abbreviated form. There may be more homework assigned outside of session. Second, the CBT provider working in primary care may provide "warm handoff" consultative services, in which the medical provider co-facilitates or acts as an add-on to the medical appointment. In these cases, the approach to CBT may be simply orienting patients to the CBT model to help them understand whether they would like to engage in treatment. There may also be brief, in-the-moment strategies (e.g., cognitive re-framing, Socratic questioning, education about differences between thoughts and feelings, assistance with behavioral action plans) applied to specific problems. Finally, the BH provider may serve as a coach or consultant to the medical provider who might be providing CBT. This often functions similarly to clinical supervision or peer consultation like other settings, with awareness of the medical viewpoint/culture, decreased availability of time and limitations in the training background of the medical provider who is seeking consultation. Driven by the practical considerations and the unique setting of primary care, sessions are often shorter, treatment is abbreviated, and the goal for treatment is increased functionality rather than full recovery.

Billing and CBT in primary care

As noted in the discussion of the PCBH, billing can often be a challenge in primary care. There is great variability in the United States and the requirements of insurers. These challenges may extend to any less-formulated billable services, such as coordination of care, outreach and CBT education, screening, or consultation with other providers. Understanding the use of the current procedural terminology (CPT) and healthcare common procedure coding system (HCPCS) is essential in the delivery of mental health treatment. Billing models typically fall into two camps: those organizations that do not individual bill for mental health services versus those that attempted to individual bill patients for services provided. Although nonprimary care settings might have well-established infrastructures and more commonly used billing codes, it is essential that the provider understand how and what to bill for, to avoid harm and unexpected costs to the patient. Given the "on the fly" nature of mental health treatment in primary care, providers may unexpectedly provide a service and not have the time capability to ensure how and what to bill for. Staying current with billing practices, identifying strategies to accurately represent and reimburse treatment provided in primary care, as well as continuing to productively advocate for policy change to the larger public is essential as the world evolves toward more comprehensive care.

Informed consent

Although not necessarily specific to CBT, it is important to highlight the importance of informed consent and provision of mental health services in primary care. Informed consent is well-embedded into the practice of mental health care, and clearly acknowledged in the American Psychological Association's ethical codes. However, for those that work in primary care, the practice of informed consent differs significantly from those working in specialty care or other siloed practices.

Often, although there may be a provision of CBT or other mental health services, mental health providers that work in primary care may serve as a consultant to the team, and their interaction with the patient may be quite indirect. In primary care, delivery of informed consent and its associated parameters may vary. Who does this, and how do they conduct informed consent? Often, this responsibility may fall on the larger practice or the primary care provider. Depending on the clinic, the clinic administration may also be responsible for presenting patients with their rights and responsibilities, and it is understood that HIPAA is a legal standard to abide by. Very often, patients are presented their rights and responsibilities as part of the appointment registration process, including signing paperwork. Most behavioral health providers who work in primary care do not have their own separate documentation of informed consent.

Given the nature of informed consent in primary care, mental health providers who are providing CBT or other mental health treatments might consider a brief explanation of their role within the team, the limits of confidentiality (especially important given the nature of collaborative and integrated care), as well as highlighting relevant documentation processes. This can be done at the outset of treatment, but often might need to be revisited through the process of treatment, especially as medical issues that require contact with the primary care provider emerge. It is prudent to document these conversations as part of medical record.

Documentation

When practicing in primary care, the mental health provider will document within the medical record. For those that work within the PCBH model, many may use the traditional subjective, objective, assessment, plan (SOAP) note, to adhere to documentation practices in primary care. No matter the format, it is important for the CBT provider to understand the audience, who might be viewing the notes, and adjust what might be included and excluded within the note beyond basic legal requirements of notes. Documentation should be sure to avoid use of jargon that may not be understood by the medical team, should highlight specific recommendations and a plan moving forward.

Brief CBT: A general approach in primary care

Modified or brief CBT allows for the compression of traditional CBT (12–20 sessions) into four to eight sessions. The initial session involves orienting the patient to CBT, identifying patient concerns, and setting therapy goals. This might involve a targeted approach and a narrower focus that results in a focus on function improvement. The second session might include a continued assessment of the patient's concerns and/or introduction of techniques, including but not limited to addressing maladaptive thoughts, behavioral activation, relaxation training, and/or problem-solving techniques.

Brief CBT is most amenable to targeted problems, like adjustment difficulties, anxiety or depression, or other lifestyle changes that might be addressed in primary care. In primary care, this targeted approach is often helpful for those that are experiencing difficulties related to medical conditions. This might include behavioral changes such as those necessary for weight management, diabetes, or other medical conditions, coping with the onset of a new illness, or even preventative behaviors, such as increasing exercise or improving diet. This approach to therapy may not be suitable for those who struggle with numerous comorbid conditions, and/or personality disorders.

Therapist skills

Given the abbreviated nature of treatment in primary care, providers who are strong in specific skills may be more suited to CBT in primary care. These include a strong ability to quickly build rapport with patients, efficiently collaborate to narrow in on goals for short-term treatment, and present material clearly and concisely. As with other therapies, nonspecific factors are critical to successful therapy. Providers who can be nonjudgmental and collaborative will be more suited to this work, however, these qualities are best paired with the ability to be assertive and direct, given the time-limited, and targeted nature of these interventions.

Patient characteristics

As with many interventions, it is important to determine who is a suitable candidate for CBT in primary care. As previously mentioned for brief CBT, those that do not have numerous co-morbid conditions or psychosocial crises may be more appropriate, given the necessity to focus on specific problems without numerous complicating factors. Patients who are highly motivated, willing to engage in activities outside of session, and can manage the cognitive tasks associated with CBT are often better candidates for therapy in primary care.

Specific considerations in therapy

Insomnia

Evidence has suggested that insomnia can be addressed in primary care using both individual (Goodie & Hunter, 2014; Hunter et al., 2009; Pigeon et al., 2019) and group approaches (Davidson, Dawson, & Krsmanovic, 2019). One adaptation of CBT-I includes a four-session protocol that was used with veterans who presented with co-occurring insomnia and depression (Pigeon et al., 2019). Sessions were shortened to 15- to 30-min sessions, and two of the sessions were conducted by telephone. A review of RCTs that have examined the efficacy of CBT-I in primary care indicated that the number of sessions ranged from two to eight, many using self-help manuals (Cheung, Jarrin, Ballot, Bharwani, & Morin, 2019; Davidson, Dickson, & Han, 2019). Interventions that were at least four or more sessions and included stimulus control and sleep restriction had the largest impact on sleep onset latency, wake time after sleep onset and the Insomnia Severity Index. Several studies have used brief telephone follow-up. Interventions were delivered by registered nurses, general practitioners, social workers and behavioral health providers.

Depression and anxiety

Although there are lower treatment effect sizes of CBT applied within primary care settings, treatment sessions are less frequent (eight sessions or less) and applying CBT in primary care for depression and anxiety has been shown to be efficacious (Zhang et al., 2019). RCTs examining CBT for anxiety and depression have been tested using individual, group, and couple's treatment. A metaanalysis of 57 RCTs indicated that the number of treatment sessions typically ranged from 2 to 20 sessions, 30–90 min each. These also included booster sessions, telephone sessions, with promising evidence of online modules and treatment. Results also suggested that use of technology-based interventions might further reduce stigma and increase accessibility. Other evidence has suggested that CBT for depression in primary care might be most effective for those with mild to moderate depression, regardless of delivery modality, including individual, group, or therapist guided (Santoft et al., 2019). In addition, CBT outside of primary care had greater treatment effects, as did individual treatment (as compared to technology-assisted and group treatment; Zhang et al., 2019).

Summary

As the landscape of medicine shifts and changes, it has become increasingly clear that primary care is a setting that can readily access and address the many mental health needs of patients within our communities. Despite the barriers, the existing models of care would allow for more seamless integration that would improve access to care and enhance patient outcomes. This has and can be done. The availability of modified treatments for CBT and the feasibility of the application of evidence-based treatment in primary care are evident. As time and research evolve it is essential to consider this setting as one way to increase access to essential services needed within our communities.

Mini-dictionary of terms

Integrated care: An approach to care that allows for improved access to care in a setting (like primary care) where there is a high access to patients and allows for collaboration between providers.

Collaborative care model (CCM): A team-based approach to integrated care. It includes three aspects, including collaboration between providers, stepped-care approaches, and outcomes-driven improvements.

Screening, brief intervention, and referral to treatment (SBIRT): A model of treatment originally used for alcohol and substance use disorders. Involves a screening process to identify at-risk individuals, and then those that are considered to be higher risk are referred for specialty treatment by the medical provider.

Primary care behavioral health model (PCBH): A model of treatment that integrates behavioral health providers within the primary care treatment and allows for a broader range of services to be provided in this setting. Involves co-location of the medical record.

Key facts

- Among the 45 million American adults with mental illness, primary care physicians are the main source of behavioral health care.
- Primary care physicians manage some type of psychiatric disorder for roughly one-third of their entire patient panels.
- Individuals with chronic illnesses such as heart disease, diabetes or cancer are more likely to suffer from depression.
- While there are many different models and approaches to integrated care, the three most prominent models of integrated care are the collaborative care model (CCM), screening, brief intervention, and referral to treatment (SBIRT) and the primary care behavioral health (PCBH) model.
- There is evidence that suggests that CBT can be effectively used for the treatment of a variety of conditions often seen in primary care, including low back pain, sleep difficulties, and irritable bowel syndrome.

Applications to other areas

Not applicable to this chapter.

Summary points

- Coexisting psychological conditions can worsen health outcomes in common chronic diseases such as heart disease, cancer and chronic respiratory disease (American Psychological Association, 2014).
- Collaborative treatment between PCPs and behavioral health providers creates a multidisciplinary infrastructure better prepared to address mental health concerns.
- Given the abbreviated nature of treatment in primary care, providers who have a strong ability to quickly build rapport with patients, efficiently collaborate to narrow in on goals for short-term treatment, present material clearly and concisely may be better suited to this work.
- Although there are lower treatment effect sizes of CBT applied within primary care settings, treatment sessions are generally less frequent and are eight sessions or less and applying CBT in primary care for depression and anxiety has been shown to be efficacious.

References

Ahmedani, B. K., Simon, G. E., Stewart, C., Beck, A., Waitzfelder, B. E., Rossom, R., … Solberg, L. I. (2014). Health care contacts in the year before suicide death. *Journal of General Internal Medicine, 29*(6), 870–877. https://doi.org/10.1007/s11606-014-2767-3.

American Academy of Family Physicians. (2018). *Mental health care services by family physicians (position paper)*. https://www.aafp.org/about/policies/all/mental-health-services.html.

American Psychological Association. (2014). *Briefing series on the role of psychology in health care*. https://www.apa.org/health/briefs/primary-care.pdf.

Archer, J., Bower, P., Gilbody, S., Lovell, K., Richards, D., Gask, L., … Coventry, P. (2012). Collaborative care for depression and anxiety problems. *Cochrane Database of Systematic Reviews, 10*. https://doi.org/10.1002/14651858.CD006525.pub2.

Beck, A. J., Page, C., Buche, J., Schoebel, V., & Wayment, C. (2019). *Behavioral service provision by primary care physicians*. https://www.behavioralhealthworkforce.org/wp-content/uploads/2019/12/Y4-P10-BH-Capacityof-PC-Phys_Full.pdf.

Camacho, E. M., Davies, L. M., Hann, M., Small, N., Bower, P., Chew-Graham, C., … Coventry, P. (2018). Long-term clinical and cost-effectiveness of collaborative care (versus usual care) for people with mental–physical multimorbidity: Cluster-randomised trial. *The British Journal of Psychiatry, 213*(2), 456–463. https://doi.org/10.1192/bjp.2018.70.

Camacho, E. M., Ntais, D., Coventry, P., Bower, P., Lovell, K., Chew-Graham, C., … Davies, L. M. (2016). Long-term cost-effectiveness of collaborative care (vs usual care) for people with depression and comorbid diabetes or cardiovascular disease: A Markov model informed by the COINCIDE randomised controlled trial. *BMJ Open, 6*(10). https://doi.org/10.1136/bmjopen-2016-012514, e012514.

Centers for Disease Control and Prevention. (2011). Mental illness surveillance among adults in the United States. *Morbidity and Mortality Weekly Report, 60*, S1–S29.

Centers for Disease Control and Prevention. (2012). *Mental health and chronic disease*. https://www.cdc.gov/workplacehealthpromotion/tools-resources/pdfs/issue-brief-no-2-mental-health-and-chronic-disease.pdf.

Centers for Disease Control and Prevention. (2018). *Mental health-related physician office visits by adults aged 18 and over: United States 2012–2014*. https://www.cdc.gov/nchs/products/databriefs/db311.htm.

Cheung, J. M. Y., Jarrin, D. C., Ballot, O., Bharwani, A. A., & Morin, C. M. (2019). A systematic review of cognitive behavioral therapy for insomnia implemented in primary care and community settings. *Sleep Medicine Reviews, 44*, 23–36. https://doi.org/10.1016/j.smrv.2018.11.001.

Collins, C., Hewson, D. L., Munger, R., & Wade, T. (2010). *Evolving models of behavioral health integration in primary care.* Milbank Memorial Fund. https://www.milbank.org/publications/evolving-models-of-behavioral-health-integration-in-primary-care/.

Daré, L. O., Bruand, P.-E., Gérard, D., Marin, B., Lameyre, V., Boumédiène, F., & Preux, P.-M. (2019). Co-morbidities of mental disorders and chronic physical diseases in developing and emerging countries: A meta-analysis. *BMC Public Health, 19*(1), 304. https://doi.org/10.1186/s12889-019-6623-6.

Davidson, J. R., Dawson, S., & Krsmanovic, A. (2019). Effectiveness of group cognitive behavioral therapy for insomnia (CBT-I) in a primary care setting. *Behavioral Sleep Medicine, 17*(2), 191–201. https://doi.org/10.1080/15402002.2017.1318753.

Davidson, J. R., Dickson, C., & Han, H. (2019). Cognitive behavioural treatment for insomnia in primary care: A systematic review of sleep outcomes. *British Journal of General Practice, 69*(686). https://doi.org/10.3399/bjgp19X705065, e657.

Freeman, D. S., Manson, L., Howard, J., & Hornberger, J. (2018). Financing the primary care behavioral health model. *Journal of Clinical Psychology in Medical Settings, 25*(2), 197–209. https://doi.org/10.1007/s10880-017-9529-4.

Goodie, J. L., & Hunter, C. L. (2014). Practical guidance for targeting insomnia in primary care settings. *Cognitive and Behavioral Practice, 21*(3), 261–268. https://doi.org/10.1016/j.cbpra.2014.01.001.

Green, C., Richards, D. A., Hill, J. J., Gask, L., Lovell, K., Chew-Graham, C., ... Barkham, M. (2014). Cost-effectiveness of collaborative care for depression in UK primary care: Economic evaluation of a randomised controlled trial (CADET). *PLoS One, 9*(8). https://doi.org/10.1371/journal.pone.0104225, e104225.

Heath, B., Wise Romero, P., & Reynolds, K. A. (2013). *Standard framework for levels of integrated healthcare.* Washington, DC: SAMHSA-HRSA Center for Integrated Health Solutions.

Hunter, C. L., Goodie, J. L., Oordt, M. S., & Dobmeyer, A. C. (2009). *Integrated behavioral health in primary care: Step-by-step guidance for assessment and intervention* (p. xiii). American Psychological Association. 291 https://doi.org/10.1037/11871-000.

Luoma, J. B., Martin, C. E., & Pearson, J. L. (2002). Contact with mental health and primary care providers before suicide: A review of the evidence. *American Journal of Psychiatry, 159*(6), 909–916. https://doi.org/10.1176/appi.ajp.159.6.909.

Miller, B. F., Teevan, B., Phillips, R. L., Jr., Petterson, S. M., & Bazemore, A. W. (2011). The importance of time in treating mental health in primary care. *Families, Systems & Health, 29*(2), 144–145. https://doi.org/10.1037/a0023993.

National Alliance on Mental Illness. (2021). *Identity and cultural dimensions: Black/African American.* https://www.nami.org/Your-Journey/Identity-and-Cultural-Dimensions/Black-African-American.

Peek, C. J. (2009). Integrating care for persons, not only diseases. *Journal of Clinical Psychology in Medical Settings, 16*(1), 13–20. https://doi.org/10.1007/s10880-009-9154-y.

Peek, C. J. (2013). Integrated behavioral health and primary care: A common language. In M. R. Talen, & A. Burke Valeras (Eds.), *Integrated behavioral health in primary care: Evaluating the evidence, identifying the essentials* (pp. 9–31). New York: Springer. https://doi.org/10.1007/978-1-4614-6889-9_2.

Pigeon, W. R., Funderburk, J. S., Cross, W., Bishop, T. M., & Crean, H. F. (2019). Brief CBT for insomnia delivered in primary care to patients endorsing suicidal ideation: A proof-of-concept randomized clinical trial. *Translational Behavioral Medicine, 9*(6), 1169–1177. https://doi.org/10.1093/tbm/ibz108.

Reynolds, K. M., Chesney, B. K., & Capobianco, J. (2006). A collaborative model for integrated mental and physical health care for the individual who is seriously and persistently mentally ill: The Washtenaw community health organization. *Families, Systems & Health, 24*(1), 19–27. https://doi.org/10.1037/1091-7527.24.1.19.

Rui, P., Hing, E., & Okeyode, T. (2014). *National Ambulatory Medical Care Survey: 2014 state and national summary tables.* Centers for Disease Control and Prevention. http://www.cdc.gov/nchs/ahcd/ahcd_products.htm.

Santoft, F., Axelsson, E., Öst, L.-G., Hedman-Lagerlöf, M., Fust, J., & Hedman-Lagerlöf, E. (2019). Cognitive behaviour therapy for depression in primary care: Systematic review and meta-analysis. *Psychological Medicine, 49*(8), 1266–1274. https://doi.org/10.1017/S0033291718004208.

Seaburn, D. B., Lorenz, A. D., Gunn, W. B., Jr., Gawinski, B. A., et al. (1996). *Models of collaboration: A guide for mental health professionals working with health care practitioners.* pp. xi, 350 Basic Books.

Singh, S., Subodh, B., Mehra, A., & Mehdi, A. (2017). Reactions to psychiatry referral in patients presenting with physical complaints to medical and surgical outpatient services. *Indian Journal of Psychological Medicine, 39*(5). https://doi.org/10.4103/IJPSYM.IJPSYM_402_16. Health Research premium collection; publicly available content database.

Siu, A. L., & US Preventive Services Task Force. (2016). Screening for depression in adults: Us preventive services task force recommendation statement. *JAMA, 315*(4), 380–387. https://doi.org/10.1001/jama.2015.18392.

Strosahl, K. (1998). Integrating behavioral health and primary care services: The primary mental health care model. In *Integrated primary care: The future of medical and mental health collaboration* (pp. 139–166). W. W. Norton & Company.

Substance Abuse and Mental Health Services Administration. (2016). *Rural behavioral health: Telehealth challenges and opportunities.* https://store.samhsa.gov/sites/default/files/d7/priv/sma16-4989.pdf.

Thota, A. B., Sipe, T. A., Byard, G. J., Zometa, C. S., Hahn, R. A., McKnight-Eily, L. R., ... Williams, S. P. (2012). Collaborative care to improve the management of depressive disorders: A community guide systematic review and meta-analysis. *American Journal of Preventive Medicine, 42*(5), 525–538. https://doi.org/10.1016/j.amepre.2012.01.019.

Zhang, A., Borhneimer, L. A., Weaver, A., Franklin, C., Hai, A. H., Guz, S., & Shen, L. (2019). Cognitive behavioral therapy for primary care depression and anxiety: A secondary meta-analytic review using robust variance estimation in meta-regression. *Journal of Behavioral Medicine, 42*(6), 1117–1141. https://doi.org/10.1007/s10865-019-00046-z.

Chapter 3

Mechanisms of change in cognitive-behavioral therapy for weight loss

Loana Comșa[a] and Oana David[a,b]
[a]Department of Clinical Psychology and Psychotherapy, Babeș-Bolyai University, Cluj-Napoca, Romania, [b]Data Lab: Digital Affective Technologies in Therapy and Assessment, Babeș-Bolyai University, Cluj-Napoca, Romania

Abbreviations

BMI   body mass index
CBTWL   cognitive-behavioral therapy for weight loss
SBT   standard behavioral therapy
SCT   social cognitive theory
SDT   self-determination theory
WHO   World Health Organization

Introduction

Obesity and being overweight are major risk factors for serious health problems. The World Health Organization (WHO) defines obesity on its website as "abnormal or excessive fat accumulation that presents a health risk." Body max index (BMI) is commonly used to measure obesity and is calculated by dividing the body weight by the square of the height (kg/m^2). The WHO defines overweight and obese adults, according to BMI (see Table 1).

An important fact is that over the world, in 2016, more than 1.9 billion adults were overweight, and of these, over 650 million adults were classified as having obesity. This disease is growing worldwide and has been declared a global epidemic (WHO, 2000). Overweight and obesity are associated with an increased risk of diseases: cardiovascular problems, type 2 diabetes, hypertension, vascular attack, and some forms of cancers (Cooper & Fairburn, 2001). Besides physical problems, in an unpublished work, we (Comșa & David, 2022) found that people with obesity and overweight also think differently from those with healthy weight: they have higher levels of irrationality toward food and higher levels of two irrational styles of thinking, called demanding fairness and need for comfort. Also, they behave differently, tending to eat more uncontrolled and emotionally and restraining cognitively less. People with obesity and overweight also reported lower levels of self-efficacy.

Treatments in weight loss

In treating obesity, we have three main evidence-based approaches: surgical treatment, pharmacotherapy, and psychological treatment. Surgical treatment leads to a weight loss of about 20% and is very well maintained, but it is recommended only for individuals with a BMI of over 40. Pharmacotherapy has an effect on weight loss of about 5%–10% of the initial weight. However, after the drug treatment is stopped, weight regain is a common phenomenon (Byrne, Cooper, & Fairburn, 2003; Cooper & Fairburn, 2001). Behavioral interventions are the most commonly used psychological treatments for weight loss and weight loss maintenance, but they also have modest results. The weight loss is about 10% of the initial weight, but afterward, the loss is almost always regained (Cooper & Fairburn, 2001; Teixeira et al., 2004, 2015). More specifically, after no more than a year, about half of the weight lost during the program is regained, and this rebound continues (Byrne et al., 2003). With all this, research showed that a 5%–10% reduction in weight is associated with essential health benefits: a decrease in cholesterol, blood pressure, blood glucose, and other health indices (Wing et al., 2011). Thus, a successful long-term weight loss maintenance is considered an intentional loss of 10% of initial body weight, kept for at least 1 year (Wing & Hill, 2001).

TABLE 1 WHO classification of weight.

Weight category	BMI
Underweight	<18.5
Healthy Weight	≥18.5<25
Overweight	≥25<30
Obese	≥30

In standard behavioral treatment (SBT), the work is on changing eating behaviors, reducing calorie intake, and increasing exercise. The results in improving weight loss and weight maintenance are better when cognitive techniques are added to behavioral treatments (Cooper & Fairburn, 2001). Because of their better results in a higher decrease and a more sustainable weight loss, the use of cognitive-behavioral interventions (CBT) is increasing (Shaw, O'Rourke, Del Mar, & Kenardy, 2005).

CBT for weight loss

When we are talking about CBT for weight loss, we refer to interventions in which we can identify a CBT component. In this chapter, we considered the taxonomy used in the metaanalysis by Jacob et al. (2018). According to this taxonomy, CBT interventions have CBT strategies delivered to a group or an individual. These strategies are defined by Jacob et al. (2018) as interventions that addressed cognitive and behavioral aspects of conceptualizations of weight changes. When a CBT intervention for weight loss is described, it has also to describe the techniques used for applying it in detail. These strategies had to be more than those used in SBTs, such as self-monitoring, goal settings, or stimulus control. Accepted strategies used in CBT for weight loss are those used to help patients identify and change targeting alleged mechanisms: cognitive restructuring, problem-solving, body image acceptance, self-efficacy, weight expectations, and working on maintaining the weight loss. From the studies included in the cited metaanalysis, we know that CBT interventions usually have the length of treatment ranging from 12 to 52 weeks and the number of sessions from 6 to 30 (Comşa, David, & David, 2020; Jacob et al., 2018).

CBT techniques are mostly delivered together with nutrition and exercise counseling, CBT alone, CBT together with only exercise, or CBT together with nutrition (Comşa et al., 2020).

Characteristics of CBT for weight loss interventions

Regarding the characteristics of the interventions, treatment length is a significant predictor of the results (Comşa et al., 2020; Shaw et al., 2005), which means that the longer the treatment, the higher its effect. However, the number of sessions is not associated with weight loss (Comşa et al., 2020; Jacob et al., 2018). According to our findings (Comşa et al., 2020), there is no difference in efficacy if CBT is delivered in-group or in one-to-one sessions.

Another important fact of CBT intervention is that the comparison type is a statistically significant moderator, the largest effect sizes being registered in the studies where Education represents the comparison condition, followed by No intervention, Self-help and SBT. Our results (Comşa et al., 2020), showed that it is also important who delivers the CBT interventions and point out this as a significant moderator. Larger effect sizes were found in studies where a multi-disciplinary team delivered the intervention compared to those where the intervention was delivered by a psychologist or by a counselor. Review of Shaw et al. (2005), also points out that CBT is more effective when it is delivered with exercise and diet.

The type of delivery (to group or individual), the mean age of the participants, the level of expertise of the person who delivers the intervention, and its cognitive component are not having any statistical impact on CBT interventions, implying that what is essential in CBT is to apply the protocols. Also, these results mean that we can efficiently apply CBT to all adult ages.

Regarding the participants, there is a significant negative association between weight loss and the mean BMI of the participants in CBT interventions, which suggests that the higher the BMI, the smaller the effect size of weight change.

Theories behind CBT for weight loss

The CBT intervention addresses psychological processes that interfere with weight loss and can be focused on beliefs about eating, body image acceptance, self-image, and problem-solving.

The most commonly used theories in CBT for weight loss are social cognitive theory (SCT) and self-determination theory (SDT) (Teixeira et al., 2010). The concept in SDT (Deci & Ryan, 1985) is that successful maintenance of weight loss would occur when people choose eating and exercise behaviors because they consider weight loss maintenance and its health benefits as a personal value. In other words, they will lose weight more comfortably and maintain changes successfully if their motivation for this is intrinsic (Silva et al., 2008). SCT (Bandura, 1986) suggests a triadic, reciprocal relationship between beliefs of self-competence (self-efficacy), environmental influences (social support), and behaviors. These elements serve as a foundation for interventions that address psychological factors implicated in overeating, adherence to healthy nutrition, and physical activity.

The research on CBT for weight loss

The review of Foreyt and Goodrick (1994) has identified that initial weight, self-efficacy, CBT attendance, self-monitoring, length of treatment, and social support are mediators and moderators of weight loss. In 2005, Elfhag and Rössner identified cognitive restraint as a mediator and internal motivation as a moderator of weight maintenance. The same study also identified that self-efficacy and self-confidence are both mediators and moderators of weight maintenance. These reviews delineated the mechanisms of CBT in weight loss but did not document the magnitude of their effects.

Shaw et al. (2005), assessed the effect of psychological interventions for weight loss and found that CBT significantly reduces weight when it is combined with diet and exercise, compared to diet and exercise alone. Also, they found significant differences in efficacy when CBT was compared with behavioral therapy.

Another important work is that of Jacob et al. (2018), who examined the effects of CBT for weight loss on weight, eating behaviors, and two emotional components (depression and anxiety), and found that CBT is an efficacious psychological intervention for increasing cognitive restraint and decreasing emotional eating. Regarding the depressive symptoms, CBT for weight loss did not perform better than control. In their metaanalysis, Jacob et al. (2018) examined the efficacy of CBT interventions for weight loss, eating behaviors and emotional factors such as anxiety and depression. They considered that CBT has proven to be effective for many treatments because it targets all the psychological factors (behavioral, emotional, and cognitive), not only the behavioral ones; therefore, it is essential to synthesize the evidence assessing these factors.

Our metaanalysis (Comşa et al., 2020) nicely complements the work of Jacob et al. (2018) and examines CBT's efficacy, including the cognitive factors together with the emotional and behavioral outcomes. The novelty of our work is that besides focusing on outcome analyses alone, and the efficacy or effectiveness of the intervention on weight or some psychological outcomes, we specifically looked at the psychological mechanisms of change for the outcomes. It is essential to know that an intervention is effective, but it is equally important to understand why. Our metaanalysis results were in line with all others and further documented that CBT had a significant effect at the end of the intervention on weight and also on cognitive outcomes, emotional outcomes and behavioral ones. At one-year follow-up, the results were maintained significant on all psychological outcomes, except for the emotional ones.

Association between proposed mechanisms of change and outcomes effect sizes

We found (Comşa et al., 2020) that the effect sizes for changes in cognitive factors are significantly associated with weight outcomes. From all the proposed cognitive mechanisms of change, results indicated a significant association only between motivation changes and self-efficacy changes and weight outcomes at the end of the intervention. These results are in line with other findings (Clifford, Tan, & Gorsuch, 1991; Jeffery et al., 1984; Linde, Rothman, Baldwin, & Jeffery, 2006; Palmeira et al., 2007; Teixeira et al., 2002, 2004, 2010; Williams, Grow, Freedman, Ryan, & Deci, 1996). We did not find other significant associations between other cognitive (body image and self-regulation) factors and weight outcomes.

Mechanisms of change

When we analyze weight-loss interventions, it is essential to identify psychological factors that can influence the results, called mechanisms of change (Teixeira et al., 2015). Because many studies identified motivation and self-efficacy as mediators or predictors of weight control in psychological interventions for weight management, we considered them as mechanisms of change in weight loss.

Many studies (Elfhag & Rossner, 2005; Teixeira et al., 2010, 2015) show psychological factors, such as autonomous motivation, self-efficacy, self-regulation skills, flexible eating restraint, and positive body image are mediators of weight loss. Although there are reviews that summarize findings of psychological weight-loss interventions, only a few studies reported cognitive components (e.g., motivation, self-efficacy, self-regulation, body image acceptance) as a primary or secondary outcome, besides quantitative assessments of weight change.

Because of this variety of outcomes reported by the studies, it is complicated to study them individually; therefore, we classified all the psychological outcomes into three general clusters, representing the psychological factors (behavioral, cognitive, and emotional) considered in psychological interventions (e.g., CBT).

In the Cognitive cluster, we included variables that describe cognitions. In the Emotional cluster, we included variables that describe the emotional affect, and the last cluster, the Behavioral one, represents variables that refer to lifestyle and eating behaviors. Then, we grouped the alleged cognitive mechanisms into three theoretically coherent groups: motivation, self-efficacy, and body image. This grouping was made based on the constructs measured and the description of the instruments used for their measurement. According to Bandura's SCT (Bandura, 1977), a person's self-efficacy expectations are the beliefs that he/she can successfully act. As a result, in the Self-efficacy group, we included outcomes that represent these beliefs. In the Motivation group, we put the outcomes that describe cognitions related to the activation and maintenance of a specific behavior. SDT (Ryan & Deci, 2000) describes the motivational constructs through people's needs to evolve and be part of a social scenario. Therefore, specific elements of SDT, like self-determination, self-motivation, intrinsic motivation, locus of causality, and social support, are factors they included in the Motivation group. Also, SDT addresses the factors that regulate the behavior that a person does to achieve their goals. These factors, we referred to as Self-regulation ones. The body image group contains variables that represent cognitions and concerns about their body and weight.

Protocols in CBTWL

In this chapter, we will describe the characteristics of one of the most used CBT for weight loss protocols in studies and clinical practice, namely that developed by Cooper and Fairburn (2001) at Oxford University. The authors designed this protocol considering the psychological obstacles to weight loss and its maintenance. The length of this protocol is 11 months, and the key issues are helping patients accept and value the achieved weight loss, encouraging the adoption of a goal of weight stability and not weight loss, and helping them acquire and then use the behavioral skills and cognitive responses required for successful weight control. We describe the phases and modules of the protocol in the table below (Table 2).

Limitations and conclusions

The field of CBT for weight loss has several limitations. First, because in the CBT interventions for weight loss is a great variety of instruments used to measure the psychological outcomes, it is hard to compare one intervention with another. Second, only a few studies analyze the mechanisms of change to include them in future protocols for increasing the efficiency of future CBT interventions for weight loss.

Research on CBT for weight loss showed that it is an efficient psychological treatment. Its efficacy lies in focusing on the cognitive factors together with the behavioral ones, more specifically increasing motivation and self-efficacy. We also know that CBT can be more effective when delivered by a multidisciplinary team in more extended programs since these two factors are significant moderators.

Practice and procedures: Outcome measures

Weight loss is not always the primary outcome of CBT interventions. Some of the interventions focus on changing beliefs about weight, shape, food, or weight loss and have weight loss as a secondary outcome. Weight loss, as an outcome of the intervention, can be defined as the change in weight, BMI, or waist circumference. Targeted psychological outcomes are very varied and are measured with many instruments. In Tables 3–6, we describe the most used ones in CBT interventions.

Mini-dictionary of terms

- *Self-efficacy.* Self-beliefs about one's capability to engage and produce a specific behavior.
- *Motivation.* It is a multidimensional construct that describes the psychological forces that activate a person toward a specific goal.

TABLE 2 Description of a CBT protocol (Cooper & Fairburn, 2001).

Phases	Modules
Phase one: 24–30 weeks The objectives are: 1. Weight loss 2. Addressing potential barriers to future weight maintenance	Module I—"Starting Treatment": - patient's assessment - the intervention is described
	Module II—"Establishing and Maintaining Weight Loss": - the introduction of the energy intake restriction to about 1500 kcal/day
	Module III—"Addressing Barriers to Weight Loss": - it runs in parallel with Module II - it focuses on identifying and addressing the possible problems which may interfere with adherence to the energy-restricted diet (motivational issues, inaccurate monitoring of food intake, poor food choice, excessive alcohol intake, frequent snacking, emotional eating, and binge eating)
	Module IV—"Eating Well": - the introduction of healthy eating while losing weight
	Module V—"Increasing Activity": - the establishment of a more active lifestyle in the context of weight maintenance.
	Module VI—"Body Image": - the assessment of concerns about shape - the promotion of a greater self-acceptance
	Module VII—"Weight Goals" - helping patients to consider the origins, significance, and possible arbitrariness of their weight goals, distinguishing these from their primary goals - helping them to value any changes in weight that are occurring
	Module VIII—"Primary Goals": - it focuses on helping patients address their primary goals directly
Phase Two: min. 14 weeks It is introduced at any point between weeks 24–30, according to the patient's progress **Objective:** Addressing weight stability and helping patients acquire the strategies and skills needed for long-term weight control	Module IX—"Weight Maintenance": - it addresses to create a new routine of regularly monitor weight - learning to use the appropriate cognitive responses related to vigilance at any significant weight fluctuation, increase self-efficacy, adopting flexible guidelines regarding eating - the practice of behavioral skills to act at any significant weight fluctuations

- **Body image.** A psychological construct that represents cognitions and concerns about body and weight.
- **Cognitive factors.** Any psychological factor related to cognitions.
- **Behavioral factors.** Any psychological factor related to behavior.
- **Emotional factors.** Any psychological factor related to emotional states.
- **Metaanalysis** Is a statistical analysis performed to assess the results of multiple scientific studies that address the same research question.

Key facts

Key facts of obesity

- The World Health Organization (WHO) defines obesity on its website as "abnormal or excessive fat accumulation that presents a health risk."
- BMI is commonly used to measure obesity and is calculated by dividing the body weight by the square of the height (kg/m^2).

TABLE 3 Measurement instruments for self-efficacy and motivation in weight loss.

Outcome	Assessment instrument
Self-efficacy	Weight efficacy lifestyle short form, WEL-SF (Ames, Heckman, Grothe, & Clark, 2012) measures self-efficacy in controlling eating behaviors under challenging situations. It has eight self-reported items, which are rated on a Likert scale rated from 0 = Not confident at all that I can resist overeating to 10 = Very confident that I can resist overeating). The total score of the measure is the sum of all items. A higher score reveals higher confidence in one's ability to control their eating behavior
	Exercise self-efficacy scale, ESE (Bandura, 2006), is an 18-item that measures the confidence in one's ability to exercise regularly (three or more times per week). The response is in a format ranging from 0 to 100. It has a high internal consistency with a Cronbach's alpha of 0.89
Motivation	Self-motivation inventory, SMI (Dishman & Ickes, 1981), was developed to assess self-motivation conceptualized as a behavioral tendency to persevere independently of situational reinforcements. Self-motivation is a stable trait and relatively insensitive to situational influence. It has 40 items, from which 19 are positively keyed ones, and 21 are negatively keyed. The answers are on a 5 point Likert format from 1 = Extremely uncharacteristic of me to 5 = Extremely characteristic of me. A final score is made by summing all the items, and a high score indicates high self-motivation. The instrument has high internal consistency and has Cronbach's alpha of 0.91 (Dishman & Ickes, 1981)
	Intrinsic motivation inventory, IMI (McAuley, Duncan, & Tammen, 1989), is a multidimensional measure that assesses the participants' subjective experience related to an exercise in four dimensions: interest/enjoyment, perceived competence, effort/importance, and pressure/tension. All of these dimensions are measured, each with four items. Before a total score is calculated, it has to calculate each of the four subscale scores and the average of all 16 items to obtain a single score indicating the overall level of exercise motivation (range 1–5), with higher scores indicating a more internal, self-regulated type of motivation

TABLE 4 Measurement instruments for other cognitive outcomes in weight loss.

Outcome	Assessment instrument
Body image	Body image acceptance and action questionnaire, BI-AAQ (Sandoz, Wilson, Merwin, & Kate Kellum, 2013), is a 12 item questionnaire that measures body image inflexibility/flexibility, which is defined as the capacity to experience the body-related ongoing perceptions, sensations, feelings, thoughts, and beliefs. It has a 7-point Likert scale assessing from 1 never true to 7 always true. Internal consistency is high (Cronbach's alpha = 0.93)
	Body image assessment questionnaire, BIA (Williamson, Davis, Bennett, Goreczny, & Gleaves, 1989) consists of nine cards with nine female silhouettes ranging from very thin to obese size, from which participants are asked to select the figures corresponding to their current (i.e., perceived actual body size) and their ideal body size. Body size dissatisfaction score is calculated by the difference between the perceived body size and the ideal body size rating. A higher discrepancy indicates higher levels of body size dissatisfaction
	Body shape questionnaire, BSQ (Cooper, Taylor, Cooper, & Fairbum, 1987) is a 34-item questionnaire that measures body shape preoccupations. A Likert-type scale from 1—never to 6—always is used to assess every item. The overall score is a total sum of 34. A higher score indicates more discomfort and dissatisfaction with body appearance. This instrument assesses desire to lose weight, body dissatisfaction, feelings of low self-worth in connection with weight and shape, feelings of fatness, self-consciousness about weight in public, and distressing thoughts about weight and shape. It is used for the identification and treatment of anorexia nervosa and bulimia nervosa (Cooper et al., 1987)
Self-esteem	Rosenberg self-esteem scale, RSE (Rosenberg, 1965) is a 10 item scale that evaluates global self-worth by measuring the positive and negative attitudes about oneself. It consists of 10 statements with Likert-style responses ranging from 1 (strongly disagree) to 4 (strongly agree). Item scores are summed and range from 10 to 40; Lower scores represent high negative self-esteem
Self-compassion	Self-compassion scale, SCS (Neff, 2003) is a 26 items questionnaire assessing self-compassion. The instrument has six subscales that measure three self-compassion components (self-kindness, self-judgment, common humanity, isolation, mindfulness, and over-identification). Items are rated on a 5 point Likert format scale from 1 = Almost never, to 5 = Almost always. It has good internal consistency ($\alpha = 0.92$; Neff, 2003)
Negative thoughts	Automatic thought questionnaire, ATQ (Hollon & Kendall, 1980), measures the frequency of the negative automatic thoughts about self-associated with depression. The items are assessed on a 5 points Likert scale, the frequency they experienced some negative thoughts in the last four weeks. The total score represents the frequency of these thoughts. A higher score means a higher frequency

TABLE 5 Measurement instruments for behavioral outcomes in weight loss.

Outcome	Assessment instrument
Eating behaviors	Three factor eating questionnaire, TFEQ-R21 (Cappelleri et al., 2009; Stunkard & Messick, 1985), is a 21-item instrument that measures three types of eating behavior: the cognitive restraint scale (6 items), the uncontrolled eating scale (9 items), and the emotional eating scale (6 items) Higher scores indicate higher levels/frequency of cognitive restraint, disinhibition, and emotional eating. The items are rated on a 4-point scale from1 being Completely true to 4 being Completely false. Item 21 is answered through an 8-point scale (1 = I eat everything I want, and when I want and 8 = I constantly confine my food intake. Higher scores indicate a higher tendency to engage in those eating behaviors.
	Dutch eating behavior questionnaire, DEBQ (van Strien, Frijters, Bergers, & Defares, 1986), assesses three dimensions of eating behavior: restrained, emotional, and external eating. It consists of 33 questions such as "Do you have a desire to eat when you are irritated/ smell delicious foods/etc...?" Answers were asses a 5-point Likert like scale, from 0 = never to 5 = very often
	Eating disorder examination-questionnaire, EDE-Q (Fairburn & Beglin, 1994). The EDE-Q identifies symptoms of eating disorders. The Restraint Scale (4 items) is a rough subjective measure of restraint over eating and the tendency to avoid meals and certain high-calorie foods. Higher scores indicate more efforts to restrict eating. The Eating Concern Scale (5 items) measures guilt and preoccupation with eating, fear of losing control over eating, concern about eating in front of other people, and secretive eating. Higher scores indicate more dysfunctional eating behaviors and attitudes

TABLE 6 Measurement instruments for emotional and quality of life outcomes in weight loss.

Outcome	Assessment instrument
Anxiety	The state-trait anxiety inventory, STAI (Spielberger, Gorsuch, & Lushene, 1970), is an instrument that measures traits and a state of anxiety. It has 20 items for measuring trait anxiety and 20 for state one. Answers are on a 4-point Likert scale from 1 = almost never to 4 = almost always. High scores indicate high anxiety. Internal consistency ranged from 0.86 to 0.95
Depression	Beck depression inventory 2, BDI-II (Beck, Steer, Ball, & Ranieri, 1996), is a self-administrated 21 items instrument built to assess the severity of depression in adults and children over 13 years. A total score is made by summing all the items' scores and the result can indicate a minimum, light, moderate or severe degree of depression
QOL	Obesity-related well-being questionnaire, ORWELL-97 (Mannucci et al., 1999), is an 18 items instrument that measures obesity-related quality-of-life. Items are rated on a four-point scale from 0 = not at all to 3 = much. Higher scores are indicating diminished quality of life. Orwell-97 has revealed good internal consistencies (Cronbach alpha was 0.83)

- According to WHO, an adult with a BMI between 25 and 30 is overweight, and over 30 is obese.
- In 2016, more than 1.9 billion adults were classified as having overweight, and of these, over 650 million adults were classified as having obesity.
- Obesity is a growing disease worldwide and has been declared a global epidemic (WHO, 2000).
- Overweight and obesity are associated with an increased risk of the following diseases: cardiovascular problems, type 2 diabetes, hypertension, vascular attack, and some forms of cancers (Cooper & Fairburn, 2001).

Key facts in CBTWL

- When cognitive techniques are added to behavior therapy, they improve intervention success and reduce weight regain (Cooper & Fairburn, 2001).
- Accepted strategies used in CBT for weight loss are cognitive restructuring, problem-solving, body image acceptance, self-efficacy, weight expectations, and working on maintaining the weight loss.
- CBTWL is an effective psychological intervention in weight management.

Applications to other areas

In this chapter, we have reviewed the efficacy, and the alleged mechanisms of change together with the characteristics of CBT interventions for weight loss in adults. The focus of these interventions is at the same time on behavioral changes and on cognitive ones. As we described, the cognitive aspects of CBT for weight loss are those considered to influence body weight, such as self-monitoring of negative thoughts, problem-solving, cognitive restructuring of negative beliefs about self, enhancing motivation, and self-efficacy. This work is transdiagnostic in CBT and is also common in interventions for other health problems such as depression, bulimia nervosa, binge eating, etc. So, CBTWL can also be used together with the interventions for other health problems as long as obesity is a comorbidity. We (Comşa et al., 2020) included in our metaanalysis studies that reported effects on emotional outcomes and found that CBTWL had a significant positive impact on them at the end of the interventions. Carels, Darby, Cacciapaglia, and Douglass (2004) used CBTWL to reduce cardiovascular risk factors in postmenopausal women. Bacon et al. (2002); Bacon, Stern, Van Loan, and Keim (2005) addressed their intervention to obese, chronic dieters, and Hales et al. (2016) to overweight and obese.

Regarding the weight of the participants, this intervention is also designed for healthy weight, overweight or obese (Bacon et al., 2002, 2005; Hales et al., 2016) with different health problems such as chronic dieters (Bacon et al., 2002, 2005), premenopausal women (Carels et al., 2004) to reduce the risk induced by these problems. CBTWL can be delivered in several ways: online via mobile applications (Hales et al., 2016), face-to-face in the workplace (Jamal, Moy, Azmi Mohamed, & Mukhtar, 2016), or in general practice (Munsch, Biedert, & Keller, 2003). Also, CBTWL can be delivered to children and adolescents, but this is not the subject of this chapter.

Summary points

- Results of the research on CBTWL indicated a significant association between motivation changes and self-efficacy changes and weight outcomes.
- Motivation and self-efficacy are considered alleged mechanisms of change in weight loss.
- Current evidence suggests that CBT is effective in weight loss.
- CBTWL interventions are more effective with components of self-efficacy and motivation
- CBTWL has a greater effect if a multidisciplinary team delivers it

References

Ames, G. E., Heckman, M. G., Grothe, K. B., & Clark, M. M. (2012). Eating self-efficacy: Development of a short-form WEL. *Eating Behaviors*, *13*(4). https://doi.org/10.1016/j.eatbeh.2012.03.013.

Bacon, L., Keim, N., Van Loan, M., Derricote, M., Gale, B., Kazaks, A., & Stern, J. (2002). Evaluating a 'non-diet' wellness intervention for improvement of metabolic fitness, psychological well-being and eating and activity behaviors. *International Journal of Obesity*, *26*(6), 854–865. https://doi.org/10.1038/sj.ijo.0802012.

Bacon, L., Stern, J. S., Van Loan, M. D., & Keim, N. L. (2005). Size acceptance and intuitive eating improve health for obese, female chronic dieters. *Journal of the American Dietetic Association*, *105*(6), 929–936. https://doi.org/10.1016/j.jada.2005.03.011.

Bandura, A. (1977). Self-efficacy: Toward a unifying theory of behavioral change. *Psychological Review*, *84*(2), 191–215. https://doi.org/10.1037/0033-295X.84.2.191.

Bandura, A. (1986). *Social foundations of thought and action: A social cognitive theory*. New Jersey: Prentice-Hall. 16(1).

Bandura, A. (2006). Guide for constructing self-efficacy scales. In *Self-efficacy beliefs of adolescents*. https://doi.org/10.1017/CBO9781107415324.004.

Beck, A. T., Steer, R. A., Ball, R., & Ranieri, W. F. (1996). Comparison of Beck depression inventories-IA and -II in psychiatric outpatients. *Journal of Personality Assessment*, *67*(3). https://doi.org/10.1207/s15327752jpa6703_13.

Byrne, S., Cooper, Z., & Fairburn, C. (2003). Weight maintenance and relapse in obesity: A qualitative study. *International Journal of Obesity*, *27*(8), 955–962. https://doi.org/10.1038/sj.ijo.0802305.

Cappelleri, J. C., Bushmakin, A. G., Gerber, R. A., Leidy, N. K., Sexton, C. C., Lowe, M. R., & Karlsson, J. (2009). Psychometric analysis of the three-factor eating questionnaire-R21: Results from a large diverse sample of obese and non-obese participants. *International Journal of Obesity*, *33*(6). https://doi.org/10.1038/ijo.2009.74.

Carels, R. A., Darby, L. A., Cacciapaglia, H. M., & Douglass, O. M. (2004). Reducing cardiovascular risk factors in postmenopausal women through a lifestyle change intervention. *Journal of Women's Health*, *13*(4), 412–426. https://doi.org/10.1089/154099904323087105.

Clifford, P. A., Tan, S. Y., & Gorsuch, R. L. (1991). Efficacy of a self-directed behavioral health change program: Weight, body composition, cardiovascular fitness, blood pressure, health risk, and psychosocial mediating variables. *Journal of Behavioral Medicine*, *14*(3), 303–323. https://doi.org/10.1007/BF00845457.

Comşa, L., & David, O. (2022). Relevant psychological factors in weight management. How to think and behave to lose weight and maintain it for good. *Journal of Rational Emotive and Cognitive-Behavior Therapy*. In press.

Comşa, L., David, O., & David, D. (2020). Outcomes and mechanisms of change in cognitive-behavioral interventions for weight loss: A meta-analysis of randomized clinical trials. *Behaviour Research and Therapy*, *132*. https://doi.org/10.1016/j.brat.2020.103654, 103654.

Cooper, Z., & Fairburn, C. G. (2001). A new cognitive behavioural approach to the treatment of obesity. *Behaviour Research and Therapy*, *39*(5), 499–511. https://doi.org/10.1016/S0005-7967(00)00065-6.

Cooper, P. J., Taylor, M. J., Cooper, Z., & Fairburn, C. G. (1987). The development and validation of the body shape questionnaire. *International Journal of Eating Disorders, 6*(4). https://doi.org/10.1002/1098-108X(198707)6:4<485::AID-EAT2260060405>3.0.CO;2-O.

Deci, E. L., & Ryan, R. M. (1985). Intrinsic motivation and self-determination in human behavior. In *Intrinsic motivation and self-determination in human behavior* Springer. https://doi.org/10.1007/978-1-4899-2271-7.

Dishman, R. K., & Ickes, W. (1981). Self-motivation and adherence to therapeutic exercise. *Journal of Behavioral Medicine, 4*(4). https://doi.org/10.1007/BF00846151.

Elfhag, K., & Rossner, S. (2005). Who succeeds in maintaining weight loss? A conceptual review of factors associated with weight loss maintenance and weight regain. *Obesity Reviews, 6*(1), 67–85. https://doi.org/10.1111/j.1467-789X.2005.00170.x.

Fairburn, C. G., & Beglin, S. J. (1994). Assessment of eating disorders: Interview or self-report questionnaire? *International Journal of Eating Disorders, 16*(4), 363–370. EATING DISORDER EXAMINATION QUESTIONNAIRE (EDE-Q 6.0)—Appendix.

Foreyt, J. P., & Goodrick, G. K. (1994). Attributes of successful approaches to weight loss and control. *Applied and Preventive Psychology, 3*(4), 209–215. https://doi.org/10.1016/S0962-1849(05)80095-2.

Hales, S., Turner-McGrievy, G. M., Wilcox, S., Fahim, A., Davis, R. E., Huhns, M., & Valafar, H. (2016). Social networks for improving healthy weight loss behaviors for overweight and obese adults: A randomized clinical trial of the social pounds off digitally (social POD) mobile app. *International Journal of Medical Informatics, 94*, 81–90. https://doi.org/10.1016/j.ijmedinf.2016.07.003.

Hollon, S. D., & Kendall, P. C. (1980). Cognitive self-statements in depression: Development of an automatic thoughts questionnaire. *Cognitive Therapy and Research, 4*(4). https://doi.org/10.1007/BF01178214.

Jacob, A., Moullec, G., Lavoie, K. L., Laurin, C., Cowan, T., Tisshaw, C., ... Bacon, S. L. (2018). Impact of cognitive-behavioral interventions on weight loss and psychological outcomes: A meta-analysis. *Health Psychology, 37*(5), 417–432. https://doi.org/10.1037/hea0000576.

Jamal, S. N., Moy, F. M., Azmi Mohamed, M. N., & Mukhtar, F. (2016). Effectiveness of a group support lifestyle modification (GSLiM) programme among obese adults in workplace: A randomised controlled trial. *PLoS One, 11*(8). https://doi.org/10.1371/journal.pone.0160343, e0160343.

Jeffery, R. W., Bjornson-Benson, W. M., Rosenthal, B. S., Lindquist, R. A., Kurth, C. L., & Johnson, S. L. (1984). Correlates of weight loss and its maintenance over two years of follow-up among middle-aged men. *Preventive Medicine, 13*(2), 155–168. https://doi.org/10.1016/0091-7435(84)90048-3.

Linde, J. A., Rothman, A. J., Baldwin, A. S., & Jeffery, R. W. (2006). The impact of self-efficacy on behavior change and weight change among overweight participants in a weight loss trial. *Health Psychology, 25*(3), 282–291. https://doi.org/10.1037/0278-6133.25.3.282.

Mannucci, E., Ricca, V., Barciulli, E., Di Bernardo, M., Travaglini, R., Cabras, P. L., & Rotella, C. M. (1999). Quality of life and overweight: The obesity related well-being (ORWELL 97) questionnaire. *Addictive Behaviors, 24*(3). https://doi.org/10.1016/S0306-4603(98)00055-0.

McAuley, E. D., Duncan, T., & Tammen, V. V. (1989). Psychometric properties of the intrinsic motivation inventory in a competitive sport setting: A confirmatory factor analysis. *Research Quarterly for Exercise and Sport, 60*(1). https://doi.org/10.1080/02701367.1989.10607413.

Munsch, S., Biedert, E., & Keller, U. (2003). Evaluation of a lifestyle change programme for the treatment of obesity in general practice. *Swiss Medical Weekly, 133*(9–10), 148–154. https://doi.org/2003/09/smw-10109.

Neff, K. D. (2003). The development and validation of a scale to measure self-compassion. *Self and Identity, 2*(3). https://doi.org/10.1080/15298860309027.

Palmeira, A. L., Teixeira, P. J., Branco, T. L., Martins, S. S., Minderico, C. S., Barata, J. T., ... Sardinha, L. B. (2007). Predicting short-term weight loss using four leading health behavior change theories. *International Journal of Behavioral Nutrition and Physical Activity, 4*(1), 14. https://doi.org/10.1186/1479-5868-4-14.

Rosenberg, M. (1965). Rosenberg self-esteem scale. *Journal of Religion and Health*. https://doi.org/10.1037/t01038-000.

Ryan, R. M., & Deci, E. L. (2000). Self-determination theory and the facilitation of intrinsic motivation, social development, and well-being. *American Psychologist*. https://doi.org/10.1037/0003-066X.55.1.68.

Sandoz, E. K., Wilson, K. G., Merwin, R. M., & Kate Kellum, K. (2013). Assessment of body image flexibility: The body image-acceptance and action questionnaire. *Journal of Contextual Behavioral Science, 2*(1–2). https://doi.org/10.1016/j.jcbs.2013.03.002.

Shaw, K. A., O'Rourke, P., Del Mar, C., & Kenardy, J. (2005). Psychological interventions for overweight or obesity. In K. A. Shaw (Ed.), *Cochrane database of systematic reviews* John Wiley & Sons, Ltd. https://doi.org/10.1002/14651858.CD003818.pub2.

Silva, M. N., Markland, D., Minderico, C. S., Vieira, P. N., Castro, M. M., Coutinho, S. R., ... Teixeira, P. J. (2008). A randomized controlled trial to evaluate self-determination theory for exercise adherence and weight control: Rationale and intervention description. *BMC Public Health, 8*(1), 234. https://doi.org/10.1186/1471-2458-8-234.

Spielberger, C. D., Gorsuch, R. L., & Lushene, R. E. (1970). *STAI manual for the state-trait anxiety inventory. Self-evaluation questionnaire*. Lushene Consulting Psychologists Press.

van Strien, T., Frijters, J. E. R., Bergers, G. P. A., & Defares, P. B. (1986). The Dutch eating behavior questionnaire (DEBQ) for assessment of restrained, emotional, and external eating behavior. *International Journal of Eating Disorders, 5*(2). https://doi.org/10.1002/1098-108X(198602)5:2<295::AID-EAT2260050209>3.0.CO;2-T.

Stunkard, A. J., & Messick, S. (1985). The three-factor eating questionnaire to measure dietary restraint, disinhibition and hunger. *Journal of Psychosomatic Research, 29*(1). https://doi.org/10.1016/0022-3999(85)90010-8.

Teixeira, P. J., Carraça, E. V., Marques, M. M., Rutter, H., Oppert, J.-M., De Bourdeaudhuij, I., ... Brug, J. (2015). Successful behavior change in obesity interventions in adults: A systematic review of self-regulation mediators. *BMC Medicine, 13*(1), 84. https://doi.org/10.1186/s12916-015-0323-6.

Teixeira, P., Going, S. B., Houtkooper, L. B., Cussler, E. C., Martin, C. J., Metcalfe, L. L., ... Lohman, T. G. (2002). Weight loss readiness in middle-aged women: Psychosocial predictors of success for behavioral weight reduction. *Journal of Behavioral Medicine, 25*(6), 499–523. https://doi.org/10.1023/A:1020687832448.

Teixeira, P. J., Palmeira, A. L., Branco, T. L., Martins, S. S., Minderico, C. S., Barata, J. T., ... Sardinha, L. B. (2004). Who will lose weight? A reexamination of predictors of weight loss in women. *International Journal of Behavioral Nutrition and Physical Activity, 1*(12). https://doi.org/10.1186/1479-5868-1-12.

Teixeira, P. J., Silva, M. N., Coutinho, S. R., Palmeira, A. L., Mata, J., Vieira, P. N., ... Sardinha, L. B. (2010). Mediators of weight loss and weight loss maintenance in middle-aged women. *Obesity, 18*(4), 725–735. https://doi.org/10.1038/oby.2009.281.

WHO. (2000). Obesity: Preventing and managing the global epidemic. In *World Health Organization: Technical report series*. WHO technical report series, no. 894.

Williams, G. C., Grow, V. M., Freedman, Z. R., Ryan, R. M., & Deci, E. L. (1996). Motivational predictors of weight loss and weight-loss maintenance. *Journal of Personality and Social Psychology, 70*(1), 115–126. https://doi.org/10.1037/0022-3514.70.1.115.

Williamson, D. A., Davis, C. J., Bennett, S. M., Goreczny, A. J., & Gleaves, D. H. (1989). Development of a simple procedure for assessing body image disturbances. *Behavioral Assessment, 11*(4), 433–446.

Wing, R. R., & Hill, J. O. (2001). Successful weight loss maintenance. *Annual Review of Nutrition, 21*(1), 323–341. https://doi.org/10.1146/annurev.nutr.21.1.323.

Wing, R. R., Lang, W., Wadden, T. A., Safford, M., Knowler, W. C., Bertoni, A. G., ... Wagenknecht, L. (2011). Benefits of modest weight loss in improving cardiovascular risk factors in overweight and obese individuals with type 2 diabetes. *Diabetes Care, 34*(7), 1481–1486. https://doi.org/10.2337/dc10-2415.

Chapter 4

Ethno-cognitive behavioral therapy and ethnopsychotherapy: A new narrative

Farooq Naeem
Department of Psychiatry, University of Toronto, Toronto, ON, Canada

Many countries in the Global North are becoming culturally diverse as a result of globalization. This lays a considerable onus on healthcare systems to address health disparities by providing equitable, culturally sensitive, appropriate, and effective clinical services relevant to the different population's cultural backgrounds. On the other side, in many countries in the Global South, such as Asia and Latin America, the improved economic condition and internet access to knowledge have expanded awareness of modern medicines and, as a result, of significant cultural difficulties (United Nations, 2018). In the Global South, increased interest in modern psychotherapies has resulted from economic progress and increased knowledge of mental health problems and related problems via social and electronic media and the internet. Earlier attempts to use modern psychotherapies with "cultural minorities" (our preferred term) in the Global North and the general population in the Global South have led to a growing awareness that modern psychotherapies require some adjustments when used with people in the Global South or cultural minorities in the Global North, because, unlike medical interventions, psychosocial interventions are based on the cultural values of those who develop them.

The ethnopsychotherapy and ethno-CBT

The idea that "inability to provide culturally competent and culturally adapted healthcare services leads to a disparity in services for people from minority cultures, which in turn leads to poor access to services, poor mental health outcomes, and increasing costs to society" underpins the drive to modify psychotherapies for the specific needs of cultural minorities (Kirmayer, 2012). An increasing population of cultural minorities, rising political awareness, and a persistent drive to enhance mental health outcomes have led to a push in healthcare systems to improve equity and equality in the Global North (Culyer & Wagstaff, 1993).

As a result of the aforementioned considerations, there has been a considerable push to research cultural variables in the delivery of psychological therapies. The study of psychotherapy across cultures and subcultures is known as ethnopsychotherapy (Naeem, Phiri, et al., 2015). As a result, ethno-CBT (cognitive behavior therapy) will be the branch of ethnopsychotherapy concerned with culture and CBT. This is a brand-new field that is still in its early phases of development.

Culturally adapted psychotherapies-current evidence

So far, more than 10 review articles and metaanalyses of culturally adapted psychosocial interventions have been published. The reviews reported effect sizes ranging from 0.23 to 0.75, with the majority reporting moderate to large effect sizes.

The majority of the evaluations looked at a wide range of mental health conditions and populations. One focused on schizophrenia (Degnan, Baker, Edge, & Drake, 2016), while five focused on depressive symptoms (Anik, West, Cardno, & Mir, 2021; Chowdhary et al., 2014; Rojas-García, Ruíz-Pérez, Gonçalves, Rodríguez-Barranco, & Ricci-Cabello, 2014; Rojas-García et al., 2015; van Loon, van Schaik, Dekker, & Beekman, 2013). One review focused on online and internet-based interventions (Spanhel et al., 2021). One review focused on Latinos (Sutton, 2015). The rest of the reviews included studies of different ethnic and cultural groups. Two reviews focused on youth (≤18 years) (Hodge, Jackson, & Vaughn, 2010; Huey Jr. & Polo, 2008). One study specifically focused on postnatal depression among women (Rojas-García et al., 2014). While most reviews attempted to characterize cultural adaptation, only two used a systematic method to define the nature and process in-depth (Chowdhary et al., 2014; Degnan et al., 2016). Two metaanalyses

(Rojas-García et al., 2014, 2015) focused on culturally adapted interventions for depressed persons from low socioeconomic status (Rojas-García et al., 2014, 2015).

These evaluations contained several flaws, including methodological heterogeneity, which included research with various designs and contexts (for example, intervention model, adaption process, ethnic/cultural group, tailored for majority versus minority groups, kind of control). Most crucially, the approach for selecting candidates varied from one assessment to the next. As a result, it's not unexpected that, despite the short period between the two evaluations, the number of studies fluctuated considerably from one to the next. Metaanalyses of research including subjects from various ethnic backgrounds with numerous issues who lived in different countries were frequently undertaken in reviews. Only one review reported outcomes in one ethnic group, Latinos (Sutton, 2015). Only two reviews (Chowdhary et al., 2014; Degnan et al., 2016) reported different psychotherapies in detail and grouped them based on their theoretical backgrounds together. Almost all of the papers, except for a few, combined publications that employed a variety of experimental designs.

In the same way, studies with different control conditions were combined together. At least some of these variables could be controlled for through sensitivity, subgroup, and moderator analyses. Unfortunately, subgroup analyses were not reported in the majority of metaanalyses. Part of the reason for the wide range of results and effect sizes indicated by these reviews (i.e., 0.23–0.75) could be due to methodological difficulties. The metaanalyses used a number of methodologies (e.g., aggregated mean, fixed-, and random-effect size models). It's not unexpected that most metaanalyses included the same papers because the literature in this field is scarce and the reviews aren't highly targeted. Finally, no economic evaluations were presented in any of the studies.

Most authors included a selection of studies published only in the English language. The authors reported a number of biases, such as the effect of risk biases, including publication bias, blinding of outcome assessors, and attrition bias. Publication bias was reported in some studies, but its influence on effect size was not calculated. Other factors reported influencing the effect were the choice of the control group and measurements by therapists versus self-reports and external observers.

Types and processes of cultural adaption

The limited evidence suggests that the most adaptations were made in the dimensions of language, context, concepts, family, communication, content, cultural norms and practices, context and delivery, therapeutic alliance, and treatment goals, even though not all studies focused on content and process of adaptation. (Chowdhary et al., 2014; Degnan et al., 2016). For example, in their review of adapted interventions for mental health problems, Griner and Smith (2006) cultural adaptation was classified based on whether the treatments included elements such as an explicit statement of culture, matched race or ethnicity between the client and the therapist, use of the client's preferred language, incorporation of cultural values and worldview into sessions, collaboration with cultural others, appropriately localized services, and relevant spirituality discussion. Chowdhary et al. (2014) evaluated cultural adaptation using the Medical Research Council framework (Craig et al., 2008) and Bernal and Saez-Sanriago's model (Bernal & Sáez-Santiago, 2006) to describe the nature of adaptation, while Degnan et al. (2016) used qualitative methodology to evaluate the nature of cultural adaptations. They assessed the process and components of adaption using sound methodology. The authors assessed the nature of cultural adaptations reported in the included research using a qualitative approach. The authors' nine themes after the adaptations were thematically examined were language, concepts, family, communication, content, cultural norms and practices, setting and delivery, therapeutic partnership, and treatment goals. However, both research took data from RCTs, which may have hampered the identification of themes because authors did not always record adaptations or characterized the adaptation process separately. This could have made it impossible to create a full picture.

Effect of cultural factors on therapy outcomes

Some metaanalyses tried to figure out what factors predicted a more substantial effect of culturally adapted therapies. These include aspects of the intervention, design, sample, and environment that could have influenced its effectiveness. For example, Rojas-García et al. (2015) reported that Individually administered interventions, interventions delivered in hospitals/clinics and at home, psychoeducation and interpersonal therapy-based interventions, and culturally appropriate interventions all had a more significant impact. Degnan et al. (2016) reported minimal variation in symptomatic effect depending on the type of intervention. However, interventions were more successful when patients attended with relatives rather than alone.

Multiple reviews (Griner & Smith, 2006; Smith, Rodríguez, & Bernal, 2011; Sutton, 2015) found age as a predictor of the effect of culturally adapted interventions. Of these, two reviews (Griner & Smith, 2006; Smith et al., 2011) reported that older age moderates the effect of interventions. Other studies (Griner & Smith, 2006; Hall, Ibaraki, Huang, Marti, & Stice, 2016) reported language match to influence the effectiveness of the interventions. However, Sutton (2015) failed to find the effect of language on effectiveness.

Several reviews (Griner & Smith, 2006; Hall et al., 2016; Rojas-García et al., 2015; Sutton, 2015) that explored the effect of ethnicity as a moderator found it to be an influential factor. Interestingly, some of the reviews (Griner & Smith, 2006; Rojas-García et al., 2015; Smith et al., 2011) also found that culturally adapted therapies are delivered to a homogeneous group of participants, they are more effective than when delivered to people of mixed ethnic or cultural backgrounds. One review (Rojas-García et al., 2015) also reported that culturally adapted interventions were twice as effective with Asian Americans compared to other cultural minority groups in the USA. However, another review (Degnan et al., 2016) reported a reverse effect in that these authors found interventions adapted for Chinese and majority populations to be less effective for schizophrenia than non-Chinese and minority populations.

Two metaanalyses (Benish & Quintana, 2011; van Loon et al., 2013) have delved deeper into the nature of adaptation and its impact on moderating the intervention effects. Benish and Quintana (2011) found that In comparison to culturally appropriate treatments without illness myth adaptation, illness myth adaptation exhibited a considerable moderating effect.. Similarly, van Loon et al. (2013) reported when cultural adaptation is combined with a focus on patients' cultural values, beliefs, and symptom presentation, it is more effective.. However, Sutton (Sutton, 2015) reported that The sort of cultural adaptation (surface, deep, or combination structure) appears to have no bearing on the outcome.

Three studies found no evidence of any factors having a moderating effect. Huey Jr. and Polo (2008) reported that ethnicity, problem type, clinical severity, diagnostic status, and culture-responsive treatment had no moderating influence. Similarly, Hodge et al. (2010) found no moderating effect of race or ethnicity. Finally, Sutton (2015) reported that acculturation, the study participants' immigration status, the major language of intervention, treatment style and format, and mental health construct categories did not affect the effectiveness of adapted interventions.

Need for ethnopsychotherapy

There is now sufficient evidence in the literature to show that cultural differences can influence the process of psychosocial interventions and thus the outcomes (Barrera, Castro, Strycker, & Toobert, 2013; Bhugra & Bhui, 1998; Edge et al., 2018; Sue, Zane, Nagayama Hall, & Berger, 2009). This isn't surprising, given that "most psychotherapy ideas were created by white males from the West," which "may collide with the cultural values and beliefs of clients from non-Western backgrounds." (Scorzelli & Reinke-Scorzelli, 1994).

Earlier authors on the cultural issues related to therapy, such as Laungani (2004) argued that individualism-communalism, cognitivism-emotionalism, free will-determinism, and materialism-spiritualism are four key value characteristics that distinguish Eastern and Western cultures. Asians, he noticed, are more likely to be community-oriented, employ logic less frequently, lean toward spiritual explanations, and have a deterministic view of life. While one can argue against this dichotomous view of the distinctions between eastern and western cultures, especially since globalization, improved economic situations in the east, and social and electronic media have narrowed the divide, cultural aspects remain important.

Culture and religion interact in a complicated way. Religion does, however, play a significant influence, as some religious people believe that a mental health problem is caused by a religious reason (Naeem, Phiri, Rathod, & Ayub, 2019). Some aspects of existence are predetermined in monotheistic religions, and this idea influences how people think. Hindus believe in a similar concept known as "karma." Buddhism, on the other hand, views life nonlinearly. It has been reported (Li et al., 2017) that Confucianism's teachings—filial piety, discouragement of self-centeredness, emphasis on academic achievement, and the importance of interpersonal harmony—have a huge impact on Chinese cultural values. Similarly, Taoism's teachings on a simple life, being connected with nature, and noninterference in the course of natural events, have a significant impact on the mental and emotional health of Chinese people.. Similarly, it has been pointed out that (Fierros & Smith, 2006) Many people of Latino descent hold strong religious views, which can have a significant impact on how they understand mental illness and its treatment, as well as how they perceive therapists. Fatalism, for example, is the concept that divine powers govern the world and that no human being can control or avert adversity. Similar beliefs have been reported among racialized communities in the Global North; A qualitative study was undertaken in Montreal to investigate the health beliefs of Caribbean people indicated a belief in the curative power of nonmedical interventions, most notably God and, to a lesser extent, traditional folk medicine (Whitley, 2016). Karenga's Nguzo Saba, which is made up of seven African humanistic principles (unity, self-determination, collective work and responsibility, cooperative economics, purpose, creativity, and faith), are African cultural values that contribute to the formation and reinforcement of community among African-Americans, and thus can have a significant impact on therapy. (Karenga & Karenga, 2007).

Culture

It is hard to define culture. There is a great deal of uncertainty about just how the word itself, let alone the phenomenon to which it refers, should be defined, understood and described. Terminology in this area is varied and confusing, and The terms "culture," "ethnic group," and "race," for example, are sometimes used interchangeably.

As a result, there are too many definitions of culture around. Triandis, for example, likes a broad-brush approach. Thus his definition encompasses both the physical and subjective parts of the environment in which we exist (Triandis & Brislin, 1984). The term culture is applied to all qualities of an individual's environment in a broad sense. However, according to Fernando, it usually refers to the person's nonmaterial features with other persons forming a social group. (Fernando, 2010). Others, such as Reber, describe culture as "the information system that codes how members of a structured organization, civilization, or nation interact with their social and physical environment," in contrast to this all-encompassing view (Reber, 1985).

According to anthropologists, "culture" comprises three elements: items or artifacts, ideas and knowledge, and behavioral patterns (Hendry & Underdown, 2012). However, we prefer the UNESCO definition of culture, which defines culture as "the set of distinctive spiritual, material, intellectual, and emotional features of society or a social group, that encompasses not only art and literature, but lifestyles, ways of living together, value systems, traditions, and beliefs." (UNESCO, 2001). This definition is accompanied by a UNESCO proclamation that welcomes cultural diversity, encourages cultural pluralism, defines cultural diversity, and emphasizes that cultural diversity entails human rights respect. Most crucially, this definition goes beyond race, religion, or nationality from the definition of culture. Minority cultures within a larger dominant culture are referred to as subcultures.

As a result, ethnicity, color, religion, gender, age, refugees and immigrants, and disabilities can all be used to create cultural or subcultural groups. The principles of cultural adaptation of evidence-based psychosocial therapies to people of other cultures are founded on the aforementioned values.

The ethno-CBT: Cultural adaptation of CBT

Cultural adaptation of CBT was initially defined as "making adaptations in how therapy is delivered, through the acquisition of awareness, information, and skills appropriate to a given culture, without compromising the theoretical underpinnings of CBT." (Naeem, 2012). As a result, cultural adaptation's primary goal is to "increase engagement with a client who does not share the therapist's cultural background." This definition, however, is limited to CBT. Bernal and colleagues (Bernal, Jiménez-Chafey, & Domenech Rodríguez, 2009) defined cultural adaptation of psychotherapies as "the systematic modification of an evidence-based treatment (EBT) or intervention protocol to consider language, culture, and the context in such a way that it is compatible with the client's cultural patterns, meanings, and values. Well-documented, systematic, and evaluated adaptations can help enhance research and practice. Similarly, Hall (2001) defined cultural adaptions as "the tailoring of psychotherapy to specific cultural contexts."

Some therapists in the United States have created recommendations for adapting psychosocial interventions based on their work with non-Western cultures, assuming that individuals from non-Western cultures may have distinct sets of beliefs, values, and perceptions. There are a few known guides or models of cultural adaptation: ecological validity model (Bernal, Bonilla, & Bellido, 1995; Bernal & Sáez-Santiago, 2006); Cultural Accommodation Model (Leong & Lee, 2006); model of essential elements (Podorefsky, Mcdonald-dowdell, & Beardslee, 2001); cultural adaptation process model (Domenech-Rodríguez & Wieling, 2005); data-driven adaptation; heuristic framework (Barrera Jr. & Castro, 2006); psychotherapy adaptation and modification model (Hwang, 2006); and adaptation model for American Indians (Whitbeck, 2006).

On the other hand, these first recommendations explain therapists' own experiences dealing with patients from racialized populations, provide therapy suggestions in general, and address broader philosophical, theoretical, and clinical concerns. Furthermore, none of the frameworks focused solely on CBT, and the guidelines did not directly derive from research addressing cultural variables.

Southampton adaptation framework for CBT

We developed the Southampton Adaptation Framework for CBT (Naeem, Ayub, Gobbi, & Kingdon, 2009) to culturally adapt CBT in 2009. Over the years, the framework evolved and has been described in detail in our recent publications in its current form (Naeem et al., 2016, 2019; Naeem, Phiri, et al., 2015). This framework has also been used to adapt CBT for depression and anxiety in Saudi Arabia (Alqahtani et al., 2019) and Morocco (Rhermoul, Naeem, Kingdon, Hansen,

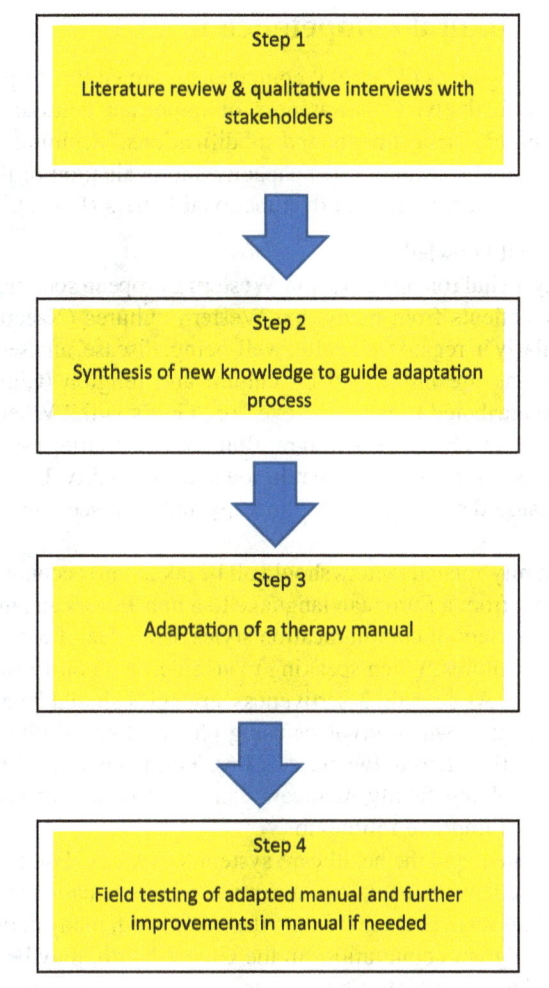

```
        ┌─────────────────────────────┐
        │          Step 1             │
        │ Literature review &         │
        │ qualitative interviews with │
        │ stakeholders                │
        └─────────────────────────────┘
                      ↓
        ┌─────────────────────────────┐
        │          Step 2             │
        │ Synthesis of new knowledge  │
        │ to guide adaptation process │
        └─────────────────────────────┘
                      ↓
        ┌─────────────────────────────┐
        │          Step 3             │
        │ Adaptation of a therapy     │
        │ manual                      │
        └─────────────────────────────┘
                      ↓
        ┌─────────────────────────────┐
        │          Step 4             │
        │ Field testing of adapted    │
        │ manual and further          │
        │ improvements in manual      │
        │ if needed                   │
        └─────────────────────────────┘
```

FIG. 1 Process of cultural adaptation here.

& Toufiq, 2017) and psychosis in China (Li et al., 2017) and for emotional dysregulation in learning disability in Canada (McQueen et al., 2018). Currently, the framework is being used in Canada to adapt CBT for depression and anxiety for the South Asian population (Naeem et al., 2021). We have used the framework to adapt and test CBT using an RCT design in various areas, such as depression (Naeem, Gul, et al., 2015), schizophrenia (Naeem, Saeed, et al., 2015), OCD (Aslam, Irfan, & Naeem, 2015), and self-harm (Husain et al., 2014) (Fig. 1).

Methodology for adaptation

We used a mixed-methods approach. We started with qualitative studies using open-ended interviews (Naeem, Gobbi, Ayub, & Kingdon, 2009; Naeem, Gobbi, Ayub, Kingdon, et al., 2010; Naeem, Ayub, Kingdon, & Gobbi, 2012) to gather the information that can be used to develop more or less precise guidelines for adaptation. Through time, these interviews have evolved into semistructured interviews that can be carried out by less experienced researchers with supervision (Li et al., 2017). Patients, their caregivers' therapists or mental health experts, and community leaders have all been involved in the past. The qualitative research looked into issues such as patients', carers', and community leaders' beliefs about a condition, its origins, and treatment, particularly nonmedical therapies, as well as patients' experiences with any nonpharmacological assistance they received. Professionals were interviewed about their experiences, obstacles, and, if any, solutions. A name-the-title technique (Naeem, Gobbi, et al., 2009) was used to find equivalent terminology rather than using literal translations. Following the adaptation of CBT in each area, a feasibility pilot study was often done to see if the modified therapy was acceptable. Finally, to determine the therapy's efficacy, a larger RCT was done.

The fundamental areas of cultural competence in CBT

According to our findings, the following areas of cultural competency termed the "triple-A principle" must be covered to adapt CBT for a specific culture effectively: (1) awareness of important cultural issues and treatment preparation; (2) assessment and engagement; and (3) "adjustments and modifications." Cultural awareness, in turn, refers to three important areas: (a) culture, religion, and spirituality; (b) capacity and evaluation of the healthcare system's capabilities and features; (c) the cultural context of cognitions and dysfunctional beliefs (Fig. 2).

1. Awareness of cultural and spiritual knowledge.
   A. Religion and spirituality play a vital role in many non-Western European societies. A biopsychosocial-spiritual paradigm of illness is used by patients from many non-Western cultures (Naeem, Phiri, et al., 2015). This model impacts their beliefs, particularly in regards to health, well-being, disease, and seeking help when they are in trouble. Beliefs on the cause-effect link are influenced by culture and religion (Cinnirella & Loewenthal, 1999). For example, a mishap could be attributed to the "evil eye" or "God's will." When faced with adversity, people frequently turn to religious coping mechanisms. (Bhugra, Bhui, & Rosemarie, 1999). On the other hand, many beliefs and stigmas around mental illness may be rooted in religion and spirituality. Understanding ideas about the causes of mental illness is critical because they can influence therapy and help-seeking options and pathways (Lloyd et al., 1998) (Fig. 3).

   Language, gender, and family-related factors should all be taken into account. Language obstacles must be taken into account. When translating from a European language to a non-European language, literal translations are ineffective. Cultures have a vast variety of communication styles. The idea of assertiveness, for example, is absent in many non-Western European cultures. When speaking to an elder in a non-European culture, assertiveness may be unacceptable in some instances. As a result, assertiveness strategies should be applied culturally sensitive, such as learning the "apologies technique," which involves using phrases like "With a big apology, before a person disagrees," etc. (Farooq Naeem, 2013; Hays & Iwamasa, 2006). You're dealing with the family in communal societies, not simply the individual. Involving family members can boost treatment involvement, ensure that homework assignments are completed, and improve follow-up.
   B. It is critical to evaluate the context and the healthcare system's capacity. For example, distance from the treatment facility may be a problem for patients in many low- and middle-income nations where psychiatric clinics are limited to larger cities (Li et al., 2017; Naeem et al., 2010). On the other hand, many patients in the Global South, as well as some from racialized and immigrant communities in the Global North, may be unaware of the healthcare system, mental health problems, available psychiatric treatments, and their likely outcomes, which can lead to underutilization of culturally adapted interventions if due consideration is not paid to these factors.

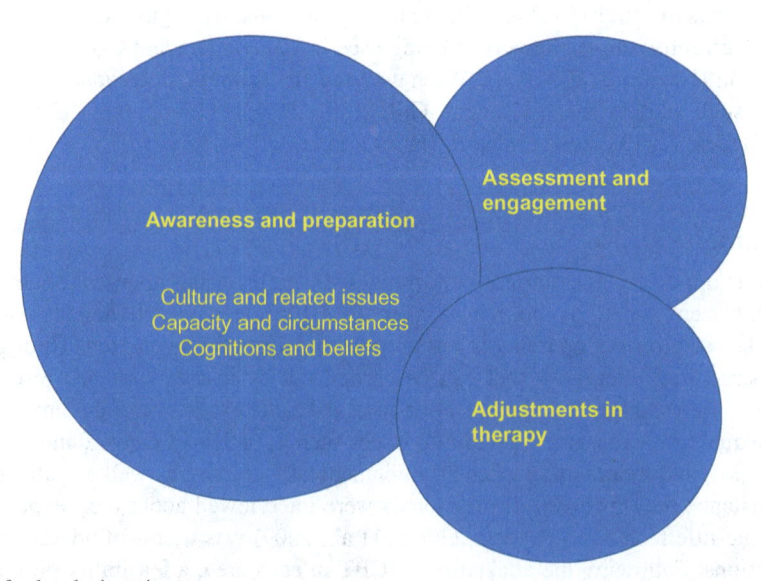

FIG. 2 Fundamental areas of cultural adaptation.

South Asians (Naeem)	South Asians in UK (Bheeka)	Chinese population in China (Li)
Psycho social stress or worry ++++ Poverty + loss of balance of mind ++ Overthinking ++ personality + **Biological** Hereditary ++ chemicals in brain ++ childbirth + phlegm + increased heat in liver + **Spiritual or religious causes** spirits, magic, taweeds, fear of hawai things (ghosts etc) ++ learning of spiritualism + evile eye + Gods will + **Other causes** Masturbation + **Don't know ++**	**Spiritual/religious causes** +++ **Psycho social** Stress ++ Interpersonal conflicts + **Biological** (4.4%). Dual explanatory models of psychosis (77.7%), combining prescribed medication and seeing a traditional faith healer as a treatment method **Don't know ++**	**Psycho social** Stress + problems in the society + Family related problems + Work or education related problems + Trauma + being mocked + **Biological** Over stimulation + Outbreak of emotions + Genetic factors + **Spiritual or religious causes** Air pollution + communication of ground of earth + PS. All patients had received spiritual treatment before coming to medical professionals **Don't know ++**

FIG. 3 Bio psycho socio spiritual model.

C. The content of Cognitive errors and dysfunctional beliefs vary from culture to culture.

There is at least some evidence from research from Turkey and Hong Kong, for example provides evidence of such variations (Sahin & Sahin, 1992; Tam et al., 2007). Typical beliefs that a Western European therapist could deem dysfunctional include the belief that one must rely on others, please others, submit to the demands of those in authority, and sacrifice one's own needs for the benefit of family; however, these beliefs are normal and in some instances even admirable in persons from the communal societies (Chen & Davenport, 2005; Farooq Naeem, 2013; Laungani, 2004) (Fig. 4).

2. Assessment and engagement

Being aware of the aforementioned cultural and religious challenges can aid in conducting a culturally sensitive evaluation and engaging customers from other cultures. The therapist can begin by starting with the patient's ideas about mental illness, healing, and healers. "What do you think is the cause of the illness?" "Who can treat these symptoms?" and "What is the optimal treatment for these symptoms?" are some of the questions that might help the therapist get understanding. Next, the therapist should inquire about the patient's spiritual and religious basis of symptoms. The therapist can then ask about the patient's experience and expectations of the healthcare system and their interactions with any healers (e.g., faith healers, religious healers, magicians, and herbalists). Because some patients with anxiety or depression also have bodily concerns, assessing somatic concerns is critical. Shame, guilt and stigma attached to mental illness within the clients'

Culture & related issues
- Cause and effect relationship (Bio-psycho-socio-spiritual model)
- Language and communication
- Family related issues

Cognitions & Beliefs
- Beliefs about health & illness
- Beliefs about treatment/healers
- Cognitive errors & dysfunctional beliefs

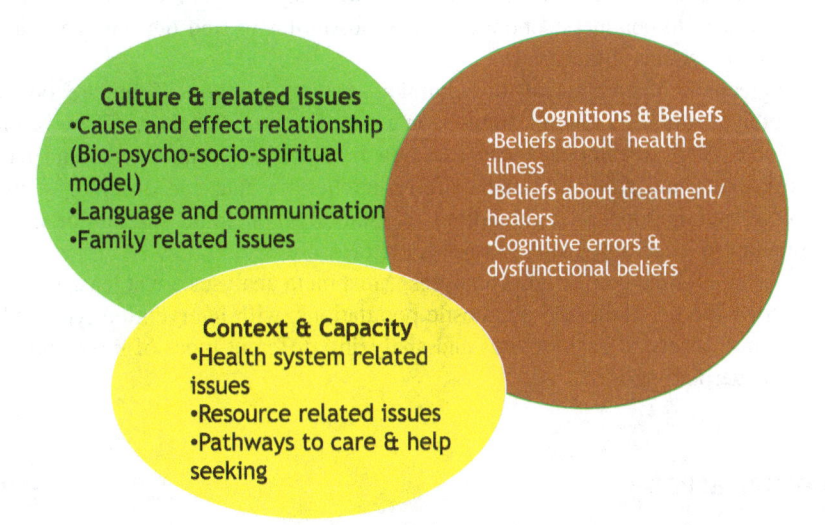

Context & Capacity
- Health system related issues
- Resource related issues
- Pathways to care & help seeking

FIG. 4 Awareness related issues.

context should be explored during the initial examination. Finally, structured assessments using validated tools can be used to assess the patient's beliefs about illness and its treatment (Lloyd et al., 1998) and their level of acculturation (Frey & Roysircar, 2004; Wallace, Pomery, Latimer, Martinez, & Salovey, 2010). Therapists should make themselves familiar with the biopsychosocial (Kinderman, 2014) and the spiritual models of illness (Verghese, 2008) to formulate clients' problems. Finally, teaching the patient and their family on the biopsychosocial causes of illness is an important aspect of the evaluation process.

Patients from non-Western cultures have been found to drop out of therapy at high rates (Rathod & Kingdon, 2014). Therefore, the initial few sessions are always crucial. Because some patients expect instant relief from bothersome symptoms, emphasizing symptom management at the start of therapy can increase participation and raise the patient's trust in the therapist.

3. Adjustments to the therapy

CBT techniques, in our opinion, do not require large modifications to function for people from non-Western European cultures. Minor changes, however, will be required. Problem solving, activity scheduling and behavioral experiments are examples of procedures that require just minor alterations. Non-Western patients prefer muscle relaxation and breathing techniques. Breathing exercises are popular in non-Western cultures and are part of many religious and spiritual traditions. Because most clients with depression and anxiety focus on physical symptoms, thought diaries can be employed with a column for physical symptoms. It takes a lot of effort to assist clients in recognizing their thoughts and emotions. Thus the therapist should take the time to explain these to them.

To begin, a more directive counseling style might be beneficial. A collaborative approach might be adopted as therapy progresses. A saint or guru who provides sermons is the non-Western form of spiritual and emotional healing, instead of professional teaching through "Socratic dialogue," which is valued in individualistic Western countries. If Socratic discussion is employed without adequate preparation, patients from non-Western cultures often feel uncomfortable.

In non-Western cultures, the use of stories to explain a point is frequent. Stories are used by experienced healers from non-Western cultural backgrounds to communicate their messages. When combined with visuals in handouts, stories can be quite engaging. Similarly, due to low literacy levels, homework compliance is low in many non-Western societies. The patient could be provided audio cassettes or bibliographic materials from the session and encouraged to count negative thoughts with beads or counters, commonly used in Asia and Africa, instead of pen and paper. Family members can also play an important part in assisting their homework and ensuring that they attend therapy sessions.

Future directions

Current evidence suggests that culturally adapted interventions are effective. However, this is an emerging field and therefore a lot of work needs to be done. There are a number of recommendations for future work in the field. There needs to be further research into the process of adaptation to find out what works. There is also a need for further studies evaluating the effectiveness of adapted interventions when compared with nonadapted interventions as opposed to usual care. Most importantly, there is a need to conduct economic evaluation studies to find out how beneficial culturally adapted interventions are compared with standard therapies.

Currently, there is no agreement on components of cultural adaptation that work. There are too many frameworks, and none expect the Southampton adaption framework focuses on a specific type of therapy. There is a need for a framework that is universally acceptable. Such a framework can use data from existing frameworks and can provide a common evidence-based framework that could guide clinicians and researchers in adapting different types of therapies. Future studies should also focus on potential moderators of cultural adaptation on outcome. It might be worth considering adding a checklist for culturally adapted interventions in consort guidelines.

There is also a need to conduct high quality meta analyses. Most meta analyses are at least 10 years old. Future metaanalyses should also focus on specific ethnic and diagnostic populations, with intervention types subanalyzed, instead of combining participants from a variety of backgrounds and analyzing different types of interventions together regardless of their varied theoretical underpinning.

Applications to other areas

Lessons learned from the study of cultural adaptation of CBT can be applied to other psychotherapies.

Key facts

1. There is sufficient evidence to suggest that culturally adapted CBT is effective.
2. Culturally adapted CBT can improve engagement and thus therapy outcomes.
3. CBT can be culturally adapted using qualitative methods and by engaging stakeholders.

Summary points

1. The ethnopsychotherapy or study of psychotherapy across cultures is an emerging field.
2. Modern psychotherapies were developed in western Europe and north America and therefore are underpinned by the dominant values of these cultures.
3. There is evidence to suggest that adapted psychotherapies have better outcomes.
4. A few framework for adapting psychotherapies exist, but none provides specific guidelines, and none is based in research.
5. The Southampton adaptation framework for CBT specifically focuses on CBT, is evidence based, and has been tested in numerous RCTs. However, it might not be applicable to other therapies.
6. There is a need for more research in this area, with fully powered RCTs, economic evaluations, and implementation data.

References

Algahtani, H. M. S., Almulhim, A., AlNajjar, F. A., Ali, M. K., Irfan, M., Ayub, M., & Naeem, F. (2019). Cultural adaptation of cognitive behavioural therapy (CBT) for patients with depression and anxiety in Saudi Arabia and Bahrain: A qualitative study exploring views of patients, carers, and mental health professionals. *The Cognitive Behaviour Therapist, 12*. https://doi.org/10.1017/S1754470X1900028X.

Anik, E., West, R. M., Cardno, A. G., & Mir, G. (2021). Culturally adapted psychotherapies for depressed adults: A systematic review and meta-analysis. *Journal of Affective Disorders, 278*, 296–310. https://doi.org/10.1016/j.jad.2020.09.051.

Aslam, M., Irfan, M., & Naeem, F. (2015). Brief culturally adapted cognitive behaviour therapy for obsessive compulsive disorder: A pilot study. *Pakistan Journal of Medical Sciences, 31*(4), 874–879. https://doi.org/10.12669/pjms.314.7385.

Barrera, M., Castro, F. G., Strycker, L. A., & Toobert, D. J. (2013). Cultural adaptations of behavioral health interventions: A progress report. *Journal of Consulting and Clinical Psychology, 81*(2), 196–205. https://doi.org/10.1037/a0027085.

Barrera, M., Jr., & Castro, F. G. (2006). A heuristic framework for the cultural adaptation of interventions. *Clinical Psychology: Science and Practice, 13*(4), 311–316. https://doi.org/10.1111/j.1468-2850.2006.00043.x.

Benish, S. G., & Quintana. (2011). Culturally adapted psychotherapy and the legitimacy of myth: A direct-comparison meta-analysis. *Journal of Counseling Psychology, 58*(3), 279–289. https://doi.org/10.1037/a0023626.

Bernal, G., Bonilla, J., & Bellido, C. (1995). Ecological validity and cultural sensitivity for outcome research: Issues for the cultural adaptation and development of psychosocial treatments with hispanics. *Journal of Abnormal Child Psychology, 23*(1), 67–82. https://doi.org/10.1007/BF01447045.

Bernal, G., Jiménez-Chafey, M. I., & Domenech Rodríguez, M. M. (2009). Cultural adaptation of treatments: A resource for considering culture in evidence-based practice. *Professional Psychology: Research and Practice, 40*(4), 361–368. https://doi.org/10.1037/a0016401.

Bernal, G., & Sáez-Santiago, E. (2006). Culturally centered psychosocial interventions. *Journal of Community Psychology, 34*(2), 121–132. https://doi.org/10.1002/jcop.20096.

Bhugra, D., & Bhui, K. (1998). Psychotherapy for ethnic minorities: Issues, context and practice. *British Journal of Psychotherapy, 14*(3), 310–326. https://doi.org/10.1111/j.1752-0118.1998.tb00385.x.

Bhugra, D., Bhui, K., & Rosemarie, M. D. (1999). Cultural identity and its measurement: A questionnaire for Asians. *International Review of Psychiatry, 11*(2–3), 244–249. https://doi.org/10.1080/09540269974438.

Chen, S. W.-H., & Davenport, D. S. (2005). Cognitive-behavioral therapy with Chinese American clients: Cautions and modifications. *Psychotherapy: Theory, Research, Practice, Training, 42*(1), 101–110. https://doi.org/10.1037/0033-3204.42.1.101.

Chowdhary, N., Jotheeswaran, A. T., Nadkarni, A., Hollon, S. D., King, M., Jordans, M. J. D., … Patel, V. (2014). The methods and outcomes of cultural adaptations of psychological treatments for depressive disorders: A systematic review. *Psychological Medicine, 44*(06), 1131–1146. https://doi.org/10.1017/S0033291713001785.

Cinnirella, M., & Loewenthal, K. M. (1999). Religious and ethnic group influences on beliefs about mental illness: A qualitative interview study. *British Journal of Medical Psychology, 72*(4), 505–524. https://doi.org/10.1348/000711299160202.

Craig, P., Dieppe, P., Macintyre, S., Michie, S., Nazareth, I., & Petticrew, M. (2008). Developing and evaluating complex interventions: The new Medical Research Council guidance. *BMJ, 337*. https://doi.org/10.1136/bmj.a1655, a1655.

Culyer, A. J., & Wagstaff, A. (1993). Equity and equality in health and health care. *Journal of Health Economics, 12*(4), 431–457.

Degnan, A., Baker, S., Edge, D., & Drake, R. (2016). The nature and efficacy of culturally-adapted psychosocial interventions for schizophrenia: A systematic review and meta-analysis. *Psychological Medicine, 48*, 714–727.

Domenech-Rodríguez, M., & Wieling, E. (2005). Developing culturally appropriate, evidence-based treatments for interventions with ethnic minority populations. In *Voices of color: First-person accounts of ethnic minority therapists* (pp. 313–333). Sage Publications, Inc. https://doi.org/10.4135/9781452231662.n18.

Edge, D., Degnan, A., Cotterill, S., Berry, K., Baker, J., Drake, R., & Abel, K. (2018). *Culturally adapted family intervention (CaFI) for African-Caribbean people diagnosed with schizophrenia and their families: A mixed-methods feasibility study of development, implementation and acceptability.* NIHR Journals Library. http://www.ncbi.nlm.nih.gov/books/NBK525363/.

Farooq Naeem, M. A. (2013). *Culturally adapted CBT (CaCBT) for depression, therapy manual for use with South Asian Muslims [Kindle Edition]*.

Fernando, S. (2010). *Mental health, race and culture* (3rd ed.). Red Globe Press.

Fierros, M., & Smith, C. (2006). The relevance of hispanic culture to the treatment of a patient with posttraumatic stress disorder (PTSD). *Psychiatry (Edgmont), 3*(10), 49–56.

Frey, L. L., & Roysircar, G. (2004). Effects of acculturation and worldview for white American, south American, south Asian, and southeast Asian students. *International Journal for the Advancement of Counselling, 26*(3), 229–248. https://doi.org/10.1023/B:ADCO.0000035527.46652.d2.

Griner, D., & Smith, T. B. (2006). Culturally adapted mental health intervention: A meta-analytic review. *Psychotherapy: Theory, Research, Practice, Training, 43*(4), 531–548. https://doi.org/10.1037/0033-3204.43.4.531.

Hall, G. C. N. (2001). Psychotherapy research with ethnic minorities: Empirical, ethical, and conceptual issues. *Journal of Consulting and Clinical Psychology, 69*(3), 502–510. https://doi.org/10.1037/0022-006X.69.3.502.

Hall, G. C. N., Ibaraki, A. Y., Huang, E. R., Marti, C. N., & Stice, E. (2016). A Meta-analysis of cultural adaptations of psychological interventions. *Behavior Therapy, 47*(6), 993–1014. https://doi.org/10.1016/j.beth.2016.09.005.

Hays, P. A., & Iwamasa, G. Y. (Eds.). (2006). *Culturally responsive cognitive-behavioral therapy: Assessment, practice, and supervision* American Psychological Association.

Hendry, J., & Underdown, S. (2012). *Anthropology: A beginner's guide*. Oneworld Publications.

Hodge, D. R., Jackson, K. F., & Vaughn, M. G. (2010). Culturally sensitive interventions and health and behavioral health youth outcomes: A meta-analytic review. *Social Work in Health Care, 49*(5), 401–423. https://doi.org/10.1080/00981381003648398.

Huey, S., Jr., & Polo, A. J. (2008). Evidence-based psychosocial treatments for ethnic minority youth. *Journal of Clinical Child & Adolescent Psychology, 37*(1), 262–301. https://doi.org/10.1080/15374410701820174.

Husain, N., Afsar, S., Ara, J., Fayyaz, H., Rahman, R. U., Tomenson, B., … Chaudhry, I. B. (2014). Brief psychological intervention after self-harm: Randomised controlled trial from Pakistan. *The British Journal of Psychiatry, 204*(6), 462–470. https://doi.org/10.1192/bjp.bp.113.138370.

Hwang, W.-C. (2006). The psychotherapy adaptation and modification framework: Application to Asian Americans. *American Psychologist, 61*(7), 702–715. https://doi.org/10.1037/0003-066X.61.7.702.

Karenga, M., & Karenga, T. (2007). The Nguzo Saba and the black family: Principles and practices of well-being and flourishing. In *Black families* (4th ed., pp. 7–28). SAGE Publications, Inc. https://doi.org/10.4135/9781452226026.

Kinderman, P. (2014). Get the message right: A psychosocial model of mental health and well-being. In P. Kinderman (Ed.), *A prescription for psychiatry: Why we need a whole new approach to mental health and wellbeing* (pp. 30–47). UK: Palgrave Macmillan. https://doi.org/10.1057/9781137408716_2.

Kirmayer, L. J. (2012). Rethinking cultural competence. *Transcultural Psychiatry, 49*(2), 149–164. https://doi.org/10.1177/1363461512444673.

Laungani, P. (2004). *Asian perspectives in counselling and psychotherapy*. Psychology Press.

Leong, F. T., & Lee, S.-H. (2006). A cultural accommodation model for cross-cultural psychotherapy: Illustrated with the case of Asian Americans. *Psychotherapy: Theory, Research, Practice, Training, 43*(4), 410–423. https://doi.org/10.1037/0033-3204.43.4.410.

Li, W., Zhang, L., Luo, X., Liu, B., Liu, Z., Lin, F., … Naeem, F. (2017). A qualitative study to explore views of patients', carers' and mental health professionals' to inform cultural adaptation of CBT for psychosis (CBTp) in China. *BMC Psychiatry, 17*. https://doi.org/10.1186/s12888-017-1290-6.

Lloyd, K. R., Jacob, K. S., Patel, V., St Louis, L., Bhugra, D., & Mann, A. H. (1998). The development of the short explanatory model interview (SEMI) and its use among primary-care attenders with common mental disorders. *Psychological Medicine, 28*(5), 1231–1237.

van Loon, A., van Schaik, A., Dekker, J., & Beekman, A. (2013). Bridging the gap for ethnic minority adult outpatients with depression and anxiety disorders by culturally adapted treatments. *Journal of Affective Disorders, 147*(1–3), 9–16. https://doi.org/10.1016/j.jad.2012.12.014.

McQueen, M., Blinkhorn, A., Broad, A., Jones, J., Naeem, F., & Ayub, M. (2018). Development of a cognitive behavioural therapy-based guided self-help intervention for adults with intellectual disability. *Journal of Applied Research in Intellectual Disabilities, 31*(5), 885–896. https://doi.org/10.1111/jar.12447.

Naeem, F. (2012). *Adaptation of cognitive behaviour therapy for depression in Pakistan*. Lap Lambert Academic Publishing GmbH KG.

Naeem, F., Ayub, M., Gobbi, M., & Kingdon, D. (2009, December). Development of Southampton adaptation framework for CBT (SAF-CBT): A framework for adaptation of CBT in non-western culture. *Journal of Pakistan Psychiatric Society*. http://www.pakmedinet.com/15940.

Naeem, F., Ayub, M., Kingdon, D., & Gobbi, M. (2012). Views of depressed patients in Pakistan concerning their illness, its causes, and treatments. *Qualitative Health Research*. https://doi.org/10.1177/1049732312450212.

Naeem, F., Gobbi, M., Ayub, M., & Kingdon, D. (2009). University students' views about compatibility of cognitive behaviour therapy (CBT) with their personal, social and religious values (a study from Pakistan). *Mental Health, Religion and Culture, 12*(8), 847–855. https://doi.org/10.1080/13674670903115226.

Naeem, F., Gobbi, M., Ayub, M., Kingdon, D., et al. (2010). Psychologists experience of cognitive behaviour therapy in a developing country: A qualitative study from Pakistan. *International Journal of Mental Health Systems, 4*, 2.

Naeem, F., Gul, M., Irfan, M., Munshi, T., Asif, A., Rashid, S., … Ayub, M. (2015). Brief culturally adapted CBT (CaCBT) for depression: A randomized controlled trial from Pakistan. *Journal of Affective Disorders, 177*, 101–107. https://doi.org/10.1016/j.jad.2015.02.012.

Naeem, F., Phiri, P., Munshi, T., Rathod, S., Ayub, M., Gobbi, M., & Kingdon, D. (2015). Using cognitive behaviour therapy with South Asian Muslims: Findings from the culturally sensitive CBT project. *International Review of Psychiatry*, *27*(3), 233–246. https://doi.org/10.3109/09540261.2015.1067598.

Naeem, F., Phiri, P., Nasar, A., Munshi, T., Ayub, M., & Rathod, S. (2016). An evidence-based framework for cultural adaptation of cognitive behaviour therapy: Process, methodology and foci of adaptation. *World Cultural Psychiatry Research Review*, *11*, 67–70.

Naeem, F., Phiri, P., Rathod, S., & Ayub, M. (2019). Cultural adaptation of cognitive–behavioural therapy. *BJPsych Advances*, 1–9. https://doi.org/10.1192/bja.2019.15.

Naeem, F., Saeed, S., Irfan, M., Kiran, T., Mehmood, N., Gul, M., … Kingdon, D. (2015). Brief culturally adapted CBT for psychosis (CaCBTp): A randomized controlled trial from a low income country. *Schizophrenia Research*. https://doi.org/10.1016/j.schres.2015.02.015.

Naeem, F., Tuck, A., Mutta, B., Dhillon, P., Thandi, G., Kassam, A., … McKenzie, K. (2021). Protocol for a multi-phase, mixed methods study to develop and evaluate culturally adapted CBT to improve community mental health services for Canadians of south Asian origin. *Trials*, *22*(1), 600. https://doi.org/10.1186/s13063-021-05547-4.

Podorefsky, D. L., Mcdonald-dowdell, M., & Beardslee, W. R. (2001). Adaptation of preventive interventions for a low-income, culturally diverse community. *Journal of the American Academy of Child & Adolescent Psychiatry*, *40*(8), 879–886. https://doi.org/10.1097/00004583-200108000-00008.

Rathod, S., & Kingdon, D. (2014). Case for cultural adaptation of psychological interventions for mental healthcare in low and middle income countries. *BMJ*, *349*. https://doi.org/10.1136/bmj.g7636, g7636.

Reber, A. S. (1985). *The penguin dictionary of psychology*. Penguin Books.

Rhermoul, F.-Z. E., Naeem, F., Kingdon, D., Hansen, L., & Toufiq, J. (2017). A qualitative study to explore views of patients, carers and mental health professionals' views on depression in Moroccan women. *International Journal of Culture and Mental Health*, 1–16. https://doi.org/10.1080/17542863.2017.1355397.

Rojas-García, A., Ruíz-Pérez, I., Gonçalves, D. C., Rodríguez-Barranco, M., & Ricci-Cabello, I. (2014). Healthcare interventions for perinatal depression in socially disadvantaged women: A systematic review and meta-analysis. *Clinical Psychology: Science and Practice*, *21*(4), 363–384. https://doi.org/10.1111/cpsp.12081.

Rojas-García, A., Ruiz-Perez, I., Rodríguez-Barranco, M., Gonçalves Bradley, D. C., Pastor-Moreno, G., & Ricci-Cabello, I. (2015). Healthcare interventions for depression in low socioeconomic status populations: A systematic review and meta-analysis. *Clinical Psychology Review*, *38*, 65–78. https://doi.org/10.1016/j.cpr.2015.03.001.

Sahin, N. H., & Sahin, N. (1992). How dysfunctional are the dysfunctional attitudes in another culture? *The British Journal of Medical Psychology*, *65*(Pt 1), 17–26.

Scorzelli, J. F., & Reinke-Scorzelli, M. (1994). Cultural sensitivity and cognitive therapy in India. *The Counseling Psychologist*, *22*(4), 603–610. https://doi.org/10.1177/0011000094224006.

Smith, T. B., Rodríguez, M. M. D., & Bernal, G. (2011). Culture. In *Psychotherapy relationships that work* (2nd ed.). Oxford University Press. https://doi.org/10.1093/acprof:oso/9780199737208.003.0016.

Spanhel, K., Balci, S., Feldhahn, F., Bengel, J., Baumeister, H., & Sander, L. B. (2021). Cultural adaptation of internet- and mobile-based interventions for mental disorders: A systematic review. *NPJ Digital Medicine*, *4*(1), 1–18. https://doi.org/10.1038/s41746-021-00498-1.

Sue, S., Zane, N., Nagayama Hall, G. C., & Berger, L. K. (2009). The case for cultural competency in psychotherapeutic interventions. *Annual Review of Psychology*, *60*, 525–548. https://doi.org/10.1146/annurev.psych.60.110707.163651.

Sutton, C. (2015). *Culturally adapted interventions for Latinos: A meta-analysis*. School of Graduate Psychology. http://commons.pacificu.edu/spp/1134.

Tam, P. W. C., Wong, D. F. K., Chow, K. K. W., Ng, F. S., Ng, R. M. K., Cheung, M. S. M., … Mak, A. D. P. (2007). Qualitative analysis of dysfunctional attitudes in Chinese persons suffering from depression. *Hong Kong Journal of Psychiatry*, *17*(4), 109.

Triandis, H. C., & Brislin, R. W. (1984). Cross-cultural psychology. *American Psychologist*, *39*(9), 1006–1016. https://doi.org/10.1037/0003-066X.39.9.1006.

UNESCO. (2001). *UNESCO universal declaration on cultural diversity*. http://portal.unesco.org/en/ev.php-URL_ID=13649&URL_DO=DO_TOPIC&URL_SECTION=-471.html.

United Nations. (2018, January 8). *January 2018 briefing on the world economic situation and prospects*. Development Policy & Analysis Division, Dept of Economic & Social Affairs, United Nations. https://www.un.org/development/desa/dpad/publication/world-economic-situation-and-prospects-january-2018-briefing-no-110/.

Verghese, A. (2008). Spirituality and mental health. *Indian Journal of Psychiatry*, *50*(4), 233–237. https://doi.org/10.4103/0019-5545.44742.

Wallace, P. M., Pomery, E. A., Latimer, A. E., Martinez, J. L., & Salovey, P. (2010). A review of acculturation measures and their utility in studies promoting Latino health. *Hispanic Journal of Behavioral Sciences*, *32*(1), 37–54.

Whitbeck, L. B. (2006). Some guiding assumptions and a theoretical model for developing culturally specific preventions with native American people. *Journal of Community Psychology*, *34*(2), 183–192. https://doi.org/10.1002/jcop.20094.

Whitley, R. (2016). Ethno-racial variation in recovery from severe mental illness: A qualitative comparison. *Canadian Journal of Psychiatry*, *61*(6), 340–347.

Chapter 5

Cognitive-behavioral therapy and cancer survival

Špela Miroševič[a] and Zalika-Klemenc Ketiš[a,b,c]

[a]*Department of Family Medicine, Faculty of Medicine Ljubljana, University of Ljubljana, Ljubljana, Slovenia,* [b]*Department of Family Medicine, Faculty of Medicine Maribor, University of Maribor, Maribor, Slovenia,* [c]*Primary Healthcare Research and Development Institute, Community Health Centre, Ljubljana, Slovenia*

Abbreviations

CBT   cognitive-behavioral therapy
MBSR   mindfulness-based therapy
RCT   randomized-controlled trial
SEGT   supportive-expressive group therapy

Introduction

"Does living better also mean living longer?" This is a difficult question that addresses the psychosocial aspect of humanity and a question that pertains to a better quality of life and longer survival. Substantial interest in the topic has been generated since Spiegel, Bloom, Kraemer, and Gottheil (1989) reported on an unexpected survival benefit of group psychosocial support in women with metastatic breast cancer. Intensive group psychotherapy was suggested to result not only in improved mood, better coping, and reduced pain, but also in longer survival (Spiegel et al., 1989). Moreover, the same suggestion was observed in cancer patients themselves. The study reported that 85% of cancer patients believe that psychological functioning predicts disease progression (Lemon, Zapka, & Clemow, 2004). Therefore, it became essential to explore these ideas and figure out the conclusions of such interventions, especially in terms of what works and what does not work on the psychosocial and survival outcomes of such investigations.

This section begins with exploring the possible biological pathways with which psychosocial intervention can prolong survival. Furthermore, we will briefly describe which psychosocial interventions are available for cancer patients and then focus on CBT and its effects on cancer survival. We raise questions such as—does every cancer patient benefit from the intervention? In what stages is specific intervention more beneficial? Do timing and length of intervention influence prolonged survival and psychological factors in cancer patients?

Linking psychosocial morbidity to cancer progression

Up to one third of all newly diagnosed cancer patients experience depression or other psychiatric morbidity related to diagnosis and treatment (Singer et al., 2009) and remain symptomatic up to 6 years later (Grassi & Rosti, 1996). The prevalence of psychological morbidity in cancer patients increases with cancer severity and symptoms such as pain, fatigue, and malaise (Massie, 2004). However, the determination of clinically significant morbidity can be difficult to assess, especially with advanced cancers, due to similar somatic and psychological symptoms and the characteristics of advanced disease (Bottomley, 1998; Lloyd-Williams & Friedman, 1999).

A biopsychosocial model of cancer survival suggests that modifiable psychosocial factors, such as depressive symptoms, low social support, and psychological stress may increase cancer risk and/or progression (Aizer et al., 2013; Kroenke, Kubzansky, Schernhammer, Holmes, & Kawachi, 2006; Spiegel & Giese-Davis, 2003). There is divided, but strong evidence that depression predicts cancer progression and mortality (Dalton, Mellemkjaer, Olsen, Mortensen, & Johansen, 2002; Penninx et al., 1998; Shekelle et al., 1981; Stommel, Given, & Given, 2002; Watson & Richardson, 1999). Furthermore, psychosocial support has been reported to reduce depression, anxiety, and pain, and may even increase the survival time with cancer; however, studies of the latter have not been converged yet (Fawzy et al., 1993; Goodwin et al., 2001; Kuchler et al., 1999; Richardson, Zarnegar, Bisno, & Levine, 1990; Spiegel et al., 1989).

Psychophysiological mechanisms linking depression and cancer progression include dysregulation of the HPA axis, especially daily imbalance in cortisol (Sephton et al., 2009) and melatonin (Hansen et al., 2012). Depression also affects different parts of the immune function that may affect cancer surveillance (Grivennikov, Greten, & Karin, 2010). Moreover, depressed patients tend to be less assertive in demanding more thorough medical evaluation and maintain worse self-care. The bidirectional relationship between cancer and depression can, therefore, make new options for therapeutic intervention on psychophysiological outcomes (Spiegel & Giese-Davis, 2003).

Psychosocial interventions for cancer patients

Psychosocial intervention is defined as any intervention that focuses on the treatment of psychological or social factors rather than biological ones (Ruddy & House, 2005). The definition is broad and allows for an inclusion of various methods, including learning coping skills, promoting social support, encouraging emotional expression, education and methods focusing on existential questions and spirituality. We have reviewed the currently available psychosocial interventions (see below) and have identified the following therapeutical methods (see Fig. 1) supporting the patient, teaching coping skills and psychoeducation, encouraging emotion expression and interpersonal exploration, and exploring existential questions.

Supportive-expressive group therapy (SEGT) is a long-term group therapy, usually lasting 90 min every week. It involves 3–15 participants and is led by two therapists. It is defined as a semistructured treatment and involves the creation of supportive and emotionally expressive therapy and mostly includes existential questions. Moreover, participants are encouraged to learn coping skills (to confront their problems and decrease anxiety and pain, mostly by hypnosis), to promote and improve social support, enhance meaning (confrontation of dying and death, reordering priorities) and to enhance social support and improve communication with physicians (Spiegel et al., 2007).

A more standardized treatment is mindfulness-based therapy (MBSR), whose purpose is to enhance support and help patients in distress. It is especially successful in preventing depression and helping people who are in pain and suffer from a range of conditions and life issues that were initially difficult to treat in hospital settings (Carlson, Speca, Patel, & Goodey, 2003; Speca, Carlson, Goodey, & Angen, 2000; Wurtzen et al., 2013).

Supportive intervention focuses on discussing feelings, concerns and problems. The group usually has neither specific, preplanned agenda nor a set of structured exercises. The therapist serves as a facilitator, pointing out common issues that underlie individual problems (e.g., helplessness, sense of loss of control) and encourages participation by all group members (Telch & Telch, 1986; Temel et al., 2010).

FIG. 1 Identified methods used in the psychosocial interventions for cancer patients. This figure presents careful identification of each method that has been reported in psychosocial interventions used in studies, exploring psychosocial interventions and psychosocial benefits/survival of cancer patients.

	SEGT	CBT	MBT	PE	Supportive
Support	+	+/-	+	+/-	+
Coping skills	+/-	+/-	+	+/-	-
Psychoeducation	+/-	+/-	-	+	+/-
Emotion expression	+	-	-	-	-
Interpersonal exploration	+/-	+/-	-	-	+/-
Existential questions	+/-	-	+/-	-	-

Notes. Figure 1: + means specific method is present; +/- means specific method may be present or not; - means specific method is not present.

In the majority of the reviewed studies, psychoeducation is reported as a "control group," or it is added to the more known interventions (CBT, SEGT), and only a few studies used it as an independent intervention (Dolbeault et al., 2009; Jacobs, Ross, Walker, & Stockdale, 1983; Salzer et al., 2010). Cancer patients would learn how to cope with the illness and also about the changes that occur during and after treatment. In addition, psychoeducation improves treatment adherence and efficacy (Golant, Altman, & Martin, 2003).

A reasonably brief cognitive-behavioral therapy (CBT) is defined as an action-oriented form of psychosocial therapy. Important modules used in CBT are psychoeducation, relaxation and stress management, enhanced communication, assertiveness, problem-solving and constructive thinking (Edelman, Bell, & Kidman, 1999; Fawzy et al., 1993; Telch & Telch, 1986). Now, we will look at this approach more thoroughly.

Description of methods

CBT includes learning about new coping skills, problem-solving therapies, and cognitive restructuring methods. The main areas of focus are cognition and behavior. The goal of CBT is to correct the patient's behavior and the maladaptive coping responses and to develop healthy ones (Dobson, 2010). With cancer patients, this is particularly important because of the many stressful situations that they generally face—especially in the early stages.

Therefore, one of the key methods in CBT for cancer patients is to introduce new coping mechanisms. Teaching coping skills helps the patient to cope with the side effects of the physical (pain, vomiting), as well as psychological (anxiety, depression) and social (loneliness, assertiveness) components of the disease. The studies report on several different ways to enhance learning coping mechanisms, such as relaxation, stress management, assertive communication (Andersen et al., 2008), cognitive restructuring and problem-solving (Fawzy et al., 1990), and pleasant activity planning (Telch & Telch, 1986). Coping skills training is designed to give patients the feeling that there is something in the course of the disease that they can control. It can be especially beneficial when cancer patients learn about their diagnosis, as well as going through difficult treatment. At the beginning of the diagnosis, patients with better coping mechanisms can gain more knowledge and choose better treatment (Fawzy, 1999). Patients can be more assertive in expressing their needs. Appropriate care at the beginning can predict the progress of the disease since it seems that patients can have the highest impact on the disease at the very beginning.

In CBT it is also important that the patient and their families are educated about the disease including its course and progress, which seems to have a calming effect as the patient and their family realize that things are not as bad as they might have seemed at the beginning. Education seems to be especially beneficial at the beginning of the diagnosis (Dolbeault et al., 2009). Nowadays, there is so much information available about cancer that patients and their families should strive to receive relevant information.

Duration and setting of the CBT

CBT is usually shorter and has predefined modules in most cases. This gives a patient a sense of control that is taken from them with the diagnosis. Time-limited and relatively short and well-organized psychotherapies are especially appropriate for newly diagnosed cancer patients (Fawzy et al., 1993; Fukui, Koike, Ooba, & Uchitomi, 2003). Short interventions are more accessible and may contribute to reducing distress among recently diagnosed breast cancer patients, especially those with high initial distress. The length of the therapy depends in part on the goal of the therapy. It is suggested that 8 to 10 weeks seems to be sufficient time to consolidate group cohesion and conduct psychotherapeutic work (Stagl et al., 2015). Moreover, how long the therapy lasts also depends on what kind of method will be used in the intervention.

One study (Reardon, Cukrowicz, Reeves, & Joiner, 2002) showed that it is not the duration of the therapy that benefits most, but the regularity and intensity of the treatment. However, other studies (Howard, Kopta, Krause, & Orlinsky, 1986; Howard, Lueger, Maling, & Martinovich, 1993) suggest that the number of sessions needed for a significant improvement of symptoms varies as a function of the outcome. Thus, even if the dose of therapy (number of sessions) is associated with clinical benefits (the effects), improvements in short-term therapy have been observed as well. Short-term therapies seem to be more successful when they are goal-oriented, more structured, and directive, such as CBT.

How does CBT in cancer survival works?

There is no doubt that CBT is effective in decreasing the symptoms of depression and anxiety. A group of cancer patients who undertook CBT reported on fewer depressive symptoms and better emotional and physical well-being (Stagl et al., 2015; Vargas et al., 2014). The study showed that group-based CBT enhanced physiological and psychological adaptation when women recover from surgery or undergo cancer-related treatments (Antoni et al., 2012; Phillips et al., 2011).

TABLE 1 Descriptive summary of randomized-controlled trials exploring CBT and cancer-survival.

	Sample				Intervention				Outcome
Author	No. of patients in I and C	Type and stage of cancer	Age, years (M±SD or %)	Married (%)	Format	Dose (hours)	Duration	Follow-up (M)	Survival = HR (95% CI)
Cunningham et al. (1998)	I=30 C=36	Breast, metastatic	49.5±13	74	Group	84	35 weeks	5 years	0.76 (0.43, 1.35)
Edelman et al. (1999)	I=43 C=49	Breast, metastatic	50±8.5	70.5	Group	50	6 months	5 years	1.32 (0.84, 2.08)
Andersen et al. (2008)	I=114 C=113	Breast, stage II and III	≥50=51%	73.5	Group	39	1 year	11 years	0.44 (0.22, 0.86)
Stagl et al. (2015)	I=120 C=120	Breast, 0–IIIB	49.7±9.0	62.5	Group	15	10 weeks	11 years	0.21 (0.05, 0.93)

Notes: I, intervention group; C, control group; CBT, cognitive-behavioral therapy.
This table presents pooled data from the RCTs, information about the included patients (number of patients in each group, type and stage of cancer, age, and marital status), information about the intervention (format, dose, duration and follow-up), and the outcome (survival measured with hazard ratio).

There are a large number of studies exploring CBT in cancer patients, reporting the effect of CBT on the QoL, fatigue, insomnia, menopausal symptoms, and pain (Getu et al., 2021).

To explore whether the psychological impact can also have an effect on cancer survival, we reviewed randomized-controlled studies exploring CBT and cancer survival (see Table 1; after (Mirosevic et al., 2019). Only four studies were included in this review and the results showed that CBT can increase survival only in the early-stage cancer patients (Edelman, Bell, & Kidman, 1999; Stagl et al., 2015). Two studies that included patients with metastatic cancer did not confirm the same results (Cunningham et al., 1998; Edelman, Lemon, Bell, & Kidman, 1999). Mirosevic et al. (2019) performed a metaanalysis of early- versus late-stage cancer patients. Results showed that studies sampling cancer patients in an early stage have a significant effect on survival with a large effect size (NNT of 2.6), while studies sampling patients with later stages showed no statistically significant effect on survival (Mirosevic et al., 2019). Both groups were compared with a z-test and showed a significant survival effect favoring studies with a CBT that included patients with an early-stage diagnosis. These results suggest that there may be a possibility to modify psychological functioning and consequently reduce the risk of cancer progression. Specifically, CBT improved stress management skills, which was found to decrease serum cortisol levels (Andersen et al., 2008). By learning new coping skills, cancer patients may lower distress and inflammatory and metastatic processes via stress pathways (Lutgendorf, Sood, & Antoni, 2010). Importantly, there are other nondirect relations by which cancer patients participating in CBT can benefit. Other cancer patients in the group may influence women's behaviors to decrease cancer risk with less alcohol consumption, more physical activity, and a healthy diet (Danaei et al., 2005).

Although results look promising, they should be considered with caution. In the majority of the included studies, survival was not a primary endpoint. Thus, sample sizes were often not considered for this analysis and were, therefore, small. We also observed a high number of missing data. While it is accurate that CBT can improve psychological functioning, further studies are needed to confirm whether improving psychological functioning can also improve survival. Because survival can improve only if psychological functioning improves, further studies should evaluate the underlying mechanisms of such improvements by which survival benefit is possible in nonmetastatic cancer patients. It would also be interesting to address changes in inflammatory, immune, neuroendocrine, and other tumor-promoting processes.

Conclusion

Assuming there is no single intervention format that is ideal for every patient at every point in time, it is important to know what components of any particular intervention are important with specific cancer patients at a specific time of the cancer trajectory. Good education is certainly better than bad therapy. For designing appropriate intervention, some of the facts appeared to be more efficient than others. Those patients who came to the intervention sooner not only showed better acceptance of the intervention, but showed longer survival as well. Perhaps the phase of the disease is the most critical factor in determining which specific method and intervention should be used. Teaching patients how to cope with their anxiety, depression, pain, thoughts, insomnia and how to demand more thorough medical help appeared to be one of the main things that could prolong survival, especially in the newly-diagnosed patients. However, for the patients in the later phase of the disease, it is especially important to teach them that no feeling is too frightening to face. Many cancer patients do not want to feel negative feelings about their disease. Furthermore, outbursts of emotions are in general not welcome in medical settings. Patients might also fear that if they talk about their sadness, the crying will not stop. However, the opposite has been observed: The more patients try to suppress feelings, the more powerful these feelings become. Supporting patients to safely express their emotions in a safe environment should be a priority.

Practice and procedures

When considering CBT for the earlier stages of cancer patients, we suggest that the below evidence-based practice is followed:

- Keep the group homogenous (only early-stage cancer patients, only female/male).
- Do the interview prior to the start of the program and explore the motivation of the participants.
- CBT should be performed with a licensed clinical psychologist and CBT therapist.
- Each session should be predefined and last 90 min for at least 8 weeks, preferably 10 weeks.
- Important sessions to include: coping and psychological adaptation, cognitive reframing, stress reappraisal, assertiveness training, learning coping skills, and anger management.

Mini-dictionary of terms

Psychosocial intervention—Any nonpharmaceutical treatment focused on addressing psychological or social factors.

Psychological morbidity—Any psychological state that is considered to be clinically relevant, usually with symptoms of anxiety and depression.

Supportive-expressive group therapy—Psychosocial intervention with a semistructured group approach. It involves supportive and emotionally expressive therapy with existential questions.

Mindfulness-based therapy—Psychosocial intervention that focuses on coping with stress and pain in the present moment.

Supportive intervention—Psychosocial intervention with no prepared agenda. It allows patients to discuss their feelings, concerns and problems.

Psychoeducational intervention—Psychosocial intervention that focuses on education about the disease, changes in the body and emotions.

Key facts of CBT in cancer survival

- CBT is a short-term form of a very focused psychosocial intervention.
- CBT in the treatment of cancer survival can be either conducted individually or in a group, but is mostly conducted individually.
- CBT's predominantly used methods with cancer patients are stress regulation via learning about the coping mechanisms and relaxation techniques, practicing assertiveness, psychoeducation, and cognitive restructuring methods.
- CBT rarely uses emotional expression, interpersonal exploration, and existential questioning.
- CBT for cancer patients mainly focuses on decreasing psychological morbidity, i.e., symptoms of anxiety and depression.
- CBT was found to be effective only when given in the earlier stages of cancer patients.

Application to other areas

In this section, we review studies that have attempted to answer the question whether psychosocial interventions, specifically CBT, increase cancer survival. This knowledge is specifically important for CBT therapists working with cancer patients; however, it is also valuable for therapists working with other modalities. For example, this review indicates that

CBT is especially valuable when given to cancer patients in the earlier phases of the diagnosis (nonmetastatic cancer patients). Importantly, there are other approaches that may be more important for other phases of cancer survivorship (e.g., supportive-expressive therapy for patients with metastatic cancer).

Summary points

- CBT is effective in decreasing symptoms of anxiety and depression.
- CBT was evaluated for cancer survivors in four studies.
- Two studies found that CBT improves cancer survivorship.
- Based on the moderator analysis, treatment time is of key importance when deciding which psychosocial treatment a clinician should use.
- CBT was found to be more effective when used in the earlier stages of the disease.

References

Aizer, A. A., Chen, M. H., McCarthy, E. P., Mendu, M. L., Koo, S., Wilhite, T. J., … Nguyen, P. L. (2013). Marital status and survival in patients with cancer. *Journal of Clinical Oncology*, *31*(31), 3869–3876. https://doi.org/10.1200/JCO.2013.49.6489.

Andersen, B. L., Yang, H. C., Farrar, W. B., Golden-Kreutz, D. M., Emery, C. F., Thornton, L. M., … Carson, W. E., 3rd. (2008). Psychologic intervention improves survival for breast cancer patients: A randomized clinical trial. *Cancer*, *113*(12), 3450–3458. https://doi.org/10.1002/cncr.23969.

Antoni, M. H., Lutgendorf, S. K., Blomberg, B., Carver, C. S., Lechner, S., Diaz, A., … Cole, S. W. (2012). Cognitive-behavioral stress management reverses anxiety-related leukocyte transcriptional dynamics. *Biological Psychiatry*, *71*(4), 366–372.

Bottomley, A. (1998). Depression in cancer patients: A literature review. *European Journal of Cancer Care*, *7*(3), 181–191. Retrieved from: http://www.ncbi.nlm.nih.gov/pubmed/9793010.

Carlson, L. E., Speca, M., Patel, K. D., & Goodey, E. (2003). Mindfulness-based stress reduction in relation to quality of life, mood, symptoms of stress, and immune parameters in breast and prostate cancer outpatients. *Psychosomatic Medicine*, *65*(4), 571–581. Retrieved from: http://www.ncbi.nlm.nih.gov/pubmed/12883107.

Cunningham, A. J., Edmonds, C., Jenkins, G., Pollack, H., Lockwood, G., & Warr, D. (1998). A randomized controlled trial of the effects of group psychological therapy on survival in women with metastatic breast cancer. *Psycho-Oncology*, *7*(6), 508–517.

Dalton, S. O., Mellemkjaer, L., Olsen, J. H., Mortensen, P. B., & Johansen, C. (2002). Depression and cancer risk: A register-based study of patients hospitalized with affective disorders, Denmark, 1969–1993. *American Journal of Epidemiology*, *155*(12), 1088–1095. Retrieved from: http://www.ncbi.nlm.nih.gov/pubmed/12048222.

Danaei, G., Vander Hoorn, S., Lopez, A. D., Murray, C. J., Ezzati, M., & Group, C. R. A. C. (2005). Causes of cancer in the world: Comparative risk assessment of nine behavioural and environmental risk factors. *The Lancet*, *366*(9499), 1784–1793.

Dobson, K. S. (2010). *Handbook of cognitive-behavioral therapies*. New York: Guilford Press.

Dolbeault, S., Cayrou, S., Bredart, A., Viala, A. L., Desclaux, B., Saltel, P., & Dickes, P. (2009). The effectiveness of a psycho-educational group after early-stage breast cancer treatment: Results of a randomized French study. *Psychooncology*, *18*(6), 647–656. https://doi.org/10.1002/pon.1440.

Edelman, S., Bell, D. R., & Kidman, A. D. (1999). A group cognitive behaviour therapy programme with metastatic breast cancer patients. *Psychooncology*, *8*(4), 295–305. https://doi.org/10.1002/(SICI)1099-1611(199907/08)8:4<295::AID-PON386>3.0.CO;2-Y.

Edelman, S., Lemon, J., Bell, D. R., & Kidman, A. D. (1999). Effects of group CBT on the survival time of patients with metastatic breast cancer. *Psycho-Oncology*, *8*(6), 474–481.

Fawzy, F. I. (1999). Psychosocial interventions for patients with cancer: What works and what doesn't. *European Journal of Cancer*, *35*(11), 1559–1564. Retrieved from: https://www.ncbi.nlm.nih.gov/pubmed/10673962.

Fawzy, F. I., Cousins, N., Fawzy, N. W., Kemeny, M. E., Elashoff, R., & Morton, D. (1990). A structured psychiatric intervention for cancer patients. I. Changes over time in methods of coping and affective disturbance. *Archives of General Psychiatry*, *47*(8), 720–725. Retrieved from: http://www.ncbi.nlm.nih.gov/pubmed/2378543.

Fawzy, F. I., Fawzy, N. W., Hyun, C. S., Elashoff, R., Guthrie, D., Fahey, J. L., & Morton, D. L. (1993). Malignant melanoma. Effects of an early structured psychiatric intervention, coping, and affective state on recurrence and survival 6 years later. *Archives of General Psychiatry*, *50*(9), 681–689. Retrieved from: http://www.ncbi.nlm.nih.gov/pubmed/8357293.

Fukui, S., Koike, M., Ooba, A., & Uchitomi, Y. (2003). The effect of a psychosocial group intervention on loneliness and social support for Japanese women with primary breast cancer. *Oncology Nursing Forum*, *30*(5), 823–830. https://doi.org/10.1188/03.ONF.823-830.

Getu, M. A., Chen, C., Panpan, W., Mboineki, J. F., Dhakal, K., & Du, R. (2021). The effect of cognitive behavioral therapy on the quality of life of breast cancer patients: A systematic review and meta-analysis of randomized controlled trials. *Quality of Life Research*, *30*(2), 367–384.

Golant, M., Altman, T., & Martin, C. (2003). Managing cancer side effects to improve quality of life: A cancer psychoeducation program. *Cancer Nursing*, *26*(1), 37–44. quiz 45-36 https://doi.org/10.1097/00002820-200302000-00005.

Goodwin, P. J., Leszcz, M., Ennis, M., Koopmans, J., Vincent, L., Guther, H., … Hunter, J. (2001). The effect of group psychosocial support on survival in metastatic breast cancer. *The New England Journal of Medicine*, *345*(24), 1719–1726. https://doi.org/10.1056/NEJMoa011871.

Grassi, L., & Rosti, G. (1996). Psychosocial morbidity and adjustment to illness among long-term cancer survivors. A six-year follow-up study. *Psychosomatics*, *37*(6), 523–532. https://doi.org/10.1016/S0033-3182(96)71516-5.

Grivennikov, S. I., Greten, F. R., & Karin, M. (2010). Immunity, inflammation, and cancer. *Cell*, *140*(6), 883–899. https://doi.org/10.1016/j.cell.2010.01.025.

Hansen, M. V., Madsen, M. T., Hageman, I., Rasmussen, L. S., Bokmand, S., Rosenberg, J., & Gogenur, I. (2012). The effect of MELatOnin on depression, anxietY, cognitive function and sleep disturbances in patients with breast cancer. The MELODY trial: Protocol for a randomised, placebo-controlled, double-blinded trial. *BMJ Open*, *2*(1), e000647. https://doi.org/10.1136/bmjopen-2011-000647.

Howard, K. I., Kopta, S. M., Krause, M. S., & Orlinsky, D. E. (1986). The dose-effect relationship in psychotherapy. *The American Psychologist*, *41*(2), 159–164. Retrieved from: http://www.ncbi.nlm.nih.gov/pubmed/3516036.

Howard, K. I., Lueger, R. J., Maling, M. S., & Martinovich, Z. (1993). A phase model of psychotherapy outcome: Causal mediation of change. *Journal of Consulting and Clinical Psychology*, *61*(4), 678–685. Retrieved from: http://www.ncbi.nlm.nih.gov/pubmed/8370864.

Jacobs, C., Ross, R. D., Walker, I. M., & Stockdale, F. E. (1983). Behavior of cancer patients: A randomized study of the effects of education and peer support groups. *American Journal of Clinical Oncology*, *6*(3), 347–353. Retrieved from: http://www.ncbi.nlm.nih.gov/pubmed/6342360.

Kroenke, C. H., Kubzansky, L. D., Schernhammer, E. S., Holmes, M. D., & Kawachi, I. (2006). Social networks, social support, and survival after breast cancer diagnosis. *Journal of Clinical Oncology*, *24*(7), 1105–1111. https://doi.org/10.1200/JCO.2005.04.2846.

Kuchler, T., Henne-Bruns, D., Rappat, S., Graul, J., Holst, K., Williams, J. I., & Wood-Dauphinee, S. (1999). Impact of psychotherapeutic support on gastrointestinal cancer patients undergoing surgery: Survival results of a trial. *Hepato-Gastroenterology*, *46*(25), 322–335. Retrieved from: http://www.ncbi.nlm.nih.gov/pubmed/10228816.

Lemon, S. C., Zapka, J. G., & Clemow, L. (2004). Health behavior change among women with recent familial diagnosis of breast cancer. *Preventive Medicine*, *39*(2), 253–262. https://doi.org/10.1016/j.ypmed.2004.03.039.

Lloyd-Williams, M., & Friedman, T. (1999). Depression in terminally ill patients. *The American Journal of Hospice & Palliative Care*, *16*(6), 704–705. Retrieved from: http://www.ncbi.nlm.nih.gov/pubmed/11094906.

Lutgendorf, S. K., Sood, A. K., & Antoni, M. H. (2010). Host factors and cancer progression: Biobehavioral signaling pathways and interventions. *Journal of Clinical Oncology*, *28*(26), 4094.

Massie, M. J. (2004). Prevalence of depression in patients with cancer. *Journal of the National Cancer Institute Monographs*, (32), 57–71. https://doi.org/10.1093/jncimonographs/lgh014.

Mirosevic, S., Jo, B., Kraemer, H. C., Ershadi, M., Neri, E., & Spiegel, D. (2019). Not just another meta-analysis: Sources of heterogeneity in psychosocial treatment effect on cancer survival. *Cancer Medicine*, *8*(1), 363–373.

Penninx, B. W., Guralnik, J. M., Pahor, M., Ferrucci, L., Cerhan, J. R., Wallace, R. B., & Havlik, R. J. (1998). Chronically depressed mood and cancer risk in older persons. *Journal of the National Cancer Institute*, *90*(24), 1888–1893. Retrieved from: http://www.ncbi.nlm.nih.gov/pubmed/9862626.

Phillips, K. M., Antoni, M. H., Carver, C. S., Lechner, S. C., Penedo, F. J., McCullough, M. E., … Blomberg, B. B. (2011). Stress management skills and reductions in serum cortisol across the year after surgery for non-metastatic breast cancer. *Cognitive Therapy and Research*, *35*(6), 595–600.

Reardon, M. L., Cukrowicz, K. C., Reeves, M. D., & Joiner, T. E. (2002). Duration and regularity of therapy attendance as predictors of treatment outcome in an adult outpatient population. *Psychotherapy Research*, *12*(3), 273–285.

Richardson, J. L., Zarnegar, Z., Bisno, B., & Levine, A. (1990). Psychosocial status at initiation of cancer treatment and survival. *Journal of Psychosomatic Research*, *34*(2), 189–201. Retrieved from: http://www.ncbi.nlm.nih.gov/pubmed/2325003.

Ruddy, R., & House, A. (2005). Psychosocial interventions for conversion disorder. *Cochrane Database of Systematic Reviews*, (4), CD005331. https://doi.org/10.1002/14651858.CD005331.pub2.

Salzer, M. S., Palmer, S. C., Kaplan, K., Brusilovskiy, E., Ten Have, T., Hampshire, M., & Coyne, J. C. (2010). A randomized, controlled study of internet peer-to-peer interactions among women newly diagnosed with breast cancer. *Psychooncology*, *19*(4), 441–446. https://doi.org/10.1002/pon.1586.

Sephton, S. E., Dhabhar, F. S., Keuroghlian, A. S., Giese-Davis, J., McEwen, B. S., Ionan, A. C., & Spiegel, D. (2009). Depression, cortisol, and suppressed cell-mediated immunity in metastatic breast cancer. *Brain, Behavior, and Immunity*, *23*(8), 1148–1155. https://doi.org/10.1016/j.bbi.2009.07.007.

Shekelle, R. B., Raynor, W. J., Jr., Ostfeld, A. M., Garron, D. C., Bieliauskas, L. A., Liu, S. C., & Paul, O. (1981). Psychological depression and 17-year risk of death from cancer. *Psychosomatic Medicine*, *43*(2), 117–125. Retrieved from: http://www.ncbi.nlm.nih.gov/pubmed/7267935.

Singer, S., Gotze, H., Mobius, C., Witzigmann, H., Kortmann, R. D., Lehmann, A., & Hauss, J. (2009). Quality of care and emotional support from the inpatient cancer patient's perspective. *Langenbeck's Archives of Surgery*, *394*(4), 723–731. https://doi.org/10.1007/s00423-009-0489-5.

Speca, M., Carlson, L. E., Goodey, E., & Angen, M. (2000). A randomized, wait-list controlled clinical trial: The effect of a mindfulness meditation-based stress reduction program on mood and symptoms of stress in cancer outpatients. *Psychosomatic Medicine*, *62*(5), 613–622. Retrieved from: http://www.ncbi.nlm.nih.gov/pubmed/11020090.

Spiegel, D., Bloom, J. R., Kraemer, H. C., & Gottheil, E. (1989). Effect of psychosocial treatment on survival of patients with metastatic breast cancer. *Lancet*, *2*(8668), 888–891. https://doi.org/10.1016/s0140-6736(89)91551-1.

Spiegel, D., Butler, L. D., Giese-Davis, J., Koopman, C., Miller, E., DiMiceli, S., & Kraemer, H. C. (2007). Effects of supportive-expressive group therapy on survival of patients with metastatic breast cancer: A randomized prospective trial. *Cancer*, *110*(5), 1130–1138. https://doi.org/10.1002/cncr.22890.

Spiegel, D., & Giese-Davis, J. (2003). Depression and cancer: Mechanisms and disease progression. *Biological Psychiatry*, *54*(3), 269–282. Retrieved from: https://www.ncbi.nlm.nih.gov/pubmed/12893103.

Stagl, J. M., Bouchard, L. C., Lechner, S. C., Blomberg, B. B., Gudenkauf, L. M., Jutagir, D. R., & Antoni, M. H. (2015). Long-term psychological benefits of cognitive-behavioral stress management for women with breast cancer: 11-year follow-up of a randomized controlled trial. *Cancer*, *121*(11), 1873–1881.

Stommel, M., Given, B. A., & Given, C. W. (2002). Depression and functional status as predictors of death among cancer patients. *Cancer*, *94*(10), 2719–2727. Retrieved from: http://www.ncbi.nlm.nih.gov/pubmed/12173342.

Telch, C. F., & Telch, M. J. (1986). Group coping skills instruction and supportive group therapy for cancer patients: A comparison of strategies. *Journal of Consulting and Clinical Psychology*, *54*(6), 802–808. Retrieved from: http://www.ncbi.nlm.nih.gov/pubmed/3794024.

Temel, J. S., Greer, J. A., Muzikansky, A., Gallagher, E. R., Admane, S., Jackson, V. A., … Lynch, T. J. (2010). Early palliative care for patients with metastatic non-small-cell lung cancer. *The New England Journal of Medicine*, *363*(8), 733–742. https://doi.org/10.1056/NEJMoa1000678.

Vargas, S., Antoni, M. H., Carver, C. S., Lechner, S. C., Wohlgemuth, W., Llabre, M., … DerHagopian, R. P. (2014). Sleep quality and fatigue after a stress management intervention for women with early-stage breast cancer in southern Florida. *International Journal of Behavioral Medicine*, *21*(6), 971–981.

Watson, R. J., & Richardson, P. H. (1999). Identifying randomized controlled trials of cognitive therapy for depression: Comparing the efficiency of Embase, Medline and PsycINFO bibliographic databases. *The British Journal of Medical Psychology*, *72*(Pt 4), 535–542. Retrieved from: http://www.ncbi.nlm.nih.gov/pubmed/10616135.

Wurtzen, H., Dalton, S. O., Elsass, P., Sumbundu, A. D., Steding-Jensen, M., Karlsen, R. V., … Johansen, C. (2013). Mindfulness significantly reduces self-reported levels of anxiety and depression: Results of a randomised controlled trial among 336 Danish women treated for stage I-III breast cancer. *European Journal of Cancer*, *49*(6), 1365–1373. https://doi.org/10.1016/j.ejca.2012.10.030.

Section B

Cognitive behavioral therapy in adults

Chapter 6

Acrophobia and consumer-based automated virtual reality cognitive behavior therapy

Tara Donker and Markus Heinrichs

Department of Psychology, Laboratory of Biological and Personality Psychology, Albert-Ludwigs University of Freiburg, Freiburg im Breisgau, Germany

Abbreviations

AR	augmented reality
ARET	augmented reality exposure therapy
CAVE	cave automatic virtual environment
CBT	cognitive behavior therapy
HMD	head-mounted displays
HRV	heart rate variability
RCT	randomized controlled trial
VR	virtual reality
VRET	virtual reality exposure therapy

Introduction

A specific phobia is characterized by an excessive, persistent fear of a specific object or a certain situation (e.g., animals, thunder, flying, driving a car), which interferes significantly in daily life (American Psychiatric Association, 2013). The life time prevalence of a specific phobia is around 15% (Kessler, Petukhova, Sampson, Zaslavsky, & Wittchen, 2012). Having a specific phobia can raise the risk of developing another anxiety disorder or a depressive disorder (Trumpf, Margraf, Vriends, Meyer, & Becker, 2010). Acrophobia, or the fear of heights, is the most prevalent type of specific phobia with an estimated life time prevalence rate of 3%–6% (Becker et al., 2007; Curtis, Magee, Eaton, & Wittchen, 1998; Depla, ten Have, van Balkom, & de Graaf, 2008; Kapfhammer, Huppert, Grill, Fitz, & Brandt, 2015; LeBeau et al., 2010; Oosterink, de Jongh, & Hoogstraten, 2009). Typical fear-evoking situations are, for example, looking down from a tower, hiking and mountaineering, climbing ladders, and walking over a bridge (Huppert, Wuehr, & Brandt, 2020). The most common symptoms people with acrophobia suffer from in those situations include anxiety, vertigo, unsteadiness, weak knees, rapid heartbeat, sweating, and panic attacks (Huppert et al., 2020). To prevent these symptoms, one coping strategy applied by most acrophobics is simply to avoid situations involving heights [Huppert, Grill, & Brandt, 2013; Schäfer et al., 2014). This is an effective short term strategy, but in the long term, such avoidance leads to even higher levels of anxiety (Clark & Beck, 2009), restrictions in daily activities, and reduced quality of life (Huppert et al., 2013; Schäfer et al., 2014). This is why the current evidence-based treatment of choice is cognitive behavioral therapy (CBT) (Nathan & Gorman, 2007), in which exposure is the most essential ingredient (Antony & Barlow, 2002; Botella et al., 2016).

Cognitive behavioral therapy for specific phobia

In exposure treatment, patients repeatedly and gradually expose themselves to the object or situation they fear in order to learn that their feared expectations fail to come true, thus enabling a novel adaptive learning mechanism (Craske, Treanor, Conway, Zbozinek, & Vervliet, 2014). The mechanism believed to underlie exposure therapy is called inhibitory learning, meaning that through exposure, the familiar association of a feared (conditional) stimulus (e.g., a dog) and a (conditional) response (e.g., dog bites) is inhibited by a new association (the dog will not bite) (Craske et al., 2014). One of the most important strategies to strengthen such a new association is to design exposures so that the patient's expectations are maximally contradicted with regards to what actually happens (expectancy violation) (e.g., Davey, 1992). Exposure exists in several variations, including imaginary exposure (to a feared object or situation in your imagination), interoceptive exposure (to internal body queues such as increased heart rate, sweating), or in vivo exposure (exposure in real life). There is no doubt that exposure therapy is effective for anxiety disorders (e.g., Hofmann & Smits, 2008; Norton & Price, 2007). Yet despite available and effective evidence-based therapy, less than 8% of people with a specific phobia seek treatment (Botella et al., 2016; Mackenzie, Reynolds, Cairney, Streiner, & Sareen, 2012), and even less than 1% actually undergoes specific phobia treatment (Stinson et al., 2007). There are several possible reasons for these low numbers, including the lack of evidence-based treatment offered by health care providers, long waiting lists, shortage of trained therapists and the reluctance of therapists and patients to undergo exposure therapy as some consider it cruel or unethical (Botella et al., 2016). Expensive treatments also contribute to the limited uptake of evidence-based therapy (Kessler, Berglund, Bruce, et al., 2001). Efforts to overcome these hurdles are found in digital technology, such as virtual reality (VR).

Virtual reality

VR is a technology able to create computer-generated "analogs" of our world. VR makes it possible to create artificial experiences in real time, making the user feel "immersed" and able to interact as if it were the real world (Botella, Fernández-Álvarez, Guillén, García-Palacios, & Baños, 2017). This so-called immersive VR is capable of producing the sensation of actually being in life-sized new environments (Freeman et al., 2017), and it is made possible by integrating computers, and head-mounted displays (HMDs that occlude users' view of the outside world), body-tracking sensors, specialized interface devices, and 3D graphics (Rizzo & Koenig, 2017). Using this equipment, the computer-generated simulated world changes naturally with head and body movements (Rizzo & Koenig, 2017). In fact, VR is capable of making users believe the experience is real, and is thus capable of evoking real life emotions (Botella, Perpiñá, Baños, & García-Palacios, 1998; McLuhan, 1964).

VR, and in particular VR exposure therapy (VRET) has revealed tremendous promise for alleviating mental health disorders (Rizzo & Koenig, 2017). Instead of exposure in vivo, that is, exposure to real life settings, in VRET, an individual is exposed to the feared object or situation in an immersive virtual environment (Slater & Sanchez-Vives, 2016). VR has several advantages over in vivo exposure. One is having 100% control over the exposure environment (Botella et al., 2017). For example, instead of being exposed to a dog in real life, where the dog would decide when to bark or move, the dog in VR is programmed to bark or move in a controlled way. This offers researchers the opportunity to manipulate and tailor key variables associated with stimulus presentation (e.g., color, size, movement), context, and the intensity of exposure according to the patient's needs (Miloff et al., 2016). Another advantage of VR is its great flexibility. Whereas first-generation software and hardware enabled only simple simulations mimicking real world (mostly exposure) situations, activities can be carried out in VR nowadays that are impossible in the real world (Geraets, van der Stouwe, Pot-Kolder, & Veling, 2021). For instance, new VR techniques include body swapping, which is the illusion of a virtual body experienced as one's own (Slater & Sanchez-Vives, 2016). Finally, with VR technology involving HMDs and body motion software, researchers are able to extract unprecedented amounts of data, such as gaze focus (Emmelkamp, 2005; Miloff et al., 2016).

Virtual reality cognitive behavioral therapy for specific phobias

The first studies using VR to replace real life settings for psychological disorders date back to mid-1990, in which the first clinical trial was employing VR for acrophobia (Rothbaum et al., 1995). Since then, numerous studies have been carried out, and metaanalyses have provided evidence of VRET in reducing specific phobias and other anxiety disorders (Chou et al., 2021; Freeman et al., 2017; Geraets et al., 2021). Recent studies have begun to target other disorders such as eating and substance disorders, depression, and even psychosis (Geraets et al., 2021). There is also a growing body of evidence that VR yields effects comparable to in vivo exposures (Carl et al., 2019; Fodor et al., 2018; Morina, Ijntema, Meyerbröker, & Emmelkamp, 2015). For example, Emmelkamp et al. (2002) investigated the effectiveness of VRET vs exposure in vivo in patients suffering from acrophobia. The VR environments were exact replicas of the genuine environments that had been experienced in the in vivo exposure, e.g., a mall (MagnaPlaza) in Amsterdam that has four floors with escalators and a balustrade, and a roof garden at the top of a 65-ft-high university building. Results were similar between the two groups (Emmelkamp et al., 2002). However, although high quality treatment studies with adequate sample sizes are growing, they are still limited, in particular compared to active control groups (Geraets et al., 2021).

The problem of scalability and accessibility of virtual reality CBT

VR hardware emerged in the 1980s and was mostly used in specialist laboratories (Slater & Sanchez-Vives, 2016) due to its high costs and complexity (Rizzo & Koenig, 2017). For example, the cave automatic virtual environment (CAVE) projects computer images onto the walls of a room, and the participant wears tracked shutter glasses to view the scene three-dimensionally (Cruz-Neira, Sandin, & DeFanti, 1993; Freeman et al., 2017). Understandably, this was not a scalable solution for evidence-based treatment. VR's dissemination has long been hampered by high costs, technical obstacles, availability and the lack of training facilities (Geraets et al., 2021). Partly driven by investments in the entertainment and gaming industry (Rizzo & Koenig, 2017), such as Sony (Project Morpheus), Microsoft (Hololens), HTC (ViVe), and Facebook (Oculus Rift) (Miloff et al., 2016), VR has since moved beyond specialist laboratories (Freeman et al., 2017) into consumer homes. Whereas VR was a professional niche product costing as much as $35,000 (Miloff et al., 2016) or more, over the last two decades, advances in VR-based technologies (e.g., computational speed, computer graphics, tracking sensors) have now made it possible to increase the accessibility and lower the price of sophisticated, immersive, and interactive VR systems (Rizzo & Koenig, 2017). VR consumer products are now priced around $599 plus the cost of a suitable computer (Hoffman et al., 2014; Miloff et al., 2016). Moreover, thanks to recent innovations, even smartphones can now run VR software, such as the Samsung Gear VR ($99), Google Day Dream ($30) or a mobile VR viewer that can cost as little as $5–10. With penetration rates of smartphone ownership of >75% in Western countries and growing 45% in developing countries (Pew Research Center, 2019), this clearly creates the opportunity for the accessible delivery of VR mental health. In particular, as consumer VR has now made it possible to deliver VR treatment without the presence of a therapist, a new era of treatment dissemination in VR has evolved, namely automated, stand-alone, or self-guided (mobile) VR therapy. This narrative review was carried out to provide a critical overview of the evidence-based stand-alone, automated or self-guided consumer-based VR interventions (from here onward called "automated consumer-based" VR), for specific phobia, with particular focus on acrophobia.

Practice and procedures

Selection and inclusion of studies

We conducted a literature search using several electronic databases (e.g., PubMed, Scopus) in July 2021 by combining index and text words of automated, stand-alone, or self-guided, smartphone app-based virtual reality, and specific phobia. We restricted the search of papers published after June 2014 (when the first consumer VR was sold). Studies were included if they specifically mentioned that the intervention was automated, self-guided, stand-alone, or carried out remotely by participants, and if the VR hardware and software used was a consumer product (e.g., a mobile VR application downloadable on a mobile device). The control group could consist of treatment-as-usual, attention-placebo, another treatment or a wait list. Studies without a control group (pre-post design) were also included. Studies were included if they were published in peer-reviewed international journals in English, Dutch, or German. Studies were excluded if mental health symptoms/disorders were not an outcome measure, if it did not include an intervention, if the intervention targeted a medical disorder (e.g., fibromyalgia) or if the intervention did not target virtual reality but, for instance, augmented reality. Conference abstracts, protocol papers, and case studies were also excluded. In total, we examined 3101 abstracts from which 34 full-text papers were retrieved. Reference lists from systematic reviews, metaanalyses and books obtained in the search were also screened for potentially relevant references. This resulted in 43 potentially relevant abstracts. Of those, 30 were retrieved for full-text assessment, but none met our inclusion criteria. A total of eight trials did so.

Characteristics of included studies

The characteristics of the studies we included are summarized in Table 1. Five trials targeted acrophobia or the so-called fear of heights (Bentz et al., 2021 [Easy Heights], Donker et al., 2019 [ZeroPhobia], Freeman et al., 2018 [Now I can do heights], Hong et al., 2017 and Levy et al., 2016), two trials (Lindner et al., 2020; Miloff et al., 2016) describing one app for arachnophobia (fear of spiders) (Itsy) and one trial investigated the effects of nyctophobia, also known as the fear of darkness (Paulus et al., 2019). In total, 579 participants were recruited across all trials. Three trials applied a randomized controlled trial (RCT) design (Bentz et al., 2021; Donker et al., 2019; Freeman, Lister, et al., 2019; Freeman, Yu, et al., 2019), three a pre-post design (Levy et al., 2016; Lindner et al., 2020; Paulus et al., 2019), one study used a noninferiority design (Miloff et al., 2016), and another trial employed a randomized design (Hong et al., 2017). Regarding control groups, one trial used both an attention-placebo and no intervention as a control group (Bentz et al., 2021), one trial used a wait-list

TABLE 1 Characteristics of included studies targeting automated consumer-based VR-CBT or VRET for specific phobia by type of disorder.

Study, name of app	Specific phobia type	Trial	Primary outcome measure	Conditions	Sample	Sample size	VR equipment	VR environment	Setting	Delivery length	Contact with research team
Bentz, Wang, Ibach, et al., 2021; *Easy Heights*	Acrophobia	RCT	BAT	Phase 1: VRET vs. Placebo VR; Phase 2: VRET vs. no intervention	Adults from primary care institute with and without a DSM-V diagnosis	Phase 1: 77 (39/38); Phase 2: 50 (25/25)	Samsung Galaxy smartphone, Google Daydream View 2, controllers, noise canceling headphones	Rural mountain, cloudy weather, urban town	Phase 1: Lookout tower; Phase 2: own home	Phase 1: 1 day; Phase 2: 2 weeks	Yes
Donker et al., 2019; Donker, Klaveren, Cornelisz, Kok, & van Gelder, 2020; *ZeroPhobia*	Acrophobia	RCT	AQ	VR-CBT vs WL	Adults from the general population	193 (96/97)	Mobile VR viewer (Google cardboard) and participants' own Android smartphone	Inside a theater	Own home	3 weeks	No
Freeman et al., 2019; Freeman et al., 2019; *Now I can do heights*	Acrophobia	RCT	HIQ	VR-BE vs. CAU (no treatment)	Adults from the general population	100 (49/51)	Gaming PC, HTC vive, ear-phones, hand controllers	Atrium of a 10-story office complex	Oxford VR	2 weeks	Yes
Hong, Kim, Jung, Kyeong, & Kim, 2017; *unkown*	Acrophobia	RT	Unknown	High vs. Low fear	Students	48 (24/24)	Samsung Galaxy S6, Samsung Gear VR, Samsung Gear S2 smartwatch	Elevator, cliff driving, heli-skiing, rooftop walking	VR center	2 weeks	Yes
Levy, Leboucher, Rautureau, & Jouvent, 2016; *unkown*	Acrophobia	Pre-post cross-over	Unknown	e-VRET and pVRET	Adults from the hospital, with a DSM-IV diagnosis	6	Laptop, Sony HMZ-T1 HMD, Mio Alpha wristband	Subway station, 24-story tower block	Hospital	3 weeks	Yes
Lindner et al., 2020; *Itsy*	Arachno-Phobia	Pre-post	FSQ	VRET	Adults from the general population and students	25	Samsung Galaxy S6, Samsung Gear VR	Spiders with increasing levels of realness	Stockholm University	1 session	Yes
Miloff et al., 2016; *Itsy*	Arachno-Phobia	Noninferiority trial	BAT	VRET vs. OST In vivo exposure	Adults from the general population with a DSM-V diagnosis	100 (50/50)	Samsung Galaxy S6, Samsung Gear VR	Spiders with increasing levels of realness	Stockholm University	1 session	Yes
Paulus, Suryan, Wijayanti, Yusuf, & Iskandarsyah, 2019; *unkown*	Nycto-phobia	Pre-post	Fear of darkness thermometer	VR-CBT	Students	30	Android Smartphone 5.0.2 (Lollipop), Samsung Gear VR, wireless headphone, wireless Joystick	From daylight till the night	Laboratory Room, University of Padjadjaran	1 session	Yes

This table shows an overview of the characteristics of included studies targeting automated VR studies for specific phobia. AQ, acrophobia questionnaire; BAT, behavioral approach test; DSM-V, diagnostic and statistical manual of mental disorders, fifth edition; FSQ, fear of questionnaire; HIQ, heights interpretation questionnaire; OST, one session therapist-led; RCT, randomized controlled trial; RT, randomized trial; VR, virtual reality; VR-BE, VR behavioral experiments; VR-CBT, VR-cognitive behavior therapy; e-VRET, remote VRET; p-VRET, physical presence of a therapist VRET.

control group (Donker et al., 2019), one trial used care as usual (which was no intervention) as a control group (Freeman et al., 2018) and one trial made an active comparison (Miloff et al., 2016). Hong et al. (2017) randomized participants with high vs. low fear to one of two arms. Four trials (Donker et al., 2019; Freeman et al., 2018; Lindner et al., 2020; Miloff et al., 2016) recruited from the general population, two trials recruited students (Hong et al., 2017; Paulus et al., 2019) and two studies recruited from a hospital or primary care institute (Bentz et al., 2021; Levy et al., 2016). Six studies delivered the intervention through a smartphone app (Bentz et al., 2021, Donker et al., 2019; Hong et al., 2017; Lindner et al., 2020; Miloff et al., 2016; Paulus et al., 2019), while the other two studies used a PC or gaming laptop (Freeman et al., 2018; Levy et al., 2016). HMDs used in the studies ranged from very rudimentary (Mobile VR viewer [formerly known as Google Cardboard]) (Donker et al., 2019), to high-end consumer VR software (HTC vive and gaming laptop, Sony HMZ-T1 and a powerful laptop) (Freeman et al., 2018; Levy et al., 2016). Delivery length varied between one session (Lindner et al., 2020; Miloff et al., 2016; Paulus et al., 2019) and 3 weeks (Donker et al., 2019; Levy et al., 2016). Five studies assessed post-test outcomes only, whereas four others included follow-up assessments too (4 weeks, 3, 6 and 12 months) (Donker et al., 2019; Freeman et al., 2018; Lindner et al., 2020; Miloff et al., 2016). All but one trial (Donker et al., 2019) were conducted in the presence of a research team. Six studies were conducted in a VR lab at the university or hospital, while two studies were (partly) conducted in a natural setting (participants' own home) (Bentz et al., 2021; Donker et al., 2019).

Effects and quality of automated VR studies for specific phobia

Table 2 shows the results of the included VR studies. All but one (Levy et al., 2016) thereof demonstrated significant reductions ($P < 0.05$) in specific phobia symptoms and, when documented, large effect sizes at post-test and follow-up (Bentz, Wang, & Ibach, 2021; Donker et al., 2019; Freeman, Lister, et al., 2019; Freeman, Yu, et al., 2019; Lindner et al., 2020; Miloff et al., 2016). One trial (Levy et al., 2016) compared remote vs. the physical presence of a therapist

TABLE 2 Results of automated consumer-based VR-CBT or VRET.

Study, name of app	Specific phobia type	Outcomes of interest	Effect size (Cohen's d)	Dropout
Bentz et al., 2021; *Easy Heights*	Acrophobia	After phase 2, but not phase 1, participants in the Easy Heights condition showed significantly higher BAT scores compared to participants in the control condition. The app was well accepted	$d=1.3$ (between, post-test)	Phase 1: 3% Phase 2: 7%
Donker et al., 2019; *ZeroPhobia*	Acrophobia	Compared to controls, participants in the VR-CBT app group showed a significant reduction of acrophobia symptoms at posttest. The number needed to treat was 1.7. The app was rated user-friendly	$d=1.14$ (between, post-test) $d=2.68$ (within, 3-mo fu)	23%
Freeman, Lister, et al., 2019; Freeman, Yu, et al., 2019; *Now I can do heights*	Acrophobia	Compared with participants in the control group, the VR-BE treatment reduced fear of heights at the end of treatment. The number needed to treat was 1.3	$d=2.00$ (between, post-test) $d=2.00$ (between, 4-week fu)	0%
Hong et al., 2017; *unkown*	Acrophobia	Both high and low-fear groups demonstrated a significant decrease in acrophobia symptoms at post-test, while the degree of the reduction was significantly greater in the high-fear group than the low-fear group. Gaze-down percentage and subjective fear rating showed a significant group difference, but HR did not	Not provided	Not mentioned

Continued

TABLE 2 Results of automated consumer-based VR-CBT or VRET—cont'd

Study, name of app	Specific phobia type	Outcomes of interest	Effect size (Cohen's d)	Dropout
Levy et al., 2016; *unkown*	Acrophobia	Outcomes (anxiety, presence, therapeutic alliance, heart rate variability) between e-VRET and p-VRET were similar between the two conditions. The remote sessions were well accepted	Not provided	0%
Lindner et al., 2020; *Itsy*	Arachnophobia	There was a significant reduction in fear of spider symptoms at post-test. Results were maintained at a six-month follow-up. The intervention was tolerated and practical	$d = 1.26$ (within, post-test) $d = 1.98$ (within, 6-mo fu)	Not mentioned
Miloff et al., 2016; *Itsy*	Arachnophobia	Large, significant reductions in behavioral avoidance and self-reported fear in both groups at post-treatment were found, with VRET approaching the strong treatment benefits of OST over time. Noninferiority was identified at 3- and 12- months follow-up but was significantly worse until 12 months	VRET $d = 1.49$ (within, post-test) $d = 1.64$ (within, 3-mo fu) $d = 2.01$ (within, 12-mo fu) *OST* in vivo *exposure* $d = 2.39$ (within, post-test) $d = 2.68$ (within, 3-mo fu) $d = 2.27$ (within, 12-mo fu)	4% of the participants completed post-test BAT
Paulus et al., 2019; *unkown*	Fear of darkness	The fear of darkness mobile assisted VR app was significantly less at post-test. The app was acceptable and easy to use	Not provided	30%

This table summarizes the results of automated VR treatments for specific phobia: BAT, behavioral approach test; OST, one session therapist-led; VR, virtual reality; VR-BE, VR behavioral experiments; VR-CBT, VR-cognitive behavior therapy; VRET, virtual reality exposure therapy; e-VRET, remote VRET; p-VRET, physical presence of a therapist VRET.

VR and observed similar results between the two conditions, but the study's sample size was low ($N = 6$) and they do not state whether acrophobia symptoms changed significantly over time. Another study (Millof et al., 2019) conducted a noninferiority trial comparing automated VRET with one session exposure in vivo therapy. They concluded that noninferiority was achieved at the 3- and 12-month follow-ups between the two conditions, and that automated VRET performed significantly worse up until the 12-month follow-up. The drop-out rate ranged from 0% (both nonmobile VR) (Freeman, Lister, et al., 2019; Freeman, Yu, et al., 2019; Levy et al., 2016) to 30% (mobile VR) (Paulus et al., 2019). Aside from psychological outcome measures, two studies included psychophysiological outcome measures (heart rate variability [HRV], measured with a smart watch or wristband) and one study included gaze-down measurements. The gaze-down percentage differed significantly between high and low-fear acrophobia groups, but HRV revealed no differences (Hong et al., 2017; Levy et al., 2016). Automated VR treatments were rated as user-friendly, acceptable, tolerable, and practical (Bentz et al., 2021; Donker et al., 2019; Levy et al., 2016; Lindner et al., 2020; Paulus et al., 2019).

Table 3 specifies the VR environments in each included acrophobia trial (Bentz et al., 2021; Donker et al., 2019, 2020; Freeman et al., 2018; Hong et al., 2017; Levy et al., 2016). VR environments ranged from natural settings (e.g., rural mountain, cliff, snowy mountains) to urban settings (urban town, theater, 10- to 24-story tower block or office complex, elevator). Two studies included sounds in the VR environment (e.g., of the wind or birds, of an elevator going up), while three out of five studies mentioned gamifications in the VR environment for engagement purposes. Please see Figs. 1 and 2

TABLE 3 Description of VR environments of automated, VR-CBT or VRET for acrophobia.

Study, name of app	Description of the VR environment
Bentz et al., 2021; *Easy Heights*	The Easy Heights app captures three scenarios: a rural mountain, cloudy weather and an urban town. The user starts on the ground level on a virtual platform. In each of the scenarios the user can choose between 16 different height levels between 0 and 75 m. Each VR scenario is accompanied by sound according to the specific scenario. Think of the sound of wind when a user is at high altitude or the sound of birds in the rural mountain scenario. Gamified reinforcement elements include getting a balloon when a level is completed, and sound effects when the virtual platform moves upward. Users start at ground level (0 m) and continue to higher levels only if their distress according to the subjective units of distress scales (SUDS) is below 4 for two consecutive ratings. The user can select a SUD score via gaze selection. SUDS are rated continuously during VR exposure, with the first SUD being rated after 10 s at each level. The maximum VR exposure time is 20 min in study phase 1 and 30 min in study phase 2
Donker et al., 2019, 2020, *ZeroPhobia*	ZeroPhobia Fear of heights' VR environment consists of a theater in which the user is replacing a theater manager. The user needs to complete several tasks to prepare the theater for tonight's performance. These tasks are conducted at increasing altitude levels. The first level begins with changing a light bulb on a small ladder. Then, the user proceeds to the edge of the stage to connect the speakers. The third level involves repairing a small damaged platform. For this, the user needs to climb a very high ladder. Next, the spotlight on the highest balcony needs fixing, and finally a cat needs to be rescued that is standing on the end of a gangway high above the stage. A sixth level involves randomly choosing a spot on one of the balconies to look down. The user can choose low or high balconies. The VR environment is further gamified by the use of assets located on the theater floor that users need to collect, such as bags that have been left behind by people in last night's audience, tools to repair the damaged platform, or escaped kittens. In this way, participants are encouraged to look down to confront their own fears ZeroPhobia also provides four 360° videos in which users cross a high bridge, sit on a rooftop with their legs dangling off the edge, stand on the top of a high building, or stand on a high crane. According to their SUD scores, users continue to the next level (SUD score under 4) or are advised to repeat the same level. ZeroPhobia makes use of gaze selection to explore the VR environment or to select SUD scores. Users can practice with VR as often and as long as they want
Freeman et al., 2018; *Now I can do heights*	In "Now I can do heights" the users meet a virtual coach who takes him/her to the atrium of a large 10-story office complex. Many height cues are visible in the atrium, such as balls in the air, people moving. The user can choose a floor (of five in total in the beginning) to start practicing. The user is provided with tasks on each floor to find out whether his or her fear is accurate. The VR environment is gamified. For example, a cat needs to be rescued from a tree, the user plays a music instrument near the floor's edge or can throw balls over the floor's rim. The virtual coach explains each task and gives empathic encouragement. She also repeats key learning points. The participant can practice with VR for as long desired, and can decide when to move to higher floors. The virtual coach asks for belief ratings for the height threat at the end of the session when the user is back down on the atrium ground level. She also encourages the user to practice with real heights between sessions
Hong et al., 2017; *unkown*	Four different situations are presented in this VR environment: an elevator, a cliff, a helicopter over snow-covered mountains, and rooftop walking. The elevator is made of glass and ascends from the 1st to the 45th floor. Sounds of the elevator operating are included. In the second VR environment, the user is on the back seat of a car driving on a road over a steep cliff and is encouraged to look outside. In the heli-skiing environment, users look down from a helicopter flying over snow-covered mountains. They then jump out of the helicopter and ski down a mountain slope at high speed. The last environment consists of a rooftop in which participants walk on a rooftop encircling the top of a tall tower. Each of these environments takes about 5 min to complete. Users can practice with the environments twice per session (2 sessions in total)
Levy et al., 2016; *unkown*	The VR environment consists of subway stations and a 24-story tower block. No further information is provided

This table describes the VR environments for automated VR for acrophobia: SUD, subjects unit of distress; VR, virtual reality.

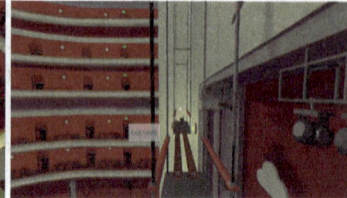

FIG. 1 Screenshots of the automated VR environment for acrophobia: ZeroPhobia. Representative automated VR environment with rudimentary VR equipment: Level 2: connecting speakers. Level 4: replacing a light bulb. Level 5: saving a cat. *(From T. Donker (own materials).)*

FIG. 2 Screenshots of the automated VR environment for acrophobia: Now I can do Heights. Representative automated VR environment with high-end VR equipment: *Now I can do Heights. (From Freeman, D., Haselton, P., Freeman, J., Spanlang, B., Kishore, S., Albery, E., Denne, M., Brown, P., Slater, M., & Nickless, A. (2018). Automated psychological therapy using immersive virtual reality for treatment of fear of heights: a single-blind, parallel-group, randomised controlled trial. Lancet Psychiatry, 5, 625–632 with permission.)*

Stills from the atrium of a large ten-story office complex

Stills from the atrium of a large ten-story office complex

for screenshots of two VR environments for acrophobia, ranging from low-end VR consumer equipment (Fig. 1, Donker et al., 2019) to high-end VR consumer equipment (Freeman et al., 2018).

As with all new domains yet to be explored, the first studies examining consumer VR carried out single-arm trials with small cohorts (Hong et al., 2017; Levy et al., 2016) while more recent studies applied RCT designs with larger sample sizes (e.g., Bentz et al., 2021; Donker et al., 2019; Freeman et al., 2018; Miloff et al., 2016). As several of the trials included in this narrative review provided insufficient information about the primary outcome measure, dropout and adherence rates, recruitment strategy, statistical analyses, or the VR environment used, their replication and/or quality is compromised.

Conclusions

The studies included in this narrative review demonstrated generally promising results for evidence-based automated consumer-based VR-CBT or VRET in reducing the symptoms of acrophobia, arachnophobia, and nyctophobia. However, their findings do need to be taken with caution because of the limited amount of studies included, and the low quality of some of them. For instance, as there were no control conditions in several studies, it is difficult to rule out whether the reported effects are attributable to the VR treatment itself or to other factors, such as regression to the mean or natural remission. In addition, no strong conclusions about other specific phobia types can be drawn because of the absence of automated consumer-based VR studies targeting other forms of specific phobia. However, it is only a matter of time, as the next automated VR-CBT study on the fear of flying is ongoing (Fehribach, Toffolo, van Straten, & Donker, 2021). Furthermore, only one study conducted automated consumer-based VR-CBT in a natural setting without the presence of therapists or research team members (Donker et al., 2019). Hence, the potential effects of human influence in the other studies cannot be ruled out. More research, and the replication of fully automated (that is: with no research team member or therapist present) consumer-based VR-CBT in real life settings is essential to demonstrate its effectiveness.

Overall, the usability and acceptability of automated, consumer-based VR CBT or VRET, if assessed, were positive, indicating that automated VR treatments may indeed help make evidence-based self-help treatments more accessible. Likewise, adherence rates were high. Interestingly, in the two studies delivered through a laptop, the adherence was 100% (Freeman et al., 2018; Levy et al., 2016) unlike in other studies that experienced dropout rates of 23% or 30% (Donker et al., 2019; Paulus et al., 2019). VR-equipment quality may account for these retention differences, but given the low number of studies, we cannot draw conclusions about what influences dropout rates yet. Some studies included follow-up measures at three or even 12 months (Donker et al., 2019; Lindner et al., 2020; Miloff et al., 2016), indicating sustainable effects, but further investigation of long-term timeframes of automated consumer-based VR-CBT or VRET is needed to be able to draw firm conclusions about the sustainability of results.

Limitations and recommendations for further research

This narrative review has several limitations. First, the number of included studies was low, thus limiting our interpretations as to whether automated consumer-based VR treatments are effective in reducing specific phobia symptoms and, for example, if the quality of the VR equipment influences results. Secondly, we had to exclude other technologies in this field such as augmented reality (AR), since VR was the focus of this chapter. Nevertheless, consumer-based AR-CBT or augmented reality exposure therapy (ARET) has great potential to facilitate the dissemination of evidence-based treatment, as its usually costs less than VR (as a person's entire environment need not be modeled (Botella et al., 2016), and AR viewers are not needed. For instance, a recent RCT targeting stand-alone mobile ARET for the fear of spiders (Zimmer et al., 2021) demonstrated high effect sizes. Third, as several studies included in this review are likely to be underpowered, their results are less robust, which also, hampers our interpretations. Finally, because of the low quality of some studies, there may have been a bias in the positive results reported (Cuijpers, Andersson, Donker, & van Straten, 2011).

More research is needed to investigate the effectiveness, user-friendliness, safety, and working mechanisms of consumer-based automated VR-CBT in real-life settings with no research team present. In addition, more investigation is required to examine long-term effects and other types of specific phobia and mental health disorders, the cost-effectiveness of automated VR vs. care as usual, and the influence of low-end vs. high-end consumer VR software with adequate sample sizes. In addition, studies addressing the uptake, dissemination, and implementation of automated VR among those in need are absolutely essential to investigate whether the potential of automated VR, that is, accessible, affordable treatment for everyone, is in fact feasible. Finally, there is a growing availability of low-cost VR readily available in the app store for alleviating mental health disorders without any evidence-base. Therefore, patients need to be warned or guided about which low-cost VR treatments actually work and for which it is unknown.

In sum, although a firm conclusion cannot yet be drawn, the findings from this narrative review suggest that automated VR can be effective in reducing specific phobia symptoms. Automated VR is not meant to replace traditional psychological treatment. Rather, given the ease of dissemination and its affordability, consumer-based automated VR has great potential to make evidence-based mental health treatment more accessible for those whose needs are not being met.

Mini-dictionary of terms

- **App-based VR**. A VR application directly downloadable on a mobile device, such as a smartphone or tablet.
- **Automated VR**. A VR treatment that is automated. When the treatment is activated, it progresses automatically until the end of the treatment, with no therapist involvement.
- **Consumer VR**. Instead of niche markets, consumer VR products are intended for consumers, e.g., the average household. These products prices are priced considerably lower, and the product requires no extensive technical knowledge or know-how to operate.
- **Head-mounted display**. A head-mounted display (HMD) is a viewing screen placed on a person's head. It can be either in the form of a goggle or helmet. It displays an image, one for each eye, thereby creating an overall stereo view. Each of the two images are computed and rendered individually with regards to the position of each eye corresponding to a mathematical manifestation of a three-dimensional (3D) virtual scene. An HMD blocks the user's view of the outside world.
- **Immersive VR**. A form of VR with the aim to immerse the user inside the VR world so convincingly that they can fully engage in the computer-created environment.
- **Self-guided**. A treatment developed to be followed without guidance by a therapist.
- **Stand-alone**. An encompassing treatment meant to function without any interference or guidance by a therapist.
- **Virtual reality (VR)**. Virtual reality (VR) is a technology able to create computer-generated "analogs" of our world. The technology enables a simulated experience that can be similar to or completely different from the real world.

Key facts

Key facts of consumer-based automated VR-CBT for specific phobia

- The first VR device prototype was called Sensorama and was developed by Morton Heilig in 1962. VR in the old days easily cost over $70,000. It was developed for a niche market (e.g., medical, automobile industry, military training).
- The first studies on VR for therapeutic purposes were conducted in the mid-1990s, yielding promising results.
- The first consumer-VR software was developed in 2014 mostly for use in the gaming industry. Prices vary from $10.
- In 2015, the first pilot study on automated VR treatment for specific phobia (acrophobia) was conducted.
- Automated consumer-based VR treatment ranges from very rudimentary (mobile VR viewer and an app on a smartphone) to high-end (gaming laptop and HTC vive HMD) VR equipment.
- Automated consumer-based VR treatment can be applied to improve the scalability and dissemination of evidence-based treatment.

Applications to other areas

In this chapter we have reviewed automated consumer-based VR-CBT and CBT components (e.g., VRET) suitable for people suffering from a specific phobia, with special focus on acrophobia. The studies included in this narrative review demonstrated promising results on reducing the symptoms of acrophobia, arachnophobia, and nyctophobia. Because of its relative low cost and easy dissemination, consumer-based automated VR treatment has great potential to increase the accessibility of evidence-based mental health treatment. Interestingly, aside from specific phobia, consumer-based automated VR treatment has the potential to become broadly applicable for other mental health disorders. Research into traditional VR treatment, which uses high-end VR equipment and is guided by a therapist, has also shown positive results for other anxiety disorders, such as post-traumatic stress disorder, social anxiety, and obsessive compulsive disorder (Freeman et al., 2017). Furthermore, VR treatment research has begun to explore its potential for substance use disorders, psychotic disorders, eating disorders and depression, again with promising results (Geraets et al., 2021). The potential of stand-alone, automated or self-guided interventions to reduce symptoms of these serious mental health conditions has been demonstrated by several studies and metaanalyses targeting internet interventions in which therapist guidance was absent (e.g., Bücker, Westermann, Kühn, & Moritz, 2019; Karyotaki, Riper, Twisk, et al., 2017; Torok et al., 2019). In addition, research has demonstrated that self-guided internet-interventions for serious mental health conditions are acceptable, user-friendly, and not harmful (e.g., Fuller-Tyszkiewicz et al., 2018; Karyotaki et al., 2018). In line with these findings, there is no reason to doubt that automated consumer-based VR treatment can also be effective in helping to alleviate other mental health disorders. For instance, Lindner et al. (2019) investigated the effectiveness of a self-guided mobile VR app for a specific form of social anxiety disorder, namely speaking anxiety, and demonstrated equal effects compared to one session of therapist-led exposure in vivo treatment. The next step targeting more complex disorders is being taken by Freeman et al. (2017), Freeman, Lister, et al. (2019) and Freeman, Yu, et al. (2019), who is currently investigating automated consumer-based VR treatments for delusions and psychosis. Automated consumer-based VR treatment also shows promise for medical conditions such as clinical pain (Donegan, Ryan, Swidrak, & Sanchez-Vives, 2020).

Summary points

- Specific phobia is one of the most prevalent mental health disorders worldwide.
- Most patients do not receive evidence-based treatment.
- Consumer-based automated VR treatment has the potential to significantly increase the accessibility of evidence-based treatment.
- Consumer-based automated VR treatment has demonstrated significant reductions in specific phobia symptoms of acrophobia, arachnophobia, and nyctophobia.
- Consumer-based automated VR treatment has been rated as user-friendly, acceptable, and tolerable.
- More research is needed to examine long-term effects, other types of (more complex) mental health disorders, the quality of consumer VR equipment, its cost-effectiveness and comparability with active control conditions, preferably applying RCT designs in natural settings without research team members and with sufficient power adhering to CONSORT guidelines.
- Implementation in clinical practice and uptake by the general population will be another important area of research to demonstrate the added value of consumer-based automated VR treatment.
- Consumer-based automated VR treatment holds promise for other types of (more complex) mental health and medical disorders.

References

American Psychiatric Association. (2013). *Diagnostic and statistical manual of mental disorders* (5th ed.). American Psychiatric Association.

Antony, M. M., & Barlow, D. H. (2002). Specific phobia. In D. H. Barlow (Ed.), *Anxiety and its disorders: The nature and treatment of anxiety and panic* (2nd ed., pp. 390–417). New York: Guilford.

Becker, E. S., Rinck, M., Türke, V., Kause, P., Goodwin, R., Neumer, S., et al. (2007). Epidemiology of specific phobia subtypes: Findings from the Dresden mental health study. *European Journal of Psychiatry, 22*, 69–74.

Bentz, D., Wang, N., Ibach, M. K., et al. (2021). Effectiveness of a stand-alone, smartphone-based virtual reality exposure app to reduce fear of heights in real-life: A randomized trial. *NPJ Digital Medicine, 4*, 16.

Botella, C., Fernández-Álvarez, J., Guillén, V., García-Palacios, A., & Baños, R. (2017). Recent Progress in virtual reality exposure therapy for phobias: A systematic review. *Current Psychiatry Reports, 19*, 42.

Botella, C., Pérez-Ara, M.Á., Bretón-López, J., Quero, S., García-Palacios, A., & Baños, R. M. (2016). In vivo versus augmented reality exposure in the treatment of small animal phobia: A randomized controlled trial. *PLoS One, 11*, e0148237.

Botella, C., Perpiñá, C., Baños, R. M., & García-Palacios, A. (1998). Virtual reality: A new clinical setting lab. *Studies in Health Technology and Informatics, 58*, 73–81.

Bücker, L., Westermann, S., Kühn, S., & Moritz, S. (2019). A self-guided internet-based intervention for individuals with gambling problems: Study protocol for a randomized controlled trial. *Trials, 20*, 74.

Carl, E., Stein, A. T., Levihn-Coon, A., Pogue, J. R., Rothbaum, B., Emmelkamp, P., et al. (2019). Virtual reality exposure therapy for anxiety and related disorders: A meta-analysis of randomized controlled trials. *Journal of Anxiety Disorders, 61*, 27–36.

Chou, P. H., Tseng, P. T., Wu, Y. C., Chang, J. P. C., Tu, Y. K., Stubbs, B., et al. (2021). Efficacy and acceptability of different interventions for acrophobia: A network meta-analysis of randomised controlled trials. *Journal of Affective Disorders, 282*, 786–794.

Clark, D. A., & Beck, A. T. (2009). *Cognitive therapy of anxiety disorders: Science and practice*. New York: Guilford Publications.

Craske, M. G., Treanor, M., Conway, C. C., Zbozinek, T., & Vervliet, B. (2014). Maximizing exposure therapy: An inhibitory learning approach. *Behaviour Research and Therapy, 58*, 10–23.

Cruz-Neira, C., Sandin, D. J., & DeFanti, T. A. (1993). Surround-screen projection-based virtual reality: The design and implementation of the CAVE. *ACM Computer Graphics (SIGGRAPH) Proceedings, 27*, 135–142.

Cuijpers, P., Andersson, G., Donker, T., & van Straten, A. (2011). Psychological treatment of depression: Results of a series of meta-analyses. *Nordic Journal of Psychiatry, 65*, 354–364.

Curtis, G. C., Magee, W. J., Eaton, W. W., & Wittchen, H. U. (1998). Specific fears and phobias. Epidemiology and classification. *British Journal of Psychiatry, 173*, 212–217.

Davey, G. C. (1992). Classical conditioning and the acquisition of human fears and phobias: A review and synthesis of the literature. *Advances in Behaviour Research and Therapy, 14*, 29–66.

Depla, M. F., ten Have, M. L., van Balkom, A. J., & de Graaf, R. (2008). Specific fears and phobias in the general population: Results from the Netherlands mental health survey and incidence study (NEMESIS). *Social Psychiatry and Psychiatric Epidemiology, 43*, 200–208. 18.

Donegan, T., Ryan, B. E., Swidrak, J., & Sanchez-Vives, M. V. (2020). Immersive virtual reality for clinical pain: Considerations for effective therapy. *Frontiers in Virtual Reality, 1*, 9.

Donker, T., Cornelisz, I., van Klaveren, C., van Straten, A., Carlbring, P., Cuijpers, P., et al. (2019). Effectiveness of self-guided app-based virtual reality cognitive behavior therapy for acrophobia: A randomized clinical trial. *JAMA Psychiatry, 76*, 682–690.

Donker, T., Klaveren, C., Cornelisz, I., Kok, R. N., & van Gelder, J.-L. (2020). Analysis of usage data from a self-guided app-based virtual reality cognitive behavior therapy for acrophobia: A randomized controlled trial. *Journal of Clinical Medicine, 9*, 1614.

Emmelkamp, P. M. (2005). Technological innovations in clinical assessment and psychotherapy. *Psychotherapy and Psychosomatics, 74*, 336–343.

Emmelkamp, P. M., Krijn, M., Hulsbosch, A. M., de Vries, S., Schuemie, M. J., & van der Mast, C. A. (2002). Virtual reality treatment versus exposure in vivo: A comparative evaluation in acrophobia. *Behaviour Research and Therapy, 40*, 509–516.

Fehribach, J., Toffolo, M., van Straten, A., & Donker, T. (2021). Virtual reality self-help treatment for aviophobia: A protocol for a randomized-controlled trial. *JMIR Research Protocols, 10*(4). https://doi.org/10.2196/2200, e22008.

Fodor, L. A., Coteț, C. D., Cuijpers, P., Szamoskozi, Ș., David, D., & Cristea, I. A. (2018). The effectiveness of virtual reality based interventions for symptoms of anxiety and depression: A meta-analysis. *Science Reports, 8*, 10323.

Freeman, D., Haselton, P., Freeman, J., Spanlang, B., Kishore, S., Albery, E., et al. (2018). Automated psychological therapy using immersive virtual reality for treatment of fear of heights: A single-blind, parallel-group, randomised controlled trial. *Lancet Psychiatry, 5*, 625–632.

Freeman, D., Lister, R., Waite, F., Yu, L. M., Slater, M., Dunn, G., et al. (2019). Automated psychological therapy using virtual reality (VR) for patients with persecutory delusions: Study protocol for a single-blind parallel-group randomised controlled trial (THRIVE). *Trials, 29*, 87.

Freeman, D., Reeve, S., Robinson, A., Ehlers, A., Clark, D., Spanlang, B., et al. (2017). Virtual reality in the assessment, understanding, and treatment of mental health disorders. *Psychological Medicine, 47*, 2393–2400.

Freeman, D., Yu, L. M., Kabir, T., Martin, J., Craven, M., Leal, J., et al. (2019). Automated virtual reality (VR) cognitive therapy for patients with psychosis: Study protocol for a single-blind parallel group randomised controlled trial (gameChange). *BMJ Open, 27*, e031606.

Fuller-Tyszkiewicz, M., Richardson, B., Klein, B., Skouteris, H., Christensen, H., Austin, D., et al. (2018). A Mobile app-based intervention for depression: End-user and expert usability testing study. *JMIR Mental Health, 5*, e54.

Geraets, C. N. W., van der Stouwe, E. C. D., Pot-Kolder, R., & Veling, W. (2021). Advances in immersive virtual reality interventions for mental disorders—A new reality? *Current Opinion in Psychology, 13*, 40–45.

Hoffman, H. G., Meyer, W. J., 3rd., Ramirez, M., Roberts, L., Seibel, E. J., Atzori, B., et al. (2014). Feasibility of articulated arm mounted oculus rift virtual reality goggles for adjunctive pain control during occupational therapy in pediatric burn patients. *Cyberpsychology, Behaviour and Social Networking, 17*, 397–401.

Hofmann, S. G., & Smits, J. A. (2008). Cognitive-behavioral therapy for adult anxiety disorders: A meta-analysis of randomized placebo-controlled trials. *Journal of Clinical Psychiatry, 69*, 621–632.

Hong, Y. J., Kim, H. E., Jung, Y. H., Kyeong, S., & Kim, J. J. (2017). Usefulness of the Mobile virtual reality self-training for overcoming a fear of heights. *Cyberpsychology, Behaviour and Social Networking, 20*, 753–761.

Huppert, D., Grill, E., & Brandt, T. (2013). Down on heights? One in three has visual height intolerance. *Journal of Neurology, 260*, 597–604.

Karyotaki, E., Kemmeren, L., Riper, H., Twisk, J., Hoogendoorn, A., Kleiboer, A., et al. (2018). Is self-guided internet-based cognitive behavioural therapy (iCBT) harmful? An individual participant data meta-analysis. *Psychological Medicine, 48*, 2456–2466.

Karyotaki, E., Riper, H., Twisk, J., et al. (2017). Efficacy of self-guided internet-based cognitive behavioral therapy in the treatment of depressive symptoms: A Meta-analysis of individual participant data. *JAMA Psychiatry, 74*, 351–359.

Huppert, D., Wuehr, M., & Brandt, T. (2020). Acrophobia and visual height intolerance: Advances in epidemiology and mechanisms. *Journal of Neurology, 267*, 231–240. https://doi.org/10.1007/s00415-020-09805-4.

Kapfhammer, H. P., Huppert, D., Grill, E., Fitz, W., & Brandt, T. (2015). Visual height intolerance and acrophobia: Clinical characteristics and comorbidity patterns. *European Archives of Psychiatry and Clinical Neuroscience, 265*, 375–385.

Kessler, R. C., Berglund, P. A., Bruce, M. L., et al. (2001). The prevalence and correlates of untreated serious mental illness. *Health Services Research, 36*, 987–1007.

Kessler, R. C., Petukhova, M., Sampson, N. A., Zaslavsky, A. M., & Wittchen, H. U. (2012). Twelve month and lifetime prevalence and lifetime morbid risk of anxiety and mood disorders in the United States. *International Journal of Methods in Psychiatric Research, 21*, 169–184.

LeBeau, R. T., Glenn, D., Liao, B., Wittchen, H. U., Beesdo-Baum, K., Ollendick, T., et al. (2010). Specific phobia: A review of DSM-IV specific phobia and preliminary recommendations for DSM-V. *Depression and Anxiety, 27*, 148–167.

Levy, F., Leboucher, P., Rautureau, G., & Jouvent, R. (2016). E-virtual reality exposure therapy in acrophobia: A pilot study. *Journal of Telemedicine and Telecare, 22*, 215–220.

Lindner, P., Miloff, A., Bergman, C., Andersson, G., Hamilton, W., & Carlbring, P. (2020). Gamified, automated virtual reality exposure therapy for fear of spiders: A single-subject trial under simulated real-world conditions. *Frontiers in Psychiatry, 3*, 116.

Lindner, P., Miloff, A., Fagernäs, S., Andersen, J., Sigeman, M., Andersson, G., et al. (2019). Therapist-led and self-led one-session virtual reality exposure therapy for public speaking anxiety with consumer hardware and software: A randomized controlled trial. *Journal of Anxiety Disorders, 61*, 45–54.

Mackenzie, C. S., Reynolds, K., Cairney, J., Streiner, D., & Sareen, J. (2012). Disorder-specific mental health service use for mood and anxiety disorders: Associations with age, sex, and psychiatric comorbidity. *Depression and Anxiety, 29*, 234–242.

McLuhan, M. (1964). *Understanding media*. New York: Signet.

Miloff, A., Lindner, P., Hamilton, W., Reuterskiöld, L., Andersson, G., & Carlbring, P. (2016). Single-session gamified virtual reality exposure therapy for spider phobia vs. traditional exposure therapy: Study protocol for a randomized controlled non-inferiority trial. *Trials, 17*. 60.

Morina, N., Ijntema, H., Meyerbröker, K., & Emmelkamp, P. M. (2015). Can virtual reality exposure therapy gains be generalized to real-life? A meta-analysis of studies applying behavioral assessments. *Behaviour Research and Therapy, 74*, 18–24.

Nathan, P. E., & Gorman, J. M. (2007). *A guide to treatments that work* (3rd ed.). New York: Oxford University Press.

Norton, P. J., & Price, E. C. (2007). A meta-analytic review of adult cognitive-behavioral treatment outcome across the anxiety disorders. *Journal of Nervous and Mental Disease, 195*, 521–531.

Oosterink, F., de Jongh, A., & Hoogstraten, J. (2009). Prevalence of dental fear and phobia relative to other fear and phobia types. *European Journal of Oral Science, 117*, 135–143.

Paulus, E., Suryan, M., Wijayanti, P. A. K., Yusuf, F. P., & Iskandarsyah, A. (2019). The use of mobile-assisted virtual reality in fear of darkness therapy. *Telkomnika, 17*, 282–290.

Pew Research Center. (2019). *Smartphone ownership is growing rapidly around the world, but not always equally*. Washington, DC: Pew Research Center.

Rothbaum, B. O., Hodges, L. F., Kooper, R., Opdyke, D., Williford, J. S., & North, M. (1995). Effectiveness of computer-generated (virtual reality) graded exposure in the treatment of acrophobia. *American Journal of Psychiatry, 152*, 626–628.

Rizzo, A. S., & Koenig, S. T. (2017). Is clinical virtual reality ready for primetime? *Neuropsychology, 31*, 877–899.

Schäfer, F., Müller, M., Huppert, D., Brandt, T., Tife, T., & Grill, G. (2014). Consequences of visual height intolerance for quality of life: A qualitative study. *Quality of Life Research, 23*, 697–705.

Slater, M., & Sanchez-Vives, M. V. (2016). Enhancing our lives with immersive virtual reality. *Frontiers in Robotics and Artificial Intelligence, 3*, 74.

Stinson, F. S., Dawson, D. S., Chou, S. P., Smith, S., Goldstein, R. B., Ruan, W. J., et al. (2007). The epidemiology of DSM-IV specific phobia in the USA: Results from the national epidemiologic survey on alcohol and related conditions. *Psychological Medicine, 37*, 1047–1059.

Torok, M., Han, J., Baker, S., Werner-Seidler, A., Wong, I., Larsen, M. E., et al. (2019). Suicide prevention using self-guided digital interventions: A systematic review and meta-analysis of randomised controlled trials. *The Lancet, 2*, E25–E36.

Trumpf, J., Margraf, J., Vriends, N., Meyer, A. H., & Becker, E. S. (2010). Specific phobia predicts psychopathology in young women. *Social Psychiatry and Psychiatric Epidemiology, 45*, 1161–1166.

Zimmer, A., Wang, N., Ibach, M., Fehlmann, B., Schicktanz, N., Bentz, D., et al. (2021). Effectiveness of a smartphone-based, augmented reality exposure app to reduce fear of spiders in real-life: A randomized controlled trial. *Journal of Anxiety Disorders, 82*, 102442.

Chapter 7

Cognitive behavioral therapy and adjustment disorder

Soledad Quero[a,c], Sara Fernández-Buendía[a], Rosa M. Baños[b,c], and Cristina Botella[a,c]

[a]Department of Basic, Clinical Psychology, and Psychobiology, Universitat Jaume I, Castellón, Spain, [b]Department of Personality, Evaluation and Psychological Treatment, Universitat de València, Valencia, Spain, [c]CIBER Fisiopatología de la Obesidad y Nutrición (CIBERON), Instituto Salud Carlos III, Madrid, Spain

Abbreviations

ADNM	adjustment disorder new module
AjD	adjustment disorder
APA	American Psychological Association
CBT	cognitive behavioral therapy
DSM	diagnostic and statistical manual of mental disorders
EMDR	eye movement desensitization and reprocessing
EMMA	engaging media for mental health applications
IADQ	International Adjustment Disorder Questionnaire
ICD	International Classification of Diseases and Related Health Problems
ICTs	information and communication technologies
MINI	mini international neuropsychiatric interview
PTSD	posttraumatic stress disorder
RCT	randomized control trial
SCAN	schedule for clinical assessment in neuropsychiatry
SCID	structured clinical interview for DSM-5
TAO	adjustment disorder online
TAU	treatment as usual
TEO	online emotional therapy
VR	virtual reality
WHO	World Health Organization

Introduction

Throughout life, people have to face multiple and diverse stressful situations (life changes, moves, divorces, breakups, loss of a job,…). In most cases, they have the necessary resources to overcome them successfully. Nevertheless, there are some occasions in which people are overwhelmed by their circumstances and show emotional and/or behavioral alterations as a consequence of the stressor or stressors. Diagnostic manuals have tried to bring together these various maladaptive reactions under the diagnostic category known as adjustment disorder (AjD). This stressor-associated disorder has been identified as the seventh most frequently diagnosed by psychiatrists (Reed, Correia, Esparza, Saxena, & Maj, 2011) and the ninth most diagnosed by psychologists (Evans et al., 2013). However, this elevated use of the diagnosis has not been accompanied by an increase in research focused on the study of its course, assessment and treatment (Zelviene & Kazlauskas, 2018). In contrast, AjD has been considered a controversial category since its inclusion in diagnostic manuals (Maercker, Einsle, & Köllner, 2007; Zelviene & Kazlauskas, 2018). Although its definition has undergone a number of modifications to reach its current status, it still remains subject to criticism today (Maercker & Lorenz, 2018).

This chapter will address various aspects related to AjD, including the diagnostic criteria of the disorder, the impact it can have on the lives of people who suffer from it, some aspects related to assessment and, more in depth, the evidence of current treatments, specifically CBT. Finally, the work developed by the authors' research group will be presented, as well as some future trends and an applied case study.

Classification and diagnostic criteria of AjD

The two main diagnostic manuals, the Diagnostic and Statistical Manual of Mental Disorders (DSM-5; American Psychiatric Association, 2013) and the International Classification of Diseases (ICD-11; World Health Organization, 2018), propose a definition of AjD which has considerable similarities but also some important differences. First of all, both manuals include AjD in the subgroup of stress-related disorders. Therefore, the presence of a stressor is a necessary condition to the diagnosis. There is a wide variety of stressful factors that can be associated with an AjD (personal, economic, work-related, relational…). The manuals also share the idea that maladaptive symptoms are originated as a consequence of the stressor(s), which must produce significant impairment in the patient's functioning. They also emphasize that symptoms usually resolve within 6 months (unless the stressful factor persists for longer) and that diagnosis requires that criteria for other mental disorders are not met.

Nevertheless, we can find some significant discrepancies in the manuals. For example, ICD-11 symptoms focus on worry, intrusive thoughts and rumination according to Maercker's proposal (Maercker et al., 2007). This differs from the DSM-5, which maintains a more general description of symptoms, only referring to "emotional and behavioral symptoms."

One of the most important differences refers to the criterion of clinical significance. The DSM-5 is not very strict and allows the criterion to be met when the person presents distress or interference as a consequence of the symptoms. Some authors consider that this may result in an over-pathologizing of stress responses and propose that both criteria (distress and functional impairment) should be met to diagnose AjD (Baumeister & Kufner, 2009; Maercker et al., 2007). In the case of ICD-11, both conditions are required in addition to diagnose AjD.

Finally, it should be noted that the DSM-5 makes a proposal for subtypes of AjD, unlike the ICD-11 which does not include any subtype of the disorder because it considers that there is a lack of evidence to support its existence (Maercker & Lorenz, 2018). The DSM-5 also distinguishes between an acute (the duration is less than 6 months) and chronic AjD (the symptoms last longer than 6 months as long as the stressor or its consequences are sustained).

The definition included in the diagnostic manuals has been controversial from the beginning and its poverty has led to the under-research of AjD. The current conceptualization suggests that the disorder is a kind of "residual category" and that it is a subthreshold disorder (Maercker & Lorenz, 2018; Zelviene & Kazlauskas, 2018). One of the reasons for this is the transitory character of the disorder and that the diagnosis requires that the criteria for no other mental disorder are met. Consequently, it is sometimes regarded as a mild disorder despite evidence to the contrary (Casey, 2014).

Prevalence

Over the years AjD has been commonly excluded from epidemiological studies of mental health (Zelviene & Kazlauskas, 2018). Furthermore, the few investigations that have included it in their analyses had serious methodological limitations or were highly inconsistent with each other (Perkonigg, Lorenz, & Maercker, 2018).

Several factors such as the heterogeneity of the samples and the lack of a standardized measurement instrument for AjD make it difficult to obtain clear results (O'Donnell, Agathos, Metcalf, Gibson, & Lau, 2019; Zelviene & Kazlauskas, 2018). In the general population the estimated prevalence is about 1%–2% according to Zelviene and Kazlauskas (2018). One of the few cohort studies that have been conducted with this type of sample is the one of Maercker et al. (2012) in Germany. The sample consisted of participants with ages between 14 and 93 years old and AjD symptoms were measured following the Maercker's model of intrusive symptoms, avoidance and maladaptive coping (Maercker et al., 2007). The prevalence was estimated at 1.4% for those individuals who did not meet the interference criterion, whereas it was 0.9% for those who showed significant impairment.

For studies with high-risk samples, prevalence figures are usually higher. For example, in patients hospitalized for injuries an estimated prevalence of 19% was found at 3 months after admission and 16% 12 months later (O'Donnell et al., 2016), in a sample of outpatients from a psychiatric clinic it was estimated at 11.5% (Yaseen, 2017) and in a group of people who had lost their job involuntarily the figure was 15.5% at 12 months after stressor took place (Perkonigg et al., 2018). More specifically, the prevalence was 13.8% for males and 17.2% for females, suggesting that women are twice more likely to develop AjD symptoms than men.

Course and impact

AjD is considered a transitory disorder, so that the symptomatology does not usually last more than 6 months once the stressor and its consequences have disappeared (WHO, 2018). Nevertheless, in some cases the symptoms can become more complicated and last longer, increasing the likelihood of developing other psychiatric disorders such as major depression and anxiety (O'Donnell et al., 2016).

AjD causes in people who suffer it a high interference (O'Donnell et al., 2016). There is a link between suffering from this disorder and developing suicidal ideation and behavior (Casey, Jabbar, O'Leary, & Doherty, 2015; Fegan & Doherty, 2019). In fact, AjD is considered one of the most frequent disorders among people who commit self-harm (Benton & Lynch, 2006), and according to Polyakova, Knobler, Ambrumova, and Lerner (1998), the time between the first symptom and the suicide attempt was shorter in patients with AjD than people with major depression.

In addition to the impact that AjD can have on quality of life, there are many economic, social and occupational costs associated with it. In Europe, the estimated cost of stress and trauma-related disorders exceeds 200 billion euros per year (Kalisch et al., 2017).

Assessment

The lack of consensus on the definition of AjD has resulted in a lack of specific assessment instruments (O'Donnell et al., 2019; Zelviene & Kazlauskas, 2018). Very few of the most commonly used interviews include the assessment of this disorder (Casey, 2014). In addition, some of the instruments that allow its diagnosis, including the Structured Clinical Interview for DSM-5 (SCID-I; First, Spitzer, Gibbon, & Williams, 1999), the Schedule for Clinical Assessment in Neuropsychiatry (SCAN; World Health Organization, 1992) and the Mini International Neuropsychiatric Interview (MINI) (Sheehan et al., 1998) do not do it comprehensively (O'Donnell et al., 2019). For example, the SCAN does not include specific questions for AjD and the SCID and MINI interviews only allow the diagnosis of AjD if criteria for any other mental disorder are not present.

Despite the remarkable lack of instruments for AjD, new diagnostic interviews have recently been published. A clear example is the Diagnostic Interview Adjustment Disorder (DIAD; Cornelius, Brouwer, de Boer, Groothoff, & Vanderklink, 2014), which is formed by 29 items following the DSM-5 criteria. It consists of several sections and, among other aspects, analyses the stressful events that have taken place in the 3 years prior to the assessment (including their start date, end date, …), as well as the symptomatology and distress, the temporal relationship between the stressor(s) and the symptoms and the degree of interference caused by the disorder. The validity of this instrument has already been analyzed by Cornelius et al. (2014) with promising results, however, more research is needed to determine reliability and validity.

Following Maercker's proposal for a new definition of AjD in ICD-11 (Maercker et al., 2007) new measures for the disorder have been developed and validated, including the Adjustment Disorder New Module Questionnaire (ADNM; Einsle, Köllner, Dannemann, & Maercker, 2010) and the International Adjustment Disorder Questionnaire (IADQ; Shevlin et al., 2019). Specifically, the ADNM-20 can be used both as a structured clinical interview and in self-report format. It consists of a list of 29 stressors that the person may have experienced in the year prior to the assessment. The instrument includes 20 items which assess intrusions, avoidance, inability to adapt, anxiety, depression, behavioral symptoms and the interference in different areas of functioning. The reliability and validity of the ADNM-20 has been tested in several studies (Bachem, Perkonigg, Stein, & Maercker, 2017). Similarly, the 8-item (ADNM-8; Kazlauskas, Gegieckaite, Eimontas, Zelviene, & Maercker, 2018; Kazlauskas, Zelviene, Lorenz, Quero, & Maercker, 2018) and 4-item (ADNM-4; Ben-Ezra, Mahat-Shamir, Lorenz, Lavenda, & Maercker, 2018) versions also present well-established validation data.

Psychological treatment

The main aim of AjD treatments is to enable the patient to cope with the stressful event (Baños et al., 2008). It is proposed that these interventions usually have 3 common components: reducing or eliminating the impact of the stressor, increasing the patient's coping skills, and reducing or eliminating symptomatology (Domhardt & Baumeister, 2018).

Currently there is no evidence-based treatment of choice for AjD (Casey, 2014). Some of the few systematic reviews analyzing treatments for AjD (Constantin, Dinu, Rogozea, Burtea, & Leasu, 2020; Domhardt & Baumeister, 2018; O'Donnell, Metcalf, Watson, Phelps, & Varker, 2018) highlight the wide variety of interventions and techniques proposed for the treatment of this disorder (brief psychodynamic therapy, relaxation-based therapy, mindfulness, bibliotherapy, interpersonal psychotherapy, cognitive behavioral therapy,…). In addition, many of the intervention studies published have methodological limitations (O'Donnell et al., 2018). Specifically, in the systematic review by O'Donnell et al. (2018) including 29 articles addressing the treatment of AjD the quality of the studies were classified from low to very low. They also found a wide diversity of psychological treatments for AjD, but the most frequent form of treatment was CBT. CBT is also the most recommended by authors (Benton & Lynch, 2006). In general terms and considering the transient nature of

AjD, it is recommended that treatment for this disorder should be focal, brief and rapid using CBT techniques which are evidence-based. Next a review of CBT interventions for AjD will be presented.

Cognitive behavioral therapy

CBT combines both cognitive (e.g., cognitive restructuring) and behavioral (e.g., exposure, behavioral activation, problem solving,…) techniques to produce changes in patients' thoughts and behaviors and, consequently, in their emotions.

Several RCTs have tested the efficacy of CBT delivered in the work setting comparing this treatment with other interventions such as a CBT-based return-to-work intervention (Salomonsson et al., 2017, 2019), a treatment as usual control group (Carta et al., 2012; Van der Klink, Blonk, Schene, & van Dijk, 2003) or a no-treatment condition (Dalgaard et al., 2014, 2017; Glasscock, Carstensen, & Dalgaard, 2018). Overall, CBT showed to be superior to all other interventions with variable effect sizes ranging from small (Glasscock et al., 2018) to moderate (Salomonsson et al., 2019).

The efficacy of CBT in group format has also been tested by 2 RCT. The treatment was effective in reducing symptoms for AjD patients, compared to patients who did not receive any intervention (Cruces et al., 2018) and no significant differences were observed with respect to the individual application of this same intervention (Shaffer, Shapiro, Sank, & Coghlan, 1981). These results are promising since comparable effects are found in both formats individual and group therapy, with the cost-efficacy advantage for the group format.

Finally other studies have analyzed the efficacy of other formats of CBT. Specifically, the bibliotherapy format has been compared with a waitlist control group (Bachem & Maercker, 2016) or with face-to-face delivery of the same treatment (Salomonsson et al., 2018). Both groups presented a clear reduction in AjD symptomatology. However, in the study by Salomonsson et al. (2018) the effect was greater for the group that included face-to-face therapy with a therapist.

These promising results should be taken with caution since that many of the aforementioned studies have serious limitations (e.g., they have very heterogeneous samples, do not report treatment effect sizes, do not include control groups or do not establish adequate follow-up of the results). More research will give light on the efficacy of CBT for AjD.

ICTs for the treatment of AjD

In the last years, new forms of treatment that are being tested by our research team and that make use of Information and Communication Technologies (ICTs) to improve the effectiveness and efficiency of current interventions have emerged.

Virtual reality: EMMA's world

A technology that is becoming increasingly important for the treatment of AjD and other trauma and stress-related disorders is virtual reality (VR) (Quero et al., 2017). This tool allows the creation of interactive virtual worlds through a computer and gives the therapists greater control over the environment. Briefly, it allows them to represent difficult situations graduating the level of difficulty to different stages of treatment (Freeman et al., 2017). According to this author VR can increase access to treatments for mental disorders and may even improve the outcomes of interventions.

Several metaanalyses and systematic reviews have already highlighted the usefulness of VR for the treatment of Post-Traumatic Stress Disorder (e.g., Deng et al., 2019; Eshuis et al., 2020). Regarding specifically AjD treatment, the only intervention identified that uses VR is the one developed by Botella, Baños, and Guillén (2008), which will be explained below.

The protocol developed by Labpsitec research group at Jaume I University (Botella et al., 2008) includes CBT and positive psychology strategies that are accompanied by the use of a VR system for the application of the elaboration/exposure component (Quero et al., 2019).

One of the limitations of the VR studies that have been conducted so far with PTSD patients is that the virtual scenarios they used were specific to certain traumatic events (e.g., a war or a sexual assault) (Baños et al., 2008). However, EMMA's World is an open VR system that adapts its features to the patients' need. It does not need to physically represent the event, but makes use of symbols to expose the patient to their emotions and cognitions whatever the stressful or traumatic event they have experienced (Baños et al., 2008; Quero, Molés, et al., 2019).

The aim is to promote the emotional processing of the stressful event, as proposed by Foa and Kozak (1986). According to these authors, it is necessary to activate the fear structure to include new information and help the person to develop a new meaning of the event (Quero, Molés, et al., 2019).

The EMMA's World consists of a virtual environment that can be manipulated by the patient and therapist via a computer according to the needs of the treatment. It does not require VR glasses for its use, but images are projected on a large

FIG. 1 EMMA's world environments.

screen. EMMA has different scenarios (a desert, an island, a forest, a snowy landscape and a meadow) representing different emotions (see Fig. 1). The environment can be changed in real time and it is also possible to adjust certain factors (time of day, weather,…) and to include new symbols in addition to those already incorporated (e.g., images, figures, music,…) to encourage emotional expression (Botella et al., 2006; Botella et al., 2006; Quero, Molés, et al., 2019).

The VR system also has a set of tools that facilitate the patient's emotional expression (Botella, Baños, Guerrero, et al., 2006; Botella, Baños, Rey, et al., 2006). One of the most important is the "Book of Life," where the patient can represent the adverse experience and update the information as the treatment progresses and constructs a new meaning for the stressful event.

Treatment components

The intervention includes 6 weekly sessions lasting 1.5 h, although it can be extended to two more sessions if the patients require it (Botella et al., 2008; Quero, Molés, et al., 2019). The components of the treatment are briefly presented below:

Psychoeducation: problems are presented as part of life and the most common reactions to a stressful situation are explained to the patients. Finally, an explanatory model on the development and maintenance of AjD is presented.

Elaboration/exposure to the stressful event: this component makes use of the EMMA's World system to enable people to elaborate the negative event and develop a new more positive meaning of it. In this way, the emotional expression of the patients is encouraged through the different symbols and options included in the system. This module is also accompanied by live exposure to those elements related to the stressful event that patients avoid (situations, places, people,…).

Positive psychology techniques: this module includes Popper's (1995) "Problem acceptance training" which tries to make the patients understand that problems are part of life and have an important function. Another exercise is the one of Peterson and Seligman's (2004) "My best virtues or strengths" in which the patients have to select from a list of virtues which ones they have and which ones they would like to develop. Finally, it is necessary to mention the "Heuristics" activity, in which the patients have to choose some statements that make them feel more able to cope with the stressful event (Quero, Molés, et al., 2019).

Strategies from Neimeyer (2000): this section includes the "Book of Life" mentioned above, the "Vital trace" activity in which the patients have to find positive aspects of the adverse event, and finally, the "Projection letter to the future" in which the patients have to write a letter to themselves from the future. This exercise tests the patients' ability to imagine themselves coping with the stressful event in a not too distant future.

Relapse prevention: this is the last component of most psychological treatments and aims to maintain the learning achieved during the intervention.

Efficacy data

The efficacy of this AjD CBT protocol has been tested by several preliminary case studies (Andreu-Mateu, Botella, Quero, Guillén, & Baños, 2012; Baños et al., 2008; Quero et al., 2017) and in a study comparing the effectiveness of CBT with and without VR support (Baños et al., 2011) for the treatment of stress-related disorders (AjD among others). Results showed comparable effects for both conditions (standard CBT and CBT supported by VR). A more recent RCT (Quero, Molés, et al., 2019) compared the application of this protocol with and without VR with a wait-list control group in a large sample of AjD patients. The results showed that, although both treatment groups produced improvements in measures and were effective for the treatment of AjD, the condition that included the use of VR resulted in greater symptom improvement after 1-year follow-up. In addition, the majority of participants expressed a preference for the VR condition (66.7%) over the non-VR condition (33.6%).

Finally, additional evidence for this AjD CBT protocol supported by VR was obtained in a pilot randomized clinical trial (Quero et al., 2019) aimed at examining the feasibility (doability, initial efficacy, and acceptability) of a digital support system to deliver homework via the Internet in the treatment of AjD. The aim was to explore how to increase homework engagement in psychotherapy. *Terapia Emocional Online* (TEO) system is a digital support system that allows therapists to create and send personalized homework sessions to patients via the Internet. The system consists of two platforms, one for the therapist and one for the patient. The first one allows the professional to create the tasks for each user combining multimedia resources (images, videos, audios, texts). The patient's platform allows the patient to answer questionnaires and access the assigned homework tasks. The AjD homework contents included in the protocol can be found in Quero, Rachyla, et al. (2019). Participants in both interventions (traditional homework condition vs. TEO system homework condition) showed statistically significant improvements with large effect sizes after treatment, with no differences between groups. The therapeutic gains were maintained at 1-year follow-up, and additional improvement was also observed at long term, with no differences between groups. This is the first study confirming that the homework component can be successfully applied in a conventional way and through the Internet.

Internet-based treatments for AjD

The need to develop brief treatments that reach as many people as possible has led to the incorporation of Internet in the treatment of psychological disorders (Kazdin & Blase, 2011; Kazlauskas, Gegieckaite, et al., 2018; Kazlauskas, Zelviene, et al., 2018). In addition, the pandemic situation caused by COVID-19 has made even more evident the urgency of developing large-scale interventions that can be administered in any circumstance or location, whatever the epidemiological situation (Kazlauskas & Quero, 2020).

The research suggests that self-help treatments via the internet may be suitable for the treatment of mental disorders, producing similar effects to those achieved with face-to-face therapy (Carlbring, Andersson, Cuijpers, Riper, & Hedman-Lagerlöf, 2018). Furthermore, using these interventions can offer multiple advantages in terms of flexibility, cost savings and confidentiality (Musiat & Tarrier, 2014).

To the authors' knowledge only 5 Randomized Controlled Trials (RCTs) have been conducted for assessing the efficacy of self-applied interventions for AjD (Bachem & Maercker, 2016; Eimontas, Rimsaite, Gegieckaite, Zelviene, & Kazlauskas, 2018; Lindsäter et al., 2018; Moser, Bachem, Berger, & Maercker, 2019; Rachyla et al., 2020). The treatments tested are brief, ranging from 4 to 6 modules, except for the study of Lindsäter et al. (2018) which includes 12 modules. All of these articles found clinical improvement in patients who received the treatment. However, in two of them where the intervention was completely self-administered (without any support or guidance) dropout rates exceeded 65% (Eimontas et al., 2018; Moser et al., 2019). Decreasing these rates remains one of the challenges of internet-based interventions.

Internet-based treatment for AjD: TAO

TAO or Adjustment Disorder Online (Trastorno Adaptativo Online in Spanish) was developed by LABPSITEC research team and it is a self-applied treatment program for AjD that has been adapted to a web platform. The intervention protocol is based on the previously describe CBT by Botella et al. (2008), but it was updated to adapt it to a Web format (Rachyla et al., 2018).

The main objective of TAO is to help the person to be able to adequately process the emotions related to the stressful event and to learn from it (Rachyla et al., 2020). One of the advantages of this program is that it can be adapted to any person, regardless of the stressful event they have experienced (Rachyla et al., 2020). The treatment is accessible via a website (https://www.psicologiaytecnologia.com/) and to log in the person needs their own username and password which are provided at the beginning of the treatment.

TAO consists of 7 sequential treatment modules (about 60 min long) which include: psychoeducation, emotional regulation techniques, exposure techniques, problem solving, mindfulness, acceptance and processing of the adverse event, positive psychology techniques, and relapse prevention (Rachyla et al., 2018). All the components make use of multimedia content, texts and activities (see Fig. 2). Some of them can be downloaded in PDF format (Rachyla et al., 2018).

FIG. 2 TAO web platform and tools.

In addition to the treatment modules, the web platform incorporates a set of tools. Some of the most relevant are: a section where the content can be reviewed at any time, a calendar to keep track of the sessions and graphs showing the patient's progress throughout the treatment (Rachyla et al., 2018, 2020).

Treatment components

The 7 modules of the programmer have a similar structure. All of them start with the presentation of the agenda, then the specific contents for that module, the practice activities, the postmodule assessment, the homework assignments and a summary of the key ideas (Rachyla et al., 2018). A brief description of the content included in the modules is presented in Table 1. A more detailed description can be found in Rachyla et al. (2018).

TABLE 1 TAO treatment modules and their content.

Module	Content
0. Starting this programmer	A description of the treatment program is included. An explanation is also presented to the patient on the importance of their active participation during evaluations and homework between sessions
1. Understanding emotional reactions	A description of the most common reactions to stressful events is presented and the explanatory model of AjD and some useful strategies for managing emotional reactions (such as behavioral activation and the slow breathing technique) are explained
2. Learning to cope with negative emotions	Exposure and problem-solving techniques are explained
3. Accepting problems	The first part of the elaboration of the event (known as "Acceptance") is introduced, making use of the "Book of Life," metaphors and acceptance procedures. This module also includes a mindfulness practice
4. Learning from problems	Problems are presented as an opportunity to learn. The practice of elaboration is continued (chapter known as "Coping" in the "Book of Life") and the concept of virtues or psychological strengths is introduced
5. Changing the meaning of problems	A new metaphorical description of the event is made, elaboration is continued (chapter known as "Change of meaning" in the "Book of Life") and the patient's ability to develop a new attitude toward problems is promoted ("Letter from the future" and "Heuristics" activities)
6. Relapse prevention	Progress and skills achieved during treatment are evaluated, techniques are reviewed, high-risk situations are worked on and follow-ups are planned

Efficacy and acceptability data

The research carried out by Rachyla et al. (2017) tested the acceptability and usability of TAO. The results showed that the program was well accepted by both patients and therapists, and both found the programmer easy to use and appreciated the clear and simple organization of the treatment modules, the presence of multimedia material and the use of easy-to-understand language. However, they also expressed the need to reduce the amount of text. In a later study, Rachyla et al. (2020) conducted a RCT to test the effectiveness of TAO (accompanied by a brief weekly phone call) in comparison to a wait-list control. Participants in TAO condition showed statistically significant improvements, compared to the control group. More specifically, the program was effective in reducing negative affect, depressive symptoms, anxiety and AjD symptoms. In addition, it was successful in enhancing patients' posttraumatic growth, positive affect and quality of life. This trend was maintained in the long-term, however, the lack of a control group at follow-up limits the generalization of the results.

One of the limitations of the research was the drop-out rate (23.55%) (Rachyla et al., 2020). Although this figure is lower than those reported in the literature for this type of interventions (30%) (Melville, Casey, & Kavanagh, 2010), more research is still needed to reduce the number of people who drop out of internet-based treatments.

Future directions

Blended treatments

One solution to the high number of drop-outs from self-administered treatments is the use of blended interventions, which combine self-applied treatment with face-to-face sessions with a therapist. This way of delivering the treatment components facilitates the learning process for patients (Cucciare, Weingardt, & Villafranca, 2008). In addition, face-to-face sessions allow a therapeutic relationship to be established with the professional, which decreases dropouts (Erbe, Eichert, Riper, & Ebert, 2017). Finally, this type of format also results in significant financial savings that can exceed 33%–57% (Schuster, Fichtenbauer, Sparr, Berger, & Laireiter, 2018).

So far, the authors are not aware of any blended psychological treatment for AjD that has been tested. Only a study with similar characteristics is that of Leterme et al. (2020), however, the face-to-face sessions were conducted by a nurse instead of a psychologist and lasted only 10 min.

Our research team is working on a feasibility study to test a blended application of TAO. Using this format, the patient will continue to do most of the treatment through the web platform but will receive a face-to-face support session from time to time. The average time spent on each module will be about 10–12 days and after each module there will be an individual session via videoconference with a therapist. During the face-to-face sessions (lasting about 30 min) the therapist will explain the main contents of each module, answer questions and motivate the patient to continue with the therapy. It will take about 12 weeks to complete the intervention (Rachyla et al., 2018, 2020). It is expected that this way combining online and face-to-face will help to reduce drop-out rates while still being a cost-effective treatment alternative.

Practical case

Marcos is a 33-year-old man who asked for psychological help because his mood had worsened since he lost his job 5 months ago. After working in a law firm for 8 years, the company started to make financial losses and decided to lay off some workers. Marcos was one of them and referred that the day he was informed of his dismissal was "very hard" and that he experienced it as "unbearable." Marcos reported that he experiences feelings of guilt about the reason for the dismissal ("what did I do wrong?" "why did it have to be me?") and that he finds it very difficult to accept his new reality. He now has many free hours in the day and spends them worrying and ruminating about whether he will be able to pay the mortgage or whether he will find a new job. He believes that he will not find a job that he likes as much as his old one. He also feels unable to look for a new job because he thinks he might not be fit to be a lawyer. Since the dismissal, Marcos has stopped doing many of the activities he used to enjoy (he no longer meets his friends, he has stopped hiking,…) and says that he is now a much "sadder" and "irritable with others" person.

Using the "Book of Life" of TAO treatment program, the elaboration of this patient's stressful event was carried out. In the different parts of the elaboration (acceptance, coping and change of meaning) he was asked to write about his thoughts, emotions, sensations and memories and to answer a series of questions about how he saw the stressful event. A brief summary of the meaning reconstruction process in this patient is presented below.

Chapter 1: Acceptance

Write down what you remember about the stressful event (thoughts, emotions, sensations,...) and choose a metaphor to represent it. Answer the following questions: What expectations of life did you have before the stressful event? What goals and objectives did you want to achieve? Do you think your life has changed since the event took place?

I have chosen the stone as a symbol to represent how I feel because this problem is like a weight that I am carrying and that I cannot get rid of. This weight prevents me from moving forward with my life. Before I was fired, I had not received any negative feedback from my boss, so I was not expecting it. When he told me "We have to take this decision for the good of the company, I am very sorry Marcos", I started to feel very anxious, my heart ached, and I couldn't help crying in front of him. Now I feel ashamed, and I think I shouldn't have reacted like that. After the dismissal I went home and told my partner, who tried to console me but was unsuccessful. I spent the day in bed and that is what I have continued to do until now. I'm not like I used to be, I've become much less cheerful and sociable. I spend the day angry or crying and I can't sleep at night. Also, I don't feel like or have the strength to look for another job, so I avoid anything that has to do with that. I feel that I am weak, that I will not be able to overcome this situation and that I will not be able to find another job.

Chapter 2: Coping

Write down what you remember about the stressful event (thoughts, emotions, sensations,...) and choose a metaphor to represent it. Answer the following questions: Do you think you may be interpreting the events in a very negative way? Do you think there are alternative ways of doing it? What has the stressful event taught you? Did it help you to value some specific aspects of your life more?

In this chapter I have chosen a storm as a metaphor because I think it represents my current state. Somehow, right now everything is complicated in my life, and I am overwhelmed, but I hope that one day the storm will pass and I will be able to feel better. I can't stop thinking about whether I did something wrong, whether I did something to get myself fired. It keeps giving me insomnia and affects my relationships with others. My partner keeps telling me that I need to encourage myself to look for another job, that I am a valid person. But the truth is that right now that doesn't comfort me, rather, it makes me feel guilty for not doing anything.

The truth is that I may be assessing the situation too negatively. Maybe there is something to learn from all this. I now know that I have a partner who supports me in bad times, and I am grateful for that. I can also see it as an opportunity to get a better job.

Chapter 3: Change of meaning

Write down what you remember about the stressful event (thoughts, emotions, sensations,...) and choose a metaphor to represent it. Answer the following questions: Do you think that since the beginning of the treatment it has changed the way you see the problem? Do you think that facing this event that has been so painful for you can have benefits in the future? Do you think that the way you see the problem will be different in 20 or 30 years?

I have decided to use the metaphor of the stone again. However, now I have a pickaxe, a tool, so that I can break it bit by bit. I know it will take a lot of effort, but I think it will be worth it. I still spend many days in bed and my motivation to look for work is still low. However, I am starting to realize that I can look at redundancy as an opportunity rather than a problem. A chance to change, to turn my life around and to choose where and how I want to work. Not everything was positive in the previous company, they were often late in paying my salary, so now I will look for a place that gives me more stability. I think that if I continue to work on these ideas, I will feel much better (…).

Mini-dictionary of terms

- ***Stressor.*** It refers to any stimulus or event that can potentially produce a stress response in the person. It can refer to social (divorces, conflicts, break-ups in love or with friends,...), economic, work or health aspects (illnesses), among others.
- ***Subthreshold disorder.*** Those disorders that, although they may course with some relevant symptomatology, do not meet the minimum diagnostic criteria necessary to make a full diagnosis of a mental disorder.
- ***Clinical significance.*** It indicates the point at which the interference produced by the symptomatology exceeds the threshold of normality.
- ***Module.*** The modules are the online equivalent of face-to-face therapy sessions.

- *Self-applied treatment.* Interventions that can be carried out by the person without the need for a therapist. They are usually via the internet.
- *Internet-based treatment.* Those treatments that make use of the internet for their application.
- *Blended format.* A treatment delivery format that involves alternating a self-delivered intervention via the internet with face-to-face sessions with a therapist.
- *Randomized Control Trial.* A type of experimental research that assesses the impact of a treatment on a randomly selected sample. The control group is also randomly selected from the population.
- *Wait-list control group.* This is a comparison group that receives the same treatment as the experimental group but after the end of the study.
- *Treatment As Usual.* It is a kind of control group in which the patient receives the treatment usually received in clinical practice.
- *Feasibility and acceptability studies.* They focus on analyzing whether applying a certain treatment is feasible in relation to different aspects (economic, technical, user acceptance, usability of the system,...).

Applications to other areas

EMMA's World: The adaptive VR system described in the present chapter can be used to treat other stress and trauma related disorders (prolonged grief disorder, acute stress disorder, PTSD) (e.g., Baños et al., 2011), other psychological disorders such as phobias (Botella, Baños, Guerrero, et al., 2006; Botella, Baños, Rey, et al., 2006) or physical conditions (fibromyalgia) where emotions play an important role (García-Palacios et al., 2015).

Key facts of adjustment disorder

- It is one of the most frequently diagnosed disorders.
- The prevalence ranges from 1% to 2% in the general population to 19% in high-risk samples.
- The prevalence, assessment, and treatment of this disorder are poorly investigated.
- Adjustment disorder is associated with an increased risk of suicidal ideation and behavior.
- This disorder carries a high social, economic, and occupational cost.

Key facts of ICTs for the treatment of adjustment disorder

- Previous studies suggest that adjustment disorder may be suitable for treatments that make use of technologies.
- Different technology-supported treatments have been tested for the treatment of AjD, such as virtual reality and Internet-based treatments.
- These interventions can offer multiple advantages, such as flexibility, easy access, cost savings, and confidentiality.

Summary points

- There is currently no treatment of choice for AjD.
- The intervention that has shown the greatest efficacy is CBT.
- The use of ICTs can improve the outcomes of current interventions.
- VR has proven to be a useful tool with promising results in the treatment of AjD.
- Using the internet to assign homework between sessions is doable and feasible and could improve patients' motivation during treatment.
- The application of an internet-based treatment for AjD has shown good acceptability and usability, resulting in significant symptom improvement compared to a wait-list control group.
- The use of a blended treatment format (combining online and face-to-face delivery) may improve adherence to treatment and decrease dropout rates, which would result also in cost savings of approximately 33%–57%.

References

American Psychiatric Association. (2013). *Diagnostic and statistical manual of mental disorders* (5th ed.). Madrid: Editorial Médica Panamericana.

Andreu-Mateu, S., Botella, C., Quero, S., Guillén, V., & Baños, R. M. (2012). El tratamiento de los trastornos adaptativos: cuando el estímulo estresante sigue presente. *Behavioral Psychology, 20*, 323–348.

Bachem, R., & Maercker, A. (2016). Self-help interventions for adjustment disorder problems: A randomized waiting-list controlled study in a sample of burglary victims. *Cognitive Behaviour Therapy*, 45(5), 397–413. https://doi.org/10.1080/16506073.2016.1191083.

Bachem, R., Perkonigg, A., Stein, D. J., & Maercker, A. (2017). Measuring the ICD-11 adjustment disorder concept: Validity and sensitivity to change of the Adjustment Disorder-New Module questionnaire in a clinical intervention study. *International Journal of Methods in Psychiatric Research*, 26(4). https://doi.org/10.1002/mpr.1545, e1545.

Baños, R. M., Guillén, V., Botella, C., García-Palacios, A., Jorquera, M., & Quero, S. (2008). Un programa de tratamiento para los trastornos adaptativos. Un estudio de caso. *Apuntes de Psicología*, 26(2), 303–316. Retrieved fro https://idus.us.es/bitstream/handle/11441/84983/12.pdf?sequence=1&isAllowed=y.

Baños, R. M., Guillen, V., Quero, S., Garcia-Palacios, A., Alcaniz, M., & Botella, C. (2011). A virtual reality system for the treatment of stress-related disorders: A preliminary analysis of efficacy compared to a standard cognitive behavioral program. *International Journal of Human-Computer Studies*, 69(9), 602–613. https://doi.org/10.1016/j.ijhcs.2011.06.002.

Baumeister, H., & Kufner, K. (2009). It is time to adjust the adjustment disorder category. *Current Opinion in Psychiatry*, 22(4), 409–412. https://doi.org/10.1097/YCO.0b013e32832cae5e.

Ben-Ezra, M., Mahat-Shamir, M., Lorenz, L., Lavenda, O., & Maercker, A. (2018). Screening of adjustment disorder: Scale based on the ICD-11 and the adjustment disorder new module. *Journal of Psychiatric Research*, 103, 91–96. https://doi.org/10.1016/j.jpsychires.2018.05.011.

Benton, T. D., & Lynch, J. (2006). *Adjustment disorders*. EMedicine.

Botella, C., Baños, R. M., Guerrero, B., García-Palacios, A., Quero, S., & Raya, M. A. (2006). Using a flexible virtual environment for treating a storm phobia. *PsychNology Journal*, 4(2), 129–144. Retrieved fro http://www.psychnology.org/File/PNJ4(2)/PSYCHNOLOGY_JOURNAL_4_2_BOTELLA.pdf.

Botella, C., Baños, R. M., & Guillén, V. (2008). Creciendo en la adversidad. Una propuesta de tratamiento para los trastornos adaptativos. In C. Vázquez, & G. Hervás (Eds.), *Psicología Positiva Aplicada* (pp. 129–154). Bilbao: DDB.

Botella, C., Baños, R. M., Rey, B., Alcañiz, M., Guillén, V., Quero, S., & García-Palacios, A. (2006). Using an adaptive display for the treatment of emotional disorders: A preliminary analysis of effectiveness. In *CHI 2006, April 22–27, 2006, Montreal, Canadá*.

Carlbring, P., Andersson, G., Cuijpers, P., Riper, H., & Hedman-Lagerlöf, E. (2018). Internet-based vs. face-to-face cognitive behavior therapy for psychiatric and somatic disorders: An updated systematic review and meta-analysis. *Cognitive Behaviour Therapy*, 47(1), 1–18. https://doi.org/10.1080/16506073.2017.1401115.

Carta, M. G., Petretto, D., Adamo, S., Bhat, K. M., Lecca, M. E., Mura, G., & Moro, M. F. (2012). Counseling in primary care improves depression and quality of life. *Clinical Practice and Epidemiology in Mental Health: CP & EMH*, 8, 152. https://doi.org/10.2174/1745017901208010152.

Casey, P. (2014). Adjustment disorder: New developments. *Current Psychiatry Reports*, 16(6), 451. https://doi.org/10.1007/s11920-014-0451-2.

Casey, P., Jabbar, F., O'Leary, E., & Doherty, A. M. (2015). Suicidal behaviours in adjustment disorder and depressive episode. *Journal of Affective Disorders*, 174, 441–446. https://doi.org/10.1016/j.jad.2014.12.003.

Constantin, D., Dinu, E. A., Rogozea, L., Burtea, V., & Leasu, F. G. (2020). Therapeutic interventions for adjustment disorder: A systematic review. *American Journal of Therapeutics*, 27(4), e375–e386. https://doi.org/10.1097/MJT.0000000000001170.

Cornelius, L. R., Brouwer, S., de Boer, M. R., Groothoff, J. W., & Vanderklink, J. J. (2014). Development and validation of the diagnostic interview for adjustment disorder (DIAD). *International Journal of Methods in Psychiatric Research*, 23(2), 192–207. https://doi.org/10.1002/mpr.1418.

Cruces, J. M. S., Cuenca, I. M. G., Lacomba-Trejo, L., Adell, M.Á. C., Navarro, I. C., Cortés, M. F., … Álvarez, E. C. (2018). Group therapy for patients with adjustment disorder in primary care. *The Spanish Journal of Psychology*, 21. https://doi.org/10.1017/sjp.2018.51.

Cucciare, M. A., Weingardt, K. R., & Villafranca, S. (2008). Using blended learning to implement evidence-based psychotherapies. *Clinical Psychology: Science and Practice*, 15(4), 299–307. https://doi.org/10.1111/j.1468-2850.2008.00141.x.

Dalgaard, L., Eskildsen, A., Carstensen, O., Willert, M., Andersen, J., & Glasscock, D. (2014). Changes in self-reported sleep and cognitive failures: A randomized controlled trial of a stress management intervention. *Scandinavian Journal of Work Environment Health*, 40(6). Retrieved fro http://www.jstor.org/stable/43188058.

Dalgaard, V. L., Andersen, L. P. S., Andersen, J. H., Willert, M. V., Carstensen, O., & Glasscock, D. J. (2017). Work-focused cognitive behavioral intervention for psychological complaints in patients on sick leave due to work-related stress: Results from a randomized controlled trial. *Journal of Negative Results in Biomedicine*, 16(1), 1–12. https://doi.org/10.1186/s12952-017-0078-z.

Deng, W., Hu, D., Xu, S., Liu, X., Zhao, J., Chen, Q., … Li, X. (2019). The efficacy of virtual reality exposure therapy for PTSD symptoms: A systematic review and meta-analysis. *Journal of Affective Disorders*, 257, 698–709. https://doi.org/10.1016/j.jad.2019.07.086.

Domhardt, M., & Baumeister, H. (2018). Psychotherapy of adjustment disorders: Current state and future directions. *The World Journal of Biological Psychiatry*, 19(1), S21–S35. https://doi.org/10.1080/15622975.2018.1467041.

Eimontas, J., Rimsaite, Z., Gegieckaite, G., Zelviene, P., & Kazlauskas, E. (2018). Internet-based self-help intervention for ICD-11 adjustment disorder: Preliminary findings. *Psychiatric Quarterly*, 89(2), 451–460. https://doi.org/10.1007/s11126-017-9547-2.

Einsle, F., Köllner, V., Dannemann, S., & Maercker, A. (2010). Development and validation of a self-report for the assessment of adjustment disorders. *Psychology, Health & Medicine*, 15(5), 584–595. https://doi.org/10.1080/13548506.2010.487107.

Erbe, D., Eichert, H. C., Riper, H., & Ebert, D. D. (2017). Blending face-to-face and internet-based interventions for the treatment of mental disorders in adults: Systematic review. *Journal of Medical Internet Research*, 19(9). https://doi.org/10.2196/jmir.6588, e306.

Eshuis, L. V., van Gelderen, M. J., van Zuiden, M., Nijdam, M. J., Vermetten, E., Olff, M., & Bakker, A. (2020). Efficacy of immersive PTSD treatments: A systematic review of virtual and augmented reality exposure therapy and a meta-analysis of virtual reality exposure therapy. *Journal of Psychiatric Research*. https://doi.org/10.1016/j.jpsychires.2020.11.030.

Evans, S. C., Reed, G. M., Roberts, M. C., Esparza, P., Watts, A. D., Correia, J. M., … Saxena, S. (2013). Psychologists' perspectives on the diagnostic classification of mental disorders: results from the WHO-IUPsyS Global Survey. *International Journal of Psychology, 48*(3), 177–193. https://doi.org/10.1080/00207594.2013.804189.

Fegan, J., & Doherty, A. M. (2019). Adjustment disorder and suicidal behaviours presenting in the general medical setting: A systematic review. *International Journal of Environmental Research and Public Health, 16*(16), 2967. https://doi.org/10.3390/ijerph16162967.

First, M., Spitzer, R., Gibbon, M., & Williams, J. (1999). *Structured clinical interview for DSM-IV Axis I disorders: SCID-I*. New York: Biometrics Research Department.

Foa, E. B., & Kozak, M. J. (1986). Emotional processing of fear: Exposure to corrective information. *Psychological Bulletin, 99*(1), 20. https://doi.org/10.1037/0033-2909.99.1.20.

Freeman, D., Reeve, S., Robinson, A., Ehlers, A., Clark, D., Spanlang, B., & Slater, M. (2017). Virtual reality in the assessment, understanding, and treatment of mental health disorders. *Psychological Medicine, 47*(14), 2393–2400. https://doi.org/10.1017/S003329171700040X.

García-Palacios, A., Herrero, H., Vizcaíno, Y., Belmonte, M. A., Castilla, D., Molinari, G., … Botella, C. (2015). Integrating virtual reality with activity management for the treatment of fibromyalgia. *The Clinical Journal of Pain, 31*(6), 564–572. https://doi.org/10.1097/AJP.0000000000000196.

Glasscock, D. J., Carstensen, O., & Dalgaard, V. L. (2018). Recovery from work-related stress: A randomized controlled trial of a stress management intervention in a clinical sample. *International Archives of Occupational and Environmental Health, 91*(6), 675–687. https://doi.org/10.1007/s00420-018-1314-7.

Kalisch, R., Baker, D. G., Basten, U., Boks, M. P., Bonanno, G. A., Brummelman, E., Chmitorz, A., Fernàndez, G., Fiebach, C., Galatzer-Levy, I., Geuze, E., Groppa, S., Helmreich, I., Hendler, T., Hermans, E., Jovanovic, T., Kubiak, T., Lieb, K., Lutz, B., … Kleim, B. (2017). The resilience framework as a strategy to combat stress-related disorders. *Nature Human Behaviour, 1*(11), 784–790. https://doi.org/10.1038/s41562-017-0200-8.

Kazdin, A. E., & Blase, S. L. (2011). Rebooting psychotherapy research and practice to reduce the burden of mental illness. *Perspectives on Psychological Science, 6*(1), 21–37. https://doi.org/10.1177/1745691610393527.

Kazlauskas, E., Zelviene, P., Lorenz, L., Quero, S., & Maercker, A. (2018). A scoping review of ICD-11 adjustment disorder research. *European Journal of Psychotraumatology, 8*(7), 1421819. https://doi.org/10.1080/20008198.2017.1421819.

Kazlauskas, E., Gegieckaite, G., Eimontas, J., Zelviene, P., & Maercker, A. (2018). A brief measure of the international classification of diseases-11 adjustment disorder: Investigation of psychometric properties in an adult help-seeking sample. *Psychopathology, 51*, 10–15. https://doi.org/10.1159/000484415.

Kazlauskas, E., & Quero, S. (2020). Adjustment and coronavirus: How to prepare for COVID-19 pandemic-related adjustment disorder worldwide? *Psychological Trauma: Theory, Research, Practice, and Policy, 12*(1), 22. https://doi.org/10.1037/tra0000706.

Leterme, A. C., Behal, H., Demarty, A. L., Barasino, O., Rougegrez, L., Labreuche, J., … Servant, D. (2020). A blended cognitive behavioral intervention for patients with adjustment disorder with anxiety: A randomized controlled trial. *Internet Interventions, 21*. https://doi.org/10.1016/j.invent.2020.100329, 100329.

Lindsäter, E., Axelsson, E., Salomonsson, S., Santoft, F., Ejeby, K., Ljotsson, B., … Hedman-Lagerlöf, E. (2018). Internet-based cognitive behavioral therapy for chronic stress: A randomized controlled trial. *Psychotherapy and Psychosomatics, 87*(5), 296–305. https://doi.org/10.1159/000490742.

Maercker, A., Einsle, F., & Köllner, V. (2007). Adjustment disorders as stress response syndromes: A new diagnostic concept and its exploration in a medical sample. *Psychopathology, 40*(3), 135–146. https://doi.org/10.1159/000099290.

Maercker, A., & Lorenz, L. (2018). Adjustment disorder diagnosis: Improving clinical utility. *The World Journal of Biological Psychiatry, 19*(1), S3–S13. https://doi.org/10.1080/15622975.2018.1449967.

Maercker, A., Forstmeier, S., Pielmaier, L., Spangenberg, L., Brähler, E., & Glaesmer, H. (2012). Adjustment disorders: Prevalence in a representative nationwide survey in Germany. *Social Psychiatry and Psychiatric Epidemiology, 47*(11), 1745–1752. https://doi.org/10.1007/s00127-012-0493-x.

Melville, K., Casey, L., & Kavanagh, D. (2010). Dropout from Internet-based treatment for psychological disorders. *British Journal of Clinical Psychology, 49*(4), 455–471. https://doi.org/10.1348/014466509X472138.

Moser, C., Bachem, R., Berger, T., & Maercker, A. (2019). ZIEL: Internet-based self-help for adjustment problems: Results of a randomized controlled trial. *Journal of Clinical Medicine, 8*(10), 1655. https://doi.org/10.3390/jcm8101655.

Musiat, P., & Tarrier, N. (2014). Collateral outcomes in e-mental health: A systematic review of the evidence for added benefits of computerized cognitive behavior therapy interventions for mental health. *Psychological Medicine, 44*(15), 3137–3150. https://doi.org/10.1017/S0033291714000245.

Neimeyer, R. A. (2000). Searching for the meaning of meaning: Grief therapy and the process of reconstruction. *Death Studies, 24*(6), 541–558. https://doi.org/10.1080/07481180050121480.

O'Donnell, M. L., Alkemade, N., Creamer, M., McFarlane, A., Silove, D., Bryant, R., … Forbes, D. (2016). A longitudinal study of adjustment disorder after trauma exposure. *The American Journal of Psychiatry, 173*(12), 1231–1238. https://doi.org/10.1176/appi.ajp.2016.16010071.

O'Donnell, M. L., Metcalf, O., Watson, L., Phelps, A., & Varker, T. (2018). A systematic review of psychological and pharmacological treatments for adjustment disorder in adults. *Journal of Traumatic Stress, 31*(3), 321–331. https://doi.org/10.1002/jts.22295.

O'Donnell, M. L., Agathos, J. A., Metcalf, O., Gibson, K., & Lau, W. (2019). Adjustment disorder: Current developments and future directions. *International Journal of Environmental Research and Public Health, 16*(14), 2537. https://doi.org/10.3390/ijerph16142537.

Peterson, C., & Seligman, M. E. (2004). *Vol. 1. Character strengths and virtues: A handbook and classification*. USA: Oxford University Press.

Perkonigg, A., Lorenz, L., & Maercker, A. (2018). Prevalence and correlates of ICD-11 adjustment disorder: Findings from the Zurich adjustment disorder study. *International Journal of Clinical and Health Psychology, 18*(3), 209–217. https://doi.org/10.1016/j.ijchp.2018.05.001.

Polyakova, I., Knobler, H. Y., Ambrumova, A., & Lerner, V. (1998). Characteristics of suicidal attempts in major depression versus adjustment reactions. *Journal of Affective Disorders, 47*(1–3), 159–167. https://doi.org/10.1016/S0165-0327(97)00137-7.

Popper, K. (1995). *Vol. 39. Karl Popper: Philosophy and problems*. Cambridge: Cambridge University Press.

Quero, S., Andreu-Mateu, S., Moragrega Vergara, I., Baños, R., Molés Amposta, M., Nebot Ibáñez, S., & Botella, C. (2017). Un programa cognitivo-conductual que utiliza la realidad virtual para el tratamiento de los trastornos adaptativos: una serie de casos. *Revista Argentina de Clínica Psicológica*, *26*(1), 5–18. Retrieved fro http://repositori.uji.es/xmlui/bitstream/handle/10234/170773/54562.pdf?sequence=1.

Quero, S., Molés, M., Campos, D., Andreu-Mateu, S., Baños, R. M., & Botella, C. (2019). An adaptive virtual reality system for the treatment of adjustment disorder and complicated grief: 1-year follow-up efficacy data. *Clinical Psychology & Psychotherapy, 26*, 204–217. https://doi.org/10.1002/cpp.2342.

Quero, S., Rachyla, I., Molés, M., Mor, S., Tur, C., Cuijpers, P., ... Botella, C. (2019). Can between-session homework be delivered digitally? A pilot randomized clinical trial of CBT for adjustment disorders. *International Journal of Environmental Research and Public Health, 16*(20), 3842. https://doi.org/10.3390/ijerph16203842.

Rachyla, I., Quero, S., Pérez Ara, M.Á., Molés Amposta, M., Campos, D., & Mira, A. (2017). Web-based, self-help intervention for adjustment disorders: Acceptance and usability. *Annual Review of Cybertherapy and Telemedicine, 15*, 207–210. Retrieved fro http://repositori.uji.es/xmlui/bitstream/handle/10234/174699/58317.pdf?sequence=1&isAllowed=y.

Rachyla, I., Pérez-Ara, M., Molés, M., Campos, D., Mira, A., Botella, C., & Quero, S. (2018). An internet-based intervention for adjustment disorder (TAO): Study protocol for a randomized controlled trial. *BMC Psychiatry, 18*(1), 1–10. https://doi.org/10.1186/s12888-018-1751-6.

Rachyla, I., Mor, S., Cuijpers, P., Botella, C., Castilla, D., & Quero, S. (2020). A guided Internet-delivered intervention for adjustment disorders: A randomized controlled trial. *Clinical Psychology & Psychotherapy*, 1–12. https://doi.org/10.1002/cpp.2518.

Reed, G. M., Correia, J. M., Esparza, P., Saxena, S., & Maj, M. (2011). The WPA-WHO global survey of psychiatrists' attitudes towards mental disorders classification. *World Psychiatry, 10*, 118–131. https://doi.org/10.1002/j.2051-5545.2011.tb00034.x.

Salomonsson, S., Santoft, F., Lindsäter, E., Ejeby, K., Ljótsson, B., Öst, L. G., ... Hedman-Lagerlöf, E. (2017). Cognitive–behavioural therapy and return-to-work intervention for patients on sick leave due to common mental disorders: A randomised controlled trial. *Occupational and Environmental Medicine, 74*(12), 905–912.

Salomonsson, S., Santoft, F., Lindsäter, E., Ejeby, K., Ljótsson, B., Öst, L. G., ... Hedman-Lagerlöf, E. (2018). Stepped care in primary care-guided self-help and face-to-face cognitive behavioural therapy for common mental disorders: A randomized controlled trial. *Psychological Medicine, 48*(10), 1644–1654. https://doi.org/10.1017/S0033291717003129.

Salomonsson, S., Santoft, F., Lindsäter, E., Ejeby, K., Ingvar, M., Ljótsson, B., ... Hedman-Lagerlöf, E. (2019). Effects of cognitive behavioural therapy and return-to-work intervention for patients on sick leave due to stress-related disorders: Results from a randomized trial. *Scandinavian Journal of Psychology, 61*(2), 281–289. https://doi.org/10.1111/sjop.12590.

Schuster, R., Fichtenbauer, I., Sparr, V. M., Berger, T., & Laireiter, A. R. (2018). Feasibility of a blended group treatment (bGT) for major depression: Uncontrolled interventional study in a university setting. *BMJ Open, 8*(3). https://doi.org/10.1136/bmjopen-2017-018412, e018412.

Shaffer, C. S., Shapiro, J., Sank, L. I., & Coghlan, D. J. (1981). Positive changes in depression, anxiety, and assertion following individual and group cognitive behavior therapy intervention. *Cognitive Therapy and Research, 5*(2), 149–157. https://doi.org/10.1007/BF01172523.

Sheehan, D., Lecrubier, Y., Sheehan, K., Amorim, P., Janavs, J., Weiller, E., ... Dunbar, G. (1998). The Mini-International Neuropsychiatric Interview (MINI): The development and validation of a structured diagnostic psychiatric interview for DSM-IV and ICD-10. *The Journal of Clinical Psychiatry*, *59*(20), 22–33. https://www.psychiatrist.com/wp-content/uploads/2021/02/15175_mini-international-neuropsychiatric-interview-mini.pdf.

Shevlin, M., Hyland, P., Ben-Ezra, M., Karatzias, T., Cloitre, M., Vallières, F., & Maercker, A. (2019). Measuring ICD-11 adjustment disorder: The development and initial validation of the International Adjustment Disorder Questionnaire. *Acta Psychiatrica Scandinavica, 141*(3), 265–274. https://doi.org/10.1111/acps.13126.

Van der Klink, J. L. J., Blonk, R. W. B., Schene, A. H., & van Dijk, F. J. (2003). Reducing long term sickness absence by an activating intervention in adjustment disorders. *Occupational and Environmental Medicine, 60*, 429–437. https://doi.org/10.1136/oem.60.6.429.

World Health Organization. (1992). *International classification of diseases* (10th). Geneva: WHO.

World Health Organization. (2018). *International classification of diseases* (11th ed.). Ginebra: OMS.

Yaseen, Y. A. (2017). Adjustment disorder: Prevalence, sociodemographic risk factors, and its subtypes in outpatient psychiatric clinic. *Asian Journal of Psychiatry, 28*, 82–85. https://doi.org/10.1016/j.ajp.2017.03.012.

Zelviene, P., & Kazlauskas, E. (2018). Adjustment disorder: Current perspectives. *Neuropsychiatric Disease and Treatment, 14*, 375. https://doi.org/10.2147/NDT.S121072.

Chapter 8

Anxiety disorders: Mindfulness-based cognitive behavioral therapy

Jennifer Apolinário-Hagen[a], Marie Drüge[b], Roy Danino[c], and Siegfried Tasseit[d]

[a]Institute of Occupational, Social and Environmental Medicine, Faculty of Medicine, Centre for Health and Society, Heinrich Heine University Düsseldorf, Düsseldorf, Germany, [b]Department of Psychology, University of Zurich, Zurich, Switzerland, [c]Institute of Psychology, Heinrich Heine University Düsseldorf, Düsseldorf, Germany, [d]Psychotherapeutic Private Practice, Bad Gandersheim, Germany

Abbreviations

ACT	acceptance and commitment therapy
CAU	care as usual
CBT	cognitive behavioral therapy
CI	confidence interval
CT	cognitive therapy
GAD	generalized anxiety disorder
iMBCT	Internet-delivered mindfulness-based cognitive therapy
MABIs	mindfulness- and acceptance-based interventions
MARS	mobile app rating scale
MBCT	mindfulness-based cognitive therapy
MBIs	mindfulness-based interventions
MBSR	mindfulness-based stress reduction
MCT	metacognitive therapy
mHealth	mobile health (apps)
PD	panic disorder
RCT	randomized controlled trial

Introduction

Anxiety disorders represent an immense global burden, with an estimated life-time prevalence of 16.6% (Somers, Goldner, Waraich, & Hsu, 2006) and up to 33.7% (Bandelow & Michaelis, 2015). Different medications such as selective serotonin reuptake inhibitors and psychological treatments have been shown to be effective in anxiety disorders (Bandelow et al., 2015). Among the psychological treatment options, cognitive behavioral therapy (CBT) is the first-line treatment for anxiety disorders (Carpenter et al., 2018). However, there is an enormous treatment gap in anxiety disorders, as international survey data from 21 countries by a World Health Organization (WHO) initiative suggests (Alonso et al., 2018): According to Alonso et al. (2018), out of all survey participants screened positive for a 12-month Diagnostic and Statistical Manual of Mental Disorders (DSM-IV) anxiety disorder, only 27.6% received any treatment at all, while 9.8% were provided with a potentially suitable treatment, such as individual CBT. While medication can involve side effects, effective psychological treatments like individual CBT may also not be the preferred choice by all patients with anxiety disorders (Bandelow et al., 2015). As nonresponse and treatment resistance can pose a severe issue in anxiety disorders, combining a first-line therapeutic option with compatible novel therapeutic options such as mindfulness-based interventions (MBIs) could be a useful strategy (Hofmann, Asnaani, Vonk, Sawyer, & Fang, 2012; Roy-Byrne, 2015).

In this chapter we aim to provide a narrative overview of several current application options and the evidence base for using mindfulness-based cognitive therapy (MBCT) as specific MBI in the treatment of anxiety disorders, with the main focus on generalized anxiety disorder (GAD). Moreover, we discuss the possibilities and limitations of recent developments like mindfulness apps and potential side effects as well as contraindications of MBCT in anxiety disorders among adult populations.

Mindfulness-based interventions

MBIs are becoming increasingly common in clinical practice, as stand-alone programs or as a supplement to other treatment approaches. Mindfulness has been defined as "… the awareness that emerges through paying attention on purpose, in the present moment, and nonjudgmentally to the unfolding of experience moment by moment" (Kabat-Zinn, 2003, p. 145). MBIs like mindfulness-based stress reduction (MBSR) and MBCT have been classified as part of the "third wave" CBT family of intervention formats (Hayes & Hofmann, 2017). "Third wave" or process-based CBT interventions are supposed to address generic psychological skills, such as dealing with the context or process of thinking rather than the content of thoughts, in terms of one's association to thoughts and feelings by promoting nonjudgmental moment-by-moment awareness, mindfulness and acceptance (Hayes & Hofmann, 2017; Hofmann, Sawyer, & Fang, 2010). "Third wave" CBTs involve a broad range of diverse intervention approaches, including MBSR (Kabat-Zinn, 1990), MBCT (Segal, Williams, & Teasdale, 2012), acceptance and commitment therapy (ACT; Hayes, Luoma, Bond, Masuda, & Lillis, 2006) and metacognitive therapy (MCT; Wells & King, 2006). Among these approaches, MBSR and MBCT are the most commonly applied MBIs. Initially, MBSR was developed as the first mindful meditation group program for the adjunctive treatment of chronic pain (Kabat-Zinn, 2003), but its scope has been extended to a wide range of health outcomes in both healthy and clinical populations (Khoury, Sharma, Rush, & Fournier, 2015; Lee et al., 2020). In contrast to MBSR, MBCT is a disorder-specific MBI that was developed for depression relapse prophylaxis, has a greater scope on formal mindfulness practice, and incorporates elements of cognitive therapy (CT) (Edenfield & Saeed, 2012).

Mindfulness-based cognitive therapy

Based on CT and MBSR, Segal et al. (2012) developed MBCT for the relapse prevention of recurrent depressive episodes. In line with the *cognitive vulnerability model*, MBCT has been grounded on the assumption in case of monitoring even minor discrepancies between desired and actual states, the habitual "driven-doing mode" (or automatic pilot) triggers and reactivates automatic (ruminative) negative thoughts linked to depressed mood, increasing the risk of relapse each time and even more with each new depressive episode (Segal et al., 2012). Therefore, the MBCT program aims at teaching the ability "… to recognize and disengage from mood states characterized by self-perpetuating patterns of ruminative negative thinking" (Segal et al., 2012, p. 74). Training mindfulness skills like decentering makes it possible to reduce one's habitual tendency to "float away" into sequences of ruminative thoughts (Teasdale, Segal, & Williams, 1995, p. 34).

Although CBT and MBCT share common principles and methods of CT, they differ in their therapeutic stance (Sipe & Eisendrath, 2012), as illustrated in Table 1. Specifically, in MBCT it is learned how to get along with negative thoughts and unpleasant feelings and to not treat thoughts as facts or value them as dysfunctional. Instead of reinforcing more functional responses, mindfulness practice intends to promote a nonjudgmental present-moment awareness and different ways of looking at one's personal relationships to own thoughts without attempting to challenge, replace or change them (Segal et al., 2012). By these means, the underlying processes of maladaptive emotion regulation strategies can be addressed with mindfulness training that is an integral part of MBCT.

MBCT group format, scope and extensions

The eight-session MBCT group course with 2h training per week includes a range of formal and informal mindful meditation exercises that are repeated throughout the course and incorporates homework assignments as well as mindful practice exercises to be trained in daily life (Segal et al., 2012), as shown in Table 2.

Prior to participating in the MBCT program, an initial interview lasting about 1h is conducted with each prospective participant, including information about MBCT, expectations and contraindications. An MBCT course usually starts with group orientation and ground rules (e.g., confidentiality), and is divided in such a way that the early sessions (1–4) focus on the recognition and bringing awareness to "autopilot" or "driven-doing mode" and the training of formal mindfulness practice to get into "being mode," while the later sessions (5–8) build upon the cultivation of independent mindful practice and aim at bringing awareness to thoughts and feelings in daily life as well as coping strategies, such as disengagement from negative thoughts (Segal et al., 2012). MBCT is thus not proposed as another smart problem-solving technique, but rather as a different way to coping. Since brooding won't solve the problem, distancing oneself from the usual mindset of goal-orientation in solving problems and struggling, for instance, by training mindfulness of the breath in daily life whenever rumination and worries occur, is advocated in MBCT (Segal et al., 2012).

Meta-analytic evidence suggests an association between home practice following MBCT and MBSR regarding mindfulness skills in daily life and positive mental health outcomes (Parsons, Crane, Parsons, Fjorback, & Kuyken, 2017).

TABLE 1 Key features and differences between cognitive behavioral therapy (CBT) and mindfulness-based cognitive therapy (MBCT) concerning the therapeutic aims stance in the treatment of anxiety and mood disorders.

CBT	MBCT
Focusing on the problem/content of thoughts, goal orientation (e.g., more adaptive responses)	Focusing on the process/context, orientation on one's relationship to thoughts and feelings (process of thinking) and body sensations
Learning to recognize and alter negative thoughts and unpleasant feelings	Adopting new ways of "being with the problem," noticing and allowing thoughts and feelings
Differentiating between dysfunctional (negative) thoughts and "healthy" responses	Viewing thoughts as thoughts (versus facts), without judgment
Debating and challenging maladaptive beliefs, formulating new interpretations, altering maladaptive thoughts and feelings	Paying attention, noticing, allowing and accepting thoughts and feelings (without adjusting or changing them)
Interventions aiming at reinforcing more adaptive responses for dealing with the problem (e.g., addressing behavioral avoidance), changing the content of thoughts	Interventions as well as mindful exercises aiming at cultivating a nonjudgmental present-moment awareness of one's experience, accepting thoughts (allowing, decentering)
Therapist instructs, supports and coaches the patients (help for self-help), helps solving the problem, role as therapist	Mindful stance of the therapist (embodying the approach, personal daily mindfulness practice), role as instructor
Professional qualification in CBT, optionally: additionally trained in other approaches	Professional qualification in an evidence-based therapeutic approach (e.g., CBT) plus in-depth mindfulness training (in case of MBCT for mental disorders)
Collaborative approach, working alliance	Collaborative approach, but not scoping on solving the problem
Asking for causes of the problem ("why" questions) in the inquiry	Not asking for causes of the problem ("how" or "what" questions) in the inquiry
Main focus on the content of thoughts and beliefs, but also on physical responses	Main focus on awareness of inner experience including body sensations
Coping with manifest symptoms (as a treatment), including relapse prophylaxis	Recognizing early warning signals and disengage (focus on relapse prevention)

This table presents key differences in the therapeutic stance between cognitive behavioral therapy (CBT) and mindfulness-based cognitive therapy (MBCT) in both group and individual treatment settings.
Adapted and extended from Sipe, W. E. B., & Eisendrath, S. J. (2012). Mindfulness-based cognitive therapy: Theory and practice. *Canadian Journal of Psychiatry. Revue Canadienne De Psychiatrie, 57*(2), 65. https://doi.org/10.1177/070674371205700202; Segal, Z. V., Williams, J. M. G., & Teasdale, J. D. (2012). *Mindfulness-based cognitive therapy for depression* (2nd ed.). The Guilford Press.

Indeed, an MBCT program provides participants with handouts, forms for recording home practice and guidance on how to maintain practice commitment. According to Sipe and Eisendrath (2012), home practice of mindfulness skills learned in MBCT should amount to 45 min per day. Beyond the commonly applied group format, elements of MBCT can be also found in individual treatments (Michalak et al., 2019), digital interventions (Nissen et al., 2020) and videoconferencing MBCT (Moulton-Perkins, Moulton, Cavanagh, Jozavi, & Strauss, 2020).

MBCT: Application to anxiety disorders

Overall, MBCT represents a viable approach in dealing with symptoms of anxiety disorders, as worries and rumination can be effectively addressed with mindfulness training and CT techniques (Sipe & Eisendrath, 2012). Furthermore, MBCT may also influence underlying mechanisms of excessive worry such as intolerance of uncertainty in anxiety disorders, for instance, in panic disorder (PD) (Kim, Lee, Kim, Choi, & Lee, 2016).

Mindfulness in MBCT for anxiety disorders can be trained in group sessions following the MBSR structure using formal mindful practice techniques like the body scan and can specifically focus on worries in anxiety disorders by incorporating cognitive strategies like practicing the skill of disengaging or decentering, providing psychoeducation and homework assignments (Evans, 2016; Evans et al., 2008). In recent years, different studies have presented MBCT protocols that have been adapted to anxiety disorders such as GAD (Wong et al., 2016), social anxiety disorder (Piet, Hougaard, Hecksher, & Rosenberg, 2010) and PD (Kim et al., 2016) or other forms of anxiety such as health anxiety (Surawy, McManus, Muse, &

TABLE 2 Sample of an 8-week group mindfulness-based cognitive therapy (MBCT) program for generalized anxiety disorder (GAD).

Session	Theme/focus	Anxiety-related specifications, content, general aspects	Mindful practice, homework and review
Session 1	Awareness and automatic pilot	Group orientation and ground rules Psychoeducation (mindfulness, the automatic pilot); disorder-specific: e.g., automatic negative thoughts in GAD, worries and rumination Introduction into home practice (mindful eating, body scan)	Body scan Mindful eating Homework: body scan, awareness of the autopilot modus Mindful eating: raisin exercise and review
Session 2	Dealing with barriers	Psychoeducation (association to thoughts and feelings) Dealing with difficulties regarding mindful practice/rumination and worries, bodily sensations, etc. Thoughts and feelings (anxiety) exercise Pleasant events calendar	Body scan Sitting meditation (or alternatives) Homework and mindful practice review
Session 3	Mindfulness of the breath/gathering the scattered mind	Psychoeducation (awareness of mind wandering, focus on the breath) Maintaining awareness using the breath as an anchor Different ways to practice awareness of the breath Unpleasant events calendar or worry diary (setting up)	5 min seeing or hearing Sitting meditation (30 min), or mindful movement (hatha yoga)/mindful stretching and review Homework and mindful practice review 3-min breathing space and review
Session 4	Staying present	Psychoeducation (staying present) Defining the "territory" of GAD/anxiety Explore common coping strategies (avoidance, safety behaviors, rumination/worry) versus taking a breathing space	Sitting meditation Homework and mindful practice review 3-min breathing space and review
Session 5	Allowing/letting be	Psychoeducation (exploring difficulties) Acceptance (not working against rumination and worry in GAD)	Sitting meditation (how to deal with difficulties) Homework and mindful practice review 3-min breathing space and review
Session 6	Thoughts are not facts	Psychoeducation: cognitive biases Anxiety-related thoughts and alternative viewpoints exercise (dealing with rumination and worries in GAD)	Sitting meditation (working with difficulties) Homework and mindful practice review 3-min breathing space and review Discuss breathing space
Session 7	How can I take best care of myself?	Psychoeducation (e.g., identifying triggers); exploring associations between activity and anxiety; identifying warning signs and actions in GAD Formulate realistic and meaningful goals Development of an action plan Generate a list of pleasant and mastery activities and plan how best to schedule pleasant activities in daily life	Sitting meditation (includes working with difficulties) Homework and mindful practice review 3-min breathing space or mindful walking
Session 8	Using what had been learned to better deal future anxiety	Relapse prevention (comorbid depression in GAD), plans for the maintenance of mindful practice Discuss plans to maintain the practice and link them to positive reasons for doing so	Body scan Homework, mindful practice and course review End of the course, with a closing meditation

This table illustrates contents of mindfulness-based cognitive therapy (MBCT) for generalized anxiety disorder (GAD). Adapted and extended from Segal et al. (2012), including GAD-specific modifications based on Evans et al. (2008), Evans (2016), Sado et al. (2018), and Wong et al. (2016).

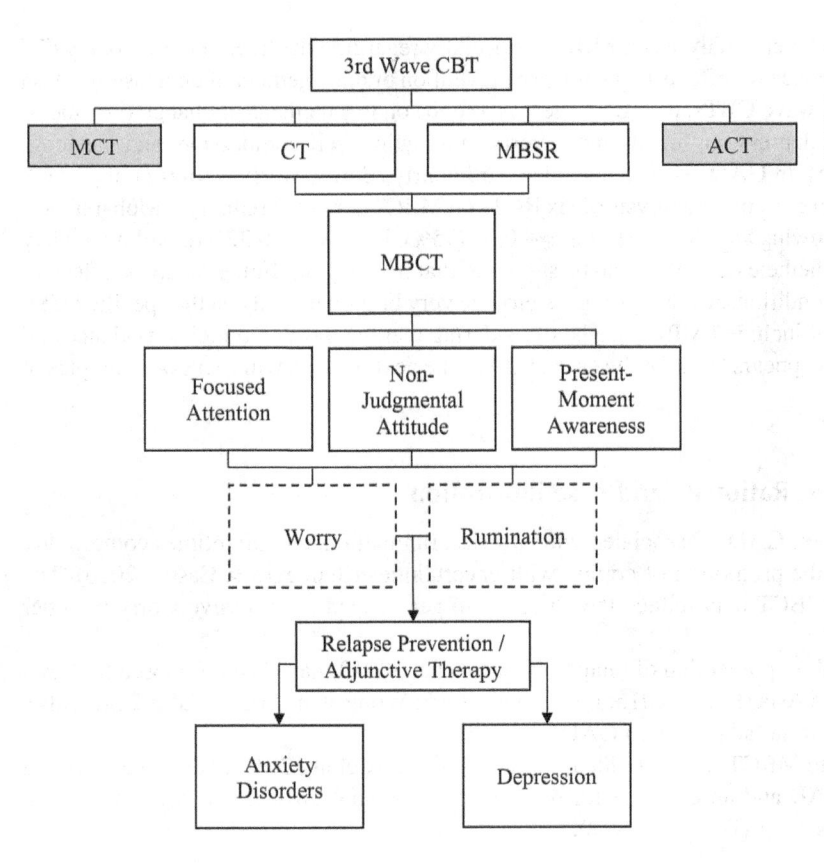

FIG. 1 Mindfulness-based cognitive therapy for anxiety disorders. *CBT*, cognitive behavioral therapy; *CT*, cognitive therapy; *MBCT*, mindfulness-based cognitive therapy; *MBSR*, mindfulness-based stress reduction; *MCT*, metacognitive therapy.

Williams, 2015) and geriatric anxiety in older patients (Hazlett-Stevens, Singer, & Chong, 2019), and to provide a MBCT targeting different anxiety disorders at the same time (Sado et al., 2018) (Fig. 1).

MBCT for anxiety: Current evidence

To date, most review articles on the efficacy of mindfulness- and acceptance-based interventions (MABIs) for dealing with anxiety have focused on a variety of intervention approaches, settings (individual, group, digital) as well as outcomes and populations, and seldom involve randomized controlled trials (RCTs) on MBCT. Most research supports MBIs for anxiety symptoms across different approaches, showing large effect sizes in pre-post-comparisons ($g = 0.89$, 10 studies) and waitlist-controlled trials ($g = 0.96$, 4 studies) (Khoury et al., 2013). A moderate effect size of MBIs/MABIs in the reduction of anxiety symptoms (Hedge's $g = 0.63$) has been identified by a meta-analysis of $n = 39$ RCTs with clinical populations (Hofmann, Sawyer, Witt, & Oh, 2010). Regarding the subset of samples involving patients with anxiety disorders, it was shown that MABIs had a large effect size ($g = 0.97$) compared to control conditions such as care-as-usual (CAU, waitlist-control). Concerning the specific effects of stand-alone MBIs (i.e., not integrated into a broader treatment setting), another meta-analysis revealed small to medium effects on anxiety (SMD = 0.39; 95% CI [0.22, 0.56]) compared to control conditions (Blanck et al., 2018). These findings are in line with another meta-analysis of 19 trials (Vøllestad, Nielsen, & Nielsen, 2012), showing a large effect size ($g = 0.83$) of different MABIs on anxiety. Other meta-analyses indicated that MBIs help reduce the symptoms of anxiety during the treatment of chronic or life-threatening medical conditions, such as cancer (Zhang et al., 2015).

In another review of pharmacological and psychological treatments for anxiety disorders, an effect of MBIs of $d = 1.56$ (95% CI [1.20, 1.92]) was found based on four studies albeit with a quite high risk of bias, that it is was not possible to compare MBIs with other treatments (Bandelow et al., 2015). Overall, these methodological shortcomings make it difficult to recommend MBIs like MBCT as an alternative to CBT, since the evidence base is inconsistent (de Abreu Costa, D'Alò de Oliveira, Tatton-Ramos, Manfro, & Salum, 2019). While a recent review of 12 systematic reviews suggested the efficacy of MBIs was comparable with CBT for anxiety disorders (Fumero, Peñate, Oyanadel, & Porter, 2020), prior reviews yielded other results (Goldberg et al., 2018; Strauss, Cavanagh, Oliver, & Pettman, 2014).

There are only few systematic reviews and meta-analyses on MBCT, especially regarding the treatment of anxiety disorders, as most overviews concentrate on evidence for effects in the relapse prevention or management of depression (Musa et al., 2020). Furthermore, as with other third wave CBTs, in MBCT there is a scope on mechanisms of change (i.e., mediation analyses), which include in MBCT for depression, for instance, worry, rumination, self-compassion, and metacognitive awareness (van der Velden et al., 2015). In GAD, MBCT may also particularly address worry (Evans et al., 2008).

Concerning specific anxiety disorders, there is a meta-analysis of six RCTs on MBCT for GAD (Ghahari, Mohammadi-Hasel, Malakouti, & Roshanpajouh, 2020) showing an effect size of $g = -0.65$ (95%CI $[-0.97, -0.32]$) regarding anxiety symptom reduction compared to controls. Nonetheless, while it can be supposed that training mindfulness may be effective in reducing anxiety symptoms across health conditions, existing reviews provide very limited insights on the specific effectiveness of MBCT due to the small number of included MBCT trials. Considering that the most convincing evidence and rationale for using MBCT in anxiety disorders appears to be for the case of GAD, the next passage will focus on examples of MBCT for GAD.

MBCT for generalized anxiety disorder: Rationale and case illustration

As the least effectively treated anxiety disorder, GAD is associated with worries, rumination, and oftentimes comorbidity with depression. CT strategies may involve the promotion of coping with uncertainties (Starcevic & Castle, 2016). The development of mindfulness skills through MBCT may reduce the tendency to get trapped in excessive worry (van der Velden et al., 2015).

Adaptations of the original MBCT manual for prevention of relapse by Segal et al. (2012) have been proposed for GAD across the world, including North America and Asia (Evans, 2016; Evans et al., 2008; Wong et al., 2016). Table 2 presents a synthesis of selected MBCT 8-week group format adaptions to GAD.

Nonetheless, it is also possible to integrate MBCT into individual therapy (Michalak et al., 2019). Here, we present a case illustration on individual MBCT for GAD and an example for dealing with potential difficulties when integrating MBCT for GAD into an individual therapy setting (Boxes 1 and 2).

Box 1 Case illustration (Maria). MBCT for GAD in an individual setting

Maria, a 58-year old flight assistant working at the airport, has been suffering from GAD after a treatment for breast cancer 2 years earlier. Although her treatment went fine, her worrying and rumination (e.g., death, accident of relative, losing her job) restrained her everyday activities such as work and leisure time activities to a minimum. Consequently, the mother of three also developed depressive symptoms (persistent low mood, and a loss of interests and pleasure), e.g., she did not want to spent time with her children anymore. Her main worry was to get sick again, including developing physical stress symptoms (sweating, muscle tension, trembling). She felt persistently nervous and restless, especially when there was an appointment for aftercare check-ups. Right before the therapeutic progress was initiated, she avoided the appointments completely. Apart from that her own avoidance behavior included touching her breasts ("what if there was another knot?"), driving a car ("What if I had a panic attack" or "I would lose control otherwise?") up to leaving the house for work ("What if I can't find a parking space?"). She also involved others in her avoidance behavior (e.g., asking her son not to use a plane for holidays).

After using the first sessions to create a sustainable therapeutic alliance while focusing on the resources of the patient and reactivating activities (e.g., Nordic walking), CBT involved psychoeducation, functional analysis of behavior and elements of MCT for GAD (Wells & King, 2006). While some of the goals were already attained (e.g., reducing the avoidance behavior so that she was able to drive a car again), some of the physical symptoms and worries about her state of health remained. Therefore, elements of MBCT were adapted for GAD following the original manual for depression (Segal et al., 2012) and specifications related to GAD (Borkovec & Sharpless, 2004) aiming at increasing a nonjudgmental awareness of emotional experience in the present moment and to build skills in terms of decentering from thoughts and feelings, enhancing acceptance, and decreasing experiential avoidance.

To engage with the present moment and explore body sensations, the *body scan* (Kabat-Zinn, 1990) was taught. Maria was very open to this method and used it five times a week. First, she reported relaxation from that ritual, which motivated her to engage further into the practice. Maria developed a sensitivity for changes of her physical sensations (e.g., sweating, feeling nervous) and realized that her fear increased rather than disappeared (on a scale from 0 to 100). Maria noticed the reactions of her mind (automatic catastrophic thoughts) and also how quickly her worries shift from one topic to another. Having noted that we practiced returning her attention gently but firmly to the present instead "This is me thinking about the future again, I am here now. Nothing happened yet. I'm accepting that I can't control the future." To anchor Maia further in the "here and now," the technique "breathing space" was introduced, a short practice which encourages present-moment awareness in everyday life.

MBCT aims at reducing the impact of negative thoughts and physical sensations by encouraging more distance to them and to the reactivity that arises in relation to them. By these means, the cycle of escalation that might otherwise lead to anxious

Box 1 Case illustration (Maria). MBCT for GAD in an individual setting—cont'd

preoccupation and avoidance behavior or worrying can be broken. This leads to the possibility of making choices about how to respond in a flexible way rather than react in a habitual way. Maria was able to go to the aftercare checkups again, because she felt able to respond to the results. We used the results of the functional analysis of behavior (triggers, thoughts, emotions, and avoidance behaviors) to compare them with arising thoughts, e.g., "Something bad will happen and I need to be aware" and feelings, e.g., feeling restless in sitting meditation. Afterward, we tried to train experiential awareness and nonjudgmental acceptance of whatever arises in each moment. Maria developed thoughts like "well, this fear is here now. These worries about catastrophes are here again, I can deal with them, they won't kill me." One main issue was to accept that there is no 100% safety, which was hard for her to deal with. It helped her letting go of controlling the future and accepting the present moment, which decreased her restlessness immensely. Also, she realized a big sadness and a fear of dying arising in sitting mediation, which she had suppressed before. The suppression had been needed to enable her to go through the process of cancer treatment and recovery. In therapy she opened up to this fear, noticed it, but did not want to be paralyzed by this fear. Therefore, some techniques of ACT (Hayes et al., 2006) were used to analyze what's truly meaningful to her and to engage into a value-oriented life. It further motivated her to strengthen her relationships though opening up about her feelings and reduce her avoidance behavior. By the end of the therapy, Maria noticed thoughts about her health faded and, although at times the worries and negative thoughts would arise in a rush, they would not lead her behavior into avoidance. Maria reduced her workload to 70% and grew more into her new role as a grandmother, as her oldest son got his first child.

Box 2 Example for potential difficulties with integrating MBCT elements in the individual treatment of GAD

Many individuals with GAD referred by their family doctors experience somatic symptoms and symptoms of autonomic hyperarousal (e.g., increased heart rate, shortness of breath). Here, we present experience with MBCT elements in a psychotherapeutic practice in Germany. MBCT has been proposed for this form of anxiety disorders within the framework of the CBT approved by the German health insurances primarily in an individual treatment setting.

The basis for integrating the MBCT elements was the German translation of the instructions of the manual by Segal et al. (2012) in the translated version (Michalak, Heidenreich, & Williams, 2012) for the first session with the exercise "Body Scan" starting with: "Lie down and get comfortable. ... Allow your eyes to close gently" (Michalak et al., 2012, p. 37). Here are some exemplary statements on potential difficulties with mindfulness practice, as expressed in the individual treatment setting.

Patients with GAD with vegetative symptoms often say: "I want to try this, but I want to keep my eyes open for a while ... and I feel now lying down, how my heart beats even stronger, ... it's getting faster, ... I know that from me in unusual situations, ... I don't think I can stand it so well ... ."

The therapist in the role as MBCT instructor continues in the first session by saying: "Take a few moments and make contact with the movements of your breath and the sensations in your body ...".

The patient observes his or her body and remarks: "It is now also tight around my chest ... I now hardly get any air at what you say with breathing ... I think I cannot stand it for long ... I think we have to stop now ...".

With this form of resistance, it could be problematic to simply continue with the exercise. It can make sense to stop here and say, for example, "Maybe today is not the right time. If you like, we can try it again next time and then maybe do something different." This may also mean to use an adapted instruction for body scan or to use another technique following the person-centered therapy approach, or combining CBT techniques with hypnotherapy.

Mindfulness-based interventions: Adverse events and contraindications

Remarkably, very little is known about potential harms and contraindications of established MBIs, as probable adverse events such as worsening of symptoms, distress, discomfort or confusion have been seldom adequately monitored and reported (Baer, Crane, Miller, & Kuyken, 2019). Increased awareness or focused attention to bodily sensations and thoughts may contribute to increases in symptoms given the interoceptive nature of mindful practice. Mindfulness exercises can involve temporary unpleasant physical responses among patients with anxiety disorders, including increased heart rate. Anecdotal data from case reports suggested different meditation-induced adverse events across MBIs (van Dam et al., 2018). However, MBCT exercises are brief compared to other forms of meditation.

In a systematic review of 36 RCTs on the safety of MBIs ($n = 25$ MBSR, $n = 11$ MBCT), none of the MBCT studies reported intervention-related adverse events, while one study mentioned intervention-unrelated adverse events (Wong,

Chan, Zhang, Lee, & Tsoi, 2018). There is furthermore no clear guidance on contraindications for MBCT apart from general considerations, such as present drug abuse, not receiving adequate treatment, mismatch with preferences, and severe acute problems like psychosis or suicidality (Segal et al., 2012). Principally, some patients may experience discomfort to open up in a group setting, while others may experience feelings of failure when they do not adhere to mindful practice recommendations, and unmet expectations may lead to disappointment and frustration (e.g., expecting mindfulness to be equivalent to relaxation) (Evans, 2016). Thus, it is crucial to discuss the setting and individual expectations prior to MBCT attendance (Segal et al., 2012). Also, the treatment with some effective medications in anxiety disorders like benzodiazepines (Bandelow et al., 2015) may be a contraindication for MBIs involving exposure techniques.

Furthermore, it appears crucial to provide a foundation for realistic treatment expectations, as the term "mindfulness" in the context of MBIs may be misleading (van Dam et al., 2018). Some patients may have negative attitudes toward labels like "mindfulness" or "meditation." In addition, the language form used by the therapist does not always correspond to the language used by the patient. In MBCT, clarifications of expectations and contraindications represent an integral part of the initial interview with each prospective participant, which may buffer some of the previously outlined issues. Therefore, MBCT for anxiety disorders should be applied by qualified therapists (e.g., trained in CBT).

Newer developments: Mindfulness apps and Internet-delivered MBCT

Mindfulness practice on a daily basis following MBCT is required to maintain intervention effects (Evans, 2016), and may be supported by using mindfulness mobile health (mHealth) apps. Most of these publicly available apps, however, focus on self-help and training of mindful meditation practice rather than on structured, professionally guided treatment support as in MBCT and CBT. To date, there is preliminary research support on the efficacy of evidence-based apps. A recent meta-analysis of 27 RCTs showed significant effects of such apps on increasing acceptance and mindfulness ($g=0.29$) (Linardon, 2020). Moreover, different systematic reviews on mindfulness apps showed small effect sizes regarding the improvement of anxiety symptoms. For example, a meta-analysis of 34 trials on mindfulness apps on several mental health outcomes identified significant effect sizes for anxiety ($g=0.28$, 15 studies) and depression ($g=0.33$, 15 studies) compared to control conditions at the postintervention assessment (Gál, Ştefan, & Cristea, 2021). The small effects of stand-alone mental health apps are in line with other reviews covering a broader range of approaches (Weisel et al., 2019).

Remarkably, there are many low-quality apps in commercial app stores, which represent a key barrier to finding a helpful mindfulness app. A systematic review (Schultchen et al., 2020) of 192 mindfulness-focused apps in European commercial app stores using the Mobile Application Rating Scale (MARS) indicated moderate quality, but only few apps ($n=7$) were tested in a RCT, the vast majority were associated with a lack of data security and privacy policy and most were not grounded on evidence-based techniques regarding mindfulness and behavior change. To date, only a few mindfulness apps can be recommended as additional support for MBCT and CBT, but not replace them.

Regarding web-based MBIs, a meta-analysis of 12 trials found significantly reduced anxiety in the total clinical sample ($g=-0.433$) and subgroup with an anxiety disorder ($g=-0.719$) (Sevilla-Llewellyn-Jones, Santesteban-Echarri, Pryor, McGorry, & Alvarez-Jimenez, 2018). In contrast to (native) mindfulness apps, Internet-based computerized programs are more commonly applied in clinical settings and adhere more to the traditional structure of MBCT, whereas they may be also used with smartphones (as web-apps; websites optimized for smartphone screens). Within the broad range of MBIs, there are a few studies on the efficacy of digital formats of MBCT (Internet-based MBCT, iMBCT) for specific patient populations with anxiety disorders. For example, an RCT on iMBCT for anxiety and depression in $n=150$ cancer survivors revealed significant effects for anxiety reduction at postintervention ($d=0.45$) that were maintained at a 6-month follow-up ($d=0.40$) compared CAU (Nissen et al., 2020). There are also some studies comparing outcomes between MBCT and iMBCT, showing no differences in anxiety (Compen et al., 2018) and therapeutic alliance (Bisseling et al., 2019) between delivery modes. Overall, clinical practice would benefit from unbiased information on digital interventions to consult patients interested in these options. In addition, recent research indicates a preference for blended therapies (i.e., combination of face-to-face contacts with the therapist with digital self-help intervention components) over complete online delivery (Phillips, Himmler, & Schreyögg, 2021). Therefore, therapists interested in iMBCT may consider using blended or stepped-care approaches. It is also possible to perform MBCT using videoconferencing, which has become an attractive option especially during the Covid-19 pandemic (Moulton-Perkins et al., 2020).

Conclusions

MBCT connects disorder-specific elements of CT with mindfulness practice from MBSR, which makes it suitable not only for depression relapse prevention but also to deal with anxiety symptoms (Hofmann, Sawyer, & Fang, 2010). To date, there is a need for more high-quality RCTs to draw conclusions on the effectiveness of traditional MBIs/MBCT (Creswell, 2017; Goldberg et al., 2018; Norton, Abbott, Norberg, & Hunt, 2015) and digitalized MBIs/iMBCT (O'Connor, Munnelly, Whelan, & McHugh, 2018). Still, the provision of MBCT by experienced behavioral therapists represents a suitable option, especially in cases when traditional CBT is not working as expected or encounters its limits. Finally, the evidence base for the application of this relatively novel approach that was not initially developed for anxiety disorders is still limited but growing, as it includes the training of mindfulness as a generic skill that is compatible with the goals of CBT.

Practice and procedures

The 8-week MBCT group program (Segal et al., 2012) provides the basis for adaptations to anxiety disorders (see, Table 2), whereas elements of MBCT can also be integrated into individual psychotherapy (Michalak et al., 2019), as illustrated in Box 1.

Mini-dictionary of terms

- *Mindfulness-based cognitive therapy (MBCT)*. Manualized program with roots in MBSR and CBT/CT that was initially developed as an 8-week group format for the relapse prevention of recurrent depression.
- *Metacognitive awareness*. Ability to observe one's own thinking processes.
- *Breathing space*. Brief mindful exercise (e.g., 3 min) applied in MBCT and MBSR, with three steps: (1) noticing what one is experiencing right now with broad attentional focus, (2) exclusively narrowing attention to one's own breath, and (3) again broaden the focus to the whole body, including thoughts and feelings.
- *Body scan*. Guided meditation exercise used in MBCT and MBSR. Mindful imaginary journey through one's whole body (noticing bodily sensations, thoughts and feelings in the "here and now"), without judging or changing these observed inner states.
- *Raisin exercise*. Brief exercise, often applied in the first session of MBCT bringing mindfulness into daily routines—in this case—eating a raisin, by paying full attention to eating with all senses; noticing how it smells, looks like, tastes, etc.
- *Sitting meditation*. Especially suitable for beginners of MBCT (reduced risk of falling asleep in a seated position, alertness), becoming aware of a specific inner event while sitting, by counting breaths or by focusing on a specific imagined picture or word.

Key facts of MBCT

- MBCT is an 8-week group intervention that was developed by Segal et al. (2012), combining techniques from MBSR and CBT/CT, especially in the relapse prevention of recurrent depressive disorders.
- MBCT target processes underlying mental health outcomes, such as focused attention, present-moment awareness, and nonjudgmental attitude.
- MBCT may augment first-line treatments like CBT for anxiety disorders, as mindfulness practice can also address rumination and worry (e.g., in GAD).
- Further MBCT developments target other fields in clinical psychology (e.g., anxiety disorders) and modes of delivery (e.g., individual treatment settings).
- Regular mindful training of present-moment awareness and acceptance is an established method to promote mental health and to reduce ruminative thinking.
- There is preliminary evidence for the efficacy of MBCT in the treatment of anxiety disorders, but further research is required as well as more data on potential adverse events and contraindications.

Applications to other areas

In this chapter, we have reviewed the current status of MBCT for anxiety disorders. Despite the promising rationale for using MBCT in anxiety disorders (Sipe & Eisendrath, 2012) and the availability of adapted MBCT protocols for anxiety disorders like GAD (Evans, 2016), it should be noted that the clinical testing is at an earlier stage compared to CBT in terms of both traditional MBCT (Creswell, 2017; Goldberg et al., 2018; Norton et al., 2015) and iMBCT (O'Connor et al., 2018). Moreover, few publicly available mindfulness apps fulfill minimal quality criteria for evidence-based and safe behavior change interventions, which makes it difficult to identify useful apps (Schultchen et al., 2020). In addition, potential adverse events of mindfulness practice should be considered and monitored (Baer et al., 2019). In MBCT, the expectations of prospective participants as well as individual contraindications are discussed individually prior to the course in an initial interview. Taken together, MBCT can be currently recommended as an adjunctive option for anxiety disorders. Besides anxiety and depression, there exists preliminary research support for MBCT for insomnia, chronic pain and cancer, making it an emerging field with further application options (Eisendrath, 2016).

Summary points

- Anxiety disorders are common, given an estimated life-time prevalence of up to 33.7%(Bandelow & Michaelis, 2015).
- Many people with anxiety disorders drop out of first-line treatments, do not receive a suitable treatment like CBT or simply remain untreated.
- MBCT—as an approach combining mindfulness practice with elements of CT—has extended its scope from relapse prevention of recurrent depressive disorders to also treat anxiety disorders.
- Despite convincing rationale for using MBCT in the treatment of anxiety disorders, the evidence base is still limited.
- Mindfulness mHealth apps may be suitable to support the daily practice of mindfulness, but should not replace (face-to-face) MBCT.

References

Alonso, J., Liu, Z., Evans-Lacko, S., Sadikova, E., Sampson, N., Chatterji, S., Abdulmalik, J., Aguilar-Gaxiola, S., Al-Hamzawi, A., Andrade, L. H., Bruffaerts, R., Cardoso, G., Cia, A., Florescu, S., de Girolamo, G., Gureje, O., Haro, J. M., He, Y., de Jonge, P., … Thornicroft, G. (2018). Treatment gap for anxiety disorders is global: Results of the World Mental Health Surveys in 21 countries. *Depression and Anxiety, 35*(3), 195–208. https://doi.org/10.1002/da.22711.

Baer, R., Crane, C., Miller, E., & Kuyken, W. (2019). Doing no harm in mindfulness-based programs: Conceptual issues and empirical findings. *Clinical Psychology Review, 71*, 101–114. https://doi.org/10.1016/j.cpr.2019.01.001.

Bandelow, B., & Michaelis, S. (2015). Epidemiology of anxiety disorders in the 21st century. *Dialogues in Clinical Neuroscience, 17*(3), 327–335.

Bandelow, B., Reitt, M., Röver, C., Michaelis, S., Görlich, Y., & Wedekind, D. (2015). Efficacy of treatments for anxiety disorders. *International Clinical Psychopharmacology, 30*(4), 183–192. https://doi.org/10.1097/YIC.0000000000000078.

Bisseling, E., Cillessen, L., Spinhoven, P., Schellekens, M., Compen, F., van der Lee, M., & Speckens, A. (2019). Development of the therapeutic alliance and its association with internet-based mindfulness-based cognitive therapy for distressed cancer patients: Secondary analysis of a multicenter randomized controlled trial. *Journal of Medical Internet Research, 21*(10). https://doi.org/10.2196/14065, e14065.

Blanck, P., Perleth, S., Heidenreich, T., Kröger, P., Ditzen, B., Bents, H., & Mander, J. (2018). Effects of mindfulness exercises as stand-alone intervention on symptoms of anxiety and depression: Systematic review and meta-analysis. *Behaviour Research and Therapy, 102*, 25–35. https://doi.org/10.1016/j.brat.2017.12.002.

Borkovec, T. D., & Sharpless, B. (2004). Generalized anxiety disorder: Bringing cognitive-behavioral therapy into the valued present. In *Mindfulness and acceptance: Expanding the cognitive-behavioral tradition* (pp. 209–242). Guilford Press.

Carpenter, J. K., Andrews, L. A., Witcraft, S. M., Powers, M. B., Smits, J. A. J., & Hofmann, S. G. (2018). Cognitive behavioral therapy for anxiety and related disorders: A meta-analysis of randomized placebo-controlled trials. *Depression and Anxiety, 35*(6), 502–514. https://doi.org/10.1002/da.22728.

Compen, F., Bisseling, E., Schellekens, M., Donders, R., Carlson, L., van der Lee, M., & Speckens, A. (2018). Face-to-face and internet-based mindfulness-based cognitive therapy compared with treatment as usual in reducing psychological distress in patients with cancer: A multicenter randomized controlled trial. *Journal of Clinical Oncology: Official Journal of the American Society of Clinical Oncology, 36*(23), 2413–2421. https://doi.org/10.1200/JCO.2017.76.5669.

Creswell, J. D. (2017). Mindfulness interventions. *Annual Review of Psychology, 68*, 491–516. https://doi.org/10.1146/annurev-psych-042716-051139.

de Abreu Costa, M., D'Alò de Oliveira, G. S., Tatton-Ramos, T., Manfro, G. G., & Salum, G. A. (2019). Anxiety and stress-related disorders and mindfulness-based interventions: A systematic review and multilevel meta-analysis and meta-regression of multiple outcomes. *Mindfulness, 10*(6), 996–1005. https://doi.org/10.1007/s12671-018-1058-1.

Edenfield, T. M., & Saeed, S. A. (2012). An update on mindfulness meditation as a self-help treatment for anxiety and depression. *Psychology Research and Behavior Management, 5*, 131–141. https://doi.org/10.2147/PRBM.S34937.

Eisendrath, S. J. (Ed.). (2016). *Mindfulness-based cognitive therapy: Innovative applications* Springer. https://doi.org/10.1007/978-3-319-29866-5.

Evans, S. (2016). Mindfulness-based cognitive therapy for generalized anxiety disorder. In S. J. Eisendrath (Ed.), *Mindfulness-based cognitive therapy: Innovative applications* (pp. 145–154). Springer.

Evans, S., Ferrando, S., Findler, M., Stowell, C., Smart, C., & Haglin, D. (2008). Mindfulness-based cognitive therapy for generalized anxiety disorder. *Journal of Anxiety Disorders, 22*(4), 716–721. https://doi.org/10.1016/j.janxdis.2007.07.005.

Fumero, A., Peñate, W., Oyanadel, C., & Porter, B. (2020). The effectiveness of mindfulness-based interventions on anxiety disorders. A systematic meta-review. *European Journal of Investigation in Health, Psychology and Education, 10*(3), 704–719. https://doi.org/10.3390/ejihpe10030052.

Gál, É., Ştefan, S., & Cristea, I. A. (2021). The efficacy of mindfulness meditation apps in enhancing users' well-being and mental health related outcomes: A meta-analysis of randomized controlled trials. *Journal of Affective Disorders, 279*, 131–142. https://doi.org/10.1016/j.jad.2020.09.134.

Ghahari, S., Mohammadi-Hasel, K., Malakouti, S. K., & Roshanpajouh, M. (2020). Mindfulness-based cognitive therapy for generalised anxiety disorder: A systematic review and meta-analysis. *East Asian Archives of Psychiatry, 30*(2), 52–56. https://doi.org/10.12809/eaap1885.

Goldberg, S. B., Tucker, R. P., Greene, P. A., Davidson, R. J., Wampold, B. E., Kearney, D. J., & Simpson, T. L. (2018). Mindfulness-based interventions for psychiatric disorders: A systematic review and meta-analysis. *Clinical Psychology Review, 59*, 52–60. https://doi.org/10.1016/j.cpr.2017.10.011.

Hayes, S. C., & Hofmann, S. G. (2017). The third wave of cognitive behavioral therapy and the rise of process-based care. *World Psychiatry, 16*(3), 245–246. https://doi.org/10.1002/wps.20442.

Hayes, S. C., Luoma, J. B., Bond, F. W., Masuda, A., & Lillis, J. (2006). Acceptance and commitment therapy: Model, processes and outcomes. *Behaviour Research and Therapy, 44*(1), 1–25. https://doi.org/10.1016/j.brat.2005.06.006.

Hazlett-Stevens, H., Singer, J., & Chong, A. (2019). Mindfulness-based stress reduction and mindfulness-based cognitive therapy with older adults: A qualitative review of randomized controlled outcome research. *Clinical Gerontologist, 42*(4), 347–358. https://doi.org/10.1080/07317115.2018.1518282.

Hofmann, S. G., Asnaani, A., Vonk, I. J., Sawyer, A. T., & Fang, A. (2012). The efficacy of cognitive behavioral therapy: A review of meta-analyses. *Cognitive Therapy and Research, 36*(5), 427–440. https://doi.org/10.1007/s10608-012-9476-1.

Hofmann, S. G., Sawyer, A. T., & Fang, A. (2010). The empirical status of the "new wave" of CBT. *The Psychiatric Clinics of North America, 33*(3), 701–710. https://doi.org/10.1016/j.psc.2010.04.006.

Hofmann, S. G., Sawyer, A. T., Witt, A. A., & Oh, D. (2010). The effect of mindfulness-based therapy on anxiety and depression: A meta-analytic review. *Journal of Consulting and Clinical Psychology, 78*(2), 169–183. https://doi.org/10.1037/a0018555.

Kabat-Zinn, J. (1990). *Full catastrophe living: Using the wisdom of your body and mind to face stress, pain, and illness*. Delacorte Press.

Kabat-Zinn, J. (2003). Mindfulness-based interventions in context: Past, present, and future. *Clinical Psychology: Science and Practice, 10*(2), 144–156. https://doi.org/10.1093/clipsy.bpg016.

Khoury, B., Lecomte, T., Fortin, G., Masse, M., Therien, P., Bouchard, V., … Hofmann, S. G. (2013). Mindfulness-based therapy: A comprehensive meta-analysis. *Clinical Psychology Review, 33*(6), 763–771. https://doi.org/10.1016/j.cpr.2013.05.005.

Khoury, B., Sharma, M., Rush, S. E., & Fournier, C. (2015). Mindfulness-based stress reduction for healthy individuals: A meta-analysis. *Journal of Psychosomatic Research, 78*(6), 519–528. https://doi.org/10.1016/j.jpsychores.2015.03.009.

Kim, M. K., Lee, K. S., Kim, B., Choi, T. K., & Lee, S.-H. (2016). Impact of mindfulness-based cognitive therapy on intolerance of uncertainty in patients with panic disorder. *Psychiatry Investigation, 13*(2), 196–202. https://doi.org/10.4306/pi.2016.13.2.196.

Lee, E. K. P., Yeung, N. C. Y., Xu, Z., Zhang, D., Yu, C.-P., & Wong, S. Y. S. (2020). Effect and acceptability of mindfulness-based stress reduction program on patients with elevated blood pressure or hypertension: A meta-analysis of randomized controlled trials. *Hypertension, 76*(6), 1992–2001. https://doi.org/10.1161/HYPERTENSIONAHA.120.16160.

Linardon, J. (2020). Can acceptance, mindfulness, and self-compassion be learned by smartphone apps? A systematic and meta-analytic review of randomized controlled trials. *Behavior Therapy, 51*(4), 646–658. https://doi.org/10.1016/j.beth.2019.10.002.

Michalak, J., Crane, C., Germer, C. K., Gold, E., Heidenreich, T., Mander, J., … Segal, Z. V. (2019). Principles for a responsible integration of mindfulness in individual therapy. *Mindfulness, 10*(5), 799–811. https://doi.org/10.1007/s12671-019-01142-6.

Michalak, J., Heidenreich, T., & Williams, J. M. G. (2012). *Achtsamkeit: Fortschritte der Psychotherapie. Vol. 48* (1st ed.). Hogrefe Verlag.

Moulton-Perkins, A., Moulton, D., Cavanagh, K., Jozavi, A., & Strauss, C. (2020). Systematic review of mindfulness-based cognitive therapy and mindfulness-based stress reduction via group videoconferencing: Feasibility, acceptability, safety, and efficacy. *Journal of Psychotherapy Integration*. https://doi.org/10.1037/int0000216.

Musa, Z. A., Kim Lam, S., Binti Mamat @ Mukhtar, F, Kwong Yan, S., Tajudeen Olalekan, O., & Kim Geok, S. (2020). Effectiveness of mindfulness-based cognitive therapy on the management of depressive disorder: Systematic review. *International Journal of Africa Nursing Sciences, 12*. https://doi.org/10.1016/j.ijans.2020.100200, 100200.

Nissen, E. R., O'Connor, M., Kaldo, V., Højris, I., Borre, M., Zachariae, R., & Mehlsen, M. (2020). Internet-delivered mindfulness-based cognitive therapy for anxiety and depression in cancer survivors: A randomized controlled trial. *Psycho-Oncology, 29*(1), 68–75. https://doi.org/10.1002/pon.5237.

Norton, A. R., Abbott, M. J., Norberg, M. M., & Hunt, C. (2015). A systematic review of mindfulness and acceptance-based treatments for social anxiety disorder. *Journal of Clinical Psychology, 71*(4), 283–301. https://doi.org/10.1002/jclp.22144.

O'Connor, M., Munnelly, A., Whelan, R., & McHugh, L. (2018). The efficacy and acceptability of third-wave behavioral and cognitive eHealth treatments: A systematic review and meta-analysis of randomized controlled trials. *Behavior Therapy, 49*(3), 459–475. https://doi.org/10.1016/j.beth.2017.07.007.

Parsons, C. E., Crane, C., Parsons, L. J., Fjorback, L. O., & Kuyken, W. (2017). Home practice in mindfulness-based cognitive therapy and mindfulness-based stress reduction: A systematic review and meta-analysis of participants' mindfulness practice and its association with outcomes. *Behaviour Research and Therapy, 95*, 29–41. https://doi.org/10.1016/j.brat.2017.05.004.

Phillips, E. A., Himmler, S. F., & Schreyögg, J. (2021). Preferences for e-mental health interventions in Germany: A discrete choice experiment. *Value in Health*. https://doi.org/10.1016/j.jval.2020.09.018.

Piet, J., Hougaard, E., Hecksher, M. S., & Rosenberg, N. K. (2010). A randomized pilot study of mindfulness-based cognitive therapy and group cognitive-behavioral therapy for young adults with social phobia. *Scandinavian Journal of Psychology, 51*(5), 403–410. https://doi.org/10.1111/j.1467-9450.2009.00801.x.

Roy-Byrne, P. (2015). Treatment-refractory anxiety; definition, risk factors, and treatment challenges. *Dialogues in Clinical Neuroscience, 17*(2), 191–206.

Sado, M., Park, S., Ninomiya, A., Sato, Y., Fujisawa, D., Shirahase, J., & Mimura, M. (2018). Feasibility study of mindfulness-based cognitive therapy for anxiety disorders in a Japanese setting. *BMC Research Notes, 11*(1), 653. https://doi.org/10.1186/s13104-018-3744-4.

Schultchen, D., Terhorst, Y., Holderied, T., Stach, M., Messner, E.-M., Baumeister, H., & Sander, L. B. (2020). Stay present with your phone: A systematic review and standardized rating of mindfulness apps in European app stores. *International Journal of Behavioral Medicine.* https://doi.org/10.1007/s12529-020-09944-y.

Segal, Z. V., Williams, J. M. G., & Teasdale, J. D. (Eds.). (2012). *Mindfulness-based cognitive therapy for depression* (2nd ed.). The Guilford Press.

Sevilla-Llewellyn-Jones, J., Santesteban-Echarri, O., Pryor, I., McGorry, P., & Alvarez-Jimenez, M. (2018). Web-based mindfulness interventions for mental health treatment: Systematic review and meta-analysis. *JMIR Mental Health, 5*(3). https://doi.org/10.2196/10278, e10278.

Sipe, W. E. B., & Eisendrath, S. J. (2012). Mindfulness-based cognitive therapy: Theory and practice. *Canadian Journal of Psychiatry, 57*(2), 63–69. https://doi.org/10.1177/070674371205700202.

Somers, J. M., Goldner, E. M., Waraich, P., & Hsu, L. (2006). Prevalence and incidence studies of anxiety disorders: A systematic review of the literature. *Canadian Journal of Psychiatry, 51*(2), 100–113. https://doi.org/10.1177/070674370605100206.

Starcevic, V., & Castle, D. J. (2016). Anxiety disorders. In G. Fink (Ed.), *Handbook of stress: Vol. 1. Stress: Concepts cognition emotion and behavior* (pp. 203–211). Elsevier. https://doi.org/10.1016/B978-0-12-800951-2.00024-8.

Strauss, C., Cavanagh, K., Oliver, A., & Pettman, D. (2014). Mindfulness-based interventions for people diagnosed with a current episode of an anxiety or depressive disorder: A meta-analysis of randomised controlled trials. *PLoS One, 9*(4). https://doi.org/10.1371/journal.pone.0096110, e96110.

Surawy, C., McManus, F., Muse, K., & Williams, J. M. G. (2015). Mindfulness-based cognitive therapy (MBCT) for health anxiety (hypochondriasis): Rationale, implementation and case illustration. *Mindfulness, 6*(2), 382–392. https://doi.org/10.1007/s12671-013-0271-1.

Teasdale, J. D., Segal, Z., & Williams, J. M. (1995). How does cognitive therapy prevent depressive relapse and why should attentional control (mindfulness) training help? *Behaviour Research and Therapy, 33*(1), 25–39.

van Dam, N. T., van Vugt, M. K., Vago, D. R., Schmalzl, L., Saron, C. D., Olendzki, A., … Meyer, D. E. (2018). Mind the hype: A critical evaluation and prescriptive agenda for research on mindfulness and meditation. *Perspectives on Psychological Science, 13*(1), 36–61. https://doi.org/10.1177/1745691617709589.

van der Velden, A. M., Kuyken, W., Wattar, U., Crane, C., Pallesen, K. J., Dahlgaard, J., … Piet, J. (2015). A systematic review of mechanisms of change in mindfulness-based cognitive therapy in the treatment of recurrent major depressive disorder. *Clinical Psychology Review, 37*, 26–39. https://doi.org/10.1016/j.cpr.2015.02.001.

Vøllestad, J., Nielsen, M. B., & Nielsen, G. H. (2012). Mindfulness- and acceptance-based interventions for anxiety disorders: A systematic review and meta-analysis. *British Journal of Clinical Psychology, 51*(3), 239–260. https://doi.org/10.1111/j.2044-8260.2011.02024.x.

Weisel, K. K., Fuhrmann, L. M., Berking, M., Baumeister, H., Cuijpers, P., & Ebert, D. D. (2019). Standalone smartphone apps for mental health—A systematic review and meta-analysis. *NPJ Digital Medicine, 2*(1), 118. https://doi.org/10.1038/s41746-019-0188-8.

Wells, A., & King, P. (2006). Metacognitive therapy for generalized anxiety disorder: An open trial. *Journal of Behavior Therapy and Experimental Psychiatry, 37*(3), 206–212. https://doi.org/10.1016/j.jbtep.2005.07.002.

Wong, S. Y. S., Chan, J. Y. C., Zhang, D., Lee, E. K. P., & Tsoi, K. K. F. (2018). The safety of mindfulness-based interventions: A systematic review of randomized controlled trials. *Mindfulness, 9*(5), 1344–1357. https://doi.org/10.1007/s12671-018-0897-0.

Wong, S. Y. S., Yip, B. H. K., Mak, W. W. S., Mercer, S., Cheung, E. Y. L., Ling, C. Y. M., … Ma, H. S. W. (2016). Mindfulness-based cognitive therapy v. Group psychoeducation for people with generalised anxiety disorder: Randomised controlled trial. *British Journal of Psychiatry: The Journal of Mental Science, 209*(1), 68–75. https://doi.org/10.1192/bjp.bp.115.166124.

Zhang, M. F., Wen, Y. S., Liu, W. Y., Peng, L. F., Wu, X. D., & Liu, Q. W. (2015). Effectiveness of mindfulness-based therapy for reducing anxiety and depression in patients with cancer: A meta-analysis. *Medicine, 94*(45). https://doi.org/10.1097/MD.0000000000000897, e0897-0.

Chapter 9

Avoidant/restrictive food intake disorder: Features and use of cognitive-behavioral therapy

P. Evelyna Kambanis, Christopher J. Mancuso, and Angeline R. Bottera
Department of Psychology, University of Wyoming, Laramie, WY, United States

Abbreviations

ARFID avoidant/restrictive food intake disorder
CBT-AR cognitive-behavioral therapy for avoidant/restrictive food intake disorder

Introduction

Prior to the release of the fifth edition of the *Diagnostic and Statistical Manual of Mental Disorders* (*DSM-5*; American Psychiatric Association, 2013), a gap in clinical diagnoses was identified, highlighting the need for an eating disorder that characterized avoidant and restrictive eating behaviors spanning beyond infancy and early childhood. Thus, as an expansion and reformulation of the *DSM-IV* (APA, 2000), diagnosis of "feeding disorder of infancy or early childhood," avoidant/restrictive food intake disorder (ARFID), was introduced in the *DSM-5* (APA, 2013). ARFID is an eating disorder spanning the lifespan and weight spectrum, characterized by avoidant or restrictive eating by volume (i.e., restriction of amount) and/or variety (i.e., avoidance of specific foods). The disorder is associated with a host of medical and/or psychological sequelae, including significant weight loss (or failure to achieve expected weight gain or faltering growth in children); significant nutritional deficiency; dependence on enteral feeding or oral nutrition supplements; and/or marked interference with psychosocial functioning (APA, 2013).

A diagnosis of ARFID cannot be due to lack of available food or associated cultural norms. Diagnostically, the core clinical feature that distinguishes ARFID from other eating disorders—such as anorexia nervosa and bulimia nervosa—is the absence of shape and weight concerns. Accordingly, ARFID cannot be conferred if an individual has a diagnosis of anorexia nervosa, bulimia nervosa or experiences a concurrent disturbance in the way in which their body shape/weight is experienced. Finally, avoidant/restrictive eating in ARFID cannot be attributable to a concurrent medical condition or be better explained by another medical condition.

The three ARFID profiles

Since the induction of ARFID in the *DSM-5* (APA, 2013), researchers have suggested a need for increased precision in defining and conceptualizing the various ARFID presentations (e.g., Eddy et al., 2019). The *DSM-5* (APA, 2013) presents three distinct ARFID phenotypes: (a) sensory sensitivity; (b) fear of aversive consequences; and (c) lack of interest in food or eating. Although prior studies conceptualized the three ARFID presentations as mutually exclusive categories (e.g., Nicholls, Chater, & Lask, 2000), growing evidence suggests that these profiles overlap (Bryant-Waugh et al., 2019). Thomas et al. (2017) proposed a three-dimensional neurobiological model of ARFID wherein the three core presentations occur along a continuum of severity and are not mutually exclusive (Fig. 1); that is, an individual with ARFID may present with difficulty across domains, ranging in severity. Importantly, not every individual who meets diagnostic criteria for ARFID will fit neatly into one of the three profiles. Using this three-dimensional model of avoidant/restrictive eating may aid in not only characterizing the cross-sectioning nature of the various ARFID phenotypes, but also in understanding risk factors relevant for treatment development and relapse prevention (Thomas et al., 2017).

FIG. 1 Thomas et al. (2017) three-dimensional model of ARFID. Figure depicting that the three ARFID profiles (i.e., sensory sensitivity, fear of aversive consequences, lack of interest in food/eating) vary in severity and are not mutually exclusive. *(Borrowed with permission from Thomas, J. J., Lawson, E. A., Micali, N., Misra, M., Deckersbach, T., & Eddy, K. T. (2017). Avoidant/restrictive food intake disorder: A three-dimensional model of neurobiology with implications for etiology and treatment.* Current Psychiatry Reports, 19(8), *1–9. https://doi.org/10.1007/s11920-017-0795-5.)*

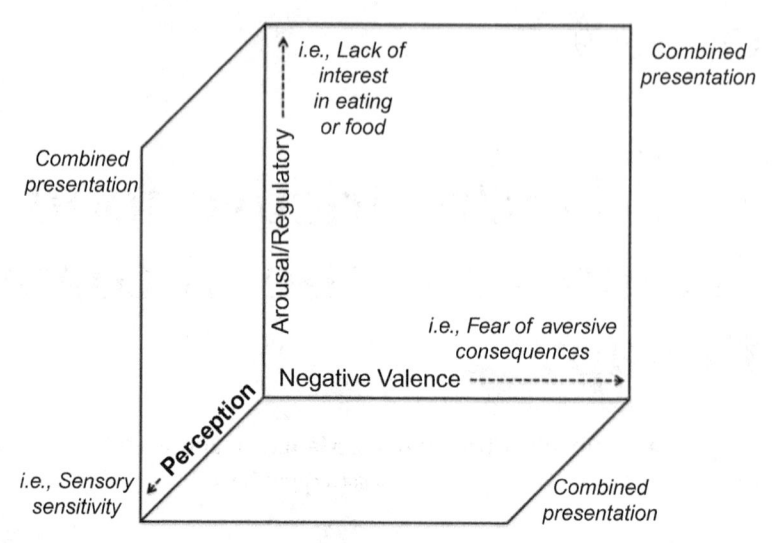

Cognitive-behavioral conceptualization of ARFID

Thomas and Eddy (2019) proposed the first conceptual model to explain core ARFID symptoms (Fig. 2; Thomas & Eddy, 2019). According to this model, individuals with ARFID are born with a biological predisposition that contributes to negative feelings and predictions about the consequences of eating that maintain avoidant/restrictive eating. For individuals with this biological predisposition, food avoidance/restriction may become chronic if they are unable to develop more adaptive eating behaviors. Chronic food restriction is associated with a host of physical (e.g., low weight, nutritional deficiencies) and psychological (e.g., social avoidance, reduced salience of hunger cues) sequelae. These consequences function to reinforce food avoidance/restriction. Thomas and Eddy (2019) suggested that though an initial, unchangeable biological predisposition or event may trigger food avoidance, "it is now the pattern of avoidance or restriction itself that serves to maintain the patient's negative predictions about eating and its attendant physical and psychological consequences. In turn, these physical and psychological factors serve to further reinforce food restriction, in a reciprocal feedback loop" (Thomas & Eddy, 2019, p. 20). Accordingly, by directly targeting food avoidance/restriction, the associated negative feelings, predictions, and consequences may be reduced, in turn interrupting the reciprocal feedback loop.

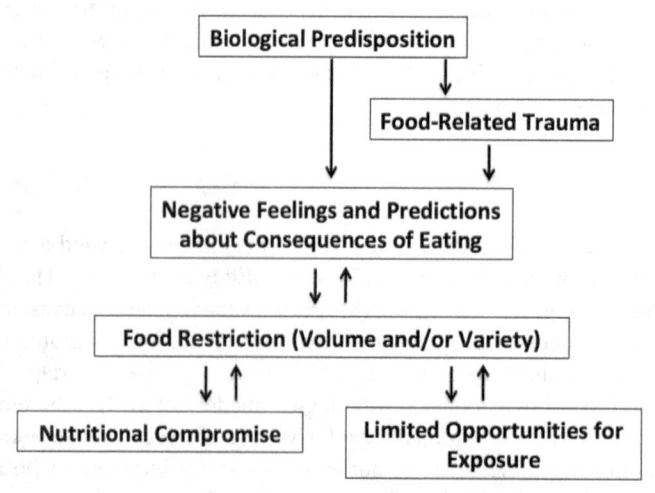

FIG. 2 General cognitive-behavioral model of ARFID. Figure depicting the general cognitive-behavioral model of ARFID. *(Borrowed with permission from Thomas, J. J., & Eddy, K. T. (2019). Cognitive-behavioral therapy for avoidant/restrictive food intake disorder: Children, adolescents, and adults. Cambridge University Press. and Cambridge University Press.)*

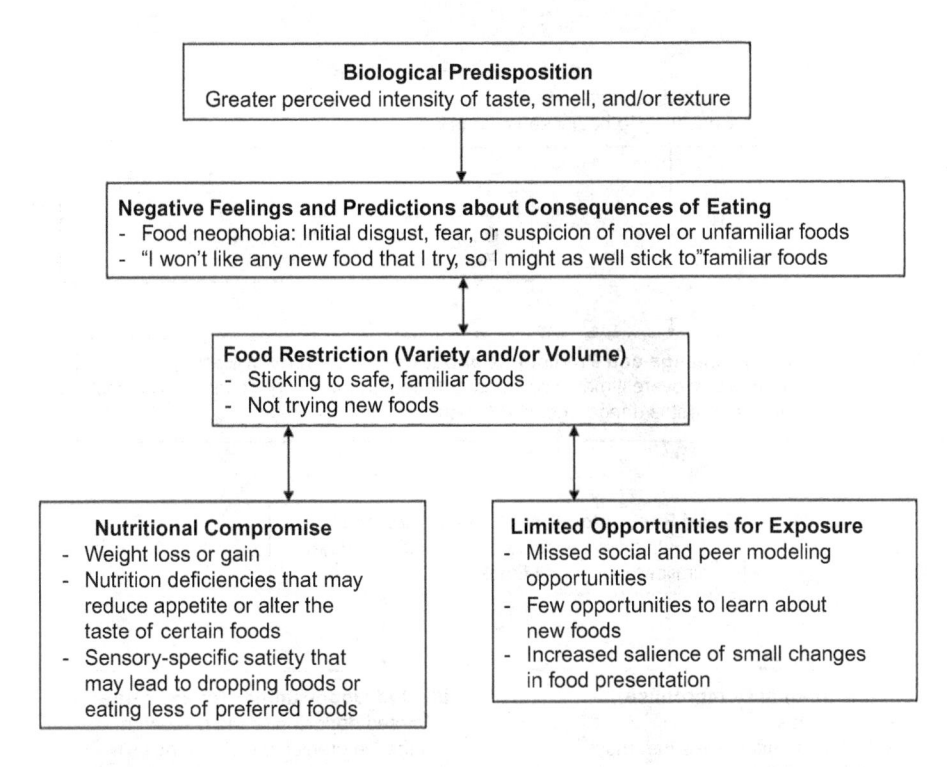

FIG. 3 Cognitive-behavioral model of ARFID with sensory sensitivity. Figure depicting the cognitive-behavioral model of ARFID for the sensory sensitivity profile. *(Borrowed with permission from Thomas, J. J., & Eddy, K. T. (2019). Cognitive-behavioral therapy for avoidant/restrictive food intake disorder: Children, adolescents, and adults. Cambridge University Press.)*

Sensory sensitivity

Individuals with ARFID characterized by the sensory sensitivity profile are hypothesized to be predisposed to experience the sensory components of food (e.g., taste, texture, smell) more intensely than others (Fig. 3; Thomas & Eddy, 2019). Accordingly, resistance to inclusion of novel or diverse foods may be more pronounced. Such individuals may present weight loss or gain and/or nutritional deficiencies resulting from their restricted range of food choices, which are often centered around palatable, processed foods. Due to the potentially high caloric content of such foods, individuals characterized by the sensory sensitivity profile may present with overweight or obesity or vitamin or mineral imbalances. Engaging in eating might be particularly challenging if preferred foods are not available, or if others are eating novel foods that induce distress (e.g., watching an individual eat a food with a novel appearance). This may perpetuate further food avoidance, contributing to a dramatically small range of "safe foods."

Fear of aversive consequences

For those presenting with the fear of aversive consequences profile, an increased tendency toward anxious temperament and focus on physiological sensations may contribute to experiencing an allergic reaction, choking, or vomiting as especially traumatizing (Fig. 4; Thomas & Eddy, 2019). Consequently, foods associated with these prior experiences—or even eating in general—may contribute to intense and chronic levels of distress. Individuals characterized by the fear of aversive consequences profile may perceive eating to be dangerous, leading them to restrict a certain food, groups of foods, or nearly all foods or beverages. Weight loss and nutritional deficiencies may occur, further disrupting typical hunger cues and taste perception. Individuals may avoid social opportunities involving food, in turn also forfeiting the opportunity to challenge cognitions related to perceived negative consequences of eating (e.g., if I eat this apple, I will choke). Further, individuals may engage in a myriad of safety behaviors (e.g., requiring sips of water between bites, taking small bites, excessively chewing foods).

Lack of interest in food or eating

Individuals presenting with an observed lack of interest in food/eating are hypothesized to find food less enjoyable (hedonic appetite) and/or experience a general absence of hunger (homeostatic appetite; Fig. 5; Thomas & Eddy, 2019). Disruptions in biologic hunger cues, predictions about how one may feel after eating (e.g., full, bloated, nauseous), and subsequent dietary restriction may partially explain an observed general disinterest in food. Similar to the other profiles, lack of interest

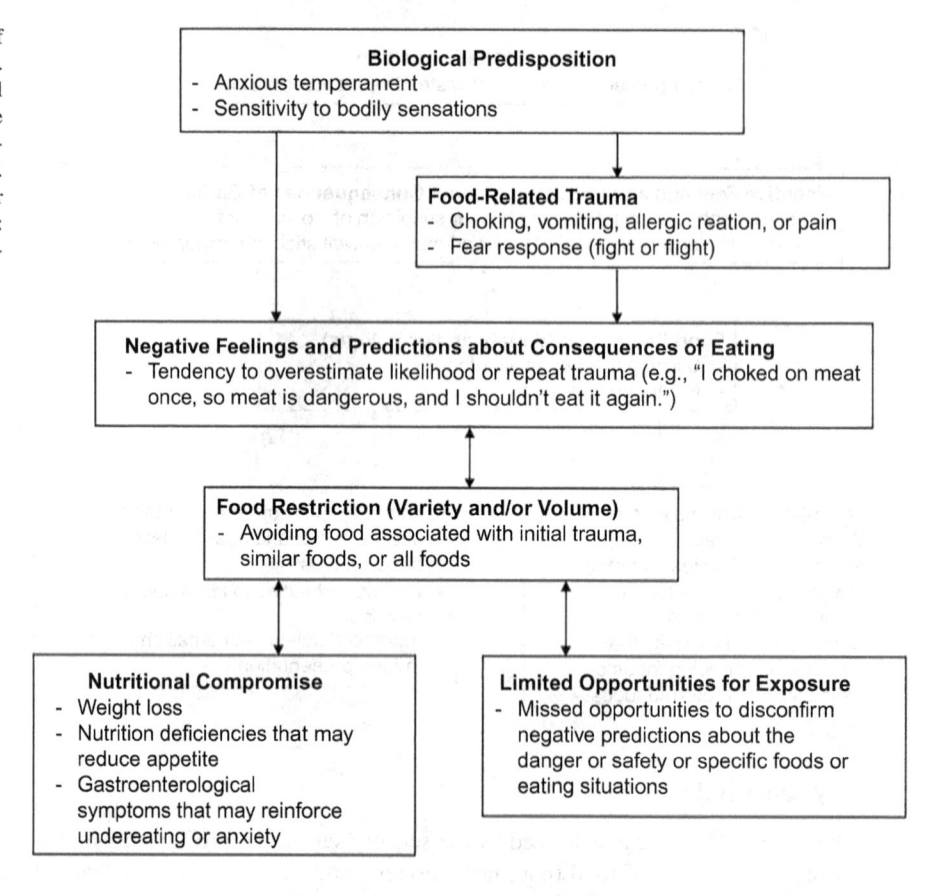

FIG. 4 Cognitive-behavioral model of ARFID with fear of aversive consequences. Figure depicting the cognitive-behavioral model of ARFID for the fear of aversive consequences profile. *(Borrowed with permission from Thomas, J. J., & Eddy, K. T. (2019). Cognitive-behavioral therapy for avoidant/restrictive food intake disorder: Children, adolescents, and adults. Cambridge University Press.)*

Biological Predisposition
- Anxious temperament
- Sensitivity to bodily sensations

Food-Related Trauma
- Choking, vomiting, allergic reation, or pain
- Fear response (fight or flight)

Negative Feelings and Predictions about Consequences of Eating
- Tendency to overestimate likelihood or repeat trauma (e.g., "I choked on meat once, so meat is dangerous, and I shouldn't eat it again.")

Food Restriction (Variety and/or Volume)
- Avoiding food associated with initial trauma, similar foods, or all foods

Nutritional Compromise
- Weight loss
- Nutrition deficiencies that may reduce appetite
- Gastroenterological symptoms that may reinforce undereating or anxiety

Limited Opportunities for Exposure
- Missed opportunities to disconfirm negative predictions about the danger or safety or specific foods or eating situations

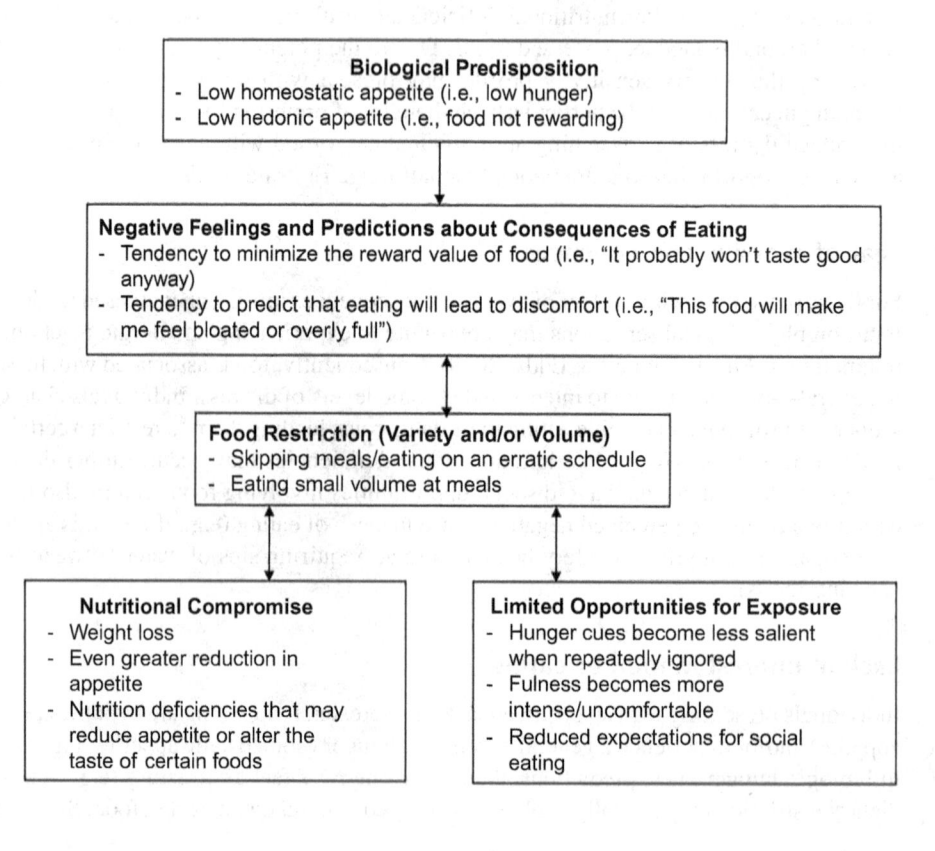

FIG. 5 Cognitive-behavioral model of ARFID with lack of interest in food or eating. Figure depicting the cognitive-behavioral model of ARFID for the lack of interest in food or eating profile. *(Borrowed with permission from Thomas, J. J., & Eddy, K. T. (2019). Cognitive-behavioral therapy for avoidant/restrictive food intake disorder: Children, adolescents, and adults. Cambridge University Press.)*

Biological Predisposition
- Low homeostatic appetite (i.e., low hunger)
- Low hedonic appetite (i.e., food not rewarding)

Negative Feelings and Predictions about Consequences of Eating
- Tendency to minimize the reward value of food (i.e., "It probably won't taste good anyway)
- Tendency to predict that eating will lead to discomfort (i.e., "This food will make me feel bloated or overly full")

Food Restriction (Variety and/or Volume)
- Skipping meals/eating on an erratic schedule
- Eating small volume at meals

Nutritional Compromise
- Weight loss
- Even greater reduction in appetite
- Nutrition deficiencies that may reduce appetite or alter the taste of certain foods

Limited Opportunities for Exposure
- Hunger cues become less salient when repeatedly ignored
- Fulness becomes more intense/uncomfortable
- Reduced expectations for social eating

in food or eating may lead to vitamin and mineral deficiencies, low-weighted clinical presentation (though normal and overweight presentations also occur), or failure to thrive (which may also coincide with low energy, mood, or disinterest in activities previously enjoyed). Often, individuals characterized by the lack of interest profile will not eat for hours between meals, which maintains appetite disruption. Avoidance of eating out may make it difficult for individuals to challenge perceptions of physical sensations of illness and thwarts opportunities in which they may otherwise evaluate the hedonic aspect of eating.

Cognitive-behavioral therapy for avoidant/restrictive food intake disorder

Overview of treatment

Cognitive-behavioral therapy for ARFID (CBT-AR; Thomas & Eddy, 2019), based on the cognitive-behavioral conceptualization of ARFID, is designed to reduce nutritional compromise and increase opportunities for exposure to novel foods, in turn reducing negative feelings and predictions about eating. CBT-AR is a flexible and modular outpatient treatment appropriate for children (>10 years old), adolescents, and adults with ARFID. Treatment spans 20–30 sessions depending on an individual's weight status (i.e., 20 sessions for individuals who are not underweight and 30 sessions for those who have significant weight to gain) over the course of approximately 6–12 months. CBT-AR is appropriate for individuals with ARFID who are medically stable, currently accepting at least some food by mouth, and not receiving enteral feeding. Though family-supported CBT-AR exists for child and early adolescent patients (10–15 years old) and young adult patients (16 years and older) who live at home and have significant weight to gain, we focus on individual CBT-AR, which is designed for late adolescent and adult patients, 16 years and older, without significant weight to gain. Importantly, though session attendees might differ, interventions are similar across the age span.

CBT-AR contains four stages (Table 1). Stage 1 involves psychoeducation and early change and spans 2–4 sessions. Stage 2 focuses on treatment planning and is typically completed in two sessions. Stage 3 is the heart of treatment (14–22 sessions) and focuses on addressing maintaining mechanisms in each ARFID domain (i.e., sensory sensitivity, fear of

TABLE 1 Stages, corresponding sessions, and interventions for CBT-AR.

Stage	Length	Intervention
Stage 1	2–4 sessions	Psychoeducation on ARFID and CBT-AR and early change • Self-monitoring and establishing a regular pattern of eating (eating preferred foods) • Increasing volume (for underweight patients) and/or variety of preferred foods • Individualized formulations of maintaining mechanism(s)
Stage 2	2 sessions	Treatment planning • Continue increasing food volume and/or variety • Psychoeducation about five basic food groups and nutritional deficiencies • Selecting new foods to learn about in Stage 3
Stage 3	14–22 sessions	Addressing maintaining mechanisms • Sensory sensitivity: Systematic desensitization to novel foods; plans for out-of-session practice • Fear of aversive consequences: Development of fear/avoidance hierarchy, graded exposure • Lack of interest in food/eating: Interoceptive exposures; in-session exposures to preferred foods
Stage 4	2 sessions	Relapse prevention • Evaluating whether treatment goals have been met • Identifying CBT-AR strategies to continue implementing at home • Continuing to learn about novel foods

Note: ARFID, avoidant/restrictive food intake disorder; CBT-AR, cognitive-behavioral therapy for avoidant/restrictive food intake disorder.

aversive consequences, lack of interest in food or eating). Finally, Stage 4, which usually occurs over two sessions, focuses on relapse prevention.

CBT is the first-line treatment for other eating disorders (Hay, Claudino, Touyz, & Abd Elbaky, 2015; Herpertz et al., 2011; National Institute of Clinical Excellence, 2017), lending credibility to its use with ARFID. To date, only one study (Thomas et al., 2021) reported on outcomes of CBT-AR for an outpatient sample of 15 adults with ARFID. All patients endorsed high ratings of treatment credibility and anticipated improvement following the first session, and a large percentage (93%) of completers indicated satisfaction at the conclusion of treatment. Patients included an average of 18 novel foods into their diets, and the subset of patients who were underweight demonstrated a significant increase in body mass index following CBT-AR. Following treatment, almost half (47%) of patients no longer met diagnostic criteria for ARFID. Thomas et al. (2020) also reported on outcomes of CBT-AR for an outpatient sample of 20 children (ages 10–17 years) with ARFID of relatively mild severity. Using intent-to-treat analyses, clinicians rated 17 patients (85%) as "much improved" or "very much improved." Patients included an average of approximately 17 novel foods into their diets, and the subset of patients who were underweight demonstrated significant weight gain, moving from the 10th to the 20th percentile for body mass index. Following treatment, 70% of patients no longer met diagnostic criteria for ARFID. Importantly, such gains occurred without a large multidisciplinary team—patients were only treated by a therapist and physician. Overall, findings from these studies suggest that CBT-AR is a promising and efficacious treatment for children and adults with ARFID.

Assessment and treatment preparation

A thorough psychiatric evaluation should be conducted at the outset of treatment to: (a) confirm the presence of ARFID; (b) determine the severity of ARFID; and (c) identify maintaining mechanisms (i.e., sensory sensitivity, fear of aversive consequences, lack of interest in food or eating) of ARFID. In addition to the Structured Clinical Interview for *DSM-5* (SCID-5; First, 2014) and the Eating Disorder Assessment for *DSM-5* (EDA-5; Sysko et al., 2015), Bryant-Waugh et al. (2019) developed the first clinical interview specifically designed to assess for the presence and severity of ARIFD—the Pica, ARFID, and Rumination Disorder Interview (PARDI). Unlike other clinical interviews for eating disorders, the PARDI confers ARFID diagnoses and assesses the severity of the three ARFID profiles. This assessment can be used to personalize CBT-AR to the individual. This is especially important given the variety of ARFID presentations and the potential for cooccurrence of phenotypic features within an individual (Thomas et al., 2017).

Prior to assessing the three ARFID phenotypes, clinicians should begin with an assessment of the patient's dietary avoidance/restriction. Collateral reports may be useful or necessary if the assessor is concerned about the patient's level of insight into their eating behaviors. Dietary avoidance/restriction may be assessed by asking the patient about their consumption of food across the five basic food groups (i.e., fruits, vegetables, grains, protein, dairy) and their daily pattern of eating (e.g., "What do you eat on a typical day?"). Such questions are included in the PARDI. By focusing early assessment within these domains, the assessor may determine if the patient is limiting their consumption in terms of volume or variety. Importantly, the assessor should rule out other eating disorder diagnoses (e.g., anorexia nervosa, bulimia nervosa), by confirming the absence of shape and weight concerns as the maintaining mechanism of dietary restriction.

In addition to a comprehensive psychiatric assessment, a medical evaluation is needed prior to initiating treatment (Aulinas et al., 2020; Brigham, Manzo, Eddy, & Thomas, 2017). Specifically, it is imperative to confirm the patient's medical stability prior to engaging in CBT-AR in an outpatient care setting. The assessment of current height and weight should utilize the patient's growth charts and estimated trajectory to determine their target weight for treatment, as the patient's growth may have been impacted by their avoidant/restrictive eating behaviors. Other critical aspects addressed in the medical evaluation include vitamin deficiencies, use of/dependence on oral nutrition supplements (e.g., Boost), partial or full dependence on enteral feeding, current medications, and other medical comorbidities that may impact eating (e.g., allergies, oral-motor difficulties).

Following psychiatric and medical assessments, a patient's treatment team and plan can be determined. As CBT-AR is an outpatient treatment, the treatment team will typically only require a primary care physician, who will continually monitor the patient's health across treatment, and a behavioral health provider (e.g., a psychologist), who will administer the CBT-AR protocol. However, other providers (e.g., psychiatrist, dietitian, gastroenterologist, endocrinologist, speech/language pathologist, occupational therapist) could be added as needed (Eddy et al., 2019).

Treatment protocol

Stage 1

The initial stage of CBT-AR generally spans 2–4 sessions. During this stage, the therapist provides psychoeducation about ARFID (and the patient's maintaining mechanism[s] of sensory sensitivity, fear of aversive consequences, and/or lack of interest in food or eating), and CBT-AR. Sessions typically begin with the clinician setting an agenda to demonstrate the time-sensitive nature of treatment, providing the patient an opportunity to add items to the session agenda (or, if more appropriate, subsequent session agendas). To ensure that weight gain goals are being met or to monitor weight changes over the course of treatment, an in-session weigh-in is conducted. During in-session weigh-ins, the clinician openly weighs their patient and discusses their weight with them.

In Session 1, following psychoeducation and the initiation of in-session weigh-ins, the clinician introduces establishing a regular pattern of eating (i.e., eating three meal and 2–3 snacks per day [for underweight patients, three snacks are nearly always necessary], eating every 3–4h; Fairburn, 2008) and self-monitoring. The clinician encourages the patient to begin with preferred foods, with minor changes and variations in the presentation of these foods (e.g., rotating preferred meals), and to reintroduce previously avoided foods to facilitate early change. In Session 2, the clinician collaboratively establishes the working conceptualization of the patient's disordered eating patterns based on the cognitive-behavioral conceptualization of ARFID. For patients who are underweight, Sessions 3 and 4 should be implemented. Session 3 incorporates a therapeutic meal in which the clinician and patient work to establish strategies to be incorporated into regular eating in support of increased dietary intake. This meal typically comprises calorie-dense preferred foods in addition to one novel food. The clinician may coach the patient by providing specific instructions to help them increase the volume of their food intake (e.g., "Take another bite" "Try to not put your fork down between bites"). After the patient has consumed an adequate volume during the in-session meal, the clinician encourages them to take one bite of the novel food item. In Session 4, the clinician and patient discuss continued increases in dietary intake through a caloric increase of 500 cal per day to support a weight gain of 1–2 pounds per week. Positive changes to daily eating should be encouraged and areas in which improvements may be made (e.g., increasing consumption of calorie-dense foods) should be addressed.

Once the following criteria are met, Stage 2 may begin: the patient (a) demonstrates understanding of ARFID (including their specific presentation) and what will happen in CBT-AR; (b) has established self-monitoring of daily food intake; (c) is eating at regular intervals; and (d) has begun increasing volume (by 500 cal per day if underweight) or variety of their food intake.

Stage 2

Stage 2 of CBT-AR comprises two sessions with a continued emphasis placed on increasing food volume and variety and psychoeducation about nutritional deficiencies. Session 1 in Stage 2 includes a review of assigned at-home practice tasks and meals. For patients who are underweight, increases in daily caloric intake are reviewed. If a patient has not gained sufficient weight, an emphasis is placed on first increasing the volume (before the variety) of foods. As nutritional deficiencies are common among those with ARFID, a review of the patient's medical assessment should be conducted, and macronutrient and micronutrient deficiencies should be discussed. When a nutritional deficiency is identified, symptoms, treatments of symptoms that may be prescribed by a doctor, and potential correction of the deficiency by diversifying intake are reviewed.

In Session 2 of Stage 2, a review of the patient's dietary intake within the five basic food groups (i.e., fruits, vegetables, grains, protein, dairy) is conducted. The therapist may identify food groups that are under- or over-represented in the patient's diet. For example, one study (Harshman et al., 2019) compared dietary intake between ARFID and healthy controls and found that vegetable and protein intake was significantly lower, and intake of added sugars and total carbohydrates was significantly higher in ARFID compared to healthy controls. In preparation for Stage 3, the clinician may begin to discuss foods from the primary food groups that the patient may have eaten recently (e.g., once in the past month and may eat if it is offered to them) or may be willing to learn about (e.g., willing to attempt to eat a small amount of). Incorporating these food groups in Stage 3 may support resolution of a patient's nutritional deficiencies, encourage further weight gain, and/or ameliorate psychosocial impairment.

Once the following criteria are met, Stage 3 may begin: the patient (a) is steadily gaining weight (if underweight); (b) has identified foods that could be added to correct any nutritional deficiencies; (c) has continued to increase eating flexibility by consuming slight variations in preferred foods or by eliminating safety behaviors; and (d) has identified several food groups to learn about in Stage 3.

Stage 3

Stage 3, the heart of treatment, is conducted over a span of 14–22 sessions during which the clinician selects the module(s) most appropriate to treat the patient's maintaining mechanisms (i.e., sensory sensitivity, fear of aversive consequences, lack of interest in food/eating). For patients with multiple maintaining mechanisms, the clinician starts with the module addressing the patient's primary, or most impairing, mechanism. While the modules differ depending upon the patient's maintaining mechanism, the common element is exposure.

Sensory sensitivity

The sensory sensitivity profile is addressed via systematic desensitization to novel foods by repeated in-session exposure to, and exploration of, appearance, feel, smell, taste, and texture. The patient is invited to bring five novel foods to each session and to nonjudgmentally describe each food. In early sessions, the patient selects small portions of foods to practice eating throughout the week to facilitate habituation. In later sessions, the patient works to incorporate larger volumes of these novel foods into their diet to meet caloric needs.

Fear of aversive consequences

For patients characterized by the fear of aversive consequences profile, the clinician provides psychoeducation about how avoidance increases anxiety and collaboratively works with the patient to develop a fear and avoidance hierarchy of foods and eating-related situations that the patient fears will lead to negative outcomes. Subsequently, the clinician conducts in-session exposures to these foods and situations in which choking, vomiting, or other feared consequences are hypothesized to occur, continuing these exposures until the patient has completed the most distressing task on their hierarchy. In order to create the patient's hierarchy, the clinician first helps the patient develop a Subjective Units of Distress (SUDS) scale which ranges from 0 (no distress whatsoever) to 100 (highest distress imaginable). For each step in the hierarchy, the clinician should ask the patient where this step falls on their SUDS scale. At the start of the first food exposure and every exposure thereafter, the clinician should ask the patient to describe the feared consequence that they are concerned will occur as a result of the exposure (e.g., "I will vomit"). The clinician then asks the patient to estimate the probability of the feared outcome, ranging from 0% to 100% (e.g., "I am 95% certain I will vomit"). The patient next rates their SUDS and begins the exposure while the clinician continues to take periodic SUDS rating as the exposure continues. After the exposure is complete, the clinician should ask then patient to reestimate the probability of the feared outcome. The clinician asks the patient to repeat these exposures for homework in order to further test the patient's predictions that their feared outcomes will occur.

Lack of interest in food or eating

For individuals presenting with an apparent lack of interest in food or eating, a series of interoceptive exposures are conducted in session to help the patient habituate to sensations associated with eating and fullness. Such exposures might include pushing one's belly out (to mimic bloating), gulping water (to mimic fullness), or spinning in a chair (to mimic nausea). In-session exposures to preferred foods are conducted to reinforce what the patient might enjoy about eating (e.g., describing the sensory properties of preferred foods). Further, self-monitoring occurs to increase awareness of the patient's hunger and fullness.

Once the following criteria are met, Stage 4 may begin: the patient (a) is no longer underweight; (b) is eating at regular intervals and has increased volume or variety within meals and snacks; (c) is regularly incorporating foods that will help resolve nutrition deficiencies; (d) has at least partially resolved their primary ARFID maintaining mechanism(s).

Stage 4

The final stage of CBT-AR is administered over a period of two sessions, during which readiness to terminate treatment is assessed and relapse prevention plans are developed. Stage 4 should be conducted over a period of 4 weeks, with Session 1 occurring 2 weeks after Stage 3 is complete and Session 2 occurring 2 weeks following Session 1. This enables the patient sufficient time to continue with progress made during session. Session 1 should focus on the symptomatic changes that have occurred over the course of treatment. The formulation of the patient's specific presentation (developed in Stage 2) should be examined, placing emphasis on evaluating the patient's progress. Additionally, the clinician should review foods from the five basic food groups that the patient has incorporated into their diet. Session 1 of Stage 4 also includes a conversation about the patient's readiness to complete treatment. The following may serve as indicators of the patient's readiness to terminate treatment: (a) no longer meeting criteria for ARFID; (b) consuming a variety of foods from the primary food groups; (c) on an expected growth trajectory; (c) corrected nutritional deficiencies; and/or (d) reduction in clinical impairment related to ARFID. Session 2 of Stage 4 focuses on the cocreation of a relapse prevention plan. Collaboratively, the clinician and patient review how the patient's eating has changed over the course of treatment, identify CBT-AR strategies for continued implementation, set goals for continued progress, and anticipate potential challenges that may arise in subsequent months. Problematic behaviors that may signal relapse should be identified and techniques used to address these behaviors should be reinforced.

CBT-AR case example

Presenting complaint

Anna was a 19-year-old girl with a prolonged history of avoidant and restrictive eating referred to us by her prior therapist for treatment for ARFID. At her intake, Anna reported that though people had commented on her low weight throughout the course of her life, she never believed she met criteria for an eating disorder and just considered herself "a picky eater." Anna shared that she was "relieved" to learn of her ARFID diagnosis and its corresponding treatment. She reported that there were a number of foods she refused to eat specifically because of their sensory characteristics (i.e., texture or taste). Anna stated that such foods included yogurt, cheese, granola, and most fruits. Anna noted that her avoidant and restrictive eating began when she was 4 years old. At the time of initial assessment, her diet consisted of a bagel for breakfast and tortellini for lunch. She stated that dinners typically varied because her mother would cook for the family. Anna provided an example of her mother making tacos. However, Anna refused to combine the taco shell and taco meat, stating that they "do not go together." Anna's diet seldom included fruits, vegetables, or dairy, and her protein intake was limited to what her mother cooked for dinner. As a consequence of her ARFID symptoms, Anna experienced psychosocial impairment at college, stating that she often avoided situations during which she would have to eat with others. Restaurant dinners with her friends and family were limited to destinations where Anna's preferred foods were available (e.g., fast food restaurants like McDonalds, where she could order French fries).

Assessment and treatment preparation

Prior to initiating treatment, the clinician referred Anna for a medical evaluation with a physician. Anna was 5'4" tall and weighed 117.21 pounds (body mass index = 20.1 kg/m^2). Anna's blood work indicated an iron, zinc, and vitamin B12 deficiency. While her physician recommended a multivitamin in conjunction with other supplements to correct these deficiencies, Anna shared that she was hesitant to take pills due to her dislike of swallowing them. Accordingly, one of her treatment goals was to expand her dietary variety so as to avoid swallowing pills. Anna shared that another treatment goal was to diversify her preferred and acceptable foods so that she would be able to eat in social situations, and at restaurants of her friends' choosing. She stated that this was particularly important to her because, as a theater major, she wanted to be able to enjoy dinners with the cast and crew following theatrical productions. Anna followed-up, reporting that these dinners often take place at a vegetarian restaurant in her college town and that this restaurant does not serve any foods that Anna is willing to eat. Anna also reported that expanding her dietary intake by volume would enable her to have more energy for her dance and stage combat classes. Because Anna was an adult, her therapist recommended individual CBT-AR. As Anna was not underweight, the therapist planned for approximately 20 sessions of treatment. Based on her presentation and case conceptualization, Stage 3 of treatment would focus exclusively on the sensory sensitivity module with the primary treatment goal of introducing novel foods and expanding dietary variety.

Stage 1

Anna was receptive to psychoeducation about ARFID. Upon intake, Anna was eating only three meals per day. Thus, the therapist intervened to support a regular pattern of eating, which included 2–3 snacks per day in addition to her three meals. Anna was able to identify foods she would be willing to introduce as snacks. These included two fruits (i.e., apple, orange) and peanut butter (either on the apple, or in a sandwich). In cocreating her individualized formulation, Anna's clinician pointed out that her long-standing reliance on processed grains (i.e., bagel, tortellini) made the other food groups increasingly novel, and thus less approachable. Anna noted how her avoidance of social activities limited her exposure to novel foods that her peers might model eating (e.g., vegetarian foods). Anna set a goal to introduce four foods—wavy

Pringles (as opposed to her typically preferred original texture Pringles), cinnamon raisin bagels (as opposed to plain) with brown sugar cream cheese, sugar snap peas at dinner and bologna slices or cheese sticks at lunch.

Stage 2

Anna was successful in meeting her goal of incorporating snacks and introducing new foods. Anna was able to identify several proteins, fruits, and vegetables to learn about in CBT-AR. Due to concerns about Anna's low intake of iron, zinc, and B12, Anna and her clinician reviewed her nutritional deficiencies and identified foods rich in iron (e.g., spinach, beef, chickpeas), zinc (e.g., cashews, chickpeas, oatmeal, fortified cereals) and vitamin B12 (e.g., poultry, fish, eggs, yogurt, cheese) she was willing to learn about. At the end of Session 4, Anna identified five foods that she was willing to learn about in the subsequent session (raisins, corn, turkey cold cuts, cashews, spinach) and planned to bring small tasting portions of each.

Stage 3

The clinician focused on a single maintaining mechanism for the entirety of Stage 3—sensory sensitivity. She began Session 5 by asking Anna to look at, touch, smell, taste, and chew each of the five foods she had brought to session. Anna was surprised to find that she did not dislike any of the foods she had brought with her. She was fond of the raisins, which she committed to adding to her morning snack 3–4 times during the upcoming week. She reported trying "ants on a log" at home given her like of raisins and peanut butter and stated that doing so enabled her to incorporate celery into her diet. She also stated that she was willing to continue learning about turkey cold cuts, cashews, and spinach, and she committed to taking small tastes of each at home prior to the next session. Of note, Anna said that she did not want to continue learning about the corn. The therapist acknowledged that though Anna might wish to learn about these foods in the future, she could decline practicing any foods that she did not initially find appealing. The therapist emphasized that it was the range of foods across the five basic food groups, rather than specific foods, that were important. Anna also shared with the therapist that she had spontaneously tried eating cherries and green beans, as well as varying her breakfast to include fortified milk and Fruit Loops cereal. Anna's therapist praised her for making these changes on her own outside of session.

Anna returned to Session 6 having successfully incorporated the raisins and having practiced eating turkey cold cuts, cashews, and spinach. Anna continued to try new foods in each session and began to feel increasingly comfortable trying novel foods outside of session as well. By the time Anna had tried 15 novel foods in session and three novel foods (fortified milk, Fruit Loops cereal, celery) spontaneously at home, the therapist encouraged Anna to consider how she might incorporate the newly tasted foods into her routine meals and snacks. Between Sessions 9 and 14, Anna continued to try fruits, vegetables, and proteins in session, incorporating these food groups into her daily diet. Anna reported that the changes she had made to her dietary intake enabled her to attend more social events. Further, she stated that she was looking forward to an upcoming cast dinner following a theater production at her university. By Session 18, Anna felt she had made sufficient progress, and together with her therapist, made the decision to move on to Stage 4.

Stage 4

In Sessions 19 and 20, the therapist and Anna reviewed the foods that she had successfully incorporated into her daily diet. Anna reported that she was surprised with how much more energy she had than at the start of treatment. They also discussed which foods that she had initially expressed interest in learning about remained to be tasted in the future. Anna reported that she had successfully incorporated 23 novel foods into her diet, including several fruits, vegetables proteins, dairy and grains. Anna was now also consistently eating several iron-, zinc-, and vitamin B12-rich foods. Anna identified several foods she would like to continue learning about posttreatment. The therapist encouraged Anna to continue to use the skills she had learned in CBT-AR to embark on a process of life-long learning and dietary expansion. Anna set a goal to try one novel food per week for the next several months. In cocreating her relapse prevention plan in the final session, Anna identified several CBT-AR strategies (e.g., taking small bites when trying a novel food for the first time) that she had found helpful and wanted to continue implementing posttreatment.

Outcome and conclusion

After completing 17 sessions of CBT-AR, Anna was highly successful at increasing dietary variety, more than doubling the absolute number of foods that she ate on a consistent basis. Importantly, the new foods she added came from the food groups that had been relatively underrepresented in her diet at the beginning of treatment (i.e., fruits, vegetables, dairy, and proteins), and included several foods that she selected to correct her nutrition deficiencies. Although Anna began by tasting novel foods in-session early in treatment, she quickly graduated to trying novel foods at home and incorporating them in larger portions into her routine meals and snacks. In the second half of the treatment, she began to report decreases in psychosocial impairment as her dietary variety increased. By session 17, her deficiencies had resolved and Anna was very pleased with this because through treatment, she was able to avoid swallowing pills. In summary, Anna presents an example of a highly motivated woman with a relatively mild form of illness, a single ARFID maintaining mechanism, and no psychiatric comorbidities. All of these factors likely facilitated her rapid progress in CBT-AR.

Though it is a relatively new diagnosis, the feeding problems and clinical impairment associated with ARFID are not. This formalization of the disorder in fifth edition of the *Diagnostic and Statistical Manual of Mental Disorders* (APA, 2013) propelled the research of this unique eating disorder and its heterogeneous presentations. According to the model proposed by Thomas and Eddy (2019), individuals with ARFID are born with a biological predisposition contributing to negative feelings and predictions about the consequences of eating that maintain avoidant and/or restrictive eating. CBT-AR was developed out of necessity to aid individuals struggling with their eating and represents a promising clinical approach to treating the disorder. Data to date suggest that CBT-AR is an effective and efficient treatment in expanding dietary intake volume and variety. Many of those treated with CBT-AR fully recover, gaining weight and experiencing a reduction in psychosocial impairment. Future dismantling studies may help to identify components of CBT-AR that are critical in eliciting change as well as those that may be superfluous. Current research is seeking to examine biological underpinnings of the disorder with the hope that understanding the pathophysiology of the disorder may lead to new and effective treatment targets.

Practice and procedures

- Pica, ARFID and Rumination Disorder Interview (PARDI; Bryant-Waugh et al., 2019): The PARDI is a comprehensive, multiinformant semistructured clinical interview to assess and diagnose pica, ARFID, and rumination disorder in children and adults. The PARDI captures clinical features and severity and provides dimensional scores for each of the three ARFID profiles (i.e., sensory sensitivity, fear of aversive consequences, lack of interest in eating).

Mini-dictionary of terms

- Avoidant/restrictive food intake disorder: a new feeding and eating disorder introduced in the fifth edition of the *Diagnostic and Statistical Manual of Mental Disorders* characterized by avoidant or restrictive eating resulting in one (or more) of the following: (a) significant weight loss or failure to achieve expected weight gain; (b) nutritional deficiency; (c) dependence on enteral feeding or oral nutrition supplements; and (d) interference with psychosocial functioning.
- ARFID profiles: Three core ARFID presentations—sensory sensitivity, fear of aversive consequences, and lack of interest in food or eating.
- Sensory sensitivity: One of the profiles of ARFID; some people with ARFID find that novel foods have strange or intense sensory characteristics (e.g., tastes, textures, smells) and they feel safer and more comfortable eating foods that they are familiar with and know well.
- Fear of aversive consequences: One of the profiles of ARFID; some people with ARFID have had scary experiences with food, like throwing up, choking, or allergic reactions, so they may avoid foods that have made them sick or stop eating altogether.
- Lack of interest in food or eating: One of the profiles of ARFID; some people with ARFID do not feel hungry very often, think eating is a chore, or get full very quickly.

Key facts of ARFID and CBT-AR

- ARFID differs from other eating disorders, such as anorexia nervosa, because individuals with ARFID do not present with shape or weight concerns.
- ARFID occurs across the weight spectrum and lifespan.
- The three mechanisms maintaining dietary avoidance/restriction in ARFID are sensory sensitivity, fear of aversive consequences, and lack of interest in food or eating.
- ARFID is associated with one (or more) of the following: (a) significant weight loss; (b) significant nutritional deficiency; (c) dependence on enteral feeding or oral nutrition supplements; (d) marked interference with psychosocial functioning.

- CBT-AR is a modular outpatient treatment that includes four stages and spans across 20–40 sessions depending on an individual's weight presentation.
- Main treatment goals in CBT-AR include achieving or maintaining a healthy weight, correcting any nutritional deficiencies, eating foods from each of the five basic food groups (i.e., fruits, vegetables, proteins, dairy, grains), and feeling more comfortable eating in social situations.

Applications to other areas

Given that ARFID was introduced as a new diagnosis in the fifth edition of the *Diagnostic and Statistical Manual of Mental Disorders*, nonmental health care providers (e.g., primary care physicians, pediatricians) should be aware of the sign and symptoms associated with this disorder. This will help with the early identification of and prevention of ARFID. It is possible that, if unfamiliar with the disorder, many providers could perceive frank ARFID as normative picky eating. Providers should also be aware that ARFID is different from other eating disorders, like anorexia nervosa and bulimia nervosa, because people with ARFID do not worry much about their weight or shape. Thus, it is possible to confer an eating disorder diagnosis among individuals who do not experience shape or weight concerns. Instead, individuals with ARFID might have one, two, or three of important concerns—sensory sensitivity, fear of aversive consequences, or lack of interest in food or eating. It is important to understand that someone with ARFID is not just being "picky" or "stubborn." Rather, those with ARFID have underlying biological traits that initially made their eating habits a logical choice. There are helpful steps patients and families can take to interrupt patterns of behavior attributable to ARFID. Given the availability and efficacy of CBT-AR, providers should consider referring patients presenting with such concerns to treatment programs for eating disorders that target ARFID.

Summary points

- ARFID is a restrictive eating disorder that spans the lifespan and weight spectrum. There are three ARFID profiles: (a) sensory sensitivity; (b) fear of aversive consequences; and (c) lack of interest in food or eating. The three profiles occur along a continuum of severity and are not mutually exclusive.
- CBT-AR is designed to reduce nutritional compromise and increase opportunities for exposure to novel foods, in turn reducing negative feelings and predictions about eating. It is a modular outpatient treatment appropriate for children (10 years and older), adolescents, and adults with ARFID.
- In Stage 1 of CBT-AR (psychoeducation and early change), the clinician provides psychoeducation about ARFID, encourages the patient to establish a regular pattern of eating and self-monitoring, and helps the patient increase their food intake by volume (if underweight) and/or variety.
- In Stage 2 of CBT-AR (treatment planning), the clinician provides psychoeducation about nutritional deficiencies and supports the patient in selecting novel foods to learn about and incorporate in Stage 3 that will support resolution of these deficiencies, encourage further weight gain (if underweight), and/or reduce psychosocial impairment.
- In Stage 3 of CBT-AR (addressing maintaining mechanisms), the therapist selects the module(s) most relevant to the patient's maintaining mechanism(s). Though the specific components of Stage 3 depend upon the maintaining mechanism(s) being targeted, the common thread underlying all is in-session exposure and between-session practice.
- In Stage 4 of CBT-AR (relapse prevention), the clinician and patient evaluate progress and cocreate a relapse prevention plan.

References

American Psychiatric Association. (2000). *Diagnostic and statistical manual of mental disorders (DSM-IV®)*. American Psychiatric Pub.

American Psychiatric Association. (2013). *Diagnostic and statistical manual of mental disorders (DSM-5®)*. American Psychiatric Pub.

Aulinas, A., Marengi, D. A., Galbiati, F., Asanza, E., Slattery, M., Mancuso, C. J., et al. (2020). Medical comorbidities and endocrine dysfunction in low-weight females with avoidant/restrictive food intake disorder compared to anorexia nervosa and healthy controls. *International Journal of Eating Disorders, 53*(4), 631–636.

Brigham, K. S., Manzo, L. D., Eddy, K. T., & Thomas, J. J. (2017). Evaluation and treatment of avoidant/restrictive food intake disorder (ARFID) in adolescents. *Current Pediatrics Report, 6*(2), 107–113.

Bryant-Waugh, R., Micali, N., Cooke, L., Lawson, E. A., Eddy, K. T., & Thomas, J. J. (2019). Development of the pica, ARFID, and rumination disorder interview, a multi-informant, semi-structured interview of feeding disorders across the lifespan: A pilot study for ages 10–22. *International Journal of Eating Disorders, 52*(4), 378–387. https://doi.org/10.1002/eat.22958.

Eddy, K. T., Harshman, S. G., Becker, K. R., Bern, E., Bryant-Waugh, R., Hilbert, A., et al. (2019). Radcliffe ARFID workgroup: Toward operationalization of research diagnostic criteria and directions for the field. *International Journal of Eating Disorders, 52*(4), 361–366.

Fairburn, C. (2008). *Cognitive behavior therapy and eating disorders*. New York, NY: Guilford Press.

First, M. B. (2014). Structured clinical interview for the DSM (SCID). *The Encyclopedia of Clinical Psychology*, 1–6.

Harshman, S. G., Wons, O., Rogers, M. S., Izquierdo, A. M., Holmes, T. M., Pulumo, R. L., et al. (2019). A diet high in processed foods, total carbohydrates and added sugars, and low in vegetables and protein is characteristic of youth with avoidant/restrictive food intake disorder. *Nutrients, 11*(9), 2013.

Hay, P. J., Claudino, A. M., Touyz, S., & Abd Elbaky, G. (2015). *Individual psychological therapy in the outpatient treatment of adults with anorexia nervosa* (p. CD003909). The Cochrane Library.

Herpertz, S., Hagenah, U., Vocks, S., von Wietersheim, J., Cuntz, U., Zeeck, A., et al. (2011). The diagnosis and treatment of eating disorders. *Deutsches Arzteblatt International, 108*, 678–685.

National Institute of Clinical Excellence. (2017). *Eating disorders: Recognition and treatment*. London, United Kingdom: National Institute of Clinical Excellence.

Nicholls, D., Chater, R., & Lask, B. (2000). Children into DSM don't go: A comparison of classification systems for eating disorders in childhood and early adolescence. *International Journal of Eating Disorders, 28*(3), 317–324.

Sysko, R., Glasofer, D. R., Hildebrandt, T., Klimek, P., Mitchell, J. E., Berg, K. C., et al. (2015). The Eating Disorder Assessment for DSM-5 (EDA-5): Development and validation of a structured interview for feeding and eating disorders. *International Journal of Eating Disorders, 48*(5), 452–463. https://doi.org/10.1002/eat.22388.

Thomas, J. J., Becker, K. R., Breithaupt, K., Murray, H. B., Jo, J. H., Kuhnle, M. C., et al. (2021). Cognitive-behavioral therapy for adults with avoidant/restrictive food intake disorder. *Journal of Behavioral and Cognitive Therapy, 31*, 47–55.

Thomas, J. J., Becker, K. R., Kuhnle, M. C., Jo, J. H., Harshman, S. G., Wons, O. B., … Eddy, K. T. (2020). Cognitive-behavioral therapy for avoidant/restrictive food intake disorder: Feasibility, acceptability, and proof-of-concept for children and adolescents. *The International Journal of Eating Disorders, 53*(10), 1636–1646. https://doi.org/10.1002/eat.23355.

Thomas, J. J., & Eddy, K. T. (2019). *Cognitive-behavioral therapy for avoidant/restrictive food intake disorder: Children, adolescents, and adults*. Cambridge University Press.

Thomas, J. J., Lawson, E. A., Micali, N., Misra, M., Deckersbach, T., & Eddy, K. T. (2017). Avoidant/restrictive food intake disorder: A three-dimensional model of neurobiology with implications for etiology and treatment. *Current Psychiatry Reports, 19*(8), 1–9. https://doi.org/10.1007/s11920-017-0795-5.

Chapter 10

Diabetes-related distress and HbA1c: The use of cognitive behavioral therapy

Peerasak Lerttrakarnnon[a], G. Lamar Robert[b], Puriwat Fakfum[b], and Kongprai Tunsuchart[c]

[a]Aging and Aging Palliative Care Research Cluster, Department of Family Medicine, Faculty of Medicine, Chiang Mai University, Chiang Mai, Thailand, [b]Aging and Aging Palliative Care Research Cluster, Faculty of Medicine, Chiang Mai University, Chiang Mai, Thailand, [c]District Public Health Assistance, Sarapee District Public Health Office, Chiang Mai, Thailand

Abbreviations

DM	diabetes mellitus
FPG	fasting plasma glucose
HbA1c	hemoglobin A1c
IGT	impaired glucose tolerance
IFG	impaired fasting glucose
USD	United States Dollar
DRD	diabetes-related distress
T2DM	type 2 diabetes mellitus
T1DM	type 1 diabetes mellitus
CBT	cognitive behavioral therapy
PAID	problems areas in diabetes
DDS	Diabetes Distress Scale
PHQ	Patient Health Question
GSES	General Self-Efficacy Scale
BMI	Body Mass Index
CPD	chronic physical diseases
RCTs	randomized controlled trials
DSNs	diabetes specialist nurse
MBSR	mindfulness-based stress reduction
MBCT	mindfulness-based cognitive therapy
ACT	acceptance and commitment therapy
BG-CBT	brief group-cognitive behavioral therapy
DOT	diabetes online therapy

Introduction

Diabetes mellitus (DM) is a serious, long-term, or chronic condition that occurs when there are raised levels of glucose in a person's blood because their body produces either none or insufficient quantities of the hormone insulin and cannot effectively use any insulin it does produce (International Diabetes Federation, 2019). Stages of DM pathology can be divided into prediabetes, diabetes, and associated complications. Diagnosis criteria for diabetes in this study were fasting plasma glucose (FPG) (no caloric intake for at least 8h) $\geq$126mg/dL, random plasma glucose >200mg/dL or hemoglobin (Hb) A1c $\geq$6.5mg/dL. Prediabetes is defined as raised blood glucose level above the normal range but below the diabetic blood level. This condition is known as impaired glucose tolerance (IGT) defined as FPG<126mg/dL and 2-h plasma glucose after 75g oral glucose load=140–200mg/dL or as impaired fasting glucose (IFG) defined as FPG=110–125mg/dL (International Diabetes Federation, 2019).

In 2019, the estimated global prevalence of diabetes was 9.3% (463 million people), estimated to rise to 10.2% (578 million) by 2030 and 10.9% (700 million) by 2045. The prevalence was higher in urban (10.8%) than rural (7.2%) areas, and in high-income (10.4%) more than low-income countries (4.0%). The prevalence of diabetes in 2019 showed an age-related increase and a similar trend was predicted to continue in 2030 and 2045 as the population of the world ages. The estimated incidence of diabetes in women aged 20 to 79 years is slightly lower than in men (9.0% vs. 9.6%) (International Diabetes Federation, 2019). Regionally, the highest prevalence in 2019 was in the North American and Caribbean Region at 27.0%. Countries with the highest number of people older than 65 years with diabetes are China, the United States of America and India (Sinclair et al., 2020). In 2019, one in two (50.1%), or 231.9 million of the 463 million adults were living with diabetes. One in two (50.1%) people living with diabetes do not know that they have diabetes. In 2019, one in two adults living with diabetes (50.1% of the world's population or 231.9 million people) did not know that they had the disease (Saeedi et al., 2019). Globally, the estimated cost of diabetes care in 2019 was USD 760 billion, projected to grow to USD 825 billion by 2030 and USD 845 billion by 2045 (Williams et al., 2020). It has been estimated that approximately 4.2 million adults aged 20 to 79 years would die as a result of diabetes and its complications in 2019, equivalent to one death every eight seconds. Diabetes is estimated to be associated with 11.3% of global deaths from all causes among people in that age group. Almost half (46.2%) of deaths associated with diabetes among the 20 to 79 age group occur in people under the age of 60 (International Diabetes Federation, 2019). It is estimated that 19.3% of people aged 65 to 99 years (135.6 million, 95% CI: 107.6–170.6 million) are living with diabetes, and it is projected that the number will reach 195.2 million by 2030 and 276.2 million by 2045. An estimated 4.2 million deaths among 20- to 79-year-old adults are attributable to diabetes. Diabetes is estimated to contribute to 11.3% of deaths globally, ranging from 6.8% in the Sub-Saharan Africa region to 16.2% in the Middle East and North Africa. About half (46.2%) of the deaths attributable to diabetes occur in people under the age of 60. The Sub-Saharan Africa region has the highest (73.1%) proportion of deaths attributable to diabetes in people under the age of 60, while the European region has the lowest (31.4%) (Saeedi et al., 2020).

Type 2 diabetes patients represent the majority (around 90%) of diabetes patients worldwide (International Diabetes Federation, 2019). Many adults with type 2 diabetes mellitus (T2DM) experience psychosocial burdens and mental health problems associated with the disease. Diabetes-related distress (DRD) also has distinct deleterious effects on self-care behaviors and disease control. Reducing DRD in adults with T2DM could enhance their psychological well-being, health-related quality of life, self-care abilities and disease control as well as reducing depressive symptoms (Chew, Vos, Metzendorf, Scholten, & Rutten, 2017). A systemic review and meta-analysis study identified psychological interventions which could potentially improve glycemic control in adults with type 2 diabetes including self-help materials, cognitive behavioral therapy (CBT) and counseling (Winkley et al., 2020). The following section includes a narrative review of the efficacy of CBT in treating DRD and achieving glycemic control in DM patients.

DRD and associated factors

Definition and prevalence of DRD

DRD is a psychosocial problem which can be identified through screening and routine monitoring for diabetes in patients, particularly when diabetes treatment targets are not met and/or at the onset of diabetes-related complications (American Diabetes Association, 2021). DRD is a syndrome comprised of multidimensional components including worry, conflict, frustration, and discouragement that can accompany living with diabetes. Negative physical and psychological effects can be directly attributable to long-term suffering resulting from diabetes-related emotional distress (Fisher, Gonzalez, & Polonsky, 2014; Thanakwang, Thinganjana, & Konggumnerd, 2014).

Two common validated diabetes-related distress assessments are Problems Areas in Diabetes (PAID) (Polonsky et al., 1995) and Diabetes Distress Scale (DDS) (Polonsky et al., 2005). A systematic review and meta-analysis by Perrin, Davies, Robertson, Snoek, and Khunti (2017) included 55 studies ($n=36,998$) and reported an overall prevalence of 36% for diabetes distress (including 25 PAID studies, 9 PAID-5 studies and 21 DDS studies) in people with T2DM. The prevalence of diabetes distress was significantly higher in samples with a higher incidence of comorbid depressive symptoms and in those with a female sample majority (Perrin et al., 2017). The prevalence of high DRD (Table 1) as measured by PAID or PAID-5 (score ≥ 40) ranged from 1.2% to 60.2% (Aikens, 2012; Al-Ozairi et al., 2020; Delahanty et al., 2007; Karlsen et al., 2012; Kasteleyn et al., 2015; Kuniss et al., 2017; Liu et al., 2020; Pintaudi et al., 2015; Stoop et al., 2014). The prevalence of moderate to high DRD by the 17-item Diabetes Distress Scale (DDS-17) (Table 2) was 6.9% to 76.2% (Geleta et al., 2021; Tunsuchart et al., 2020a; Alzughbi et al., 2020; Parsa et al., 2019; Hamed et al., 2019; Hemavathi et al., 2019; Aljuaid et al., 2018; Nanayakkara et al., 2018; Arif et al., 2018; Chew, Vos, Stellato, & Rutten, 2017; Zhou et al., 2017; Fisher et al., 2008). Most of the participants in those studies were type 2 diabetics.

TABLE 1 Comparison of diabetes-related distress prevalence by problem area using Problems Areas in Diabetes Scale (PAID) scores ≥40.

Author(s)	Country	Sample size (persons)	Mean age [±SD] or range (years)	Mean HbA1c	Incidence of diabetes related distress (%)
Delahanty et al. (2007)	America	815 T2DM	66	7.44	24.57
Aikens (2012)	America	253 T2DM	57.3±8.3 (27–88)	7.6±1.6	21
Karlsen, Oftedal, and Bru (2012)	Norway	378 T2DM	30–70	7.1±1.1	22
Stoop et al. (2014)	Netherlands	774 T2DM (primary care) 526 T2DM (secondary care)	65 (58–73)	6.8 (6.3–7.5)	10 (primary care=4, secondary care=19)
Kasteleyn et al. (2015)	Netherlands	590 T2DM	18–85	–	7.1
Pintaudi et al. (2015)	Italy	2374 T2DM	65.0±10.2	7.7±1.5	60.2
Kuniss et al. (2017)	Germany	345 T2DM	72.3±9.7	6.4±1.0	1.2
Liu et al. (2020)	China	1512 T2DM	60.63±11.29	–	55.9 (PAID-5)
Al-Ozairi, Al Ozairi, Blythe, Taghadom, and Ismail (2020)	Kuwait	465 T2DM	55.3±10.1	–	14

Factors associated with DRD

Diabetes distress levels and the prevalence of elevated diabetes distress were considerably lower in the participants treated in primary care (8±11; 4%) than those in secondary care (23±21; 19%) (Stoop et al., 2014). Previous studies have reported that DRD in diabetes patients is associated with younger age (Chew et al., 2016; Stoop et al., 2014), female gender (Fisher et al., 2008), a high incidence of complications/comorbidities (Fisher et al., 2008), anxiety (Fisher et al., 2008) and depression (Fisher et al., 2009; Jia et al., 2009; Ting et al., 2011; Zhang et al., 2013), low levels of physical activity (Fisher, Glasgow, & Strycker, 2010), obesity (Ting et al., 2011), low medication compliance (Aikens, 2012; Pandit et al., 2014), poor diet control, low self-efficacy (Fisher, Hessler, Polonsky, & Mullan, 2012), poor self-care behavior (Polonsky et al., 1995), poor quality of life (Stanković, Jasović-Gasić, & Lecić-Tosevski, 2013), Chinese ethnicity (Chew et al., 2016), higher Patient Health Question (PHQ) scores (Chew et al., 2016), lower sleep time (Zhou et al., 2017), low General Self-Efficacy Scale (GSES) scores (Zhou et al., 2017), a family history of diabetes, (Parsa et al., 2019), duration of diabetes (Parsa et al., 2019), type of treatment (Parsa et al., 2019), HbA1c (Nanayakkara et al., 2018; Stoop et al., 2014; Strandberg, Graue, Wentzel-Larsen, Peyrot, & Rokne, 2014), suboptimal self-care (Nanayakkara et al., 2018), poor coping styles (Karlsen et al., 2012), perceived support (Karlsen et al., 2012), ethnic minority status (Stoop et al., 2014), using insulin (Stoop et al., 2014), higher body mass index (BMI) (Stoop et al., 2014), and the presence of neuropathy (Stoop et al., 2014). A high level of DRD has been reported to be associated with high HbA1c levels (Al-Ozairi et al., 2020; Fisher et al., 2008, 2010; Nanayakkara et al., 2018; Pandit et al., 2014; Tsujii et al., 2012), insulin treatment (Delahanty et al., 2007; Kuniss et al., 2017; Liu et al., 2020; Nanayakkara et al., 2018), younger age (Fisher et al., 2008), female gender (Nanayakkara et al., 2018; Pintaudi et al., 2015), difficulty following dietary recommendations (Nanayakkara et al., 2018), depression (Nanayakkara et al., 2018), having a larger number of children (Chittem, Chawak, Sridharan, & Sahay, 2019), a feeling of a greater sense of control over one's illness (Chittem et al., 2019), having more illness-related worries (Chittem et al., 2019), and living alone (Pintaudi et al., 2015). Lower DDS was associated with older age, lower BMI, higher self-efficacy, higher levels of health care provider support, and a healthy diet (Wardian & Sun, 2014). HbA1c levels, body mass index and triglyceride were the major predictors of diabetes distress (Parsa et al., 2019). A phenomenological study of the elderly (≥ 65 years) by Hernandez et al. (2019) reported that the most frequent symptoms of T2DM and related distress in patients were fatigue, hypoglycemia, diarrhea, pain, loss of balance, and falling which led to substantial loss of independence, decreased quality of life, and constrained social lives due to restricted activities. Furthermore,

TABLE 2 Comparison of diabetes-related distress prevalence and components using the 17-item Diabetes Distress Scale (DDS-17).

Authors	City, Country	Sample size (persons)	Mean age [±SD], or range (years)	Prevalence of diabetes related distress by category (%)				
				Total	Interpersonal	Physician-related	Emotional	Regimen-related
Fisher et al. (2008)	America	506 T2DM	57.8±9.86	18.0	–	–	–	–
Chew, Vos, Mohd-Sidik, and Rutten (2016)	Malaysia	697 T2DM	56.9	49.2	–	–	–	–
Zhou et al. (2017)	China	363 T2DM	55.3±11.7	42.1	35.8	54.3	17.1	28.1
Arif et al. (2018)	Islamabad, Pakistan	349 T2DM	26–85	76.2	54.4	50.7	84.8	82.5
Nanayakkara, Pease, Ranasinha, et al. (2018)	Australia	2552 T2DM	63±13	6.9	6.9	3.3	12.6	11.3
Aljuaid, Almutairi, Assiri, Almalki, and Alswat (2018)	Taif, Saudi Arabia	509 T2DM		25.0	7.7	24.9	54.0	12.7
Wardian et al. (2018)	America	585 DM	55.2	–	19.5	9.1	41.9	42.9
Hemavathi, Satyavani, Smina, and Vijay (2019)	South India	400 T2DM	25–65	38.75	–	–	–	–
Hamed, Ibrahim, Ali, and Kheir (2019)	Egypt	350 T2DM	33–65	15.7	–	–	–	–
Parsa, Aghamohammadi, and Abazari (2019)	Ardabil, Iran	220 T2DM	58.82±9.41	63.6	50.9	64.1	74.1	75.5
Alzughbi et al. (2020)	Jazan, Saudi Arabia	300 T2DM	23–83	30.0 (7.3 plus depression)	12.3	11.7	10.7	7.0
Tunsuchart, Lerttrakarnnon, Srithanaviboonchai, Likhitsathian, and Skulphan (2020a)	Chiang Mai, Thailand	370 T2DM	52–69	8.9	12.4	4.3	27.1	15.4
Geleta et al. (2021)	Southwest, Ethiopia	321 T2DM	28–54	36.8	12.1	34	43.6	34.9

CBT on DRD or other outcomes in DM patients

A study by (Chew, Vos, Metzendorf, et al., 2017) proposed a conceptual framework for the influences of cognition and emotion on various aspects of diabetes management (Fig. 1). This systematic review and meta-analysis of randomized controlled trials (RCTs) of the effects of psychological interventions for DRD in adults (18 years and older) with T2DM showed that none of the psychological interventions improved DRD more than standard care (Chew et al., 2016). However, a systematic review and meta-analysis of psychological interventions in T2DM patients found the probability of intervention effectiveness is higher for self-help materials, CBT, and counseling than with standard care. Psychological interventions for adults with T2DM were found to provide minimal clinical benefit in improving glycemic control (Winkley et al., 2020).

CBT is a form of psychotherapy that can help reduce potentially detrimental emotions, behavior, and physiological responses (Beck, 2011). It is widely used to treat psychiatric disorders, psychological problems, medical problems with a psychological component, comorbid psychiatric disorders, difficulties in adjustment to illness, poor adherence to treatment and other illness-related behavioral problems (Halford & Brown, 2009).

This section reviews CBT intervention in DM patients and in DM patients with comorbidities (Table 3). Effects of CBT on DM patients and DM patients with comorbidity include reduction of HbA1c (Wei et al., 2018; Whitehead et al., 2017; Yang et al., 2020; Tunsuchart, Lerttrakarnnon, Srithanaviboonchai, Likhitsathian, & Skulphan, 2020b; Ni et al., 2020; Zuo et al., 2020), especially short-term and medium-term glycemic control (Uchendu & Blake, 2017), reduced DDS scores (Tunsuchart et al., 2020a), reduction of depressive symptoms (Yang et al., 2020; Kanapathy & Bogle, 2019), improved self-efficacy and self-concept related to successful diabetes management (Kanapathy & Bogle, 2019), improvement in short- and medium-term anxiety and depression as well as long-term depression (Uchendu & Blake, 2017) and improved quality of sleep (Zuo et al., 2020). In addition, behavioral strategies were found to have a positive effect on glycemic control, and cognitive strategies had a greater effect on depression symptoms (Yang et al., 2020). However, results of CBT from a systematic review and meta-analysis were mixed for diabetes-related distress (of 5 studies, only 3 reported decreased DRD) and for quality of life (Uchendu & Blake, 2017).

CBT intervention can be divided into individual types or into groups of types plus combinations with other strategies. An interpersonal strategy delivered via a group method was reported to have a greater effect on both HbA1c and depression symptoms (Yang et al. (2020). Types of CBT interventions include giving homework assignments, stress management, and interpersonal strategies. CBT strategies include diabetes-focused CBT using secure online real-time instant messaging delivered by diabetes specialist nurses (DSNs) (Doherty et al., 2021), a telephone-administered program (Yang et al., 2020), mindfulness-based stress reduction (MBSR) and mindfulness-based cognitive therapy (MBCT) (Bogusch & O'Brien, 2019; Ni et al., 2020), acceptance and commitment therapy (ACT) and mindfulness interventions (Pinhas-Hamiel & Hamiel, 2020), CBT plus lifestyle counseling (Cummings et al., 2019), motivational interviewing (MI) and CBT skills training provided by a nurse (Magill et al., 2018), internet cognitive behavioral therapy (iCBT)

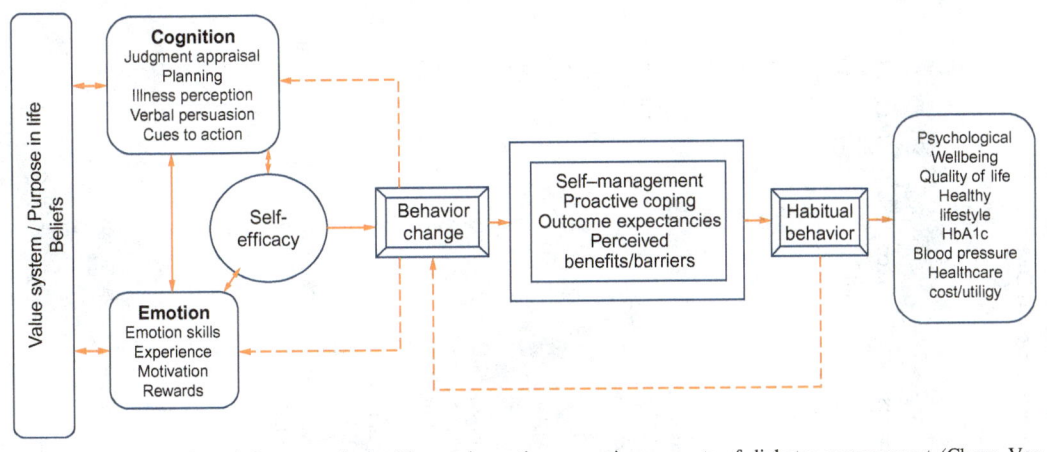

FIG. 1 Conceptual framework of the influences of cognition and emotion on various aspects of diabetes management (Chew, Vos, Metzendorf, et al., 2017).

TABLE 3 Studies of cognitive behavioral therapy (CBT) and CBT in combination with other interventions for diabetes-related distress (DRD) or HbA1c and outcomes in DM patients and in DM patients with comorbidities.

Authors	Place	Method	Mean age [± SD] or range (years)	Outcome variables	Sample size (persons)	Intervention	Outcome
Doherty et al. (2021)	UK	Pre and post uncontrolled	38.1 ± 13.1	HbA1C, PHQ 9, GAD 7, DDS (DDS-17)	71 T1DM, 27 (6months FU), 36 (12 months FU)	12 months online diabetes focus CBT-based intervention (Diabetes Online Therapy) by a diabetes specialist nurse	Feasible to conduct a full-scale text-based synchronized real-time diabetes-focused CBT as an efficacy randomized controlled trial.
Tunsuchart et al. (2020b)	Thailand	A quasi-experimental pretest/posttest design	56.04 ± 8.33	RD, HbA1c, food consumption behavior, physical activity, adherence to medication	28 intervention/ 28 control T2DM	6 brief group cognitive behavioral therapy sessions (BG-CBT)	Had a significant effect on the amelioration of diabetes distress, improvement of food consumption behavior, and reduction of HbA1c levels.
Andreae, Andreae, Richman, Cherrington, and Safford (2020)	America	Cluster-randomized clinical trial	58.9 ± 10.4	Physical activity, HbA1c, pain, QOL	96 intervention/ 99 control DM	CBT-based program delivery by trained community members (3 months, peer delivered, telephone administered program	Improved physical activity, functional status, pain, QOL (HbA1C not significant between control and intervention)
Ni, Ma, and Li (2020)	China	A systematic review and meta-analysis	18.0–68.7 intervention/ 18.5–59.8 control	Depression, quality of life, HbA1c	741 T1DM, T2DM	Mindfulness-based stress reduction (MBSR) and mindfulness-based cognitive therapy (MBCT)	MBSR and MBCT were beneficial in improving depression, the mental composite score of QoL and HbA1C in diabetes.
Zuo et al. (2020)	China	Case-control	63.9 ± 10.2 intervention/ 61.7 ± 10.4 control18	The Pittsburgh Sleep Quality Index (PSQI), the Diabetes-Specific Quality of Life Scale (DSQLS), HbA1c	94 intervention/ 93 control T2DM with poor sleep quality	CBT with aerobic exercise	CBT was effective for sleep disturbances, improving sleep quality, increasing QOL, and decreasing glycemic levels in participants with T2DM.

Yang, Li, and Sun (2020)	China	A meta-analysis of randomized controlled trial	≥18	HbA1c, depression	2619 T1DM, T2DM	Homework assignment, stress management, interpersonal strategy via group	Reduced HbA1c (−0.275%, 95% CI: −0.443 to −0.107; $P<.01$) (group and behavioral strategy), reduced depression symptoms (average reduction −2.788 to −1.207 (95% CI: −4.450) (cognitive strategy)
Winkley et al. (2020)	UK	A systematic review and meta-analysis, network meta-analysis	≥18	Mean change HbA1c	14,796 T2DM ∼70 RCTs	Psychological intervention	Self-help materials, cognitive behavioral therapy, counseling were probably the most effective interventions
Bogusch and O'Brien (2019)	America	Meta-analytic review	—	HbA1c, FBS, DRD, QOL, SMBG, self-management	673 T1DM, T2DM	Mindfulness-based interventions (MBIs) (6 Mindfulness-Based Stress Reduction (MBSR), 3 Mindfulness-Based Cognitive Therapy (MBCT), 5 others	MBIs showed promise for improving psychological well-being through reductions in DRD and increases in quality of life.
Newby et al. (2017)	Australia	A randomized controlled trial	46.7 ± 12.6	Web-based PHQ9, PAID, HbA1C, K 10, SF 12, GAD 7, PHQ 15	27 intervention/ 41 control -T1DM, T2 DM with major depressive disorder	Internet-based Cognitive Behavioral Therapy (6 automated cartoon-style web-based sessions teaching CBT skills over 10 weeks)	iCBT for depression was an efficacious, accessible treatment option for people with diabetes.
Uchendu and Blake (2017)	UK	A systematic review and meta-analysis of randomized controlled trial	≥18	Glycemic control, DRD, anxiety, QOL	1091 T1DM, T2DM	CBT	CBT was effective in reducing short-term and medium-term glycemic control, although no significant effect was found for long-term glycemic control. CBT improved short- and medium-term anxiety and depression, and long-term depression. Mixed results were found for diabetes-related distress and quality of life.

Continued

TABLE 3 Studies of cognitive behavioral therapy (CBT) and CBT in combination with other interventions for diabetes-related distress (DRD) or HbA1c and outcomes in DM patients and in DM patients with comorbidities—cont'd

Authors	Place	Method	Mean age [± SD] or range (years)	Outcome variables	Sample size (persons)	Intervention	Outcome
Wei et al. (2018)	UK	A randomized controlled trial	11–16	HbA1c	43 intervention/ 42 control T1DM	6 session of 1 to 1 CBT (control—Nondirective supportive counseling (NDC)	CBT demonstrated better maintenance of glycemic control compared with NDC.
Whitehead et al. (2017)	Australia	A randomized controlled trial	>18	HbA1c	34 education/ 39 education plus ACT/45 control in uncontrolled T2DM	- Nurse-led educational intervention alone - Nurse-led intervention using education and acceptance and commitment therapy (ACT)	At 6 months post intervention, HbA1c was reduced in both intervention groups with a greater reduction noted in the nurse-led education intervention.
Chew, Vos, Stellato, and Rutten (2017)	Netherland	A systematic review and meta-analysis of randomized controlled trials	43.2–70.7	DRD, HbA1c, Health-related all causes mortality (HRQoL), all-cause mortality	4458 intervention/ 3213 control T2DM	Psychological intervention for diabetes-related distress in adults with T2DM	Low-quality evidence showed none of the psychological interventions improved DRD better than standard care. Low-quality evidence of improved self-efficacy and HbA1c after psychological interventions.

(Newby et al., 2017), nurse-led intervention using education plus acceptance and commitment therapy (ACT) (Whitehead et al., 2017). A study by Andreae et al. (2020) found CBT-based, peer-coach-delivered behavioral intervention in adults with diabetes as well as chronic joint pain who were living in rural areas resulted in self-reports of increased levels of exercise despite pain and improvements in functioning, pain, and health-related quality of life compared with an attention control program over a 3 month period. The intervention, however, did not change HbA1c levels.

Practices and procedures

DRD assessment tools

- Problem areas in the Diabetes (PAID) scale (Table 4), have been reported a well-validated 20-item self-report questionnaire (Polonsky et al., 1995). The original version used a 6-point Likert scale where patients rated the degree to which each item was currently problematic for them, with scores ranging from 1 ("no problem") to 6 ("serious problem"). A total scale score is computed by summing the item responses, giving a possible range of from 24 to 144. However, most studies have used a modified version with score re-scaling of from 0 to 100 (Delahanty et al., 2007). A simplified version, PAID-5, which consists of items 3, 6, 12, 16, 19 of PAID, has been used to conduct questionnaire surveys. PAID-5 scores are transformed into a 0 to 100 scale, with higher scores indicating greater emotional distress. A cutoff of ≥ 40 is used to indicate the probable existence of diabetes-related distress (Liu et al., 2020; Pintaudi et al., 2015).
- The diabetes distress scale (DDS) questionnaire (Table 5) is composed of 17 items grouped into four subcomponents: five items about emotional burden (EB), three items about regimen-related distress (RD), three items about

TABLE 4 Areas evaluated in the Problem Areas in Diabetes Scale (PAID) (Delahanty et al., 2007; Polonsky et al., 1995).

No.	Problem area
1	Worrying about the future and the possibility of serious complications
2	Feeling guilty or anxious when you get off track with your diabetes management
3	Feeling scared when you think about having/living with diabetes
4	Feeling discouraged with your diabetes regimen
5	Feeling depressed when you think about having/living with diabetes
6	Feeling constantly concerned about food and eating
7	Feeling "burned out" by the constant effort to manage diabetes
8	Feeling angry when you think about having/living with diabetes
9	Coping with complications of diabetes
10	Feeling that diabetes is taking up too much mental and physical energy
11	Worrying about reactions
12	Not knowing if the mood or feelings you are experiencing are related to your blood glucose
13	Feeling overwhelmed by your diabetes regimen
14	Feeling alone with diabetes
15	Feelings of deprivation regarding food and meals
16	Not "accepting" diabetes
17	Not having clear and concrete goals for your diabetes care
18	Uncomfortable interactions around diabetes with family/friends
19	Feeling that friends/family are not supportive of diabetes management efforts
20	Feeling unsatisfied with your diabetes physician

TABLE 5 The 17-item Diabetes-related Distress Scale (DDS-17) (Fisher et al., 2008; Polonsky et al., 2005).

No.	Emotional burden (EB)
1	Feeling that diabetes is taking up too much of my mental and physical energy every day
2	Feeling angry, scared and/or depressed when I think about living with diabetes.
3	Feeling that I will end up with serious long-term complication, no matter what I do.
4	Feeling that diabetes controls my life.
5	Feeling overwhelmed by the demands of living with diabetes.
	Physician related-distress (PD)
1	Feeling that my doctor doesn't know enough about diabetes and diabetes care.
2	Feeling that my doctor doesn't give me clear enough directions on how to manage my diabetes.
3	Feeling that my doctor doesn't take my concerns seriously enough.
4	Feeling that I don't have a doctor who I can see regularly enough about my diabetes.
	Regimen related-distress (RD)
1	Feeling that I am not testing my blood sugars frequently enough.
2	Feeling that I am often failing with my diabetes routine.
3	Not feeling confident in my day-to-day ability to manage diabetes.
4	Feeling that I am not sticking closely enough to a good meal plan.
5	Not feeling motivated to keep up my diabetes self-management.
	Interpersonal distress (ID)
1	Feeling that friends or family are not supportive enough of self-care efforts (e.g., planning activities that conflict with my schedule, encouraging me to eat the "wrong" foods).
2	Feeling that friends or family don't appreciate how difficult living with diabetes can be.
3	Feeling that friends or family don't give me the emotional support that I would like.

Responses are on a 6-point scale: 1=Not a problem; 2=A slight problem; 3=A moderate problem; 4=A somewhat serious problem; 5=A serious problem; 6=A very serious problem.

interpersonal-related distress (ID), and four items about physician-related distress (PD). Each item is rated on a 6-point scale from 1 (no problem) to 6 (serious problems) (Polonsky et al., 2005). A mean score of 2.0 to 2.9 indicates moderate distress, and ≥3 indicates a high level of distress (Fisher et al., 2012).

CBT intervention programs in DM with DRD

A previous systematic review and meta-analysis study of CBT intervention groups, glycemic control and psychological outcomes in adults with DM (Uchendu & Blake, 2017) involved between 6 and 21 sessions each lasting 30 to 120 min over a period of 6 weeks to 4 months. Some intervention groups had booster sessions aimed at reinforcing lessons learned during previous CBT sessions. The formats of the CBT included face-to-face individual sessions, face-to-face group sessions, individual telephone sessions, and web-based sessions. Examples of programs that were shown to reduce DRD and/or control blood sugar level include:

- Brief group CBT programs (BG-CBT) (Tunsuchart et al., 2020b) consisting of 6 weekly sessions. Objectives of the session included (1) building relationships and understanding the connections between stimuli, thoughts, emotions, behavior, and physiology; (2) a focus on identifying stimuli that promote inappropriate behavior and methods to modify that behavior, including finding ways to manage stimuli; (3) attempts to identify inappropriate behavior affecting emotions, thoughts, and physiology, as well as the influence of external stimuli, including searching for methods to change

inappropriate behavior; (4) a focus on negative automatic thoughts; (5) selection of negative automatic thoughts to be improved or modified; (6) summarizing knowledge and understanding of what had been learned, including systems for monitoring achievement of the mutually established goals.
- Individual 6 weekly sessions of consultations with a qualified CBT therapist plus follow-up sessions at 6 and 12 months for adolescents with type 1 diabetes (Wei et al., 2018). Program strategies included provision of information sheets, assignments to be completed at home covering material discussed during the session. Objects included: (1) developing and maintaining a therapeutic relationship; (2) cognitive restructuring: identifying negative automatic thoughts, recognizing associations between thoughts, feelings, and behavior, and replacing those with more balanced thoughts; and problem solving, assertiveness training, and relaxation (Al-Ozairi et al., 2020).
- Education plus acceptance and commitment therapy (ACT) intervention (Whitehead et al., 2017). The ACT components consist of mindfulness and acceptance training in relation to difficult thoughts and feelings about diabetes, exploration of personal values related to diabetes, and a focus on the ability to act in a valued direction while confronting difficult experiences. The objectives of the intervention include increased acceptance of diabetes-related thoughts and feelings and a reduction in the extent to which thoughts and feelings interfere with valued action, an increase in understanding of diabetes, greater satisfaction with diabetes management, an increase in self-management activities and maintenance or improvement of mental health as measured by levels of anxiety and depression.
- Diabetes online therapy (DOT) (Doherty et al., 2021). DOT is delivered via a real-time instant messaging system using text communication over the internet via phone or computer using an internet-based platform. The platform was developed by a commercial company that provides internet CBT for depression and anxiety that had been commissioned by the National Health Service (NHS) for use in some parts of England. The DOT platform functions to set up appointments, provide automatic transcription of sessions which can be shared by the patient and the therapist, allowing secure communication outside regular sessions, charting progress using disease-specific measures, holding online sessions and providing downloadable material for guided self-help. Ten 50 min sessions were conducted over 3 months by CBT nurses who had had 3 days of training. At the end of each session, the therapist completed a brief summary outlining the main points of covered in the session along with the agreed-to CBT homework. A self-learning manual which included basic techniques (eliciting negative diabetes cognitions, the basic 3-systems formulation, goal setting, agenda setting and feedback, activity scheduling, behavioral goals and experiments) was used by the participants.

Mini-dictionary of terms

Fasting plasma glucose (FPG): Blood glucose concentration in millimoles per liter (mmol/l) or milligrams per deciliter (mg/dL) of an individual who had not eaten anything for at least 8 h.
Hemoglobin A1c (HbA1c): Hemoglobin bound with glucose (glycosylated hemoglobin) showing the average percentage of blood glucose over the past two to 3 months. The range is: normal <5.7%, prediabetic 5.7%–6.4% and diabetic ≥6.5%.
Term of glycemic control: The duration of follow-up HbA1 level after CBT intervention was divided into short-term (up to 4 months), medium-term (up to 8 months) and long-term (up to 12 months).
Poor diet control: The patient usually eats certain unhealthy types of food, e.g., fast food, soda, and tea.
Self-efficacy: The individual's belief in their capacity to execute behaviors necessary to achieve specific performance goals.

Key facts about diabetes mellitus epidemiology (International Diabetes Federation, 2019)

- An estimated 463 million people (9.3% of the world population) aged 20 to 79 years were diabetic in 2019.
- Type 2 diabetes is the most common type of diabetes, accounting for around 90% of diabetes worldwide.
- The estimated number of adults aged 20 to 79 years with impaired glucose tolerance is 374 million (7.5% of the world population in this age group).
- The number of people older than 65 years (65–99 years) with diabetes is 135.6 million (19.3% of the world population in this age group).
- Diabetes significantly increases the risk of disease, including coronary heart disease by 160%, ischemic heart disease by 127%, hemorrhagic stroke by 56%, and cardiovascular death by 132%.

Key facts about psychosocial issues to diabetes mellitus (American Diabetes Association, 2021)

- All diabetes patients should receive psychosocial care that is integrated with a collaborative, patient-centered approach with the goals of optimizing health outcomes and health-related quality of life.
- All diabetes patients should be assessed for symptoms of diabetes distress, depression, anxiety, disordered eating and for cognitive capacity using appropriate standardized and validated tools at the initial visit, at periodic intervals, and when there is a change in disease, treatment, or life circumstance, including caregivers and family members.
- Older diabetic adults (≥ 65 years) should be screened for cognitive impairment and depression.
- Diabetes patients should be screened for anxiety if they exhibit anxiety or worries regarding diabetes complications, insulin administration, or taking of medications, as well as fear of hypoglycemia and/or hypoglycemia, unawareness that interferes with self-management behaviors, and in those who express fear, dread, or irrational thoughts and/or show anxiety symptoms such as avoidance behaviors, excessive repetitive behaviors, or social withdrawal. Patients should be referred for treatment if any anxiety symptoms are present.
- All diabetes patients should be screened for depression annually.
- Diabetes patients with depression should be referred to mental health providers with experience in using cognitive behavioral therapy, interpersonal therapy, or other evidence-based treatment approaches in conjunction with collaborative care along with the patient's diabetes treatment team.

Applications to other areas

The BG-CBT Intervention Program has been found to have a significant effect in the amelioration of diabetes distress, improvement of food consumption behavior, and reduction of HbA1c levels (Tunsuchart et al., 2020b). Currently, the Covid-19 global pandemic is affecting populations around the world. Methods to help protect people from Covid-19 infection include vaccination, avoiding crowds and close contact with others by keeping a physical distance of at least 1 m, wearing a properly fitted mask and frequent hand cleaning with alcohol-based hand rub or soap and water. One result of the Covid-19 pandemic is that BG-CBT intervention with face-to-face group meetings of 6 to 8 participants with a therapist for 6 session might not currently be safe due to the risk of infection. An alternative intervention which avoids such close contact with others is diabetes online therapy (DOT). DOT is delivered via a real-time instant messaging system using text communication via phone or computer using an internet-based platform (Doherty et al., 2021). This alternative of BG-CBT intervention program delivers the 6 sessions online, effectively preventing Covid-19 infection. Additionally, social support from family or friends may contribute to better glycemic control by ameliolating the effect of diabetes stress (Lee, Piette, Heisler, & Rosland, 2018). Applying knowledge and skills of CBT in these group may benefit.

Summary points

- The overall prevalence of diabetes-related distress is approximately 36% with a range of 1.2%–76.2% depending on the assessment tools used and the geographic location.
- An interpersonal strategy delivered via group CBT has a greater effect on HbA1c than other strategies.
- Behavioral strategies have a greater effect on glycemic control.
- CBT has had mixed results in reducing diabetes-related distress.
- Self-management by people with diabetes is an important part of successful prevention or delaying of diabetes complications.

References

Aikens, J. E. (2012). Prospective associations between emotional distress and poor outcomes in type 2 diabetes. *Diabetes Care*, *35*(12), 2472–2478. https://doi.org/10.2337/dc12-0181.

Aljuaid, M. O., Almutairi, A. M., Assiri, M. A., Almalki, D. M., & Alswat, K. (2018). Diabetes-related distress assessment among type 2 diabetes patients. *Journal of Diabetes Research*, *2018*, 7328128. https://doi.org/10.1155/2018/7328128.

Al-Ozairi, E., Al Ozairi, A., Blythe, C., Taghadom, E., & Ismail, K. (2020). The epidemiology of depression and diabetes distress in type 2 diabetes in Kuwait. *Journal of Diabetes Research*, *2020*, 7414050. https://doi.org/10.1155/2020/7414050.

Alzughbi, T., Badedi, M., Darraj, H., Hummadi, A., Jaddoh, S., Solan, Y., et al. (2020). Diabetes-related distress and depression in Saudis with type 2 diabetes. *Psychology Research and Behavior Management*, *13*, 453–458. https://doi.org/10.2147/PRBM.S255631.

American Diabetes Association. (2021). 5. Facilitating behavior change and well-being to improve health outcomes: Standards of medical care in diabetes-2021. *Diabetes Care*, S53–S72. https://doi.org/10.2337/dc21-S005. 33298416.

Andreae, S. J., Andreae, L. J., Richman, J. S., Cherrington, A. L., & Safford, M. M. (2020). Peer-delivered cognitive behavioral training to improve functioning in patients with diabetes: A cluster-randomized trial. *Annals of Family Medicine*, *18*(1), 15–23. https://doi.org/10.1370/afm.2469.

Arif, M. A., Syed, F., Javed, M. U., Arif, S. A., Hyder, G. E., & Awais-ur-Rehman. (2018). The ADRIFT study – assessing diabetes distress and its associated factors in the Pakistani population. *The Journal of the Pakistan Medical Association*, *68*(11), 1590–1596.

Beck, J. S. (2011). *Cognitive behavior therapy: Basics and beyond*. Guilford Press.

Bogusch, L. M., & O'Brien, W. H. (2019). The effects of mindfulness-based interventions on diabetes-related distress, quality of life, and metabolic control among persons with diabetes: A meta-analytic review. *Behavioral medicine (Washington, D.C.)*, *45*(1), 19–29. https://doi.org/10.1080/08964289.2018.1432549.

Chew, B. H., Vos, R., Mohd-Sidik, S., & Rutten, G. E. (2016). Diabetes-related distress, depression and distress-depression among adults with type 2 diabetes mellitus in Malaysia. *PLoS One*, *11*(3). https://doi.org/10.1371/journal.pone.0152095, e0152095.

Chew, B. H., Vos, R. C., Metzendorf, M. I., Scholten, R. J., & Rutten, G. E. (2017). Psychological interventions for diabetes-related distress in adults with type 2 diabetes mellitus. *The Cochrane Database of Systematic Reviews*, *9*(9), CD011469. https://doi.org/10.1002/14651858.CD011469.pub2.

Chew, B. H., Vos, R. C., Stellato, R. K., & Rutten, G. (2017). Diabetes-related distress and depressive symptoms are not merely negative over a 3-year period in Malaysian adults with type 2 diabetes mellitus receiving regular primary diabetes care. *Frontiers in Psychology*, *8*, 1834. https://doi.org/10.3389/fpsyg.2017.01834.

Chittem, M., Chawak, S., Sridharan, S. G., & Sahay, R. (2019). The relationship between diabetes-related emotional distress and illness perceptions among Indian patients with type II diabetes. *Diabetes & metabolic syndrome*, *13*(2), 965–967. https://doi.org/10.1016/j.dsx.2018.12.018.

Conversano, C. (2019). Common psychological factors in chronic diseases. *Frontiers in Psychology*, *10*, 2727. https://doi.org/10.3389/fpsyg.2019.02727.

Cummings, D. M., Lutes, L. D., Littlewood, K., Solar, C., Carraway, M., Kirian, K., … Hambidge, B. (2019). Randomized trial of a tailored cognitive behavioral intervention in type 2 diabetes with comorbid depressive and/or regimen-related distress symptoms: 12-month outcomes from COMRADE. *Diabetes Care*, *42*(5), 841–848. https://doi.org/10.2337/dc18-1841.

Delahanty, L. M., Grant, R. W., Wittenberg, E., Bosch, J. L., Wexler, D. J., Cagliero, E., et al. (2007). Association of diabetes-related emotional distress with diabetes treatment in primary care patients with type 2 diabetes. *Diabetic Medicine: A Journal of the British Diabetic Association*, *24*(1), 48–54. https://doi.org/10.1111/j.1464-5491.2007.02028.x.

Doherty, A. M., Herrmann-Werner, A., Rowe, A., Brown, J., Weich, S., & Ismail, K. (2021). Feasibility study of real-time online text-based CBT to support self-management for people with type 1 diabetes: The diabetes on-line therapy (DOT) study. *BMJ Open Diabetes Research & Care*, *9*(1). https://doi.org/10.1136/bmjdrc-2020-00193, e001934.

Fisher, L., Glasgow, R. E., & Strycker, L. A. (2010). The relationship between diabetes distress and clinical depression with glycemic control among patients with type 2 diabetes. *Diabetes Care*, *33*(5), 1034–1036. https://doi.org/10.2337/dc09-2175.

Fisher, L., Gonzalez, J. S., & Polonsky, W. H. (2014). The confusing tale of depression and distress in patients with diabetes: A call for greater clarity and precision. *Diabetic Medicine: A Journal of the British Diabetic Association*, *31*(7), 764–772. https://doi.org/10.1111/dme.12428.

Fisher, L., Hessler, D. M., Polonsky, W. H., & Mullan, J. (2012). When is diabetes distress clinically meaningful?: Establishing cut points for the diabetes distress scale. *Diabetes Care*, *35*(2), 259–264. https://doi.org/10.2337/dc11-1572.

Fisher, L., Mullan, J. T., Skaff, M. M., Glasgow, R. E., Arean, P., & Hessler, D. (2009). Predicting diabetes distress in patients with type 2 diabetes: A longitudinal study. *Diabetic Medicine: A Journal of the British Diabetic Association*, *26*(6), 622–627. https://doi.org/10.1111/j.1464-5491.2009.02730.x.

Fisher, L., Skaff, M. M., Mullan, J. T., Arean, P., Glasgow, R., & Masharani, U. (2008). A longitudinal study of affective and anxiety disorders, depressive affect and diabetes distress in adults with type 2 diabetes. *Diabetic Medicine: A Journal of the British Diabetic Association*, *25*(9), 1096–1101. https://doi.org/10.1111/j.1464-5491.2008.02533.x.

Geleta, B. A., Dingata, S. T., Emanu, M. D., Eba, L. B., Abera, K. B., & Tsegaye, D. (2021). Prevalence of diabetes related distress and associated factors among type 2 diabetes patients attending hospitals, Southwest Ethiopia, 2020: A cross-sectional study. *Patient Related Outcome Measures*, *12*, 13–22. https://doi.org/10.2147/PROM.S290412.

Halford, J., & Brown, T. (2009). Cognitive-behavioural therapy as an adjunctive treatment in chronic physical illness. *Advances in Psychiatric Treatment*, *15*(4), 306–317. https://doi.org/10.1192/apt.bp.107.330731.

Hamed, M. S., Ibrahim, N. A., Ali, H. M., & Kheir, C. G. (2019). Study of the effect of glycemic control on the diabetes-related distress in a sample of Egyptian patients with diabetes mellitus. *Diabetes Updates*, *5*. https://doi.org/10.15761/DU.1000134.

Hemavathi, P., Satyavani, K., Smina, T. P., & Vijay, V. (2019). Assessment of diabetes related distress among subjects with type 2 diabetes in South India. *International Journal of Psychology and Behavioral Sciences*, *11*, 1–5.

Hernandez, L., Leutwyler, H., Cataldo, J., Kanaya, A., Swislocki, A., & Chesla, C. (2019). Symptom experience of older adults with type 2 diabetes and diabetes-related distress. *Nursing Research*, *68*(5), 374–382. https://doi.org/10.1097/NNR.0000000000000370.

International Diabetes Federation. (2019). *IDF Diabetes Atlas* (9th ed.). Brussels, Belgium: International Diabetes Federation.

Jia, W., Gao, X., Pang, C., Hou, X., Bao, Y., Liu, W., et al. (2009). Prevalence and risk factors of albuminuria and chronic kidney disease in Chinese population with type 2 diabetes and impaired glucose regulation: Shanghai diabetic complications study (SHDCS). *Nephrology, Dialysis, Transplantation: Official Publication of the European Dialysis and Transplant Association- European Renal Association*, *24*(12), 3724–3731. https://doi.org/10.1093/ndt/gfp349.

Karlsen, B., Oftedal, B., & Bru, E. (2012). The relationship between clinical indicators, coping styles, perceived support and diabetes-related distress among adults with type 2 diabetes. *Journal of Advanced Nursing*, *68*(2), 391–401. https://doi.org/10.1111/j.1365-2648.2011.05751.x.

Kasteleyn, M. J., de Vries, L., van Puffelen, A. L., Schellevis, F. G., Rijken, M., Vos, R. C., et al. (2015). Diabetes-related distress over the course of illness: Results from the Diacourse study. *Diabetic Medicine: A Journal of the British Diabetic Association, 32*(12), 1617–1624. https://doi.org/10.1111/dme.12743.

Kuniss, N., Rechtacek, T., Kloos, C., Müller, U. A., Roth, J., Burghardt, K., et al. (2017). Diabetes-related burden and distress in people with diabetes mellitus at primary care level in Germany. *Acta Diabetologica, 54*(5), 471–478. https://doi.org/10.1007/s00592-017-0972-3.

Lee, A. A., Piette, J. D., Heisler, M., & Rosland, A. M. (2018). Diabetes distress and glycemic control: The buffering effect of autonomy support from important family members and friends. *Diabetes Care, 41*(6), 1157–1163. https://doi.org/10.2337/dc17-2396.

Liu, S. Y., Huang, J., Dong, Q. L., Li, B., Zhao, X., Xu, R., et al. (2020). Diabetes distress, happiness, and its associated factors among type 2 diabetes mellitus patients with different therapies. *Medicine, 99*(11). https://doi.org/10.1097/MD.0000000000018831, e18831.

Magill, N., Graves, H., de Zoysa, N., Winkley, K., Amiel, S., Shuttlewood, E., et al. (2018). Assessing treatment fidelity and contamination in a cluster randomised controlled trial of motivational interviewing and cognitive behavioural therapy skills in type 2 diabetes. *BMC Family Practice, 19*(1), 60. https://doi.org/10.1186/s12875-018-0742-5.

Nanayakkara, N., Pease, A., Ranasinha, S., et al. (2018). Depression and diabetes distress in adults with type 2 diabetes: Results from the Australian National Diabetes Audit (ANDA) 2016. *Scientific Reports, 8*, 7846. https://doi.org/10.1038/s41598-018-26138-5.

Newby, J., Robins, L., Wilhelm, K., Smith, J., Fletcher, T., Gillis, I., et al. (2017). Web-based cognitive behavior therapy for depression in people with diabetes mellitus: A randomized controlled trial. *Journal of Medical Internet Research, 19*(5). https://doi.org/10.2196/jmir.7274, e157.

Ni, Y., Ma, L., & Li, J. (2020). Effects of mindfulness-based stress reduction and mindfulness-based cognitive therapy in people with diabetes: A systematic review and meta-analysis. *Journal of Nursing Scholarship: An Official Publication of Sigma Theta Tau International Honor Society of Nursing, 52*(4), 379–388. https://doi.org/10.1111/jnu.12560.

Pandit, A. U., Bailey, S. C., Curtis, L. M., Seligman, H. K., Davis, T. C., Parker, R. M., et al. (2014). Disease-related distress, self-care and clinical outcomes among low-income patients with diabetes. *Journal of Epidemiology and Community Health, 68*(6), 557–564. https://doi.org/10.1136/jech-2013-203063.

Parsa, S., Aghamohammadi, M., & Abazari, M. (2019). Diabetes distress and its clinical determinants in patients with type II diabetes. *Diabetes & Metabolic Syndrome, 13*(2), 1275–1279. https://doi.org/10.1016/j.dsx.2019.02.007.

Perrin, N. E., Davies, M. J., Robertson, N., Snoek, F. J., & Khunti, K. (2017). The prevalence of diabetes-specific emotional distress in people with type 2 diabetes: A systematic review and meta-analysis. *Diabetic Medicine: A Journal of the British Diabetic Association, 34*(11), 1508–1520. https://doi.org/10.1111/dme.13448.

Pinhas-Hamiel, O., & Hamiel, D. (2020). Cognitive behavioral therapy and mindfulness-based cognitive therapy in children and adolescents with type 2 diabetes. *Current Diabetes Reports, 20*(10), 55. https://doi.org/10.1007/s11892-020-01345-5.

Pintaudi, B., Lucisano, G., Gentile, S., Bulotta, A., Skovlund, S. E., Vespasiani, G., et al. (2015). Correlates of diabetes-related distress in type 2 diabetes: Findings from the benchmarking network for clinical and humanistic outcomes in diabetes (BENCH-D) study. *Journal of Psychosomatic Research, 79*(5), 348–354. https://doi.org/10.1016/j.jpsychores.2015.08.010.

Polonsky, W. H., Anderson, B. J., Lohrer, P. A., Welch, G., Jacobson, A. M., Aponte, J. E., et al. (1995). Assessment of diabetes-related distress. *Diabetes Care, 18*(6), 754–760. https://doi.org/10.2337/diacare.18.6.754.

Polonsky, W. H., Fisher, L., Earles, J., Dudl, R. J., Lees, J., Mullan, J., et al. (2005). Assessing psychosocial distress in diabetes: Development of the diabetes distress scale. *Diabetes Care, 28*(3), 626–631. https://doi.org/10.2337/diacare.28.3.626.

Saeedi, P., Petersohn, I., Salpea, P., Malanda, B., Karuranga, S., Unwin, N., et al. (2019). Global and regional diabetes prevalence estimates for 2019 and projections for 2030 and 2045: Results from the International Diabetes Federation Diabetes Atlas, 9th edition. *Diabetes Research and Clinical Practice, 157*, 107843. https://doi.org/10.1016/j.diabres.2019.107843.

Saeedi, P., Salpea, P., Karuranga, S., Petersohn, I., Malanda, B., Gregg, E. W., et al. (2020). Mortality attributable to diabetes in 20-79 years old adults, 2019 estimates: Results from the international Diabetes Federation Diabetes Atlas, 9th edition. *Diabetes Research and Clinical Practice, 162*, 108086. https://doi.org/10.1016/j.diabres.2020.1080.

Sinclair, A., Saeedi, P., Kaundal, A., Karuranga, S., Malanda, B., & Williams, R. (2020). Diabetes and global ageing among 65-99-year-old adults: Findings from the International Diabetes Federation diabetes Atlas, 9th edition. *Diabetes Research and Clinical Practice, 162*, 108078. https://doi.org/10.1016/j.diabres.2020.108078.

Stanković, Z., Jasović-Gasić, M., & Lecić-Tosevski, D. (2013). Psychological problems in patients with type 2 diabetes – clinical considerations. *Vojnosanitetski Pregled, 70*(12), 1138–1144. https://doi.org/10.2298/vsp1312138s.

Stoop, C. H., Nefs, G., Pop, V. J., Wijnands-van Gent, C. J., Tack, C. J., Geelhoed-Duijvestijn, P. H., et al. (2014). Diabetes-specific emotional distress in people with type 2 diabetes: A comparison between primary and secondary care. *Diabetic Medicine: A Journal of the British Diabetic Association, 31*(10), 1252–1259. https://doi.org/10.1111/dme.12472.

Strandberg, R. B., Graue, M., Wentzel-Larsen, T., Peyrot, M., & Rokne, B. (2014). Relationships of diabetes-specific emotional distress, depression, anxiety, and overall well-being with HbA1c in adult persons with type 1 diabetes. *Journal of Psychosomatic Research, 77*(3), 174–179. https://doi.org/10.1016/j.jpsychores.2014.06.015.

Thanakwang, K., Thinganjana, W., & Konggumnerd, R. (2014). Psychometric properties of the Thai version of the Diabetes Distress Scale in diabetic seniors. *Clinical Interventions in Aging, 9*, 1353–1361. https://doi.org/10.2147/CIA.S67200.

Ting, R. Z., Nan, H., Yu, M. W., Kong, A. P., Ma, R. C., Wong, R. Y., et al. (2011). Diabetes-related distress and physical and psychological health in Chinese type 2 diabetic patients. *Diabetes Care, 34*(5), 1094–1096. https://doi.org/10.2337/dc10-1612.

Tsujii, S., Hayashino, Y., Ishii, H., & Diabetes Distress and Care Registry at Tenri Study Group. (2012). Diabetes distress, but not depressive symptoms, is associated with glycaemic control among Japanese patients with type 2 diabetes: Diabetes distress and care registry at Tenri (DDCRT 1). *Diabetic Medicine: A Journal of the British Diabetic Association, 29*(11), 1451–1455. https://doi.org/10.1111/j.1464-5491.2012.03647.x.

Tunsuchart, K., Lerttrakarnnon, P., Srithanaviboonchai, K., Likhitsathian, S., & Skulphan, S. (2020a). Type 2 diabetes mellitus related distress in Thailand. *International Journal of Environmental Research and Public Health, 17*(7), 2329. https://doi.org/10.3390/ijerph17072329.

Tunsuchart, K., Lerttrakarnnon, P., Srithanaviboonchai, K., Likhitsathian, S., & Skulphan, S. (2020b). Benefits of brief group cognitive behavioral therapy in reducing diabetes-related distress and HbA1c in uncontrolled type 2 diabetes mellitus patients in Thailand. *International Journal of Environmental Research and Public Health, 17*(15), 5564. https://doi.org/10.3390/ijerph17155564.

Uchendu, C., & Blake, H. (2017). Effectiveness of cognitive-behavioural therapy on glycaemic control and psychological outcomes in adults with diabetes mellitus: A systematic review and meta-analysis of randomized controlled trials. *Diabetic Medicine: A Journal of the British Diabetic Association, 34*(3), 328–339. https://doi.org/10.1111/dme.13195.

Wardian, J., & Sun, F. (2014). Factors associated with diabetes-related distress: Implications for diabetes self-management. *Social Work in Health Care, 53*(4), 364–381. https://doi.org/10.1080/00981389.2014.884038.

Wardian, J. L., Tate, J., Folaron, I., Graybill, S., True, M., & Sauerwein, T. (2018). Who's distressed? A comparison of diabetes-related distress by type of diabetes and medication. *Patient Education and Counseling, 101*(8), 1490–1495. https://doi.org/10.1016/j.pec.2018.03.001.

Wei, C., Allen, R. J., Tallis, P. M., Ryan, F. J., Hunt, L. P., Shield, J. P., et al. (2018). Cognitive behavioural therapy stabilises glycaemic control in adolescents with type 1 diabetes-outcomes from a randomised control trial. *Pediatric Diabetes, 19*(1), 106–113. https://doi.org/10.1111/pedi.12519.

Whitehead, L. C., Crowe, M. T., Carter, J. D., Maskill, V. R., Carlyle, D., Bugge, C., et al. (2017). A nurse-led education and cognitive behaviour therapy-based intervention among adults with uncontrolled type 2 diabetes: A randomised controlled trial. *Journal of Evaluation in Clinical Practice, 23*(4), 821–829. https://doi.org/10.1111/jep.12725.

Williams, R., Karuranga, S., Malanda, B., Saeedi, P., Basit, A., Besançon, S., et al. (2020). Global and regional estimates and projections of diabetes-related health expenditure: Results from the international diabetes federation diabetes Atlas, 9th edition. *Diabetes Research and Clinical Practice, 162*. https://doi.org/10.1016/j.diabres.2020.108072, 108072.

Winkley, K., Upsher, R., Stahl, D., Pollard, D., Brennan, A., Heller, S. R., et al. (2020). Psychological interventions to improve glycemic control in adults with type 2 diabetes: A systematic review and meta-analysis. *BMJ Open Diabetes Research & Care, 8*(1). https://doi.org/10.1136/bmjdrc-2019-001150, e001150.

Yang, X., Li, Z., & Sun, J. (2020). Effects of cognitive behavioral therapy-based intervention on improving glycaemic, psychological, and physiological outcomes in adult patients with diabetes mellitus: A meta-analysis of randomized controlled trials. *Frontiers in Psychiatry, 11*, 711. https://doi.org/10.3389/fpsyt.2020.00711.

Zhang, J., Xu, C. P., Wu, H. X., Xue, X. J., Xu, Z. J., Li, Y., et al. (2013). Comparative study of the influence of diabetes distress and depression on treatment adherence in Chinese patients with type 2 diabetes: A cross-sectional survey in the People's Republic of China. *Neuropsychiatric Disease and Treatment, 9*, 1289–1294. https://doi.org/10.2147/NDT.S49798.

Zhou, H., Zhu, J., Liu, L., Li, F., Fish, A. F., Chen, T., et al. (2017). Diabetes-related distress and its associated factors among patients with type 2 diabetes mellitus in China. *Psychiatry Research, 252*, 45–50. https://doi.org/10.1016/j.psychres.2017.02.049.

Zuo, X., Dong, Z., Zhang, P., Zhang, P., Chang, G., Xiang, Q., et al. (2020). Effects of cognitive behavioral therapy on sleep disturbances and quality of life among adults with type 2 diabetes mellitus: A randomized controlled trial. *Nutrition, Metabolism, and Cardiovascular Diseases: NMCD, 30*(11), 1980–1988. https://doi.org/10.1016/j.numecd.2020.06.024.

Chapter 11

Dizziness: Features and the use of cognitive behavioral therapy

Masaki Kondo

Department of Psychiatry and Cognitive-Behavioral Medicine, Nagoya City University, Nagoya, Japan

Abbreviations

ACT	acceptance and commitment therapy
CBT	cognitive behavioral therapy
CSD	chronic subjective dizziness
DHI	dizziness handicap inventory
PPPD	persistent postural-perceptual dizziness
PPV	phobic postural vertigo
RCT	randomized controlled trial
SSRI	selective serotonin reuptake inhibitor
VR	vestibular rehabilitation

Introduction

In the treatment of dizziness, cognitive behavioral therapy (CBT) has been applied to a functional vestibular disorder, persistent postural-perceptual dizziness (PPPD), and plays an important role in the treatment of PPPD. This article first describes the disease concept, prevalence and clinical course, symptoms, pathophysiology, and treatment of PPPD, and then describes how each of these issues can be used in CBT. Next, previous studies on CBT for PPPD and related disorders are presented, followed by cognitive behavioral model and specific CBT techniques from a practical perspective. Finally, a pilot study on acceptance and commitment therapy (ACT), which is one of the third generations of behavioral therapy, for PPPD is mentioned briefly.

Transition of the concept of the functional vestibular disorder

In 1871, Westphal described agoraphobia as "Dizziness, spatial disorientation, and anxiety while attempting to shop in the open spaces and motion-rich environments of town squares" (Staab, 2012). In other words, agoraphobia initially included a dizziness component, but later the dizziness component disappeared and the concept changed to one related to panic disorder. Since around 1990, several disease concepts of functional dizziness have been proposed from the viewpoint of the interaction of physical and psychological factors: phobic postural vertigo (PPV) (Brandt, Huppert, & Dieterich, 1994) focused on neurotic personality tendencies and exacerbation of vertigo with upright posture, space motion discomfort (Romas, Jacob, & Lilienfeld, 1997) focused on exacerbation by motion and spatial perception, visual vertigo (Bronstein, 1995) focused on exacerbation by visual stimuli, and chronic subjective dizziness (CSD) (Staab & Ruckenstein, 2007) focused on these exacerbating factors and environmental sensitivity. As research on these diseases progressed, they came to be regarded as different aspects of the same disease, and the momentum for the development of international diagnostic criteria integrating these diseases was growing. A diagnostic criteria committee was established in the Bárány Society (the International Society for Neuro-otology) in 2010, and the diagnostic criteria for PPPD were published by the Bárány Society in 2017 (Staab et al., 2017). In 2018, the diagnostic guideline for PPPD was included in the International Classification of Diseases 11th Revision (ICD-11) (World Health Organization, 2018).

The key point here in CBT is that PPPD is defined as a functional vestibular disorder involving the interaction of both physical and psychological factors, and that it is not a structural or psychiatric condition. Therapists need to educate patients as such.

Persistent postural-perceptual dizziness

Prevalence and clinical course

The prevalence of PPPD in the general population has not been reported, however, it is considered a common disease. Using the Visual Vertigo Analogue Scale (Dannenbaum, Chilingaryan, & Fung, 2011) and Situational Characteristics Questionnaire (Jacob, Lilienfeld, Furman, Durrant, & Turner, 1989), which measures the degree of visual stimulation by visual vertigo and space motion comfort, respectively, a previous study found that 9% and 50% of the general population recruited through advertising scored above the 25 percentile of the group of patients with PPPD (Powell, Derry-Sumner, Rajenderkumar, Rushton, & Sumner, 2020). This means that visual stimuli-induced exacerbation of dizziness is common in the general population and that PPPD may not be a rare disease. In a tertiary referral dizziness clinic, PPV, a PPPD-related disease, is the second most common diagnosis following benign paroxysmal positional vertigo (Brandt, 2013) and is the most frequent diagnosis in the age group from 20 to 50 years old (Strupp et al., 2003). In CSD, which is also a PPPD-related disease, the average duration from onset to visit to a tertiary facility was 4.5 years, and some patients had symptoms lasting for decades (Staab & Ruckenstein, 2007). Patients' ages averaged in the 40s and were distributed from adolescence to old age (Bittar & von Sohsten Lins, 2014; Staab & Ruckenstein, 2007; Yan et al., 2017). A long-term follow-up study of patients with PPV revealed that symptoms resolved spontaneously in a small number of patients. Many patients with PPPD are likely to remain symptomatic with no treatment, and three quarters developed comorbidities of anxiety or depression (Huppert, Strupp, Rettinger, Hecht, & Brandt, 2005).

Knowledge of prevalence and clinical course is important in psychoeducation. Many patients with PPPD have experienced undiagnosed dizziness after visiting hospitals and undergoing various tests, and their anxiety may be heightened by thinking that their dizziness is rare or untreatable. PPPD requires appropriate education that it is a common condition but time-consuming diagnosis and that it is a functional somatic disorder with no abnormalities on physical examination.

Diagnosis and primary symptoms

The diagnostic criteria for PPPD are listed in Table 1 (Staab et al., 2017).

Criterion A describes features of primary symptoms that are dizziness, unsteadiness, or nonspinning vertigo. Primary symptoms tend to last for prolonged (hours-long) periods. It may worsen and lessen spontaneously, which can affect patient's cognition and behavior. In addition, primary symptoms may tend to increase as the day progresses (Staab et al., 2017). The therapist should ask the patient whether dizziness tends to increase in the evening. If the patient answers yes, it should be ascertained whether the patient interprets fatigue as exacerbating dizziness.

TABLE 1 Diagnostic criteria for persistent postural-perceptual dizziness.

PPPD is a chronic vestibular disorder defined by criteria A–E below. All five criteria must be fulfilled to make the diagnosis

A. One or more symptoms of dizziness, unsteadiness, or nonspinning vertigo are present on most days for 3 months or more
   1. Symptoms last for prolonged (hours-long) periods of time, but may wax and wane in severity
   2. Symptoms need not be present continuously throughout the entire day
B. Persistent symptoms occur without specific provocation, but are exacerbated by three factors:
   1. Upright posture
   2. Active or passive motion without regard to direction or position
   3. Exposure to moving visual stimuli or complex visual patterns
C. The disorder is precipitated by conditions that cause vertigo, unsteadiness, dizziness, or problems with balance including acute, episodic, or chronic vestibular syndromes, other neurologic or medical illnesses, or psychological distress
   1. When the precipitant is an acute or episodic condition, symptoms settle into the pattern of criterion A as the precipitant resolves, but they may occur intermittently at first, and then consolidate into a persistent course
   2. When the precipitant is a chronic syndrome, symptoms may develop slowly at first and worsen gradually
D. Symptoms cause significant distress or functional impairment
E. Symptoms are not better accounted for by another disease or disorder

TABLE 2 Typical cognitive appraisals and avoidance behaviors for primary symptoms of persistent postural-perceptual dizziness.

Features of primary symptoms	Typical cognitive appraisals	Typical avoidance behaviors
Waxing and waning spontaneously	"Dizziness may get worse while I am meeting someone, and I may cause him or her trouble"	Avoiding activities that require them to stay in a place for a while, or that make it difficult to take a break in between (e.g., avoiding eating out with less familiar acquaintances, or avoiding making a long oral presentation at a business conference)
Increasing as the day progresses (worsening gradually in the evening)	"If I get tired, my dizziness will get worse and I will have to lie down"	Avoiding activities that cause fatigue (e.g., avoiding room cleaning), even when not exposed to exacerbating factors, (e.g., avoiding talking on the phone for a long time with a friend)
Momentary transient flares (lasting just seconds)	"My body has gone wrong forever. It won't get better" "I will fall and get injured"	Avoiding the situation in which flares can occur if they are predictable

Another important point about primary symptoms is that momentary transient flares (lasting just seconds) may occur spontaneously or with movement in some patients (Staab et al., 2017). If the patient can anticipate the situation in which such flares occur, he or she may avoid the situation. If the situation is unpredictable, patients may keep constant vigilance for their physical sensations and may show similar anticipatory anxiety and avoidance behaviors in patients with panic disorder who experience frequent, unexpected panic attacks. Typical cognitive and behavioral patterns for these features of primary symptoms are shown in Table 2.

Exacerbation factors

The B criterion in Table 1 is an item of exacerbating factors, and all three factors (upright posture, motion, and visual stimuli) must be present at any point in the history. Because avoidance of exacerbating factors is present in almost all patients, therapists must take care it with CBT and must have a good understanding of the real-life situations in which exacerbating factors are present in individual patients (Table 3).

Exacerbation by upright posture refers to worsening of primary symptoms while standing or walking. In more severely affected patients, therapists need to check if they are sitting or lying down more often and if they are doing safety behaviors to avoid losing their balance.

TABLE 3 Typical cognitive appraisals and avoidance behaviors for exacerbating factors of persistent postural-perceptual dizziness.

Exacerbating factors	Typical cognitive appraisals	Typical avoidance behaviors and safety behaviors
Upright posture (standing or walking)	"I will fall and get injured" "If I lose my balance, people around me will be suspicious of me"	Avoiding standing or walking for a long time (e.g., increasingly sitting or lying down), or taking safety behaviors (e.g., using a cane, walking near a handrail, or having a family member accompany the patient)
Active or passive motion	"My balance organ has gone wrong and I will have to lie down"	Avoiding active motion (e.g., moving slowly, or immobilizing some joints during activities) or passive motion (e.g., avoiding getting into a car or train), or taking safety behaviors (e.g., listening to music on a train for distraction)
Exposure to moving visual stimuli or complex visual patterns	"I've got a disease that no one know what it is" "I will fall and get injured"	Avoiding the situation with moving objects or image and complex visual patterns, or taking safety behaviors (e.g., focusing as little as possible on visual stimuli)

Exacerbation by motion means worsening of primary symptoms caused by head movements (e.g., looking up, down, or sideways), other active movements (e.g., stretching, bending, turning, tidying, cleaning, washing, cooking, or exercising), or passive movements (e.g., getting in a car or train, riding an escalator or elevator). There are a wide variety of situations and behaviors in which motion can exacerbate dizziness, and it is necessary to carry out the detailed inquiry with examples of specific movements. For example, driving a car by oneself may not make dizziness worse, but riding as a passenger in a car or on a bus may aggravate it. Another example is that a patient avoids riding in a car, because dizziness may not worsen while riding in the car but may worsen after getting out of the car. In addition, many patients find it comfortable to remain motionless, but some patients report that moving at a moderate pace (e.g., walking, riding a bicycle, or rocking a chair) is more tolerable than standing or sitting motionless (Dieterich, Staab, & Brandt, 2016). Small differences in these situations need to be noted and evaluated in detail.

Exacerbations by visual stimuli include moving visual stimuli (e.g., watching a movie with a lot of motion, watching a screen scroll, or watching a moving train or car), complex visual pattern (e.g., looking at a display shelf in a grocery store, going to a shopping center, looking at a crowd, looking at a patterned wall or floor, looking at a spreadsheet with a lot of values, or watching a movie with many flashing lights), visual stimuli with fewer spatial reference points (e.g., large fields, large warehouses, open atriums, or a long corridor with unpatterned white walls), and narrow-field visual stimuli (e.g., reading a newspaper or book, sewing, or knitting).

Evaluation of avoidance behaviors and safety behaviors for motion and visual stimuli is often difficult and requires more clinical experience for therapists. For example, if a patient says, "Riding the train makes my dizziness worse," it is important to determine whether the dizziness is worsened by looking at a large number of passengers in the train (complex visual stimuli), by looking at the flowing scenery outside the train (moving visual stimuli), or by simply riding the train regardless of visual stimuli (passive motion). This clear distinction of exacerbating factors allows therapists and patients to identify dysfunctional thoughts as targets in behavioral experiments or cognitive restructuring.

Precipitant factors (onset factors)

Criterion C in Table 1 is the entry for mode of onset. In the majority of cases, the primary symptoms begin following the triggering acute vertigo or dizziness. Many patients are obsessed with the causes of acute dizziness and continue to complain, "Is there something wrong with my balance system?" or "If I cannot find out the cause of onset, my dizziness won't be cured.". Also, if the triggering acute dizziness was caused by a panic attack, nonspecific anxiety, or transient mental stress, the patient's dizziness may have been treated as psychogenic dizziness by previous medical care providers and the patient may believe that the symptoms do not improve until the "causative" anxiety or mental stress is resolved. In these cases, therapists should educate the patient that there are various causes of acute dizziness or vertigo that trigger PPPD, that they are not directly related to the maintenance factors of PPPD, and that PPPD is a functional disease, not psychogenic (Staab et al., 2017).

Exclusion of other disorders, and comorbidity

The E Criterion in Table 1 describes that primary symptoms and exacerbating factors cannot be adequately explained by other vestibular or medical disorders. Thus, diagnosis of PPPD requires neuro-otologist examination, vestibular function tests, and imaging tests as needed.

It should be noted that PPPD can coexist with other vestibular, medical, and psychiatric disorders. Vestibular disorders (e.g., Ménière's disease, benign paroxysmal positional vertigo, or vestibular migraine) or psychiatric disorders (e.g., panic disorder, generalized anxiety disorder, or major depression) that can be precipitants of PPPD may persist. In such cases, it is necessary to provide the patient with appropriate treatment for comorbid vestibular or psychiatric disorders, and education should be provided to enable patients to distinguish symptoms of PPPD from those of comorbid disorders.

Hypothesis of pathophysiology

The proposed pathological hypothesis of PPPD is a maladaptation of the equilibrium system (Staab, 2020). PPPD is typically triggered by acute balance problem. The most common triggers of acute balance problem are central and peripheral vestibular disorders, vestibular migraine attacks, panic attacks, anxiety-induced dizziness, concussive injuries of the brain, whiplash injuries of the neck, and autonomic disorders (Staab et al., 2017). Immediately after the triggering balance problem, the acute balance adaptation strategies begin in the patient's equilibrium system, including dependence on visual and somatosensory perception, postural control in high-risk situations, and environmental vigilance. Dependence on visual and somatosensory perception means an automatic physiological change that the patient is usually unaware of, which decreases the input of the vestibular sensation and, instead, increases the input of the visual and somatosensory perceptions to maintain body balance. Postural control in high-risk situations refers to postural control in situations that needs high-demand control strategy not to lose balance. For example, tensioning the antigravity muscles not to fall are included. Patients with PPV showed changes in postural control characterized by high-frequency, low-amplitude postural sway associated with co-contraction of the crural muscles during standing still (Wuehr et al., 2013). Some patients with PPPD show immobilizing some joints as much as possible or narrowing the stride length. Environmental vigilance means to always be cautious of the exacerbating factors and triggers of dizziness or unsteadiness.

In many cases, such acute balance adaptation strategies gradually become unnecessary, and patients recover by readjusting to the environment. However, if psychological characteristics such as neurotic personality tendencies or trait anxiety are present, patients enter a vicious cycle in which the acute balancing adaptive strategies are sustained by upright posture, motion, and visual stimuli, and the condition is maintained without readaptation. In the vicious circle, the acute balancing adaptive strategies increase phobic self-observation (self-monitoring of internal sensations). Increased self-monitoring disrupts automatic motor control and leads to overcompensating motion (May cause gait disturbance). At the same time, the fear of dizziness and unsteadiness confuses and distorts the processing of incoming sensory information, thereby resulting in noisy, inaccurate sensory inputs. These physiological changes cause distortion and failed integration of afferent signals. This leads to the perception of dizziness or unsteadiness, followed again by the acute balancing adaptive strategies (Popkirov, Staab, & Stone, 2018; Popkirov, Stone, & Holle-Lee, 2018).

Psychological factors involved in pathophysiology

In this model, the psychological factors involved in maintaining the vicious cycle of the acute balancing adaptive strategies is self-monitoring of interoceptive sensations and fear of dizziness and unsteadiness that are influenced by fear of falling, health anxiety, and avoidance behaviors (Popkirov, Staab, & Stone, 2018; Popkirov, Stone, & Holle-Lee, 2018). Two proposals for CBT mentioned below are consistent with this model. The cognitive behavioral model of PPPD proposed by Whalley and Cane (2017) is similar to the cognitive-behavioral formulation of health anxiety and is regarded as a specific expression of health anxiety in dizziness. Edelman, Mahoney, and Cremer (2012) noted similarities between CSD and panic disorder and performed CBT for CSD, assuming that both were accompanied by a conditioned fear response to unpleasant bodily sensations.

There are also some clinical studies that support this maladaptation model of the equilibrium system. People with personality traits associated with anxiety, such as neuroticism and introversion, are at higher risk for CSD (Staab, Rohe, Eggers, & Shepard, 2014). A report of clinical characteristics related to PPPD showed that PPPD patients had significantly more neuroticism and introverted personality than controls (Yan et al., 2017). In another previous study, compared to patients with cured vestibular disease, PPPD patients had greater body vigilance and more negative illness perception, which may be risk factors for the development of PPPD (Trinidade, Harman, Stone, Staab, & Goebel, 2021).

Treatment: Antidepressant

Antidepressants, vestibular rehabilitation, and CBT are known treatments for PPPD, but evidence is still lacking. For antidepressants, various observational studies have suggested its effectiveness, but no randomized controlled trials (RCT) have been reported. In functional diseases, as in psychiatric diseases, patient's expectation for treatment (placebo response) is considered to have a significant effect on efficacy, and RCTs are needed in the future. In prospective observational studies of selective serotonin reuptake inhibitors (SSRI) in patients with CSD, 18%–25% of patients discontinued taking SSRIs due to side effects or inadequate efficacy, and 55%–67% showed remission or treatment response (Staab & Ruckenstein, 2005; Staab, Ruckenstein, & Amsterdam, 2004). Serotonin-norepinephrine reuptake inhibitors are also used for chronic nonspecific dizziness (Horii et al., 2016). Because side effects such as dizziness and nausea can cause dropouts, a lower starting dose (e.g., 5 mg for escitalopram) is recommended than for depression. In clinical practice, CBT is often combined with SSRIs. Further research is needed to determine if combination therapy for PPPD enhances the effectiveness.

Treatment: Vestibular rehabilitation

Vestibular rehabilitation (VR) is developed as a physical therapy aimed at promoting vestibular compensation, typically for acute unilateral vestibular disorders. Vestibular compensation is defined as a neurological compensatory change in which neuroplasticity of the central vestibular system corrects the bilateral imbalance and improves vestibular symptoms. However, it has been suggested that the primary mechanism of clinical effectiveness of VR for PPPD is the promotion of behavioral habituation to persistent vestibular symptoms and motion hypersensitivity rather than physical enhancement of central vestibular compensation (Staab, 2011). VR for PPPD and related disorders aims to desensitize alarming balance control systems using habituation and relaxation techniques (Popkirov, Stone, & Holle-Lee, 2018). There are no reported RCTs of VR for PPPD including related diseases compared to controls. A retrospective case series, a telephone survey of 26 patients with PPPD who received VR, has been reported, and 14 patients (53.8%) answered that VR was helpful (Thompson, Goetting, Staab, & Shepard, 2015).

VR for persistent dizziness is considered to be more useful when combined with CBT because it is mainly effective as behavioral therapy. Several studies on CBT plus VR combination therapy for PPPD-related disease or chronic nonspecific dizziness have been conducted (Andersson, Asmundson, Denev, Nilsson, & Larsen, 2006; Holmberg, Karlberg, Harlacher, & Magnusson, 2007; Holmberg, Karlberg, Harlacher, Rivano-Fischer, & Magnusson, 2006; Jacob, Whitney, Detweiler-Shostak, & Furman, 2001; Johansson, Akerlund, Larsen, & Andersson, 2001). VR can be used as interoceptive exposure targeting dizziness and related symptoms in the context of CBT (Toshishige et al., 2020) and as acceptance and mindfulness exercise for dizziness and related distress in the context of ACT (Kuwabara et al., 2020). These are described in the later section.

Cognitive behavioral therapy for persistent postural-perceptual dizziness

Effectiveness studies

Two previous controlled studies of CBT for PPPD-related diseases have been reported. In a nonrandomized controlled study of individual CBT for PPV, 39 patients (age: 24–62, mean 45 years) were alternately assigned to self-treatment vestibular rehabilitation exercises (self-VR) alone or CBT plus self-VR (Holmberg et al., 2006). The mean duration of illness was approximately 60 months (range: 1–360 months). CBT (45–60 min per session, 8–12 sessions) was not manual-based and based on analysis of individual patients. Initially, psychoeducation was given on the effects of cognitive appraisals and avoidance behaviors, and the principles of exposure therapy. For dysfunctional cognitive appraisals of natural body sway, psychoeducation and feedback by family members and mirror gazing were provided. Breathing techniques and applied relaxation were conducted for excessive tension of the antigravity muscles. Using these methods, the patients were exposed to situations in stores, bridges, or open spaces, sometimes accompanied by a psychologist. Data from 16 patients in the CBT plus self-VR group and 15 patients in the self-VR group who completed the treatment were analyzed. Compared to the self-VR group, the combination therapy group showed a significant improvement in handicap due to dizziness, anxiety, and depression after treatment (Holmberg et al., 2006). However, the significant effect had disappeared at the 1-year follow-up (Holmberg et al., 2007). CBT in this study does not appear to include therapeutic components to directly approach hypervigilance and negative appraisals for physical sensations.

An RCT have been reported for patients with CSD (Edelman et al., 2012). 41 patients with CSD were randomly assigned to individual CBT and waiting list control. The median duration of disease was 9 months, the mean was 34.9 months, and the range was 1–240 months. CBT was performed in three weekly sessions. Psychoeducation about CSD was first offered. In order to reduce the threat perceptions of undiagnosed serious diseases, in-session behavioral experiments were conducted to examine the effects of excessive vigilance and attention-switching. Avoidance and safety behaviors were identified, and situational exposure was conducted. Patients were offered a metaphor for "surfing with the wave" (similar to the "panic surfing" metaphor used for panic disorder (Lamplugh, Berle, Milicevic, & Starcevic, 2008)), which aims to leave the sensations as they are rather than resisting them. Interoceptive exposure (e.g., intentionally moving the head) was performed if necessary. Finally, patients were encouraged to live a normal life despite their symptoms, instead of trying to monitor or control symptoms. At posttreatment, the CBT group showed a significant improvement in the handicap due to dizziness compared with the waiting group, but no significant improvement in anxiety and depression (Edelman et al., 2012). The treatment effect was maintained for 6 months after treatment, and approximately 60% of patients achieved clinically significant changes pretreatment to 6 months posttreatment (Mahoney, Edelman, & Cremer, 2012). Because of the relatively short duration of disease in patients enrolled in this study, it has been pointed out that CBT may be effective if performed relatively early after onset, but less effective in patients with longer duration of disease (Dieterich et al., 2016). However, the CBT program in this study focused on hypervigilance and negative interpretation for symptoms and adopts acceptance strategy for unpleasant sensations, which may have contributed to the long-term therapeutic effect.

An RCT titled CBT as augmentation of sertraline for PPPD has been reported (Yu, Xue, Zhang, & Zhou, 2018). However, descriptions of the treatment (gaining patients' trust, encouraging patients to communicate with others, making patients expose and check social factors such as family and work that contribute to PPPD, and educating patients about the development and treatment of PPPD) does not seem to include treatment elements specific to CBT, and this study is not discussed here.

Predictors of effectiveness

In the aforementioned study of CBT for CSD, only high anxiety at baseline predicted a higher handicap due to dizziness at 6 months posttreatment, but not duration of illness, psychiatric comorbidities, medical comorbidities, or severity of dizziness symptoms at baseline (Mahoney et al., 2012).

In a single-arm observational study of group CBT in 37 patients with CSD, only presence of comorbid anxiety disorders predicted improvement in dizziness handicap from baseline to 6 months posttreatment, but not age, sex, duration of illness, comorbid vestibular disorder, level of anxiety, level of depression, presence of semicircular canal paresis, and posturography indices (velocity of movement of the center of pressure, envelopment area, and Romberg ratio) (Toshishige et al., 2020). In this study, only patients with disease duration of 3 months or longer were enrolled, with a median duration of 18 months and an interquartile range of 10–51 months. CBT was performed in the group format of three or four patients, with one 120-min session per week for a total of five or six sessions. The treatment components conducted for all patients were psychoeducation about CSD, interoceptive exposure to dizziness and related symptoms using VR (repetitive head movements), and behavioral experiments to verify the effect of avoidance and safety behaviors with graded exposure in feared situation. Abdominal breathing, progressive muscle relaxation, and attention training were added for some patients. The effect size was similar to that in the previous study of Mahoney et al. (2012) between pretreatment and 6 months posttreatment, and significant improvements were shown regardless of the presence or absence of comorbid vestibular disorders. It has been concluded that group format and focusing on body vigilance by interoceptive exposure may have promoted the therapeutic effect, particularly in patients with comorbid anxiety disorders (Toshishige et al., 2020).

Practice and procedures

Measurement instruments

The Dizziness Handicap Inventory (DHI) (Jacobson & Newman, 1990) and the Vertigo Symptom Scale short form (VSS) (Yardley, Beech, Zander, Evans, & Weinman, 1998) are the most commonly-used patient-reported outcome measures in clinical researches on chronic dizziness, and the former is most frequently used as the main outcome (Fong, Li, Aslakson, & Agrawal, 2015).

DHI is a self-administered 25-item scale which assesses the handicap of daily life due to dizziness. The total score ranges from 0 to 100, and the higher the score, the greater the handicap. The developer of the original version recommends rating no handicap less than or equal to 14. The minimally important difference, defined as the minimum change in which a change in an individual patient can be interpreted as clinically beneficial, has been reported as 18 points for distribution-based methods (Jacobson & Newman, 1990) and 11 points for anchor-based methods (Tamber, Wilhelmsen, & Strand, 2009).

VSS is a self-administered 15-item scale which measures the frequency of vestibular balance symptoms (e.g., vertigo, dizziness, or unsteadiness) and autonomic-anxiety symptoms (e.g., palpitation, shortness of breath, or sweating) for a month before. The total score ranges from 0 to 60, and the higher the score, the more frequent the symptoms. Neither minimally important difference nor the cut-off value for asymptomatic has been reported.

The Niigata PPPD Questionnaire (Yagi et al., 2019) has been developed as a self-rating scale specific to PPPD. This questionnaire consists of 12 items that measure the responsibility to exacerbating factors (upright posture, motion, and visual stimuli), and is rated on a scale of 0–6 points for each item, for a total score of 0–72 points.

The Hospital Anxiety and Depression Scale (Zigmond & Snaith, 1983) has been used most frequently to measure anxiety and depression in patients with dizziness (Piker, Kaylie, Garrison, & Tucci, 2015). Other measures may be helpful, including measures of health anxiety, hypervigilance for physical sensations, and negative evaluation of illness.

Cognitive behavioral model (formulation)

As mentioned earlier, excessive vigilance and negative interpretation for dizziness-related physical sensations are considered to be important psychological factors contributing to the maintenance of PPPD. Thus, CBT for PPPD is similar to CBT for panic disorder and health anxiety.

Panic disorder is an anxiety disorder characterized by repeated unexpected panic attacks, anticipatory anxiety about the occurrence or consequences of panic attacks, and avoidance behaviors (American Psychiatric Association, 2013). The psychological factors involved in the development and maintenance of panic disorder have been described as follows. First, anxiety about panic is developed after the initial panic attack. This anxiety is established by two factors: interoceptive conditioning (i.e., learned anxiety conditioned by internal states) and misappraisal of bodily sensations (i.e., misinterpretation of bodily sensations to mean something catastrophic, such as a sign of loss of control). Anxiety about panic can be interpreted as the sensitization of the trait of anxiety sensitivity (similar to neuroticism) by the experience of precipitant panic attacks. Next, the development of learned anxiety about a particular bodily sensation leads to hypersensitivity to even normal somatic sensations. Then, daily activities that produce internal sensations similar to those felt during a panic attack can heighten anxiety and trigger panic attacks. Finally, internal sensations, which are difficult to escape, may give panic attacks a meaning that cannot be predicted or controlled, thereby maintaining chronic anticipatory anxiety. The vicious cycle of panic attacks and anxiety is maintained in this way (Craske & Barlow, 2007).

The aforementioned pathophysiological hypothesis and related psychological features of PPPD are very similar to this explanation of the onset and maintenance of panic disorder. Therefore, the following psychological process of the development and maintenance of PPPD may be hypothesized: First, in individuals with the trait of neuroticism, anxiety about dizziness is formed after experiencing precipitants of PPPD. This anxiety is attributed to two factors: interoceptive conditioning (i.e., learned anxiety conditioned by the internal sensation of dizziness) and misappraisal of dizziness (i.e., interpreting dizziness as a catastrophic sign of complete loss of balance control). Second, the formation of learned anxiety about dizziness can lead to hypersensitivity to even safe dizziness-like sensations (e.g., natural body sway), as well as impaired integration of visual, somatosensory, and vestibular sensations related to body balance (e.g., the automatic upweighting of visual and somatosensory). Then, motion and visual stimuli in daily life can exacerbate dizziness. Finally, because the sensation of dizziness is difficult to control, anxiety about dizziness becomes chronic.

The cognitive-behavioral model of maintenance of PPPD created based on this psychological process is shown in Fig. 1. This model refers to the CBT model diagram of panic disorder by Hofmann (2011), and may be similar to the cognitive behavioral model of CSD postulated by Edelman et al. (2012) and the cognitive behavioral model of PPPD proposed by Whalley and Cane (2017). Future research about this cognitive behavioral model is needed.

Health anxiety is a concept that explains hypochondriasis (illness anxiety disorder in Diagnostic and statistical manual of mental disorders, fifth edition (DSM-5)) (American Psychiatric Association, 2013). Because PPPD does not usually cause as severe health anxiety as hypochondriasis, CBT for PPPD uses some strategies of CBT for hypochondriasis.

Psychoeducation and sharing the cognitive behavioral model

Fig. 2 illustrates at which points in the vicious circle model the specific treatment components of CBT work. Psychoeducation that PPPD is a common and treatable disease, that it is a functional physical disorder in which test abnormalities do

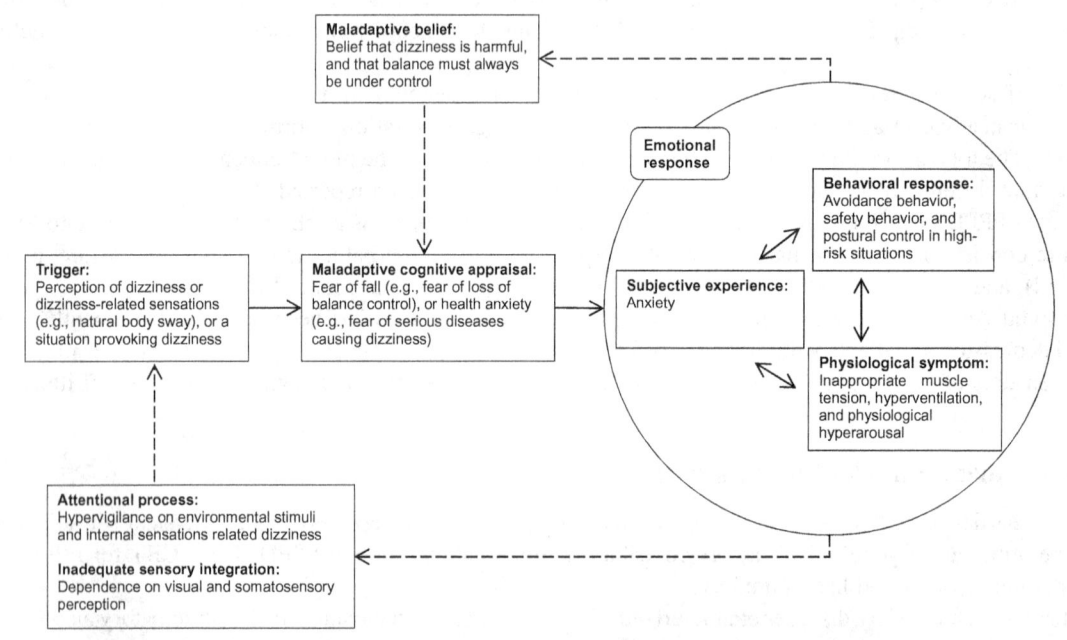

FIG. 1 Cognitive behavioral model of persistent postural-perceptual dizziness. The *solid arrows* show the relationship between the ABC model (Activating event (Trigger)—Belief (Maladaptive cognitive appraisal)—Consequence (Emotional response)), which is the principle of CBT. The *dotted arrows* indicate the backloop from Emotional response.

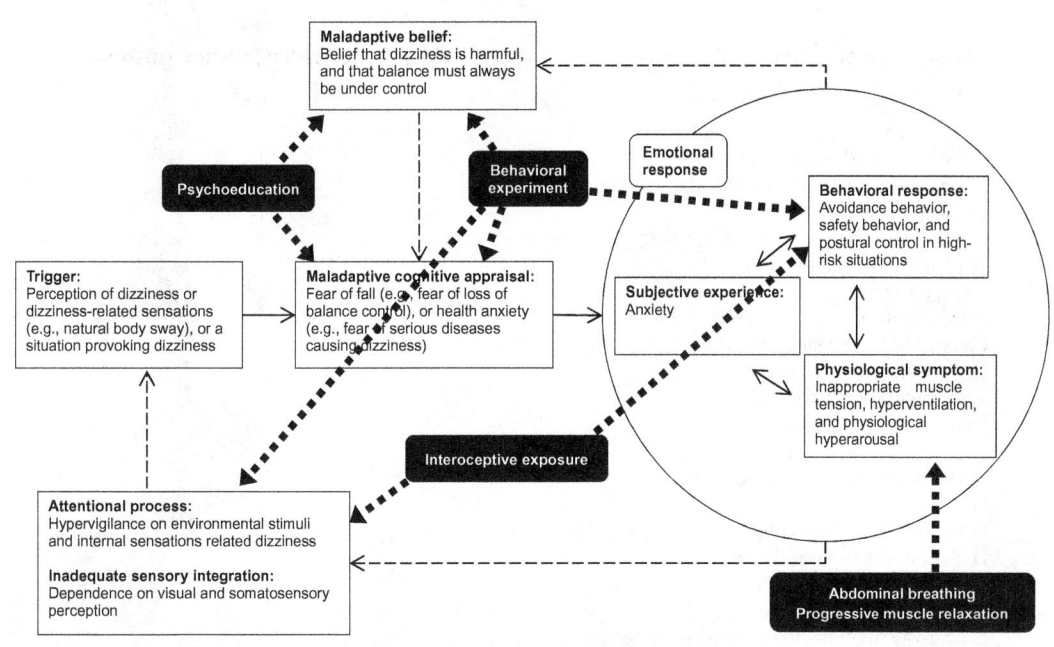

FIG. 2 Specific techniques of cognitive behavioral therapy for persistent postural-perceptual dizziness. The *white letters* on the black background indicate the specific technique of CBT, and the *dotted arrows* extending from the technique indicate the target of the intervention.

not appear, and that it usually takes a long time to be diagnosed can reduce patients' health anxiety and misappraisal for illness. By making a detailed assessment of the primary symptoms and exacerbating factors, patients understand that they do not have an unidentified disease.

Patients can learn that they can cope with their condition in a different way by understanding the psychological model of maintaining PPPD. The author helps patients understand the vicious circle model by sharing with the following four examples in Fig. 1: (1) hyperventilation, (2) inappropriate muscle tension, (3) avoidance behavior, and (4) excessive attention to dizziness. Since some patients are resistant to a psychological approach, initially the cognitive aspect should not be emphasized, and sharing the vicious circle regarding physiological symptoms, behaviors, and attentional processes may facilitate the introduction of treatment.

Relaxation techniques

Abdominal breathing and progressive muscle relaxation (Jacobson, 1925) are often used as relaxation techniques in CBT. Because hyperventilation tends to maintain panic attacks and anticipatory anxiety in panic disorder, CBT for panic disorder usually involves abdominal breathing (breathing retraining) (Craske & Barlow, 2007). It is well known that hyperventilation can also induce dizziness and unsteadiness, and many patients with PPPD have a tendency to hyperventilation, and abdominal breathing often relieves symptoms of PPPD. As described above, patients with PPPD have inappropriate excessive muscle tension as postural control in high-risk situations (Wuehr et al., 2013). Since it is input to the central equilibrium system as an afferent stimulus and may form an abnormal loop of posture control, relaxation techniques such as the progressive muscle relaxation are considered to be effective.

Relaxation techniques can be expected to correct directly hyperventilation and excessive muscle tension, as well as enhance the ability to self-monitor breathing, muscle tension, and other bodily sensations. However, it is important to note that some patients use such relaxation techniques as safety behaviors. Because attempts to avoid dizziness with relaxation techniques often result in high levels of hypervigilance and anxiety about dizziness and may maintain PPPD, therapists should carefully assess whether relaxation techniques are working effectively or adversely in individual patients. To determine this, it may be useful to conduct behavioral experiments in the next section.

Cognitive techniques

Commonly used techniques for cognitive interventions are cognitive restructuring and behavioral experiments (Beck, 1970). Because some patients with PPPD believe that they have a physical disorder and do not require cognitive

TABLE 4 List of behaviors that can be safety behaviors in patients with persistent postural-perceptual dizziness.

Move slowly
Move your head as little as possible
Do not turn your head in any particular direction
Leaning on something
Holding on to a shopping cart, a railing, or someone
Walking near a wall or railing
Using a cane
Asking someone to follow you
Carry as little luggage as possible
Carry medication with you
Take medication before going out
Plant your feet firmly on the ground
Strengthen your body muscles
Widen your legs so as not to fall
Narrow your stride
Don't look too much at the shelves or products
Look only at the desired item
Look blankly without focus
Try not to move your eyes
Look away from moving objects or unpleasant sights
Focus your attention on something else
Distract yourself with something else
Tell yourself over and over again, "I'm okay, I'm okay"
Take a deep breath
Do abdominal breathing
Do progressive muscle relaxation

If the patient is engaging in these behaviors to avoid dizziness or anxiety about dizziness, then they are considered safety behaviors.

intervention, they may not prefer the cognitive restructuring that targets thoughts without actual behaviors. Although behavioral experiment is also designed to correct maladaptive cognitive appraisals, it is constructed as hypothesis-testing approach by actually performing them and is easily accepted by such patients.

Common themes of behavioral experiments in PPPD are as follows. (1) Stopping acute balance control strategies including avoidance and safety behaviors in situations where catastrophic consequences are feared; (2) exposure to external stimuli that disturb balance including exacerbating factors (upright posture, motion, and visual stimuli); and (3) changing attention processes for internal conditions and bodily sensations (e.g., distraction) to confirm their effects. Table 4 lists behaviors that can be safe behaviors in PPPD patients, which can be used when discussing safety behaviors with patients. The article of Whalley and Cane (2017) provided useful examples of behavioral experiments. Behavioral experiments are more likely to be accepted by patients if they are initiated in the form of graded exposure which starts with a low anxiety task. As the behavioral experiment is repeated, the patient becomes interested in the behavioral experiment on the topic that he or she wants to examine.

Behavioral techniques

As noted earlier, interoceptive exposure targeting dizziness and related symptoms may play an important role in CBT for PPPD. Interoceptive exposure in PPPD is similar to that in panic disorder (Craske & Barlow, 2007). Simple VR exercises (e.g., shaking the head from side to side, or up and down) can be used to induce dizziness. The therapist instructs the patient to fully experience the induced dizziness and to observe a gradual decrease in anxiety. Repeating this process in homework typically reduces in anxiety about dizziness and hypervigilance.

Acceptance and commitment therapy for persistent postural-perceptual dizziness

Thus, CBT for PPPD that targets anxiety about dizziness is a promising treatment, however, the author hypothesized that CBT for PPPD may have limited therapeutic effects. In some PPPD patients comorbid with refractory organic vestibular disease (e.g., refractory vestibular neuritis with difficulty in vestibular compensation), behavioral experiments and interoceptive exposure were unsuccessful because of vertigo or dizziness caused by organic vestibular diseases. It is also possible to interpret the acute phase balance adjustment strategy that maintains PPPD as a psychological and physiological process aimed at controlling dizziness. Therefore, the author hypothesized that acceptance and commitment therapy (ACT) is more effective because it aims to increase valued actions while accepting distress without controlling it (Hayes, Strosahl, & Wilson, 2011). A pilot study of ACT combined with VR was conducted in 27 patients with PPPD. ACT+VR program included six of 2-h weekly sessions for three patients, consisting of psychoeducation for PPPD, six core processes of ACT (acceptance, cognitive defusion, present moment, self-as-context, value, and committed action) and VR as a mindfulness exercise for the purpose of accepting dizziness. Three out of four patients showed significant improvement (remission or treatment response) between baseline and 6 months posttreatment (Kuwabara et al., 2020). Further research on ACT for PPPD is warranted.

Mini-dictionary of terms

- *Persistent postural-perceptual dizziness*. A functional somatic disease that presents with persistent dizziness, unsteadiness, or nonspinning vertigo exacerbated by upright posture, motion, and visual stimuli.
- *Dizziness*. Nonmotion sensations that disrupt or impair spatial orientation.
- *Unsteadiness*. Feeling unsteady when standing or walking.
- *Nonspinning vertigo*. False or distorted sensations that one's body is shaking, swaying, rocking or bouncing, or similar sensations of movement in the surroundings.
- *Behavioral experiment*. A CBT technique in which patients actually act to verify how well the content of their automatic thoughts (cognitive appraisals) fits into reality.
- *Interoceptive exposure*. A CBT technique that aims to eliminate conditioned anxiety through repeated exposure to unpleasant physical sensations evoked by stimuli.
- *Avoidance behaviors*. Behaviors to avoid staying in feared situations in order to reduce anxiety.
- *Safety behaviors*. Behaviors carried out in feared situations to reduce anxiety.

Key facts of persistent postural-perceptual dizziness

- Persistent postural-perceptual dizziness (PPPD) is defined as a functional somatic disorder in 2017.
- The primary symptoms of PPPD are persistent dizziness, unsteadiness, or nonspinning vertigo that are exacerbated by upright posture, motion, and visual stimuli.
- The prevalence of PPPD in the general population is still unknown but is considered a common disease.
- The pathological hypothesis of PPPD is the maladaptive persistence of acute balance control strategy.
- Targeting hypervigilance and negative appraisals for dizziness and related sensations may be the key factor in CBT for PPPD.
- A typical CBT for PPPD is similar to CBT for panic disorder, and consists of psychoeducation, relaxation techniques, cognitive techniques such as behavioral experiment, and behavioral techniques such as interoceptive exposure.
- Vestibular rehabilitation is assumed to work as habituation exercise for PPPD and can be conducted as interoceptive exposure in the context of CBT for PPPD.
- Acceptance and commitment therapy has been applied recently for PPPD.

Applications to other areas

In this chapter, I have reviewed CBT for PPPD that similar to CBT for panic disorder. The diagnostic criteria of PPPD by the Bárány Society requires all three exacerbating factors (upright posture, motion, and visual stimuli) to be present in the history, but CBT can be applied to cases with one or two exacerbating factors. Kuwabara et al. (2020) reported that the effect of ACT on patients with PPPD with one to three exacerbating factors was similar to the effect on only patients with PPPD with three exacerbating factors. Toshishige et al. (2020) also reported CBT for CSD patients with —one to three exacerbating factors. CBT may be feasible regardless of the number of exacerbating factors if the patient's dizziness is consistent with the pathophysiological hypothesis of PPPD (inappropriate persistence of acute balance control strategy). It is unclear whether CBT can be applied to patients with functional nonpersistent dizziness or functional episodic vertigo, but it may be applied if hypervigilance and maladaptive perception of dizziness or vertigo are present in the patient, and further research is needed. It is also not at all clear whether CBT can be applied to organic vestibular diseases (e.g.,

Ménière's disease) in which mental stress may be related to worsening of symptoms, but even if it can, its cognitive behavioral model is likely to be different from that of PPPD.

Summary points

- Persistent postural-perceptual dizziness (PPPD) is a functional somatic disorder.
- PPPD is considered a common disease.
- The primary symptoms of PPPD are persistent dizziness, unsteadiness, or nonspinning vertigo.
- The pathogenesis of PPPD is hypothesized to be the inappropriate persistence of acute balance control strategies.
- Targeting hypervigilance and negative appraisals for dizziness may be the key factor in CBT for PPPD.
- CBT for PPPD is similar to CBT for panic disorder.
- Vestibular rehabilitation can be conducted as interoceptive exposure in CBT for PPPD.
- Acceptance and commitment therapy has been applied recently for PPPD.

References

American Psychiatric Association. (2013). *Diagnostic and statistical manual of mental disorders, Fifth Edition (DSM-5)*. American Psychiatric Association.

Andersson, G., Asmundson, G. J., Denev, J., Nilsson, J., & Larsen, H. C. (2006). A controlled trial of cognitive-behavior therapy combined with vestibular rehabilitation in the treatment of dizziness. *Behaviour Research and Therapy, 44*(9), 1265–1273.

Beck, A. T. (1970). Cognitive therapy: Nature and relation to behavior therapy. *Behavior Therapy, 1*(2), 184–200.

Bittar, R. S., & von Sohsten Lins, E. M. (2014). Clinical characteristics of patients with persistent postural and perceptual dizziness. *Brazilian Journal of Otorhinolaryngology, 81*, 276–282.

Brandt, T. (2013). *Vertigo: Its multisensory syndromes*. Springer Science & Business Media.

Brandt, T., Huppert, D., & Dieterich, M. (1994). Phobic postural vertigo: A first follow-up. *Journal of Neurology, 241*(4), 191–195.

Bronstein, A. M. (1995). The visual vertigo syndrome. *Acta Oto-Laryngologica. Supplementum, 520*(Pt 1), 45–48.

Craske, M. G., & Barlow, D. H. (2007). *Mastery of your anxiety and panic: Therapist guide. Vol. 2.* Oxford University Press.

Dannenbaum, E., Chilingaryan, G., & Fung, J. (2011). Visual vertigo analogue scale: An assessment questionnaire for visual vertigo. *Journal of Vestibular Research, 21*(3), 153–159.

Dieterich, M., Staab, J., & Brandt, T. (2016). Functional (psychogenic) dizziness. *Handbook of Clinical Neurology, 139*, 447–468.

Edelman, S., Mahoney, A. E., & Cremer, P. D. (2012). Cognitive behavior therapy for chronic subjective dizziness: A randomized, controlled trial. *American Journal of Otolaryngology, 33*(4), 395–401.

Fong, E., Li, C., Aslakson, R., & Agrawal, Y. (2015). Systematic review of patient-reported outcome measures in clinical vestibular research. *Archives of Physical Medicine and Rehabilitation, 96*(2), 357–365.

Hayes, S. C., Strosahl, K. D., & Wilson, K. G. (2011). *Acceptance and commitment therapy: The process and practice of mindful change.* Guilford Press.

Hofmann, S. G. (2011). *An introduction to modern CBT: Psychological solutions to mental health problems.* John Wiley & Sons.

Holmberg, J., Karlberg, M., Harlacher, U., & Magnusson, M. (2007). One-year follow-up of cognitive behavioral therapy for phobic postural vertigo. *Journal of Neurology, 254*(9), 1189–1192.

Holmberg, J., Karlberg, M., Harlacher, U., Rivano-Fischer, M., & Magnusson, M. (2006). Treatment of phobic postural vertigo. A controlled study of cognitive-behavioral therapy and self-controlled desensitization. *Journal of Neurology, 253*(4), 500–506.

Horii, A., Imai, T., Kitahara, T., Uno, A., Morita, Y., Takahashi, K., & Inohara, H. (2016). Psychiatric comorbidities and use of milnacipran in patients with chronic dizziness. *Journal of Vestibular Research, 26*(3), 335–340.

Huppert, D., Strupp, M., Rettinger, N., Hecht, J., & Brandt, T. (2005). Phobic postural vertigo—A long-term follow-up (5 to 15 years) of 106 patients. *Journal of Neurology, 252*(5), 564–569.

Jacob, R. G., Lilienfeld, S. O., Furman, J. M., Durrant, J. D., & Turner, S. M. (1989). Panic disorder with vestibular dysfunction: Further clinical observations and description of space and motion phobic stimuli. *Journal of Anxiety Disorders, 3*(2), 117–130.

Jacob, R. G., Whitney, S. L., Detweiler-Shostak, G., & Furman, J. M. (2001). Vestibular rehabilitation for patients with agoraphobia and vestibular dysfunction: A pilot study. *Journal of Anxiety Disorders, 15*(1–2), 131–146.

Jacobson, E. (1925). Progressive relaxation. *The American Journal of Psychology*, 73–87.

Jacobson, G. P., & Newman, C. W. (1990). The development of the Dizziness Handicap Inventory. *Archives of Otolaryngology – Head and Neck Surgery, 116*(4), 424–427.

Johansson, M., Akerlund, D., Larsen, H. C., & Andersson, G. (2001). Randomized controlled trial of vestibular rehabilitation combined with cognitive-behavioral therapy for dizziness in older people. *Otolaryngology and Head and Neck Surgery, 125*(3), 151–156.

Kuwabara, J., Kondo, M., Kabaya, K., Watanabe, W., Shiraishi, N., Sakai, M., ... Akechi, T. (2020). Acceptance and commitment therapy combined with vestibular rehabilitation for persistent postural-perceptual dizziness: A pilot study. *American Journal of Otolaryngology, 41*(6), 102609.

Lamplugh, C., Berle, D., Milicevic, D., & Starcevic, V. (2008). A pilot study of cognitive behaviour therapy for panic disorder augmented by panic surfing. *Clinical Psychology & Psychotherapy, 15*(6), 440–445.

Mahoney, A. E., Edelman, S., & Cremer, P. D. (2012). Cognitive behaviour therapy for chronic subjective dizziness: Longer-term gains and predictors of disability. *American Journal of Otolaryngology, 34*(2), 115–120.

Piker, E. G., Kaylie, D. M., Garrison, D., & Tucci, D. L. (2015). Hospital anxiety and depression scale: Factor structure, internal consistency and convergent validity in patients with dizziness. *Audiology and Neuro-Otology, 20*(6), 394–399.

Popkirov, S., Staab, J. P., & Stone, J. (2018). Persistent postural-perceptual dizziness (PPPD): A common, characteristic and treatable cause of chronic dizziness. *Practical Neurology, 18*(1), 5–13.

Popkirov, S., Stone, J., & Holle-Lee, D. (2018). Treatment of persistent postural-perceptual dizziness (PPPD) and related disorders. *Current Treatment Options in Neurology, 20*(12), 50.

Powell, G., Derry-Sumner, H., Rajenderkumar, D., Rushton, S. K., & Sumner, P. (2020). Persistent postural perceptual dizziness is on a spectrum in the general population. *Neurology, 94*(18), e1929–e1938.

Romas, R. T., Jacob, R. G., & Lilienfeld, S. O. (1997). Space and motion discomfort in Brazilian versus American patients with anxiety disorders. *Journal of Anxiety Disorders, 11*(2), 131–139.

Staab, J. P. (2011). Behavioral aspects of vestibular rehabilitation. *NeuroRehabilitation, 29*(2), 179–183.

Staab, J. P. (2012). Chronic subjective dizziness. *Continuum, 18*(5 Neuro-otology), 1118–1141.

Staab, J. P. (2020). Persistent postural-perceptual dizziness. *Seminars in Neurology, 40*(1), 130–137.

Staab, J. P., Eckhardt-Henn, A., Horii, A., Jacob, R., Strupp, M., Brandt, T., & Bronstein, A. (2017). Diagnostic criteria for persistent postural-perceptual dizziness (PPPD): Consensus document of the committee for the Classification of Vestibular Disorders of the Barany Society. *Journal of Vestibular Research, 27*(4), 191–208.

Staab, J. P., Rohe, D. E., Eggers, S. D., & Shepard, N. T. (2014). Anxious, introverted personality traits in patients with chronic subjective dizziness. *Journal of Psychosomatic Research, 76*(1), 80–83.

Staab, J. P., & Ruckenstein, M. J. (2005). Chronic dizziness and anxiety: Effect of course of illness on treatment outcome. *Archives of Otolaryngology – Head and Neck Surgery, 131*(8), 675–679.

Staab, J. P., & Ruckenstein, M. J. (2007). Expanding the differential diagnosis of chronic dizziness. *Archives of Otolaryngology – Head and Neck Surgery, 133*(2), 170–176.

Staab, J. P., Ruckenstein, M. J., & Amsterdam, J. D. (2004). A prospective trial of sertraline for chronic subjective dizziness. *Laryngoscope, 114*(9), 1637–1641.

Strupp, M., Glaser, M., Karch, C., Rettinger, N., Dieterich, M., & Brandt, T. (2003). The most common form of dizziness in middle age: Phobic postural vertigo. *Nervenarzt, 74*(10), 911–914.

Tamber, A. L., Wilhelmsen, K. T., & Strand, L. I. (2009). Measurement properties of the Dizziness Handicap Inventory by cross-sectional and longitudinal designs. *Health and Quality of Life Outcomes, 7*, 101.

Thompson, K. J., Goetting, J. C., Staab, J. P., & Shepard, N. T. (2015). Retrospective review and telephone follow-up to evaluate a physical therapy protocol for treating persistent postural-perceptual dizziness: A pilot study. *Journal of Vestibular Research, 25*(2), 97–103.

Toshishige, Y., Kondo, M., Kabaya, K., Watanabe, W., Fukui, A., Kuwabara, J., … Akechi, T. (2020). Cognitive-behavioural therapy for chronic subjective dizziness: Predictors of improvement in Dizziness Handicap Inventory at 6 months posttreatment. *Acta Oto-Laryngologica, 140*(10), 827–832.

Trinidade, A., Harman, P., Stone, J., Staab, J. P., & Goebel, J. A. (2021). Assessment of potential risk factors for the development of persistent postural-perceptual dizziness: A case-control pilot study. *Frontiers in Neurology, 11*, 601883.

Whalley, M. G., & Cane, D. A. (2017). A cognitive-behavioral model of persistent postural-perceptual dizziness. *Cognitive and Behavioral Practice, 24*(1), 72–89.

World Health Organization. (2018). *Persistent postural-perceptual dizziness.* Retrieved from https://icd.who.int/browse11/l-m/en#/http://id.who.int/icd/entity/2005792829.

Wuehr, M., Pradhan, C., Novozhilov, S., Krafczyk, S., Brandt, T., Jahn, K., & Schniepp, R. (2013). Inadequate interaction between open- and closed-loop postural control in phobic postural vertigo. *Journal of Neurology, 260*(5), 1314–1323.

Yagi, C., Morita, Y., Kitazawa, M., Nonomura, Y., Yamagishi, T., Ohshima, S., … Horii, A. (2019). A validated questionnaire to assess the severity of persistent postural-perceptual dizziness (PPPD): The Niigata PPPD Questionnaire (NPQ). *Otology & Neurotology, 40*(7), e747–e752.

Yan, Z., Cui, L., Yu, T., Liang, H., Wang, Y., & Chen, C. (2017). Analysis of the characteristics of persistent postural-perceptual dizziness: A clinical-based study in China. *International Journal of Audiology, 56*(1), 33–37.

Yardley, L., Beech, S., Zander, L., Evans, T., & Weinman, J. (1998). A randomized controlled trial of exercise therapy for dizziness and vertigo in primary care. *British Journal of General Practice, 48*(429), 1136–1140.

Yu, Y. C., Xue, H., Zhang, Y. X., & Zhou, J. (2018). Cognitive behavior therapy as augmentation for sertraline in treating patients with persistent postural-perceptual dizziness. *BioMed Research International, 2018*, 8518631.

Zigmond, A. S., & Snaith, R. P. (1983). The hospital anxiety and depression scale. *Acta Psychiatrica Scandinavica, 67*(6), 361–370.

Chapter 12

Epilepsy, sexual function, and mindfulness-based cognitive therapy

Zainab Alimoradi[a], Mark D. Griffiths[b], and Amir H. Pakpour[c]
[a]Social Determinants of Health Research Center, Research Institute for Prevention of Non-Communicable Diseases, Qazvin University of Medical Sciences, Qazvin, Iran, [b]International Gaming Research Unit, Psychology Department, Nottingham Trent University, Nottingham, United Kingdom, [c]Department of Nursing, School of Health and Welfare, Jönköping University, Jönköping, Sweden

Abbreviations

AEDs	antiepileptic drugs
APA	American Psychiatric Association
CBT	cognitive-behavioral therapy
HHcy	hyperhomocysteinemia
ICD	International Classification of Diseases
MBCT	mindfulness-based cognitive therapy
MBCT-S	mindfulness-based cognitive therapy for sexual problems
Ox-LDL	oxidized low-density lipoprotein
PWE	people with epilepsy
SHBG	sex hormone-binding globulin
vWF	von Willebrand factor
WHO	World Health Organization

Introduction

Epilepsy is a brain disease that requires ongoing treatment, characterized by abnormal brain activity that causes seizures or unusual behaviors and feelings, and sometimes loss of consciousness (Perrotta, 2020). This disease has various neurological, cognitive, psychological, and social consequences and covers a significant part of the burden of diseases in the world (World Health Organization, 2019). The economic consequences of epilepsy-related morbidity and mortality are substantial (O'Donohoe, Choudhury, & Callander, 2020).

Approximately 46 million people worldwide are affected by epilepsy (Beghi et al., 2019). Moreover, 80% of people with epilepsy (PWE) live in low- and middle-income countries and do not have adequate access to treatment. Despite the effectiveness and low cost of anticonvulsant drugs, the treatment gap is 75% in most low-income countries and more than 50% in most middle-income countries (WHO, 2019). The number of individuals with epilepsy is expected to increase worldwide due to increased life expectancy. The risk of premature death among individuals with epilepsy is up to three times that of the general population. In addition, epilepsy has a high risk of disability, psychological problems (more commonly depression and anxiety), and social isolation (Feigin et al., 2019). All over the world, individuals with epilepsy and their families suffer from stigma and discrimination, and many children with epilepsy do not go to school. Adults are barred from work, driving, or marriage. Therefore, the lives of individuals with epilepsy are often affected by stigma, discrimination, and human rights violations (Fiest, Birbeck, Jacoby, & Jette, 2014).

In a recent systematic review (Owolabi et al., 2020), the overall prevalence of lifelong epilepsy and active epilepsy was 9 and 16 per 1000, respectively. Its prevalence is estimated to be highest in African countries (30.2 per 1000 population) (Owolabi et al., 2020). The prevalence of active epilepsy in general, in low- and high-income countries is 6.38, 6.68, and 5.49 per 1000, respectively. The reason for the difference between low- and high-income countries can be explained by the prevalence of selective risk factors (mostly infection and trauma), the structure of the high-risk population, and the treatment distance. In addition, methodological issues such as more accurate confirmation of cases and elimination of separate and acute seizures are another reason for these reported differences (Feigin et al., 2019; Fiest et al., 2017).

Effects of epilepsy on various aspects of life

The physical, psychological, and social consequences of epilepsy impose significant burdens among individuals with the disease and their families (WHO, 2019). Around the world, individuals with epilepsy and their families suffer from stigma and discrimination and often face serious problems in quality of life, education, employment, marriage and reproduction (Hamedi-Shahraki et al., 2019; Penn Miller et al., 2021). Individuals with the stigma of epilepsy have lower self-esteem and quality of life, more social isolation, poorer mental health, and worse disease control (Akyol & Nehir, 2021). However, the stigma burden is higher for individuals living in low-income and less developed countries, and as a result, the stigma of epilepsy is higher among those experiencing social and economic problems (Fiest et al., 2014).

The family of an individual with epilepsy may be isolated from the community for fear of transmitting the disease or feeling different from others, or forced to live in a separate bedroom from the rest of the family (Njamnshi et al., 2010). In many Asian and African cultures, women with epilepsy are not good wives because they are unable to properly care for children, cook, or participate in household chores. Single adult women can be exposed to sexual abuse, physical abuse and/or extreme poverty. In a Zambian study, the rate of rape for women with epilepsy was 20% vs. 3% compared to women with other chronic diseases (Birbeck, Chomba, Atadzhanov, Mbewe, & Haworth, 2007). Women with epilepsy in Nigeria face a number of social and economic challenges, with one-third reporting physical abuse of family members and 10% reporting rape (Komolafe et al., 2012). Therefore, epilepsy affects various aspects of individuals' lives, including their marital and sexual life.

Effects of epilepsy on sexual function among individuals with epilepsy

Epilepsy, like many common medical disorders such as diabetes, high blood pressure, and depression, can cause sexual dysfunction. However, the prevalence and nature of sexual dysfunction among PWE, its causes, and optimal management strategies are unclear. Many factors contribute to this relative lack of data on sexual dysfunction among PWE (Rathore, Henning, Luef, & Radhakrishnan, 2019). Both physicians and patients are often reluctant to discuss sexual health in clinical practice. There is considerable discomfort among physicians in the diagnosis and treatment of sexual dysfunction among PWE, which often requires a multidisciplinary approach. In addition, many patients, especially in developing countries, consider discussing sexual dysfunction a taboo and, in turn, accept it as part of their illness (Rathore et al., 2019). The natural human sexual response can be divided into four stages (Khan & Gunasekaran, 2019), each of which may be disrupted:

(1) Sexual desire in which desire usually includes imagination and desire to have sexual activity.
(2) Excitement or sexual arousal comprising an individual's sense of sexual pleasure. This involves physiological changes such as swelling and erection of the penis in men, and pelvic congestion, inflammation of the external genitalia, and vaginal lubrication in women.
(3) Orgasm that is accompanied by the release of sexual tension and rhythmic contraction of perineal muscles and genitals.
(4) Resolution that is accompanied by a feeling of comfort and muscle relaxation. Men are physiologically unable to regain an erection following ejaculation for different periods of time, while women may be able to respond to additional stimuli.

According to the International Classification of Mental and Behavioral Disorders, IX-10 (World Health Organization 1992), "sexual dysfunction" is the inability of an individual to enjoy sexual intercourse. Sexual dysfunction is a problem that an individual or a couple experiences at any stage of normal sexual activity that causes distress and tension in interpersonal relationships. Several different systems have been proposed for the classification of sexual disorders. The two most widely used classification systems for classifying sexual dysfunction are the eleventh revision of the International Classification of Diseases (ICD-11; World Health Organization, 2019) and the fifth edition of the Diagnostic and Statistical Manual of Mental Disorders (DSM-5; American Psychiatric Association, 2013). These two diagnostic manuals divide sexual disorders into four broad categories: sexual desire disorders, sexual arousal disorders, orgasmic disorders, and sexual pain disorders. These disorders are mostly classified as organic or functional in origin based on four areas of sexual function (APA, 2013; Reed et al., 2016).

A major limitation in identifying sexual disorders is the mental nature of the disorder and its impact on interpersonal relationships with the partner and socio-cultural factors. Hyposexuality is defined as the decrease in sexual desire and sexual activity less than once a month. However, sexual desire is often influenced by social, cultural, and environmental factors. The diagnosis of sexual dysfunction also depends on the degree of anxiety and stress. Individuals might experience different kinds of sexual dysfunctions simultaneously. These factors should be considered when diagnosing and classifying sexual disorders among PWE (Rathore et al., 2019).

Prevalence of sexual dysfunction among patients with epilepsy

Epilepsy-related sexual dysfunction is not yet fully understood, but 30%–66% of men with epilepsy and 14%–50% of women with epilepsy experience some form of sexual dysfunction (Atif, Sarwar, & Scahill, 2016). The results of a recent metaanalysis showed that compared to control groups, sexual dysfunction among women and men with epilepsy were 2.69 times and 4.85 times higher, respectively (Zhao et al., 2019).

Among women with epilepsy, there are four types of sexual dysfunction. However, the majority of women have impaired sexual desire and arousal, while sexual problems caused by orgasm and pain are less common. Men with epilepsy are 1.5–2 times more likely to have sexual dysfunction than healthy men. While erectile dysfunction is the most common sexual disorder among men with epilepsy, approximately 10%–20% of patients experience decreased libido. Sexual dysfunction is more common among patients with uncontrolled epilepsy and those with anxiety and related depression, while patients with controlled epilepsy are less likely to have sexual dysfunctions (Rathore et al., 2019). Also, individuals with lesional epilepsy report more sexual dysfunctions than nonlesional ones (Ogunjimi, Yaria, Makanjuola, & Ogunniyi, 2018). Risk factors identified for sexual dysfunction among patients with epilepsy include depression, type of antiepileptic drug (AED), lateralization of epilepsy, endocrine dysfunction, and temporal lobe involvement (Harden & Pennell, 2013; Wulsin, Solomon, Privitera, Danzer, & Herman, 2016; Zelená, Kuba, Soška, & Rektor, 2011).

Mechanisms of sexual dysfunction in epilepsy

The mechanism of sexual dysfunction among patients with epilepsy is not fully understood and most of the available evidence indicates that sexual dysfunction in epilepsy is caused by several factors including disease-related factors and drug treatment, psychiatric factors, and social factors (Markoula et al., 2020).

Direct effects of the disease: Sexual dysfunction may be due to the direct effects of epilepsy. Sexual dysfunction usually begins after the onset of seizures and even independently of the effects of medications and accompanying psychiatric illnesses and is present in untreated patients with epilepsy (Bhugra & Colombini, 2018). Similarly, patients with focal epilepsy experience four times more sexual dysfunction than individuals with generalized epilepsy. Sexual dysfunctions are also more common among patients with uncontrolled epilepsy, especially right temporal lobe epilepsy. These factors suggest that central mechanisms of interference with pituitary-hypothalamic function, independent of the use of AEDs or in association with other mental illnesses, may lead to sexual dysfunction among patients with epilepsy. Impairment of pulsatile gonadotropin and dopamine secretion can lead to impaired sexual function by causing hypogonadism and hyperprolactinemia (Yogarajah & Mula, 2017). In addition, menopausal women start menopause earlier, which may affect their sex life due to the effects of menopausal changes in sex hormones (Dennerstein, Alexander, & Kotz, 2003).

Antiepileptic drugs: Despite the ambiguity concerning the direct effects of epilepsy on sexual function, evidence suggests there are consequential effects of AEDs on sex hormone levels and sexual dysfunction by various mechanisms (Calabrò, Marino, & Bramanti, 2011). Conventional AEDs, especially those that interfere with cytochrome $p450$ metabolism, are associated with sexual dysfunction (Calabrò, 2017). AEDs, especially enzyme stimulants such as carbamazepine and phenytoin, can cause sexual dysfunction by accelerating the metabolism of sex hormones, increasing sex hormone-binding globulin (SHBG) levels, and consequently lowering active testosterone levels. However, AEDs that inhibit liver enzymes (such as valproate) increase estrogen levels (Calabrò, 2017; Calabrò, Grisolaghi, Quattrini, Bramanti, & Magaudda, 2013). AEDs increase liver metabolism, sex steroids, gonads, and adrenals. In addition, AEDs suppress the pituitary-hypothalamic axis and affect serotonergic pathways, causing hypogonadotropic hypogonadism and sexual dysfunction. Common AEDs include carbamazepine and phenytoin, which are associated with low levels of free testosterone and high levels of SHBG (Yogarajah & Mula, 2017). There is growing evidence that newer AEDs also have a negative effect on sexual function by being able to modulate neurotransmission in the brain and spinal cord and negatively affecting monoamine pathways (Calabrò, 2017).

Vascular changes associated with risk factors or atherosclerotic changes due to epilepsy: The association of vascular disease with epilepsy is not uncommon. The possibility of erectile dysfunction due to epilepsy-related vascular changes is not well known, but it may be due to chronic epilepsy and its long-term treatment, especially with enzymatic inhibitory AEDs, and unfavorable vascular profiles including von Willebrand factor (vWF), fibrinogen levels, oxidized low-density lipoprotein (Ox-LDL), hypercholesterolemia/dyslipidemia, increased lipoprotein (a) (Lpa), hyperhomocysteinemia (HHcy), hyperuricemia, hyperinsulinemia, and increased insulin resistance, hyperlipidemia, and hyperlip metabolic syndrome increases. Chronic epilepsy is also associated with increased internal carotid artery thickness, premature death from ischemic heart disease, and stroke (Moussa, Papatsoris, Abou Chakra, Dabboucy, & Fares, 2020). All of these changes can be a cause for erectile dysfunction among male patients.

Comorbid psychiatric conditions: Individuals with epilepsy have significant accompanying psychiatric illnesses, and are 2–5 times more likely to develop any mental disorder, and 1 in 3 patients with epilepsy is diagnosed with a psychiatric illness in their lifetime. Psychiatric comorbidities have a poor prognosis because they have been associated with poor response to treatment (drugs and surgery), increased morbidity, and increased mortality (Lin & Pakpour, 2017; Mula, Kanner, Jetté, & Sander, 2021). Psychiatric illnesses (e.g., depression, anxiety, and psychosis), the use of antipsychotic, and antidepressant drugs in association with epilepsy can increase the chances of developing sexual dysfunction (Pellinen et al., 2021; Zelená et al., 2011).

Social factors: In addition, various psychosocial factors can play an important role in causing sexual dysfunction among individuals with epilepsy. Such individuals have low self-esteem, experience stigma, have lower social development, and often feel social isolation. These factors may lead to increased rejection, feelings of sexual inadequacy, and lack of sexual attractiveness among these patients. In addition, anxiety and fear of seizures during sexual intercourse can lead to avoidance of sexual activity, which can lead to feelings of rejection and dissatisfaction in the partner (Rathore et al., 2019).

Strategies for the treatment of sexual disorders in patients with epilepsy

Issues concerning sexual activity should be raised by healthcare professionals as a routine part of the management of patients with epilepsy and should consequently be referred to a psychiatric clinic (Suleyman, Yitayih, Fanta, Kebede, & Madoro, 2021). Initiation of appropriate treatment for sexual dysfunction among patients with epilepsy requires a thorough evaluation. Treatment should be considered according to the patient's needs, epilepsy status, and comorbidities, and available medications for epilepsy management. Various methods may be useful for the treatment of sexual dysfunction among epileptic patients, including: behavioral approaches to improve sexual function, waiting to achieve adjustment, reducing the dose of current AEDs, delaying medication until after sex, adjunctive therapy, and changing the diverse range of AEDs available to individual patients.

Several adjuvant drugs, such as buspirone, yohimbine, neostigmine, cyproheptadine, mesansrine, amantadine, and dexamphetamine, are used to treat AED-related sexual disorders (Atif, Azeem, & Sarwar, 2016). Although AEDs may increase the risk of sexual problems due to a wide range of drug-related side effects, such as changes in sex hormones, the benefits of treating AEDs are likely to outweigh the negative consequences (Lin, Updegraff, & Pakpour, 2016). AEDs can reduce anxiety and disease-related stigmas by improving symptoms. By reducing anxiety and minimizing stigma, women's sexual functioning may be indirectly improved. On the other hand, choosing AEDs with less effect on sex hormones can help improve sexual function among patients (Bradford & Meston, 2006).

Quality of life and adherence to treatment also increase as sexual functioning improves (Lin, Burri, Fridlund, & Pakpour, 2017). Despite the high rate of sexual problems among individuals with epilepsy, few actively seek treatment for their problems. Patients do not seek treatment for a variety of reasons, such as negative beliefs about seeking help, shame, perceived barriers such as time constraints, and social norms (Lin, Oveisi, Burri, & Pakpour, 2017). Given that sexual health is an important component in quality of life, attention to healthy sexual function among patients with epilepsy should be considered by healthcare providers (Lin, Potenza, Broström, Blycker, & Pakpour, 2019). Due to the significant prevalence of sexual dysfunction among patients with epilepsy, the synergistic effect of sexual dysfunction and mental disorders, design and implementation of psychological interventions including mindfulness-based cognitive therapy (MBCT) for sexual disorders among patients with epilepsy have been proposed (Paterson, Handy, & Brotto, 2017).

Mindfulness-based cognitive therapy

MBCT is an approach to psychotherapy that uses cognitive-behavioral therapy (CBT) techniques alongside mindfulness meditation techniques and similar psychological strategies. This method was originally developed to prevent recurrence among individuals with major depressive disorder (Sommers-Flanagan & Sommers-Flanagan, 2018). Mindfulness can be briefly described as unconscious and present consciousness (Hanh, 2016). Mindfulness has been used for the treatment of many medical conditions including pain disorders, depression, anxiety, borderline personality disorder, substance abuse, eating disorders, psychosis, and behavioral problems in children (Brotto, Krychman, & Jacobson, 2008).

Psychological therapies are classified into different groups of approaches. The common factors approach suggests that the effectiveness of different approaches in treatment and counseling is activated by factors that are present in all evidence-based therapies. These factors include therapeutic relationship, empathy, and active listening skills, but the factor that undoubtedly exists in effective therapies is learning. It has been suggested that mindfulness is also a common factor in various therapeutic approaches. Mindfulness is a core process of psychotherapy, according to which the development of mindfulness leads to the acquisition of new perspectives and is not related to the usual set of responses, including automatic thoughts and behaviors. In the technical eclecticism "approach" to treatment, the therapist chooses appropriate and comfortable techniques from different approaches, such as mindfulness skills, based on understanding the client's needs. The "theoretical integration" approach in therapy works with the goal of bringing together diverse theoretical systems under a metatheoretical framework. In particular, MBCT reinforces learning theory through the coemergence model of reinforcement, cognitive and coping techniques, emotional and social neuroscience, mindfulness, and empathy (Shires, Cayoun, & Francis, 2019). Focusing on cognitive processes distinguishes MBCT from other mindfulness-based therapies (Hayes, Villatte, Levin, & Hildebrandt, 2011). CBT uses it as training for the desired situation and the role of cognition in it (Manicavasgar, Parker, & Perich, 2011).

Practice and procedures of mindfulness-based cognitive therapy

In MBCT, mindfulness is based on four principles including body awareness (body posture and movement/physical activity), body emotions (including emotion-related items), mental states (including emotional states), and mental content training (thoughts, images). MBCT comprises four stages of learning, grouped into two stages: "internalization" and "externalization," which allows change at the system level. These steps are designed to develop mindfulness, cognitive and behavioral skills in key areas of function including intrapersonal ("personal stage"), situational ("exposure stage"), interpersonal stage, and "empathy stage." Stage 1 is for internalization skills and Stages 2–4 are for externalization skills, usually hierarchical. The purpose of this hierarchical integration is first to train clients to internalize attention in order to regulate attention and emotion, and then to develop these skills in the context of their mental state (Shires et al., 2019). In summary, each of the MBCT steps has two categories of behavioral training and meditation training as follows (Table 1):

TABLE 1 Summary of MBCT steps.

Stages at system level	Stage 1: Internalization		Stage 2: externalization	
	Personal stage	*Exposure stage*	*Interpersonal stage*	*Empathy stage*
Main focus of each stage	Focus on mindfulness training for deep levels of metacognitive awareness and mutual understanding, accepting the instability of phenomena, increasing relaxation and a sense of self-efficacy, and an emphasis on commitment to exercise	The exposure process is performed as "bipolar exposure" guided imagery with interoceptive exposure to subjective units of distress targets while remaining equanimous followed by in vivo exposure using equanimity	Behavioral tasks include experiential ownership with interpersonal exposure to avoid interpersonal avoidance and conflict. Not reacting to the reactions of others (seeing suffering), and being bold and forthright	Behavioral tasks are being aware of moral boundaries and commitment to ethics, compassion for oneself and others, preventing recurrence, and preservation of achievements
Interventional session content	Session 1—progressive muscle relaxation; Session 2—breathing mindfulness; Session 3—one-sided body scan; Session 4—one-sided body scan without voice and functional exercise	Session 5—bilateral body scans for symmetrical and faster involvement of the wider sensory-psychological networks; and Session 6—partial movement with constant attention, to establish flow in the body while avoiding the desire for pleasant emotions	Session 7—sweeping *en masse* with "free flow" through the entire body in a single pass while remaining equanimous with pleasure; and Session 8—transversal scanning by passing attention transversally through the body to feel the interior of the body with equanimity	Session 9—sweeping in-depth by passing attention with vertical free flow on the inside of the body with equanimity (loving-kindness meditation; mindful practice of five ethical precepts); and Session 10—maintenance exercise of respiratory awareness, body scan, and affection. Finally, there is a review the program

Step 1 (personal stage): In behavioral training, there is a focus on mindfulness training for deep levels of metacognitive awareness and mutual understanding, accepting the instability of phenomena, increasing relaxation and a sense of self-efficacy, and an emphasis on commitment to exercise. Meditation exercises consist of four sessions: Session 1—progressive muscle relaxation (14 min twice a day); Session 2—breathing mindfulness (30 min twice a day from here to there); Session 3—one-sided body scan; Session 4—one-sided body scan without voice and functional exercise.

Step 2 (exposure stage): The exposure process is performed as "bipolar exposure" guided imagery with interoceptive exposure to subjective units of distress targets while remaining equanimous followed by in vivo exposure using equanimity. Meditation exercises at this stage consist of two sessions: Session 5—bilateral body scans for symmetrical and faster involvement of the wider sensory-psychological networks (practicing bipolar exposure for 11 min after 30 min of meditation); and Session 6—partial movement with constant attention, to establish flow in the body while avoiding the desire for pleasant emotions (practicing bipolar exposure for 11 min after 30 min of meditation).

Step 3 (interpersonal stage): In this step, behavioral tasks include experiential ownership with interpersonal exposure to avoid interpersonal avoidance and conflict. Not reacting to the reactions of others (seeing suffering), and being bold and forthright. Meditation exercises in this stage consist of two sessions: Session 7—sweeping *en masse* with "free flow" through the entire body in a single pass while remaining equanimous with pleasure; and Session 8—transversal scanning by passing attention transversally through the body to feel the interior of the body with equanimity.

Stage 4 (empathy stage): In this stage, behavioral tasks are being aware of moral boundaries and commitment to ethics, compassion for oneself and others, preventing recurrence, and preservation of achievements. At this stage, two meditation practice sessions are held: Session 9—sweeping in-depth by passing attention with vertical free flow on the inside of the body with equanimity (loving-kindness meditation for 8 min; mindful practice of five ethical precepts); and Session 10—maintenance exercise once a day for 45 min, including 10 min of respiratory awareness, 25 min of body scan, 10 min of affection. Finally, there is a review the program (Shires et al., 2019).

Almost every meditation session lasts a week. It should be noted that in order to maximize the effectiveness of training, clients should adhere to three basic principles, including adequate exercise frequency (usually twice a day), adequate duration (usually 30 min per session), and adequate exercise practice (conscious effort to reduce, identify, and respond to emerging experiences). MBCT research shows that clients who follow this protocol benefit the most (Scott-Hamilton, Schutte, & Brown, 2016). In addition, clients are taught to relax in their daily lives. They must learn to control the body's emotions as much as possible in everyday situations, identify the normal patterns of emotion they have experienced during stressful events, and at the same time increase their capacity by increasing calmness to avoid learned responses. Consequently, mutual awareness, which is created during the formal practice of meditation, becomes a skillful tool to prevent the strengthening of useful reactionary habits in daily life (Shires et al., 2019). Mindfulness, one of the essential components of thoughtful meditation, can be done at any time and in conjunction with daily and mundane tasks (Hanh, 2016).

Application of mindfulness-based cognitive therapy to promote sexual function

Sexual dysfunctions are related to distraction, anxiety, inhibition, self-criticism about sexual function, and a lack of attention to the present and sexual stimuli (Bitzer & Kirana, 2021). Therefore, it is posited that mindfulness approaches may be useful for this particular group of individuals, as they try to experience the "here and now" through meditation, paving the way for meaning and happiness (Brotto & Heiman, 2007). Given the overlap in psychological processing and the mechanisms of depression and libido reduction, Paterson et al. (2017) adapted the MBCT protocol for depression to address problems with libido and arousal (Paterson et al., 2017). They used mind-focused sex therapy exercises to create MBCT-S to treat female sexual dysfunction of general population (Adam, De Sutter, Day, & Grimm, 2020; Hucker & McCabe, 2014). MBCT-S to promote sexual function in different groups such as patients with gynecological cancers (Brotto et al., 2012), provoked vestibulodynia (Brotto et al., 2019), and epilepsy (Lin et al., 2019) have been used.

Evidence-based experiences of mindfulness-based cognitive therapy interventions to enhance sexual function

The first study to use MBCT to improve sexual function in women with gynecological cancer was conducted by Brotto et al. (2012). In this study, 31 individuals with a history of endometrial or cervical cancer with a mean age of 54 years with sexual dysfunction or sexual arousal concerns were randomly assigned to the intervention group with three 90-min sessions of MBCT or the control group was assigned to a 2-month waiting list. In the first session, the multifactorial causes of women's sexual problems were taught, an introduction to the cognitive challenges of maladaptive sexual beliefs, the prevalence of sexual disorders after cancer, and familiarity with exercises related to body image and mindfulness. In the second session, homework was reviewed, cognitive challenges with intellectual backgrounds, mindfulness exercises in the session, training on the relationship between mindfulness, body image and gender, and motivation training methods were taught. In the third session, training on the relationship, mindfulness, and gender, pelvic floor health training, sex-focused sex technique and the inclusion of mindfulness in sexual exercises were performed. Sexual response, sexual discomfort and mood, as well as physiological and mental sexual arousal were assessed before, 1 month, and 6 months after treatment. The status of individuals on the waiting list did not change, and the treatment led to a significant improvement in all areas of sexual response and a tendency to give importance to reducing sexual discomfort. Perception of sexual arousal during viewing of an erotic film also increased significantly after the intervention. It was concluded that a brief mindfulness-based intervention was effective in improving sexual function (Brotto et al., 2012).

Adam et al. (2020) compared the effectiveness of mindfulness-based behavioral techniques with conventional video-based CBT in the treatment of women with orgasmic disorder. In this study, 65 women with a mean age of 33 years were randomly assigned to MBCT ($N = 35$) or CBT group ($N = 30$). Sexual function and sexual distress and sexual awareness were assessed as outcomes of this study before and after treatment and 2 months after its completion. The intervention program was that the women received seven videos per week and watched them as much as they wanted. Each video taught next week's sexual exercises that participants had to do alone and/or with their partner. The MiBCT team also received and practiced additional mindfulness exercises. The results of the study showed that women in both groups significantly improved sexual function and reduced sexual anxiety, as well as improved desire, arousal, orgasm and sexual satisfaction. They suggested that these results should guide physicians' decisions with respect to evaluating the relevance and real added value of suggesting mindfulness exercises to their patients with such problems (Adam et al., 2020).

Brotto et al. (2019) compared the effect of MBCT versus CBT for the treatment of stimulated vestibulodynia. In this study, 130 participants were divided into two groups: CBT ($n = 63$) and MiBCT ($n = 67$) with a mean age of 31 and 33 years. The main outcome was self-reported pain during penetration assessed at baseline, and then reassessed after immediate treatment and at 6-month follow-up. Secondary outcomes included pain rating with a valvesometer, severe pain, increased pain awareness, pain acceptance, sexual function, and sexual distress. There was a significant relationship between group and time for their reported pain. Recovery with MiBCT was greater than in CBT cases. For all secondary outcomes, both groups resulted in similar significant improvements and the benefits and were maintained for 6 months. It was concluded that mindfulness was a promising approach to improving self-reported vaginal pain (Brotto et al., 2019).

Farajkhoda, Sohran, Molaeinezhad, and Fallahzadeh (2019) investigated the effectiveness of mindfulness-based therapeutic counseling on improving sexual satisfaction among women of childbearing age. In this study, 20 women of reproductive age were randomly assigned to two groups of mindfulness and waiting list. The intervention was presented across eight counseling sessions with a MBCT approach to improve sexual satisfaction. The intervention program comprised automatic guidance (first session), encountering obstacles (second session), presence of mind (third session), remaining in the present moment (fourth session), acceptance (fifth session), thoughts are not real (sixth session), self-care (seventh session), and applying the education (eighth session). Sexual satisfaction was measured at baseline, 8 weeks, and 12 weeks after treatment. The results showed that MBCT increased sexual satisfaction among women of childbearing age compared to the control group (Farajkhoda et al., 2019).

Hucker and McCabe (2014) examined the effect of online mindfulness-CBT on women's sexual problems. In the first study, 26 women completed treatment and the changes were compared with the waiting list control group (31 patients). In Study 2, 16 women in the control group completed treatment. The authors did not use the control group in the second study. The results showed that both treatment groups observed a significant improvement in intimacy and sexual relationship and emotional intimacy was significantly improved in the first study treatment group. Most positive changes were maintained in follow-up. The intervention program consisted of six advanced online modules, the first lasting at least a week and the rest at least every 2 weeks. All women started the program at the same time and were encouraged to complete each course within 2 weeks, although the program allowed flexible scheduling depending on the needs of each couple. Each section included psychological training and related CBT exercises designed to help women identify and challenge negative self-negative thoughts and beliefs about gender. Each module included texts for partners that covered information about women's gender, women's sexual problems, and relationship issues, as well as descriptions of therapeutic exercises. The texts also included phrases aimed at normalizing the process for men and validating their treatment experiences. Each module also included communication exercises and mindfulness exercises, both of which required the participation of a sexual partner. In addition, mindfulness exercises were used throughout the program to help women cultivate instantaneous awareness, and the use of mindfulness during mindfulness to help women manage difficult thoughts and feelings and be more present and aware (Hucker & McCabe, 2014).

Mindfulness was first introduced through the main exercises of sexual mindfulness focusing on momentary experience (e.g., self-meditation, mindfulness). Participants were encouraged to do the exercise for 5 min each day during the program. Then more mindfulness exercises were introduced, such as conscious eating, conscious movement, and exercises that included sensory awareness of the body, taking a shower or bath in later periods, more sexually mindfulness exercises, and sensory concentration sessions. Online chat groups were run for 1 h each, every 2 weeks, consisting of approximately 4–8 women in each group. All groups were facilitated by a therapist specializing in the treatment of women's sexual problems and disorders. Loose-structured chat groups were conducted including a review of program exercises and experiences over the past 2 weeks, discussion of challenges and barriers to change, specific intervention suggestions (if needed), and end of session.

In addition, the final sessions focused more on reflection and relapse prevention. During chat groups, the facilitator guided women to focus on the causes and perpetuators of their sexual problems and to explore barriers to change during treatment as well as potential solutions. The facilitator used CBT and mindfulness to reinforce the concepts of the pursuing pleasure program. Participants were also offered unlimited electronic contact with the therapist (see Hucker & McCabe, 2014 for more information on the content of chat groups and the topics covered in these discussions). Each module ended with an obstacle requirement, which included a list of questions that completed the exercises before moving on to the next module. Women had to agree that they had achieved these goals and were prepared to move forward with the question mark electronically before receiving the password to access the next pursuing pleasure module (Hucker & McCabe, 2014).

Evidence of the effectiveness of mindfulness-based cognitive therapy intervention to improve sexual function among epileptic patients

In a review of related literature, Lin et al. (2019) studied the effect of MBCT program on sexual function and intimacy among elderly women with epilepsy. In this multicenter randomized controlled trial, 660 women with epilepsy were randomly assigned to one of the groups: (i) patient and their sexual partner; (ii) patient, their sexual partner and healthcare provider; or (iii) routine treatment. The intervention treatment for the first two groups was 8 weekly MBCT-S sessions. In the second group of patients, their sexual partner and healthcare provider, three sex counseling sessions were added for healthcare providers. Self-assessment scales were used initially, 1 month and 6 months after the intervention. The results of the study showed improvement in sexual function, sexual distress, and intimacy in both interventions groups at 1-month and 6-month follow-up. The second group (patient, their sexual partner and healthcare provider) had a greater improvement in intimacy compared to the intervention group consisting patient and their sexual partner both follow-ups. In terms of sexual function, the second group performed better than the first group at 6-month follow-up. MBCT-S improved sexual function and intimacy while reducing sexual distress (Lin et al., 2019). The intervention program in the study of Lin et al. is introduced as a MBCT-S to promote sexual function in women with epilepsy.

Practice and procedures for MBCT-S

The first session provided sexual information about the prevalence of sexual dysfunction in the women with epilepsy and the relationship between epilepsy and sexual dysfunction was presented. Mindfulness was then introduced by providing a rationale for using mindfulness to treat sexual problems. In addition, information on the anatomy and physiology of sexual responses was provided to couples during a long-term relationship. Homework included meditation exercises and conscious nutrition exercises for 10 min a day.

The second session began with an in-depth review of the assignments. A body scan was introduced to couples to help them become more aware of their body emotions. Couples were encouraged to complete a worksheet that focused on potential participants (or so-called "protective agents") in their sexual problems. Facilitators encouraged couples to focus on their genitals, lovingly and without judgment. Couples were asked to imagine that they had chosen a goal for their partner to first explore and describe with their eyes closed. Conscious listening was performed by couples to practice listening without judgment for a few minutes (40 min of body scan exercises daily was recommended as homework for couples).

The third session invited the couple to listen to each other for 5 min (conscious listening). Sitting meditation (including awareness of breathing and bodily responses to bodily emotions) was then practiced. Pictures were provided to the couple to explain the reason chosen and its meaning to their partners. The couple were asked to use a hand mirror to observe their genitals. In addition, they completed two worksheets to examine sexual beliefs and body image. The homework included stretching and breathing exercises for 40 min a day.

The fourth session began with conscious listening for 5 min. The session continued with sitting meditation with a focus on nonjudgment and acceptance of awareness of breathing, body, sounds and thoughts. Two worksheets describing a cognitive model assumed that biases were presented to couples. At the end of the session, couples were asked to imagine walking in pairs, one guiding, the other blindfolded or blindfolded. Sitting meditation was considered as homework (40 min daily).

The fifth session emphasized sitting meditation, including conscious awareness of breathing. Observation and tactile practice were performed by the couple. The couples were then asked to do the first "concentration exercise" by the presenters. In focusing on exercise, women and their partners were encouraged to pay attention to their bodies while taking a shower and afterward. They were then encouraged to describe their feelings about their bodies in nonjudgmental ways. A handout on their individual sexual response cycles was completed by couples. Alternative meanings of sexual/relationship situations were discussed by the couple. Sitting meditation was considered as homework (40 min daily).

The sixth session included self-observation exercises and touching exercises. The session also explored approaches to couples' problems or concerns. The presenters asked the couple to deliberately bring up a problem in order to be aware of the related thoughts and experiences in their body. Couples were then asked to listen to their partner's pulse. Homework included 25 min of daily hard work.

In the seventh session, the presenters invited the couple to become aware of any physical sensations in their body. Exercises including sexual arousal with sexual assistance, awareness of sexual feelings and pleasurable touch were performed by couples. At the end of the session, the long-term intent worksheet was completed by the couple. In the last (eighth) session, the couple repeated all the mindfulness exercises (Lin et al., 2019).

Application of the aforementioned MBCT-S program led to improvements in sexual function, sexual distress, and intimacy at 1-month and 6-month follow-ups. Therefore, MBCT-S appears to be efficacious in enhancing sexual function and reducing sex-related distress among women with epilepsy (Lin et al., 2019).

Conclusion

MBCT is an 8-week, evidence-based program that combines mindfulness skills and cognitive therapy techniques for working with different problems including sexual dysfunction. Mindfulness, or awareness of present-moment experience, is developed through exercises designed to strengthen attentional capacity. MBCT incorporates many CBT principles and techniques for working with thoughts, emotions, body sensations, and behaviors. MBCT-S has been developed to improve sexual health in different female groups such as women with concerns regarding sexual desire and/or arousal, patients with gynecological cancers, provoked vestibulodynia, and epilepsy. Based on the findings in the extant literature, MBCT-S programs lead to improvements in sexual function, decreased sexual distress, and increased intimacy among women, but there is little evidence regarding effectiveness of MBCT-S for sexual problems among men. Therefore, further studies are needed to investigate the effectiveness of MBCT-S programs among both male and female individuals with different health conditions as well as their effect on different aspects of sexual life.

Mini-dictionary of terms

Sexual dysfunction: The natural human sexual response can be divided into four stages (desire, excitement, orgasm, and resolution). Each of these stages may be disrupted and lead to sexual dysfunction.

CBT: Cognitive-behavioral therapy (CBT) is a type of psychotherapeutic treatment that helps people learn how to identify and change destructive or disturbing thought patterns that have a negative influence on behavior and emotions. CBT focuses on changing the automatic negative thoughts that can contribute to and worsen emotional difficulties, depression, and anxiety. These spontaneous negative thoughts have a detrimental influence on mood. Through CBT, these thoughts are identified, challenged, and replaced with more objective, realistic thoughts.

MBCT: Mindfulness-based cognitive therapy (MBCT) is an approach to psychotherapy that uses cognitive-behavioral therapy (CBT) techniques alongside mindfulness meditation techniques and similar psychological strategies.

Key facts

- Epilepsy, like many common medical disorders can cause sexual dysfunction.
- Sexual dysfunction in epilepsy is caused by several factors including disease-related factors and drug treatment, psychiatric factors, and social factors.
- Prevalence of sexual dysfunction among PWE is significant.

- Contribution of psychiatric comorbidity and social factors to increasing sexual dysfunction among PWE is considerable.
- Design and implementation of psychological interventions including MBCT for sexual disorders among patients with epilepsy have been proposed.
- Issues concerning sexual activity should be raised by healthcare professionals as a routine part of the management of patients with epilepsy and MBCT can be considered as an effective promoting strategy.

Summary points

- Mindfulness-based cognitive therapy (MBCT) can be adapted to treat sexual dysfunction for PWE.
- Based on current evidence, MBCT for sexual problems programs lead to improvements in sexual function, sexual distress, and intimacy among women with epilepsy.
- There is little evidence regarding effectiveness of MBCT for sexual problems among men with epilepsy.
- Further studies are needed to investigate the effectiveness of MBCT for sexual problems programs on different aspects of sexual life.

References

Adam, F., De Sutter, P., Day, J., & Grimm, E. (2020). A randomized study comparing video-based mindfulness-based cognitive therapy with video-based traditional cognitive behavioral therapy in a sample of women struggling to achieve orgasm. *The Journal of Sexual Medicine*, *17*(2), 312–324. https://doi.org/10.1016/j.jsxm.2019.10.022.

Akyol, T., & Nehir, S. (2021). Effects of attitudes of patients with epilepsy towards their disease on mental health and quality of life. *Erciyes Medical Journal*, *43*(1), 47–54.

American Psychiatric Association. (2013). *Diagnostic and statistical manual of mental disorders: DSM-5*. Arlington, VA: American Psychiatric Association.

Atif, M., Azeem, M., & Sarwar, M. R. (2016). Potential problems and recommendations regarding substitution of generic antiepileptic drugs: A systematic review of literature. *Springerplus*, *5*(1), 182.

Atif, M., Sarwar, M. R., & Scahill, S. (2016). The relationship between epilepsy and sexual dysfunction: A review of the literature. *Springerplus*, *5*(1), 2070.

Beghi, E., Giussani, G., Nichols, E., Abd-Allah, F., Abdela, J., Abdelalim, A., ... Murray, C. J. L. (2019). Global, regional, and national burden of epilepsy, 1990–2016: A systematic analysis for the global burden of disease study 2016. *Lancet Neurology*, *18*(4), 357–375. https://doi.org/10.1016/S1474-4422(18)30454-X.

Bhugra, D., & Colombini, G. (2018). Sexual dysfunction: Classification and assessment. *Advances in Psychiatric Treatment*, *19*(1), 48–55. https://doi.org/10.1192/apt.bp.112.010884.

Birbeck, G., Chomba, E., Atadzhanov, M., Mbewe, E., & Haworth, A. (2007). The social and economic impact of epilepsy in Zambia: A cross-sectional study. *Lancet Neurology*, *6*(1), 39–44.

Bitzer, J., & Kirana, P.-S. E. (2021). Female sexual dysfunctions. In M. Lew-Starowicz, A. Giraldi, & T. Krüger (Eds.), *Psychiatry and sexual medicine* (pp. 109–134). Cham: Springer.

Bradford, A., & Meston, C. M. (2006). The impact of anxiety on sexual arousal in women. *Behaviour Research and Therapy*, *44*(8), 1067–1077. https://doi.org/10.1016/j.brat.2005.08.006.

Brotto, L. A., Bergeron, S., Zdaniuk, B., Driscoll, M., Grabovac, A., Sadownik, L. A., ... Basson, R. (2019). A comparison of mindfulness-based cognitive therapy vs cognitive behavioral therapy for the treatment of provoked vestibulodynia in a hospital clinic setting. *The Journal of Sexual Medicine*, *16*(6), 909–923. https://doi.org/10.1016/j.jsxm.2019.04.002.

Brotto, L. A., Erskine, Y., Carey, M., Ehlen, T., Finlayson, S., Heywood, M., ... Miller, D. (2012). A brief mindfulness-based cognitive behavioral intervention improves sexual functioning versus wait-list control in women treated for gynecologic cancer. *Gynecologic Oncology*, *125*(2), 320–325. https://doi.org/10.1016/j.ygyno.2012.01.035.

Brotto, L. A., & Heiman, J. R. (2007). Mindfulness in sex therapy: Applications for women with sexual difficulties following gynecologic cancer. *Sexual and Relationship Therapy*, *22*(1), 3–11.

Brotto, L. A., Krychman, M., & Jacobson, P. (2008). Eastern approaches for enhancing women's sexuality: Mindfulness, acupuncture, and yoga (CME). *The Journal of Sexual Medicine*, *5*(12), 2741–2748.

Calabrò, R. S. (2017). Sexual dysfunction and topiramate: What does lie beneath the tip of the iceberg? *Epilepsy & Behavior*, *73*, 281–282. https://doi.org/10.1016/j.yebeh.2017.05.028.

Calabrò, R. S., Grisolaghi, J., Quattrini, F., Bramanti, P., & Magaudda, A. (2013). Prevalence and clinical features of sexual dysfunction in male with epilepsy: The first southern Italy hospital-based study. *International Journal of Neuroscience*, *123*(10), 732–737. https://doi.org/10.3109/00207454.2013.798783.

Calabrò, R. S., Marino, S., & Bramanti, P. (2011). Sexual and reproductive dysfunction associated with antiepileptic drug use in men with epilepsy. *Expert Review of Neurotherapeutics*, *11*(6), 887–895. https://doi.org/10.1586/ern.11.58.

Dennerstein, L., Alexander, J. L., & Kotz, K. (2003). The menopause and sexual functioning: A review of the population-based studies. *Annual Review of Sex Research*, *14*(1), 64–82.

Farajkhoda, T., Sohran, F., Molaeinezhad, M., & Fallahzadeh, H. (2019). The effectiveness of mindfulness-based cognitive therapy consultation on improving sexual satisfaction of women in reproductive age: A clinical trial study in Iran. *Journal of Advanced Pharmacy Education & Research*, *9*(S2), 151–160.

Feigin, V. L., Nichols, E., Alam, T., Bannick, M. S., Beghi, E., Blake, N., ... Ellenbogen, R. G. (2019). Global, regional, and national burden of neurological disorders, 1990–2016: A systematic analysis for the global burden of disease study 2016. *Lancet Neurology*, *18*(5), 459–480.

Fiest, K. M., Birbeck, G. L., Jacoby, A., & Jette, N. (2014). Stigma in epilepsy. *Current Neurology and Neuroscience Reports*, *14*(5), 436–440.

Fiest, K. M., Sauro, K. M., Wiebe, S., Patten, S. B., Kwon, C.-S., Dykeman, J., ... Jetté, N. (2017). Prevalence and incidence of epilepsy: A systematic review and meta-analysis of international studies. *Neurology*, *88*(3), 296–303.

Hamedi-Shahraki, S., Eshraghian, M. R., Yekaninejad, M. S., Nikoobakht, M., Rasekhi, A., Chen, H., & Pakpour, A. (2019). Health-related quality of life and medication adherence in elderly patients with epilepsy. *Neurologia i Neurochirurgia Polska*, *53*(2), 123–130. https://doi.org/10.5603/PJNNS.a2019.0008.

Hanh, T. N. (2016). *The miracle of mindfulness: An introduction to the practice of meditation*. Beacon Press.

Harden, C. L., & Pennell, P. B. (2013). Neuroendocrine considerations in the treatment of men and women with epilepsy. *Lancet Neurology*, *12*(1), 72–83. https://doi.org/10.1016/S1474-4422(12)70239-9.

Hayes, S. C., Villatte, M., Levin, M., & Hildebrandt, M. (2011). Open, aware, and active: Contextual approaches as an emerging trend in the behavioral and cognitive therapies. *Annual Review of Clinical Psychology*, *7*, 141–168.

Hucker, A., & McCabe, M. P. (2014). An online, mindfulness-based, cognitive-behavioral therapy for female sexual difficulties: Impact on relationship functioning. *Journal of Sex & Marital Therapy*, *40*(6), 561–576. https://doi.org/10.1080/0092623X.2013.796578.

Khan, S. D., & Gunasekaran, K. (2019). The human sexual response. In K. Gunasekaran, & S. Khan (Eds.), *Sexual medicine* (pp. 1–9). Springer.

Komolafe, M. A., Sunmonu, T. A., Afolabi, O. T., Komolafe, E. O., Fabusiwa, F. O., Groce, N., ... Olaniyan, S. O. (2012). The social and economic impacts of epilepsy on women in Nigeria. *Epilepsy & Behavior*, *24*(1), 97–101.

Lin, C.-Y., Burri, A., Fridlund, B., & Pakpour, A. H. (2017). Female sexual function mediates the effects of medication adherence on quality of life in people with epilepsy. *Epilepsy & Behavior*, *67*, 60–65.

Lin, C.-Y., Oveisi, S., Burri, A., & Pakpour, A. H. (2017). Theory of planned behavior including self-stigma and perceived barriers explain help-seeking behavior for sexual problems in Iranian women suffering from epilepsy. *Epilepsy & Behavior*, *68*, 123–128.

Lin, C.-Y., & Pakpour, A. H. (2017). Using hospital anxiety and depression scale (HADS) on patients with epilepsy: Confirmatory factor analysis and Rasch models. *Seizure*, *45*, 42–46.

Lin, C.-Y., Updegraff, J. A., & Pakpour, A. H. (2016). The relationship between the theory of planned behavior and medication adherence in patients with epilepsy. *Epilepsy & Behavior*, *61*, 231–236. https://doi.org/10.1016/j.yebeh.2016.05.030.

Lin, C. Y., Potenza, M. N., Broström, A., Blycker, G. R., & Pakpour, A. H. (2019). Mindfulness-based cognitive therapy for sexuality (MBCT-S) improves sexual functioning and intimacy among older women with epilepsy: A multicenter randomized controlled trial. *Seizure*, *73*, 64–74. https://doi.org/10.1016/j.seizure.2019.10.010.

Manicavasgar, V., Parker, G., & Perich, T. (2011). Mindfulness-based cognitive therapy vs cognitive behaviour therapy as a treatment for non-melancholic depression. *Journal of Affective Disorders*, *130*(1–2), 138–144.

Markoula, S., Siarava, E., Keramida, A., Chatzistefanidis, D., Zikopoulos, A., Kyritsis, A. P., & Georgiou, I. (2020). Reproductive health in patients with epilepsy. *Epilepsy & Behavior*, *113*, 107563.

Moussa, M., Papatsoris, A. G., Abou Chakra, M., Dabboucy, B., & Fares, Y. (2020). Erectile dysfunction in common neurological conditions: A narrative review. *Archivio Italiano di Urologia, Andrologia*, *92*(4), 371–385.

Mula, M., Kanner, A. M., Jetté, N., & Sander, J. W. (2021). Psychiatric comorbidities in people with epilepsy. *Neurology: Clinical Practice*, *11*(2), e112–e120. https://doi.org/10.1212/cpj.0000000000000874.

Njamnshi, A. K., Bissek, A.-C. Z.-K., Yepnjio, F. N., Tabah, E. N., Angwafor, S. A., Kuate, C. T., ... Kepeden, M.-N. Z. (2010). A community survey of knowledge, perceptions, and practice with respect to epilepsy among traditional healers in the Batibo health district, Cameroon. *Epilepsy & Behavior*, *17*(1), 95–102.

O'Donohoe, T. J., Choudhury, A., & Callander, E. (2020). Global macroeconomic burden of epilepsy and the role for neurosurgery: A modelling study based upon the 2016 global burden of disease data. *European Journal of Neurology*, *27*(2), 360–368. https://doi.org/10.1111/ene.14085.

Ogunjimi, L., Yaria, J., Makanjuola, A., & Ogunniyi, A. (2018). Sexual dysfunction among Nigerian women with epilepsy. *Epilepsy & Behavior*, *83*, 108–112.

Owolabi, L. F., Adamu, B., Jibo, A. M., Owolabi, S. D., Isa, A. I., Alhaji, I. D., & Enwere, O. O. (2020). Prevalence of active epilepsy, lifetime epilepsy prevalence, and burden of epilepsy in sub-Saharan Africa from meta-analysis of door-to-door population-based surveys. *Epilepsy & Behavior*, *103*. https://doi.org/10.1016/j.yebeh.2019.106846, 106846.

Paterson, L. Q. P., Handy, A. B., & Brotto, L. A. (2017). A pilot study of eight-session mindfulness-based cognitive therapy adapted for women's sexual interest/arousal disorder. *Journal of Sex Research*, *54*(7), 850–861. https://doi.org/10.1080/00224499.2016.1208800.

Pellinen, J., Chong, D. J., Elder, C., Guinnessey, P., Wallach, A. I., Devinsky, O., & Friedman, D. (2021). The impact of medications and medical comorbidities on sexual function in people with epilepsy. *Epilepsy Research*, *172*. https://doi.org/10.1016/j.eplepsyres.2021.106596, 106596.

Penn Miller, I., Hecker, J., Fureman, B., Meskis, M. A., Roberds, S., Jones, M., Grabenstatter, H., ... Lubbers, L. (2021). Epilepsy community at an inflection point: Translating research toward curing the epilepsies and improving patient outcomes. *Epilepsy Currents*, *21*(5), 385–388.

Perrotta, G. (2020). Epilepsy: From pediatric to adulthood. Definition, classifications, neurobiological profiles and clinical treatments. *Journal of Neurology, Neurological Science and Disorders, 6*(1), 14–29.

Rathore, C., Henning, O. J., Luef, G., & Radhakrishnan, K. (2019). Sexual dysfunction in people with epilepsy. *Epilepsy & Behavior, 100*. https://doi.org/10.1016/j.yebeh.2019.106495, 106495.

Reed, G. M., Drescher, J., Krueger, R. B., Atalla, E., Cochran, S. D., First, M. B., … Saxena, S.. (2016). Disorders related to sexuality and gender identity in the ICD-11: Revising the ICD-10 classification based on current scientific evidence, best clinical practices, and human rights considerations. *World Psychiatry, 15*(3), 205–221. https://doi.org/10.1002/wps.20354.

Scott-Hamilton, J., Schutte, N. S., & Brown, R. F. (2016). Effects of a mindfulness intervention on sports-anxiety, pessimism, and flow in competitive cyclists. *Applied Psychology. Health and Well-Being, 8*(1), 85–103.

Shires, A., Cayoun, B., & Francis, S. (2019). *The clinical handbook of mindfulness-integrated cognitive behavior therapy: A step-by-step guide for therapists*. Wiley.

Sommers-Flanagan, J., & Sommers-Flanagan, R. (2018). *Counseling and psychotherapy theories in context and practice: Skills, strategies, and techniques*. Wiley.

Suleyman, B., Yitayih, S., Fanta, T., Kebede, A., & Madoro, D. (2021). Sexual dysfunction and its predictors among male patients with epilepsy attending public hospitals, East Ethiopia: A cross-sectional study. *Health Science Journal, 15*(5), 841.

World Health Organization. (2019). *Epilepsy: A public health imperative*. World Health Organization.

Wulsin, A. C., Solomon, M. B., Privitera, M. D., Danzer, S. C., & Herman, J. P. (2016). Hypothalamic-pituitary-adrenocortical axis dysfunction in epilepsy. *Physiology & Behavior, 166*, 22–31. https://doi.org/10.1016/j.physbeh.2016.05.015.

Yogarajah, M., & Mula, M. (2017). Sexual dysfunction in epilepsy and the role of anti-epileptic drugs. *Current Pharmaceutical Design, 23*(37), 5649–5661.

Zelená, V., Kuba, R., Soška, V., & Rektor, I. (2011). Depression as a prominent cause of sexual dysfunction in women with epilepsy. *Epilepsy & Behavior, 20*(3), 539–544. https://doi.org/10.1016/j.yebeh.2011.01.014.

Zhao, S., Tang, Z., Xie, Q., Wang, J., Luo, L., Liu, Y., … Zhao, Z.. (2019). Association between epilepsy and risk of sexual dysfunction: A meta-analysis. *Seizure, 65*, 80–88.

Chapter 13

Female sexual dysfunction: Applications of cognitive-behavioral therapy

Françoise Adam and Elise Grimm
Psychological Sciences Research Institute, Faculty of Psychological and Educational Sciences, UCLouvain, Ottignies-Louvain-la-Neuve, Belgium

Abbreviations

APA	American Psychiatric Association
CBT	cognitive-behavioral therapy
DSM-5	diagnostic and statistical manual—fifth edition
DSM-IV-TR	diagnostic and statistical manual—fourth edition—revised
FDA	Food and Drug Administration
FOD	female orgasmic disorder
FSD	female sexual disorder
FSDS-R	Female Sexual Distress Scale–revised
FSFI	female sexual function index
FSIAD	female sexual interest/arousal disorder
MSD	male sexual disorder

Introduction

The World Health Organization (2021) defines sexual health as "an integral part of overall health, well-being, and quality of life," which changes over the lifespan following exposure to specific life events (e.g., medical interventions, menopause, reproductive experiences, or relationships) and should be assessed in routine medical and psychological examinations in women. According to a study by Bachmann (2006), 60% of professionals were aware that one- to three-quarters of their patients suffered from female sexual disorders (FSDs). Despite this, most were reluctant to openly discuss their patients' sexual functioning or to perform a full assessment.

Patients are also reluctant to discuss their sexual difficulties to avoid generating a certain discomfort in their practitioners. Less than 25% of sexually active men and women suffering from sexual difficulties actually consult for these (Laumann, Glasser, Neves, & Moreira, 2009). Patients with sexual difficulties are often hesitant to consult and will not be sufficiently informed as to which professional to turn to. Consequently, sexual topics remain taboo for the majority of patients and practitioners.

However, approximately 43% to 44% of women suffer from FSDs (Laumann, Paik, & Rosen, 1999; Shifren, Monz, Russo, Segreti, & Johannes, 2008). FSDs refer to the difficulty in one or more phases of the sexual response, which includes desire, arousal, lubrication, and orgasm. One should note that 12% of women with FSDs also experience emotional distress and negative emotions such as guilt, frustration, stress, anger, and embarrassment with respect to their sex life (Shifren et al., 2008). FSDs negatively affect women's sexual and marital satisfaction, as well as quality of life (Nappi et al., 2016). Despite diagnosis being the first step for women to access appropriate treatment, FSDs remain often unidentified. In light of the current limitations discussed, the goal of this chapter is to provide clinicians and practitioners with a better understanding of assessment and empirically validated treatments for FSDs.

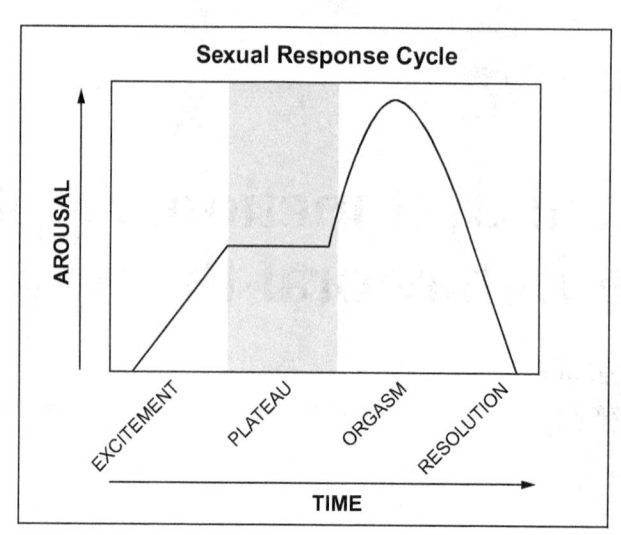

FIG. 1 The human sexual response. This figure shows the human sexual response. *(From Masters, W. H., & Johnson, V. E. (1966). Human sexual response. Little, Brown and Company: Boston with permission.)*

The female sexual response

The first sexual response model was linear and based on the physiological reactions observed during coitus (Masters & Johnson, 1966). Fig. 1 shows the four successive phases originally identified as arousal, plateau, orgasm, and resolution. The authors observed that the two physiological reactions of women's sexual response of women are vasocongestion of the genitals and increased neuromuscular tension in the entire body. Kaplan (1979) later completed this model by adding the phase of desire, essential for the increase of sexual excitation in women.

The circular model (Basson, 2005; Basson, Wierman, Van Lankveld, & Brotto, 2010) proposes a more complex expression of the female sexual response, incorporating emotional and relationship elements, such as the emotional intimacy context (i.e., relationship with partner, person's state of mind). This model emphasizes that spontaneous desire (i.e., sexual drive) is not always the driving force behind women's engagement in sexual activities. Desire can also be triggered by a context of emotional intimacy, which precedes physical arousal and includes the desire to be close or to experience sexual pleasure.

According to Basson and colleagues (e.g., Basson et al., 2000, 2010), female arousal is the culmination of both genital (i.e., biological) and subjective (i.e., psychological) arousal. Sexual stimuli will enable the reaching of a certain degree of sexual excitement, which the woman can perceive both psychologically (subjective excitement) and physiologically (genital excitement). If the woman evaluates this increase in sexual arousal as pleasurable, erotic and/or sexual, this will reinforce both types of arousal and the desire to pursue the sexual activity. If the woman evaluates the rise in sexual arousal in an unpleasant manner (e.g., with anxiety or guilt), this may inhibit the female sexual response. Finally, both physical and emotional satisfaction, observed through orgasm attainment, are important for developing emotional intimacy between partners and for increasing the desire to engage in future sexual activity (Basson et al., 2010). This model provides a better understanding of the interplay between emotional, interpersonal, and physical factors of the female sexual response. Several authors (e.g., Nowosielski, Wróbel, & Kowalczyk, 2016), however, claim that no single model truly captures the complexity of the female sexual experience. Clinicians are therefore advised to thoroughly assess each patient's own experience.

Sexual dysfunction model

In striving to reach an understanding of the precipitating and maintaining factors of sexual dysfunctions, a recent systematic review by Tavares, Moura, and Nobre (2020) exposed the important role of cognitive processing factors. Negative thoughts related to sexual performance or a negative body image can severely interfere with women's sexual arousal (Dove & Wiederman, 2000). Individuals with sexual dysfunction are more likely to activate negative sexual cognitive schemas, which in turn are fueled by dysfunctional sexual beliefs (Nobre, Gouveia, & Gomes, 2003; Nobre & Pinto-Gouveia, 2006, 2008). This can generate negative automatic thoughts and unpleasant emotions, including anxiety, sadness, guilt, and anger. These will prevent the person from focusing upon the erotic and sexual stimuli and thereby hinder the sexual response (Nobre & Pinto-Gouveia, 2006). Otherwise known as cognitive distraction, this process is one of the dysfunctional processes most often encountered by men and women with sexual dysfunctions (Brotto et al., 2016), as it decreases both subjective and physiological arousal (Dove & Wiederman, 2000).

Barlow's model of cognitive interference (1986) provides a better understanding of cognitive distraction. The latter explains that "negative affective responses may contribute to the avoidance of erotic cues and thus facilitate a kind of cognitive interference produced by focusing on non-erotic cues" (p. 144). People with sexual dysfunctions tend to focus on performance-related concerns and nonerotic stimuli during sexual activities. This reduces sexual arousal and increases performance anxiety. In contrast, women without sexual dysfunctions are more likely to focus their attention upon erotic and sexual stimuli, thereby increasing sexual arousal. For example, when a woman experiences a decrease in sexual desire during sexual activities, she may experience this as a failure and consequently develop anxious apprehension about sexual activities.

Wiegel, Scepkowski, and Barlow (2007) adapted Barlow's (1986) model to focus upon the dysfunctional cognitive and emotional processes that result from anxious apprehensions. They identified four principal consequences, namely (1) a sense of loss of control over the sexual situation, accompanied by unpleasant thoughts and emotions; (2) hypervigilance toward signs of sexual arousal that are judged unsatisfactory; (3) a shift in self-focused attention (i.e., self-evaluative focus) with a focus toward the perception of one's inability to cope with the situation; and (4) significant physiological activation (i.e., arousal). As a result, anxious apprehensions regarding sexual activities and the avoidance of unpleasant sexual situations increase and maintain FSDs. Evidence suggests that such cognitive processing factors should be systematically taken into account in the assessment and the treatment during psychological interventions (Tavares et al., 2020). These processes form a vicious circle displayed in Fig. 2, the latter of which can be presented to the patient during assessment.

Practice and procedures

Diagnosing female sexual dysfunctions

According to the American Psychiatric Association's (APA) Diagnostic and Statistical Manual Fifth Edition (DSM-5, 2013), sexual dysfunctions are defined as "a clinically significant disturbance in a person's ability to respond sexually

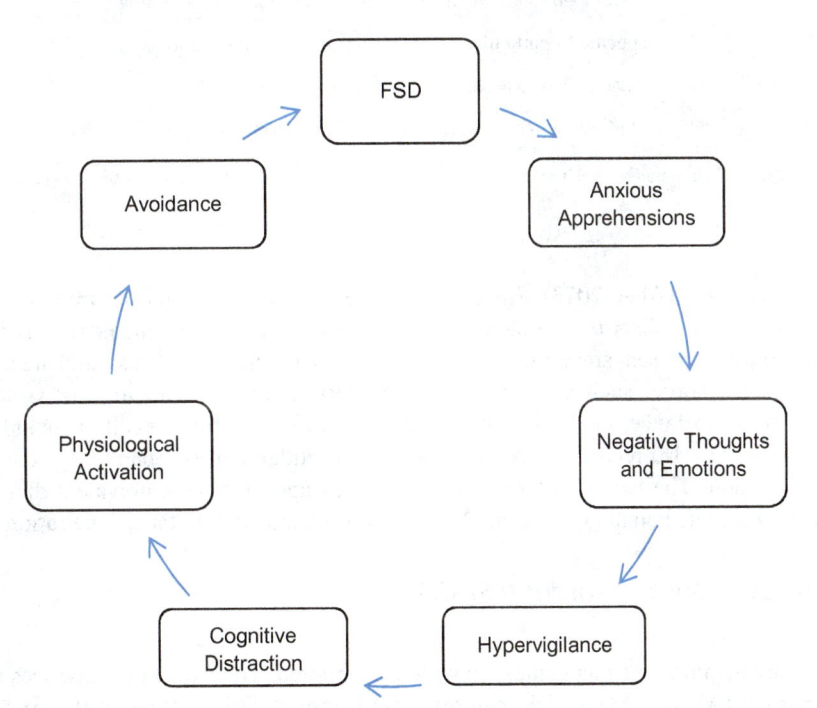

FIG. 2 Vicious circle implicated in female sexual dysfunctions. This figure represents the mechanisms involved in the appearance and maintenance of FSDs. *(The figure is adapted from Wiegel, M., Scepkowski, L. A., & Barlow, D. H. (2007). Cognitive-affective processes in sexual arousal and sexual dysfunction. In* Kinsey Institute Conference, 1st, Jul, 2003, Bloomington, IN, US; This Work Was Presented at the Aforementioned Conference, *cognitive-emotional model of sexual dysfunctions with permission.)*

TABLE 1 DSM-5 diagnostic criteria for female sexual dysfunctions and specifiers.

Female sexual dysfunctions	Diagnostic criteria
Female sexual interest/arousal disorder	*At least three of the six following symptoms*: Absence of or reduction in (a) interest in sexual activity; (b) sexual thoughts or fantasies; (c) sexual activity initiation; (d) sexual excitement or pleasure during sexual activities in (almost) all sexual encounters; (e) sexual interest/arousal following internal or external sexual or erotic cues; and/or (f) genital or nongenital sensations during sexual activity during sexual activities in (almost) all sexual encounters
Female orgasmic disorder	*Presence of one or two symptoms:* (a) marked delay in –, infrequency of –, or absence of orgasm and (b) markedly reduced intensity of orgasmic sensations almost all or all the time during sexual activity
Genito-pelvic pain/penetration disorder	Presence of "persistent or recurrent difficulties" with respect to at least one of four symptoms: (a) vaginal penetration during intercourse; (b) vulvovaginal or pelvic pain during vaginal intercourse or attempted penetration; (c) important fear regarding vulvovaginal/pelvic pain either in anticipation of—or resulting from—vaginal penetration; or (d) tension or tightening of the pelvic floor muscles during (attempted) vaginal penetration
Substance/medication-induced sexual dysfunction	Sexual dysfunctions resulting from (a) substance intoxication or medication withdrawal, and (b) involving medication or substance known to produce such disturbances in sexual functioning. The disturbance must not be exclusively present during delirium, and clinicians must specify with a mild, moderate, severe or without a substance use disorder. The diagnosis also requires the specification with respect to the temporal relationship between medication or substance intake and subsequent sexual dysfunction, namely (1) during intoxication, (2) during withdrawal, or (3) after medication use
Other specified/unspecified sexual dysfunction	Presence of several sexual-related symptoms causing clinical and significant distress, though the symptoms do not meet all diagnostic criteria for a specific disorder class. In the former category, the clinician chooses to reveal the reason for which a diagnostic cannot be established, while they will point to insufficient information in the latter
Specifiers	
Lifelong	Present since the first sexual experiences
Acquired	Developed following a period of normal sexual functioning
Situational	Specific to particular partners, stimulations, or situations
Generalized	Generalized to a number of situations
Severity	Mild, moderate, or severe distress linked to the symptoms

Table containing all DSM-5 diagnostic criteria and specifiers required to be taken into consideration when evaluating female sexual dysfunctions.

or to experience sexual pleasure" (APA, 2013). Since the new DSM-5, sexual diagnoses are sex-specific. FSD diagnoses require one to establish whether the disorder is lifelong, acquired, situational, or generalized (see Table 1 for an overview of criteria). The symptoms must be present since 6 months, cause significant clinical distress, and are not better attributable to another reason (i.e., mental disorder such as depression or anxiety, relationship distress or violence, other significant stressors, or the effects of a substance, medication or medical condition). A final specifier includes the current severity (Table 1). Additionally, the DSM-5 recommends adequate clinical judgment to separate dysfunction from the absence of adequate sexual stimulation. The next chapter section will focus upon the most prevalent diagnoses, namely Female Sexual Interest/Arousal Disorder, Female Orgasmic Disorder, and Genito-Pelvic Pain/Penetration Disorder.

Female sexual interest/arousal disorder (FSIAD)
Diagnosis

FSIAD is characterized by the interpersonal context in which it is rooted. The woman experiences a "desire discrepancy" with respect to their partner (APA, 2013), which manifests itself through at least three of the six symptoms presented in Table 1. It is important to note that if a lifelong absence of sexual interest or arousal is better explained by one's identification as "asexual," the current diagnosis cannot be made. This most recent criteria has considerably changed in two ways since APA's Fourth Edition of the DSM (DSM-IV-TR, APA, 2000; for a full description of changes between diagnoses, see Clayton & Juarez, 2019). First, sexual desire no longer represents a prerequisite for a healthy sexual response. It is well accepted that desire may result from and/or accompany a healthy arousal response following sexual stimuli. Second, the DSM-5 also considers arousal as a subjective experience in addition to the physical phenomenon previously described. The desire and arousal disorders previously observed in the DSM-IV-TR have thus merged under one category. With the new diagnosis requiring further validation, it is not uncommon for women to show different symptom profiles.

Prevalence and etiology

The lack of desire and/or arousal is the most frequent sexual-related complaints (Shifren et al., 2008). Statistics indicate low sexual interest in 22% of U.S. women (Laumann et al., 1999), and that 21%–31% of active women report low desire (McCormick, Lewis, Somley, & Kahan, 2007). The lack of frequent sexual activity, difficulties experiencing orgasm, pain during sexual activity, unrealistic expectations regarding the appropriate level of sexual interest/arousal, poor sexual techniques, and reduced sexual knowledge represent additional, important variables to identify and acknowledge during diagnosis (Parish et al., 2016). In addition, changes in hormones (e.g., estradiol, androgen), neurotransmitters (e.g. dopamine), and medical disorders (e.g., diabetes, cancer, depression) have been cited as prominent risk factors for the disorder (Parish et al., 2016). Regarding the psychological factors, daily stressor occurrence or more general current life situation may also negatively impact arousal and/or desire (Meston & Stanton, 2017). On a relationship level, a woman's feelings for her partner represent an important predictor of sexual desire, even when accounting for hormonal levels (e.g., Guthrie, Dennerstein, Taffe, Lehert, & Burger, 2004). With respect to societal factors, variables such as religion and culture may hinder arousal and/or desire due to the guilt and shame they both generate in relation to sexual experiences (Meston & Stanton, 2017).

Female orgasmic disorder (FOD)

Diagnosis

The criteria for FOD requires the self-reported presence of one or two symptoms linked to the experience and intensity of orgasm (see Table 1). It is additionally necessary to specify whether the female patient has never experienced an orgasm in any other situation before. Oftentimes, the diagnosis is accompanied by difficulties in sexual interest and/or arousal and can be guided by the observation of physiological changes (APA, 2013).

Prevalence and etiology

According to the DSM-5, around 10%–42% of women will report a diagnosis of FOD. Wincze and Weisberg (2015) estimate the prevalence rate as much higher, given the DSM-5 fails to account for women who are unable to experience orgasm, yet are not distressed by it. Akin to all female sexual dysfunctions, the etiology of FOD is multifold (for a review, see Parish et al., 2016). First, the lack of sexual excitatory activation from the central neuroendocrine or peripheral nervous systems (i.e., dopamine, oxytocin, melanocortin, and norepinephrine), and/or the increased sexual excitatory activation from central neuroendocrine sexual inhibitory processes (i.e., opioids, serotonin) may hinder orgasm attainment. Second, FOD may result from low hormone levels, medical disorders, or complications from medical procedures. Third, psychological factors such as negative emotions (e.g., guilt, shame, emotional distress), negative schemas and dysfunctional sex-related beliefs, mood disorders, excessive fatigue, or overthinking during sex are also known to be associated with orgasmic disorder (Adam, Day, De Sutter, & Brasseur, 2017; Parish et al., 2016). As result of sex-related ruminations, women may also engage in cognitive distraction, which, consequently, may inhibit them from staying focused upon sexual and erotic stimuli during sexual activities and further hinder orgasm (for a review, see Adam et al., 2017). Fourth, psychosocial problems may also cause orgasmic disorder, including communication problems within the couple, traumatic past couple experience or sexual abuse, or cultural and religious restrictions. The presence of sexual dysfunctions in the partner may also inhibit oneself from reaching orgasm.

Genito-pelvic pain/penetration disorder

Diagnosis

The diagnosis for genito-pelvic pain/penetration disorder requires the patient to experience "persistent or recurrent difficulties" in at least one of the four symptoms presented in Table 1 (APA, 2013). A frequent observation associated with the disorder is the avoidance of not only sexual activities but also of gynecological examinations. In addition, it is common for women to report relationship difficulties or feelings of inadequate femininity. The previous, separate classifications of dyspareunia and vaginismus found in the DSM-IV-TR (APA, 2000) have been combined to form the current diagnostic.

Dyspareunia (pain during intercourse) is a "recurrent acute pain located between the vaginal introitus to the uterus and adnexae" (Bergeron et al., 2001). Vaginismus is characterized by marked tension or tightness of the pelvic floor musculature during attempts at vaginal penetration (APA, 2013).

Prevalence and etiology

Studies report that 14%–34% of younger women and 45% of older women are affected by penetration disorder, respectively (Van Lankveld et al., 2010). At the biological level, it is well understood that the majority of sexual pain difficulties linked to the genital skin or mucus membranes result from inflammation caused by acute general infections (Bergeron, Corsini-Munt, Aerts, Rancourt, & Rosen, 2015). In contrast, the risk factors underlying vulvodynia are not clear. Pain may be the result of multiple physiological changes including malignant vulva lesions, transient conditions (e.g., infections such as herpes), dermatological disease-related lesions in tissues, hypersensitivity in the tissues around and within the vestibule, and age- or event-related hormonal changes (for reviews, see Bergeron et al., 2015; Conforti, 2017; Dias-Amaral & Marques-Pinto, 2018). Psychological responses such as cognitive catastrophizing, fear, and hypervigilance to pain, subsequently leads to avoidance of sexual interactions and contribute to the maintenance of the disorder (Conforti, 2017). Low self-efficacy and cognitive schemas (e.g., incompetence, rejection) also contribute, as women will stay focused on negative interpretations of their sexual events (Dias-Amaral & Marques-Pinto, 2018). Significantly stressful events and emotional disorders, such as depression and anxiety, are also thought to sustain muscle contractions. Finally, relationship-related factors, with a focus on both the woman's communication about her chronic pain and the partner's response, may also play an important role (Conforti, 2017).

Assessing female sexual dysfunctions

Setting and format

According to the American College of Obstetricians and Gynecologists (2017), sexuality should be approached through open dialogues and open-ended questions using gender-neutral terminology. Questions should be progressive, ranging from less to more intimate, as such to avoid sudden patient discomfort. The PLISSIT model (Annon, 1976) recommends several steps for FSD assessment: (1) asking the patient's permission (*P*) to address the area of sexuality; (2) providing limited information (*LI*) about sexual functioning; (3) offering specific suggestions (*SS*) for improving sexual functioning, and (4) initiating intensive therapy (*IT*).

Assessment lasts about 3–5 sessions depending on whether the patient is consulting alone or as a couple. Half of the time, patients will attend the consultations with their partner. According to Brotto et al. (2016), there is an interdependent link between both partner's sexual functioning. If one partner presents a sexual dysfunction, the emergence of sexual dysfunctions and decreased sexual satisfaction in the other is frequent (up to a three-fold probability: Chew et al., 2020). Should the patient consults alone, their partner should be invited to at least one session for evaluation. Based on the case conceptualization, the clinician may suggest individual, couple, or mixed formats (e.g., individual sessions with one partner and occasional consultations as a couple). The choice of therapy format is highly dependent upon the sexual dysfunction type, its presence in one or both partners, and the patients' objectives.

Assessment goals

The first assessment goal is to evaluate the presence of sexual dysfunctions according to the DSM-V and biopsychosocial model (Brotto et al., 2016). The second assessment goal is to exclude any biological causes behind the patient's sexual difficulties. An interdisciplinary approach (e.g., between gynecologist, physical therapist, psychologist) is therefore essential (Chew et al., 2020). The third goal is to determine the patients' expectations for therapy. Remission is often the main priority. However, therapy may also improve other aspects of the marital and sexual relationship, such as communication and relationship satisfaction, without necessary resolving the sexual difficulties. Furthermore, the absence of sexual dysfunctions does not guarantee that patients are sexually satisfied (Brotto et al., 2016). Some patients report sexual distress without meeting the diagnostic criteria for FSD, and they may therefore still benefit from sex therapy. Finally, the fourth goal is to provide information on sexual functioning, not only to reassure patients, but also to formulate explanatory hypotheses for the difficulties encountered. Psychoeducation promotes a better understanding of the predisposing, precipitating, and maintaining factors of sexual difficulties.

TABLE 2 Biopsychosocial factors required for female sexual dysfunction assessment.

Biological factors
Medical and surgical conditions: lower urinary tract problems, endometriosis, uterine fibroids, breast and ovarian cancer, spinal cord injury, hormonal changes, and medications (e.g., antidepressants and antianxiety)

Sociocultural factors
Age, education, ethnicity, income

Lifestyle factors
Unhealthy diet, lack of exercise, smoking, alcohol and drug abuse

Life stressors
Infertility, postpartum period, aging and menopause

Interpersonal and relational factors
Intimacy, lower relationship satisfaction, partner sexual function, poor sexual communication, partner illness, partner discrepancies in level of sexual desire between partner, negative partner response, partner violence

Psychological factors
Developmental factors
Problematic attachment, childhood abuse or neglect, history of sexual abuse and trauma, early sexual experiences (e.g., masturbation), personality traits (e.g., neuroticism, introversion and low positive traits affects), negative cognitive schema, sexual beliefs (e.g., sexual myths, body image beliefs)

Psychological processing factors
Causal attribution to sexual problems, performance anxiety, efficacy expectations, cognitive distraction, content of negative thoughts during sexual activity, and negative emotions during sexual activity (e.g., Anxiety and low mood, and anxious apprehension of sexual activity)

Comorbid mental health issues
Stress, depression, physical illness (e.g., multiple sclerosis, diabetes, breast cancer survivors), anxiety disorders (e.g., social phobia, panic disorder), posttraumatic stress disorder, substance use disorder, and medication

The table shows the main factors which must be accounted for during assessment, primarily based on Brotto et al. (2016) and Khajehei et al. (2015).

Assessment interview

The first assessment sessions presented in the current chapter are based upon a biopsychosocial approach and Brotto and colleagues' (2016) systematic review, with questions evaluating biological, psychological, interpersonal, and sociocultural factors, as well as life events and stressors. The clinician must question lifestyle choices such as an unhealthy diet, a lack of exercise, smoking, and alcohol and drug abuse, in addition to other factors that may influence sexual functioning, namely negative body image, negative attitude toward sex, types of sexual practices (Khajehei, Doherty, & Tilley, 2015), sexual self-esteem, and sexual satisfaction (Brotto et al., 2016). Table 2 provides an exhaustive overview of the different factors to consider for a biopsychosocial assessment of FSD. Sexual dysfunctions are assessed both diachronically (i.e., understanding the onset and evolution of the difficulties) and synchronically (i.e., evaluating the interacting effects of cognitive, emotional, behavioral, and environmental dimensions of the sexual difficulties). FSDs should be consistently evaluated at the beginning, middle, and end of therapy to evaluate treatment efficacy.

First session

The American College of Obstetricians and Gynecologists (2017) recommends for clinicians to gather information concerning physical health and sexual history, after which a DSM-5 diagnosis (APA, 2013) may be proposed. Beginning with medical questions will help put patients at ease before addressing details that are more intimate. The clinician will then address interpersonal and relational factors, as well as life stressors. Finally, the clinician explores psychological factors. All assessment questions can be found in Table 3 and can be sequentially discussed with the client in the presented order.

Second and third session

The second session hones in on the maintaining psychological factors responsible for the patient's sexual difficulties (Table 3). The clinician should carry out the diachronic evaluation of the patient's affective and sexual history (e.g., education, sexual experiences, traumas, etc.), and sexual functioning. Should the patient be consulting with his or her partner, the clinician should hear each individual independently, oftentimes during the third session. Moreover, this individual

TABLE 3 Assessment interview questions.

Assessment session number	Important questions
1	**Biological, sociocultural, and lifestyle factors** • What is your date of birth? • Do you currently work? • Do you have any disorders, even if they are not directly linked to the current consultation? • Are you taking any medications? • Have you had any operations in the past? • Do you smoke? Do you drink alcohol? • Do you exercise? **Interpersonal and relational factors** and **life stressors** • Are you currently in a relationship? Are you married? Length of relationship? Cohabitation? • Do you have children? Do you have a desire to become pregnant? • How is the marital relationship with your partner outside of sex? • Are there any stressful events that you relate to your sexual difficulties? **Psychological factors** • Could I ask you some more intimate and sexual questions? • Could you explain to me how it is going at the intimate and sexual level? • When was the last time you had sex? How did it go? • Are you still having sex? How often? • How long have you been experiencing these sexual difficulties? In what situation(s)? What was it like before? • What motivated you to come for a consultation now? • What are your expectations from the consultation?
2	**Affective and sexual history** • At what age did you start your emotional and sexual life? • What was your first sexual experience like? • Have you ever used self-stimulation/masturbation? • What kind of education did you receive about intimacy and sex? Do you think this might have had an impact on the development of your emotional and sexual life? • Have you had any negative experiences related to sexuality? • Have you had other partners? How did it go sexually? **Current sexual functioning** • What is the atmosphere in which you have sex? Is it conducive to the development of intimacy and sexuality? • Do you practice foreplay (emotional, erotic, and sexual touching)? Do you find it stimulating/exciting? • At the beginning of sexual activity, do you feel sexual desire? Is it spontaneous or does it come on gradually during sexual stimulation? • Do you lubricate? • Is there penetration during sexual activity? Is penetration stimulating/exciting or painful? • Do you ever reach orgasm? • How do you feel physically and emotionally after sexual activities?
3 or 4	**Sexual dysfunctions** • Could you explain to me how it went the last time you had sexual difficulties?

The table shows the specific and sequential questions that should be addressed by the clinicians during assessment. It is recommended to follow the order and the respective sessions to which these questions belong.

session allows the clinician to meet the patient outside the marital dynamics. However, it is essential to note that the clinician cannot conceal secrets that may hinder therapeutic goals (e.g., infidelity). It is therefore essential to inform both patients that anything said in the individual interview could be discussed during the couple consultation.

Next, the clinician will assess the patient's current sexual functioning by asking specific questions about sexual desire, lubrication, arousal, and orgasm, in order to highlight possible sexual dysfunction that is causing sexual distress (American College of Obstetricians and Gynecologist, 2017). It is also necessary to assess whether there is sexual pain related to sexual activities (before/during/after). To facilitate the discussion, the clinician can draw the Masters and Johnson (1966) model (see Fig. 1) to target the sexual difficulties the patient has experienced over the past four weeks. At the end of the interview, the clinician can assess the FSD using standardized and validated questionnaires (see the Validated Questionnaires section for further information). The clinician should also identify the presence of FSD based on the DSM-V (APA, 2013) diagnostic criteria (see Table 1 for the full specifications). In order to synchronically consider all dimensions of the sexual difficulties (emotional, cognitive, behavioral, and interpersonal), patients should be invited to answer, in session or as homework, the questions presented in Fig. 3. The synchronic and diachronic analyses will enable the identification of any reinforcing mechanisms.

The third session will allow the clinician to give a first feedback regarding the sexual dysfunctions encountered and to analyze the precipitating and maintaining mechanisms. If the individuals are consulting as a couple, this will be done during the fourth session. If they are consulting as a couple, each partner will have the opportunity to explain their own difficulties.

Fourth or fifth sessions

The fourth session is designed to present the case conceptualization and to agree upon both therapeutic goals and format. If the patient consults with their partner, this will occur during the fifth session. The clinician will begin with presenting the history and progression of the sexual difficulties (i.e., the result of the diachronic analysis), highlighting the predisposing factors (e.g., sexual myths, childhood sexual abuse), precipitating factors (e.g., postpartum period), and maintaining factors (e.g., negative partner response, avoidance strategies). The clinician will then present the synchronic analysis (see Fig. 3) and formulate hypotheses regarding the sexual difficulties encountered. Conceptualization will show the patient that the clinician understands the difficulties, which could strengthen their therapeutic alliance. Understanding one's sexual functioning and underlying mechanisms, which strengthen motivation in the individuals and assigns them a more active role in therapy.

More importantly, case conceptualization will guide therapeutic goals and the corresponding choice of intervention. Individual, couple, or mixed-formats will be selected based on the clients' best interests. To reinforce a biopsychosocial

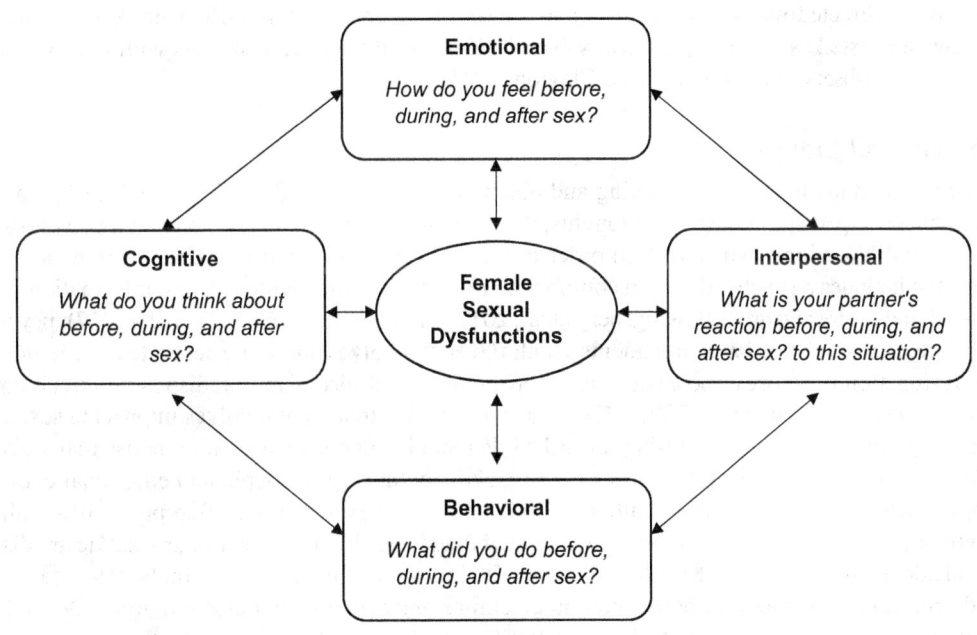

FIG. 3 Synchronic analysis of sexual dysfunctions. Figure representing the cognitive, emotional, behavioral, and interpersonal questions that should be systematically investigated with patients.

treatment approach, the clinician may also suggest for the patient consult other relevant professionals (e.g., gynecologist, psychologist, or urologist).

Validated questionnaires

Questionnaires allow for the objective and standardized assessment of the sexual dysfunctions. The female sexual function index (FSFI, Rosen et al., 2000) and the Female Sexual Distress Scale-revised (FSDS-R, DeRogatis, Clayton, Lewis-D'Agostino, Wunderlich, & Fu, 2008) are both self-report questionnaires and most widely used among clinicians and researchers. These should be administered during assessment, as well as during mid- and post-treatment evaluations.

The female sexual function index (FSFI)

The FSFI (Rosen et al., 2000) assesses female sexual functioning along six dimensions: desire, arousal, lubrication, orgasm, sexual satisfaction, and pain. It is composed of 19 items, assessed on a 6-point Likert scale. The total score ranges from 0 to 36, with a score of 26.55 or lower indicating FSD presence (Wiegel, Meston, & Rosen, 2005). Given this tool's high internal consistency (i.e., Cronbach's alpha of 0.86), it represents a reliable measure for assessing FSDs.

The Female Sexual Distress Scale—revised (FSDS-R)

The FSDS-R (DeRogatis et al., 2008) assesses personal sexual distress. This questionnaire is composed of 13 items assessed on a 5-point Likert scale. The total score ranges from 0 to 52, with a score of 11 or higher specifying the presence of sexual distress. Given the tool's high internal consistency (Cronbach's alpha of 0.86), it represents a reliable measure of FSD.

Treating female sexual dysfunctions

FSDs can be either treated medically or psychologically. In the current section, we only review the latter approach. It is important to note that embarrassment, stigma, lack of options for evidence-based treatments, distance, and cost all represent barriers for women to seek psychological treatment (Stephenson, Zippan, & Brotto, 2021). Currently, no single treatment has been established as the prime treatment for FSDs (for a recent systematic review, see Weinberger, Houman, Caron, & Anger, 2019). A full review of available treatments has recently been summarized in a systematic review by Marchand (2020). Among these, cognitive-behavioral therapy (CBT) is the only approach having shown sufficient evidence for both the treatment and its underlying theory for both female and male sexual dysfunctions (Emanu, Avildsen, & Nelson, 2018; ter Kuile, Both, & van Lankveld, 2010; Wheeler & Guntupalli, 2020). Difficult sex-related cognitions, behaviors, and emotions should therefore be addressed using CBT in combination with other techniques (Laan, Rellini, & Barnes, 2013), which include sensate focus, directed masturbation, systematic desensitization, and pelvic-floor exercises. Mindfulness-based CBT has also been suggested as a recent empirically-based treatment, though clinical trials with rigorous designs are still required to confirm the observed effects (Pyke & Clayton, 2015).

Cognitive-behavioral therapy (CBT)

CBT aims to identify and modify the precipitating and maintaining factors of FSDs. These include the individuals' cognitions (e.g., unrealistic expectations, negative thoughts, dysfunctional schemas) and behaviors (e.g., exposure exercises to target experiential and behavioral avoidance), in order to reduce fear of penetration, pain, or sex-related anxiety (Kane et al., 2019). It often includes a psychoeducation component, providing women with accurate information regarding erotic stimulation, and sexual desire and arousal (Kingsberg et al., 2017). As stated by Kingsberg et al. (2017), psychoeducation is evidently at the core of all evidence-based treatments for all FSDs. It involves providing accurate information regarding the female sexual dysfunction diagnoses, increasing sexual awareness, and discussing predisposing, precipitating, perpetuating, and protective factors (Brotto et al., 2008). The next steps of CBT treatment involves improving sexual skills in both the woman and her partner, as well as increasing reward while reducing punishments within nonsexual aspects of the relationship (ter Kuile et al., 2010). CBT also includes training skills pertaining to couple and emotional communication, as well as developing sensual fantasy. As per traditional CBT, homework exercises are often prescribed within the couple.

Specific components are modified according to the treated diagnoses. In the context of arousal/desire disorder, orgasm consistency is an added component (ter Kuile et al., 2010). It aims to educate the couple on how to enable the woman to reach orgasm during sexual interactions, before both male climax and actual intercourse initiation. Depending on the relevance, it also incorporates coital alignment techniques, to allow direct clitoral stimulation by the penis during sexual intercourse. CBT for orgasmic disorder incorporates sensate focus, directed masturbation, systematic desensitization, and pelvic-floor exercises, which will be discussed further below. Treatment adaptations for sexual pain include targeting not only pain, muscle tightness, and vaginal penetration, but also anxiety (Al-Abbadey, Liossi, Curran, Schoth, & Graham, 2016). Beyond its original format, CBT also exists under group (Bergeron, Khalifé, Dupuis, & McDuff, 2016), couple (Corsini-Munt, Bergeron, Rosen, Mayrand, & Delisle, 2014), bibliotherapy (Van Lankveld et al., 2006), and online formats (Stephenson et al., 2021).

A more recent and promising therapeutic endeavor is mindfulness-based CBT. In the context of FSDs, mindfulness emphasizes the acceptance of sex-related thoughts and emotions. Based on Brotto and colleagues' work (e.g., Paterson, Handy, & Brotto, 2017), an 8-session group treatment option has been developed. Similarly to CBT, homework exercises include the monitoring of sexual beliefs, desire/arousal complaints, and body image through appropriate worksheets. Individuals are also expected to complete daily mindfulness exercises. Moreover, an online self-guided mindfulness treatment program has also been proposed (Adam, De Sutter, Day, & Grimm, 2020). The intervention consists of seven videos, which proposes increasing mindfulness exercises and combined elements of psychoeducation, diaphragmatic breathing, pelvic movement exercises, foreplay exercises, erotic imagination, directed masturbation, cognitive restructuring, and acceptance of one's womanhood (for the full program, see Adam et al., 2020).

Sensate focus

Sensate focus is considered a validated and important first-line technique for sexual dysfunctions (Laan et al., 2013), with a specific interest for interest/arousal disorder (ter Kuile et al., 2010). Originally coined by Masters and Johnson (1966), sensate focus therapy is a couple-based approach, which consists of gradual touching exercises with the aim of reducing avoidance and sexual-related anxiety, and improving communication within the relationship. Typically, couples will be invited to move from with nongenital touching, to genital touching and, ultimately, intercourse (Kingsberg et al., 2017). Individuals are invited to focus on the sexual acts in which they can directly control the locus of attention, including the voluntary behavioral action of redirecting their attention toward specific physical sensations rather than distracting (e.g., focusing on the partner's experience) or spectatoring. The latter refers to "the tendency to evaluate oneself from a third-person perspective during sex" (Laan et al., 2013). Ultimately, this will help reduce expectations and anxiety related to sexual activities. A full description of the step-by-step treatment can be found in Weiner and Avery-Clark (2014). At each steps, the authors emphasize the need for the partners to hone in on their sensations.

Directed masturbation

Directed masturbation training, incorporating elements of both CBT and mindfulness, has received the most empirical support for orgasmic disorder (Laan et al., 2013). Based on operant and classical conditioning, the objective of the technique is to abandon nonconstructive behaviors that impede on reaching orgasm, with more useful behavior choices (Both & Laan, 2004). This exposure exercise involves progressive, stepwise exposure exercises, beginning with visual and tactile exploration of one's body to genital stimulation (Marchand, 2020). The program incorporates role-playing orgasm, imagining sexual excitement increases, developing sexual fantasy, discovering masturbation techniques, and using vibrators. Similarly to CBT and sensate focus, this technique aims to reduce anxiety and spectatoring, increasing attentional to sexual cues, and challenging sex-related beliefs. A full description of the treatment can be found in Both and Laan's (2004) chapter. Efficacy rates in reaching orgasm following directed masturbation range from 60% to 90% (Both & Laan, 2004; Laan et al., 2013; for a review, see Marchand, 2020). It is especially recommended for women with particular aversion of touching her genitals.

Systematic desensitization

Based on exposure therapy, systematic desensitization aims to reduce symptomatology linked to the sexual difficulty whilst targeting any sex-related feelings of anxiety, shame, or guilt associated with engaging in social activities (Both & Laan, 2004). In the first sessions, the link between orgasm difficulties and anxiety in sexual situations is explained, exposing the rationale for exposure exercises. Individuals work with the clinician to create a hierarchy of feared anxiety-provoking sex-related situations, which can later be arranged according to increasing intimacy. Gradually, patients will expose themselves to the anxiety created by each sexual situation. One will begin with less fear-provoking situations, increasing difficulty with each session. These exposure exercises can be achieved both through imagination and in vivo (live exercises). Exercises are repeated so long as they generate significant distress for the woman. Oftentimes, the exposure exercises are accompanied by muscle relaxation techniques, as well as assertiveness and self-affirmation training. Patients will be encouraged to practice these techniques as homework assignments.

The systematic review by Marchand (2020) highlights that efficacy ranges from 10% to 56% for participants with anorgasmia, with higher success achieved in combination with anxiety medication or inclusion of the partner in the therapy. Recent recommendations warn against the use of systematic desensitization alone as a first line treatment (Laan et al., 2013; Marchand, 2020). While sex-related anxiety may be significantly reduced by the technique, a lack of carry-over effects to more general sexual function has been observed.

Pelvic-floor exercises

Besides pharmacological agents for the treatment of pain in sexual dysfunctions, expert consensus guidelines recommend CBT and pelvic floor physical therapy as a first-line treatment (e.g., Rosen, Dawson, Brooks, & Kellogg-Spadt, 2019), with an emphasis on their cumulative benefits when combined (Rosen et al., 2019). Pelvic floor exercises involve the prescription of specific exercises to relax the pelvic floor muscle, restore its proper function, and retrain the pain receptors. Techniques include stretching, massage, and myofascial trigger points (Rosenbaum & Owens, 2008). This can be optimally treated by a physical therapist specially trained in providing treatment for this sexual dysfunction. The exercises should increase blood circulation and decrease myalgia, improving therefore muscle awareness and muscle relaxation (Padoa, McLean, Morin, & Vandyken, 2021). Evidence for these exercises is multifold, with up to 71% to 80% of women expressing significant decreases in pain (for a review, see Padoa et al., 2021). Further research is warranted to distinguish the superiority of pelvic floor rehabilitation compared to CBT, as both are often combined.

Mini-dictionary of terms

- *Sexual health*. It is a state of physical, emotional, mental and social wellbeing in relation to sexuality. It is not merely the absence of disease, dysfunction or infirmity. Sexual health requires a positive and respectful approach to sexuality and sexual relationships, as well as the possibility of having pleasurable and safe sexual experiences, free of coercion, discrimination and violence.
- *Subjective arousal*. Subjective evaluation of one's sexual experience.
- *Genital arousal*. The woman's own physical perception of her bodily arousal.
- *Vasocongestion of the genitals*. The increase in blood flow to the genitals (e.g., clitoris, labia minora), which will then lead to vaginal lubrication.
- *Negative sexual cognitive schemas*. Cognitive generalities consisting of a set of ideas that people cultivate about sexuality, themselves, as well as how they perceive themselves as a sexual person.

Key facts of female sexual dysfunctions

- Sexual difficulties are encountered by approximately 43%–44% of women and 30% of men.
- Only 25% of women and men who suffer from sexual dysfunctions will dare to consult.
- Female sexual experience is very complex. Clinicians are therefore advised to thoroughly assess each patient's own experience.
- To optimize patient care, assessment should be performed during routine examinations (gynecologist, physician, psychologist).
- While there is not one established empirical treatment for all FSDs, CBT is the only intervention for which both the theory and practice have sufficient evidence.

Applications to other areas

In this chapter, we presented the initial assessment sessions and treatments for FSD. However, partners' sexual functioning is interdependent (Brotto et al., 2016), with women with FSDs being three times more likely to be in relationships with men who suffer from sexual dysfunction themselves (Chew et al., 2020). Male sexual dysfunctions (MSDs) are encountered by approximately 30% of men (Laumann et al., 1999). The DSM-V (APA, 2013) specifies erectile disorder, male hypoactive sexual desire disorder, premature (early) ejaculation, and delayed ejaculation. Masters and Johnson's (1966) sexual response model and Barlow's (1986) cognitive interference model presented earlier are also used to understand MSDs, as men with MSDs more often focus on performance concerns and nonerotic stimuli compared to men without MSDs. This decreases sexual arousal and increases anxious apprehension of sexual activities. Assessment is also based on a biopsychosocial model (Brotto et al., 2016) and the initial assessment sessions presented in this chapter could be applied to MSDs.

As previously mentioned, CBT is the only approach having shown sufficient evidence for both the treatment and its underlying mechanisms for both female and MSDs (Emanu et al., 2018). Other treatments used for males suffering with sexual dysfunctions include sex therapy (e.g., sensate focus), behavioral approaches (e.g., systematic desensitization or pelvic floor muscle rehabilitation), and psychoeducational therapy (Jaderek & Lew-Starowicz, 2019). However, the choice of a given therapy must rest on empirical findings, and the lack thereof in the domain of sexual dysfunctions means clinicians must be aware of such limits when implementing these interventions.

Summary points

- According to the DSM-5 (APA, 2013), female sexual dysfunctions include female sexual interest/arousal disorder, female orgasmic disorder, genito-pelvic pain/penetration disorder, substance/medication-induced sexual dysfunction, and other specified/unspecified sexual dysfunction.
- Assessment must be conducted in accordance with a biopsychosocial and interdisciplinary approach.
- Both pharmacological and psychological approaches have been devised for treating female sexual dysfunctions.
- Cognitive-behavioral therapy is the recommended first-line of treatment for the majority of the female sexual dysfunctions.
- Other adjunct sexual therapy techniques can be implemented, including sensate focus, directed masturbation, systematic desensitization, and pelvic-floor exercises.

References

Adam, F., Day, J., De Sutter, P., & Brasseur, C. (2017). Analyse processuelle des facteurs cognitifs du trouble de l'orgasme féminin. *Sexologies, 26*(3), 153–160.

Adam, F., De Sutter, P., Day, J., & Grimm, E. (2020). A randomized study comparing video-based mindfulness-based cognitive therapy with video-based traditional cognitive behavioral therapy in a sample of women struggling to achieve orgasm. *The Journal of Sexual Medicine, 17*(2), 312–324.

Al-Abbadey, M., Liossi, C., Curran, N., Schoth, D. E., & Graham, C. A. (2016). Treatment of female sexual pain disorders: A systematic review. *Journal of Sex & Marital Therapy, 42*(2), 99–142.

American College of Obstetricians and Gynecologist. (2017). Committee opinion no 706: Sexual health. *Obstetrics & Gynecology, 130*(1), e42.

Annon, J. S. (1976). The PLISSIT model: A proposed conceptual scheme for the behavioral treatment of sexual problems. *Journal of Sex Education Therapy, 2*(1), 1–15.

American Psychiatric AssociationPlaceholder Text (2000). Quick reference to the diagnostic criteria from DSM-IV-TR. APA, Washington, DC.

APA. (2013). *Diagnostic and statistical manual of mental disorders: DSM-5*. Vol. 10. Washington, DC: American Psychiatric Association.

Bachmann, G. (2006). Female sexuality and sexual dysfunction: Are we stuck on the learning curve? *The Journal of Sexual Medicine, 3*(4), 639–645.

Barlow, D. H. (1986). Causes of sexual dysfunction: The role of anxiety and cognitive interference. *Journal of Consulting and Clinical Psychology, 54*(2), 140.

Basson, R. (2005). Women's sexual dysfunction: Revised and expanded definitions. *Canadian Medical Association Journal, 172*(10), 1327–1333.

Basson, R., Berman, J., Burnett, A., Derogatis, L., Ferguson, D., Fourcroy, J., et al. (2000). Report of the international consensus development conference on female sexual dysfunction: Definitions and classifications. *The Journal of Urology, 163*(3), 888–893.

Basson, R., Wierman, M. E., Van Lankveld, J., & Brotto, L. (2010). Summary of the recommendations on sexual dysfunctions in women. *The Journal of Sexual Medicine, 7*(1), 314–326.

Bergeron, S., Binik, Y. M., Khalifé, S., Pagidas, K., Glazer, H. I., Meana, M., et al. (2001). A randomized comparison of group cognitive-behavioral therapy, surface electromyographic biofeedback, and vestibulectomy in the treatment of dyspareunia resulting from vulvar vestibulitis. *Pain, 91*(3), 297–306.

Bergeron, S., Corsini-Munt, S., Aerts, L., Rancourt, K., & Rosen, N. O. (2015). Female sexual pain disorders: A review of the literature on etiology and treatment. *Current Sexual Health Reports, 7*(3), 159–169.

Bergeron, S., Khalifé, S., Dupuis, M.-J., & McDuff, P. (2016). A randomized clinical trial comparing group cognitive-behavioral therapy and a topical steroid for women with dyspareunia. *Journal of Consulting and Clinical Psychology, 84*(3), 259.

Both, S., & Laan, E. (2004). Directed masturbation: A treatment of female orgasmic disorder. In *Cognitive behavior therapy: Applying empirically supported techniques in your practice* (p. 144). Wiley.

Brotto, L. A., Heiman, J. R., Goff, B., Greer, B., Lentz, G. M., Swisher, E., et al. (2008). A psychoeducational intervention for sexual dysfunction in women with gynecologic cancer. *Archives of Sexual Behavior, 37*(2), 317–329.

Brotto, L., Atallah, S., Johnson-Agbakwu, C., Rosenbaum, T., Abdo, C., Byers, E. S., et al. (2016). Psychological and interpersonal dimensions of sexual function and dysfunction. *The Journal of Sexual Medicine, 13*(4), 538–571.

Chew, P. Y., Choy, C. L., Bin Sidi, H., Abdullah, N., Roos, N. A. C., Sahimi, H. M. S., et al. (2020). The Association between female sexual dysfunction and sexual dysfunction in the male partner: A systematic review and meta-analysis. *The Journal of Sexual Medicine, 18*(1), 99–112.

Clayton, A. H., & Juarez, E. M. V. (2019). Female sexual dysfunction. *Medical Clinics of North America, 103*(4), 681–698.

Conforti, C. (2017). Genito-pelvic pain/penetration disorder (GPPPD): An overview of current terminology, etiology, and treatment. *University of Ottawa Journal of Medicine, 7*(2), 48–53.

Corsini-Munt, S., Bergeron, S., Rosen, N. O., Mayrand, M.-H., & Delisle, I. (2014). Feasibility and preliminary effectiveness of a novel cognitive-behavioral couple therapy for provoked vestibulodynia: A pilot study. *The Journal of Sexual Medicine, 11*(10), 2515–2527.

DeRogatis, L., Clayton, A., Lewis-D'Agostino, D., Wunderlich, G., & Fu, Y. (2008). Validation of the female sexual distress scale-revised for assessing distress in women with hypoactive sexual desire disorder. *The Journal of Sexual Medicine, 5*(2), 357–364.

Dias-Amaral, A., & Marques-Pinto, A. (2018). Female genito-pelvic pain/penetration disorder: Review of the related factors and overall approach. *Revista Brasileira de Ginecologia e Obstetrícia, 40*(12), 787–793.

Dove, N. L., & Wiederman, M. W. (2000). Cognitive distraction and women's sexual functioning. *Journal of Sex & Marital Therapy, 26*(1), 67–78.

Emanu, J. C., Avildsen, I., & Nelson, C. J. (2018). Psychotherapeutic treatments for male and female sexual dysfunction disorders. In *Evidence-Based Psychotherapy: The State of the Science and Practice* (pp. 253–270). Wiley.

Guthrie, J. R., Dennerstein, L., Taffe, J. R., Lehert, P., & Burger, H. G. (2004). The menopausal transition: A 9-year prospective population-based study. The Melbourne Women's midlife health project. *Climacteric, 7*(4), 375–389.

Jaderek, I., & Lew-Starowicz, M. (2019). A systematic review on mindfulness meditation-based interventions for sexual dysfunctions. *The Journal of Sexual Medicine, 16*(10), 1581–1596.

Kane, L., Dawson, S. J., Shaughnessy, K., Reissing, E. D., Ouimet, A. J., & Ashbaugh, A. R. (2019). A review of experimental research on anxiety and sexual arousal: Implications for the treatment of sexual dysfunction using cognitive behavioral therapy. *Journal of Experimental Psychopathology, 10*(2).

Kaplan, H. S. (1979). *Disorders of sexual desire and other new concepts and techniques in sex therapy*. New York: Brunner/Hazel Publications.

Khajehei, M., Doherty, M., & Tilley, P. M. (2015). An update on sexual function and dysfunction in women. *Archives of Women's Mental Health, 18*(3), 423–433.

Kingsberg, S. A., Althof, S., Simon, J. A., Bradford, A., Bitzer, J., Carvalho, J., et al. (2017). Female sexual dysfunction—Medical and psychological treatments, committee 14. *The Journal of Sexual Medicine, 14*(12), 1463–1491.

Laan, E., Rellini, A. H., & Barnes, T. (2013). Standard operating procedures for female orgasmic disorder: Consensus of the International Society for Sexual Medicine. *The Journal of Sexual Medicine, 10*(1), 74–82.

Laumann, E. O., Glasser, D. B., Neves, R. C. S., & Moreira, E. D. (2009). A population-based survey of sexual activity, sexual problems and associated help-seeking behavior patterns in mature adults in the United States of America. *International Journal of Impotence Research, 21*(3), 171–178.

Laumann, E. O., Paik, A., & Rosen, R. C. (1999). Sexual dysfunction in the United States: Prevalence and predictors. *JAMA, 281*(6), 537–544.

Marchand, E. (2020). Psychological and behavioral treatment of female orgasmic disorder. *Sexual Medicine Reviews, 9*(2), 194–211.

Masters, W. H., & Johnson, V. E. (1966). *Human sexual response*. Boston: Little, Brown and Company.

McCormick, C. M., Lewis, E., Somley, B., & Kahan, T. A. (2007). Individual differences in cortisol levels and performance on a test of executive function in men and women. *Physiology & Behavior, 91*(1), 87–94.

Meston, C. M., & Stanton, A. M. (2017). Evaluation of female sexual interest/arousal disorder. In *The textbook of clinical sexual medicine* (pp. 155–163). Springer.

Nappi, R. E., Cucinella, L., Martella, S., Rossi, M., Tiranini, L., & Martini, E. (2016). Female sexual dysfunction (FSD): Prevalence and impact on quality of life (QoL). *Maturitas, 94*, 87–91.

Nobre, P., Gouveia, J. P., & Gomes, F. A. (2003). Sexual dysfunctional beliefs questionnaire: An instrument to assess sexual dysfunctional beliefs as vulnerability factors to sexual problems. *Sexual and Relationship Therapy, 18*(2), 171–204.

Nobre, P. J., & Pinto-Gouveia, J. (2006). Emotions during sexual activity: Differences between sexually functional and dysfunctional men and women. *Archives of Sexual Behavior, 35*(4), 491–499.

Nobre, P. J., & Pinto-Gouveia, J. (2008). Differences in automatic thoughts presented during sexual activity between sexually functional and dysfunctional men and women. *Cognitive Therapy and Research, 32*(1), 37–49.

Nowosielski, K., Wróbel, B., & Kowalczyk, R. (2016). Women's endorsement of models of sexual response: Correlates and predictors. *Archives of Sexual Behavior, 45*(2), 291–302.

Padoa, A., McLean, L., Morin, M., & Vandyken, C. (2021). The overactive pelvic floor (OPF) and sexual dysfunction. Part 2. Evaluation and treatment of sexual dysfunction in OPF patients. *Sexual Medicine Reviews, 9*(1), 76–92.

Parish, S. J., Goldstein, A. T., Goldstein, S. W., Goldstein, I., Pfaus, J., Clayton, A. H., et al. (2016). Toward a more evidence-based nosology and nomenclature for female sexual dysfunctions—Part II. *The Journal of Sexual Medicine, 13*(12), 1888–1906.

Paterson, L. Q., Handy, A. B., & Brotto, L. A. (2017). A pilot study of eight-session mindfulness-based cognitive therapy adapted for women's sexual interest/arousal disorder. *The Journal of Sex Research, 54*(7), 850–861.

Pyke, R. E., & Clayton, A. H. (2015). Psychological treatment trials for hypoactive sexual desire disorder: A sexual medicine critique and perspective. *The Journal of Sexual Medicine, 12*(12), 2451–2458.

Rosen, N. O., Dawson, S. J., Brooks, M., & Kellogg-Spadt, S. (2019). Treatment of vulvodynia: Pharmacological and non-pharmacological approaches. *Drugs, 79*(5), 483–493.

Rosen, R., Brown, J., Heiman, S., Leiblum, C., Meston, R., Shabsigh, D., et al. (2000). The female sexual function index (FSFI): A multidimensional self-report instrument for the assessment of female sexual function. *Journal of Sex & Marital Therapy, 26*(2), 191–208.

Rosenbaum, T. Y., & Owens, A. (2008). Continuing medical education: The role of pelvic floor physical therapy in the treatment of pelvic and genital pain-related sexual dysfunction (CME). *The Journal of Sexual Medicine, 5*(3), 513–523.

Shifren, J. L., Monz, B. U., Russo, P. A., Segreti, A., & Johannes, C. B. (2008). Sexual problems and distress in United States women: Prevalence and correlates. *Obstetrics & Gynecology*, *112*(5), 970–978.

Stephenson, K. R., Zippan, N., & Brotto, L. A. (2021). Feasibility of a cognitive behavioral online intervention for women with sexual interest/arousal disorder. *Journal of Clinical Psychology*, *77*, 1877–1893.

Tavares, I. M., Moura, C. V., & Nobre, P. J. (2020). The role of cognitive processing factors in sexual function and dysfunction in women and men: A systematic review. *Sexual Medicine Reviews*, *8*(3), 403–430.

ter Kuile, M. M., Both, S., & van Lankveld, J. J. (2010). Cognitive behavioral therapy for sexual dysfunctions in women. *Psychiatric Clinics*, *33*(3), 595–610.

Van Lankveld, J. J., Granot, M., Schultz, W. C. W., Binik, Y. M., Wesselmann, U., Pukall, C. F., et al. (2010). Women's sexual pain disorders. *The Journal of Sexual Medicine*, *7*(1), 615–631.

Van Lankveld, J. J., ter Kuile, M. M., de Groot, H. E., Melles, R., Nefs, J., & Zandbergen, M. (2006). Cognitive-behavioral therapy for women with lifelong vaginismus: A randomized waiting-list controlled trial of efficacy. *Journal of Consulting and Clinical Psychology*, *74*(1), 168–178.

Weinberger, J. M., Houman, J., Caron, A. T., & Anger, J. (2019). Female sexual dysfunction: A systematic review of outcomes across various treatment modalities. *Sexual Medicine Reviews*, *7*(2), 223–250.

Weiner, L., & Avery-Clark, C. (2014). Sensate focus: Clarifying the Masters and Johnson's model. *Sexual and Relationship Therapy*, *29*(3), 307–319.

Wheeler, L. J., & Guntupalli, S. R. (2020). Female sexual dysfunction: Pharmacologic and therapeutic interventions. *Obstetrics & Gynecology*, *136*(1), 174–186.

Wiegel, M., Meston, C., & Rosen, R. (2005). The female sexual function index (FSFI): Cross-validation and development of clinical cutoff scores. *Journal of Sex & Marital Therapy*, *31*(1), 1–20.

Wiegel, M., Scepkowski, L. A., & Barlow, D. H. (2007). Cognitive-affective processes in sexual arousal and sexual dysfunction. In *Kinsey Institute Conference, 1st, Jul, 2003, Bloomington, IN, US; This work was presented at the Aforementioned Conference*.

Wincze, J. P., & Weisberg, R. B. (2015). *Sexual dysfunction: A guide for assessment and treatment*. Guilford Publications.

World Health Organization. (2021). *Sexual health throughout life: Definition*. Retrieved 13 July 2021, from: https://www.euro.who.int/en/health-topics/Life-stages/sexual-and-reproductive-health/news/news/2011/06/sexual-health-throughout-life/definition.

Chapter 14

Cognitive behavior therapy for insomnia in adults

Susmita Halder[a] and Akash Kumar Mahato[b]
[a]Department of Psychology, St. Xavier's University, Kolkata, India [b]Amity Institute of Behavioural Health and Allied Sciences, Amity University Kolkata, Kolkata, West Bengal, India

Abbreviations

CBT cognitive behavior therapy
DSM diagnostic and statistical manual of mental disorders
ICD International Classification of Diseases
REM rapid eye movement

The last refuge of the insomniac is a sense of superiority to the sleeping world.

—Leonard Cohen

Introduction

Humans spend around one third of their life sleeping. It is an important automated process for maintenance of healthy functioning of the brain. The process of sleep is not merely a resting phase in a day, but an important bodily function aiding in rejuvenation of bodily systems and immune functions. Chronic deprivation of sleep is known to be associated with heightened risk of several disorders including diabetes, cardiovascular disease as well as mental health issues. Across cultures good sleep has always been considered as an indicator of one having peace of mind, being free of worries and good mental health. On the contrary, losing sleep is synonymous with being worrisome and being stressed.

How prevalent it is: Insomnia typically develops with growing age and corresponding increase in level of familial and occupation related stress and worries, decreasing bodily functions and physical and mental morbidities. Insomnia is seen in larger proportion in middle aged and older adults. It is not a thumb rule, but a significant proportion of population worldwide experience sleep disturbances with their growing age. The prevalence estimates of insomnia in general population vary from 10% to 33% (AASM, 2008; Bhaskar, Hemavathy, & Prasad, 2016; Panda et al., 2012) depending on type of insomnia and methodology adopted. By an estimate roughly 10% of adults do have chronic insomnia. There has been an increasing trend in reports of sleep disturbances and insomnia even in adolescents and younger adult population (Ford, Cunningham, Giles, & Croft, 2015; Pallesen, Sivertsen, Nordhus, & Bjorvatn, 2014) which is a point of concern. Individuals differ in terms of daily requirement of sleep hours, which vary across people depending primarily on their age, where toddlers and children typically require more hours of sleep compared to adults. Usually after the age of 18 years, healthy individuals do need an average of 7–8 h of sleep. Social and local environmental influences may bring some differences in these patterns but they are more or less uniform across the globe. Children, adolescents and young adults usually enjoy good sleep and are not bothered about sleep disturbances. Behavioral insomnia's though are reported in children with an estimated prevalence of 10%–30% (Burnham, Goodlin-Jones, Gaylor, & Anders, 2002; Carter, Hathaway, & Lettieri, 2014). Acute forms of insomnia could be common in younger adults too, but are mostly transient in nature.

The ICD-10 (World Health Organization, 1992) lists nonorganic insomnia under "nonorganic sleep disorders" as a condition of unsatisfactory quantity and/or quality of sleep, which persists for a considerable period of time, including difficulty falling asleep, difficulty staying asleep, or early final wakening. It specifies insomnia as a common symptom of several mental and physical disorders, and to be classified in addition to the basic disorder only if it dominates the clinical picture. However, it separated nonorganic insomnia with insomnia due to organic conditions listed in another chapter.

To make things clearer and less confusing, the ICD-11 has a separate chapter for sleep-wake disorders that includes all relevant sleep-related diagnoses. This is also in line with the changes in diagnostic criteria for insomnia in the Diagnostic Statistical Manual of Mental Disorders, version 5 (DSM-5).

Why targeting insomnia is important

Sleep disturbances, primarily insomnia has been considered indicator of an underlying or overt psychiatric morbidity, so much so that mental health professionals essentially include queries on sleep disturbances if any present in their clients. Enquiry items on sleep disturbance are part of gold standard screening tools, e.g., GHQ 12 (Goldberg & Williams, 1988) for mental disorders or psychological distress. They also have been incorporated in the official diagnostic criteria for many mental disorders, including major depression, generalized anxiety disorder and substance-related disorders. Survey studies have demonstrated significantly higher rate (40%) of psychiatric morbidities in individuals with insomnia compared to around 16% in those not having sleep complaints (Ford & Kamerow, 1989) and the figures are consistent with findings globally (Szelenberger & Soldatos, 2005).

Recent research have indicated a two way relationship of insomnia with major mental disorders where insomnia is not mere a co-morbid or secondary condition but also a precursor of major mental disorders like anxiety and depression (Johnson, Roth, & Breslau, 2006; Roth, Franklin, & Bramley, 2007). Thus it becomes imperative to intervene early in clients with insomnia to reduce risk of other psychiatric manifestation and improve treatment outcome.

Etiological factors of insomnia and its nonpharmacological management

It is important to know the etiological perspectives when making therapeutic plan for insomnia. Insomnia can be caused by organic as well as nonorganic psychosocial factors and a combination of both. Medical conditions like diabetes, chronic pain, cardiovascular disease, obesity are known risk factors for insomnia. Psychosocial and environmental risk factors predominantly include high presumptive stress (Morin, Rodrigue, & Ivers, 2003), presence of anxiety and depressive symptoms. Occupational stress is a known risk factor for insomnia. Certain occupations, e.g., those involved in shift duties in transportation, health services, and manufacturing are known for its proneness for sleep disturbances, but it is mostly related to sleep deprivation and frequently changing sleep schedule and in most cases it auto remits upon settling down of the occupational demand. Certain studies also suggest higher insomnia rates in individuals with conjugal difficulties, individuals with lesser domestic income, lacking social support and in unemployed youth (Dollander, 2002; Grandner et al., 2015; Kent, Uchino, Cribbet, Bowen, & Smith, 2015). A significant percentage of individuals with insomnia do engage in sleep hindering behavior before bedtime which is pretty much amenable with good sleep hygiene practices. Considering vivid manifestations, co morbidities and high concurrent prevalence of major psychiatric illnesses in insomnia, management of insomnia is almost never the management of insomnia alone.

Management of insomnia in past and present: Since ages, people have tried different remedies to get good sleep. Some of the remedies prescribed by physicians in medieval and early modern periods are still in practice which might be the prelude for present day sleep hygiene. Pharmacological management of sleep disturbance and insomnia has been in existence since early 1800; however modern day hypnotic drugs, the benzodiazepines were introduced in mid-20th century only. There have been regular refinement in these hypnotic and sedative drugs, however clients might hesitate to resort to medications, especially in cases of nonorganic insomnia due to certain level of stigma attached with "sleeping pills." Sleeping pills have typically been associated with risk of addiction, increased tolerance to the medications with time, unwanted side effects and even increased risk of mortality for long term users (Kripke, Langer, & Kline, 2012). Even though there is high prevalence of insomnia in individuals with anxiety and depression; individuals taking sleeping pills might be stigmatized of suffering from mental health issues irrespective of the real situation. Studies do suggest that prescription of hypnotics in primary care and consumption of these medicines in daily life are highly moralized issues where certain population look into taking sleeping pills as an unnatural interference into a natural state' and associated with loss of control and addiction (Venn & Arber, 2012). Similar themes are prevalent across different populations and patients are often ambivalent to use medications due to negative feelings associated with hypnotic medications (Gabe & Thorogood, 1986).

In view of these, nonpharmacological management do become the first choice of treatment for clients having mild to moderate level of sleep disturbances and insomnia. Within the array of nonpharmacological management of insomnia, both evidence based techniques as well as popular techniques in local culture are in practice. Use of specific herbs or massage before bedtime is common in several cultures. Culture specific use of acupressure, acupuncture, yoga and meditation has been documented in studies (Gooneratne, 2008). However, the evidence base of these practices is mixed and often these practices remain confined to specific country populations only. Use of CBT in insomnia has gradually gained popularity and gradually it has become one of the most evidence based treatment for insomnia (Edinger & Means, 2005; Wu, Appleman, Salazar, & Ong, 2015) and suggested as a first-line treatment option for insomnia to be considered by primary care providers (Mitchell, Gehrman, Perlis, & Umscheid, 2012).

Cognitive behavior therapy for insomnia (CBT—I) refers to combinations of behavioral techniques and conventional cognitive restructuring and has evolved as a multicomponent treatment approach (Sharma & Andrade, 2012). CBT in insomnia has been found to have good outcome primarily due to its simplistic structure, no compulsory addition of pharmacotherapy that removes the risk of any addiction and rebound of symptoms and relatively lesser number of sessions. Cognitive Behavior therapy for insomnia (CBT—I) is a multicomponent therapy primarily consisting of **Stimulus control, Sleep restriction, Cognitive therapy, Counter arousal strategies and Sleep hygiene**. Individual components of CBT-I may be used as suitable to clients. A brief of the major components of CBT-I are as follows:

Stimulus control: This component of CBT-I was developed by Richard Bootzin (1977), and comprise of set of instructions to address conditioned arousal. The technique focuses on identifying behavioral cues that weakens the association of bed and bedroom with sleep and strengthen those which promotes sleep. Major emphasis is on limiting the bed and the bedroom for sleeping and sex only. Majority of the review studies on evidence base of CBT-I include stimulus control therapy as one of its core components. In broader terms the techniques of stimulus control therapy overlaps with that of sleep hygiene.

Sleep restriction: Sleep restriction is another component of CBT-I developed by Spielman, Saskin, and Thorpy (1987). Its primary goal is to restrict and regulate the time spend by the client on bed to match with actual sleep need. Using sleep logs, actual sleep time of client against time spent on bed is calculated to arrive on sleep efficacy score, which is preferred to be more than 85%. Sleep restriction is commonly done in combination of stimulus control and has been found effective in regulating sleep wake cycle. The aim of sleep restriction is not to restrict actual sleep time but to restrict the time spent in bed initially and gradually increase with increasing efficacy. The restricted bed time ideally should not go less than 5.5 h. For some clients the process appears very difficult in the beginning and in absence of monitoring by caregivers at home, sleep restriction may appear difficult to comply.

Sleep hygiene: Sleep hygiene aims to psycho-educate clients regarding good and bad sleep habits. Giving a generic overview of daily habits which can affect onset and latency of sleep, clients are asked to identify any sleep unhygienic behavior they might have and to correct them. The most common components include psycho educating clients about effect of screen time, type and time of food intake, substances, untimely exercise; the physical environment of the bedroom etc. Sleep hygiene is invariably an essential component of CBT—I, but hardly ever the sole component of CBT—I. Sleep hygiene in itself helps improving client's insight and promoting healthy activities promoting sleep.

Whether evidence base of CBT-I is universal?

While literature suggests a good evidence base of Cognitive behavior therapy in management of insomnia (Morin et al., 1999; Okajima, Komada, & Inoue, 2011; Smith, Huang, & Manber, 2005), especially in comparison to hypnotics; large scale studies with rigorous designs to reduce detection and performance bias are advised to improve the quality of the evidence (Wu et al., 2015 b). Studies contributing to evidence base of CBT-I are mostly from western countries. Randomized controlled trails are very less from India and neighboring countries.

Data on specific use pattern of CBT in insomnia by professionals across countries could be less. Asian countries have taken their own time to adapt to psychotherapies of western origin and CBT-I is no exception. There are various studies from India that mention use of CBT—I; however there are several factors that may limit the optimum efficacy of CBT-I in local population, that include lack of formal training of professionals; less number of clinics offering both pharmacological and psychotherapeutic services under one roof; lack of liaison among professionals; and client's own ambiguity on their choice of treatment considering their own lack of knowledge and high expectations of quick results. Insomnia often is not a single manifest and comes along with set of other symptoms. Studies suggest high prevalence of co-morbid mental illness conditions in insomnia and vice versa, thus making consultation for insomnia a tricky one. Customizing psychotherapeutic models to adjust cultural barriers always helps in maximizing treatment outcome and holds true for CBT in insomnia. Without changing the basic frame of CBT in insomnia, the approach might need certain modifications to suit client's individual limitations and required cultural adaptations.

Practice and procedures

A generic perspective on practice of CBT in insomnia in Indian context is discussed further. Even though the procedures are within the framework of CBT, certain adaptations to overcome cultural barriers are recommended. The perspective is largely based on experience of the authors in their clinical practice and resonates with more or less similar practice scenario for CBT-I in India and neighboring countries.

Recommended adaptations to CBT-I in Indian context: It's a well-known fact that individuals decide upon their health seeking behavior and professional consultations depending on their own perception of the need of change, need of treatment, efficacy of the treatment modality as well as treatment cost. This applies to evidence based psychotherapies too, which face certain level of resistance from both professionals as well as clients owing to their misperceptions. In Indian context where professional psychotherapy practice is primarily limited to urban and semi urban population, treatment cost is a big factor, and local population is not that aware of psychotherapy in general, it becomes important to modify and customize the standard delivery of CBT in insomnia for better treatment outcome.

Points to consider

- **Taking a thorough medical history is important:** As insomnia is highly concurrent and co morbid with medical as well as psychiatric conditions, it is important to take a detailed case history exploring other clinical condition of the client including history of diabetes, cardiovascular disease, hypertension, overactive thyroid, chronic pain, body mass index, stress perception, any history of psychoactive substance, anxiety, depression as well as family history. Taking a note of current medications for other clinical conditions is important which might be interfering with sleep.
- **Patients may fail to recognize their symptoms:** Studies (Bhaskar et al., 2016) suggest that a significant percentage of patients fail to recognize their insomnia despite having the symptoms. This might be present relatively in young and middle aged adults. Though empirical data lack in this domain, but it is hypothecated that young and middle aged adults might be ignoring their sleep disturbances to meet their pressing occupational demands. Such clients could be more at risk of engaging in unhygienic sleep practices, which may continue the sleep disturbances as conditioned behavior even when occupational demands are not pressing.
- **Inclusion of Psychoeducation:** Psychoeducation has been suggested to be an integral part of psychotherapy in Indian context, irrespective of the clinical condition (Halder & Mahato, 2012, 2019). It not only helps in enhancing insight but also aids in better treatment compliance and reduce dropout rate. Treatment cost is a major issue, as psychotherapies are not covered under medical insurance in many Asian countries including India. Several clients expecting quick recovery look at the "talking sessions" futile, thus Psychoeducation is must. The Psychoeducation sessions preferably should stretch up to 2–3 sessions. Taking a cue from the bio-psycho-social model, the client need to be explained what the situation is, the tentative etiology of the condition, why it is being maintained and what is required for its remission. The Psychoeducation sessions are ideal for exploring and correcting client's misperceptions regarding insomnia, if any. Clients also need to be explained about the individual differences in the requirement of hours of sleep. If somebody feels refreshed after having 5h of sleep, he need not worry to get extra 2h of sleep.
- **Using aids for Psychoeducation:** A customized pictorial description (Fig. 1) of the circadian rhythm along with the characteristic brain waves during different sleep stages and its progression has been found useful by the authors in the Psychoeducation process, which clients can relate with their own sleep pattern and what is hindering their sleep. Clients also need to be psycho-educated in simple language about the different stages of sleep in a healthy individual. People generally do have the notion that sleep is a single process that should continue after onset till morning without any hindrance or breaks. Incorporating the stages in Psychoeducation helps client correct their perception and understand that intensity of sleep varies as per stages which repeats several times in a single night. It also helps nullifying the worries of having lighter sleep in early morning hours or minor awakenings at around 3.00–4.00 AM.
- **Utility of lab test:** A sizeable number of clients, especially those coming directly for psychotherapeutic intervention may not understand the need of lab test like blood test or polysomnography. As these tests would require additional consultation, clients need to be explained regarding these through Psychoeducation.
- **Analysis of dreams:** People across cultures have always been intrigued by the vivid dreams they see and different meanings are attached to specific types of dream content. Indian population is no exception and therapists are not surprised by queries or request for interpretation of their dreams. Authors do not intent to include any psychoanalytic component in the CBT model, but in event clients do experience same type of dream frequently, it is advised to explore underlying conflicts and address them separately. The mechanism of dreams can also be explained in simplistic way happening during the REM sleep stage, facilitating Psychoeducation.
- **Family counseling:** Apart from Psychoeducation, family counseling is equally important. Studies have found lack of social support as risk factor for insomnia (Krause & Rainville, 2020; Stafford, Bendayan, Tymoszuk, & Kuh, 2017), thus counseling family members, especially bed partners and immediate family members is of importance to resolve

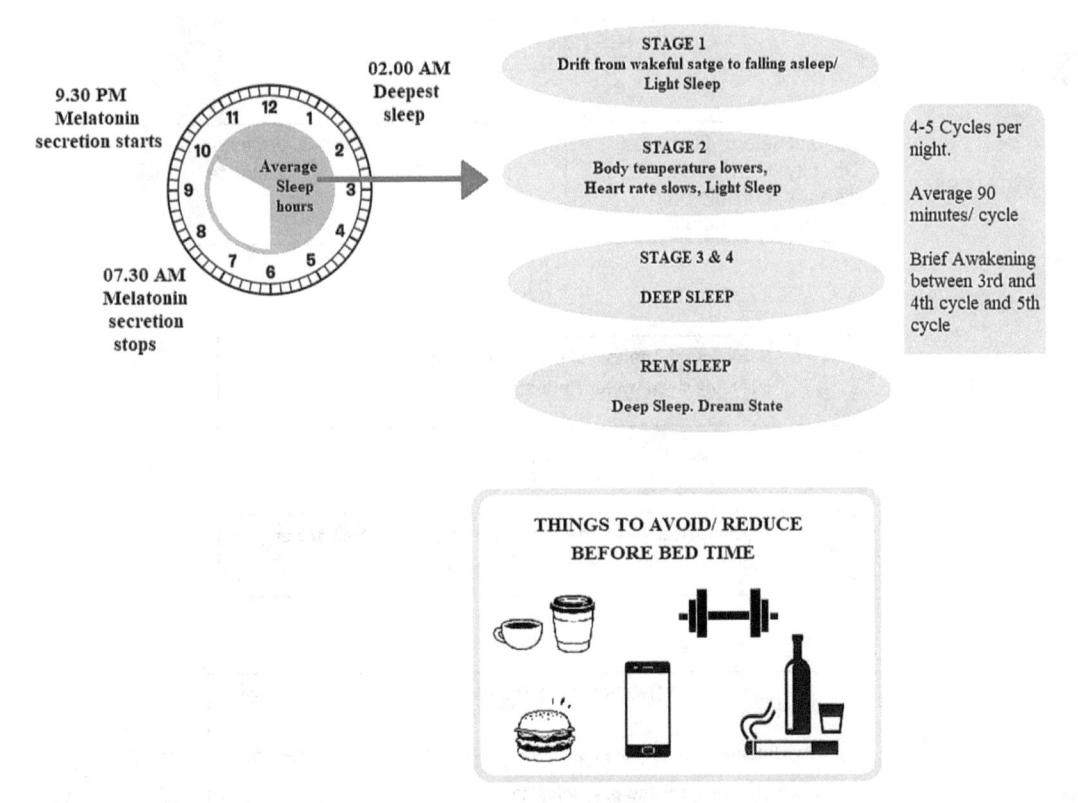

FIG. 1 Pictorial aid for Psychoeducation for patients with insomnia. A customized pictorial aid for psycho-educating clients regarding circadian rhythm, sleep wake cycle and role of sleep hygiene habits in good sleep. Image is representative only.

- conflicts that could be the perpetuating factor for insomnia. A supportive familial environment also helps in better estimation of sleep quantity and quality (Jackowska, Dockray, Hendrickx, & Steptoe, 2011).
- **Relaxation training:** Considering high co morbidity of anxiety with insomnia, relaxation training should be actively integrated with CBT in insomnia (Williams, Roth, Vatthauer, & McCrae, 2013). It is an important adjunct in the CBT process for insomnia. However, therapists need to be careful for clients with significant depressive symptoms where it could be contraindicative. Some clients may not be very comfortable with relaxation training but are okay with indigenous techniques of Yoga, which have equally good indication in insomnia. Efficacy of yoga in insomnia is established through randomized controlled trails and meta-analytical studies (Mustian et al., 2013; Wang, Chen, Pan, et al., 2020) and therapists may suggest their clients for appropriate yoga techniques.
- **All symptoms are important:** Insomnia shares an intricate relationship with anxiety and depressive symptoms. It could be risky to be complacent and ignore sleep issues when they appear as co morbid or secondary to depressive symptoms. Even if depressive symptoms appear to have remitted, nonresolution of sleep disturbance may perpetuate or be a risk factor for relapse of anxiety and depressive symptoms (Baglioni et al., 2011).
- **Customizing sleep diaries:** Sleep dairies are essential in the cognitive behavioral treatment of insomnia. They are structured and regular log of the sleep pattern, sleep onset time, rating of quality of sleep, fatigue experienced during wake hours which are noted from the outset and throughout treatment by the clients. Sleep diaries are helpful in monitoring sleep continuity variables. Despite their usefulness, many clients may not possess the tenacity to get involved into these activities, are impatient and rigid not to engage in additional behavioral tasks. It is advised to keep the structure of the sleep diary minimal so that clients can fill the logs with ease.
- **Barriers in sleep hygiene:** Insomnia is often maintained by unhygienic sleep environment and requires correction in terms of controlling room temperature, noise levels, removing clutter and unwanted articles in the bedroom. In several cases these environmental manipulations may not be possible considering individual background and limitations of the client. In these cases effort should be made to involve the immediate caregiver or family members to explain the situation and to make best possible efforts to create an environment conducive for sleep for the client. Also, considering the very simplistic framework of sleep hygiene, clients may take it very lightly. The underlying thought could be,

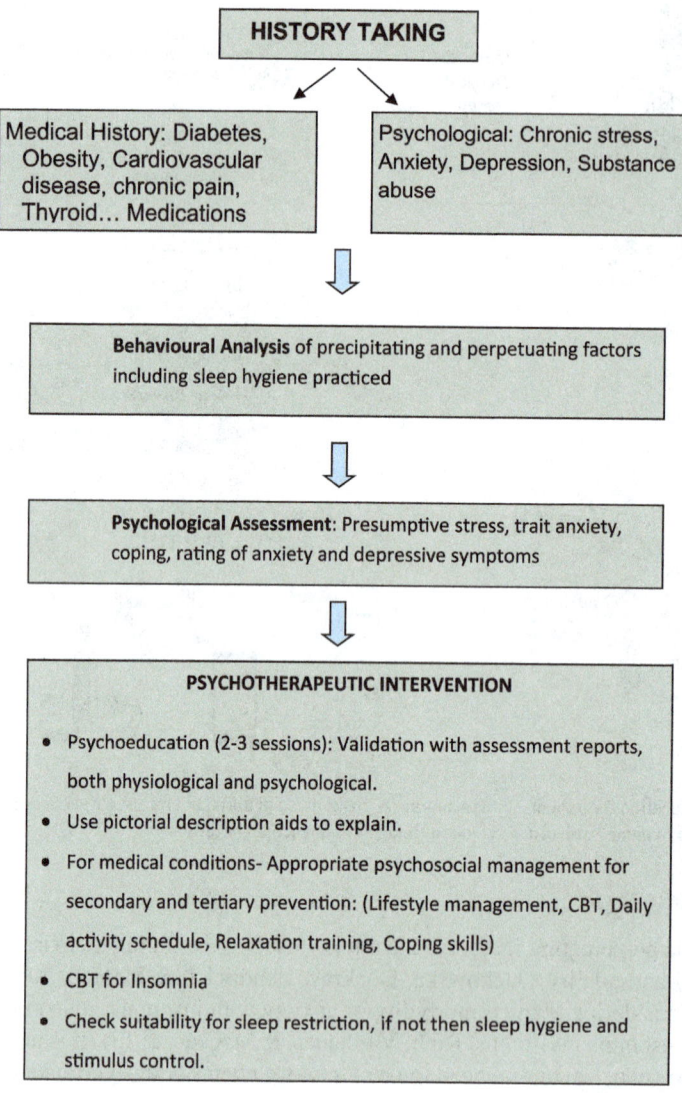

```
                          ┌─────────────────┐
                          │  HISTORY TAKING │
                          └─────────────────┘
                    ↙                          ↘
   ┌─────────────────────────────┐   ┌──────────────────────────────┐
   │ Medical History: Diabetes,  │   │ Psychological: Chronic stress,│
   │ Obesity, Cardiovascular     │   │ Anxiety, Depression, Substance│
   │ disease, chronic pain,      │   │ abuse                         │
   │ Thyroid... Medications      │   │                               │
   └─────────────────────────────┘   └──────────────────────────────┘
                                ⇩
   ┌───────────────────────────────────────────────────────────────┐
   │ Behavioural Analysis of precipitating and perpetuating factors│
   │ including sleep hygiene practiced                             │
   └───────────────────────────────────────────────────────────────┘
                                ⇩
   ┌───────────────────────────────────────────────────────────────┐
   │ Psychological Assessment: Presumptive stress, trait anxiety,  │
   │ coping, rating of anxiety and depressive symptoms             │
   └───────────────────────────────────────────────────────────────┘
                                ⇩
   ┌───────────────────────────────────────────────────────────────┐
   │            PSYCHOTHERAPEUTIC INTERVENTION                     │
   │                                                               │
   │ • Psychoeducation (2-3 sessions): Validation with assessment  │
   │   reports, both physiological and psychological.              │
   │ • Use pictorial description aids to explain.                  │
   │ • For medical conditions- Appropriate psychosocial management │
   │   for secondary and tertiary prevention: (Lifestyle           │
   │   management, CBT, Daily activity schedule, Relaxation        │
   │   training, Coping skills)                                    │
   │ • CBT for Insomnia                                            │
   │ • Check suitability for sleep restriction, if not then sleep  │
   │   hygiene and stimulus control.                               │
   └───────────────────────────────────────────────────────────────┘
```

FIG. 2 Framework for psychotherapeutic intervention for patients with insomnia.

"*I know about all these. Please tell me something which is of your expertise.*" Use of sleep diary and psycho-educational aids help in encountering these situations.

- **Use of mobile apps:** Some clients may be proactive regarding their sleep issues and looking for self-help through mobile apps for improving sleep related issues. A scrutiny of the app being used, if any by the client should be done, if the modules and activities are in sync with therapists CBT model. Clients should not be dissuaded directly, but advised accordingly.

A compact framework for undertaking a client with insomnia for psychotherapeutic intervention is depicted below (Fig. 2):

We discuss the indications of CBT in insomnia and the challenges are explained through following three case studies. The case studies are of different clients consulting for insomnia, but differed in their clinical profile. The application of CBT along with additional techniques in these cases will establish the need of certain customization in the therapeutic plan to make it more effective.

Case 1

Client was 17 years old male, studying in 12th grade hailing from an urban locality. He was brought by his parents with complaints of sleep disturbances, feeling drowsy during day hours, unable to concentrate on studies, and deteriorating academic progress. There were also complaints from his school teachers. The client had difficulty falling asleep and he used to get sleep for maximum 4–5 h. He was diagnosed with Insomnia by a general physician and referred for psychotherapeutic intervention. In initial interview the client failed to identify any significant antecedent or triggering factor for his sleep issues. However, further detailed history revealed the client was ambiguous regarding the choice of his academic stream he was pursuing. He worried a lot about his academic progression, whether he was doing it right or not and subsequently he lost interest in his studies and had poor concentration and was irritable in mood. He stressed that his sleep problem had nothing to do with his academic issues. He was engaged in psychotherapy that primarily included activity scheduling, academic plan and career counseling, CBT, and sleep hygiene. He was not very compliant on maintaining sleep diary and preferred reporting verbally. The client was skeptical about the therapy sessions initially but gradually could be engaged for eight sessions with a booster session and one follow up session.

In this case, the client had difficulty falling asleep. Subsequently he had automatic negative thoughts like: "*I am wasting my time... how I can sleep? ... I will struggle with my studies tomorrow morning.*" Subsequent emotional manifestations were in terms of irritability.

Case 2

The client was a 58 years old male, businessperson hailing from an urban locality. He had undergone cardiac surgery 6 months prior to consultation and was on medications. He had been prescribed with sleeping pills along with his regular medications which he was continuing. The sleeping pills were supposed to be stopped and used only on need basis. However, he reported difficulty falling asleep if he did not took the pills. He was having difficulty in falling asleep and spent a considerable period of time worrying being unable to sleep and its possible impact on his health. He also had palpitations at night occasionally. Detailed clinical history and psychological evaluations revealed a predominantly anxious temperament of the client. Unlike the first case, this client relatively had better acceptance for CBT for his conditions. He constantly worried about his disturbed sleep; the effect of sleeping pills he was continuing but was unable to discontinue them. He was engaged in psychotherapy that primarily included relaxation training, CBT, sleep restriction and sleep hygiene spanning 12 sessions with follow up sessions for next 3 months. Relaxation training included deep breathing only and muscle contraction and holding breath were skipped considering cardiac issues of the client. CBT focused not only on his sleep issues, but also his underlying anxiety and its role in perpetuating sleep issues. Significant improvement was found in sleep disturbances as well as anxiety levels of the patient at follow up.

In this case, the automatic negative thoughts identified were: "*I will not be able to sleep without sleeping pills*" ... "*my health is going to deteriorate complicating my medical condition.*" Subsequent emotional manifestations were in terms of worries, anxiety and low mood.

Case 3

The client was a 32 years old unemployed female, with previous history of depressive episodes and treated with antidepressants and CBT. She was relatively maintaining well and sessions were terminated 7 month back. She consulted again primarily for being unable to sleep for past 1 month. Client mentioned that she was having difficulty in initiation of sleep which was bothering her. She mentioned her mood was relatively stable but she was apprehensive of relapse or further worsening of her depressive symptoms. As her own devised strategy she had started spending more time on her mobile during night. As per the client, being engaged with the mobile phone helped her to avoid her negative thoughts or apprehensions regarding not getting sleep. Further clinical interview revealed low mood, ideas of helplessness and irritability in the client. The client was unwilling to continue with medications apprehending side effects and continue with CBT only considering her benefit from CBT earlier.

In this case, the automatic negative thoughts identified were: "*I am not able to sleep because my symptom of depression is relapsing*" ... "*What will happen if my symptoms relapse*" ... "*I will be suffering from depression for whole of my life*" ... "*Whether I will be fully cured ever*"? Subsequent emotional manifestations were in terms of low mood, feelings of helplessness and despair. She was engaged in psychotherapy, which primarily included CBT, stimulus control and sleep hygiene. Sleep issues were resolved after 8 sessions; however she was suggested to continue CBT for her recurrent depression.

The core beliefs, subsequent automatic negative thoughts and associated emotions of the patients are depicted in figure below (Fig. 3).

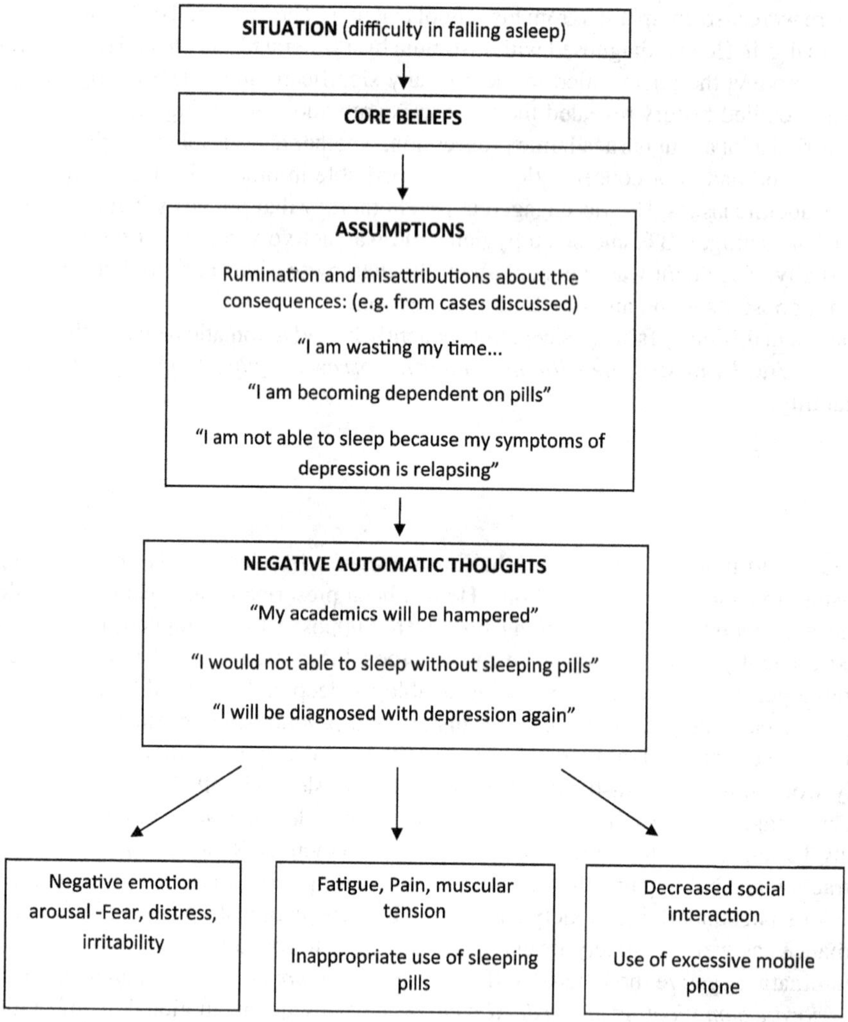

```
┌─────────────────────────────────────────┐
│ SITUATION (difficulty in falling asleep)│
└─────────────────────────────────────────┘
                    ↓
            ┌───────────────┐
            │ CORE BELIEFS  │
            └───────────────┘
                    ↓
┌─────────────────────────────────────────┐
│              ASSUMPTIONS                │
│ Rumination and misattributions about the│
│ consequences: (e.g. from cases discussed)│
│                                         │
│ "I am wasting my time...                │
│                                         │
│ "I am becoming dependent on pills"      │
│                                         │
│ "I am not able to sleep because my symptoms of │
│ depression is relapsing"                │
└─────────────────────────────────────────┘
                    ↓
┌─────────────────────────────────────────┐
│        NEGATIVE AUTOMATIC THOUGHTS      │
│                                         │
│ "My academics will be hampered"         │
│                                         │
│ "I would not able to sleep without sleeping pills" │
│                                         │
│ "I will be diagnosed with depression again" │
└─────────────────────────────────────────┘
         ↙              ↓              ↘
┌──────────────┐ ┌──────────────┐ ┌──────────────┐
│Negative emotion│ │Fatigue, Pain,│ │Decreased social│
│arousal -Fear, │ │muscular      │ │interaction    │
│distress,      │ │tension       │               │
│irritability   │ │              │ │Use of excessive│
│               │ │Inappropriate │ │mobile phone   │
│               │ │use of sleeping│ │              │
│               │ │pills         │ │              │
└──────────────┘ └──────────────┘ └──────────────┘
```

FIG. 3 The core beliefs, automatic negative thoughts, and emotions of the Insomnia clients.

Psychotherapeutic intervention

All the above cases of insomnia were primarily intervened with CBT. We discuss the common components of the psychotherapeutic intervention for all these three cases, which differed in terms of their manifestation of insomnia, co morbid physical illness and psychiatric illness.

- Educating clients: As mentioned earlier, Psychoeducation is deemed integral part of CBT, and was effectively used to enhance the insight levels of the clients regarding their insomnia. In the initial sessions after the symptoms were identified and clinical history taking and behavioral analysis was done; the need for appropriate therapeutic measures was discussed with the clients. Clients were explained the basic tenet of CBT and its efficacy in insomnia. This also helped the clients to decide on stopping/restarting their medications. The Psychoeducation also helped on correcting client's perception of their insomnia in relation to their co-morbid conditions. In case of Client 1, Psychoeducation focused on explaining association of his stress related to his academic issues which he perceived as separate issues till then. In case of Client 2, the tentative formulation of his failure to quit the hypnotics was explained. The possibility of having good sleep without hypnotics using CBT was explained. How his co-morbid cardiac condition could be a possible risk for a better treatment outcome was also explained. For client 3, who had terminated her pharmacotherapy as well as psychotherapy sessions after certain improvement of her depressive symptoms was explained the risk of improper

compliance and the need of continuing medications as required and prescribed by the psychiatrist to minimize risk of full blown relapse. The role of unhygienic sleep practices in insomnia was explained to her.
- Behavioral analysis to explore the triggering and maintaining factor for insomnia was done for all clients. Sleep diary was used to assess the sleep pattern, its quantity and quality of clients. Client 1 was reluctant on maintaining sleep diaries, but other two clients were very receptive for it. Associated and underlying thoughts were explored through Socratic dialogue.
- Maintaining sleep hygiene: Sleep hygiene is one of the easiest but very effective components of CBT in insomnia. Sleep hygiene was stressed for each of the client by adhering to fixed schedule to go to bed and waking up. Activities hindering sleep were identified through behavioral analysis in each of the client. The role of these hindering activities in insomnia through conditioning was explained to clients and was strictly advised to cut on them. Certain environmental manipulation was done for each client.
- Relaxation therapy: Relaxation techniques were introduced, demonstrated to the clients and asked to practice. It was modified for each client considering their co-morbid conditions. For client 2, the focus was given on deep breathing, skipping muscular contraction. Yoga and meditation were additionally included considering clients own inclination. For other clients, deep diaphragmatic breathing and muscular relaxation was advised.
- Cognitive therapy: Major techniques used

    Identifying the negative automatic thoughts related to insomnia: Clients were engaged in Socratic dialogue to identify their negative automatic thoughts related to insomnia. Dysfunctional thought record was made and explained to clients regarding their automatic negative thoughts, its relation to their subsequent emotions and behavior. The negative emotions were stressed how they were becoming conditioned with bed and bedroom and how controlling the stimulus will benefit them.

    Challenging and modifying beliefs: Insomnia associated beliefs and thoughts were categorized through guided discovery and thought challenging techniques. Dysfunctional thoughts were confronted through repetitive exposure to evidences and identifying the discrepancies between the thought patterns. The clients became more conscious and considered the discrepancies regarding their own thoughts in the sessions.

    In different occasions alteration of core beliefs was done through alternate and more plausible explanations.

    Cognitive restructuring: The preceding steps helped client recognize how their dysfunctional thoughts were a factor in perpetuating their sleep issues as well as their mental ill health condition. Clients were suggested to adapt healthier and better adaptive thoughts identified and relate with better emotional and behavioral outcome.

Eventually, the management of insomnia through CBT had two broad components, cognitive and behavioral. The cognitive component of CBT for insomnia involved exploring and assessing thoughts, feelings and behavior of clients regarding their sleep. The dysfunctional and unhelpful thoughts were corrected. The behavioral components of CBT for Insomnia addressed problematic and unhygienic habits to promote better sleep. Techniques included were: maintaining sleep hygiene, relaxation, stimulus control and sleep restrictions. Depending on individual client's suitability and co-morbid conditions a customized CBT was drafted for each patient. The clients though differed in their clinical profile, co-morbid conditions and their perception of therapeutic indication of CBT for them; CBT was effective in bringing improvement in insomnia and related symptoms.

Mini-dictionary of terms

- Rapid eye movement: Rapid eye movement (REM) is one of the phases of sleep in which the human brain is characteristically active like an awake individual. It is very deep phase of sleep and majority of dreaming happens in this phase.
- Stimulus control: One of the prominent techniques of CBT for insomnia, that focuses on identifying behavioral cues that weakens the association of bed and bedroom with sleep and strengthen those which promotes sleep.
- Sleep restriction: It's a method to restrict and regulate the time spend by the client on bed to match with his actual sleep need. It is commonly done in combination of stimulus control and has been found effective in regulating sleep wake cycle.
- Psychoeducation: Psychoeducation constitutes semistructured informative sessions to make clients acquaint with their mental health condition, to understand its basic nature, causes and treatment options. It has been found useful in improving treatment compliance and treatment outcome.
- Relaxation training: Relaxation training refers to practices aiming to relaxed state in wake of stress and anxiety provoking conditions. Deep breathing exercises, progressive muscular relaxation are some of the popular techniques.

Key points

- Insomnia is prevalent up to 33% in general population.
- Insomnia could be precursor of major mental illness like anxiety and depression.
- Improvement in insomnia is associated with better mental health treatment outcome.
- Evidence base of CBT-I need further support with culture specific adaptations

Applications to other areas

The entwined nature of insomnia with major mental illnesses including anxiety, depression as well as psychosis makes insomnia an important treatment goal not only in psychiatric clinics but also in general medical setups. The high prevalence of insomnia in stress related disorders and its further association with medical conditions like cardiovascular disease highlights the priority of insomnia management. With a gradual increase in the aging population, which often is troubled with insomnia, CBT could be effectively used in effective amelioration of insomnia and improving quality of life in geriatric population too. The scope of CBT in insomnia has broadened its horizon to reduce risk of major psychiatric illness symptoms, exacerbation of subclinical symptoms as well as prevention of relapse of such symptoms. While CBT in itself is an evidence based treatment, the protocols might need certain modifications to adjust with cultural demands. The advent of mobile based apps and self-monitoring devices have made the task of therapist easier, though evidence base of the mobile apps and their incorporation in CBT module is yet to be explored. Further studies on larger population with stricter methodology are required to see the efficacy of customized CBT modules in insomnia and related conditions.

Summary points

- Insomnia is highly prevalent and contrary to perception also troubles younger population.
- Chronic insomnia could be a serious risk for major mental illness considering its high concurrent presence in mental ill health conditions.
- Pharmacological management of insomnia comes with risk of side effects and addiction.
- The simplistic structure and relatively less intense nature of CBT in insomnia nullifies the risk of addiction and side effects that pharmacotherapy might have; making it a treatment of choice in insomnia with comparative or even superior efficacy.
- Individual perceptions of the client, cultural influences that may induce certain level of hesitancy to use evidence based practices and presence of co-morbid conditions might be a hindrance in application of CBT in insomnia and related conditions.
- CBT-I enjoys good evidence base even in nonwestern cultures, however, certain customization and tailoring of modules are required to ensure better compliance and outcome.
- In addition to standard procedures, Psychoeducation deem to be an important component in CBT for insomnia.

References

American Academy of Sleep Medicine. (2008). *Insomnia factsheet*. Retrieved from: https://aasm.org/resources/factsheets/insomnia.pdf.

Baglioni, C., Battagliese, G., Feige, B., Spiegelhalder, K., Nissen, C., Voderholzer, U., … Riemann, D. (2011). Insomnia as a predictor of depression: A meta-analytic evaluation of longitudinal epidemiological studies. *Journal of Affective Disorders*, 135(1–3), 10–19. https://doi.org/10.1016/j.jad.2011.01.011.

Bhaskar, S., Hemavathy, D., & Prasad, S. (2016). Prevalence of chronic insomnia in adult patients and its correlation with medical comorbidities. *Journal of Family Medicine and Primary Care*, 5(4), 780–784. https://doi.org/10.4103/2249-4863.201153.

Bootzin, R. (1977). Effects of self-control procedures for insomnia. In R. Stuart (Ed.), *Behavioral self-management: Strategies and outcomes* (pp. 176–195). New York, NY: Brunner/Mazel.

Burnham, M. M., Goodlin-Jones, B. L., Gaylor, E. E., & Anders, T. F. (2002). Nighttime sleep-wake patterns and self-soothing from birth to one year of age: A longitudinal intervention study. *Journal of Child Psychology and Psychiatry*, 43(6), 713–725.

Carter, K. A., Hathaway, N. E., & Lettieri, C. F. (2014). Common sleep disorders in children. *American Family Physician*, 89(5), 368–377.

Dollander, M. (2002). Etiologies de l'insomnie chez l'adulte [Etiology of adult insomnia]. *L'Encephale*, 28(6 Pt 1), 493–502.

Edinger, J. D., & Means, M. K. (2005). Cognitive-behavioral therapy for primary insomnia. *Clinical Psychology Review*, 25(5), 539–558. https://doi.org/10.1016/j.cpr.2005.04.003.

Ford, D. E., & Kamerow, D. B. (1989). Epidemiologic study of sleep disturbances and psychiatric disorders. An opportunity for prevention? *JAMA*, 262(11), 1479–1484. https://doi.org/10.1001/jama.262.11.1479.

Ford, E. S., Cunningham, T. J., Giles, W. H., & Croft, J. B. (2015). Trends in insomnia and excessive daytime sleepiness among US adults from 2002 to 2012. *Sleep Medicine, 16*, 372–378.

Gabe, J., & Thorogood, N. (1986). Prescribed drug use and the management of everyday life: The experiences of black and white working class women. *The Sociological Review, 34*(4), 737–772.

Goldberg, D. P., & Williams, P. (1988). *A users' guide to the general health questionnaire.* London: GL Assessment.

Gooneratne, N. S. (2008). Complementary and alternative medicine for sleep disturbances in older adults. *Clinics in Geriatric Medicine, 24*(1), 121. viii https://doi.org/10.1016/j.cger.2007.08.002.

Grandner, M. A., Jackson, N. J., Izci-Balserak, B., Gallagher, R. A., Murray-Bachmann, R., Williams, N. J., ... Jean-Louis, G. (2015). Social and behavioral determinants of perceived insufficient sleep. *Frontiers in Neurology, 6*, 112. https://doi.org/10.3389/fneur.2015.00112.

Halder, S., & Mahato, A. K. (2012). CBT: The Nepalese experience. In F. Naeem, & D. E. Kingdon (Eds.), *Cognitive behavior therapy in non western cultures* (pp. 79–91). New York: Nova Science.

Halder, S., & Mahato, A. K. (2019). Cognitive behavior therapy for children and adolescents: Challenges and gaps in practice. *Indian Journal of Psychological Medicine, 41*(3), 279–283. https://doi.org/10.4103/ijpsym.ijpsym_470_18.

Jackowska, M., Dockray, S., Hendrickx, H., & Steptoe, A. (2011). Psychosocial factors and sleep efficiency: Discrepancies between subjective and objective evaluations of sleep. *Psychosomatic Medicine, 73*(9), 810–816. https://doi.org/10.1097/PSY.0b013e3182359e77.

Johnson, E. O., Roth, T., & Breslau, N. (2006). The association of insomnia with anxiety disorders and depression: Exploration of the direction of risk. *Journal of Psychiatric Research, 40*(8), 700–708. https://doi.org/10.1016/j.jpsychires.2006.07.008.

Kent, R. G., Uchino, B. N., Cribbet, M. R., Bowen, K., & Smith, T. W. (2015). Social relationships and sleep quality. *Annals of Behavioral Medicine: A Publication of the Society of Behavioral Medicine, 49*(6), 912–917. https://doi.org/10.1007/s12160-015-9711-6.

Krause, N., & Rainville, G. (2020). Exploring the relationship between social support and sleep. *Health Education & Behavior, 47*(1), 153–161. https://doi.org/10.1177/1090198119871331.

Kripke, D. F., Langer, R. D., & Kline, L. E. (2012). Hypnotics' association with mortality or cancer: A matched cohort study. *BMJ Open, 2*(1). https://doi.org/10.1136/bmjopen-2012-000850, e000850.

Mitchell, M. D., Gehrman, P., Perlis, M., & Umscheid, C. A. (2012). Comparative effectiveness of cognitive behavioral therapy for insomnia: A systematic review. *BMC Family Practice, 13*, 40. https://doi.org/10.1186/1471-2296-13-40.

Morin, C. M., Hauri, P. J., Espie, C. A., Spielman, A. J., Buysse, D. J., & Bootzin, R. R. (1999). Nonpharmacologic treatment of chronic insomnia. An American Academy of Sleep Medicine review. *Sleep, 22*(8), 1134–1156. https://doi.org/10.1093/sleep/22.8.1134.

Morin, C. M., Rodrigue, S., & Ivers, H. (2003). Role of stress, arousal, and coping skills in primary insomnia. *Psychosomatic Medicine, 65*(2), 259–267. https://doi.org/10.1097/01.psy.0000030391.09558.a3.

Mustian, K. M., Sprod, L. K., Janelsins, M., Peppone, L. J., Palesh, O. G., Chandwani, K., ... Morrow, G. R. (2013). Multicenter, randomized controlled trial of yoga for sleep quality among cancer survivors. *Journal of Clinical Oncology: Official Journal of the American Society of Clinical Oncology, 31*(26), 3233–3241. https://doi.org/10.1200/JCO.2012.43.7707.

Okajima, I., Komada, Y., & Inoue, Y. (2011). A meta-analysis on the treatment effectiveness of cognitive behavioral therapy for primary insomnia. *Sleep and Biological Rhythms, 9*, 24–34.

Pallesen, S., Sivertsen, B., Nordhus, I. H., & Bjorvatn, B. (2014). A 10-year trend of insomnia prevalence in the adult Norwegian population. *Sleep Medicine, 15*, 173–179.

Panda, S., Taly, A. B., Sinha, S., Gururaj, G., Girish, N., & Nagaraja, D. (2012). Sleep-related disorders among a healthy population in South India. *Neurology India, 60*, 68–74.

Roth, T., Franklin, M., & Bramley, T. J. (2007). The state of insomnia and emerging trends. *The American Journal of Managed Care, 13*, S117–S120. 2007. Retrieved from: https://www.ajmc.com/view/nov07-2643ps117-s120.

Sharma, M. P., & Andrade, C. (2012). Behavioral interventions for insomnia: Theory and practice. *Indian Journal of Psychiatry, 54*(4), 359–366. https://doi.org/10.4103/0019-5545.104825.

Smith, M. T., Huang, M. I., & Manber, R. (2005). Cognitive behavior therapy for chronic insomnia occurring within the context of medical and psychiatric disorders. *Clinical Psychology Review, 25*(5), 559–592. https://doi.org/10.1016/j.cpr.2005.04.004.

Spielman, A. J., Saskin, P., & Thorpy, M. J. (1987). Treatment of chronic insomnia by restriction of time in bed. *Sleep, 10*(1), 45–56.

Stafford, M., Bendayan, R., Tymoszuk, U., & Kuh, D. (2017). Social support from the closest person and sleep quality in later life: Evidence from a British birth cohort study. *Journal of Psychosomatic Research, 98*, 1–9. https://doi.org/10.1016/j.jpsychores.2017.04.014.

Szelenberger, W., & Soldatos, C. (2005). Sleep disorders in psychiatric practice. *World Psychiatry: Official Journal of the World Psychiatric Association (WPA), 4*(3), 186–190.

Venn, S., & Arber, S. (2012). Understanding older peoples' decisions about the use of sleeping medication: Issues of control and autonomy. *Sociology of Health & Illness, 34*(8), 1215–1229. https://doi.org/10.1111/j.1467-9566.2012.01468.x (Epub 2012 Apr 4) 22471794.

Wang, W. L., Chen, K. H., Pan, Y. C., et al. (2020). The effect of yoga on sleep quality and insomnia in women with sleep problems: A systematic review and meta-analysis. *BMC Psychiatry, 20*, 195. https://doi.org/10.1186/s12888-020-02566-4.

Williams, J., Roth, A., Vatthauer, K., & McCrae, C. S. (2013). Cognitive behavioral treatment of insomnia. *Chest, 143*(2), 554–565. https://doi.org/10.1378/chest.12-0731.

World Health Organization. (1992). *The ICD-10 classification of mental and behavioural disorders: Clinical descriptions and diagnostic guidelines.* Geneva: World Health Organization.

Wu, J. Q., Appleman, E. R., Salazar, R. D., & Ong, J. C. (2015). Cognitive behavioral therapy for insomnia comorbid with psychiatric and medical conditions: A meta analysis. *JAMA Internal Medicine, 175*(9), 1461–1472. https://doi.org/10.1001/jamainternmed.2015.3006.

Further reading

American Academy of Sleep Medicine. (2005). *International classification of sleep disorders: Diagnostic and coding manual* (2nd ed.). Westchester, IL: American Academy of Sleep Medicine.

Nowakowski, S., Garland, S., Grandner, M., & Cuddihy, L. (Eds.). (2021). *Adapting cognitive behavioral therapy for insomnia* Academic Press.

Chapter 15

Internet-based cognitive behavioral therapy for loneliness

Anton Käll and Gerhard Andersson
Department of Behavioral Sciences and Learning, Linköping University, Campus Valla, Linköping, Sweden

Abbreviations

CBT   cognitive behavioral therapy
ICBT internet-based cognitive behavioral therapy
SST   social skills training

Loneliness: Introduction and prevalence

Humans are considered to be a species who both want and benefit from social relationships (Baumeister & Leary, 1995). When this need for belonging is not met, it can lead to the psychological state known as loneliness. Loneliness refers to an adverse emotional reaction rooted in a discrepancy between the actual social circumstances and one's wanted social situation (Peplau & Perlman, 1982). It is the subjective experience, rather than any objective characteristics, that is important in this conceptualization. One can feel lonely while being part of a large social network, but also perfectly content with limited social interaction (and even no interaction for some). This is in part due to the fact that feelings of loneliness are intimately tied to how the social world is perceived and the cognitive processes that help us make sense of it (Cacioppo & Hawkley, 2009).

Even if loneliness is a common experience at times during the life-span, long-lasting and frequent forms of loneliness are increasingly viewed as a problem in society (Cacioppo, Grippo, London, Goossens, & Cacioppo, 2015). One study reported that 5.6% of the general population endorsed feeling moderately or severely distressed due to loneliness (Beutel et al., 2017). Another study conducted in the U.K. reported that 6% of the population felt lonely often or always (Victor & Yang, 2012). For older adults over the age of 70 years, the prevalence of frequent loneliness has been found to range between 11% and 14% (Dahlberg, Agahi, & Lennartsson, 2018). While older adults frequently are portrayed as the most typical lonely persons, Barreto et al. (2021) found a significant negative impact of age on loneliness ratings indicating that younger adults were seemingly more prone to experiencing significant loneliness. Overall, the literature suggests that loneliness is a problem that can occur across the lifespan, including among children.

Connection to physical and mental health

Experiencing loneliness often or always has been linked to multiple adverse somatic outcomes, including coronary disease and stroke (Valtorta, Kanaan, Gilbody, Ronzi, & Hanratty, 2016), dementia (Sundström, Adolfsson, Nordin, & Adolfsson, 2019), and all-cause mortality (Holt-Lunstad, Smith, Baker, Harris, & Stephenson, 2015). In addition, the relationship between loneliness and mental health problems is well-established. Feeling lonely has been linked to a seven-fold increase in the risk of meeting the criteria for a mental health disorder (Meltzer et al., 2013), and a substantially increased risk of suicide attempts (Solmi et al., 2020). Results from longitudinal studies also suggest that loneliness can influence, and in turn be influenced by, symptoms of depression and social anxiety (Cacioppo, Hawkley, & Thisted, 2010; Danneel et al., 2019; Lim, Rodebaugh, Zyphur, & Gleeson, 2016). Though little is known about the psychological characteristics of populations with frequent and chronic loneliness, a recent latent profile analysis found that about 50% of the participants seeking treatment for their loneliness exhibited clinical levels of common psychiatric disorders such as major depressive disorder and social anxiety disorder (Käll, Shafran, & Andersson, 2021). In sum, the experience of loneliness has been linked to multiple psychiatric disorders and has been suggested to potentially lead to increase depressive symptoms and social anxiety. Recent findings also indicate the presence of clinical symptoms levels of psychiatric disorder among people experiencing frequent and chronic loneliness.

Reducing loneliness: Previous research and current directions

While there have been a number of different approaches to alleviating loneliness, there is currently no gold-standard approach for helping people with the kind of persistent and frequent feelings of loneliness that can be seen in clinical practice. For example, Masi, Chen, Hawkley, and Cacioppo (2011) divided the interventions included in their metaanalysis into four broad categories based on their targeted mechanism: (1) providing direct social support to the participants, (2) increasing the participants' social skills, (3) providing increased social opportunities through, for example, group activities, and (4) interventions aimed at changing social cognitions and reducing the impact of maladaptive social cognitive processes. Only this last category showed a substantial average reduction of loneliness with a moderate effect size compared to a control condition. The authors concluded that interventions with this focus were the most promising, and that CBT in particular could be an option for reducing loneliness. Later attempts at synthesizing findings from available studies have found slightly reduced estimates for interventions for adults (Hickin et al., 2021) and among adolescents and young adults (Eccles & Qualter, 2020), though they in large support the claim that psychological treatments in general, and CBT more specifically, can reduce loneliness.

Even with these general findings as a guideline, there is a lack of standardized protocols as few interventions have been tested on more than one occasion. To our knowledge, only a few such interventions exist. Our internet-based SOLUS protocol has been tested in two studies (Käll et al., 2020, 2021), showing similar effects for both studies in comparison to a waitlist control group (Cohen's d of 0.77 and 0.71, respectively). This intervention is built on CBT principles and incorporates the use of techniques and strategies commonly found in other manualized CBT protocols, such as behavioral experiments. The GROUPS 4 HEALTH program has also been tested and found to be effective in one randomized trial (Haslam et al., 2019), in addition to an earlier non-randomized study (Haslam, Cruwys, Haslam, Dingle, & Chang, 2016). This concept differs from other trials in that it is based on a social identity approach, utilizing concepts and theories from social psychology. Additionally, reductions have also been observed in studies in which loneliness was not the primary target of the intervention. This includes the PEERS social skills program aimed at adolescents and young adults with an autism spectrum disorder (Matthews et al., 2018; McVey et al., 2016), and a mindfulness protocol (Creswell et al., 2012; Lindsay, Young, Brown, Smyth, & Creswell, 2019). Three of these (SOLUS, PEERS, and the mindfulness interventions) contain elements frequently found in CBT protocols; cognitive restructuring techniques for SOLUS, social skills training (SST) for PEERS and mindfulness and acceptance exercises in the intervention developed by Creswell et al. (2012) and Lindsay et al. (2019). In addition to these studies in which the intervention protocols have been tested twice, some new findings could be interpreted as support for the suggestions made by Masi et al. (2011) 10 years ago. For example, Choi, Pepin, Marti, Stevens, and Bruce (2020) found that a remotely administered behavioral activation helped reduce loneliness significantly more than friendly visits in a randomized trial.

Loneliness: A CBT framework

Understanding how to deal with loneliness requires insight into how loneliness emerges and is maintained over time. In an earlier effort of synthesizing available findings from the intervention literature and psychological factors related to loneliness, we arrived at a model used to explain how it is perpetuated (Käll et al., 2020). The model is presented in Fig. 1 and includes both intrapersonal factors, interpersonal circumstances, and contextual triggers. In this model, the transient form of loneliness can come about as a result of a major change in the social network, for example moving to a new city, or a minor event, such as a movie scene inadvertently drawing attention to a recent break-up. This trigger does not necessarily result in loneliness as this is dependent on the first part of the equation described by Peplau and Perlman (1982); namely one's current social needs. Spending time alone is not inherently aversive, and solitude may be sought out at times. However, should this trigger draw attention to a discrepancy between the actual situation and one's preferences, the difference can result in the emotional state referred to as loneliness. As mentioned above, this is by no means a pathological or rare phenomenon. On the contrary, loneliness can be viewed as an adaptive evolutionary response prompting attempts to reconnect with others (Cacioppo, Cacioppo, & Boomsma, 2014). However, in cases where loneliness persists over time and is considered distressing cognitive and behavioral mechanisms important to consider. For example, studies on social cognition suggest that people who frequently experience loneliness has increased vigilance toward potential threats in social interactions and may remember negative outcomes of social contact to a greater extent than successful ones (Cacioppo & Hawkley, 2009; Spithoven, Bijttebier, & Goossens, 2017). Similarly, the cognitive content and attributions of persons

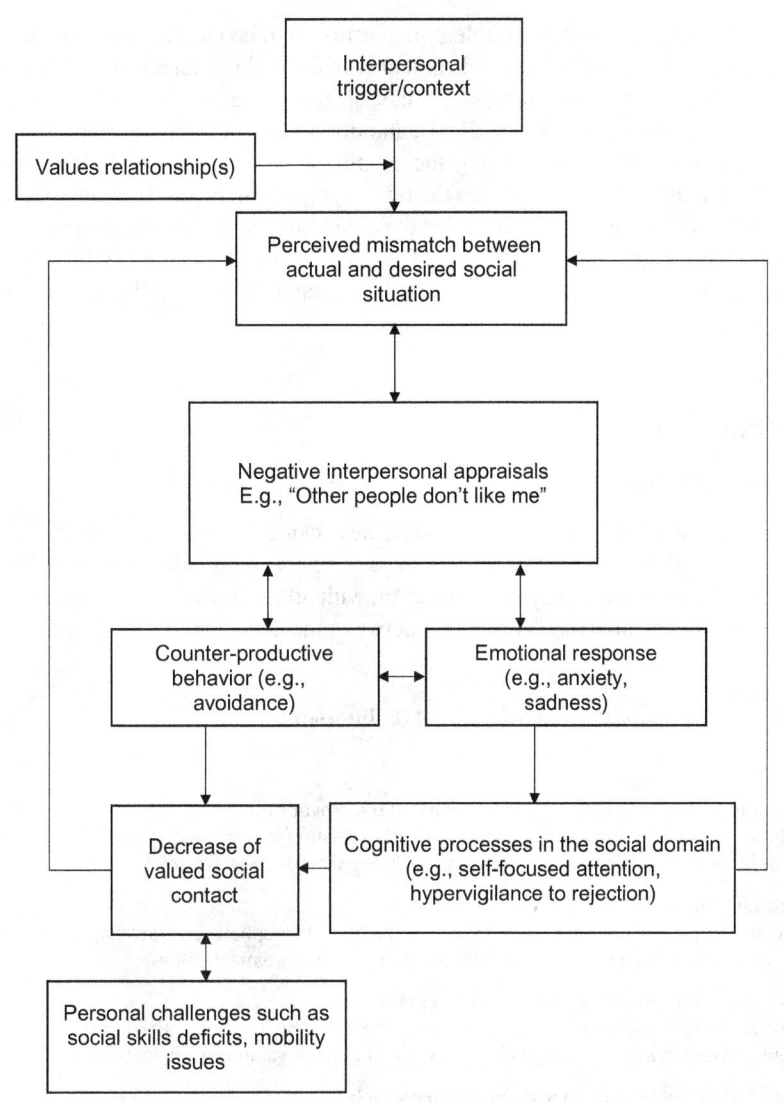

FIG. 1 A cognitive behavioral model of the maintenance of loneliness over time as seen in Käll, Shafran, et al. (2020). *(Reprinted with permission* from Käll, A., Shafran, R., Lindegaard, T., Bennett, S., Cooper, Z., Coughtrey, A., & Andersson, G. (2020). A common elements approach to the development of a modular cognitive behavioral theory for chronic loneliness. Journal of Consulting and Clinical Psychology, 88, 269–282. https://doi.org/10.1037/ccp0000454.*)*

who experience loneliness has been found to be excessively negative, which in turn may influence the probability of seeking out and maintaining social contacts that could reduce loneliness in the long-term (Vanhalst et al., 2015). There seems to be a substantial overlap with the social-cognitive processes associated with social anxiety disorder (Hofmann, 2007) and major depressive disorder (LeMoult & Gotlib, 2019). This is also the case for certain behavioral tendencies including avoidance (Nurmi, Toivonen, Salmela-Aro, & Eronen, 1997) and social withdrawal (Watson & Nesdale, 2012), both of which have been shown to have implications for CBT, for example in exposure therapies (Treanor & Barry, 2017). Assessment of these cognitive and behavioral tendencies is an important part of providing CBT for loneliness. For persons with such cognitive and behavioral characteristics associated with loneliness (e.g., avoidance and negative beliefs) it may be beneficial to target these with similar methods as is usually done in manualized CBT treatments for psychiatric conditions like anxiety and depression. Two examples are behavioral experiments and thought records commonly used to change maladaptive beliefs (McManus, Van Doorn, & Yiend, 2012). These have been found to be useful for the reduction of experienced loneliness (Käll, Jägholm, et al., 2020; Theeke et al., 2016).

CBT via the internet: ICBT

During the last two decades the Internet has been increasingly used as a way to deliver CBT which is often referred to as ICBT (Andersson, 2018). Briefly, ICBT provides the client with psychoeducational texts, pictures, and video clips that provides a rationale and explanation for how the problem in question is to be tackled (Andersson, 2015). Programs tend to be mirror face-to-face treatments in terms of length and content with the main difference being that therapist time is reduced or even automated. Guidance from a clinician/therapist has been associated with better outcomes and adherence to treatment (Baumeister, Reichler, Munzinger, & Lin, 2014), and often takes the form of monitoring progress in completing homework assignments, feedback and support during the treatment. The content of ICBT is often divided into parts, modules/lessons, that each contains a focus on a certain theme or component, much like how a CBT intervention is divided into separate sessions. ICBT provides the ability to tailor these modules to the needs of clients, as well as possibility of working with multiple problems at once or sequentially. There are many versions of ICBT which also can include text chat, video conversations, smartphone applications delivered in a secure manner (Vlaescu, Alasjö, Miloff, Carlbring, & Andersson, 2016).

Practice and procedure

Part 1: Assessment and activation

The most recent version of the SOLUS intervention for loneliness can be viewed in Table 1.

One core feature of the SOLUS intervention is *valued social contact*. As loneliness does not rely on objective characteristics of the situation, the subjective experience of connection with others is proposed to be instrumental in understanding both how loneliness comes about and how the discrepancy between the actual and wanted social situation can be resolved.

TABLE 1 Content of the most recent iteration of the SOLUS intervention.

SOLUS
Module 1: Psychoeducation regarding loneliness and how it is maintained over time
The initial module contains psychoeducation regarding loneliness and its relation to our thoughts and behaviors. Also, an introduction to a functional behavioral model (*The vicious circle of loneliness*) used during the coming modules
Module 2: Clarifying goals and values
Homework assignments related to goals (short- and long-term) and values, with an emphasis on the social realm. Also, a behavioral task, *Take the first step toward your values*, aimed at putting the principles of the module into practical use
Module 3: Behavioral techniques for increasing valued social contact
Introduction to the concept of valued social contact and a modified rationale of behavioral activation as a means of achieving this. Homework assignments centered around planning and executing behaviors meant to create valued social contact
Module 4: Techniques for challenging behavioral and emotional obstacles
Continued modified behavioral activation and a rationale for exposure exercises in cases where social anxiety may play a role in the maintenance of loneliness over time
Module 5: Cognitive restructuring
Psychoeducation about the thoughts and dysfunctional thinking patterns. Homework assignments aimed at challenging dysfunctional cognitive content and beliefs
Module 6: Behavioral experiments as a tool to overcoming loneliness
Introduction to behavioral experiments and two homework assignments related to this technique
Module 7: Increasing social skills
Psychoeducation about social skills and ways of communicating more effectively. Contains a wide range of subjects such as active listening, assertiveness, and making small talk. The participants are asked to choose the ones thought to be relevant and try them out as a homework assignment
Module 8: Evaluation of past and continued
Continued modified behavioral activation and a structured evaluation of the techniques and exercises found in previous modules. Planning for which techniques to continue using and how
Module 9: Preventing future bouts of loneliness
Information and exercises referred to as relapse prevention used to aid the continued work of reducing loneliness. Homework assignments centered around making a contingency plan should loneliness return as a concern

An important step is the identification of valued social contact that are missing. Not all social contact is equal and identifying what is missing in the current situation is important in order to identify what is needed to reduce feelings of loneliness and isolation. Some people seeking help for their loneliness have an impoverished social network in terms of quantity (Domènech-Abella et al., 2017), which requires that they expand their social circles. While not necessarily a goal on its own, increasing the number of social contacts is also thought to be an important building block for developing relationships which can eventually serve other functions, such as providing emotional support and disclosing personal troubles. However, we need to note that it is also common to have social contacts but still feeling lonely. For example, at work and even in the family it is possible to experience substantial loneliness.

Having spent time identifying what may be missing in the current situation, the efforts are then turned to realizing this valued social contact. The SOLUS intervention make use of the principles of behavioral activation (Martell, Dimidjian, & Herman-Dunn, 2010) as one of the means of achieving this. The emphasis is on producing engagement in the (social) world and is thought to be important to promote social contacts that are lacking. Furthermore, the incorporation of problem solving is also believed to be useful in the management of loneliness. The principles of behavioral activation also deals directly with the potential influence of unhelpful tendencies such as avoidance and withdrawal which are likely to maintain feelings of loneliness over time (Käll, Shafran, et al., 2020).

While behavioral activation promotes pleasurable activities in general, the focus in the SOLUS intervention is on social interactions. With this, the intervention is meant to promote social habits that increase the likelihood of achieving the valued social contact that is currently lacking. While other activities may help to reduce loneliness, the emphasis is directly on the social domain which is assumed to be of greater importance in the long term. Much like behavioral activation, the version used in SOLUS is tailored for the individual. For some, the social activities required may focus on existing relationships, such as a taking time to improve a strained romantic relationship. It may also require seeking out new social arenas and opportunities, such as taking up a new hobby with others.

Part 2: Dealing with obstacles and improving chances of success

The aim of increasing valued social contact is crucial, but obstacles are likely to arise while going through this process. These may relate to external circumstances, such as finding the time to attend meetings of an organization or other social gatherings. In these cases, incorporating problem solving described in the behavioral activation rationale (Martell et al., 2010) can be an good way to handle the obstacles. This includes defining the problem in behavioral terms, brainstorming solutions and consistently evaluating outcomes. Should the obstacles be related to intrapersonal factors such as the presence of social anxiety or other psychopathological processes that prevent the client from actively seeking out social opportunities, other strategies might be helpful. In SOLUS, we have used different versions of cognitive restructuring techniques, such as thought diaries and challenging negative automatic thoughts as well as behavioral experiments to test maladaptive beliefs regarding social situations (Käll, Jägholm, et al., 2020). We have also included a rationale and exercises derived from an exposure rationale in cases for which symptoms of anxiety prevent access to social arenas. The main point behind these is to make clients aware of the role that maladaptive cognitive and behavioral tendencies may play in interfering with attempts to seek out valued social contact and provide tools to manage these situations in a more adaptive way.

Other than these strategies to foster valued social contact, other efforts may be beneficial in increasing the chances of success when doing so. A lack of social skills has been found to be associated with loneliness, even if the findings are somewhat inconsistent regarding whether these are primarily self-perceived difficulties (e.g., Knowles, Lucas, Baumeister, & Gardner, 2015) or deficiencies that are somewhat agreed upon by different parties (e.g., Lodder, Goossens, Scholte, Engels, & Verhagen, 2016). Regardless of whether the client's social skills may actually be lacking or not, promoting effective interpersonal behaviors when socializing is believed to provide the best conditions for reducing loneliness both in the short-term (e.g., by improving the odds of a pleasurable conversation) and in the long-term (e.g., by improving one's ability to create and maintain fulfilling friendships). A study comparing regular group-based CBT for social anxiety disorder with a version with additional SST also found that adding SST lead to a greater reduction in social anxiety symptoms (Herbert et al., 2005), which may be relevant for loneliness due to its connection with symptoms of social anxiety (Lim et al., 2016). In our second research trial (Käll, Bäck, et al., 2021), the included module on social skills had a broad focus, including information and exercises about assertiveness, engaging in small talk, and active listening. We expected a large degree of heterogeneity in the population seeking help for their loneliness and therefore tried to provide a wide range of materials. Some of the more generic strategies also aimed to make the client a more skilled conversational partner in general as a means to develop social relationships. As with the behavioral activation component, the use of SST will need to be grounded in the needs of the patient.

Part 3: Maintaining gains and modular considerations

As with CBT in general, the later sessions are aimed toward preparing the client for continued work with the techniques and strategies acquired in the treatment. The long-term benefits of interventions for loneliness have not yet been studied in more extensively. We found the reductions in loneliness to be maintained at 4 months in the second trial (Käll, Bäck, et al., 2021) and after 2 years in the pilot trial (Käll, Backlund, Shafran, & Andersson, 2020). With the exception of these two studies, information about gains beyond the treatment phase is limited. The long-term maintenance of the strategies used during the treatment is thought to be important in loneliness due to the fact that relationships tend to take time to develop. Thus, the use of booster sessions or similar solutions may be advantageous even if we did not include that in the SOLUS studies.

As we have already mentioned when describing our model (Käll, Shafran, et al., 2020), it is of great importance to individualize treatment as much as possible. The steps outlined above can serve as a framework but removing or adding additional content might be essential to create a treatment that is acceptable and effective for a particular patient. In this way, the treatment is to be considered modular, with the clinician adding modules deemed to be relevant to reduce feelings of loneliness and increase the quality (and potentially quantity) of relationships. The idea is that the strategies help the client to deal with some of the factors that might be maintaining loneliness over time, though additions may be needed. As an example, some recent findings indicate the importance of an additional focus on regulating emotional experience of loneliness itself. Eres, Lim, Lanham, Jillard, and Bates (2021) found loneliness to be uniquely related to difficulties in emotional regulation as compared to non-lonely participants in a sample of persons diagnosed with social anxiety disorder. This finding builds on a previous study in which the use of dysfunctional strategies for emotional regulation predicted increased loneliness ratings (Kearns & Creaven, 2017). The same study also reported a negative relationship between more adaptive strategies (e.g., staying present in the moment) and loneliness, suggesting that a focus on emotion regulation may be useful. In line with this, some studies have reported benefits of using mindfulness as a means of reducing loneliness (Creswell et al., 2012; Zhang, Fan, Huang, & Rodriguez, 2018), though this benefit was not found for global loneliness measure in a later study (the study did find a benefit of the intervention on daily ratings of loneliness; Lindsay et al., 2019). In situations where focusing on expanding the social circle is not feasible, an emphasis on employing more functional strategies for emotional regulation could be an option. Emotional regulation strategies could thus be considered an example of a modular addition to the SOLUS intervention described above.

Mini dictionary of terms

- Loneliness: An adverse emotional reaction resulting from a perceived discrepancy between one's wanted social situation and one's actual social situation.
- ICBT: Internet-based cognitive behavioral therapy, an asynchronous form of CBT in which the content is delivered via a web page or a smartphone app.
- Valued social contact: A term used to operationalize the kind of social contact and circumstances thought to promote a sense of belonging, thus counteracting feeling of loneliness.
- Social skills training: psychoeducation and exercises meant to promote the use of effective and context-appropriate communication.

Key facts about loneliness

- Estimates suggest that around 5% of all adults and 11%–14% of adults over the age of 70 report feeling lonely often or always.
- Loneliness has been linked to negative somatic outcomes such as cardiovascular disease and all-purpose mortality.
- Loneliness is also associated with psychological and psychiatric factors such as the prevalence of psychiatric disorders.
- Longitudinal studies suggest a reciprocal influence between loneliness and symptoms of major depressive disorder and social anxiety disorder.
- Loneliness has been associated with maladaptive social cognitive processes and behavioral patterns that may prevent reconnection with others.

Applications to other areas

Even if the outlined SOLUS treatment is based on our trials on ICBT, we believe that the material has the potential to be effective when delivered in a more traditional, face- to-face synchronous setting as well. For example, similar techniques and strategies have been successfully disseminated via video call (Choi et al., 2020) and in a group setting (Theeke et al., 2016), though further research is needed. In cases where lacking social skills are thought to be a factor maintaining loneliness over time, the face-to-face format could potentially increase the feasibility by offering a better chance at assessing the participants and opportunities for intervening that are beyond the scope of ICBT, such as the possibility of role-playing scenarios found difficult at present.

Secondly, while developed as a stand-alone intervention, we believe that the previous points and ideas could be valuable when encountering loneliness in the context of treating mental health problems. The techniques and rationale used in our intervention are often based on previously developed and evaluated CBT treatments (e.g., Martell et al., 2010), which ought to facilitate incorporation. Little is known about the connection between loneliness and psychiatric disorders over time but given that loneliness has been seen to have an impact on social anxiety and depressive symptoms (Danneel et al., 2019; Lim et al., 2016) it may be beneficial to focus specifically on loneliness when encountering this problem in clinical settings.

Lastly, while our studies up to this point have been delivered in a research setting, ICBT has been successfully disseminated within regular clinical practice (Williams, O'Moore, Mason, & Andrews, 2014). Loneliness does not in of itself serve as a problem for receiving health care, but has been reported to be a common concern in primary care patients (Mullen et al., 2019). Offering the kind of help outlined above in a regular clinical setting, whether via the internet or in another format, could be a feasible way of reaching an underserved population.

Summary points

- Loneliness results from a discrepancy between the current social situation and the person's standards and expectations.
- Loneliness is closely related to both physical and mental health, with studies suggesting a reciprocal relationship with symptoms of depression and social anxiety.
- CBT has the potential to alleviate loneliness due to its proven record of dealing with maladaptive cognitive and behavioral tendencies.
- According to our rationale, loneliness stems from a lack of valued social contact. Identifying how to achieve this, creating opportunities and dealing with obstacles are thought to be the recipe for reducing feelings of loneliness.
- Our SOLUS approach has been found to be effective in two trials of ICBT.
- The outlined intervention is flexible and should be tested in a wide range of formats and settings.

References

Andersson, G. (2015). *The internet and CBT: A clinical guide.* Boca Raton: CRC Press.

Andersson, G. (2018). Internet interventions: Past, present and future. *Internet Interventions, 12,* 181–188. https://doi.org/10.1016/j.invent.2018.03.008.

Barreto, M., Victor, C., Hammond, C., Eccles, A., Richins, M. T., & Qualter, P. (2021). Loneliness around the world: Age, gender, and cultural differences in loneliness. *Personality and Individual Differences, 169.* https://doi.org/10.1016/j.paid.2020.110066, 110066.

Baumeister, H., Reichler, L., Munzinger, M., & Lin, J. (2014). The impact of guidance on internet-based mental health interventions—A systematic review. *Internet Interventions, 1,* 205–215. https://doi.org/10.1016/j.invent.2014.08.003.

Baumeister, R. F., & Leary, M. R. (1995). The need to belong: Desire for interpersonal attachments as a fundamental human motivation. *Psychological Bulletin, 117,* 497.

Beutel, M. E., Klein, E. M., Brähler, E., Reiner, I., Jünger, C., Michal, M., ... Tibubos, A. N. (2017). Loneliness in the general population: Prevalence, determinants and relations to mental health. *BMC Psychiatry, 17,* 1–7. https://doi.org/10.1186/s12888-017-1262-x.

Cacioppo, J. T., Cacioppo, S., & Boomsma, D. I. (2014). Evolutionary mechanisms for loneliness. *Cognition & Emotion, 28.* https://doi.org/10.1080/02699931.2013.837379.

Cacioppo, J. T., & Hawkley, L. C. (2009). Perceived social isolation and cognition. *Trends in Cognitive Sciences, 13,* 447–454. https://doi.org/10.1016/j.tics.2009.06.005.

Cacioppo, J. T., Hawkley, L. C., & Thisted, R. A. (2010). Perceived social isolation makes me sad: 5-year cross-lagged analyses of loneliness and depressive symptomatology in the Chicago health, aging, and social relations study. *Psychology and Aging, 25,* 453–463. https://doi.org/10.1037/a0017216.

Cacioppo, S., Grippo, A. J., London, S., Goossens, L., & Cacioppo, J. T. (2015). Loneliness: Clinical import and interventions. *Perspectives on Psychological Science, 10,* 238–249. https://doi.org/10.1177/1745691615570616.

Choi, N. G., Pepin, R., Marti, C. N., Stevens, C. J., & Bruce, M. L. (2020). Improving social connectedness for homebound older adults: Randomized controlled trial of tele-delivered behavioral activation versus tele-delivered friendly visits. *The American Journal of Geriatric Psychiatry, 28*, 698–708. https://doi.org/10.1016/j.jagp.2020.02.008.

Creswell, J. D., Irwin, M. R., Burklund, L. J., Lieberman, M. D., Arevalo, J. M. G., Ma, J., … Cole, S. W. (2012). Mindfulness-based stress reduction training reduces loneliness and pro-inflammatory gene expression in older adults: A small randomized controlled trial. *Brain, Behavior, and Immunity, 26*, 1095–1101. https://doi.org/10.1016/j.bbi.2012.07.006.

Dahlberg, L., Agahi, N., & Lennartsson, C. (2018). Lonelier than ever? Loneliness of older people over two decades. *Archives of Gerontology and Geriatrics, 75*, 96–103. https://doi.org/10.1016/j.archger.2017.11.004.

Danneel, S., Nelemans, S., Spithoven, A., Bastin, M., Bijttebier, P., Colpin, H., … Goossens, L. (2019). Internalizing problems in adolescence: Linking loneliness, social anxiety symptoms, and depressive symptoms over time. *Journal of Abnormal Child Psychology, 47*, 1691–1705. https://doi.org/10.1007/s10802-019-00539-0.

Domènech-Abella, J., Lara, E., Rubio-Valera, M., Olaya, B., Moneta, M. V., Rico-Uribe, L. A., … Haro, J. M. (2017). Loneliness and depression in the elderly: The role of social network. *Social Psychiatry and Psychiatric Epidemiology: The International Journal for Research in Social and Genetic Epidemiology and Mental Health Services, 52*, 381–390. https://doi.org/10.1007/s00127-017-1339-3.

Eccles, A. M., & Qualter, P. (2020). Alleviating loneliness in young people–A meta-analysis of interventions. *Child and Adolescent Mental Health, 26*, 17–33.

Eres, R., Lim, M. H., Lanham, S., Jillard, C., & Bates, G. (2021). Loneliness and emotion regulation: Implications of having social anxiety disorder. *Australian Journal of Psychology*, 1–11. https://doi.org/10.1080/00049530.2021.1904498.

Haslam, C., Cruwys, T., Chang, M. X. L., Bentley, S. V., Haslam, S. A., Dingle, G. A., & Jetten, J. (2019). GROUPS 4 HEALTH reduces loneliness and social anxiety in adults with psychological distress: Findings from a randomized controlled trial. *Journal of Consulting and Clinical Psychology, 87*, 787–801. https://doi.org/10.1037/ccp0000427.

Haslam, C., Cruwys, T., Haslam, S. A., Dingle, G., & Chang, M. X.-L. (2016). Groups 4 health: Evidence that a social-identity intervention that builds and strengthens social group membership improves mental health. *Journal of Affective Disorders, 194*, 188–195. https://doi.org/10.1016/j.jad.2016.01.010.

Herbert, J. D., Gaudiano, B. A., Rheingold, A. A., Myers, V. H., Dalrymple, K., & Nolan, E. M. (2005). Social skills training augments the effectiveness of cognitive behavioral group therapy for social anxiety disorder. *Behavior Therapy, 36*, 125–138.

Hickin, N., Käll, A., Shafran, R., Sutcliffe, S., Manzotti, G., & Langan, D. (2021). The effectiveness of psychological interventions for loneliness: A systematic review and meta-analysis. *Clinical Psychology Review, 88*. https://doi.org/10.1016/j.cpr.2021.102066, 102066.

Hofmann, S. G. (2007). Cognitive factors that maintain social anxiety disorder: A comprehensive model and its treatment implications. *Cognitive Behaviour Therapy, 36*, 193–209. https://doi.org/10.1080/16506070701421313.

Holt-Lunstad, J., Smith, T. B., Baker, M., Harris, T., & Stephenson, D. (2015). Loneliness and social isolation as risk factors for mortality: A meta-analytic review. *Perspectives on Psychological Science, 10*, 227–237. https://doi.org/10.1177/1745691614568352.

Kearns, S. M., & Creaven, A.-M. (2017). Individual differences in positive and negative emotion regulation: Which strategies explain variability in loneliness? *Personality and Mental Health, 11*, 64–74. https://doi.org/10.1002/pmh.1363.

Knowles, M. L., Lucas, G. M., Baumeister, R. F., & Gardner, W. L. (2015). Choking under social pressure: Social monitoring among the lonely. *Personality and Social Psychology Bulletin, 41*, 805–821. https://doi.org/10.1177/0146167215580775.

Käll, A., Backlund, U., Shafran, R., & Andersson, G. (2020). Lonesome no more? A two-year follow-up of internet-administered cognitive behavioral therapy for loneliness. *Internet Interventions, 19*. https://doi.org/10.1016/j.invent.2019.100301, 100301.

Käll, A., Bäck, M., Welin, C., Åman, H., Bjerkander, R., Wänman, M., … Andersson, G. (2021). Therapist-guided internet-based treatments for loneliness: A randomized controlled three-arm trial comparing cognitive behavioral therapy and interpersonal psychotherapy. *Psychotherapy and Psychosomatics, 90*, 351–358. https://doi.org/10.1159/000516989.

Käll, A., Jägholm, S., Hesser, H., Andersson, F., Mathaldi, A., Norkvist, B. T., … Andersson, G. (2020). Internet-based cognitive behavior therapy for loneliness: A pilot randomized controlled trial. *Behavior Therapy, 51*, 54–68. https://doi.org/10.1016/j.beth.2019.05.001.

Käll, A., Shafran, R., & Andersson, G. (2021). Exploring latent profiles of psychopathology in a sample of lonely people seeking treatment. *Journal of Psychopathology and Behavioral Assessment*. https://doi.org/10.1007/s10862-021-09870-7.

Käll, A., Shafran, R., Lindegaard, T., Bennett, S., Cooper, Z., Coughtrey, A., & Andersson, G. (2020). A common elements approach to the development of a modular cognitive behavioral theory for chronic loneliness. *Journal of Consulting and Clinical Psychology, 88*, 269–282. https://doi.org/10.1037/ccp0000454.

LeMoult, J., & Gotlib, I. H. (2019). Depression: A cognitive perspective. *Clinical Psychology Review, 69*, 51–66. https://doi.org/10.1016/j.cpr.2018.06.008.

Lim, M. H., Rodebaugh, T. L., Zyphur, M. J., & Gleeson, J. F. M. (2016). Loneliness over time: The crucial role of social anxiety. *Journal of Abnormal Psychology, 125*, 620–630. https://doi.org/10.1037/abn0000162.

Lindsay, E. K., Young, S., Brown, K. W., Smyth, J. M., & Creswell, J. D. (2019). Mindfulness training reduces loneliness and increases social contact in a randomized controlled trial. *Proceedings of the National Academy of Sciences of the United States of America, 116*, 3488–3493. https://doi.org/10.1073/pnas.1813588116.

Lodder, G. M. A., Goossens, L., Scholte, R. H. J., Engels, R. C. M. E., & Verhagen, M. (2016). Adolescent loneliness and social skills: Agreement and discrepancies between self-, meta-, and peer-evaluations. *Journal of Youth and Adolescence, 45*, 2406–2416. https://doi.org/10.1007/s10964-016-0461-y.

Martell, C. R., Dimidjian, S., & Herman-Dunn, R. (2010). *Behavioral activation for depression: A clinician's guide* (p. c2010). New York: Guilford Press.

Masi, C. M., Chen, H.-Y., Hawkley, L. C., & Cacioppo, J. T. (2011). A meta-analysis of interventions to reduce loneliness. *Personality and Social Psychology Review*, *15*, 219–266. https://doi.org/10.1177/1088868310377394.

Matthews, N. L., Orr, B. C., Warriner, K., DeCarlo, M., Sorensen, M., Laflin, J., & Smith, C. J. (2018). Exploring the effectiveness of a peer-mediated model of the PEERS curriculum: A pilot randomized control trial. *Journal of Autism and Developmental Disorders*, *48*, 2458–2475. https://doi.org/10.1007/s10803-018-3531-z.

McManus, F., Van Doorn, K., & Yiend, J. (2012). Examining the effects of thought records and behavioral experiments in instigating belief change. *Journal of Behavior Therapy and Experimental Psychiatry*, *43*, 540–547. https://doi.org/10.1016/j.jbtep.2011.07.003.

McVey, A. J., Dolan, B. K., Willar, K. S., Pleiss, S., Karst, J. S., Casnar, C. L., … Van Hecke, A. V. (2016). A replication and extension of the PEERS® for young adults social skills intervention: Examining effects on social skills and social anxiety in young adults with autism spectrum disorder. *Journal of Autism and Developmental Disorders*, *46*, 3739–3754. https://doi.org/10.1007/s10803-016-2911-5.

Meltzer, H., Bebbington, P., Dennis, M., Jenkins, R., McManus, S., & Brugha, T. (2013). Feelings of loneliness among adults with mental disorder. *Social Psychiatry and Psychiatric Epidemiology*, *48*, 5–13. https://doi.org/10.1007/s00127-012-0515-8.

Mullen, R. A., Tong, S., Sabo, R. T., Liaw, W. R., Marshall, J., Nease, D. E., Jr., … Frey, J. J., III. (2019). Loneliness in primary care patients: A prevalence study. *Annals of Family Medicine*, *17*, 108–115. https://doi.org/10.1370/afm.2358.

Nurmi, J. E., Toivonen, S., Salmela-Aro, K., & Eronen, S. (1997). Social strategies and loneliness. *The Journal of Social Psychology*, *137*, 764–777. https://doi.org/10.1080/00224549709595497.

Peplau, L. A., & Perlman, D. (1982). *Loneliness: A sourcebook of current theory, research and therapy*. New York, NY, US: Wiley.

Solmi, M., Veronese, N., Galvano, D., Favaro, A., Ostinelli, E. G., Noventa, V., … Trabucchi, M. (2020). Factors associated with loneliness: An umbrella review of observational studies. *Journal of Affective Disorders*, *271*, 131–138. https://doi.org/10.1016/j.jad.2020.03.075.

Spithoven, A. W. M., Bijttebier, P., & Goossens, L. (2017). It is all in their mind: A review on information processing bias in lonely individuals. *Clinical Psychology Review*, *58*, 97–114. https://doi.org/10.1016/j.cpr.2017.10.003.

Sundström, A., Adolfsson, A. N., Nordin, M., & Adolfsson, R. (2019). Loneliness increases the risk of all-cause dementia and Alzheimer's disease. *The Journals of Gerontology: Series B*, *75*, 919–926. https://doi.org/10.1093/geronb/gbz139.

Theeke, L. A., Mallow, J. A., Moore, J., McBurney, A., Rellick, S., & VanGilder, R. (2016). Effectiveness of LISTEN on loneliness, neuroimmunological stress response, psychosocial functioning, quality of life, and physical health measures of chronic illness. *International Journal of Nursing Sciences*, *3*, 242–251. https://doi.org/10.1016/j.ijnss.2016.08.004.

Treanor, M., & Barry, T. J. (2017). Treatment of avoidance behavior as an adjunct to exposure therapy: Insights from modern learning theory. *Behaviour Research and Therapy*, *96*, 30–36. https://doi.org/10.1016/j.brat.2017.04.009.

Valtorta, N. K., Kanaan, M., Gilbody, S., Ronzi, S., & Hanratty, B. (2016). Loneliness and social isolation as risk factors for coronary heart disease and stroke: Systematic review and meta-analysis of longitudinal observational studies. *Heart*, *102*, 1009–1016. https://doi.org/10.1136/heartjnl-2015-308790.

Vanhalst, J., Luyckx, K., Soenens, B., Van Petegem, S., Weeks, M. S., & Asher, S. R. (2015). Why do the lonely stay lonely? Chronically lonely adolescents' attributions and emotions in situations of social inclusion and exclusion. *Journal of Personality and Social Psychology*, *109*, 932–948. https://doi.org/10.1037/pspp0000051.

Victor, C. R., & Yang, K. (2012). The prevalence of loneliness among adults: A case study of the United Kingdom. *The Journal of Psychology*, *146*, 85–104. https://doi.org/10.1080/00223980.2011.613875.

Vlaescu, G., Alasjö, A., Miloff, A., Carlbring, P., & Andersson, G. (2016). Features and functionality of the Iterapi platform for internet-based psychological treatment. *Internet Interventions*, *6*, 107–114. https://doi.org/10.1016/j.invent.2016.09.006.

Watson, J., & Nesdale, D. (2012). Rejection sensitivity, social withdrawal, and loneliness in young adults. *Journal of Applied Social Psychology*, *42*, 1984–2005. https://doi.org/10.1111/j.1559-1816.2012.00927.x.

Williams, A. D., O'Moore, K., Mason, E., & Andrews, G. (2014). The effectiveness of internet cognitive behaviour therapy (iCBT) for social anxiety disorder across two routine practice pathways. *Internet Interventions*, *1*, 225–229. https://doi.org/10.1016/j.invent.2014.11.001.

Zhang, N., Fan, F.-M., Huang, S.-Y., & Rodriguez, M. A. (2018). Mindfulness training for loneliness among Chinese college students: A pilot randomized controlled trial. *International Journal of Psychology*, *53*, 373–378. https://doi.org/10.1002/ijop.12394.

Chapter 16

Mild traumatic brain injury, cognitive behavioral therapy, and psychological interventions

Karen A. Sullivan

School of Psychology and Counselling, Queensland University of Technology, Brisbane, QLD, Australia

Abbreviations

mTBI mild traumatic brain injury
PCS postconcussion syndrome
PPCS persistent postconcussion symptoms
TBI traumatic brain injury

Introduction

Traumatic brain injury (TBI) is a leading cause of death and disability worldwide (Dewan et al., 2019). A TBI occurs when an external force or blow to the head or body results in a disruption of brain function (Gómez-de-Regil, Estrella-Castillo, & Vega-Cauich, 2019; Lefevre-Dognin et al., 2020). A TBI is clinically classified according to its severity as mild, moderate, or severe. This classification formally considers features such as the unconsciousness duration and the injured person's level of responsiveness (see Fig. 1). However, the use of the term "mild" to describe the severity of a TBI is discussed as potentially unhelpful; for example, if it is taken to mean that the injury is trivial. Others have criticized this label, arguing that the term mTBI masks a spectrum of injury (Lefevre-Dognin et al., 2020).

Approximately 70%–90% of all TBIs are "mild" (mTBI). mTBIs can affect people from all age groups. Common causes include falls, sport, physical violence (assault), and transport-related incidents. In a first-ever injury, most adults will recover fully within 7–10 days without requiring significant intervention (a *typical* recovery), but the recovery from mTBI can also be *atypical*. In such instances, the mTBI symptoms are experienced as persisting, and psychological or other interventions needed to support the recovery. The focus of this chapter is mTBI and persisting symptoms. It is beyond the scope of this chapter to address the psychological care of people who have experienced *repetitive* brain trauma or a *moderate or severe* brain injury as people's needs and outcomes may differ in such circumstances. Nevertheless, when providing care for a person with a suspected mTBI, and in research applications, it is critical to consider the injury severity and any history of repetitive brain trauma.

It is difficult to estimate the number of adults who will experience an atypical mTBI recovery; however, it is commonly reported that between 10% and 15% of adults who have a mTBI will experience persistent postconcussion symptoms (PPCS). Every year, 10 million people will have a mTBI globally; thus, in absolute terms, a staggering number of individuals will be affected. An atypical recovery from a mTBI has been described as a "poor" (Iverson, 2012), "difficult" (Makdissi, Cantu, Johnston, McCrory, & Meeuwisse, 2013), or "slow" recovery (Gagnon, Galli, Friedman, Grilli, & Iverson, 2009), characterized by "persistent" (Hadanny & Efrati, 2016) or "prolonged" mTBI symptoms (Quinn, Mayer, Master, & Fann, 2018). These symptoms include physical, affective, and cognitive features, such as headache, irritability, and difficulty concentrating, respectively, which impact on functioning for weeks and months, or even years. The functional effects include disruption to work and personal relationships and decreased quality of life. The PPCS symptom profile is heterogeneous (Lange, Brickell, Ivins, Vanderploeg, & French, 2013), dependent on the method of elicitation (Edmed & Sullivan, 2014; Karaliute et al., 2021), may change over the course of recovery (Fordal et al., 2022; Lishman, 1988), and is characterized by nonspecific, nonpathognomonic symptoms (Meares et al., 2008). This highlights

FIG. 1 A schematic of timeline for the clinical recovery from mTBI. The TBI is caused by an event, such as a fall. This event produces a disruption in usual brain functions because of an external mechanical force applied to the head or body. The acute injury effects can include loss of consciousness (LOC) and posttraumatic amnesia (PTA). The duration of the LOC and the PTA and the person's score on the Glasgow Coma Scale is used to classify the TBI as mild, moderate, or severe. The postacute period is characterized by symptoms such as headaches, dizziness, nausea, irritability, difficulty concentrating and so on. Within 7–10 days, these symptoms will usually resolve, without intensive intervention. If they persist, then they may be described as persistent postconcussion symptoms (PPCS). The timeframe for this designation is between 1 and 3 months postinjury. The factors that sustain these symptoms for this duration are considered multifactorial and include psychological processes.

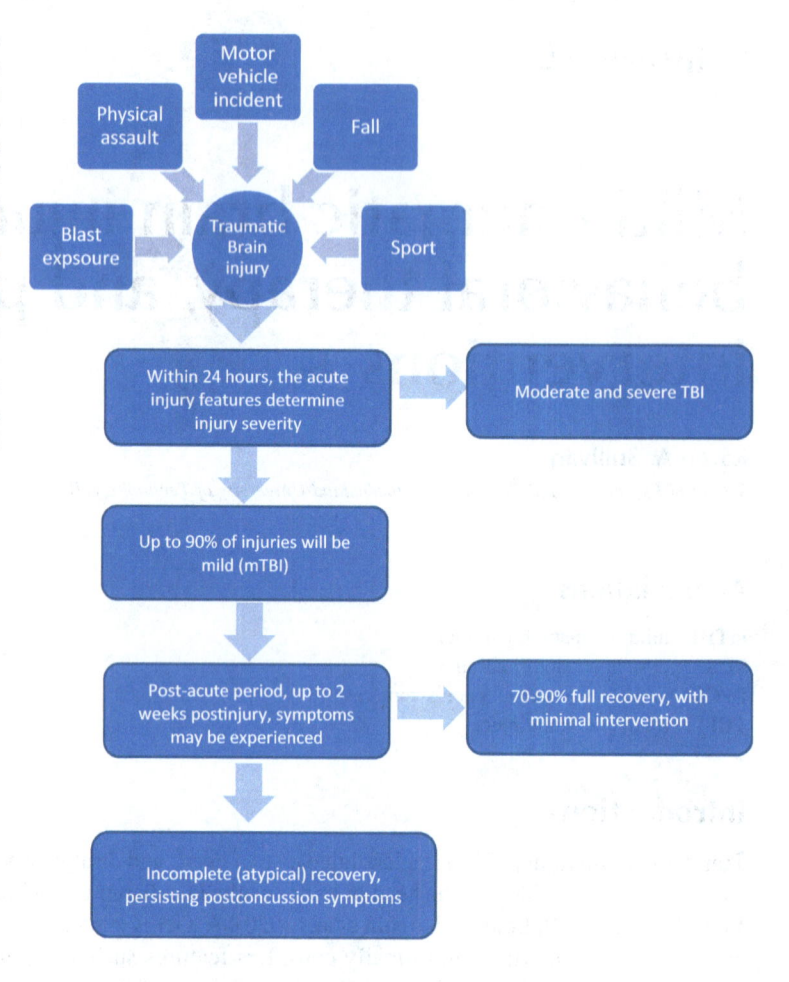

the importance of considering when and how the symptoms were assessed and the possible clinical interpretations, even before the treatment options are formulated.

Up until relatively recently, it was not uncommon to see a new diagnosis of the *postconcussion syndrome* (PCS) formally applied if a person's functioning did not return to usual within the expected time frame following the mTBI. Diagnostic criteria for the PCS were developed and appeared in major diagnostic systems, such as the Diagnostic and Statistical Manual of Mental Disorders (DSM) and the International Classification of Diseases (Voormolen et al., 2018). However, the notion of a postconcussion *syndrome* has been controversial for some time, and this remains so (Young, 2020). Further, in the 2013 revision of the DSM the relevant criteria were dropped (McIntyre, Amiri, & Kumbhare, 2020). Although alternative DSM diagnoses and diagnostic codes have been proposed for a PCS presentation (Young, 2020), in this chapter the term PCS and these alternate diagnoses are not used. Instead this chapter is framed around a discussion of the psychological approaches that might be useful for people who report persistent mTBI symptoms, or PPCS. This framing is intentionally broader than a discussion of the PCS per se.

Models of mTBI recovery

Considering models of typical and atypical mTBI recovery is important to inform decisions about whether psychological approaches could support the care of people with PPCS. Animal models of mTBI show that the acute injury effects includes significant physiological change; a chain reaction of events known as the neurometabolic cascade (Giza & Hovda, 2014). These changes can persist beyond the timeframe during which acute clinical features are typically identified in humans (Marklund et al., 2019), but they are not present weeks to months after injury and when symptoms would be described as *persistent*.

When a person is injured, the mTBI treatment is largely aimed at addressing any postacute symptoms that remain once stabilized. The standard clinical advice for recovery is to rest in the first 24–48 h postinjury, and then gradually resume usual activities as tolerated (Centers for Disease Control and Prevention, 2016; Marshall, Bayley, McCullagh, Velikonja, & Berrigan, 2012; McCrory et al., 2017). A worsening of symptoms or any new symptoms within the 7–10 day recovery period should prompt urgent medical attention (Centers for Disease Control and Prevention, 2016).

In models of mTBI recovery, the brain injury and the associated neurometabolic cascade are considered central. However, when symptoms are reported beyond the expected recovery period (atypical recovery or PPCS), there is an increased contribution to the presentation from other pre-, peri-, and postinjury factors. Collectively, these factors depict a biopsychosocial model of mTBI recovery (Iverson, 2012). Some mTBI models propose dynamic interactions between such factors (Kenzie et al., 2017; Polinder et al., 2018). For example, in addition to the "cellular level" injury response (such as the neurometabolic cascade that occurs acutely), some symptoms might be triggered or exacerbated by the wider injury context, such as if the injured occurred in particularly traumatic circumstances (e.g., in a physical assault). Another example would be if the index event (mTBI) subsequently becomes the subject of a personal injury claim, then the recovery in the weeks and months afterwards might seem to be slowed because simply engaging in legal processes can evoke PCS-like symptoms (Dunn, Lees-Haley, Brown, Williams, & English, 1995). One's premorbid ability to cope with illness and adversity can affect mTBI recovery, and psychological processes such as one's thoughts and beliefs about recovery, can all be conflated in the clinical picture, especially with increased time since injury. In some models these factors are presented as predisposing, precipitating and perpetuating the mTBI symptoms (Hou et al., 2012; Rickards, Cranston, & McWhorter, 2020), and in this way, they offer an explanation for PPCS.

In some of the earliest models of mTBI outcomes, the recovery process was considered susceptible to *physiologic* versus *psychogenic* factors; with a gradual transition to the latter, and a weaker contribution of the former with the passage of time (Lishman, 1988; Silverberg & Iverson, 2011). However, as the prior discussion of the contemporary models for mTBI recovery highlights, a wider array of interacting factors over an extended timeframe is now proposed as important to mTBI recovery. Many of the factors identified in these models have empirical support, and are noted as significant in systematic reviews of the factors that predict mTBI outcomes (Silverberg et al., 2015).

The identification of specific psychosocial factors in the atypical mTBI response invites consideration of the use of CBT as a therapeutic strategy. In particular, the formal identification of relevant factors and processes such as all-or-nothing behavior (Hou et al., 2012), fear avoidance (Silverberg, Panenka, & Iverson, 2018), poor coping skills (Ali, Mahoney, Dance, & Silverberg, 2019), motivational factors (Iverson, 2006; Sherer et al., 2020), and the cognitive perceptions and attributions of the injury, symptom origin, and recovery prospects (Gunstad & Suhr, 2001; Hou et al., 2012; Mittenberg, DiGiulio, Perrin, & Bass, 1992) are all potential targets for psychological therapies, and specifically, CBT.

This discussion highlights that there is a clear role for psychological interventions in mTBI care, particularly for atypical recovery. Further, given the timeline considerations, the therapy could be offered to *prevent* mTBI symptoms from becoming persistent, or to *manage* symptoms that have become persistent. Thus, if offered early post mTBI (within the first few weeks of injury), the therapy goal may be prevention; this approach is illustrated by Silverberg and colleague's randomized controlled trial of CBT to prevent *PPCS* (Silverberg et al., 2013). If offered late post-mTBI (e.g., 3 months or more post injury), the therapy goal may be to manage the now persistent pattern of symptoms (e.g., Belanger et al., 2020) or an individual complaint, such as insomnia. An example of the latter approach is the use of CBT-i for the treatment of insomnia following mTBI (Spencer, Collings, & Bloor, 2019). In the research on psychological approaches for mTBI and PPCS, time since injury is often poorly controlled, adding a complication to the interpretation of this literature. However, time since mTBI is clearly a very important consideration for both clinicians and researchers who are considering using CBT for mTBI.

Psychological approaches for mTBI recovery

There have been several systematic reviews of psychological interventions for the prevention and management of persistent symptoms post mTBI (Sullivan et al., 2020; Teo, Fong, Chen, & Chung, 2020). In addition to this, the literature includes several randomized controlled trials, meta-analyses, published treatment manuals, trial protocols for emerging programs, and case illustrations of relevance to this topic. There is a strong call for multidisciplinary management models (McCarty et al., 2019; Snyder & Giza, 2019) and trials of such approaches (Jaganathan & Sullivan, 2020); however, this chapter is focused solely on the evaluation of psychological therapies. Psychological approaches can be contrasted with other approaches for persistent mTBI symptoms, such as pharmacological, "device," or physical therapy (Sullivan, Hills, & Iverson, 2018); but it is important to note that nonpsychological approaches are also being actively investigated in the search for a solution to this problem (Burke, Fralick, Nejatbakhsh, Tartaglia, & Tator, 2015).

In the prevention or management of persistent symptoms after mTBI, the definition of a psychological therapy or approach, and the specific therapeutic subtypes, is both clearly fundamental and open to debate. For example, in the published reviews in this area, *counseling* is considered a psychological therapy, as is *psychoeducation* (Burke et al., 2015). Further, it is possible that some pharmacological therapies may include elements of psychoeducation; and some psychological programs might encourage physical (or cognitive) activity. For the purposes of this chapter, to qualify as a psychological therapy—and CBT specifically—it was considered that the therapy must aim to improve outcomes by changing a psychological state or process through an exploration of the links between thoughts, feelings, and behavior. Under this definition, an eligible therapy could aim to reduce the psychological distress associated with an atypical mTBI recovery, or promote cognitive self-efficacy, if challenged by negative stereotypes about brain injury. In the review by Bergersen and colleagues, their criteria for defining a psychological approach for PPCS was that the therapy should be consistent with the American Psychological Association's definition of psychotherapy, and if not from an identifiable school of thought, then it should be derived from psychological principles (Bergersen, Halvorsen, Tryti, Taylor, & Olsen, 2017). This discussion shows that the definition of a psychological therapy for PPCS, and even the characterization of some therapies as "based" on CBT, introduces a challenge for the evaluation of these therapies and their effectiveness in mTBI and PPCS.

Despite this challenge, in past reviews and scoping documents of "psychological" interventions for mTBI and PPCS, four approaches are typically identified. These include counseling, psychoeducation (or "education and reassurance"), psychotherapy interventions (e.g., CBT, mindfulness therapy, etc.), and cognitive retraining which may be aimed at (a) teaching strategies to *restore or retrain* cognitive functions or (b) *mitigating* cognitive load. Moore and colleagues provide a further description of these types of interventions and how they might work to improve PPCS (Moore, Mawdsley, Jackson, & Atherton, 2017). For example, psychoeducation is considered potentially effective as it can provide individuals with information about their symptoms and whether these are typical after mTBI, why atypical symptoms are sometimes experienced, and the multitude of factors that can contribute to symptom persistence, sometimes indirectly. Thus, by a building understanding of the injury and its effects, a new understanding can be reached, including about the potentially modifiable factors that could be addressed to improve mTBI recovery. Psychoeducation is often discussed as a standalone therapy, but it can also be part of a CBT approach. In some but not all reviews, the breakdown isolates a particular psychological therapy (Al Sayegh, Sandford, & Carson, 2010; Teo et al., 2020), such as CBT or mindfulness, but when this does not occur it is not possible to draw conclusions about the effects of therapeutic components.

CBT for mTBI recovery

Of the psychological approaches that are both plausible and tested, CBT is the most extensively researched. CBT has been used in randomized controlled trials for PPCS in children and adults (McNally et al., 2018; Silverberg et al., 2013). Some studies have applied additional selection criteria, such as targeting individuals with high early symptom burden (Ali et al., 2019). Group and individualized programs have been used (Anson & Ponsford, 2006), and the delivery has involved face to face and technology-enabled methods as well as brief (3 weeks; Ali et al., 2019) or long format programs (e.g., 12 weeks; Potter & Brown, 2012).

Several authors have published or prepared CBT treatment manuals or materials for use with people with PPCS (Ali et al., 2019; Ferguson & Mittenberg, 1996; Potter & Brown, 2012). Potter and Brown recently outlined the principles and practice of CBT for this purpose (Potter & Brown, 2012). These authors devised a 12-session program that incorporates adapted materials from an early pioneer of this approach (Mittenberg et al., 1999; Mittenberg, Zielinski, & Fichera, 1993). The Potter and Brown (2012) program is consistent with the now widely endorsed practice of encouraging a gradual and incremental return to usual activities, including sport (McCrory et al., 2017) and study (Marklund et al., 2019). A notable point about the Potter and Brown (2012) program is that it explicitly recognizes that the response to mTBI can be heterogeneous; the symptoms that trouble one person can differ from those that trouble another person, thus Potter and Brown argue against a highly prescriptive approach (Potter & Brown, 2012). Instead, the program advocated by Potter and Brown (2012) commences with establishing the foundation of the program, including that CBT is a collaborative, goal-oriented therapy. It is stressed that the program's focus is on *current* concerns, and is especially interested in the connections between a person's thoughts, feelings, and actions; and how these connections can be considered (or reconsidered) so that the problem can be addressed. The sessions in this program utilize the specific tools and strategies of problem-solving, homework, and relapse prevention.

Potter and Brown (2012) identify some key issues and themes that are relevant for CBT programs for people with PPCS such as symptom attributions and people's appraisal of the injury and their recovery potential Potter & Brown, 2012. As noted previously, empirical mTBI studies have shown that cognitive processes can distort symptom attributions, such that an everyday symptom, like headache, is experienced as worse if it occurs after a negative landmark event, like mTBI (Gunstad & Suhr, 2001). Thus, a process of exploring and facilitating cognitive re-attributions could be used, and if this reveals a modifiable alternate source for the complaint (e.g., the headache is more likely to be a consequence of caffeine withdrawal rather than the prior mTBI), this can allow this symptom to be understood differently (changed attributions), and addressed via new avenues.

If negative stereotypes about brain injury are affecting performance or expectations for recovery, the therapeutic strategy could draw attention to the discrepancy in actual versus perceived functional limitations, and aim for change by building cognitive self-efficacy. Similarly, if there is evidence of anxiety in relation to the performance of a cognitive task, the underpinning cognitive processes can be elicited and explored. CBT techniques such as thought records and Socratic questioning can be used to highlight the links between such thoughts and the subsequent feelings and actions prior to testing strategies that might result in more functional cognitive attributions (Potter & Brown, 2012).

CBT effectiveness and efficacy for mTBI and PPCS

A CBT program for PPCS was first published over 25 years ago (Ferguson & Mittenberg, 1996). This approach is theoretically plausible as already discussed, and in the literature, it is often described as *promising* (Al Sayegh et al., 2010; Brent & Max, 2017; Jaber, Hartwell, & Radel, 2019; Potter & Brown, 2012). Despite this, some authors have urged caution when drawing strong conclusions about the potential benefits of CBT for PPCS (Sullivan et al., 2020). For example, some of the most rigorous trials in this area have not shown a significant benefit of CBT for the prevention of PPCS (Silverberg et al., 2013). Further, in a recent metaanalysis PPCS symptom improvement following CBT was not shown, but other patient gains were evident (e.g., improved mood and social integration; Chen, Lin, Huda, & Tsai, 2020). Thus, CBT may be promising, but the limits of this promise must be noted. A cautious promise is needed because of the mixed results and mixed quality of the foundation studies, some of which are acknowledged as poor (Snell, Surgenor, Hay-Smith, & Siegert, 2009) or containing a high risk of bias (Sullivan et al., 2020). Other reasons for caution include the number and size of studies, which is often quite small (Ali et al., 2019), the failure to include an active control group (Arbabi et al., 2020), and problems with the use of nonstandard definitions and terms when describing the study entry criteria, outcomes, injury-intervention intervals, and the therapy itself (Karaliute et al., 2021; Snell et al., 2009). Some of the issues, such as the retention of participants throughout the trial (Silverberg et al., 2013) and potential recruitment and selection biases (Luoto et al., 2013), are not unique to CBT studies for mTBI; they are relevant for all intervention trials for PPCS.

Client considerations

Much of this discussion has focused on the likely treatment response to CBT of adults who report ongoing difficulties after mTBI (i.e., atypical recovery or PPCS). However, there is also a growing body of research into the effective CBT treatments for children and adolescents who report PPCS (Brent & Max, 2017; Committee on Sports-Related Concussions in Youth, 2014) and subpopulations, such as athletes (Elbin, Schatz, Lowder, & Kontos, 2014). Like the approach used by Potter and Brown (2012), McNally and colleagues' (2018) program was also based on early pioneer work in this area (Ferguson & Mittenberg, 1996). In the McNally et al. (2018) program, four treatment modules, each lasting between 45 and 60 min are delivered, depending on individual clinical needs. Each participant (and a guardian) receives between two and five sessions. The McNally et al. (2018) treatment modules comprise psychoeducation, activity and sleep scheduling, relaxation training, and cognitive restructuring, respectively. The module "goal" and examples of the relevant therapy activities are described in detail by McNally et al. (2018). McNally and colleagues have shown that their program reduces symptoms, and improves functional outcomes and quality-of-life; but unfortunately, their evaluation was not a randomized controlled trial; thus, a cautious interpretation is warranted.

Women have an increased risk of a worse outcome from mTBI (atypical recovery or PPCS), as do those individuals who report a higher symptom burden or greater psychological distress at the time of injury, people with comorbid or premorbid psychiatric conditions, people who are engaged in injury-related medicolegal processes, and people who may have limited social support or are unable to seek work or study accommodations in the acute period postinjury. These factors must all be considered when planning the therapy.

The role of the injury context, for example, whether the mTBI occurred during sport or as a result of intimate partner violence, can theoretically add to the traumatic stress of the injury, and thus affect the outcomes; but, unfortunately, in many treatment mTBI trials and prognostic studies this factor is not separately examined (Mathias, Harman-Smith, Bowden, Rosenfeld, & Bigler, 2014). There is some evidence from a systematic review (Mathias et al., 2014) and a simulation study (Sullivan & Wade, 2017) that supports higher reported symptoms following mTBI when the injury cause is assault as opposed to another cause. Further, in posttraumatic stress disorder, which is a chronic reaction to a trauma exposure, the outcomes are known to be related to such distinctions (i.e., a worse outcome is more likely if the index event involved interpersonal violence; Luthra et al., 2009). Future prevention and management trials must consider these factors in their design. Meanwhile, the available evidence supports the recommendation that these client considerations must be considered in the framing of CBT research and therapy for PPCS.

Recommendations for future CBT studies

Not everyone agrees that the field will be advanced by simply making improvements in the design of CBT intervention studies for PPCS, although this suggestion is commonly aired. Given the mixed findings to date, some authors have urged further consideration of the underlying illness models, and if these require adjustments prior to further investment in intervention studies (Snell et al., 2009). This might include reframing the presentation and potential diagnoses when a person presents with problems or concerns that they or others have linked to the earlier mTBI (Snell et al., 2009; Young, 2020). For example, an alternate framing of this presentation as a form of somatic-symptom disorder has been proposed (Young, 2020). Whether or not this point is agreed, it is a reminder that we must improve the quality of our trial design, carefully consider the diagnosis, and continue the search for better explanatory models. These are important considerations given that the number of registered interventional trials for PPCS has seen rapid expansion (Burke et al., 2015).

Some authors have observed that management approaches for PPCS are "evolving" from a "one-sized fits all" approach to a more targeted and specific rehabilitation plan formulated to address individual needs (Elbin et al., 2014). This may be true in some contexts. However, since mTBI is a global health issue, the availability of such treatments clearly depends on wider social and economic considerations, including variations in health care delivery. Even within a country such as Australia, there are identifiable differences in the care and support of elite or professional athletes who have had a mTBI, as compared to the support available to amateur athletes and community sport participants, and other members of the community. For example, there are specialist guidelines to improve the identification, assessment, and management of mTBI in some professional sports, and the injured athlete will often have access to a dedicated team of health and medical staff to support their recovery (Australian Rugby League Commission, 2019). CBT programs for PPCS, even if scientifically proven, may not be sufficiently accessible or affordable to support the global demand for them. Taken together, this discussion highlights that although the CBT programs discussed in this chapter lend themselves to the ideal of individualized treatments and are a theoretically plausible means of facilitating better outcomes in people who either have or are at risk of PPCS, there are some significant wider geo-political issues that are pertinent to their translation.

Conclusion

It is recognized that CBT is one of the most widely investigated psychological therapies for PPCS. This model of care is very well suited to addressing the psychological factors that are known to contribute to mTBI recovery. Despite this, continued efforts are needed to quantify the benefits of CBT for PPCS, and potentially, how this approach can be translated into strategies that local communities can adopt. At the same time, it is imperative to explore and explain the limits of the CBT promise for PPCS.

Practice and procedures

The Rivermead Postconcussion Symptoms Questionnaire (RPQ; King, Crawford, Wenden, Moss, & Wade, 1995) can be used to assess PPCS. It is the most commonly used outcome measure in intervention trials for PPCS, and in much of the research discussed in this chapter, the RPQ was used to measure PPCS. The RPQ asks the respondent to rate individual symptoms—such as headache. The ratings indicate how much of a problem a particular symptom has been over the past 24-h (from *not at all* to *severe*), compared to before the injury. The RPQ is endorsed by several international agencies for this use, and it is used in major outcome studies. Despite this, discussions continue about the choice of PPCS measure and specific applications of the RPQ, and how such choices may affect study interpretation(Karaliute et al., 2021; Voormolen et al., 2018). While further consensus on such matters is clearly needed, the informed use of the RPQ—as a standardized measure of PPCS—will add rigour to future research. Guidelines on how to optimize RPQ use in multicenter trials have also been published (Bodien et al., 2018), and such efforts are important to help build the consensus.

Mini-dictionary of terms

- *Rivermead Postconcussion Symptoms Questionnaire*. A commonly used outcome measure in CBT interventions for PPCS.
- *Postconcussion symptoms*. Symptoms such as headache, irritability, and difficulty concentrating that occur after a mTBI and are reported as persisting for weeks, months, and even years post-mTBI.

Key facts of mTBI

- Globally, TBI is estimated to affect 10 million people annually and between 70% and 90% of these injuries are mild.
- If mTBI symptoms persist beyond the typical recovery period of 1–3 months, a diagnosis of the postconcussion syndrome (PCS) may be given, or the person can be said to have persistent postconcussion symptoms (PPCS).
- More than 1 in 10 people who have had a mTBI are estimated to experience PPCS.
- Contemporary models of PPCS explain this phenomenon through a biopsychosocial lens—a position now widely supported by empirical research.
- The application of psychological therapies for people with PPCS is theoretically justified and there is some support for these approaches in the empirical research.
- While some benefits of CBT for PPCS are noted in studies with a high level of evidence—including systematic reviews and meta-analyses—there are mixed findings and stronger evidence is needed to fully understand the potential benefits.

Applications to other areas

In this chapter, several methodological issues were discussed that raise questions about the strength of the conclusions that can be drawn about the effectiveness of CBT for PPCS. These issues include potential selection biases affecting the study participants and the generalizability of the findings, diagnostic issues including poor initial (mTBI) and longer term (PPCS) icharacterizations, and ongoing discussion about the best outcome measures. These issues are not only relevant to the evaluation of the scientific evidence; they are relevant to the clinical care of individuals. Practitioners should consider these issues, when framing their care of patients.

Summary points

- mTBI is a very common injury globally; and although most people will experience a full recovery within 2 weeks, more than 1 in 10 people will report ongoing disability and persistent symptoms.
- Theoretical models of mTBI recovery identify psychological factors as contributing to a rapid recovery versus protracted symptoms.
- Specific psychological factors have been linked to a worse mTBI recovery, including cognitive and behavioral factors that can be addressed with CBT, such as cognitive misattributions and fear avoidance.
- Over 25 years ago, a CBT-based psychological therapy was developed to improve mTBI outcomes in people with persisting problems.
- Based on the accumulated evidence, including randomized controlled trials, CBT appears to provide some benefits for people with PPCS; however strong conclusions about these benefits are still premature.

References

Al Sayegh, A., Sandford, D., & Carson, A. J. (2010). Psychological approaches to treatment of postconcussion syndrome: A systematic review. *Journal of Neurology, Neurosurgery & Psychiatry, 81*(10), 1128. https://doi.org/10.1136/jnnp.2008.170092.

Ali, J. I., Mahoney, P., Dance, D., & Silverberg, N. D. (2019). Outcomes of a brief coping skills group intervention for adults with severe postconcussion symptoms. *Concussion, 4*(3), 67. https://doi.org/10.2217/cnc-2019-0011.

Anson, K., & Ponsford, J. (2006). Evaluation of a coping skills group following traumatic brain injury. *Brain Injury, 20*(2), 167–178. https://doi.org/10.1080/02699050500442956.

Arbabi, M., Sheldon, R. J. G., Bahadoran, P., Smith, J. G., Poole, N., & Agrawal, N. (2020). Treatment outcomes in mild traumatic brain injury: A systematic review of randomized controlled trials. *Brain Injury, 34*(9), 1139–1149. https://doi.org/10.1080/02699052.2020.1797168.

Australian Rugby League Commission. (2019). *Guidelines for the management of concussion in Rugby League*. Australia.

Belanger, H. G., Vanderploeg, R. D., Curtiss, G., Armistead-Jehle, P., Kennedy, J. E., Tate, D. F., … Cooper, D. B. (2020). Self-efficacy predicts response to cognitive rehabilitation in military service members with post-concussive symptoms. *Neuropsychological Rehabilitation, 30*(6), 1190–1203. https://doi.org/10.1080/09602011.2019.1575245.

Bergersen, K., Halvorsen, J.Ø., Tryti, E. A., Taylor, S. I., & Olsen, A. (2017). A systematic literature review of psychotherapeutic treatment of prolonged symptoms after mild traumatic brain injury. *Brain Injury, 31*(3), 279–289. https://doi.org/10.1080/02699052.2016.1255779.

Bodien, Y. G., McCrea, M., Dikmen, S., Temkin, N., Boase, K., Machamer, J., … Giacino, J. T. (2018). Optimizing outcome assessment in multicenter TBI trials: Perspectives from TRACK-TBI and the TBI Endpoints Development Initiative. *Journal of Head Trauma Rehabilitation, 33*(3), 147–157. https://doi.org/10.1097/htr.0000000000000367.

Brent, D. A., & Max, J. (2017). Psychiatric sequelae of concussions. *Current Psychiatry Reports, 19*(12), 108.

Burke, M. J., Fralick, M., Nejatbakhsh, N., Tartaglia, M. C., & Tator, C. H. (2015). In search of evidence-based treatment for concussion: Characteristics of current clinical trials. *Brain Injury, 29*(3), 300–305. https://doi.org/10.3109/02699052.2014.974673.

Centers for Disease Control and Prevention. (2016). *Updated mild traumatic brain injury guideline for adults*. Retrieved from https://www.cdc.gov/traumaticbraininjury/mtbi_guideline.html.

Chen, C.-L., Lin, M.-Y., Huda, M. H., & Tsai, P.-S. (2020). Effects of cognitive behavioral therapy for adults with post-concussion syndrome: A systematic review and meta-analysis of randomized controlled trials. *Journal of Psychosomatic Research, 136*. https://doi.org/10.1016/j.jpsychores.2020.110190, 110190.

Committee on Sports-Related Concussions in Youth, & Board of Children, Youth, and Families, Institute of Medicine, National Research Council. (2014). Treatment and management of prolonged symptoms and post-concussion syndrome. In R. F. Graham, M. A. Ford, et al. (Eds.), *Sports-related concussions in youth: Improving the science, changing the culture*. Washington, DC: National Academies Press.

Dewan, M. C., Rattani, A., Gupta, S., Baticulon, R. E., Hung, Y.-C., Punchak, M., ... Park, K. B. (2019). Estimating the global incidence of traumatic brain injury. *Journal of Neurosurgery, 130*(4), 1080. https://doi.org/10.3171/2017.10.Jns17352.

Dunn, J. T., Lees-Haley, P. R., Brown, R. S., Williams, C. W., & English, L. T. (1995). Neurotoxic complaint base rates of personal injury claimants: Implications for neuropsychological assessment. *Journal of Clinical Psychology, 51*(4), 577–584.

Edmed, S. L., & Sullivan, K. A. (2014). Method of symptom assessment influences cognitive, affective and somatic post-concussion-like symptom base rates. *Brain Injury, 28*(10), 1277–1282. https://doi.org/10.3109/02699052.2014.915988.

Elbin, R., Schatz, P., Lowder, H. B., & Kontos, A. P. (2014). An empirical review of treatment and rehabilitation approaches used in the acute, sub-acute, and chronic phases of recovery following sports-related concussion. *Current Treatment Options in Neurology, 16*(11), 320.

Ferguson, R. J., & Mittenberg, W. (1996). Cognitive-behavioral treatment of postconcussion syndrome: A therapist's manual. In *Sourcebook of psychological treatment manuals for adult disorders* (pp. 615–655). New York, NY: Plenum Press.

Fordal, L., Stenberg, J., Iverson, G., Saksvik, S., Karaliute, M., Vik, A., ... Skandsen, T. (2022). Trajectories of persistent postconcussion symptoms and factors associated with symptom reporting after mild traumatic brain injury. *Archives of Physical Medicine and Rehabilitation, 103*(2), 313–322. https://doi.org/10.1016/j.apmr.2021.09.016.

Gagnon, I., Galli, C., Friedman, D., Grilli, L., & Iverson, G. (2009). Active rehabilitation for children who are slow to recover following sport-related concussion. *Brain Injury, 23*(12), 956.

Giza, C. C., & Hovda, D. A. (2014). The new neurometabolic cascade of concussion. *Neurosurgery, 75*(Suppl. 4(0 4)), S24–S33. https://doi.org/10.1227/neu.0000000000000505.

Gómez-de-Regil, L., Estrella-Castillo, D. F., & Vega-Cauich, J. (2019). Psychological intervention in traumatic brain injury patients. *Behavioural Neurology, 2019*, 6937832. https://doi.org/10.1155/2019/6937832.

Gunstad, J., & Suhr, J. (2001). "Expectation as etiology" versus "the good old days": Postconcussion syndrome symptom reporting in athletes, headache sufferers, and depressed individuals. *Journal of the International Neuropsychological Society, 7*(3), 323.

Hadanny, A., & Efrati, S. (2016). Treatment of persistent post-concussion syndrome due to mild traumatic brain injury: Current status and future directions. *Expert Review of Neurotherapeutics, 16*(8), 875–887. https://doi.org/10.1080/14737175.2016.1205487.

Hou, R., Moss-Morris, R., Peveler, R., Mogg, K., Bradley, B. P., & Belli, A. (2012). When a minor head injury results in enduring symptoms: A prospective investigation of risk factors for postconcussional syndrome after mild traumatic brain injury. *Journal of Neurology Neurosurgery and Psychiatry, 83*(2), 217–223. https://doi.org/10.1136/jnnp-2011-300767.

Iverson, G. L. (2006). Ethical issues associated with the assessment of exaggeration, poor effort, and malingering. *Applied Neuropsychology, 13*(2), 77–90. https://doi.org/10.1207/s15324826an1302_3.

Iverson, G. L. (2012). A biopsychosocial conceptualization of poor outcome from mild traumatic brain injury. In J. J. Vasterling, R. A. Bryant, & T. M. Keane (Eds.), *PTSD and mild traumatic brain injury* (pp. 36–60). New York: The Guilford Press.

Jaber, A. F., Hartwell, J., & Radel, J. D. (2019). Interventions to address the needs of adults with postconcussion syndrome: A systematic review. *American Journal of Occupational Therapy, 73*(1). https://doi.org/10.5014/ajot.2019.028993. 7301205020p7301205021-7301205020p7301205012.

Jaganathan, K. S., & Sullivan, K. A. (2020). Moving towards individualised and interdisciplinary approaches to treat persistent post-concussion symptoms. *EClinicalMedicine, 18*. https://doi.org/10.1016/j.eclinm.2019.11.023, 100230.

Karaliute, M., Saksvik, S., Smevik, H., Follestad, T., Einarsen, C., Vik, A., ... Olsen, A. (2021). Methodology matters: Comparing approaches for defining persistent symptoms after mild traumatic brain injury. *Neurotrauma Reports, 2*(1), 603–617. https://doi.org/10.1089/neur.2021.0028.

Kenzie, E. S., Parks, E. L., Bigler, E. D., Lim, M. M., Chesnutt, J. C., & Wakeland, W. (2017). Concussion as a multi-scale complex system: An interdisciplinary synthesis of current knowledge. *Frontiers in Neurology, 8*(513). https://doi.org/10.3389/fneur.2017.00513.

King, N. S., Crawford, S., Wenden, F. J., Moss, N. E. G., & Wade, D. T. (1995). The rivermead post-concussion symptoms questionnaire: A measure of symptoms commonly experienced after head injury and its reliability. *Journal of Neurology, 242*(9), 587–592. https://doi.org/10.1007/bf00868811.

Lange, R., Brickell, T., Ivins, B., Vanderploeg, R., & French, L. (2013). Variable, not always persistent, postconcussion symptoms after mild TBI in U.S. military service members: A five-year cross-sectional outcome study. *Journal of Neurotrauma, 30*(11), 958–969. https://doi.org/10.1089/neu.2012.2743.

Lefevre-Dognin, C., Cogné, M., Perdrieau, V., Granger, A., Heslot, C., & Azouvi, P. (2020). Definition and epidemiology of mild traumatic brain injury. *Neuro-Chirurgie*. https://doi.org/10.1016/j.neuchi.2020.02.002.

Lishman, W. A. (1988). Physiogenesis and psychogenesis in the 'post-concussional syndrome'. *British Journal of Psychiatry, 153*, 460–469. https://doi.org/10.1192/bjp.153.4.460.

Luoto, T. M., Tenovuo, O., Kataja, A., Brander, A., Öhman, J., & Iverson, G. L. (2013). Who gets recruited in mild traumatic brain injury research? *Journal of Neurotrauma, 30*(1), 11–16. https://doi.org/10.1089/neu.2012.2611.

Luthra, R., Abramovitz, R., Greenberg, R., Schoor, A., Newcorn, J., Schmeidler, J., ... Chemtob, C. M. (2009). Relationship between type of trauma exposure and posttraumatic stress disorder among urban children and adolescents. *Journal of Interpersonal Violence, 24*(11), 1919–1927. https://doi.org/10.1177/0886260508325494.

Makdissi, M., Cantu, R. C., Johnston, K. M., McCrory, P., & Meeuwisse, W. H. (2013). The difficult concussion patient: What is the best approach to investigation and management of persistent (>10 days) postconcussive symptoms? *British Journal of Sports Medicine, 47*(5), 308–313. https://doi.org/10.1136/bjsports-2013-092255.

Marklund, N., Bellander, B.-M., Godbolt, A. K., Levin, H., McCrory, P., & Thelin, E. P. (2019). Treatments and rehabilitation in the acute and chronic state of traumatic brain injury. *Journal of Internal Medicine, 285*(6), 608–623. https://doi.org/10.1111/joim.12900.

Marshall, S., Bayley, M., McCullagh, S., Velikonja, D., & Berrigan, L. (2012). Clinical practice guidelines for mild traumatic brain injury and persistent symptoms. *Canadian Family Physician, 58*(3), 257–267. e128-240.

Mathias, J. L., Harman-Smith, Y., Bowden, S. C., Rosenfeld, J. V., & Bigler, E. D. (2014). Contribution of psychological trauma to outcomes after traumatic brain injury: Assaults versus sporting injuries. *Journal of Neurotrauma, 31*(7), 658–669. https://doi.org/10.1089/neu.2013.3160.

McCarty, C. A., Zatzick, D., Hoopes, T., Payne, K., Parrish, R., & Rivara, F. P. (2019). Collaborative care model for treatment of persistent symptoms after concussion among youth (CARE4PCS-II): Study protocol for a randomized, controlled trial. *Trials, 20*(1), 567. https://doi.org/10.1186/s13063-019-3662-3.

McCrory, P., Meeuwisse, W., Dvorak, J., Aubry, M., Bailes, J., Broglio, S., ... Vos, P. E. (2017). Consensus statement on concussion in sport—The 5th international conference on concussion in sport held in Berlin, October 2016. *British Journal of Sports Medicine, 51*(11), 838–847. https://doi.org/10.1136/bjsports-2017-097699.

McIntyre, M., Amiri, M., & Kumbhare, D. (2020). Post-concussion syndrome: A diagnosis of DSM past. *American Journal of Physical Medicine and Rehabilitation*. https://doi.org/10.1097/phm.0000000000001586. Publish Ahead of Print.

McNally, K. A., Patrick, K. E., LaFleur, J. E., Dykstra, J. B., Monahan, K., & Hoskinson, K. R. (2018). Brief cognitive behavioral intervention for children and adolescents with persistent post-concussive symptoms: A pilot study. *Child Neuropsychology, 24*(3), 396–412. https://doi.org/10.1080/09297049.2017.1280143.

Meares, S., Shores, E. A., Taylor, A. J., Batchelor, J., Bryant, R. A., Baguley, I. J., ... Marosszeky, J. E. (2008). Mild traumatic brain injury does not predict acute postconcussion syndrome. *Journal of Neurology, Neurosurgery and Psychiatry, 79*(3), 300. https://doi.org/10.1136/jnnp.2007.126565.

Mittenberg, W., DiGiulio, D., Perrin, S., & Bass, A. (1992). Symptoms following mild head injury: Expectation as aetiology. *Journal of Neurology, Neurosurgery & Psychiatry, 55*(3), 200.

Mittenberg, W., Ustarroz, J., Zielinski, R., Arboniés, A., Fichera, S., & Ferreras, A. (1999). Postconcussional syndrome: A treatment manual for patients. *Anales de Psiquiatria, 15*, 315–323.

Mittenberg, W., Zielinski, R., & Fichera, S. (1993). Recovery from mild head injury: A treatment manual for patients. *Psychotherapy in Private Practice, 12*, 37–52.

Moore, P., Mawdsley, L., Jackson, C. F., & Atherton, M. J. (2017). Psychological interventions for persisting postconcussion symptoms following traumatic brain injury. *Cochrane Database of Systematic Reviews*, (8). https://doi.org/10.1002/14651858.CD012755.

Polinder, S., Cnossen, M. C., Real, R. G. L., Covic, A., Gorbunova, A., Voormolen, D. C., ... von Steinbuechel, N. (2018). A multidimensional approach to post-concussion symptoms in mild traumatic brain injury. *Frontiers in Neurology, 9*(1113). https://doi.org/10.3389/fneur.2018.01113.

Potter, S., & Brown, R. G. (2012). Cognitive behavioural therapy and persistent post-concussional symptoms: Integrating conceptual issues and practical aspects in treatment. *Neuropsychological Rehabilitation, 22*(1), 1–25. https://doi.org/10.1080/09602011.2011.630883.

Quinn, D. K., Mayer, A. R., Master, C. L., & Fann, J. R. (2018). Prolonged postconcussive symptoms. *American Journal of Psychiatry, 175*(2), 103–111. https://doi.org/10.1176/appi.ajp.2017.17020235.

Rickards, T. A., Cranston, C. C., & McWhorter, J. (2020). Persistent post-concussive symptoms: A model of predisposing, precipitating, and perpetuating factors. *Applied Neuropsychology: Adult*, 1–11. https://doi.org/10.1080/23279095.2020.1748032.

Sherer, M., Sander, A. M., Ponsford, J., Vos, L., Poritz, J. M. P., Ngan, E., & Leon Novelo, L. (2020). Patterns of cognitive test scores and symptom complaints in persons with TBI who failed performance validity testing. *Journal of the International Neuropsychological Society, 26*(9), 932–938. https://doi.org/10.1017/S1355617720000351.

Silverberg, N. D., Gardner, A. J., Brubacher, J. R., Panenka, W. J., Li, J. J., & Iverson, G. L. (2015). Systematic review of multivariable prognostic models for mild traumatic brain injury. *Journal of Neurotrauma, 32*(8), 517–526. https://doi.org/10.1089/neu.2014.3600.

Silverberg, N. D., Hallam, B. J., Rose, A., Underwood, H., Whitfield, K., Thornton, A. E., & Whittal, M. L. (2013). Cognitive-behavioral prevention of postconcussion syndrome in at-risk patients: A pilot randomized controlled trial. *Journal of Head Trauma Rehabilitation, 28*(4), 313–322.

Silverberg, N. D., & Iverson, G. (2011). Etiology of the post-concussion syndrome: Physiogenesis and psychogenesis revisited. *NeuroRehabilitation, 29*(4), 317.

Silverberg, N. D., Panenka, W. J., & Iverson, G. L. (2018). Fear avoidance and clinical outcomes from mild traumatic brain injury. *Journal of Neurotrauma, 35*(16), 1864–1873. https://doi.org/10.1089/neu.2018.5662.

Snell, D. L., Surgenor, L. J., Hay-Smith, E. J. C., & Siegert, R. J. (2009). A systematic review of psychological treatments for mild traumatic brain injury: An update on the evidence. *Journal of Clinical and Experimental Neuropsychology, 31*(1), 20–38. https://doi.org/10.1080/13803390801978849.

Snyder, A. R., & Giza, C. C. (2019). The future of concussion. *Seminars in Pediatric Neurology, 30*, 128–137. https://doi.org/10.1016/j.spen.2019.03.018.

Spencer, R. J., Collings, A. S., & Bloor, L. E. (2019). Cognitive behavioral therapy for insomnia as treatment for post-concussive symptoms. *Medical Research, 1*(1), 5. https://doi.org/10.35702/mrj.10005.

Sullivan, K. A., Hills, A. P., & Iverson, G. L. (2018). Graded combined aerobic resistance exercise (CARE) to prevent or treat the persistent post-concussion syndrome. *Current Neurology and Neuroscience Reports, 18*(11), 75. https://doi.org/10.1007/s11910-018-0884-9.

Sullivan, K. A., Kaye, S.-A., Blaine, H., Edmed, S. L., Meares, S., Rossa, K., & Haden, C. (2020). Psychological approaches for the management of persistent postconcussion symptoms after mild traumatic brain injury: A systematic review. *Disability and Rehabilitation, 42*(16), 2243–2251. https://doi.org/10.1080/09638288.2018.1558292.

Sullivan, K. A., & Wade, C. (2017). Does the cause of the mild traumatic brain injury affect the expectation of persistent postconcussion symptoms and psychological trauma? *Journal of Clinical and Experimental Neuropsychology, 39*(4), 408–418. https://doi.org/10.1080/13803395.2016.1230597.

Teo, S. H., Fong, K. N. K., Chen, Z., & Chung, R. C. K. (2020). Cognitive and psychological interventions for the reduction of post-concussion symptoms in patients with mild traumatic brain injury: A systematic review. *Brain Injury, 34*(10), 1305–1321. https://doi.org/10.1080/02699052.2020.1802668.

Voormolen, D. C., Cnossen, M. C., Polinder, S., von Steinbuechel, N., Vos, P. E., & Haagsma, J. A. (2018). Divergent classification methods of post-concussion syndrome after mild traumatic brain injury: Prevalence rates, risk factors, and functional outcome. *Journal of Neurotrauma, 35*(11), 1233–1241. https://doi.org/10.1089/neu.2017.5257.

Young, G. (2020). Thirty complexities and controversies in mild traumatic brain injury and persistent post-concussion syndrome: A roadmap for research and practice. *Psychological Injury and Law.* https://doi.org/10.1007/s12207-020-09395-6.

Chapter 17

Multiple sclerosis fatigue and the use of cognitive behavioral therapy: A new narrative

Moussa A. Chalah and Samar S. Ayache

EA 4391, Excitabilité Nerveuse et Thérapeutique, Université Paris-Est Créteil, Créteil, France Service de Physiologie – Explorations Fonctionnelles, Hôpital Henri Mondor, Créteil, France

Abbreviations

CBT cognitive behavioral therapy
MBCT mindfulness-based cognitive therapy
MBSR mindfulness-based stress reduction
MS multiple sclerosis
PwMS patients with multiple sclerosis
RCT randomized controlled trial

Introduction

Cognitive behavioral therapy

Cognitive behavioral therapy (CBT) constitutes one of the main types of psychotherapies (Beck, 1979 cited in Churchill et al., 2010), along with psychoanalysis/psychodynamic (Freud, 1949; Klein, 1960; Jung,1963; cited in Churchill et al., 2010) and humanistic therapies (Maslow, 1943; Rogers, 1951; May, 1961 cited in Churchill et al., 2010). Historically, the CBT approach has undergone three waves of development. The first wave, also known as behavioral therapy, was based on the works about classical respondent conditioning (Pavlov, 1927, 1960; cited in Ruggiero, Spada, Caselli, & Sassaroli, 2018) and operant conditioning (Skinner, 1954; cited in Ruggiero et al., 2018).

With the second wave, the cognitive component of CBT mainly appeared with the works of Ellis, who conceived the rational emotive behavioral therapy (Ellis, 1955; 1962; Ellis & Grieger, 1986; cited in Ruggiero et al., 2018) and Beck, who developed the schema-based cognitive therapy (Beck, 1963, 1964, 1976; Beck et al., 1979; cited in Ruggiero et al., 2018). The cognitive model introduced by Beck gives attention to the role of schemas which are cognitive structures constructed and acquired along with the human experience. In a psychopathological context and under specific situations, some stimuli can activate maladaptive schematic representations of the self, world, and future (i.e., Beck triad), leading to an information processing bias with a preference for schema-congruent–negative information at the detriment of schema-incongruent–positive information. Consequently, individuals' perception could influence their emotions and behaviors.

The third wave emerged with a focus on the relationship of the individuals to their thoughts and emotions rather than to the latter's content (Hayes, 2004; cited in Hayes & Hofmann, 2017). Here, a particular emphasis is put on emotions, relationships, mindfulness, metacognition, cognitive defusion, acceptance, goals, and values (Dimidjian et al., 2016; Hayes & Hofmann, 2017). The third wave encompasses interventions such as dialectical behavior therapy, acceptance and commitment therapy, mindfulness-based cognitive therapy (MBCT), and mindfulness-based stress reduction (MBSR) (Dimidjian et al., 2016; Hayes & Hofmann, 2017).

Empirical studies suggest strong evidence regarding CBT's utility in specific psychiatric populations (Hofmann, Asnaani, Vonk, Sawyer, & Fang, 2012). Moreover, there is a growing interest in evaluating the effects of CBT on frequent comorbid symptoms in the field of neurology, such as in patients with multiple sclerosis (PwMS). In the present work, the authors start by a reappraisal of multiple sclerosis (MS) and a definition of fatigue, a common and debilitating symptom in PwMS. Afterward, they give a brief overview of the underlying mechanisms of MS fatigue and its management. Finally, they provide an analysis of the available literature that applied CBT in the context of MS fatigue and some suggestions that could help to guide future research.

Multiple sclerosis fatigue: Definition and impact

MS is an autoimmune disease of the central nervous system that affects young adults. Demyelination, axonal degeneration, and synaptic dysfunction constitute its main underlying pathophysiological processes (Chalah et al., 2015). It commonly adopts a relapsing-remitting course (i.e., intervals of clinical relapses separated by relapse-free intervals during which PwMS have a partial or complete recovery), but could follow a progressive course denoting a progressive accumulation of disability without relapses 15–20 years after the relapsing-remitting course (i.e., secondary progressive MS) or since the disease onset (i.e., primary progressive MS). PwMS suffer from a panel of unpleasant cognitive, behavioral, affective, sensory, motor, sphincter, and cerebellar symptoms depending on the location and extent of lesions. Up to 90% of PwMS suffer from fatigue which is a multidimensional (i.e., cognitive, physical, and psychosocial aspects) and a worrisome symptom that compromises essential life domains (e.g., quality of life, work productivity, socializing abilities) (Ayache & Chalah, 2017; Chen et al., 2019; van Kessel & Moss-Morris, 2006). Various definitions have been proposed to refer to this symptom ranging from simple terms such as "excessive tiredness" to more developed definitions such as a "failure to initiate and/or sustain attention tasks and/or physical activities requiring self-motivation in the absence of or not related to physical or cognitive dysfunction" (Chaudhuri & Behan, 2000; cited in Chalah et al., 2015). Some factors could precipitate this symptom, such as humidity, food ingestion, infections, and physical or mental efforts (Chalah et al., 2015). Here, one should note that fatigue perception refers to a subjective sensation reported by patients and differs from the objective decrement in motor or cognitive performance following exertion; the latter instead refers to fatiguability (Chalah et al., 2015). It is also essential to distinguish between primary MS fatigue, whose pathophysiology will be discussed in the following section, and secondary fatigue, which is the results of other factors such as mood or sleep disorders, infections, endocrinopathies, anemias, vitamins deficiencies, and medications side effects (Ayache & Chalah, 2017). The distinction between primary and secondary MS fatigue is illustrated in Fig. 1.

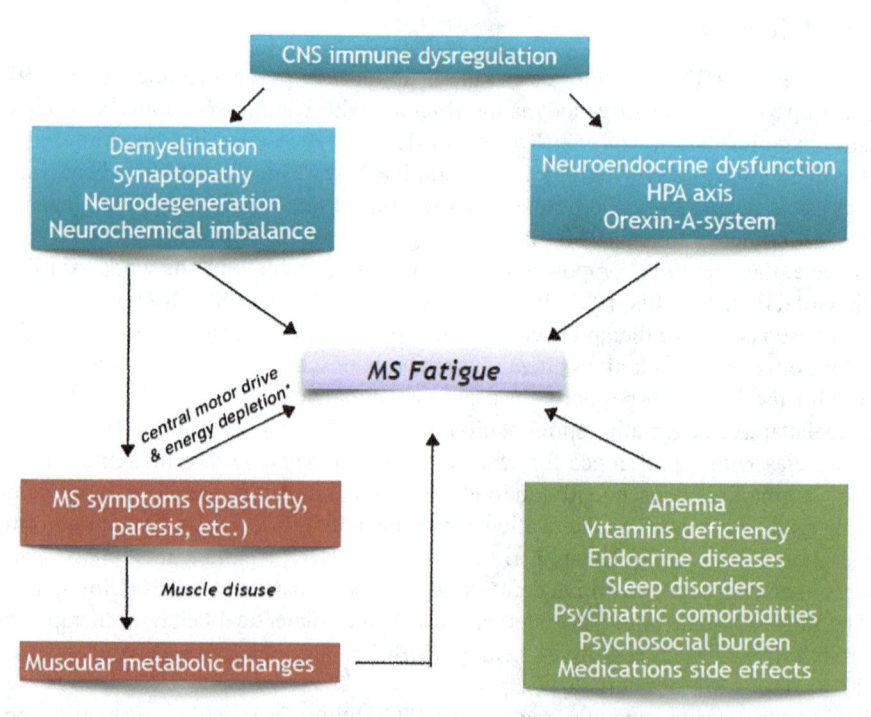

FIG. 1 The distinction between primary and secondary fatigue in multiple sclerosis. *CNS*, central nervous system; *MS*, multiple sclerosis; *, increased central motor drive and energy depletion. *(From Ayache, S. S., & Chalah, M. A. (2017). Fatigue in multiple sclerosis – Insights into evaluation and management.* Neurophysiologie clinique = Clinical Neurophysiology, 47, *139–171, with permission.)*

Nevertheless, even after excluding secondary causes, primary MS fatigue could still coexist with other psychiatric symptoms (i.e., subclinical anxiety and depression symptoms), all of which are referred to as "symptoms cluster" (Ayache & Chalah, 2020). These symptoms seem to interact with each other and share common underlying mechanisms (Ayache & Chalah, 2020).

Multiple sclerosis fatigue: Underlying mechanisms and management

MS fatigue is frequent and overwhelming, yet its underlying mechanisms remain to be explored. Its pathophysiology appears to be complex and multifactorial and incriminates neuroanatomical, neurophysiological, neuroimmune/endocrine, and neuropsychological processes (Ayache & Chalah, 2017, 2020). To start, based on anatomical and functional magnetic resonance imaging studies, accumulation of lesions/disconnections in a cortico-striato-thalamo-cortical loop has been proposed at the origin of MS fatigue (Fig. 2). Also, abnormal motor cortex excitability (e.g., GABAergic inhibitory processes) has been suggested by few works (Stampanoni Bassi et al., 2020). Moreover, few studies have suggested the involvement of inflammation (e.g., proinflammatory cytokines) in MS fatigue generation (Chalah & Ayache, 2018). Furthermore, some studies suggest a relationship between MS fatigue and some psychological factors (e.g., personality traits, coping strategies). For instance, in some works, MS fatigue has been found to be associated with alexithymia, motivational deficits, somatization behavior, high levels of neuroticism and excitability, as well as low levels of extraversion, agreeableness, and conscientiousness (Chalah & Ayache, 2017; Maggio et al., 2020).

Interestingly, a cognitive behavioral model of MS fatigue has been introduced and supposes an interaction between biological, emotional, cognitive, and behavioral factors, which aliment the vicious cycle of this symptom (van Kessel & Moss-Morris, 2006). The model assumes cerebral abnormalities at the basis of fatigue generation and considers that the ways PwMS interact with the symptom installation differ from a patient to another and might be involved in alleviating or aggravating fatigue perception. The model provides examples of unhelpful thoughts or cognitions (i.e., catastrophizing about or embarrassment from symptom occurrence, believing that the symptom reflects physical damage), unhelpful behaviors (i.e., all-or-nothing behavior, constant resting, or activity limitation), and related-emotional responses (i.e., anxiety and depressive symptoms) which, by their turn, may engender biological and physiological consequences (i.e., arousal, sleep disturbance) that contribute to maintaining or exacerbating the symptom (van Kessel & Moss-Morris, 2006). This model could help to understand the rationale behind the development of CBT protocols targeting MS fatigue.

Theoretically, CBT designed to target MS fatigue could imply psychoeducation about MS fatigue, setting and self-monitoring goals, cognitive restructuring in relation to cognitive distortions, changing the maladapted behaviors and coping strategies, and developing emotion regulation strategies. The following section provides an overview of CBT and derived interventions for which the primary outcome was MS fatigue. Trials with unspecified primary outcome or with fatigue as a secondary outcome were not considered.

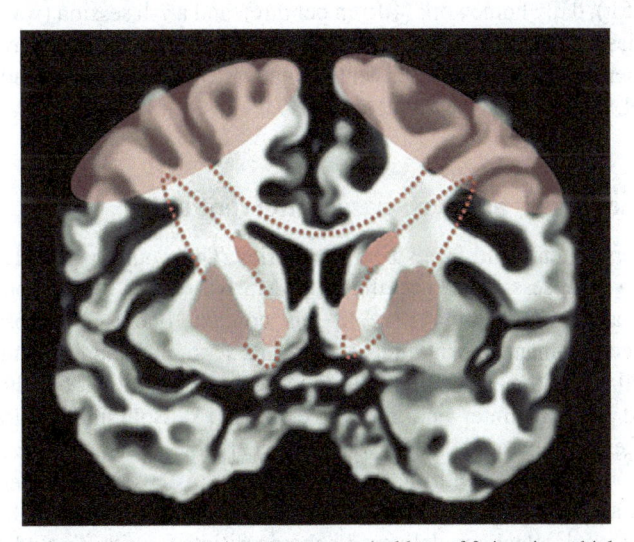

FIG. 2 A coronal view illustrating the main parts of cortico-striato-thalamo-cortical loop of fatigue in multiple sclerosis. Dysconnectivities within the loop are represented by dashed lines. *(From Ayache, S. S., & Chalah, M. A. (2017). Fatigue in multiple sclerosis – Insights into evaluation and management.* Neurophysiologie clinique = Clinical Neurophysiology, 47, *139–171, with permission.)*

Studies applying cognitive behavioral therapy in multiple sclerosis fatigue

Face-to-face individual versus group interventions

Based on the cognitive behavioral model of MS fatigue, a first randomized controlled trial (RCT) was designed to target MS fatigue (i.e., introducing MS fatigue model, targeting unhelpful thoughts and behaviors as well as negative emotions, promoting sleep-hygiene and reinstating circadian rhythm, and emphasizing the role of social support) (van Kessel et al., 2008). Compared to the control arm (i.e., relaxation training), eight individual CBT sessions (once per week, session duration: 50 min) resulted in more pronounced effects that lasted up to 6 months postintervention. In another multicenter RCT ("TREFAMS-CBT"), compared to a standardized control intervention (three consultations delivered by nurses), 12 individual patient-tailored CBT sessions (eight in the first 2 months and four over the following 2 months, session duration: 45 min, delivered by CBT therapists) resulted in a significant acute improvement in fatigue that progressively decreased until 1-year follow-up (van den Akker et al., 2017).

Besides individually applied CBT interventions, some authors assessed the effects of interventions delivered in a group format. Group CBT has been found to be a cost-effective alternative to individual therapy in some clinical settings (i.e., depression; Whitfield, 2010), and its comparability to individual CBT interventions in fatigued PwMS remains to be addressed. Other advantages of group CBT reside in the added benefits of the group cohesion, the identification with other patients which could lead to normalization effects, the within-group behavioral experiments which could enable cognitive restructuring, the possibility to act as cotherapists, and the constitution of a safe environment in which patients can experience vicarious learning and positive reinforcement (Whitfield, 2010). In fact, group CBT has been applied in the context of MS fatigue. One RCT compared a group wellness intervention (containing CBT and energy conservation elements, 2 h per week over 7 weeks) and individualized physical rehabilitation (four sessions and a phone call between sessions), both of which were followed by two phone calls following interventions (Plow, Mathiowetz, & Lowe, 2009). Both arms resulted in fatigue improvement not right after the intervention but 8 weeks later, with no group difference. In another multicenter RCT, a group-based CBT intervention with energy effectiveness techniques combined with regular local care ("FACETS" intervention; six sessions, once per week, session duration: 90 min (Thomas et al., 2010)) resulted in better fatigue improvement at 1 month (i.e., fatigue self-efficacy) and 4 months (i.e., fatigue severity and self-efficacy) follow-ups compared to the regular local care alone (Thomas et al., 2013). The latter effects persisted at 1 year (Thomas et al., 2014). Interestingly, following this trial, PwMS frequently implemented changes at the behavioral (e.g., prioritizing, pacing) and environmental (e.g., task grading, delegating) levels, as well with regards to their emotions and thoughts (e.g., no longer feeling guilty about or admitting to be a failure in case of inability to do something) (Thomas et al., 2015).

Moreover, some trials have applied third-wave CBT interventions that involved group and individual sessions. For instance, compared to usual care, an 8-week MBSR intervention resulted in fatigue improvement right after the intervention and at 6 months (Grossman et al., 2010). The program consisted of individual interviews aiming to evaluate patients' goals and attainments at the beginning and the end of the intervention, group-based sessions including mindfulness practice (once per week, session duration: 2.5 h), daily homework (40 min per day), and a 7-h session (week 6). In another non-RCT, a 10-week MBCT (eight sessions lasting 2.5 h each and homework lasting up to 1 h per day over 6 days per week) resulted in significant fatigue improvement following the intervention and at 3 months compared to a waiting list control arm (Hoogerwerf, Bol, Lobbestael, Hupperts, & van Heugten, 2017).

Face-to-face, telephone-delivered, and web-based interventions

Performing a face-to-face intervention could face some boundaries. This could be mainly related to the difficulty to frequently travel to clinics to have the sessions, especially in PwMS who could suffer from important physical disability. Traveling to clinics is also associated with a cost. In this context, one RCT assessed the effects of an 8-week telephone-delivered CBT-based self-management intervention (i.e., sessions include techniques such as setting goals, cognitive restructuring, behavioral activation, relaxation training; session duration: 45–60 min, delivered by a therapist) compared to a telephone-delivered education intervention (Ehde et al., 2015). Both interventions resulted in significant short-term fatigue improvement (after the intervention) that was still observable in the long run (at 6- and 12-month follow-up). Common factors (e.g., education, interaction with and support of a therapist, working alliance, patients' expectations) could explain the observed benefits and absence of group differences between interventions. In the same perspective, another RCT tested the effects of a 12-week telephone-delivered behavioral intervention (including patient-tailored phone calls) designed to promote fatigue self-management and physical activity in PwMS (Plow et al., 2019; Plow, Finlayson, Motl, & Bethoux, 2012; Plow, Motl, Finlayson, & Bethoux, 2020a, 2020b). Compared to contact-control and physical activity-only arms, the intervention resulted in significant fatigue improvement.

Moreover, to further overcome the difficulties (e.g., heaviness and costs) related to patients' travel to treatment centers, the face-to-face interventions and the therapists' difficulties to take in charge a large number of fatigued PwMS, using electronic health technologies, including mobile health applications, have been applied in general in PwMS, regardless of fatigue, and might have health benefits in this clinical population (Marrie et al., 2019). Self-management interventions for MS fatigue have been developed in the last decade and consisted of internet resources (e.g., websites or mobile phone applications) that incorporate self-management and adult learning principles (Akbar, Turpin, Petrin, Smyth, & Finlayson, 2018; D'hooghe et al., 2018; Finlayson, Akbar, Turpin, & Smyth, 2019; Pétrin, Akbar, Turpin, Smyth, & Finlayson, 2018) as well as CBT principles (as will be discussed later).

Internet-based CBT interventions have been developed in the context of MS fatigue (Moss-Morris et al., 2012; Pöttgen et al., 2018; van Kessel, Wouldes, & Moss-Morris, 2016). A pilot RCT has assessed the effects of a web-based CBT self-management intervention ("MSInvigor8") consisting of eight sessions accessed over 8–10 weeks combined with up to three telephone support sessions (30–60 min per phone call). Compared to standard care, MSInvigor8 resulted in significant fatigue improvement at the end of the program (Moss-Morris et al., 2012). In a later RCT, the same team documented the add-on effects of email support when combined with eight-sessions of MSInvigor8 (van Kessel et al., 2016). In a third RCT, a 12-week intervention consisting of a self-guided online interactive intervention to manage MS fatigue intervention ("ELAVIDA") based on CBT and mindfulness principles resulted in a significant fatigue reduction right after the intervention and 12 weeks later (compared to wait-list control arm) (Pöttgen et al., 2018).

Moreover, in one study performing interviews and consultations with PwMS and the health care professionals who initially delivered the abovementioned face-to-face group-based CBT intervention FACETS, the participants' views enabled proposing a web-based model of the intervention that would be important to assess in future trials (Thomas et al., 2019).

Furthermore, an ongoing multicenter RCT ("MS Fit") is testing the noninferiority of blended CBT sessions (online intervention combined with limited video consultations or face-to-face contacts) compared to face-to-face intervention (Houniet-de Gier, Beckerman, van Vliet, Knoop, & de Groot, 2020). The results of this trial will help to clarify the place of online-delivered CBT in MS fatigue.

Interestingly, a recent field trial has assessed the usability and utility of a CBT-based mobile phone application targeting MS fatigue ("MS Energize/Energise") (Babbage et al., 2019). MS Energize aims to promote self-management of MS fatigue and contains seven core topics that include an explanation of fatigue, the influence of behaviors, emotions, thoughts, internal and external factors of this symptom, and the ways to maintain benefits and overcome setbacks. Following 5–6 weeks of testing in the form of education, interactive tasks, feedbacks, quiz and vignettes, activity and sleep diaries, as well as setting goals, the patients' results support the usability of MS Energize and could help to refine the program to be tested in further trials.

Response to cognitive behavioral interventions

Some of the considered trials have included secondary analyses that enable understanding whether some variables could mediate, moderate or predict the response to a CBT intervention. For instance, the secondary analysis of the abovementioned trial by van Kessel et al. (2008) has demonstrated a significant improvement in some cognitions (focusing on fatigue, believing that the symptom reflects a damage, and having a negative representation of the symptom) and behaviors (i.e., avoidance) following CBT compared to relaxation training (Knoop, van Kessel, & Moss-Morris, 2012). Significantly, fatigue improvement was mediated by the change in the negative beliefs about the symptom (Knoop et al., 2012).

In addition, a secondary analysis of the previously seen trial by Ehde et al. (2015) has identified the patients' baseline activation (i.e., their level of knowledge, skills, and confidence in managing their health) as a moderator of the treatment response in a way that those with high baseline activation exhibited better outcomes in the self-management intervention compared to the education intervention (Ehde, Arewasikporn, Alschuler, Hughes, & Turner, 2018). Conversely, no significant group difference in fatigue improvement was observed among patients with low baseline activation. Also, other cognitions and behaviors (e.g., fatigue catastrophizing, self-efficacy) had no effects on the study outcome.

Moreover, a mediation analysis of the longitudinal patient-tailored CBT trial by van den Akker et al. (2017) has revealed that improvement in fatigue perception, physical activity, and physical functioning and the decrease in sleepiness and helplessness mediated fatigue improvement following CBT (van den Akker et al., 2018). Conversely, the reduced concentration and physical activity, as well as the increased sleepiness, mediated the increase in fatigue following the treatment (van den Akker et al., 2018).

Conversely, the few available studies that accounted for sociodemographic (i.e., age, sex, education level, and marital status) and/or clinical (MS types, disease duration, disease-modifying therapies, physical disability, cognitive deficits, anxiety or depressive symptoms) covariates do not support the effects of such variables on the studied outcomes (Ehde et al., 2015; Grossman et al., 2010; Hoogerwerf et al., 2017; Moss-Morris et al., 2012; Thomas et al., 2013; van den Akker et al., 2017; van Kessel et al., 2008).

Discussion

This narrative presents the application of CBT or its components to target fatigue in PwMS. Few studies have been published on this topic and consisted of one-to-one or group-based face-to-face interventions. To overcome the costs (travel and session costs), the number of therapists required to deliver CBT for MS fatigue, and the difficulties related to implementing face-to-face interventions in a clinical population with an important physical disability, some protocols applied telephone- or web-delivered interventions. Furthermore, a mobile phone application has been developed to help PwMS self-manage their fatigue based on CBT principles, and its effects deserve to be further explored. The available protocols suggest promising antifatigue effects of CBT in PwMS, with some but not all protocols suggesting long-term effects. Facing the heterogeneity among some of the studies outcomes, one should take into consideration the differences in the available protocols (i.e., study design [CBT alone vs combined with phone calls, email support, or energy conservation]; intervention type [i.e., CBT, MBCT, MBSR, CBT components]; sessions number [i.e., 6–12 sessions], duration, and content; difference in the control arm [i.e., education, local care, physical rehabilitation, relaxation training, waiting list]; individual vs group sessions; face-to-face vs phone- vs internet-delivered intervention). Besides the differences in the designs of the applied interventions, the cohorts might have differed with regard to clinical variables. For instance, while very few CBT studies have excluded clinically relevant anxiety (van den Akker et al., 2017) and depression (Hoogerwerf et al., 2017; van den Akker et al., 2017) in order to assess CBT effects on primary fatigue. However, in the remaining trials, some patients comorbidities (i.e., pain and moderate depressive symptoms in Ehde et al., 2015; comorbid depression allowed in Pöttgen et al., 2018), some were receiving antidepressants for longer than 3 months (Thomas et al., 2013; van Kessel et al., 2008) and others might have high anxiety and/or depression cores (Ehde et al., 2015; Grossman et al., 2010; Moss-Morris et al., 2012; Pöttgen et al., 2018; van Kessel et al., 2008, 2016) or were not assessed for anxiety and depression (Babbage et al., 2019; Plow et al., 2009; Thomas et al., 2013). In addition, while some of the available fatigue studies also documented significant improvement in anxiety (Grossman et al., 2010; Hoogerwerf et al., 2017; Moss-Morris et al., 2012; Pöttgen et al., 2018) or depressive symptoms (Ehde et al., 2015; Grossman et al., 2010; Hoogerwerf et al., 2017; Moss-Morris et al., 2012), other few studies did not document such changes (depressive symptoms in Pöttgen et al., 2018 and both symptoms in van Kessel et al., 2008, 2016; Thomas et al., 2013). Future works would benefit from assessing CBT effects on primary MS fatigue while considering the symptoms cluster (i.e., fatigue, anxiety, and depressive symptoms). Besides these affective symptoms, alexithymia seems to play a complex role in treatment response in psychiatric outcomes (Pinna, Manchia, Paribello, & Carpiniello, 2020) and merits to be assessed at baseline and following CBT interventions targeting MS fatigue.

Regarding the patients' clinical characteristics, the cohorts were predominantly constituted of women with relapsing-remitting MS (Ehde et al., 2015; Grossman et al., 2010; Hoogerwerf et al., 2017; Moss-Morris et al., 2012; Pöttgen et al., 2018; Thomas et al., 2013; van den Akker et al., 2017); some studies focused on progressive types (primary vs secondary; Ehde et al., 2015), and others did not include patients with primary progressive MS (Grossman et al., 2010; Hoogerwerf et al., 2017; van Kessel et al., 2016) or did not specify the disease (Babbage et al., 2019).

Another factor that differed among some of the available studies concerns the choice of the assessment tool, which included the Modified Fatigue Impact Scale, the Checklist Individual Strength, the Chalder Fatigue Scale, the Global Fatigue Severity subscale of the Fatigue Assessment Instrument, the Multiple Sclerosis-Fatigue Self-Efficacy scale, and the Fatigue Scale of Motor and Cognition.

Since the observed beneficial antifatigue effects could be lost progressively following the intervention (van den Akker et al., 2017), a potential strategy to maintain the effects would be by repeating the sessions. In this perspective, an ongoing multicenter trial ("MS Stay Fit") is testing the effects of internet-based blended booster sessions to maintain long-term CBT effects over 1 year (Houniet-de Gier et al., 2020). The results of this trial will help to guide the design of future CBT protocols that aim to preserve clinical improvement in time.

Another point to consider is the complexity and multifactorial nature of MS fatigue. In this context, combining several interventions with potentially different mechanisms of action might yield better effects compared to single therapies. For instance, Thomas and colleagues have combined CBT with energy effectiveness techniques (FACETS intervention; Thomas et al., 2013, 2014). In addition, Kratz and colleagues have designed the randomized "COMBO-MS" trial that aims to compare the acute and long-term effects (right after the intervention and 12 weeks later) of a 12-week intervention consisting of telephone-delivered CBT and modafinil applied as monotherapies or in combination (three treatment arms) (Kratz et al., 2019). The results of this protocol are highly awaited. Future works could explore the utility of coupling CBT with other interventions (i.e., noninvasive brain stimulation (NIBS), pharmacotherapy, exercise therapy) in optimizing MS fatigue management. For instance, NIBS and psychotherapy seem to yield better outcomes in some psychiatric disorders (i.e., anxiety and depression) (Chalah & Ayache, 2019). Combining CBT with NIBS merits to be evaluated, especially that the latter intervention (i.e., transcranial direct current stimulation) seems to yield promising antifatigue effects (Ayache & Chalah, 2018).

Practice and procedures

This chapter provides an overview of CBT protocols that targeted fatigue in patients with multiple sclerosis (MS). The authors consulted databases (PubMed/Medline, Scopus and PsychInfo) and looked for original articles—published at any time until December 2020—that applied CBT or derived interventions and focused on MS fatigue as a primary outcome. The search was restricted to articles published in English. In all the available trials, baseline and post-CBT fatigue levels were measured using subjective scales that are widely used to assess fatigue perception. Similarly, some studies used subjective tools to assess or control for anxiety or depression symptoms. The protocols were delivered in an individual or a group setting, in a face-to-face, phone-based, or web-based approach.

Mini-dictionary of terms

Autoimmune disease. A disease in which a dysfunction of the immune system results in the occurrence of an immune response (e.g., antibody production, inflammatory mediators) targeting healthy body parts.

Central nervous system. A part of the nervous system constituted of the brain (i.e., cerebrum, cerebellum, and brainstem) and the spinal cord.

Noninvasive brain stimulation techniques. Technologies that consist of modulating the excitability of brain regions via a transcranial application of a weak electric current (a technique known as transcranial electrical stimulation) or a magnetic field (a method known as transcranial magnetic stimulation) over the scalp.

Proinflammatory cytokines. Signaling molecules produced by immune cells and involved in the induction of inflammatory responses.

Synaptic dysfunction. An abnormality affecting the function or structure of the synapse which, in the context of this chapter, refers to the junction that ensure the communication between two neurons.

Key facts

- Fatigue frequently occurs in patient with multiple sclerosis (PwMS) and drastically affects their quality of life.
- The actual management of fatigue in PwMS remains challenging.
- Neuroanatomical, neurophysiological, neuroimmune/endocrine and neuropsychological substrates have been proposed for fatigue in PwMS.
- A cognitive behavioral model of fatigue in PwMS has also been proposed.
- Cognitive behavioral therapy might have its place within the therapeutic armamentarium of fatigue in PwMS.

Applications to other areas

MS fatigue is a worrisome and frequent symptom in this clinical population with limited benefits obtained with pharmaceutical management. In this chapter, we have reviewed the application of cognitive behavioral therapy (CBT) for improving fatigue in patients with multiple sclerosis (MS). Few studies have applied CBT or its components in a face-to-face format, either individually or in a group setting. Admitting the potential difficulties related to a conventional intervention (e.g., cost of travel and intervention) and the important physical disability that could occur in this clinical population, telephone-based or web-based interventions could provide an alternative solution and have been tested. The available few reports suggest promising effects of the applied protocols in improving MS fatigue. A CBT-based mobile application has recently been developed, and its effectiveness against MS fatigue will be tested. Further studies are needed to explore the optimal intervention design. Combining CBT with other interventions might further enhance the observed clinical benefits. A previous work combined CBT with energy effectiveness techniques. Moreover, an ongoing trial will explore the utility of combining a telephone-delivered CBT with a pharmacological agent (i.e., modafinil). Furthermore, the pertinence of other combinations (e.g., CBT and noninvasive brain stimulation techniques) deserves to be tested. The application of CBT in the context of MS is not only limited to fatigue; it has also been applied to target other frequent and debilitating MS symptoms (e.g., pain, affective symptoms, sleep disorders, physical disability, etc.), and its utility merits to be reviewed and analyzed.

Summary points

- This review analyzes the available literature that applied cognitive behavioral therapies (CBT) to target fatigue in Patients with Multiple Sclerosis (PwMS).
- Few studies have applied CBT in PwMS with fatigue.
- The protocols differ in their design (i.e., sessions number, duration, and content; group vs individual setting; face-to-face vs phone- or internet-delivered intervention).
- The available protocols have heterogeneous results but suggest promising antifatigue effects of CBT in PwMS.
- A CBT-based self-management mobile application for fatigue in PwMS has been developed and merits to be further studied.
- Future works could explore the utility of coupling CBT with other interventions (i.e., noninvasive brain stimulation, pharmacotherapy, exercise therapy, energy conservation) in optimizing fatigue management in PwMS.

References

Akbar, N., Turpin, K., Petrin, J., Smyth, P., & Finlayson, M. (2018). A pilot mixed-methods evaluation of MS INFoRm: A self-directed fatigue management resource for individuals with multiple sclerosis. *International Journal of Rehabilitation Research, 41*, 114–121.

Ayache, S. S., & Chalah, M. A. (2017). Fatigue in multiple sclerosis – Insights into evaluation and management. *Neurophysiologie clinique = Clinical Neurophysiology, 47*, 139–171.

Ayache, S. S., & Chalah, M. A. (2018). The place of transcranial direct current stimulation in the management of multiple sclerosis-related symptoms. *Neurodegenerative Disease Management, 8*, 411–422.

Ayache, S. S., & Chalah, M. A. (2020). Fatigue and affective manifestations in multiple sclerosis—A cluster approach. *Brain Sciences, 10*, 10.

Babbage, D. R., van Kessel, K., Drown, J., Thomas, S., Sezier, A., Thomas, P., & Kersten, P. (2019). MS Energize: Field trial of an app for self-management of fatigue for people with multiple sclerosis. *Internet Interventions, 18*, 100291.

Chalah, M. A., & Ayache, S. S. (2017). Alexithymia in multiple sclerosis: A systematic review of literature. *Neuropsychologia, 104*, 31–47.

Chalah, M. A., & Ayache, S. S. (2018). Is there a link between inflammation and fatigue in multiple sclerosis? *Journal of Inflammation Research, 11*, 253–264.

Chalah, M. A., & Ayache, S. S. (2019). Non-invasive brain stimulation and psychotherapy in anxiety and depressive disorders: A viewpoint. *Brain Sciences, 9*, 82.

Chalah, M. A., Riachi, N., Ahdab, R., Créange, A., Lefaucheur, J. P., & Ayache, S. S. (2015). Fatigue in multiple sclerosis: Neural correlates and the role of non-invasive brain stimulation. *Frontiers in Cellular Neuroscience, 9*, 460.

Chen, J., Taylor, B., Palmer, A. J., Kirk-Brown, A., van Dijk, P., Simpson, S., Jr., ... van der Mei, I. (2019). Estimating MS-related work productivity loss and factors associated with work productivity loss in a representative Australian sample of people with multiple sclerosis. *Multiple Sclerosis (Houndmills, Basingstoke, England), 25*, 994–1004.

Churchill, R., Davies, P., Caldwell, D., Moore, T. H., Jones, H., Lewis, G., & Hunot, V. (2010). Humanistic therapies versus other psychological therapies for depression. *The Cochrane Database of Systematic Reviews, 2010*, CD008700.

D'hooghe, M., Van Gassen, G., Kos, D., Bouquiaux, O., Cambron, M., Decoo, D., ... Nagels, G. (2018). Improving fatigue in multiple sclerosis by smartphone-supported energy management: The MS TeleCoach feasibility study. *Multiple Sclerosis and Related Disorders, 22*, 90–96.

Dimidjian, S., Arch, J. J., Schneider, R. L., Desormeau, P., Felder, J. N., & Segal, Z. V. (2016). Considering meta-analysis, meaning, and metaphor: A systematic review and critical examination of "third wave" cognitive and behavioral therapies. *Behavior Therapy, 47*, 886–905.

Ehde, D. M., Arewasikporn, A., Alschuler, K. N., Hughes, A. J., & Turner, A. P. (2018). Moderators of treatment outcomes after telehealth self-management and education in adults with multiple sclerosis: A secondary analysis of a randomized controlled trial. *Archives of Physical Medicine and Rehabilitation, 99*, 1265–1272.

Ehde, D. M., Elzea, J. L., Verrall, A. M., Gibbons, L. E., Smith, A. E., & Amtmann, D. (2015). Efficacy of a telephone-delivered self-management intervention for persons with multiple sclerosis: A randomized controlled trial with a one-year follow-up. *Archives of Physical Medicine and Rehabilitation, 96*, 1945–1958. e2.

Finlayson, M., Akbar, N., Turpin, K., & Smyth, P. (2019). A multi-site, randomized controlled trial of MS INFoRm, a fatigue self-management website for persons with multiple sclerosis: Rationale and study protocol. *BMC Neurology, 19*, 142.

Grossman, P., Kappos, L., Gensicke, H., D'Souza, M., Mohr, D. C., Penner, I. K., & Steiner, C. (2010). MS quality of life, depression, and fatigue improve after mindfulness training: A randomized trial. *Neurology, 75*, 1141–1149.

Hayes, S. C., & Hofmann, S. G. (2017). The third wave of cognitive behavioral therapy and the rise of process-based care. *World Psychiatry, 16*, 245–246.

Hofmann, S. G., Asnaani, A., Vonk, I. J., Sawyer, A. T., & Fang, A. (2012). The efficacy of cognitive behavioral therapy: A review of meta-analyses. *Cognitive Therapy and Research, 36*, 427–440.

Hoogerwerf, A., Bol, Y., Lobbestael, J., Hupperts, R., & van Heugten, C. M. (2017). Mindfulness-based cognitive therapy for severely fatigued multiple sclerosis patients: A waiting list controlled study. *Journal of Rehabilitation Medicine, 49*, 497–504.

Houniet-de Gier, M., Beckerman, H., van Vliet, K., Knoop, H., & de Groot, V. (2020). Testing non-inferiority of blended versus face-to-face cognitive behavioral therapy for severe fatigue in patients with multiple sclerosis and the effectiveness of blended booster sessions aimed at improving long-term outcome following both therapies: Study protocol for two observer-blinded randomized clinical trials. *Trials, 21*, 98.

Knoop, H., van Kessel, K., & Moss-Morris, R. (2012). Which cognitions and behaviours mediate the positive effect of cognitive behavioral therapy on fatigue in patients with multiple sclerosis? *Psychological Medicine, 42*, 205–213.

Kratz, A. L., Alschuler, K. N., Ehde, D. M., von Geldern, G., Little, R., Kulkarni, S., ... Braley, T. J. (2019). A randomized pragmatic trial of telephone-delivered cognitive behavioral-therapy, modafinil, and combination therapy of both for fatigue in multiple sclerosis: The design of the "COMBO-MS" trial. *Contemporary Clinical Trials, 84*, 105821.

Maggio, M. G., Cuzzola, M. F., Latella, D., Impellizzeri, F., Todaro, A., Rao, G., ... Calabrò, R. S. (2020). How personality traits affect functional outcomes in patients with multiple sclerosis: A scoping review on a poorly understood topic. *Multiple Sclerosis and Related Disorders, 46*, 102560.

Marrie, R. A., Leung, S., Tyry, T., Cutter, G. R., Fox, R., & Salter, A. (2019). Use of eHealth and mHealth technology by persons with multiple sclerosis. *Multiple Sclerosis and Related Disorders, 27*, 13–19.

Moss-Morris, R., McCrone, P., Yardley, L., van Kessel, K., Wills, G., & Dennison, L. (2012). A pilot randomised controlled trial of an Internet-based cognitive behavioral therapy self-management programme (MS Invigor8) for multiple sclerosis fatigue. *Behaviour Research and Therapy, 50*, 415–421.

Pétrin, J., Akbar, N., Turpin, K., Smyth, P., & Finlayson, M. (2018). The experience of persons with multiple sclerosis using MS INFoRm: An interactive fatigue management resource. *Qualitative Health Research, 28*, 778–788.

Pinna, F., Manchia, M., Paribello, P., & Carpiniello, B. (2020). The impact of alexithymia on treatment response in psychiatric disorders: A systematic review. *Frontiers in Psychiatry, 11*, 311.

Plow, M., Finlayson, M., Liu, J., Motl, R. W., Bethoux, F., & Sattar, A. (2019). Randomized controlled trial of a telephone-delivered physical activity and fatigue self-management interventions in adults with multiple sclerosis. *Archives of Physical Medicine and Rehabilitation, 100*, 2006–2014.

Plow, M., Finlayson, M., Motl, R. W., & Bethoux, F. (2012). Randomized controlled trial of a teleconference fatigue management plus physical activity intervention in adults with multiple sclerosis: Rationale and research protocol. *BMC Neurology, 12*, 122.

Plow, M. A., Mathiowetz, V., & Lowe, D. A. (2009). Comparing individualized rehabilitation to a group wellness intervention for persons with multiple sclerosis. *American Journal of Health Promotion: AJHP, 24*, 23–26.

Plow, M., Motl, R. W., Finlayson, M., & Bethoux, F. (2020a). Response heterogeneity in a randomized controlled trial of telerehabilitation interventions among adults with multiple sclerosis. *Journal of Telemedicine and Telecare.* https://doi.org/10.1177/1357633X20964693. 1357633X20964693. Advance online publication.

Plow, M., Motl, R. W., Finlayson, M., & Bethoux, F. (2020b). Intervention mediators in a randomized controlled trial to increase physical activity and fatigue self-management behaviors among adults with multiple sclerosis. *Annals of Behavioral Medicine, 54*(3), 213–221.

Pöttgen, J., Moss-Morris, R., Wendebourg, J. M., Feddersen, L., Lau, S., Köpke, S., ... Gold, S. M. (2018). Randomised controlled trial of a self-guided online fatigue intervention in multiple sclerosis. *Journal of Neurology, Neurosurgery, and Psychiatry, 89*, 970–976.

Ruggiero, G. M., Spada, M. M., Caselli, G., & Sassaroli, S. (2018). A historical and theoretical review of cognitive behavioral therapies: From structural self-knowledge to functional processes. *Journal of Rational-Emotive and Cognitive-Behavior Therapy: RET, 36*, 378–403.

Stampanoni Bassi, M., Buttari, F., Gilio, L., De Paolis, N., Fresegna, D., Centonze, D., & Iezzi, E. (2020). Inflammation and corticospinal functioning in multiple sclerosis: A TMS perspective. *Frontiers in Neurology, 11*, 566.

Thomas, S., Kersten, P., Thomas, P. W., Slingsby, V., Nock, A., Jones, R., ... Hillier, C. (2015). Exploring strategies used following a group-based fatigue management programme for people with multiple sclerosis (FACETS) via the Fatigue Management Strategies Questionnaire (FMSQ). *BMJ Open, 5*, e008274.

Thomas, S., Pulman, A., Thomas, P., Collard, S., Jiang, N., Dogan, H., ... Gay, M. C. (2019). Digitizing a face-to-face group fatigue management program: Exploring the views of people with multiple sclerosis and health care professionals via consultation groups and interviews. *JMIR Formative Research, 3*, e10951.

Thomas, S., Thomas, P. W., Kersten, P., Jones, R., Green, C., Nock, A., ... Hillier, C. (2013). A pragmatic parallel arm multi-centre randomised controlled trial to assess the effectiveness and cost-effectiveness of a group-based fatigue management programme (FACETS) for people with multiple sclerosis. *Journal of Neurology, Neurosurgery, and Psychiatry, 84*, 1092–1099.

Thomas, P. W., Thomas, S., Kersten, P., Jones, R., Slingsby, V., Nock, A., ... Hillier, C. (2014). One year follow-up of a pragmatic multi-centre randomised controlled trial of a group-based fatigue management programme (FACETS) for people with multiple sclerosis. *BMC Neurology, 14*, 109.

Thomas, S., Thomas, P. W., Nock, A., Slingsby, V., Galvin, K., Baker, R., ... Hillier, C. (2010). Development and preliminary evaluation of a cognitive behavioral approach to fatigue management in people with multiple sclerosis. *Patient Education and Counseling, 78*, 240–249.

van den Akker, L. E., Beckerman, H., Collette, E. H., Knoop, H., Bleijenberg, G., Twisk, J. W., ... TREFAMS-ACE Study Group. (2018). Cognitive behavioral therapy for MS-related fatigue explained: A longitudinal mediation analysis. *Journal of Psychosomatic Research, 106*, 13–24.

van den Akker, L. E., Beckerman, H., Collette, E. H., Twisk, J. W., Bleijenberg, G., Dekker, J., ... TREFAMS-ACE Study Group. (2017). Cognitive behavioral therapy positively affects fatigue in patients with multiple sclerosis: Results of a randomized controlled trial. *Multiple Sclerosis (Houndmills, Basingstoke, England), 23*, 1542–1553.

van Kessel, K., & Moss-Morris, R. (2006). Understanding multiple sclerosis fatigue: A synthesis of biological and psychological factors. *Journal of Psychosomatic Research, 61*, 583–585.

van Kessel, K., Moss-Morris, R., Willoughby, E., Chalder, T., Johnson, M. H., & Robinson, E. (2008). A randomized controlled trial of cognitive behavior therapy for multiple sclerosis fatigue. *Psychosomatic Medicine, 70*, 205–213.

van Kessel, K., Wouldes, T., & Moss-Morris, R. (2016). A New Zealand pilot randomized controlled trial of a web-based interactive self-management programme (MSInvigor8) with and without email support for the treatment of multiple sclerosis fatigue. *Clinical Rehabilitation, 30*, 454–462.

Whitfield, G. (2010). Group cognitive–behavioral therapy for anxiety and depression. *Advances in Psychiatric Treatment, 16*, 219–227.

Chapter 18

In-patient/residential treatment for obsessive-compulsive disorder

Madhuri H. Nanjundaswamy, Lavanya P. Sharma, and Shyam Sundar Arumugham

Department of Psychiatry, National Institute of Mental Health and Neurosciences (NIMHANS), Bangalore, India

Abbreviations

CBT	cognitive behavioral therapy
ERP	exposure and response prevention
OCD	obsessive-compulsive disorder
SSRI	selective serotonin reuptake inhibitors
Y-BOCS	Yale-Brown Obsessive-Compulsive Scale
SUDS	Subjective Unit of Distress Scale

Introduction

Obsessive-compulsive disorder (OCD) is a chronic psychiatric illness characterized by the presence of obsessions and/or compulsions. Obsessions are repetitive intrusive mental phenomena that occur in the form of thoughts, impulses, or images and are usually perceived as unwanted. They are often associated with anxiety or distress. Compulsions are recurrent behaviors (or mental acts) that are often performed in response to obsessions. They may be performed according to rigidly applied rules, or to achieve a sense of "completeness" (Stein et al., 2019).

OCD has a lifetime prevalence of 2% to 2.5% (Torres & Lima, 2005) and is among the most disabling psychiatric conditions worldwide (Collins et al., 2011). Evidence gathered over the past few decades have shown that OCD responds to medications with a preferential serotonin reuptake inhibiting property, as compared to nonselective antidepressants (Arumugham & Reddy, 2014). Similarly, specific models of cognitive-behavior therapy, in particular exposure and response prevention (ERP), have been found to be effective in the treatment of OCD (Fineberg et al., 2020). Thus, selective serotonin reuptake inhibitors (SSRIs) and CBT are recommended as first-line treatments for OCD. A substantial proportion of patients do not respond adequately to these first-line treatments. A variety of pharmacological, psychological, and neuromodulatory interventions have been attempted as augmenting agents in partial/nonresponders (Arumugham & Reddy, 2013). Again, CBT has proven to be the most effective augmenter for nonresponders to SSRIs (Simpson et al., 2013).

Behavioral therapy

Psychological interventions attempted in the early 20th century, based on psychodynamic and psychoanalytic theories, had limited benefit in the treatment of OCD (Foa, 2010). With the advent of behavior therapy, therapists started focusing on the "here and now," rather than past experiences and unconscious conflicts. Earlier behavioral interventions including systematic desensitization, relaxation techniques, aversion therapy and thought-stopping were not highly successful either (Foa, 2010; Reddy, Sudhir, Manjula, Arumugham, & Narayanaswamy, 2020). These interventions primarily targeted obsessive thoughts and/or the associated anxiety, with little focus on compulsions.

The introduction of the behavioral technique of exposure and response prevention (ERP) by Meyer in 1966 was a major breakthrough in the treatment of OCD (Meyer, 1966). This intervention involves gradual and prolonged exposure to the obsession/anxiety provoking stimuli and prevention of rituals/compulsions. ERP enables habituation of the fear response, which was prevented earlier by compulsive and avoidance behaviors. It may also help in disconfirmation of underlying dysfunctional beliefs/assumptions and promote extinction learning. Recent models suggest that Inhibitory learning principles, with a focus on tolerating anxiety rather than habituation, varying exposures, and replacing feared outcomes with new learning, may be more effective in fear extinction (Craske et al., 2008). Randomized controlled trials (RCT) and meta-analyses have consistently demonstrated the efficacy of ERP and hence, it remains the psychological treatment of choice in OCD (Foa et al., 2005; Reddy et al., 2020; Skapinakis et al., 2016; Stein et al., 2019).

Cognitive therapy for OCD

Cognitive therapy has been proposed as an alternate or adjunct to ERP, for those with poor response/tolerability to ERP (Abramowitz, Taylor, & McKay, 2005). Salkovskis proposed one of the most influential cognitive models for OCD (Salkovskis, 1985). These models posit that OCD patients give undue significance to normal cognitive intrusions (which are often ignored by the general population), and misinterpret them as potentially threatening due to their underlying assumption and beliefs (such as inflated responsibility, need for perfection, need to control thoughts). Obsessive beliefs questionnaire may be helpful is evaluating for these underlying beliefs (Obsessive Compulsive Cognitions Working Group, 2003). Therapists help the patients identify and modify these misinterpretations and underlying beliefs, through techniques such as cognitive restructuring and behavioral experiments. These behavioral experiments are aimed to disprove the cognitive distortions. For example, a patient avoiding black-colored clothes may be advised to wear black clothes to evaluate whether the feared consequence occurs.

It is sometimes debated whether such behavioral experiments act as a form of exposure, thus encouraging habituation to anxiety. Similarly, certain cognitive therapists argue that exposure tasks during ERP function as behavioral experiments enabling disconfirmation of dysfunctional beliefs (Huppert & Franklin, 2005). Meta-analyses of efficacy studies have found similar effect sizes for cognitive therapy and ERP (Olatunji, Davis, Powers, & Smits, 2013; Skapinakis et al., 2016). Therapists, often integrate cognitive interventions with ERP to improve the acceptability of therapy, especially in those with poor insight and intolerable anxiety (Stein et al., 2019). Cognitive restructuring helps the patient understand the excessiveness/absurdity of the obsessions and may facilitate exposure. In practice, both interventions are commonly integrated as cognitive-behavior therapy (CBT) (Fig. 1).

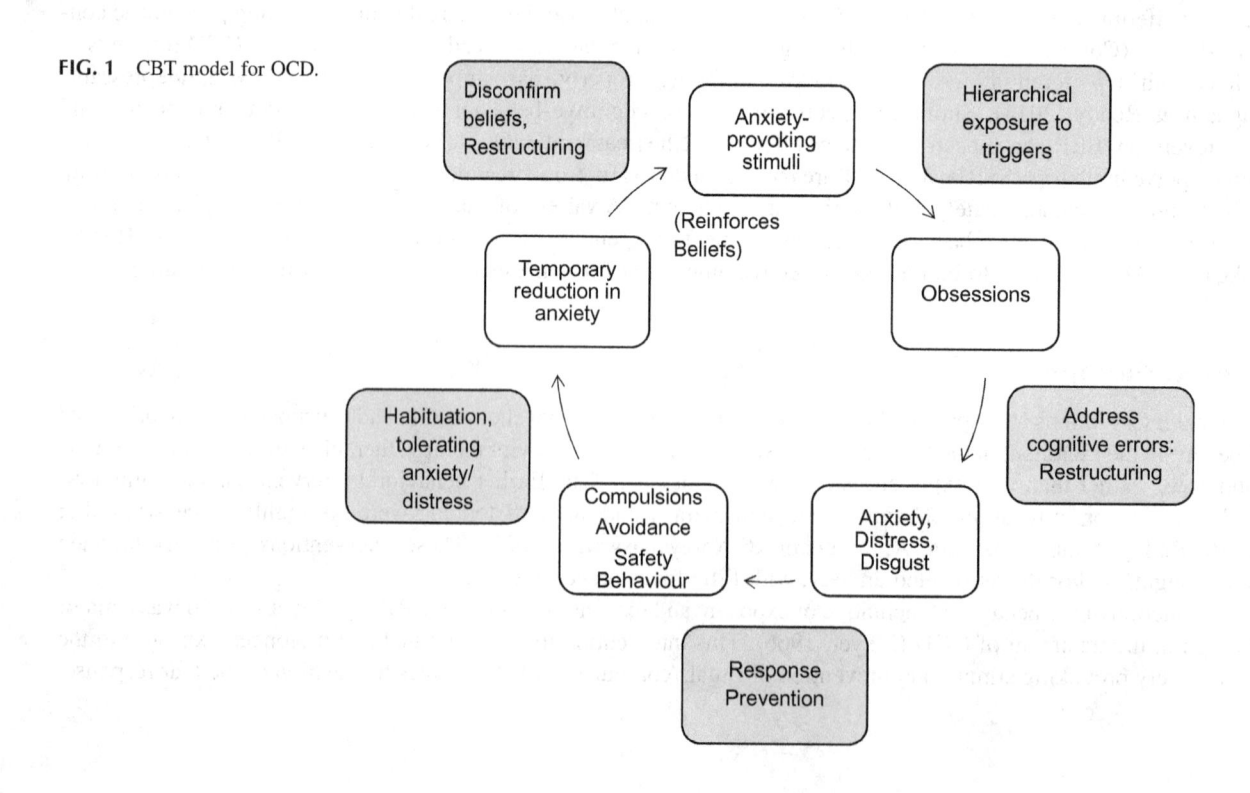

FIG. 1 CBT model for OCD.

Components of CBT

CBT for OCD usually involves the following components:

1. Assessment and establishing therapeutic relationship: The initial sessions focus on detailed assessment including confirmation of diagnosis and evaluation of comorbidity including personality disorders. It is imperative to evaluate various aspects of the OCD including nature of symptoms, severity, insight, avoidance and family accommodation. Structured instruments would be helpful in this regard (Table 1).
2. Psychoeducation: Psychoeducation should focus on nature, course of illness and rationale for choosing CBT as a treatment modality. It is imperative to discuss the principles behind treatment, structure of treatment, need for homework assignments, tolerating anxiety as a part of exposure sessions and involvement of family members, if required.
3. Sharing a personalized cognitive-behavioral formulation: A collaborative case conceptualization should be arrived based on the patient's symptoms and underlying beliefs.
4. Collaboratively build a hierarchy of triggers for obsessions/anxiety: A hierarchy of triggers based on subjective units of distress (SUDS), with lower scores representing lower anxiety/distress. This may be updated during the course of therapy based on anxiety experienced during exposures.
5. Cognitive restructuring: Cognitive restructuring may be attempted along with behavioral interventions. After eliciting faulty appraisals and cognitive distortions, restructuring is performed by gathering evidence for and against beliefs and challenging the assumptions through Socratic questioning and behavioral experiments.
6. Graded exposure: Expose the patient to triggers of obsessions/anxiety, starting with those with low SUDS. Exposure should be accompanied by abstaining from compulsive rituals. Each exposure task should continue till reduction/cessation of anxiety/distress. Therapist assisted exposure should be accompanied by homework ERP tasks. Exposure should progress gradually toward triggers with higher SUDS.
7. Address avoidance behaviors and family accommodation: Patients with OCD often demonstrate subtle avoidance behaviors (e.g., not interacting with specific people to prevent sexual/ aggressive obsessions). Family accommodation (discussed below) also helps in avoidance and hence should be gradually decreased to facilitate exposure.
8. Handling obsessive ruminations: Some patients may present with predominantly obsessions, with minimal overt compulsions (e.g., chain of thoughts regarding day-to-day mundane activities). It is important to rule out mental/cognitive compulsions which function to relieve anxiety transiently (e.g., reassuring oneself, mental prayers). If compulsions are minimal, the patient may be advised not to resist or control the obsessions and carry on with daily activities as a part of exposure.
9. Evaluate and address barriers to therapy: Therapy may not progress as expected due to various factors such as poor insight, family accommodation, personality issues, comorbid depression, anxiety etc. Such factors may have to be addressed before proceeding with therapy

TABLE 1 Tools for structured Assessment of Obsessive-Compulsive Disorder.

Diagnosis and comorbidity	Mini International Neuropsychiatric Interview Structured Clinical Interview for DSM-5 (SCID-5)
Nature and severity of symptoms	**Clinician rated:** Yale-Brown Obsessive Compulsive Scale (Y-BOCS) Dimensional Yale-Brown Obsessive Compulsive Scale (DY-BOCS) **Patient-rated:** Obsessive Compulsive Inventory-Revised (OCI-R) Dimensional Obsessive Compulsive Scale (DOCS)
Insight	Item-11 of Y-BOCS Brown Assessment of Beliefs Scale (BABS) Overvalued Ideas Scale (OVIS)
Other phenomena Family accommodation Obsessive beliefs Sensory phenomena Metacognitions	Family Accommodation Scale (FAS) Obsessive Beliefs Questionnaire (OBQ) University of São Paulo Sensory Phenomena Scale Metacognitions Questionnaire-30 (MCQ-30)

MINI (Sheehan et al., 1998), SCID-5 (MD, Williams, Karg, & Spitzer, 2016), Y-BOCS(Goodman et al., 1989), DY-BOCS (Rosario-Campos et al., 2006), OCI-R (Foa et al., 2002), DOCS(Abramowitz et al., 2010), BABS (Eisen et al., 1998), OVIS (Neziroglu, McKay, Yaryura-Tobias, Stevens, & Todaro, 1999), FAS (Calvocoressi et al., 1999), OBQ (Obsessive Compulsive Cognitions Working Group, 2003), University of São Paulo Sensory Phenomena Scale (Rosario et al., 2009), MCQ-30 (Wells & Cartwright-Hatton, 2004).

10. Plan termination and booster sessions: Termination should be planned after discussion with patient, when significant progress has been achieved or when there is plateauing of response. Termination sessions should also focus on need for continued self-guided exposure and education regarding early signs of relapse. Booster sessions may be provided as required, to consolidate the gains and prevention of relapse.

Family interventions in OCD

OCD causes significant burden on caregivers and has a negative impact on their quality of life, occupational functioning and relationships (Cicek, Cicek, Kayhan, Uguz, & Kaya, 2013; Gururaj, Math, Reddy, & Chandrashekar, 2008). The family atmosphere, in turn, affects symptoms and course of OCD. Family members may sometimes be hostile or critical toward individuals suffering from OCD (Renshaw, Chambless, & Steketee, 2003). Alternatively, family members may be accommodative by either participating in OCD rituals or modify their daily routine to decrease the anxiety/distress associated with obsessions (Calvocoressi et al., 1999). Family accommodation interferes with habituation and ERP. It is seen in majority of OCD patients across different cultures and negatively influences the course and outcome of OCD (Cherian, Pandian, Bada Math, Kandavel, & Janardhan Reddy, 2014; Lebowitz, Panza, & Bloch, 2016). Interventions addressing family accommodation have been found to be helpful in OCD (Stewart, Sumantry, & Malivoire, 2020). Supportive family members can be invited as co-therapists to supervise exposure in the home atmosphere. Such interventions may especially play a major role, especially, in collective cultures.

Treatment setting

Although the pragmatics of CBT delivery vary between centers, sessions are often conducted in an out-patient setting with 2 to 3 sessions per week over 15 to 20 sessions (Reddy et al., 2020). Availability and access to therapists remains a major constraint. Emerging evidence also support the evidence for technology-assisted therapy (Dèttore, Pozza, & Andersson, 2015). Despite these innovations, not all patients respond to out-patient CBT. CBT trials are associated with dropout and refusal rates of around 15% each (Leeuwerik, Cavanagh, & Strauss, 2019). Further, a substantial proportion of patients do not respond adequately to CBT (Fisher, Cherry, Stuart, Rigby, & Temple, 2020). Intensive supervised therapy provided in an in-patient/residential setting may be helpful in a subset of these patients. In-patient/residential treatment is recommended particularly in the context of treatment resistance, or when out-patient therapy is not feasible. It is often provided as a package, which provides an opportunity to address other important facets including familial interactions, milieu, insight, comorbidity, medication adherence etc.

In-patient/residential treatment for OCD

The earliest reports on ERP were based on therapies conducted in an in-patient setting (Meyer, 1966; Meyer, Levy, & Schnurer, 1974; Rachman, Hodgson, & Marks, 1971). Foa and Goldstein (1978) reported on the successful implementation of therapy in out-patients setting (Foa & Goldstein, 1978). Since then, multiple RCTs demonstrated the efficacy of ERP conducted as an out-patient treatment (Foa, 2010; Foa et al., 2005). Due to easier accessibility and reduction in treatment cost, CBT is currently provided primarily as an out-patient service. However, a subset of patients who do not respond adequately to out-patient CBT may require a more supervised intensive CBT program. Based, on a stepped care approach, intensive CBT provided under various degrees of supervision may be helpful for this population. Veale et al. classified the settings that provide such intensive CBT as follows (Veale et al., 2016):

1. **In-patient setting:** This provides the highest level of supervision, where treatment is provided in a hospital setup with round-the-clock nursing care. Comorbidities such as severe depression, psychosis etc. can be effectively managed in an in-patient setting. Further, in-patient facilities provide opportunity for medication management, supervision of adherence and neuromodulatory interventions. Severely ill patients such as those with suicidality, poor food intake, self-neglect, and aggressive behavior may especially be benefitted. These centers have ancillary facilities such as group therapy, family therapy, occupational therapy and vocational rehabilitation.
2. **Residential service:** These are a step-down from in-patient services. Therapists and support staff may be available during the day, but not at night. Hence, these are suggested for people who are not actively suicidal and require less supervision for daily activities and intake of medications. Patients who do not benefit from out-patient CBT and those with severely impaired daily routine may especially benefit from such services. Ancillary interventions such as occupational therapy, group therapy and family therapy are often available.

3. **Partial hospitalization:** These are similar to day care services were patients stay in the facility during the day for CBT with no facilities for overnight stay. These settings are suitable for patients planned for therapy alone and those who do not have emergent issues like suicidality.
4. **Home-based treatment:** Therapists visit the patients' home for supervision of CBT and exposure sessions. These services are optimal for subjects who have symptoms primarily in the home context. For example, some patients may not have contamination obsessions outside their home as they feel that it is "not their place." Similar facilities are provided for those with hoarding disorder. Not many centers provide such services.

In this chapter, we discuss treatment in the former two settings under the umbrella term "In-patient/residential treatment." Majority of reports on in-patient/residential treatment come from settings which provide daily intensive therapist-supervised CBT along with other treatments including self-directed exposure, occupational therapy, recreational therapy, medication management, family intervention etc. ERP remains the cornerstone of treatment. Cognitive restructuring is often provided in addition. Staying in such treatment facilities would help in planning and supervising exposure tasks on a daily basis with multiple hours of exposure sessions every day. CBT sessions are more frequent and intensive, involve a greater number of therapist hours, and conducted under close supervision.

Individual centers differ in the services provided. While some centers provide some form of round the clock staff supervision (Drummond, 1993), others provide such facilities during day alone (Thornicroft, Colson, & Marks, 1991). Hence, the nature of patients that can be admitted may also vary between such centers. For example, it may not be possible to admit people with propensity for aggressions and suicidality in the latter setting. Some centers have specialized in-patient behavior therapy setup (sometimes exclusively for those with OCD and related disorders) (Björgvinsson et al., 2013; Osgood-Hynes, Riemann, & Björgvinsson, 2003; Stewart, Stack, Farrell, Pauls, & Jenike, 2005). OCD treatment is amalgamated with general psychiatric in-patient services in other setups. As the staff are well-trained and consistent in the former, patients may feel better understood and supported. However, very few centers have such specialized facilities. Hence, CBT is often provided along with general psychiatric in-patient services, where patients with various psychiatric disorders (such as personality disorders, mood disorders, OCD) are admitted for long-term treatment. Although these centers may not have OCD-specific facilities, they may have adequate systems for supervision of CBT, provide round the clock supervision of behavioral issues and facilities available for ancillary treatments (Balachander et al., 2020).

Treating team

The treatment is often lead by a psychiatrist/psychologist specializing in OCD and CBT. Senior psychiatrists/psychologists often perform a supervisory role by training manpower, supervising treatment and formulating a management plan for each patient. Medication management is done by the psychiatrists, specialized in OCD treatment. Individual sessions of CBT are conducted by either psychologists or resident psychiatric trainees, depending on the setting. Centers often have nurses who may assist in ensuring exposure exercises, dispensing medications and handling behavioral/medical emergencies. Social workers are often involved in handling family interventions, after-care planning/follow-up and vocational rehabilitation. Some setups have counselors who assist in exposure tasks and activities of daily living. Other specialists may be called in to provide additional support on aspects such as substance dependence, neuromodulation, neuropsychological assessment, and evaluation for neurosurgical interventions.

Components of in-patient/residential treatment

Cognitive-behavior therapy

The major goal of in-patient/residential treatment is to provide supervised and intensive CBT. Most centers focus especially on ERP as the primary strategy. In-patient/residential care provides the opportunity to plan and follow through exposure tasks on a day-to-day basis. It also helps in providing intensive therapy sessions lasting 3 to 6 h of exposure per day. The milieu can be modified to facilitate exposure (e.g., monitoring/controlling water usage, staying among people to facilitate exposure to aggressive/sexual obsessions etc.). Usually, it is provided as a combination of therapist guided and self-directed exposure. Therapist assisted exposure tasks help in supervising the tasks and help in monitoring for within-session habituation.

Self-directed exposure tasks are important to promote independence and prevent shifting of vicarious responsibility during exposure tasks. For example, a patient with obsessive doubts and repeating/checking compulsions may feel reassured/satisfied with the therapist was supervising their activities and may not feel the need to check. Early morning or bedtime rituals have to be handled through self-directed exposure. Supervision by ward staff or family members may also be planned. Cognitive restructuring may also be helpful for modifying dysfunctional beliefs and misinterpretations.

Group therapy

Most programs provide group therapy in addition to individual therapy. Due to the idiosyncratic nature of obsessions/compulsions, planning of exposure tasks are generally conducted in an individual format. The admitted patients being in different stages of treatment makes group therapy more challenging. In some centers, the ERP sessions occur primarily in the group format (Osgood-Hynes et al., 2003). Such groups focus on homework review, contract setting and treatment planning. Symptom specific groups may also be conducted to decrease heterogeneity. In other centers, group therapy is provided as an add-on treatment to facilitate exposure exercises or toward other ends. For example, groups may be conducted for cognitive restructuring, providing support to caregivers, planning rehabilitation, improving coping strategies etc. Irrespective of the goals, peer-lead group sessions help in providing support, feedback and motivate members for therapy.

Medication management

A detailed evaluation conducted at intake includes evaluation of severity, nature, course of OCD and assessments for comorbidities as well as treatment history. This would help the psychiatrist in planning pharmacotherapy. Adherence to pharmacotherapy can be supervised in the in-patient/residential setting. Obsessions may interfere with medication compliance (e.g., tablets may be "contaminated" or need to take medications only on particular dates). Supervision of medication intake by nursing staff and planning exposure tasks for medication intake may ensure compliance in such scenarios. Frequent evaluation helps in monitoring for adverse effects and titration of the regimen.

Family education and support

Family members may be involved in care to various degrees. For example, in the in-patient treatment program of the OCD Clinic at National Institute of Mental Health and NeuroSciences (NIMHANS), Bangalore, India, a family member stays with the patient throughout the entire course of in-patient stay. This provides an opportunity to address familial factors that maintain OCD such as family accommodation, expressed emotions and encourage involvement of family members as cotherapists (Balachander et al., 2020). In other centers, separate individual or group sessions are held with family members to address such issues.

Discharge planning

The treatment goals have to be discussed from the day of admission, which may need modification during the course of therapy. Plans for discharge and after discharge care should thus be discussed from the beginning. Discharge may be planned when treatment goals are achieved or when it becomes clear that further stay is unlikely to yield any further advantage. In such cases, plans for alternate treatments including medication management, readmission at a later date, neuromodulatory or neurosurgical interventions may be made. It should be emphasized that discharge doesn't usually mean end of treatment. Most patients receiving such intensive treatment are treatment resistant and hence may have significant residual symptoms. Booster CBT sessions and regular follow-up care is required. Plan for after care and rehabilitation should be part of the discharge planning.

Other interventions

Other ancillary treatments are often provided as a part of the in-patient/residential care. For example, occupational therapy or vocational rehabilitation may be inculcated into the program. Noninvasive neuromodulatory interventions such as repetitive transcranial magnetic stimulation (rTMS) or transcranial direct current stimulation (tDCS) may be attempted as experimental interventions in resistant patients during the stay as they require daily sessions. Electroconvulsive therapy is not indicated in OCD unless there are comorbidities such as severe depression, catatonia etc. Interventions for comorbidities such as personality disorders, substance use disorders, anxiety, mood, psychotic disorders etc. may also be required.

Evidence for in-patient/residential treatment

Studies from different countries, including Canada, Germany, India, Italy, Norway, UK, USA evaluating the efficacy of in-patient/residential CBT have shown encouraging outcomes with significant decrease in OCD symptom severity following in-patient or residential treatment (Balachander et al., 2020; Björgvinsson et al., 2013; Dèttore, Pozza, & Coradeschi, 2013; Grøtte et al., 2018; Nowak, Osen, & Kröger, 2020; Taube-Schiff, Rector, Larkin, Mehak, & Richter, 2020; Veale et al., 2016). This is especially encouraging as improvement has been noticed in treatment-resistant patients as well. A meta-analysis of 19 studies ($N = 2306$) found an overall reduction of 10.7 (9.8–11.5, $P < .001$) point reduction Y-BOCS at discharge compared to the scores at admission (Veale, Naismith, Miles, Gledhill, et al., 2016). This meta-analysis included studies conducted both in-patient as well as residential setting. Studies have shown response rates (defined as $\geq 35\%$ reduction in Y-BOCS scores) ranging between 50% and 80% (Balachander et al., 2020; Grøtte et al., 2018). The intervention has shown promising results from adolescent population too (Björgvinsson et al., 2008; Leonard et al., 2016).

Long-term outcome

Studies have shown that the improvement is maintained at 1- to 2-year follow-up (Kordon et al., 2005; Veale, Naismith, Miles, Childs, et al., 2016). A 2-year naturalistic follow-up data from India analyzed with latent growth class modeling found the following categories of course—"remitters" (14.5%), "responders" (36.5%), "minimal responders" (34.7%), and "nonresponders" (14.6%) (Balachander et al., 2020). A German study evaluated 30 patients with OCD treated with in-patient CBT 8 to 10 years after treatment. Significant improvements were observed at follow-up with medium to large effect size. Continuation of exposure exercises at follow-up was the only significant predictor of outcome (Külz et al., 2020).

However, the evidence should be interpreted cautiously due to the following factors:

1. Although most studies included CBT as an active component, a plethora of other interventions were provided as well. For example, medication management, family interventions, vocational rehabilitation etc. were often provided. Hence the relative efficacy of individual components is not clear.
2. The studies differed with respect to the nature of interventions. For example, CBT was delivered in group format in some centers and individual format in other centers. Similarly, some studies were conducted in residential setting and others in an hospital in-patient setting (Balachander et al., 2020; Osgood-Hynes et al., 2003).
3. In the absence of control group, the role of nonspecific factors such as attention, placebo response, regression to mean etc. cannot be ruled out.
4. As the intervention is resource intensive, it is important to evaluate its cost-effectiveness.

Predictors of response

Identifying predictors would help in personalization of treatment and improving the treatment services for those showing poor outcome. There is scant literature on predictors of response from prospective systematic studies. A study from the Massachusetts General Hospital/McLean OCD Institute (OCDI), USA, ($n = 476$) found that lower baseline OCD severity, better psychosocial functioning and female gender was associated with lower OCD severity at discharge (Stewart, Yen, Stack, & Jenike, 2006). A study from the residential treatment program of the Rogers Memorial Hospital, USA ($n = 379$), evaluating potential predictors with structured assessments found baseline OCD severity to be a predictor of poor outcome (Siwiec, Riemann, & Lee, 2019). A prospective study from the OCD clinic of NIMHANS, India ($n = 58$) employing structured clinical assessment for various psychiatric diagnoses and other OCD relevant phenomena, found that poor insight at baseline was the only significant predictor of nonresponse (Nanjundaswamy, Arumugham, Narayanaswamy, & Reddy, 2020). A retrospective study with a larger sample ($n = 420$) from the same center found better insight, shorter duration of illness and lesser contamination/washing symptoms to be associated with better response both at short-term and long-term follow-up (Balachander et al., 2020). Studies have shown heterogeneous findings with respect to predictors, possibly due to the varying methodology, sample and nature of interventions. A recent systematic review found that being married or cohabiting was a consistent predictor of better outcome, while hoarding and comorbid alcohol misuse was consistently associated with poor outcome (Veale, Naismith, Miles, Gledhill, et al., 2016).

Indications for in-patient residential treatment

In-patient/residential has a specific role in stepped care approach for treatment of OCD. Table 2 summarizes the advantages and limitations of in-patient/residential therapy as compared to out-patient based CBT.

TABLE 2 Advantages/limitations of in-patient/residential treatment.

Advantages	Limitations
• Closer supervision of therapy • Higher frequency of sessions • Involvement of a Multidisciplinary team • Found effective across different settings	• More expensive than out-patient based CBT • No controlled trials available • Therapy gains may not be generalized to real-life circumstances. This is especially relevant when obsessions are triggered primarily in the home atmosphere. • Due to the involvement of multiple components, the relative efficacy of each component is not clear • The components of intervention differ between centres. Hence there is a heterogeneity in the available data. • Cost-effectiveness has not been evaluated

TABLE 3 Indications for in-patient/residential treatment.

1. Patients with treatment refractory OCD—who have not benefited from out-patient CBT and other first-line treatments
2. Patients who have not been able to access out-patient CBT.
3. Those with complex rituals and severe avoidance, which are difficult to handle on out-basis.
4. Those with comorbid personality disorders or mood disorders which interfere with out-patient exposure therapy

Given the above limitations, in-patient/residential treatment is recommended for specific indications as shown in Table 3.

B4DT: Revolutionizing ERP for OCD

The Bergen 4-day treatment (B4DT) is a novel approach for providing concentrated ERP, developed by the OCD-team from Bergen, Norway (Hansen, Hagen, Öst, Solem, & Kvale, 2018). It consists of an intense 4-day therapy format, delivered to 3 to 6 patients at a time, with equal number of therapists. The concept was borne out of evidence for potentially better outcomes with concentrated ERP as compared to weekly sessions (Abramowitz, Foa, & Franklin, 2003). It has been described as an "individual therapy in a group setting" with a therapist: patient ratio of 1:1. After an initial assessment and confirmation of the diagnosis, patients are introduced to the treatment using written material as well as instructional videos. Patients are requested to suggest relevant exposure tasks on the principle that they should attempt "exposures that their OCD would appreciate the least." The main feature of the 4-day treatment is to teach the patients to seek out anxiety-generating cues and use this anxiety and discomfort as a cue to "LEan into The anxiety" (LET-technique) instead of resorting to avoidance or compulsions. Initial exposures are therapist-assisted. The first of the 4 days (approximately 3 h) is allocated to psychoeducation and preparing individual exposure tasks. The 2nd and 3rd days involve several hours of individual and therapist-assisted exposure interspersed with brief group meetings. Caregivers also receive a psychoeducation session on day 3. The 4th day is used for summarizing and relapse prevention. Self-exposures are planned over the next three weeks. Patients are required to log homework tasks over the next 3 weeks and follow-up for an individual session to discuss their experiences and refresh treatment principles.

Studies in Norway have reported excellent response rates of around 90% and remission rates of around 76% (Kvale et al., 2018). Improvement has been found to be maintained up to 1 year of follow-up (Hansen et al., 2018). Although B4DT does not involve in-patient/residential treatment, therapy is provided over several hours each day over four days. The exposure is more concentrated than that provided in in-patient/residential setting. Exposure is not based on hierarchy. Rather, exposures that are likely to be associated with the most change are encouraged. This is different from the traditional approach, where gradual reduction in distress is emphasized. B4DT is currently attempted as a replacement for traditional out-patient based weekly CBT. It has to be seen whether this model would be helpful in treatment resistant population, where in-patient/residential treatment is indicated. The efficacy and cost-effectiveness of the intervention across different settings has to be evaluated.

Practice and procedures

We describe below the course of an in-patient treatment for OCD at our center with the following vignette:

Ms. B, a 45-year-old woman, presented with a history of over 20 years of continuous illness, characterized by fears of contamination with dirt and fecal matter. As a result, obsessions peaked around bathroom-related activities, leading to repeated washing and cleaning, using large amounts of water, using harsh detergents on herself and on most surfaces, and substantial avoidance. These compulsions would persist till she was "satisfied," an arbitrary endpoint, with her often remaining in the bathroom for days together, because of persisting doubts about whether she/her clothes were adequately clean or whether she had been re-contaminated on her way out. She had lost a significant amount of weight, had eczematous lesions over her arms and legs, and would avoid food and water to reduce her visits to the bathroom. She sometimes spends up to 5 to 6 days inside the bathroom, performing various compulsions. At admission, her YBOCS was 40/40, with good insight that faltered when confronted by obsessional thoughts. She had not responded to adequate trials of two SSRIs. She had not received out-patient CBT because of unavailability of therapists in her hometown. She was admitted in our center for in-patient CBT.

After an initial assessment, goals were set for treatment. Immediate goals were to improve oral intake and reduce time spent in the bathroom. Mealtimes were set and caloric and fluid intake gradually increased. Her treatment history was reviewed and she was started on fluoxetine (gradually increased up to 80 mg), along with clonazepam to bring down anxiety. During the initial days of hospitalization, she was unable to participate in therapy due to overwhelming anxiety. A course of anodal Transcranial Direct Current Stimulation (tDCS) to the pre-supplementary motor was provided as an experimental intervention for immediate decrease in her symptoms. Following 15 sessions of tDCS, she had notable improvement and was more cooperative for starting CBT.

After psychoeducation regarding OCD and treatment options, she was educated about the CBT model of OCD. Daily CBT sessions were conducted by a psychiatry resident, closely supervised by the senior resident and consultant attached to the OCD clinic. A hierarchy of exposure tasks was generated, collaboratively with the patient, which was modified when it was noticed that her anticipatory anxiety was higher than actual anxiety experienced during exposure tasks. Exposure was graded, moving from moderately-anxiety inducing tasks to high anxiety-inducing tasks. During the initial two weeks of therapy, exposure tasks were primarily therapist-assisted, while patient was encouraged to practice self-exposure for the same tasks. Therapist-assisted sessions of 60 to 90 min were conducted every day for 5 days a week. Exposure tasks were monitored to ensure that compulsions were not performed or replaced with other avoidance strategies. Reassurance seeking from both family members and treating team was discouraged. Self-guided exposure tasks were gradually increased.

On evaluation of her family interactions, it was found that her younger son was particularly critical of her symptoms, while her older son was excessively accommodative. Family intervention was provided to address these issues and her sons were included as co-therapists in treatment. Her motivation and understanding were assessed at regular intervals. Cognitive restructuring was also performed to address her beliefs related to contamination and threat. Her treatment progress was reviewed every week, by a multi-disciplinary team involving psychiatrists, psychologists, social workers and nursing personnel.

After six weeks of ERP, her symptoms improved remarkably and Y-BOCS score decreased to 21/40. She was able to maintain her activities of daily living independently. Due to plateauing of improvement and difficulty in further engagement in therapy, a plan was made to take a break from intensive therapy and she was discharged from the hospital. She maintained improvement until 1 year of follow-up. Booster sessions were held, revisiting the concept of habituation and suggesting tasks for self-guided exposure.

Mini-dictionary of terms

Obsessions: Repetitive intrusive mental phenomena that occur in the form of thoughts, impulses, or images and are usually perceived as unwanted. They are often associated with anxiety or distress.

Compulsions: Recurrent behaviors (or mental acts) that are often performed in response to obsessions to decrease the anxiety/distress or according to rigidly applied rules.

Exposure and response prevention: A behavioral treatment technique involving exposure of the patient to the anxiety/obsession provoking stimuli, while making a conscious decision to avoid compulsions till the anxiety reduces.

Habituation: Reduction in anxiety/distress over time after exposure to fear/obsession inducing stimuli.

Graded exposure: A behavioral technique where a patient is gradually exposed hierarchically to anxiety/obsession triggering stimuli, with the goal of habituation to stimuli lower in the hierarchy before progressing to more difficult ones.

Cognitive restructuring: A cognitive therapeutic technique where the therapist helps the patient identify and modify dysfunctional or maladaptive thoughts.

In-patient/residential treatment: Multimodal treatment involving intensive cognitive-behavior therapy, medication management, family intervention, rehabilitation etc. provided in a supervised setting, where the patient stays till the end of treatment.

Key facts

- Obsessive-compulsive has a lifetime prevalence of 2% to 3% and is among the most disabling psychiatric conditions.
- Around 20% to 30% of patient with OCD do not respond to pharmacotherapy and out-patient based cognitive-behavior therapy (CBT).
- Around 50% to 70% of patients, including those with treatment resistant OCD, respond to in-patient/residential treatment.
- Patients with good insight, lower baseline illness severity, shorter illness duration, female gender, and less contamination-themed symptoms show better response to in-patient/residential treatment.
- The Bergen 4-day treatment, a novel form of concentrated out-patient CBT, has shown response and remission rate as high as 90% and 76%, respectively.

Application to other areas

In this chapter, we have discussed in-patient/residential treatment for OCD. This intervention being resource intensive, may be applicable only to those with severe and treatment-resistant illness. Patients with poor insight into their symptoms may also benefit from intensive CBT. The role of the novel out-patient based intensive B4DT intervention has to be evaluated in this population. CBT is often recommended as first-line treatment for patients with other obsessive-compulsive related disorders including body dysmorphic disorder (BDD), trichotillomania, excoriation disorder and hoarding disorder (Reddy et al., 2020). The treatment for BDD has some overlap with that of OCD in the form of SSRIs and exposure-based interventions, but also involves other CBT techniques including cognitive restructuring and perceptual/mirror retraining (Castle et al., 2021). Patients with BDD often have poor insight as compared to OCD. Severe and treatment resistant BDD may also benefit from in-patient residential treatment. Due to partial overlap in treatment components, a similar setup would be helpful. Habit reversal therapy is found to be helpful in patients with trichotillomania and excoriation disorder (Reddy et al., 2020). It has to be evaluated whether intensive interventions are helpful in patients with these disorders, who do not respond to out-patient CBT. In-patient therapy may have limited role in hoarding disorder as it often presents exclusively in the home atmosphere.

Summary points

- A substantial minority of patients of patients do not respond to out-patient CBT and pharmacotherapy.
- In-patient/residential treatment is indicated in this population along with those who are not able to access out-patient based CBT
- In-patient/residential treatment involves a multimodal treatment which involves multiple components including intensive CBT, medication management, family interventions, group interventions, rehabilitation etc.
- Around 50% to 70% of patients respond to in-patient/residential, with improvement maintained over long-term follow-up.
- However, the evidence should be interpreted cautiously due to lack of controlled trials, cost-effectiveness evaluation and trials on relative efficacy of individual components of intervention.

References

Abramowitz, J. S., Deacon, B. J., Olatunji, B. O., Wheaton, M. G., Berman, N. C., Losardo, D., ... Hale, L. R. (2010). Assessment of obsessive-compulsive symptom dimensions: Development and evaluation of the Dimensional Obsessive-Compulsive Scale. *Psychological Assessment*, 22(1), 180–198. https://doi.org/10.1037/a0018260.

Abramowitz, J. S., Foa, E. B., & Franklin, M. E. (2003). Exposure and ritual prevention for obsessive-compulsive disorder: Effects of intensive versus twice-weekly sessions. *Journal of Consulting and Clinical Psychology*, 71(2), 394–398.

Abramowitz, J. S., Taylor, S., & McKay, D. (2005). Potentials and limitations of cognitive treatments for obsessive-compulsive disorder. *Cognitive Behaviour Therapy, 34*(3), 140–147. https://doi.org/10.1080/16506070510041202.

Arumugham, S. S., & Reddy, J. Y. C. (2013). Augmentation strategies in obsessive-compulsive disorder. *Expert Review of Neurotherapeutics, 13*(2), 187–202. quiz 203 https://doi.org/10.1586/ern.12.160.

Arumugham, S. S., & Reddy, Y. C. J. (2014). Commonly asked questions in the treatment of obsessive-compulsive disorder. *Expert Review of Neurotherapeutics, 14*(2), 151–163. https://doi.org/10.1586/14737175.2014.874287.

Balachander, S., Bajaj, A., Hazari, N., Kumar, A., Anand, N., Manjula, M., … Reddy, Y. C. J. (2020). Long-term outcomes of intensive inpatient care for severe, resistant obsessive-compulsive disorder. *Canadian Journal of Psychiatry. Revue Canadienne de Psychiatrie, 65*(11), 779–789. https://doi.org/10.1177/0706743720927830.

Björgvinsson, T., Hart, A. J., Wetterneck, C., Barrera, T. L., Chasson, G. S., Powell, D. M., … Stanley, M. A. (2013). Outcomes of specialized residential treatment for adults with obsessive-compulsive disorder. *Journal of Psychiatric Practice, 19*(5), 429–437. https://doi.org/10.1097/01.pra.0000435043.21545.60.

Björgvinsson, T., Wetterneck, C. T., Powell, D. M., Chasson, G. S., Webb, S. A., Hart, J., … Stanley, M. A. (2008). Treatment outcome for adolescent obsessive-compulsive disorder in a specialized hospital setting. *Journal of Psychiatric Practice, 14*(3), 137–145. https://doi.org/10.1097/01.pra.0000320112.36648.3e.

Calvocoressi, L., Mazure, C. M., Kasl, S. V., Skolnick, J., Fisk, D., Vegso, S. J., … Price, L. H. (1999). Family accommodation of obsessive-compulsive symptoms: Instrument development and assessment of family behavior. *The Journal of Nervous and Mental Disease, 187*(10), 636–642. https://doi.org/10.1097/00005053-199910000-00008.

Castle, D., Beilharz, F., Phillips, K. A., Brakoulias, V., Drummond, L. M., Hollander, E., … Fineberg, N. A. (2021). Body dysmorphic disorder: A treatment synthesis and consensus on behalf of the International College of Obsessive-Compulsive Spectrum Disorders and the Obsessive Compulsive and Related Disorders Network of the European College of Neuropsychopharmacology. *International Clinical Psychopharmacology, 36*(2), 61–75. https://doi.org/10.1097/YIC.0000000000000342.

Cherian, A. V., Pandian, D., Bada Math, S., Kandavel, T., & Janardhan Reddy, Y. C. (2014). Family accommodation of obsessional symptoms and naturalistic outcome of obsessive-compulsive disorder. *Psychiatry Research, 215*(2), 372–378. https://doi.org/10.1016/j.psychres.2013.11.017.

Cicek, E., Cicek, I. E., Kayhan, F., Uguz, F., & Kaya, N. (2013). Quality of life, family burden and associated factors in relatives with obsessive-compulsive disorder. *General Hospital Psychiatry, 35*(3), 253–258. https://doi.org/10.1016/j.genhosppsych.2013.01.004.

Collins, P. Y., Patel, V., Joestl, S. S., March, D., Insel, T. R., & Daar, A. S. (2011). Grand challenges in global mental health. *Nature, 475*(7354), 27–30. https://doi.org/10.1038/475027a.

Craske, M. G., Kircanski, K., Zelikowsky, M., Mystkowski, J., Chowdhury, N., & Baker, A. (2008). Optimizing inhibitory learning during exposure therapy. *Behaviour Research and Therapy, 46*(1), 5–27. https://doi.org/10.1016/j.brat.2007.10.003.

Dèttore, D., Pozza, A., & Andersson, G. (2015). Efficacy of technology-delivered cognitive behavioural therapy for OCD versus control conditions, and in comparison with therapist-administered CBT: Meta-analysis of randomized controlled trials. *Cognitive Behaviour Therapy, 44*(3), 190–211. https://doi.org/10.1080/16506073.2015.1005660.

Dèttore, D., Pozza, A., & Coradeschi, D. (2013). Does time-intensive ERP attenuate the negative impact of comorbid personality disorders on the outcome of treatment-resistant OCD? *Journal of Behavior Therapy and Experimental Psychiatry, 44*(4), 411–417. https://doi.org/10.1016/j.jbtep.2013.04.002.

Drummond, L. M. (1993). The treatment of severe, chronic, resistant obsessive-compulsive disorder: An evaluation of an in-patient programme using behavioural psychotherapy in combination with other treatments. *The British Journal of Psychiatry, 163*, 223–229. https://doi.org/10.1192/bjp.163.2.223.

Eisen, J. L., Phillips, K. A., Baer, L., Beer, D. A., Atala, K. D., & Rasmussen, S. A. (1998). The Brown Assessment Of Beliefs Scale: Reliability and validity. *The American Journal of Psychiatry, 155*(1), 102–108. https://doi.org/10.1176/ajp.155.1.102.

Fineberg, N. A., Hollander, E., Pallanti, S., Walitza, S., Grünblatt, E., Dell'Osso, B. M., Albert, U., Geller, D. A., Brakoulias, V., Janardhan Reddy, Y. C., Arumugham, S. S., Shavitt, R. G., Drummond, L., Grancini, B., De Carlo, V., Cinosi, E., Chamberlain, S. R., Ioannidis, K., Rodriguez, C. I., … Menchon, J. M. (2020). Clinical advances in obsessive-compulsive disorder: A position statement by the International College of Obsessive-Compulsive Spectrum Disorders. *International Clinical Psychopharmacology, 35*(4), 173–193. https://doi.org/10.1097/YIC.0000000000000314.

Fisher, P. L., Cherry, M. G., Stuart, T., Rigby, J. W., & Temple, J. (2020). People with obsessive-compulsive disorder often remain symptomatic following psychological treatment: A clinical significance analysis of manualised psychological interventions. *Journal of Affective Disorders, 275*, 94–108. https://doi.org/10.1016/j.jad.2020.06.019.

Foa, E. B. (2010). Cognitive behavioral therapy of obsessive-compulsive disorder. *Dialogues in Clinical Neuroscience, 12*(2), 199–207.

Foa, E. B., & Goldstein, A. (1978). Continuous exposure and complete response prevention in the treatment of obsessive-compulsive neurosis. *Behavior Therapy, 9*(5), 821–829. https://doi.org/10.1016/S0005-7894(78)80013-6.

Foa, E. B., Huppert, J. D., Leiberg, S., Langner, R., Kichic, R., Hajcak, G., & Salkovskis, P. M. (2002). The Obsessive-Compulsive Inventory: Development and validation of a short version. *Psychological Assessment, 14*(4), 485–496.

Foa, E. B., Liebowitz, M. R., Kozak, M. J., Davies, S., Campeas, R., Franklin, M. E., … Tu, X. (2005). Randomized, placebo-controlled trial of exposure and ritual prevention, clomipramine, and their combination in the treatment of obsessive-compulsive disorder. *The American Journal of Psychiatry, 162*(1), 151–161. https://doi.org/10.1176/appi.ajp.162.1.151.

Goodman, W. K., Price, L. H., Rasmussen, S. A., Mazure, C., Fleischmann, R. L., Hill, C. L., … Charney, D. S. (1989). The Yale-Brown Obsessive Compulsive Scale. I. Development, use, and reliability. *Archives of General Psychiatry, 46*(11), 1006–1011.

Grøtte, T., Hansen, B., Haseth, S., Vogel, P. A., Guzey, I. C., & Solem, S. (2018). Three-week inpatient treatment of obsessive-compulsive disorder: A 6-month follow-up study. *Frontiers in Psychology, 9*. https://doi.org/10.3389/fpsyg.2018.00620.

Gururaj, G. P., Math, S. B., Reddy, J. Y. C., & Chandrashekar, C. R. (2008). Family burden, quality of life and disability in obsessive compulsive disorder: An Indian perspective. *Journal of Postgraduate Medicine, 54*(2), 91–97.

Hansen, B., Hagen, K., Öst, L.-G., Solem, S., & Kvale, G. (2018). The Bergen 4-day OCD treatment delivered in a group setting: 12-month follow-up. *Frontiers in Psychology, 9*, 639. https://doi.org/10.3389/fpsyg.2018.00639.

Huppert, J. D., & Franklin, M. E. (2005). Cognitive behavioral therapy for obsessive-compulsive disorder: An update. *Current Psychiatry Reports, 7*(4), 268–273. https://doi.org/10.1007/s11920-005-0080-x.

Kordon, A., Kahl, K. G., Broocks, A., Voderholzer, U., Rasche-Räuchle, H., & Hohagen, F. (2005). Clinical outcome in patients with obsessive-compulsive disorder after discontinuation of SRI treatment: Results from a two–year follow–up. *European Archives of Psychiatry and Clinical Neuroscience, 255*(1), 48–50. https://doi.org/10.1007/s00406-004-0533-y.

Külz, A. K., Landmann, S., Schmidt-Ott, M., Zurowski, B., Wahl-Kordon, A., & Voderholzer, U. (2020). Long-term follow-up of cognitive-behavioral therapy for obsessive-compulsive disorder: Symptom severity and the role of exposure 8–10 years after inpatient treatment. *Journal of Cognitive Psychotherapy, 34*(3), 261–271. https://doi.org/10.1891/JCPSY-D-20-00002.

Kvale, G., Hansen, B., Björgvinsson, T., Børtveit, T., Hagen, K., Haseth, S., … Öst, L.-G. (2018). Successfully treating 90 patients with obsessive compulsive disorder in eight days: The Bergen 4-day treatment. *BMC Psychiatry, 18*(1), 323. https://doi.org/10.1186/s12888-018-1887-4.

Lebowitz, E. R., Panza, K. E., & Bloch, M. H. (2016). Family accommodation in obsessive-compulsive and anxiety disorders: A five-year update. *Expert Review of Neurotherapeutics, 16*(1), 45–53. https://doi.org/10.1586/14737175.2016.1126181.

Leeuwerik, T., Cavanagh, K., & Strauss, C. (2019). Patient adherence to cognitive behavioural therapy for obsessive-compulsive disorder: A systematic review and meta-analysis. *Journal of Anxiety Disorders, 68*. https://doi.org/10.1016/j.janxdis.2019.102135, 102135.

Leonard, R. C., Franklin, M. E., Wetterneck, C. T., Riemann, B. C., Simpson, H. B., Kinnear, K., … Lake, P. M. (2016). Residential treatment outcomes for adolescents with obsessive-compulsive disorder. *Psychotherapy Research: Journal of the Society for Psychotherapy Research, 26*(6), 727–736. https://doi.org/10.1080/10503307.2015.1065022.

MD, M. B. F., Williams, J. B. W., Karg, R. S., & Spitzer, R. L. (2016). *Structured clinical interview for DSM-5® disorders—Clinician version* (Ppk edition). American Psychiatric Association Publishing.

Meyer, V. (1966). Modification of expectations in cases with obsessional rituals. *Behaviour Research and Therapy, 4*(4), 273–280. https://doi.org/10.1016/0005-7967(66)90023-4.

Meyer, V., Levy, R., & Schnurer, A. (1974). The behavioural treatment of obsessive-compulsive disorders. In *Obsessional states* Methuen & Co. pp. viii, 352–viii, 352.

Nanjundaswamy, M., Arumugham, S., Narayanaswamy, J., & Reddy, Y. (2020). A prospective study of intensive in-patient treatment for obsessive-compulsive disorder. *Psychiatry Research, 291*. https://doi.org/10.1016/j.psychres.2020.113303, 113303.

Neziroglu, F., McKay, D., Yaryura-Tobias, J. A., Stevens, K. P., & Todaro, J. (1999). The overvalued ideas scale: Development, reliability and validity in obsessive-compulsive disorder. *Behaviour Research and Therapy, 37*(9), 881–902.

Nowak, S., Osen, B., & Kröger, C. (2020). Remission, response and its prediction after cognitive-behavioral therapy for obsessive-compulsive disorder in an inpatient setting. *Psychotherapie, Psychosomatik, Medizinische Psychologie, 70*(5), 197–204. https://doi.org/10.1055/a-0975-9628.

Obsessive Compulsive Cognitions Working Group. (2003). Psychometric validation of the obsessive beliefs questionnaire and the interpretation of intrusions inventory: Part I. *Behaviour Research and Therapy, 41*(8), 863–878.

Olatunji, B. O., Davis, M. L., Powers, M. B., & Smits, J. A. J. (2013). Cognitive-behavioral therapy for obsessive-compulsive disorder: A meta-analysis of treatment outcome and moderators. *Journal of Psychiatric Research, 47*(1), 33–41. https://doi.org/10.1016/j.jpsychires.2012.08.020.

Osgood-Hynes, D. J., Riemann, B. C., & Björgvinsson, T. (2003). *Short-term residential treatment for obsessive-compulsive disorder*. https://doi.org/10.1093/brief-treatment/mhg028.

Rachman, S., Hodgson, R., & Marks, I. M. (1971). The treatment of chronic obsessive-compulsive neurosis. *Behaviour Research and Therapy, 9*(3), 237–247. https://doi.org/10.1016/0005-7967(71)90009-x.

Reddy, Y. C. J., Sudhir, P. M., Manjula, M., Arumugham, S. S., & Narayanaswamy, J. C. (2020). Clinical practice guidelines for cognitive-behavioral therapies in anxiety disorders and obsessive-compulsive and related disorders. *Indian Journal of Psychiatry, 62*(8), 230. https://doi.org/10.4103/psychiatry.IndianJPsychiatry_773_19.

Renshaw, K. D., Chambless, D. L., & Steketee, G. (2003). Perceived criticism predicts severity of anxiety symptoms after behavioral treatment in patients with obsessive-compulsive disorder and panic disorder with agoraphobia. *Journal of Clinical Psychology, 59*(4), 411–421. https://doi.org/10.1002/jclp.10048.

Rosario, M. C., Prado, H. S., Borcato, S., Diniz, J. B., Shavitt, R. G., Hounie, A. G., … Miguel, E. (2009). Validation of the University of São Paulo Sensory Phenomena Scale: Initial psychometric properties. *CNS Spectrums, 14*(6), 315–323. https://doi.org/10.1017/S1092852900020319.

Rosario-Campos, M. C., Miguel, E. C., Quatrano, S., Chacon, P., Ferrao, Y., Findley, D., … Leckman, J. F. (2006). The Dimensional Yale-Brown Obsessive-Compulsive Scale (DY-BOCS): An instrument for assessing obsessive-compulsive symptom dimensions. *Molecular Psychiatry, 11*(5), 495–504. https://doi.org/10.1038/sj.mp.4001798.

Salkovskis, P. M. (1985). Obsessional-compulsive problems: A cognitive-behavioural analysis. *Behaviour Research and Therapy, 23*(5), 571–583.

Sheehan, D. V., Lecrubier, Y., Sheehan, K. H., Amorim, P., Janavs, J., Weiller, E., … Dunbar, G. C. (1998). The mini-international neuropsychiatric interview (M.I.N.I.): The development and validation of a structured diagnostic psychiatric interview for DSM-IV and ICD-10. *The Journal of Clinical Psychiatry, 59*(Suppl 20), 22–33. quiz 34-57.

Simpson, H. B., Foa, E. B., Liebowitz, M. R., Huppert, J. D., Cahill, S., Maher, M. J., … Campeas, R. (2013). Cognitive-behavioral therapy vs risperidone for augmenting serotonin reuptake inhibitors in obsessive-compulsive disorder: A randomized clinical trial. *JAMA Psychiatry, 70*(11), 1190–1199. https://doi.org/10.1001/jamapsychiatry.2013.1932.

Siwiec, S. G., Riemann, B. C., & Lee, H.-J. (2019). Predictors of acute outcomes for intensive residential treatment of obsessive-compulsive disorder. *Clinical Psychology & Psychotherapy, 26*(6), 661–672. https://doi.org/10.1002/cpp.2389.

Skapinakis, P., Caldwell, D. M., Hollingworth, W., Bryden, P., Fineberg, N. A., Salkovskis, P., ... Lewis, G. (2016). Pharmacological and psychotherapeutic interventions for management of obsessive-compulsive disorder in adults: A systematic review and network meta-analysis. *The Lancet. Psychiatry, 3*(8), 730–739. https://doi.org/10.1016/S2215-0366(16)30069-4.

Stein, D. J., Costa, D. L. C., Lochner, C., Miguel, E. C., Reddy, Y. C. J., Shavitt, R. G., ... Simpson, H. B. (2019). Obsessive-compulsive disorder. *Nature Reviews. Disease Primers, 5*(1), 52. https://doi.org/10.1038/s41572-019-0102-3.

Stewart, K. E., Sumantry, D., & Malivoire, B. L. (2020). Family and couple integrated cognitive-behavioural therapy for adults with OCD: A meta-analysis. *Journal of Affective Disorders, 277*, 159–168. https://doi.org/10.1016/j.jad.2020.07.140.

Stewart, S. E., Stack, D. E., Farrell, C., Pauls, D. L., & Jenike, M. A. (2005). Effectiveness of intensive residential treatment (IRT) for severe, refractory obsessive-compulsive disorder. *Journal of Psychiatric Research, 39*(6), 603–609. https://doi.org/10.1016/j.jpsychires.2005.01.004.

Stewart, S. E., Yen, C.-H., Stack, D. E., & Jenike, M. A. (2006). Outcome predictors for severe obsessive-compulsive patients in intensive residential treatment. *Journal of Psychiatric Research, 40*(6), 511–519. https://doi.org/10.1016/j.jpsychires.2005.08.007.

Taube-Schiff, M., Rector, N. A., Larkin, P., Mehak, A., & Richter, M. A. (2020). Effectiveness of intensive treatment services for obsessive compulsive disorder: Outcomes from the first Canadian residential treatment program. *International Journal of Psychiatry in Clinical Practice, 24*(1), 59–67. https://doi.org/10.1080/13651501.2019.1676450.

Thornicroft, G., Colson, L., & Marks, I. (1991). An in-patient behavioural psychotherapy unit: Description and audit. *The British Journal of Psychiatry, 158*(3), 362–367. https://doi.org/10.1192/bjp.158.3.362.

Torres, A. R., & Lima, M. C. P. (2005). Epidemiology of obsessive-compulsive disorder: A review. *Revista Brasileira De Psiquiatria (São Paulo, Brazil: 1999), 27*(3), 237–242.

Veale, D., Naismith, I., Miles, S., Gledhill, L. J., Stewart, G., & Hodsoll, J. (2016). Outcomes for residential or inpatient intensive treatment of obsessive–compulsive disorder: A systematic review and meta-analysis. *Journal of Obsessive-Compulsive and Related Disorders, 8*, 38–49. https://doi.org/10.1016/j.jocrd.2015.11.005.

Veale, D., Naismith, I., Miles, S., Childs, G., Ball, J., Muccio, F., & Darnley, S. (2016). Outcome of intensive cognitive behaviour therapy in a residential setting for people with severe obsessive compulsive disorder: A large open case series. *Behavioural and Cognitive Psychotherapy, 44*(3), 331–346. https://doi.org/10.1017/S1352465815000259.

Wells, A., & Cartwright-Hatton, S. (2004). A short form of the metacognitions questionnaire: Properties of the MCQ-30. *Behaviour Research and Therapy, 42*(4), 385–396. https://doi.org/10.1016/S0005-7967(03)00147-5.

Chapter 19

Postpartum depression and the role and position of cognitive behavioral therapy

Rachel Buhagiar[a] and Elena Mamo[b]
[a]Department of Psychiatry, Mount Carmel Hospital, Attard, Malta, [b]Psychology Department, Mater Dei Hospital, Msida, Malta

Abbreviations

APA	American Psychiatric Association
CBT	Cognitive Behavioral Therapy
DSM	Diagnostic and Statistical Manual for Mental Disorders
EPDS	Edinburgh Postnatal Depression Scale
gCBT	Group CBT
iCBT	Internet CBT
ICD	International Classification of Diseases
IPT	Interpersonal Therapy
MDD	Major Depressive Disorder
NICE	National Institute for Health and Clinical Excellence
PMH	Perinatal Mental Health
PPD	Postpartum Depression
RCT	Randomized Controlled Trial
WHO	World Health Organization

Introduction: The perinatal period & perinatal mental health

The perinatal period refers to the time of conception, through pregnancy to the end of the first year after birth. It is a time of profound change in a family's life, shaped by a multitude of determinants, including biological, psychological, social, behavioral, and environmental (Gavin, Meltzwer-Brody, Glover, & Gaynes, 2015). Each of these factors can cause significant stress and impact on the emotional well-being of the mother and father.

This transitional life stage is a time of great vulnerability for the onset and relapse of a range of mental health disorders in both men and women. Austin et al. (2017) state that for women, this risk is "higher than at many other times" (p. 13) in their life. Similarly, new fathers' depression rates were found to be twice the national average for men in the same age group (Burgess, 2011).

Postpartum depression, or PPD, features among the most common complications of the perinatal period. Importantly, perinatal psychiatric disorders can be associated with significant morbidity and mortality, including parental suicide. Notably, for perinatal mothers, this remains the leading cause of direct deaths occurring within the first postnatal year (Knight et al., 2019). Likewise, the incidence of paternal suicide was found to be 4.8% higher in postnatally depressed fathers, compared to suicide rates in fathers without depression (Quevedo et al., 2011).

This chapter will provide an overview of maternal and paternal PPD using available scientific literature and clinical experience. The importance of the role of CBT in the treatment of PPD will be discussed, and a guiding framework for its adaptation to this population will be provided.

Maternal postpartum depression

Definition & prevalence

PPD, also known as puerperal depression, postnatal depression and perinatal depression, is a clinical syndrome which is commonly defined as the onset of a major depressive episode within the first year after childbirth (Gavin et al., 2015). It is one of the three common forms of mood disturbances in women which can occur after childbirth. The two other forms are the mild, transient and self-limiting maternity blues, and the more rare but serious puerperal psychosis (Robertson, Celasun, & Stewart, 2003).

This depressive disorder is recognized as being the "most common complication of childbearing," affecting approximately 10% to 15% of women (Robertson et al., 2003, p. 12; Howard & Khalifeh, 2020). The overall prevalence varies between different studies, with statistical differences noted between different geographical regions (Shorey et al., 2018). Women in lower and middle income countries face a higher burden of perinatal depression (Fisher et al., 2011; Woody, Ferrari, Siskind, Whiteford, & Harris, 2017). Additionally, a recent meta-analysis identified an increasing prevalence beyond six months postnatally (Shorey et al., 2018). In another review, around 19% of women were found to be experiencing a depressive episode during the first three months postpartum (Gavin et al., 2005). PPD may occur for the first time in a healthy woman without a previous psychiatric history, or more commonly, in women with a prior history of depression or PPD (Kettunen, Koistinen, & Hintikka, 2014; Shorey et al., 2018).

Clinical presentation

According to Kettunen et al. (2014), PPD is "not a homogenous disorder" (p. 8) in terms of clinical course, symptom profile and severity. The signs and symptoms of PPD are identical to those during any other times (Table 1); however, the content may also encompass perinatal matters, for example, the birth experience or worries about the baby (Robertson et al., 2003). While symptoms may be potentially disabling and distressing, subthreshold symptoms may also arise and are considered

TABLE 1 Major depressive disorder.

DSM-V Diagnostic Criteria for Major Depressive Disorder and the "Peripartum Onset Specifier"

A. 5 (or more) of the following symptoms have been present during the same 2-week period and represent a change from previous functioning; at least one of the symptoms is either (1) depressed mood or (2) loss of interest or pleasure. NOTE: Symptoms that are clearly attributable to another medical condition should not be included.
  a. Depressed mood most of the day, nearly every day, as indicated by either subjective report (e.g., feels sad, empty, hopeless) or observation made by others (e.g., appears tearful).
  b. Markedly diminished interest or pleasure in all, or almost all, activities most of the day, nearly every day (as indicated by either subjective account or observation).
  c. Significant weight loss when not dieting or weight gain, or decrease or increase in appetite nearly every day.
  d. Insomnia or hypersomnia nearly every day.
  e. Psychomotor agitation or retardation nearly every day (observable by others, not merely subjective feelings of restlessness or being slowed down).
  f. Fatigue or loss of energy nearly every day.
  g. Feelings of worthlessness or excessive or inappropriate guilt (which may be delusional) nearly every day (not merely self-reproach or guilt about being sick).
  h. Diminished ability to think or concentrate, or indecisiveness, nearly every day (either by subjective account or as observed by others).
  i. Recurrent thoughts of death (not just fear of dying), recurrent suicidal ideation without a specific plan, or a suicide attempt or a specific plan for committing suicide.
B. The symptoms cause clinically significant distress or impairment in social, occupational, or other important areas of functioning.
C. The episode is not attributable to the psychological effects of a substance or to another medical condition.
NOTE: Criteria A-C represent a major depressive episode.

With peripartum onset: This specifier can be applied to the current or, if the full criteria are not currently met for a mood disorder, if onset of mood symptoms occurs during pregnancy or in the 4 weeks following delivery.
DSM-V diagnostic criteria for major depressive disorder with peripartum onset specifier (American Psychiatric Association (APA), 2017). With permission from APA (2013).

important by clinicians and researchers (Gavin et al., 2015). Symptoms may include low mood, sadness, weepiness, appetite and/or sleep disturbance, fatigue, impaired concentration, feelings of guilt, inadequacy and inability to cope.

Diagnosis & diagnostic systems

There is ongoing controversy as to how best define the onset of symptoms for PPD (Gavin et al., 2015). The most widely used classification systems in psychiatry, the Diagnostic and Statistical Manual for Mental Disorders (DSM) and the International Classification of Diseases (ICD) differ slightly in their explicit criteria for PPD. Nonetheless, both systems assume that perinatal mood disorders are not separate nosological entities, and classify them in the same class as episodes that do not occur in relation to childbirth (Di Florio, Seeley, & Jones, 2015).

The DSM, Fifth Edition (DSM-V) classifies major depressive disorders (MDD) "with peripartum onset" to comprise major depressive episodes with symptom onset in pregnancy or within the first 4 weeks after delivery (American Psychiatric Association (APA), 2017) (Table 1). In other words, DSM-V does not differentiate between depressive episodes happening in pregnancy and those which have their onset postnatally. In contrast, the ICD, Tenth Edition (ICD-10) specifies the postpartum onset as commencing within the first six weeks of delivery (World Health Organisation (WHO), 1992).

These defined time windows for a postnatal onset can be regarded as being "too narrow" (Di Florio et al., 2015, p. 110). Given that depressive symptoms can happen for several months postnatally, even after the fourth or sixth week (Bobo & Yawn, 2014), these onset specifiers are not really supported by clinical practice. In fact, many perinatal experts and clinical guidelines define PPD as happening at any time during the first postnatal year (Scottish Intercollegiate Guidelines Network (SIGN), 2012). An alternative and broader term which is often used is "perinatal depression" which includes the onset of mood symptoms during pregnancy and through one year after childbirth (Gavin et al., 2015).

Screening

It is imperative that women are screened for possible depressive disorders in the perinatal period using validated measures. The 10-item Edinburgh Postnatal Depression Scale (EPDS) is the main screening tool used worldwide for the detection of PPD (Austin et al., 2017) (Fig. 1). Although this is a self-report measure and not diagnostic, it allows for the early and timely detection of PPD sufferers and their appropriate referral to specialized services.

Adverse effects of maternal PPD

Some depressed women may feel detached from their child and may display decreased interest in holding or interacting with the baby (Gavin et al., 2015). In fact, maternal PPD has been associated with decreased maternal sensitivity, a key factor for attachment security (Siegel, 1999), and less adequate caregiving responses (Slomiana, Honvo, Emonts, Reginster, & Bruyere, 2019) . This may in turn influence the infant's temperament, socioemotional and cognitive development, and psychological adaptation over the longer term, which can persist into adolescence (Austin et al., 2017; Slomiana et al., 2019). Although a range of factors can be implicated in the association between parental psychiatric disorders and child outcomes (Fig. 2), a body of evidence now suggests that the quality of parenting is the most important potential mediator in this association (Stein et al., 2014).

Besides the mother-child relationship, depressed mothers may experience difficulties in their social relationships. They may feel isolated and less supported, for instance by their partner (Slomiana et al., 2019). Thus, maternal PPD may critically impact on the mother's psychological well-being, quality of life, and interactions with her child and partner.

Severe PPD

More severe forms of postnatal depressive episodes may present with or be associated with psychotic features, such as command hallucinations to harm the infant or delusional beliefs that the child is possessed (APA, 2017). This presentation has in fact been linked to infanticide (Gavin et al., 2015), as well as to an increased risk of self-harm in the woman (Di Florio et al., 2015).

Another feature of severe PPD is suicidality. In the recent 2016–19 report of maternal deaths in the UK, maternal suicide was found to be the second most common direct cause of death in women during or within 42 days postnatally, and the leading cause from six weeks up to one year after pregnancy (Fig. 3) (Knight et al., 2019). Additionally, 15-year findings from a UK national enquiry concluded that maternal suicides were more likely in women with depression. In this study,

Edinburgh Postnatal Depression Scale[1] (EPDS)

Name: _____   Address: _____

Your Date of Birth: _____

Baby's Date of Birth: _____   Phone: _____

As you are pregnant or have recently had a baby, we would like to know how you are feeling. Please check the answer that comes closest to how you have felt **IN THE PAST 7 DAYS**, not just how you feel today.

Here is an example, already completed.

I have felt happy:
- ☐ Yes, all the time
- ☒ Yes, most of the time   This would mean: "I have felt happy most of the time" during the past week.
- ☐ No, not very often   Please complete the other questions in the same way.
- ☐ No, not at all

In the past 7 days:

1. I have been able to laugh and see the funny side of things
   - ☐ As much as I always could
   - ☐ Not quite so much now
   - ☐ Definitely not so much now
   - ☐ Not at all

2. I have looked forward with enjoyment to things
   - ☐ As much as I ever did
   - ☐ Rather less than I used to
   - ☐ Definitely less than I used to
   - ☐ Hardly at all

*3. I have blamed myself unnecessarily when things went wrong
   - ☐ Yes, most of the time
   - ☐ Yes, some of the time
   - ☐ Not very often
   - ☐ No, never

4. I have been anxious or worried for no good reason
   - ☐ No, not at all
   - ☐ Hardly ever
   - ☐ Yes, sometimes
   - ☐ Yes, very often

*5 I have felt scared or panicky for no very good reason
   - ☐ Yes, quite a lot
   - ☐ Yes, sometimes
   - ☐ No, not much
   - ☐ No, not at all

*6. Things have been getting on top of me
   - ☐ Yes, most of the time I haven't been able to cope at all
   - ☐ Yes, sometimes I haven't been coping as well as usual
   - ☐ No, most of the time I have coped quite well
   - ☐ No, I have been coping as well as ever

*7 I have been so unhappy that I have had difficulty sleeping
   - ☐ Yes, most of the time
   - ☐ Yes, sometimes
   - ☐ Not very often
   - ☐ No, not at all

*8 I have felt sad or miserable
   - ☐ Yes, most of the time
   - ☐ Yes, quite often
   - ☐ Not very often
   - ☐ No, not at all

*9 I have been so unhappy that I have been crying
   - ☐ Yes, most of the time
   - ☐ Yes, quite often
   - ☐ Only occasionally
   - ☐ No, never

*10 The thought of harming myself has occurred to me
   - ☐ Yes, quite often
   - ☐ Sometimes
   - ☐ Hardly ever
   - ☐ Never

Administered/Reviewed by _____   Date _____

[1]Source: Cox, J.L., Holden, J.M., and Sagovsky, R. 1987. Detection of postnatal depression: Development of the 10-item Edinburgh Postnatal Depression Scale. *British Journal of Psychiatry* 150:782-786.

[2]Source: K. L. Wisner, B. L. Parry, C. M. Piontek, Postpartum Depression N Engl J Med vol. 347, No 3, July 18, 2002, 194-199

Users may reproduce the scale without further permission providing they respect copyright by quoting the names of the authors, the title and the source of the paper in all reproduced copies.

FIG. 1 The Edinburgh Postnatal Depression Scale. A 10-item self-report questionnaire used to assess for possible depression and anxiety in the perinatal period (Cox, Holden, & Sagovsky, 1987; Wisner, Parry, & Piontek, 2002).

depressed mood and emotional distress were the most common symptoms at last contact for perinatal women who died by suicide (Khalifeh, Hunt, Appleby, & Howard, 2016).

In light of all these negative consequences for the mother, infant and the partner, the dire need for robust screening programs and effective and acceptable treatments for PPD is accentuated (Sockol, Epperson, & Barber, 2011). In fact, the management of perinatal depression remains an active area of investigation and research (Gavin et al., 2015).

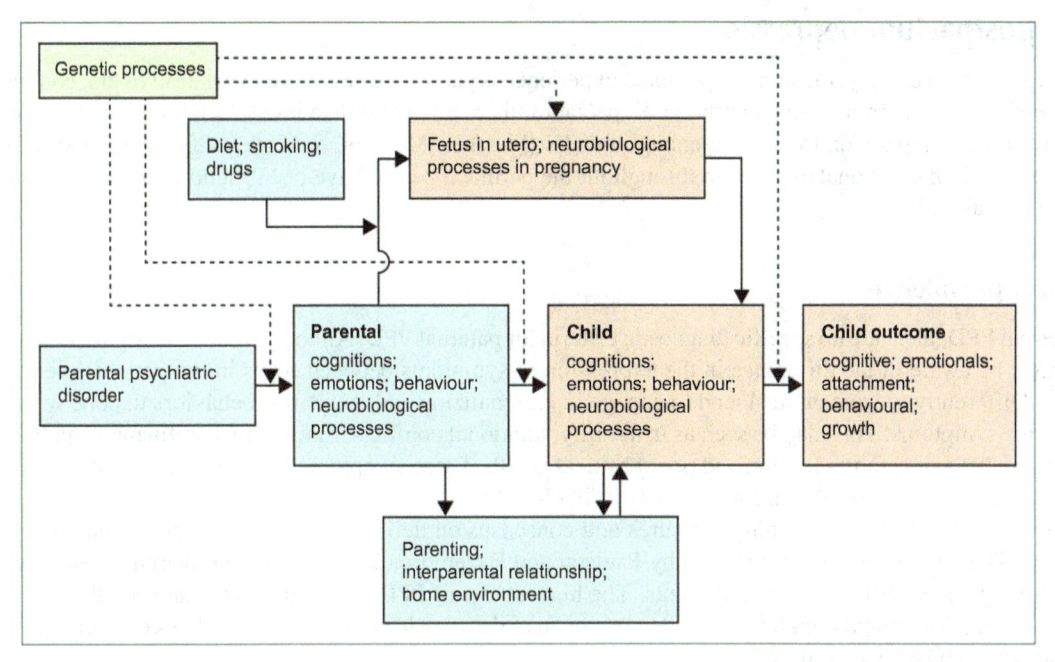

FIG. 2 Parental psychiatric disorders and child outcomes. Possible mechanisms underlying the association between parental psychopathology and child outcomes (*Dotted lines* show genetic processes. *Solid lines* show interactions. *Orange* colors refer to the child. *Blue* color refers to the parents. *Green* represents genetic processes) (Stein et al., 2014).

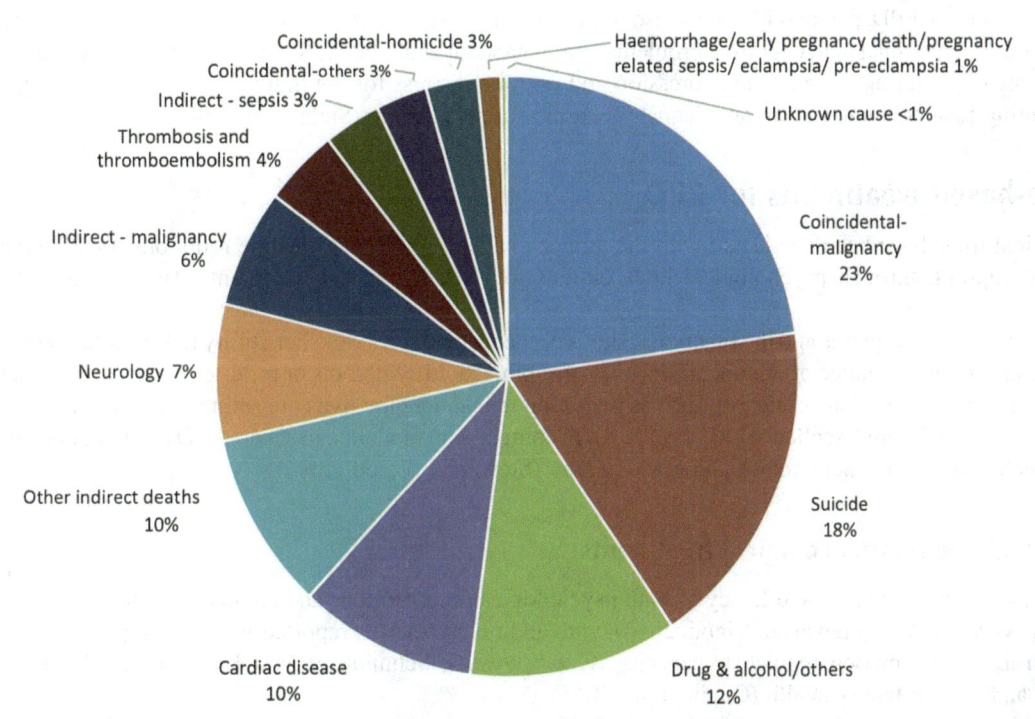

FIG. 3 Causes of death among postpartum women. Among other causes, 18% of women died by suicide between 6 weeks and one year after pregnancy in the UK between 2015 and 2017 (Knight et al., 2019).

Paternal postpartum depression

Despite being still at its infancy, the fathers' perinatal experience is part of the new inclusive narrative that complements the current societal changes surrounding parenthood. Expectant fathers often feel dismissed, lack a role or structure, and face uncertainties in their preparation to be good enough fathers (Fletcher, Matthey, & Marley, 2006). The experience of this distress and the lack of emotional expression throughout the perinatal period have been found to be predictors of paternal PPD (O'Brien et al., 2017).

Definition & prevalence

Unlike maternal PPD, there are no specific diagnostic criteria for paternal PPD. Although the same criteria for MDD may be used, one must be aware of the difference in the expression of symptoms between sexes in the perinatal period (O'Brien et al., 2017). Differently to women, men tend to engage in externalizing and avoidance behaviors to cope with or to mask their depressive symptoms. This may be seen as irritability, relational conflicts, ruminations on financial and/or child concerns, impulsive behavior, substance use, and poor physical health. These symptoms may distract from the possibility of an underlying depression (Paulson & Bazemore, 2010; Wilhelm, 2009).

Due to the lack of validated screening measures and consensus on definitions, the prevalence of paternal PPD varies across studies. The widely cited metasynthesis by Paulson and Bazemore (2010) reports an overall prevalence of 10.4%, with rates varying across different perinatal points. The highest rate of PPD was found to be between three to six-months postnatally (25.6%). Importantly, high levels of depression in males may lead to suicide, which is commonly overlooked in the perinatal period (Quevedo et al., 2011).

Paternal PPD and the family system

Research and clinical experience show a bidirectional link between maternal and paternal PPD in pregnancy and the postpartum period (Vismara et al., 2016). One in ten will experience significant distress when their partner suffers from an affective disorder (Paulson & Bazemore, 2010), possibly because of increased difficulty to meet their families' needs in such circumstances. Additionally, fathers have a 24% to 50% likelihood of suffering from perinatal mood disorders themselves when their partner is depressed; however, this can also occur independently of the woman's psychopathology (Paulson, Dauber, & Leiferman, 2006).

Similar to maternal PPD, paternal PPD may also impact negatively on the child's developmental outcomes, doubling the risk of behavioral, emotional, and cognitive problems (Ramchandani et al., 2008; Stein et al., 2014). These adverse effects can be mitigated by treating the parental depression and limiting distress for the individual, addressing potentially modifiable parenting factors, and ensuring an adequate external support system (Stein et al., 2014).

Evidence-based treatments for PPD

Several clinical trials have been conducted with the aim of evaluating different treatment options for PPD ranging from psychosocial support, through psychological interventions to pharmacological treatment (Austin et al., 2017; Sockol et al., 2011).

The severity of the symptoms and the functional status of the woman, including her ability to look after her baby, are key factors in determining the choice of treatment. In cases of subthreshold symptoms or mild-to-moderate PPD, self-help and psychosocial methods that enhance support, such as peer support and nondirective counseling, are often adequate (National Institute for Health and Care Excellence (NICE), 2014). In comparison, moderate to severe PPD often requires formal high-intensity psychological treatment and/or pharmacotherapy (Stewart & Vigod, 2016).

Psychological and pharmacological treatments

Robust evidence now exists for the efficacy of both psychological interventions and medications in the treatment of postnatal depressive symptoms (Stewart & Vigod, 2016), with additional benefits reported for other important secondary outcomes, such as infant emotion regulatory capacity (Krzeczkowski, Schmidt, & Van Lieshout, 2021), maternal-infant interactions and overall family health (Gavin et al., 2015).

In a recent meta-analysis on psychological and drug therapies for perinatal depression, symptomatic improvement from pre- to post-treatment periods was achieved for both treatment modalities, independent of one another. Among the psychological interventions included in this review were Cognitive Behavioral Therapy (CBT), Interpersonal Therapy (IPT), nondirective counseling, psychoeducational and mother-infant therapy groups. The reported effect size of 0.65 (95% CI 0.45–0.86, $P < .001$) for these interventions was comparable to that identified for psychological treatments for adult depression (Sockol et al., 2011).

Within the limited research on paternal PPD treatment, CBT remains the most researched intervention but its effectiveness still needs to be scientifically evaluated (O'Brien et al., 2017). As will be discussed later, CBT interventions need to be modified according to the fathers' specific needs and challenges when experiencing PPD.

Preferred treatment in maternal PPD

Psychotherapeutic treatment is often viewed as a more attractive and a safer option for the breastfeeding mother, who might not wish to expose her child to pharmacotherapy. Indeed, the former is considered "first-choice treatment" by many perinatal mothers (Stuart & Koleva, 2014). This higher acceptability for psychotherapy over drug therapy was also observed by van Schaik et al. (2004), who reported that "psychotherapy was assumed to solve the cause of depression" (p. 184), unlike psychotropic drugs, which in addition may have addictive potential.

Nonetheless, it is imperative that an individual risk-benefit analysis regarding the use of psychotropic medications during breastfeeding is performed, and the risks of both treated and untreated maternal illness for both the woman and the child are taken into account (Howard & Khalifeh, 2020). In addition to the woman's breastfeeding intentions, the level of distress and/or impairment caused by the disorder are equally significant aspects to be considered when making such a decision.

Indeed, for more severe illness, medication may be "an essential component of effective treatment," even in lactating women (McAllister-Williams et al., 2017, p. 10). Most selective serotonin reuptake inhibitors, the first-line pharmacological agents used in the perinatal period pass into the breast milk at a dose that is less than 10% of the maternal dose, and hence are generally viewed as compatible with lactation of healthy, full-term infants (Stewart & Vigod, 2016).

Medications may also be combined with therapy, known as combination therapy (NICE, 2014). While there is mixed evidence regarding the superiority of combination therapy over monotherapy (drug or psychological therapy) (Milgrom et al., 2015; Misri, Reebye, Corral, & Milis, 2004), the former is still recommended in moderate to severe PPD where there is limited or no response to psychological intervention or medication alone (NICE, 2014).

Barriers to care for mothers

PPD remains frequently underdiagnosed and under-treated (Thomson & Sharma, 2017). Despite their emotional distress, fewer than 50% of depressed women seek assistance or take up treatment (Milgrom et al., 2016). Contributing factors to this treatment gap encompass maternal, professional and health system factors (Gavin et al., 2015). In addition to lack of human resources, training and time constraints for health-care providers, women also report lack of motivation, difficulty trusting their provider, fear of being judged and of losing custody of their child, and lack of transportation and child care services (Gavin et al., 2015). It is important that these factors are taken into account when treatments and pathways to care are planned for perinatal women.

Barriers to care for fathers

Self and societal stigma in exposing one's own vulnerabilities pose the greatest challenge to providing care to fathers (Burgess, 2011). This difficulty is further increased with the lack of general awareness on the possible manifestation of PPD even in fathers, and also the latter's resistance to seek support during a time that is "supposed" to be dedicated specifically to the female counterpart. It has been found that father-inclusive practices and treatment options which are tailored specifically to fathers' needs are preferred (O'Brien et al., 2017). A further challenge to care is the lack of readily available and tailored screening tools for this paternal population (Carlberg, Edhborg, & Lindberg, 2018).

Interestingly, the perinatal period is a time of opportunity in identifying fathers with PPD. One avenue would be through the mother who is more likely to be engaged with healthcare services during this time. Often, neither the woman or her partner would know, or be aware that the father might be suffering from PPD. When the possibility of PPD is identified, perhaps from the description of his (usually changed) behavior or challenges being faced, the goal is to bring this to the couple's awareness, normalize it through psychoeducation, and to offer support. Since father-inclusive services are scarce, avenues for support may include going through the family doctor, a therapist, family or friends, and providing resources (websites and books) that would allow for further normalization and skill acquisition.

The role of CBT in the treatment of PPD

Over the past years, there has been a surge of research focused on the use of CBT within the perinatal population. CBT is now recognized to be an empirically validated treatment for PPD with the "problem-focused" and "time-sensitive" characteristics making it a good match for PPD (Sockol, 2015; Wenzel & Kleiman, 2015).

The effectiveness of CBT in PPD

A recent meta-analysis by Huang, Zhao, Qiang, and Fan (2018), which included a total of 20 RCTs, concluded that CBT was clinically superior to conventional therapies such as non-guided counseling in PPD. There was a significant improvement in both the short- and long-term postnatal depressive symptomatology, measured using the EPDS and the Beck Depression Inventory, for new mothers in the CBT intervention group, compared to control conditions. Contrary to this, the meta-analysis by Li et al. (2020) concluded that while CBT can effectively relieve symptoms in the short-term (<4 months), its long-term effect (>4 months) remains un-evidenced. These contrasting findings could be due to the limited number of studies on the long-term effects of CBT in PPD, as well as other clinical and methodological moderators. Indeed, different criteria have also been used in different studies, for example, in the definition of CBT protocol used, time point of intervention, level of experience of the professionals delivering the intervention, mode of delivery, as well as the type of control condition, rendering them incomparable and possibly impacting on quality of results of meta-analyses.

Further support for the effectiveness of CBT in the treatment as well as the prevention of PPD is given in the systematic review by Sockol (2015). For the included studies, treatment was initiated at different time-points in the perinatal period—some during pregnancy and others after childbirth. A statistically significant reduction in depressive symptoms for CBT-treated women with an overall moderate effect size (0.65) (95% CI 0.54–0.76, $P < .001$) was achieved.

Although well-defined cognitive-behavioral interventions are suggested to mitigate the academic ambivalence, exploration of the CBT adaptations are essential to adequately encompass the diverse perinatal needs and address the range of parenthood challenges, such as care-giving issues (Pettman et al., 2019). This adaptation of CBT will be now explored.

Adapting CBT for PPD

When adapting CBT for the perinatal population, the foundation and key principles of CBT are still retained and adhered to. The course of treatment is divided into early, middle and late phases of therapy and the traditional session structure is maintained, as shown in Fig. 4. Although the same principles may be followed for CBT in fathers, there is no available research to support this.

Within this structure, the therapist must keep the therapeutic relationship a priority. This is one of the most important agents for change within a therapeutic setting, irrespective of the modality and/or the therapist's approach (Gelso, 2014). During the perinatal period, this relationship is regarded as a most trusted relationship offering a space of empathy, warmth and genuineness, during a time of significant societal and personal expectations and profound negative automatic thoughts, self-criticism and guilt (Hardy, Cahill, & Barkham, 2007; Wenzel & Kleiman, 2015). The balance between support and structure offers the mother the opportunity to step back from the whirlwind of emotions and changes, and to experience the sense of containment and focus which characterise the CBT sessions.

Early phase of treatment and case conceptualization for PPD

The early phase is dedicated to setting the foundation of CBT and to building a strong working alliance, while gathering information, assessing the clients' level of motivation and commitment, and collaboratively drawing up a treatment plan. Throughout this work, the therapist formulates an understanding of the mother's unique psychological framework, needs, challenges and strengths.

An initial assessment is performed to understand the predisposing, precipitating and perpetuating factors that are contributing to the experience of PPD. This would include assessing for previous personal or familial episodes of depression, including PPD, and/or other mental health issues, the birth experience (current and previous), and the mother's internal working models. Additionally, it is important to explore protective factors and strengths for an in-depth conceptualization. Adopting a strengths-based approach and getting to know the client beyond her distress is also meaningful for mothers (O'Mahen et al., 2012).

> **BRIEF MOOD CHECK:** The therapist obtains a quantitative estimate of the client's mood state, which will allow the therapist and client to monitor the client's progress in treatment.
>
> **BRIDGE FROM THE PREVIOUS SESSION:** The therapist asks the client what she took away from the previous session in order to orient her to the current session and ensure that a thread runs across sessions.
>
> **AGENDA:** The therapist and client collaboratively agree on the issues to address during the current session.
>
> **DISCUSSION OF AGENDA ITEMS:** The therapist and client address the items on the agenda, taking care to balance attention to the therapeutic relationship with cognitive and behavioral change strategies. During this discussion, the therapist makes *periodic summaries* to ensure that she understands what the client is communicating and that they are in agreement with the direction that the session is taking.
>
> **HOMEWORK:** The therapist and client review the homework that the client completed in the time since the previous session, and they collaboratively develop new homework to be completed before the next session.
>
> **FINAL SUMMARY AND FEEDBACK:** The therapist invites the client to summarize what she is taking away from the current session and provide other feedback.

FIG. 4 Outline of session structure in CBT. Each session in CBT is structured in the same traditional format. While adherence to this structure is important, the flow from one part to the next must be fluid, collaborative and flexible (Wenzel & Kleiman, 2015).

This information supports the therapist in formulating a comprehensive case conceptualization of the interplay of thoughts, behaviors, and emotions pertaining to that particular client. A sample conceptualization can be seen through the model of PPD as shown below in Fig. 5.

Obstacles in the early phase

Most women in therapy for PPD have an urgency to regain their health and sense of self. However, this urgency may be accompanied by feelings of hopelessness and helplessness. This ambivalence, along with other perinatal obstacles and challenges, need to be addressed through validation and empowerment. At this point, motivational interviewing may be useful to support this process of commitment, while also helping to encourage the mother, and therefore making CBT feel more manageable for the client (Miller & Rollnick, 2013).

This therapeutic foundation built within the first few sessions gives the mother a sense of structure and encourages her to collaboratively draw up and commit to a treatment plan. The treatment plan consists of main goals and targets which will then guide the interventions in the middle phase of treatment.

It has to be said that if the client's experience of distress is too overwhelming, open discussions with the client on the possible benefits of psychotropic medication may be needed.

Middle phase of treatment: Modifying CBT interventions for PPD

Once a strong therapeutic relationship and alliance is established between the therapist and the client, the process of learning new CBT tools begins. Traditional CBT interventions used for the treatment of depression also apply to the perinatal population (Wenzel & Kleiman, 2015).

In their CBT-based program for PPD, Milgrom et al. (1999) describe key aspects to be addressed when working with this population. They outline the following: increasing pleasant and decreasing unpleasant activities (behavioral interventions), teaching self-control through self-monitoring and self-evaluation (cognitive interventions), learning problem-solving and coping skills such as relaxation, social and communication skills training. The description and methodology of each of these interventions is beyond the scope of this chapter.

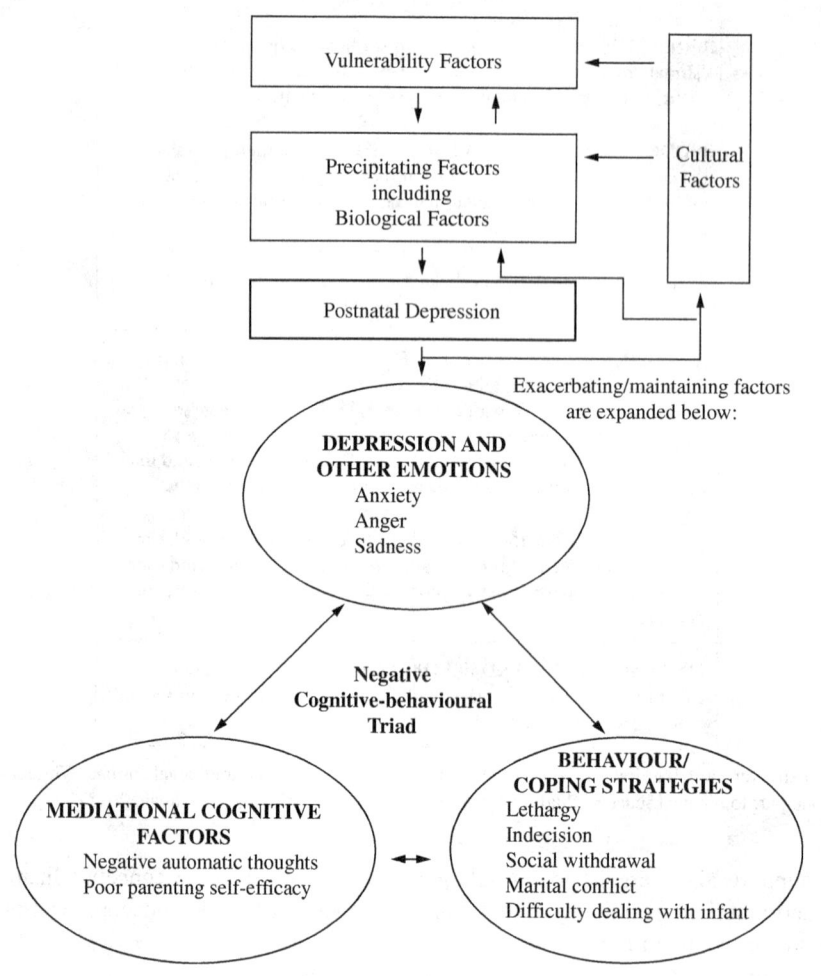

FIG. 5 Model of Postnatal Depression. A holistic model of PPD; conceptualizing the role of cognitive and behavioral mechanisms in maintaining the symptoms of PPD and the challenges that are faced in the postpartum period (Milgrom, Martin, & Negri, 1999).

The interventions and homework need to match the treatment plan and be specifically adapted to the mother's unique postpartum demands and level of support. Fundamentally, the rationale for homework is not just to comply with protocol, but the belief that it will improve the mothers' wellbeing and quality of life. As the mother practices these tasks, she will become more aware of the connection between her own thoughts, feelings and behaviors, and have the skills to make conscious choices to feel better (Wenzel & Kleiman, 2015).

Negative cognitions and behavioral challenges in PPD

The themes described in the qualitative study by O'Mahen et al. (2012) compliment the clinical experience, where the mother often places unrealistic expectations on herself. These idealistic motherhood expectations are often impossible to meet, creating a cycle of negative thoughts, such as "not being a good enough parent," which in turn trigger feelings of inadequacy and failure. These negative cognitions may be addressed through cognitive restructuring interventions, supporting the mother to identify them as unhelpful and to evaluate them against what may be achievable.

Additionally, mothers often feel stuck in the cycle of internalizing behaviors; ruminating, avoiding and hiding themselves from others. This often leaves them feeling unaccomplished, drained, isolated and at a loss on how to change these behaviors (O'Mahen et al., 2012). The use of behavioral activation interventions can help them become actively aware of how they are spending their time in order to determine what they want and what they need to engage in to improve their wellbeing. Mothers start to experience the power of behavioral activation as they feel more satisfied and accomplished (Wenzel & Kleiman, 2015). The use of skill training interventions, such as relaxation, social and problem-solving skills (Milgrom et al., 1999) are also helpful in mitigating these cognitive and behavioral challenges.

These interventions also contribute to the mother-child relationship. Specific areas of difficulty, such as interacting and playing with the baby may be incorporated into the interventions. Furthermore, CBT has shown to help mothers to self-regulate and therefore, supports them to be more present with their child, who is in turn more able to model similar emotional regulatory behaviors (Krzeczkowski et al., 2021). Having said that, if the parent-child relationship is very strained, additional interventions may be needed (Forman et al., 2007).

Obstacles in homework compliance

With the demands of parenthood, obstacles are expected in the compliance with homework tasks ("too tired; house work takes over; I prefer to take a nap when baby is asleep"). By identifying and discussing ways of overcoming these obstacles, and by offering appropriate adaptations, mothers will feel more confident. These measures will also aid treatment compliance and influence positive outcomes (O'Mahen et al., 2012; Pettman et al., 2019).

Late phase of treatment and relapse prevention

Once the therapist and the client come to a mutual agreement that the needed skills have been acquired, and the main targets in the treatment plan have been reached, the therapeutic process transitions into the late phase of treatment. Here, the focus is on maintenance and consolidation, with the aim of generalizing the learnt skills to other areas of the parent's life. A relapse prevention plan is formulated collaboratively with the client, summarizing the main insights and tools gained within the sessions. Furthermore, a plan of how to reach out for formal and informal support is drawn up, in case of future need.

Beyond traditional individual CBT treatment

Research evidence for the effectiveness and implementation of group CBT (gCBT), app-based CBT and internet-based CBT (iCBT) in the treatment of PPD is increasing. This shift to non-traditional delivery modalities helps mitigate some of the obstacles to care in the postnatal period (Sockol, 2015), thereby aiding client outreach, treatment compliance and minimization of dropouts (Huang et al., 2018). Furthermore, their utility was even more welcomed in context of the limitations posed by the COVID-19 pandemic (Chen, Selix, & Nosek, 2021).

Internet-based and application-based CBT

The feasibility and effectiveness of the use of iCBT treatment was demonstrated in a parallel 2-group RCT comparing a six session iCBT program named "MumMoodBooster" to treatment as usual (Milgrom et al., 2016). At the end of the program, 79% of the women in the intervention group no longer met diagnostic criteria for depression, in contrast to the 18% in the control group. Moreover, approximately 85% of users completed all sessions, and satisfaction with the program was rated 3.1 out of 4. In another recently published RCT by Jannati, Mazhari, Ahmadian, and Mirzaee (2020), which examined the effect of an interactive app-based CBT program for reducing PPD, the use of this mobile application resulted in a significant difference in the decrease in EPDS scores in mothers in the intervention group as compared to the control group. In addition to these promising results, mobile health interventions have the added advantage that they can be used at a time and space which is convenient to the user. On the other hand, future work focusing on comparing online psychological treatment with face-to-face CBT is needed.

Group-based CBT

Many of the reviews on the cost effectiveness and efficacy of gCBT are inconclusive due to the lack of homogeneity in the implementation of treatment groups (Gillis & Parish, 2019). Nevertheless, there are numerous benefits of gCBT, including positive therapeutic outcomes (for example validation, empowerment and healing), higher outreach (more women treated simultaneously) and importantly, social support which is missing from individual-based treatment (Gillis & Parish, 2019; Wenzel & Kleiman, 2015). This group support directly impacts on the sense of isolation and loneliness that is particularly felt in the perinatal period, above the delivered interventions aimed at utilization of support and communication skills. A well supported 12-week program can be found in the work by Milgrom et al. (1999) (Milgrom et al., 2015).

Conclusion

In summary, maternal and paternal PPD may adversely affect the whole family system; the mother, the father, the development of the child, and the relationships between them. Effective and timely treatment of PPD is central in mitigating these negative outcomes.

Different treatment options for PPD have been evidenced, ranging from psychotropic medications to psychosocial and psychological interventions, with the latter being the preferred choice of treatment for most perinatal clients. Despite the increasing awareness and the widely available research data on PPD and its treatment, it is compelling that this disorder in mothers, and evermore in fathers, remains underdiagnosed and undertreated.

Research on the psychological treatment for PPD has focused mainly on the maternal population, with the most effective treatment modalities being IPT and CBT. Implementation of CBT for PPD largely follows that for depression in the general adult population; however, important adaptations are needed to encompass the unique parental needs and challenges of the postnatal period. The therapeutic relationship, validation and holding, together with modified CBT interventions are fundamental in the effective treatment of PPD.

Practice and procedures

As an example, within national PMH services in Malta, the identification of those women experiencing PPD is facilitated through the use of the self-report EPDS questionnaire (Austin et al., 2017), as well as other validated measures to screen for other possible co-existing psychosocial and psychiatric issues. At the current time, debates are ongoing as to whether national screening should be implemented within healthcare services for all perinatal women and their partners.

Completion of the questionnaires is followed by a one-to-one individualised assessment by the specialist midwife. This allows for case formulation and holistic care planning, including signposting to other PMH professionals in the team, according to the mother's and her family's needs. Pharmacological treatment, when indicated, follows a careful risk-benefit analysis discussion between the clinician and the mother (Howard & Khalifeh, 2020; NICE, 2014).

The psychology team which forms part of the same PMH service receives frequent and regular training on up-to-date evidence-based research and interventions for all forms of PMH issues. Furthermore, regular intervision sessions between the multidisciplinary team members are held to ensure an ethical and evidence-based practice.

Mini-dictionary of terms

Perinatal period: This is the period from conception, through pregnancy to the end of the twelve months postnatally.
Perinatal mental health (PMH): This refers to the emotional and psychological well-being of the expectant or new parent during the perinatal period.
Postpartum depression (PPD): A common PMH disorder which involves a depressive episode in pregnancy or after childbirth, of varying severity, level of distress and functional impairment.
Postpartum (or baby) blues: This is the commonest form of postpartum mood disturbance. It is mild and self-limiting, characterized by mood swings which typically have their onset within a few days of giving birth and which tend to resolve within a few weeks.
Infanticide: This is the killing of a child below the age of one. Although rare, infanticide may be associated with more severe forms of PPD.
Attachment security: An emotional and physical attachment between a child and an adult/care-giver, which enables the child to feel safe and secure to explore his surroundings, knowing that there is someone reliably available to return to when distressed.
Therapeutic relationship: The genuine therapist-client relationship which is built on trust, empathy and warmth. It is the strongest predictor for change in the therapeutic process, irrespective of modality.

Key facts of postpartum depression (PPD)

- PPD is a common, disabling, yet treatable mental health disorder, which may occur during the first 12 months after childbirth in both the woman and/or her partner.
- Maternal PPD affects approximately 10% to 15% of women (Robertson et al., 2003), while current literature estimates an overall cumulative prevalence of 10.4% in men across the perinatal period (Paulson & Bazemore, 2010).

- PPD remains frequently underdiagnosed and undertreated, with fewer than 50% of women seeking assistance or deciding to take up treatment.
- Untreated maternal PPD can result in infanticide and suicide. One in six (18%) women die by suicide during the first postnatal year (Knight et al., 2019).
- In addition to causing maternal distress and poor functionality, PPD in women can impact on mother-child interactional quality, and therefore influence infant outcomes.
- Depending on the severity of PPD, treatment may include psychosocial, psychological and/or pharmacological interventions, either as mono- or combination therapy.
- CBT is an empirically validated treatment and preventative option for PPD, with several studies confirming its effectiveness in women suffering from this disorder (small to moderate effect size).

Applications to other areas

Other areas of application for CBT in the perinatal period include CBT during pregnancy and CBT for perinatal anxiety. Antenatal CBT has been shown to be an effective preventative treatment for PPD (Cho, Kwon, & Lee, 2008). Similarly, systematic reviews on CBT for perinatal anxiety have also found it to be effective (Camacho & Shields, 2018; Green et al., 2020). Wenzel and Kleiman (2015) adequately address both areas in their work on perinatal distress.

Further to this, CBT has been found to be an effective intervention, even when applied in conjunction with other schools of thought. This is especially helpful for more complex cases where there is comorbid psychopathology and/or social issues. Examples of this include the application of CBT in conjunction with family therapy (Hou et al., 2014), and mindfulness based cognitive therapy (Dimidjian et al., 2015).

Summary points

- Both psychological and/or pharmacological interventions, alone or in combination, are effective treatment options for PPD.
- Perinatal women, especially if breastfeeding, tend to prefer psychotherapeutic interventions over pharmacological interventions.
- CBT interventions used for depressive disorders outside the perinatal period can also be applied in the treatment of PPD but with adaptations to encompass the unique perinatal issues.
- A strong therapeutic relationship is fundamental in sustaining a good working alliance and in achieving positive outcomes, including treatment compliance, when working with mothers with PPD.
- More research on the adaptation of CBT protocols and interventions to encompass perinatal needs and parenthood challenges for mothers and fathers is needed.
- Besides the traditional face-to-face delivery of CBT, other modes of delivery such as internet CBT and group CBT show promising results for the field of perinatal mental health.

References

American Psychiatric Association (APA). (2017). *Diagnostic and statistical manual of mental disorders* (5th Edition). Arlington, VA: American Psychiatric Association.

Austin, M. P., Highet, N., & Expert Working Group. (2017). *Mental health care in the perinatal period: Australian clinical practice guideline*. Melbourne, Australia: Centre of Perinatal Excellence.

Bobo, W. V., & Yawn, B. P. (2014). Concise review for physicians and other clinicians: Postpartum depression. *Mayo Clinic Proceedings, 89*(6), 835–844.

Burgess, A. (2011). Fathers' roles in perinatal mental health: Causes, interactions and effects. *New Digest, 53*, 24–29.

Camacho, E. M., & Shields, G. E. (2018). Cost-effectiveness of interventions for perinatal anxiety and/or depression: A systematic review. *BMJ Open, 8*(8), e022022.

Carlberg, M., Edhborg, M., & Lindberg, L. (2018). Paternal perinatal depression assessed by the Edinburgh Postnatal Depression Scale and the Gotland Male Depression Scale: Prevalence and possible risk factors. *American Journal of Men's Health, 12*(4), 720–729.

Chen, H., Selix, N., & Nosek, M. (2021). Perinatal anxiety and depression during Covid-19. *The Journal for Nurse Practitioners, 17*(1), 26–31.

Cho, H. J., Kwon, J. H., & Lee, J. J. (2008). Antenatal cognitive-behavioral therapy for prevention of postpartum depression: A pilot study. *Yonsei Medical Journal, 49*(4), 553–562.

Cox, J. L., Holden, J. M., & Sagovsky, R. (1987). Detection of postnatal depression: Development of the 10-item Edinburgh Postnatal Depression Scale. *British Journal of Psychiatry, 150*, 782–786.

Di Florio, A., Seeley, J., & Jones, I. (2015). Diagnostic assessment of depression, anxiety, and related disorders. In J. Milgrom, & A. W. Gemmill (Eds.), *Identifying perinatal depression and anxiety. Evidence-based practice in screening, psychosocial assessment, and management* (pp. 108–120). West Sussex, UK: John Wiley & Sons.

Dimidjian, S., Goodman, S. H., Felder, J. N., Gallop, R., Brown, A. P., & Beck, A. (2015). An open trial of mindfulness-based cognitive therapy for the prevention of perinatal depressive relapse/recurrence. *Archives of Women's Mental Health, 18*(1), 85–94.

Fisher, J., Cabral de Mello, M., Patel, V., Rahman, A., Tran, T., Holton, S., & Holmes, W. (2011). Prevalence and determinants of common perinatal mental disorders in women in low- and lower-middle-income countries: A systematic review. *Bulletin of the World Health Organization, 90*, 139–149H.

Fletcher, R. J., Matthey, S., & Marley, C. G. (2006). Addressing depression and anxiety among new fathers. *Medical Journal of Australia, 185*(8), 461.

Forman, D. R., O'Hara, M. W., Stuart, S., Gorman, L. L., Larsen, K. E., & Coy, K. C. (2007). Effective treatment for postpartum depression is not sufficient to improve the developing mother–child relationship. *Development and Psychopathology, 19*(2), 585–602.

Gavin, N., Gaynes, B., Lohr, K., Meltzer-Brody, S., Gartlehner, G., & Swinson, T. (2005). Perinatal depression. A systematic review of prevalence and incidence. *Obstetrics & Gynecology, 106*(5), 1071–1083.

Gavin, N., Meltzwer-Brody, S., Glover, V., & Gaynes, B. (2015). Is population-based identification of perinatal depression and anxiety desirable. In J. Milgrom, & A. W. Gemmill (Eds.), *Identifying perinatal depression and anxiety* (pp. 11–23). West Sussex, UK: John Wiley & Sons, Ltd.

Gelso, J. C. (2014). A tripartite model of the therapeutic relationship: Theory, research, and practice. *Psychotherapy Research, 24*, 117–131. https://doi.org/10.1080/10503307.2013.845920.

Gillis, B. D., & Parish, A. L. (2019). Group-based interventions for postpartum depression: An integrative review and conceptual model. *Archives of Psychiatric Nursing, 33*(3), 290–298.

Green, S. M., Donegan, E., McCabe, R. E., Streiner, D. L., Agako, A., & Frey, B. N. (2020). Cognitive behavioral therapy for perinatal anxiety: A randomized controlled trial. *Australian & New Zealand Journal of Psychiatry, 54*(4), 423–432.

Hardy, G., Cahill, J., & Barkham, M. (2007). Active ingredients of the therapeutic relationship that promote client change: A research perspective. In P. Gilbert, & R. L. Leahy (Eds.), *The therapeutic relationship in the cognitive behavioral psychotherapies* (pp. 24–42). New York, NY: Routledge.

Hou, Y., Hu, P., Zhang, Y., Lu, Q., Wang, D., Yin, L., ... Zou, X. (2014). Cognitive behavioral therapy in combination with systemic family therapy improves mild to moderate postpartum depression. *Brazilian Journal of Psychiatry, 36*(1), 47–52.

Howard, L. M., & Khalifeh, H. (2020). Perinatal mental health: A review of progress and challenges. *World Psychiatry, 19*(3), 313–327.

Huang, L., Zhao, Y., Qiang, C., & Fan, B. (2018). Is cognitive behavioral therapy a better choice for women with postnatal depression? A systematic review and meta-analysis. *PLoS One, 13*(10). https://doi.org/10.1371/journal.pone.0205243, e0205243.

Jannati, N., Mazhari, S., Ahmadian, L., & Mirzaee, M. (2020). Effectiveness of an app-based cognitive behvaioural therapy program for postpartum depression in primary care: A randomised controlled trial. *International Journal of Medical Informatics, 141*, 104145.

Kettunen, P., Koistinen, E., & Hintikka, J. (2014). Is postpartum depression a homogenous disorder: Time of onset, severity, symptoms and hopelessness in relation to the course of depression. *BMC Pregnancy and Childbirth, 14*(402), 1–9.

Khalifeh, H., Hunt, I. M., Appleby, L., & Howard, L. M. (2016). Suicide in perinatal and non-perinatal women in contact with psychiatric services: 15 year findings from a UK national inquiry. *Lancet Psychiatry, 3*(3), 233–242.

Knight, M., Bunch, K., Tuffnell, D., Jayakody, H., Shakespeare, J., Kotnis, R., ... On behalf of MBRRACE-UK. (2019). *Saving lives, improving mothers' care. Lessons learned to inform maternity care from the UK and Ireland confidential enquiries into maternal deaths and morbidity 2014–2016*. Oxford, UK: National Perinatal Epidemiology Unit, University of Oxford.

Krzeczkowski, J. E., Schmidt, L. A., & Van Lieshout, R. J. (2021). Changes in infant emotion regulation following maternal cognitive behavioural therapy for postpartum depression. *Depression & Anxiety, 38*(4), 412–421.

Li, Z., Liu, Y., Wang, J., Liu, J., Zhang, C., & Liu, Y. (2020). Effectiveness of cognitive behavioural therapy for perinatal depression: A systematic review and meta-analysis. *Journal of Clinical Nursing, 29*(17–18), 3170–3182.

McAllister-Williams, R. H., Baldwin, D. S., Cantwell, R., Easter, A., Gilvarry, E., Glover, V., ... Young, A. H. (2017). British association for psychopharmacology consensus guidance on the issue of psychotropic medication preconception, in pregnancy and postpartum 2017. *Journal of Psychopharmacology*, 1–34.

Milgrom, J., Danaher, B. G., Holt, C., Holt, C., Seeley, J. R., Tyler, M. S., ... Ericksen, J. (2016). Internet cognitive behavioural therapy for women with postnatal depression: A randomized controlled trial of MumMoodBooster. *Journal of Medical Internet Research, 18*(3), 28–37.

Milgrom, J., Gemmill, A. W., Ericksen, J., Burrows, G., Buist, A., & Reece, J. (2015). Treatment of postnatal depression with cognitive behavioural therapy, sertraline and combination therapy: A randomised controlled trial. *Australian & New Zealand Journal of Psychiatry, 49*(3), 236–245.

Milgrom, J., Martin, P. R., & Negri, L. M. (1999). *Treating postnatal depression: A psychological approach for health care practitioners*. Chichester: Wiley.

Miller, W. R., & Rollnick, S. (2013). *Motivational interviewing: Helping people change* (3rd ed.). New York, NY: Guilford Press.

Misri, S., Reebye, P., Corral, M., & Milis, L. (2004). The use of paroxetine and cognitive-behavioral therapy in postpartum depression and anxiety: A randomised controlled trial. *Journal of Clinical Psychiatry, 65*(9), 1236–1241.

National Institute for Health and Care Excellence (NICE). (2014). *Antenatal and postnatal mental health: Clinical management and service guidance*. Retrieved from https://www.nice.org.uk/guidance/cg192/resources/antenatal-and-postnatal-mental-health-clinical-management-and-service-guidance-pdf-35109869806789.

O'Brien, A. P., McNeil, K. A., Fletcher, R., Conrad, A., Wilson, A. J., Jones, D., & Chan, S. W. (2017). New fathers' perinatal depression and anxiety—Treatment options: An integrative review. *American Journal of Men's Health, 11*(4), 863–876.

O'Mahen, H., Fedock, G., Henshaw, E., Himle, J. A., Forman, J., & Flynn, H. A. (2012). Modifying CBT for perinatal depression: What do women want?: A qualitative study. *Cognitive and Behavioral Practice, 19*(2), 359–371.

Paulson, J. F., & Bazemore, S. D. (2010). Prenatal and postpartum depression in fathers and its association with maternal depression: A meta-analysis. *JAMA, 303*(19), 1961–1969. https://doi.org/10.1001/jama.2010.605.

Paulson, J. F., Dauber, S., & Leiferman, J. A. (2006). Individual and combined effects of postpartum depression in mothers and fathers on parenting behaviour. *Pediatrics, 118*(2), 659–668. https://doi.org/10.1542/peds.2005-2948.

Pettman, D., O'Mahen, H., Svanberg, A. S., von Essen, L., Axfors, C., Blomberg, O., & Woodford, J. (2019). Effectiveness and acceptability of cognitive-behavioural therapy based interventions for maternal peripartum depression: A systematic review, meta-analysis and thematic synthesis protocol. *BMJ Open, 9*, e032659. https://doi.org/10.1136/bmjopen-2019-032659.

Quevedo, L., da Silva, R. A., Coelho, F., Pinheiro, K. A. T., Horta, B. L., Kapczinski, F., & Pinheiro, R. T. (2011). Risk of suicide and mixed episode in men in the postpartum period. *Journal of Affective Disorders, 132*(1–2), 243–246.

Ramchandani, P. G., Stein, A., O'Connor, T. G., Heron, J. O. N., Murray, L., & Evans, J. (2008). Depression in men in the postnatal period and later child psychopathology: A population cohort study. *Journal of the American Academy of Child & Adolescent Psychiatry, 47*(4), 390–398.

Robertson, E., Celasun, N., & Stewart, D. E. (2003). Risk factors for postpartum depression. In D. E. Stewart, E. Robertson, C. L. Dennis, S. L. Grace, & T. Wallington (Eds.), *Postpartum depression: Literature review of risk factors and interventions* (pp. 9–70). Toronto, Canada: University Health Network Women's Health Program.

Scottish Intercollegiate Guidelines Network (SIGN). (2012). *Management of perinatal mood disorders*. Edinburgh, Scotland: SIGN.

Shorey, S., Chee, C. Y. I., Ng, E. D., Chan, Y. H., Tam, W. W. S., & Chong, Y. S. (2018). Prevalence and incidence of postpartum depression among healthy mothers: A systematic review and meta-analysis. *Journal of Psychiatric Research, 104*, 235–248.

Siegel, D. (1999). Chapter 3: Attachment. In *The developing mind. How relationships and the brain interact to shape who we are*. Guilford Press: New York.

Slomiana, J., Honvo, G., Emonts, P., Reginster, J., & Bruyere, O. (2019). Consequences of maternal postpartum depression: A systematic review of maternal and infant outcomes. *Women's Health, 15*, 1–55.

Sockol, L. E. (2015). A systematic review of the efficacy of cognitive behavioural therapy for treating and preventing perinatal depression. *Journal of Affective Disorders, 177*, 7–21.

Sockol, L. E., Epperson, C. N., & Barber, J. P. (2011). A meta-analysis of treatments for perinatal depression. *Clinical Psychology Review, 31*, 839–849.

Stein, A., Pearson, R. M., Goodman, S. H., Rapa, E., Rahman, A., McCallum, M., … Pariante, C. M. (2014). Effects of perinatal health disorders on the fetus and child. *The Lancet, 384*, 1800–1819.

Stewart, D. E., & Vigod, S. (2016). Postpartum depression. *The New England Journal of Medicine, 375*, 2177–2186.

Stuart, S., & Koleva, H. (2014). Psychological treatments for perinatal depression. *Best Practice & Research Clinical Obstetrics & Gynaecology, 28*, 61–70.

Thomson, M., & Sharma, V. (2017). Therapeutics of postpartum depression. *Expert Review of Neurotherapeutics, 17*(5), 495–507.

van Schaik, D. J. F., Klijn, A. F. J., van Hout, H. P. J., van Marwijk, H. W. J., Beekman, A. T. F., de Haan, M., & van Dyck, R. (2004). Patients' preferences in the treatment of depressive disorder in primary care. *General Hospital Psychiatry, 26*(3), 184–189.

Vismara, L., Rollè, L., Agostini, F., Sechi, C., Fenaroli, V., Molgora, S., … Polizzi, C. (2016). Perinatal parenting stress, anxiety, and depression outcomes in first-time mothers and fathers: A 3-to 6-months postpartum follow-up study. *Frontiers in Psychology, 7*(938). https://doi.org/10.3389/fpsyg.2016.00938.

Wenzel, A., & Kleiman, K. (2015). *Cognitive behavioural therapy for perinatal distress*. New York, NY: Routledge.

Wilhelm, K. (2009). Men and depression. *Australian Family Physician, 38*, 102–105.

Wisner, K. L., Parry, B. L., & Piontek, C. M. (2002). Postpartum depression. *The New England Journal of Medicine, 347*(3), 194–199.

Woody, C. A., Ferrari, A. J., Siskind, D. J., Whiteford, H. A., & Harris, M. G. (2017). A systematic review and meta-regression of the prevalence and incidence of perinatal depression. *Journal of Affective Disorder, 219*, 86–92.

World Health Organisation (WHO). (1992). *The ICD-10 classification of mental and Behavioral disorders: Clinical descriptions and diagnostic guidelines*. Geneva, Switzerland: Author.

Chapter 20

Applications of cognitive-behavioral therapy to posttraumatic stress disorder: A focus on sleep disorders

Morohunfolu Akinnusi[a,b,c,d] and Ali A. El-Solh[a,b,c,d]

[a]VA Western New York Healthcare System, Buffalo, NY, United States, [b]Division of Pulmonary, Critical Care, and Sleep Medicine, Department of Medicine, Jacobs School of Medicine and Biomedical Sciences, Buffalo, NY, United States, [c]Department of Epidemiology and Environmental Health, School of Public Health and Health Professions, Buffalo, NY, United States, [d]University at Buffalo, Buffalo, NY, United States

Abbreviations

CBT	cognitive-behavioral therapy
CBT-I	cognitive-behavioral therapy for insomnia
DSM	diagnostic and statistical manual of mental disorders
ERRT	exposure, relaxation, and rescripting therapy
IRET	imagery rehearsal and exposure therapy
IRT	imagery rehearsal therapy
PSQI	Pittsburgh sleep quality index
PTSD	posttraumatic stress disorder
SE	sleep efficiency
TF-CBT	trauma-focused cognitive-behavioral therapy
TIB	total time in bed

Definition and pathophysiology of posttraumatic stress disorder

Posttraumatic stress disorder (PTSD) is a serious, pervasive, and distressing condition that affects both civilians and war veterans. According to the diagnostic and statistical manual (DSM-5) of Mental Disorders (5th ed.; DSM-5; American Psychiatric Association; 2013) (American Psychiatric Association, 2003) and the International Classification of Diseases (ICD-11) (Hansen, Hyland, Armour, Shevlin, & Elklit, 2015; Reed et al., 2019), PTSD is a condition characterized by recurrent distressing thoughts, intrusive aversive memories, and marked physiologic arousal following an exposure to a traumatic or stressful event.

Due to confusing nosology over the years, numerous attempts have been made to streamline and ease PTSD diagnostic criteria utilization by clinicians (O'Donnell et al., 2014) These criteria have resulted in disparate clinical constructs (Friedman, Resick, Bryant, & Brewin, 2011; Reed et al., 2019). Regardless of iterations and debates over PTSD concepts such as broad versus narrow constructs (Thompson, Gottesman, & Zalewski, 2006) and the various clinical subtypes that were proposed, the core of PTSD phenotypes comprises classes of symptoms categorized on the basis of exposure, recurrence of symptoms, avoidance, mood and reactivity alteration.

The pathophysiology of PTSD is based on the concept of fear conditioning. The model surmises that the release of stress hormone associated with the frightening experience invoked during the traumatic event promotes a strong conditional learning between cues present at the time of trauma and the fear responses. When these cues are triggered by future events, the individual exhibits a fear response not unlike the one experienced during the initial traumatic experience (Pitman et al., 2012; Yehuda, 1997). To overcome these learned associations, treatment of PTSD involves behavioral interventions deigned to foster extinction learning whereby repeated exposure to aversive cues do not simulate unpleasant memories (Foa, Rothbaum, Riggs, & Murdock, 1991; Rothbaum, Hembree, Rauch, & Foa, 2007).

Although the focus of this chapter is aimed at describing the standard psychotherapies applied for treatment of sleep-related PTSD, a brief review of PTSD treatment is warranted.

Current treatment modalities for PTSD

Available treatment modalities for PTSD comprise both psychological and pharmacological interventions. Psychotherapies are considered the first line of treatment for PTSD and are classified as trauma-focused and nontrauma-focused interventions. Trauma-focused cognitive-behavioral therapy (TF-CBT) examines memories and feelings related to the traumatic event in attempt to overcome intrusive thoughts and incapacitating fear. Examples of TF-CBT include exposure therapy (e.g., prolonged exposure), cognitive processing therapy, eye movement desensitization, and reprocessing (Bisson & Andrew, 2007; Coventry et al., 2020). In contrast, non–trauma-focused psychological interventions aim at overcoming PTSD symptoms without directly targeting thoughts and feelings related to the trauma. Relaxation, stress inoculation training, and interpersonal therapy are examples of such treatment.

Trauma-focused cognitive-behavioral therapy

Trauma-focused CBT involves helping PTSD sufferers to identify distorted thinking patterns regarding themselves, the traumatic incident and the world. It focuses on challenging and changing cognitive distortions and behaviors, improving emotional regulation, and developing personal coping strategies that target solving current problems (Kline, Cooper, Rytwinksi, & Feeny, 2018; Schoenfeld, Deviva, & Manber, 2012). In clinical terms, it is a form of psychotherapeutic measure that helps individuals learn how to identify and modify destructive or disturbing thought patterns that have a negative influence on behavior and emotions.

TF-CBT protocols draw on four core key components with varying degrees of emphasis: (1) psychoeducation; (2) anxiety management; (3) exposure; and (4) cognitive restructuring (Kubany et al., 2004; Marks, Lovell, Noshirvani, Livanou, & Thrasher, 1998). TF-CBT is offered as individual or group TF-CBT administered over 12 sessions with each session lasting between 40 and 60 min although shorter sessions of 20 min have been described with comparable efficacy (Bryant et al., 2019; Nacasch et al., 2015). Initial awareness of the diagnosis is usually considered the impetus for early intervention to limit further progression of PTSD (Kornor et al., 2008).

TF-CBT intervention has demonstrated its superior effectiveness in various groups including victims of sexual assault, terrorist attacks, war and natural disasters compared either to sham (Bryant et al., 2008, 2019) or supportive counseling alone (Blanchard et al., 2003; Bryant, Harvey, Dang, Sackville, & Basten, 1998; Duffy, Gillespie, & Clark, 2007; Foa et al., 1991; McDonagh et al., 2005). Despite the promising strategy, not all patients demonstrate sustained improvement with TF-CBT (Bradley, Greene, Russ, Dutra, & Westen, 2005; Shalev et al., 2016). Approximately 30% of those receiving behavioral interventions continue to experience PTSD symptoms (Bradley et al., 2005). In these cases, consideration of adjunctive and combination therapy with other behavioral or pharmacological treatment modalities is recommended (Lebois, Seligowski, Wolff, Hill, & Ressler, 2019).

A major limiting factor for TF-CBT is the rate of dropout from PTSD treatment. The estimate of dropout varies across studies ranging from 18% to 36% depending on the behavioral modalities used for therapy (Imel, Laska, Jakupcak, & Simpson, 2013; Rauch, Eftekhari, & Ruzek, 2012). Another challenge, which is common to other psychological treatments of PTSD, is poor availability of trained health specialists, limiting timely access to care.

Pharmacologic treatment of PTSD

While this is not the focus of this chapter, medication treatment is mentioned briefly as it offers an alternative therapy for patients with PTSD who are unable or who have not experienced any benefit from behavioral interventions. There are currently two FDA-approved serotonin reuptake inhibitors, sertraline and paroxetine, for the treatment of PTSD (Davidson et al., 2001; Marshall, Beebe, Oldham, & Zaninelli, 2001). Other agents (e.g., escitalopram, prazosin) have been used to target specific symptoms of PTSD like anxiety, depression, and nightmares (Raskind et al., 2013; Suliman et al., 2015). Some of the medications have the potential to treat several symptoms simultaneously. However, due in part, to heterogeneity of PTSD phenotypes and less fully understood phenomena such as drug adverse effects and treatment resistance, pharmacologic therapy as standalone regimen is frequently ineffective. Furthermore, the positive therapeutic response is often nonsustainable (Fonzo, Federchenco, & Lara, 2020). Invariably, patient preference plays a significant role in the prescribed and/or adopted treatment modality.

PTSD and sleep disorders

In the last decade, several studies have documented a broad array of comorbid sleep disorders in subjects with PTSD including insomnia, nightmares, sleep disordered breathing (SDB), and dream enactment behavior (El-Solh, Riaz, & Roberts, 2018; Williams, Collen, Orr, Holley, & Lettieri, 2015). Sleep impairment accounts for a significant portion of the variance in physical health complaints even after controlling for other PTSD symptoms and depression (Clum, Nishith, & Resick, 2001). It is beyond the scope of this chapter to discuss the various theoretical models responsible for the myriad of sleep manifestations observed in patients with PTSD. Instead, this section will discuss the role of CBT in the management of specific PTSD-related sleep disorders.

CBT for insomnia (CBT-I) in PTSD

Insomnia represents the predominant sleep disturbances in PTSD and is linked to poor physical and mental health (Clum et al., 2001; Gilbert, Kark, Gehrman, & Bogdanova, 2015). Difficulty falling asleep and maintaining sleep are the most frequent complaints reported to health providers (Ohayon & Shapiro, 2000) and may portend progression or worsening of PTSD symptoms (Wright et al., 2011).

The VA/DoD guidelines recommend cognitive-behavioral therapy (CBT) as the standard treatment modality for managing insomnia in patients with PTSD (VA/DOD, 2017). The basic premise of CBT-I is to alter maladaptive behaviors developed over time that contribute to poor sleep. The components of CBT-I include stimulus control to reestablish the association of bed with sleepiness instead of arousal; sleep restriction to increase sleep homeostatic drive in order to match perceived sleep duration; cognitive therapy to substitute misguided beliefs in sleep; and sleep hygiene to foster appropriate behaviors and prevent patients from engaging in stimulating activities prior to bed time (Morin, 1993).

Both observational and randomized studies have established that CBT-I is associated with objective changes in sleep architecture including increased and consolidated REM (Cervena et al., 2004). CBT-I is superior to waitlist control condition in sleep diary-derived sleep onset latency, wake after sleep onset, and higher sleep efficiency (Margolies, Rybarczyk, Vrana, Leszczyszyn, & Lynch, 2013; Talbot et al., 2014; Ulmer, Edinger, & Calhoun, 2011). CBT-I participants report improved subjective sleep and less disruptive nocturnal behaviors, with lower scores on both the Insomnia Severity Index and the Pittsburgh Sleep Quality Index (Ulmer et al., 2011). More importantly, PTSD symptoms and nightmares are reduced to a greater extent in those receiving CBT-I compared to usual care (Talbot et al., 2014). Yet, it is not uncommon for patients to still experience residual insomnia despite remission of their PTSD symptoms (Pruiksma et al., 2016). In these conditions, standard evaluation for concomitant nightmares and/or depression should be sought by healthcare providers for a combination of pharmacologic and behavioral interventions may be more effective than either modality administered alone (Margolies et al., 2013; Swanson, Favorite, Horin, & Arnedt, 2009). CBT-I components may be tailored to components that are targeting features uniquely accentuated in patients with PTSD. These may encompass in depth analysis of fear of sleep, nocturnal vigilance, or suicidal thoughts.

The safety of CBT-I treatment is a clear advantage over long-term pharmacotherapy, as are the lower risks of side effects and potential drug interactions. CBT-I is usually structured around 8–12 sessions given individually or as group therapy. However, more abbreviated versions of CBT-I with 5 sessions or less have emerged in the past two decades that have shown comparable efficacy (Buysse et al., 2011; Edinger, Wohlgemuth, Radtke, Coffman, & Carney, 2007; Troxel, Germain, & Buysse, 2012). These shorter interventions emphasize the two-process model of stimulus control and sleep restriction. The other elements of cognitive restructuring of dysfunctional beliefs and sleep hygiene are either omitted or briefly discussed.

CBT for nightmares in PTSD

The prevalence of nightmares in patients with PTSD is estimated at 50%–70% compared to 2%–7% in the general population (Spoormaker & Montgomery, 2008). Similar to insomnia, nightmares are considered predictors and potential risk factor for PTSD (Bryant, Creamer, O'Donnell, Silove, & McFarlane, 2010). Unlike idiopathic nightmares, PTSD associated nightmares are described as replicative in nature where the afflicted individual reexperiences the themes and sensory input of the traumatic event. Frequent awakenings and prolonged sensation of fear and anxiety lead to sleep deprivation, irritability and diminished quality of life (Davis, Byrd, Rhudy, & Wright, 2007).

Imagery rehearsal therapy (IRT) is the most studied form of CBTs for nightmares in PTSD (Table 1) (Hansen, Hofling, Kroner-Borowik, Stangier, & Steil, 2013; Marks, 1978). Over the years, several iterations of this therapy have been described such as exposure, relaxation, and rescripting therapy (ERRT) and imagery rescripting and exposure therapy (IRET).

TABLE 1 Summary of psychotherapy components for the treatment of trauma related nightmares.

Imagery rehearsal therapy

- Identify the recurring nightmare
- Modify the events into a new scenario with a different ending
- Rehearse the new dream daily while awake

Exposure, rescripting, and relaxation therapy

- Rescript the nightmare toward a more desirable outcome
- Practice relaxation skills at least twice daily using diaphragmatic breathing for 10 min
- Review the rescripted nightmare using imagery for approximately 15 min during the day while engaging in progressive muscle relaxation

Lucid dreaming therapy

- Emphasize that recurrent features of the nightmares can trigger lucidity
- Set the mind to awaken from dreams and recall them by writing these down
- While returning to sleep, concentrate single-mindedly on intention to remember to recognize that you are dreaming. This process can be facilitated by instituting reality checks
- Search for a familiar dream sign that recur in these dreams
- Once the mind is set to recognize these "cues," lucid dreaming can be achieved
- Rescript nightmares during these lucid dreams to a desirable outcome

Eye movement desensitization and reprocessing

- Recall distressing images while activating one type of bilateral sensory input such as hand tapping or side to side eye movement
- After each stimulation, the subject is asked to describe whatever feeling, image, or sensation comes to mind which becomes the next focus of attention
- The process is repeated numerous times until the distress related to the targeted memory is attenuated
- Redirect the distressing images toward more pleasant ones while practicing the bilateral sensory input

The intervention is designed to dissociate the maladaptive conditioned behaviors generated by nightmares from regular sleep routine. In brief, the patient selects a positive outcome to a particular distressing nightmare and rehearses the script daily until eventually the alternative scenario displaces the memory of the distressing event. IRT has additional components including psychoeducation about PTSD and nightmares as well as relaxation exercises (Forbes, Phelps, & McHugh, 2001). Treatment can be delivered in one to six sessions either individually or in group format. Overall, IRT has been shown to decrease post-traumatic nightmares frequency, reduce PTSD symptoms and improve sleep quality (Casement & Swanson, 2012). Similarly, the efficacy of EERT and IRET have been demonstrated in multitude of randomized controlled trials (RCTs) involving combat veterans and victims of sexual assault (Davis et al., 2011; Davis & Wright, 2007).

Because of the pervasiveness of sleep disturbances in PTSD, it is often that these patients may exhibit symptoms of insomnia and nightmares concomitantly. To address this scenario, treatment protocols have incorporated CBT for insomnia (CBT-I) with CBT for nightmares (Swanson et al., 2009; Ulmer et al., 2011). The combination of IRT with CBT-I have resulted in significant improvement of sleep quality, nightmares and to a lesser degree in severity of PTSD symptoms (Harb, Cook, Gehrman, Gamble, & Ross, 2009; Margolies et al., 2013). As a whole, combined treatment appears to result in larger improvements than IRT alone (Casement & Swanson, 2012) although these findings have not been uniformly established across the board (Harb et al., 2019).

CBT for obstructive sleep apnea

The relationship between obstructive sleep apnea (OSA) and PTSD has been extensively described in the literature (Fig. 1) (Colvonen et al., 2015; Jaoude, Vermont, Porhomayon, & El-Solh, 2015). In brief, the repetitive cycles of obstructive events followed by short arousals exacerbate sleep fragmentation experienced by most patients with PTSD. The hyerarousal status induces upper airway instability that furthers lower the threshold for upper airway collapsibility, predisposing to sleep apnea. Continuous positive airway pressure (CPAP) remains the primary and most effective therapy for OSA. Yet, poor CPAP adherence continues to be a major limiting factor for its effectiveness both in the civilian population and veterans afflicted by PTSD (El-Solh, Ayyar, Akinnusi, Relia, & Akinnusi, 2010; Zhang, Weed, Ren, Tang, & Zhang, 2017). The reasons for low CPAP adherence include discomfort related to mask interface and psychological barriers to behavioral

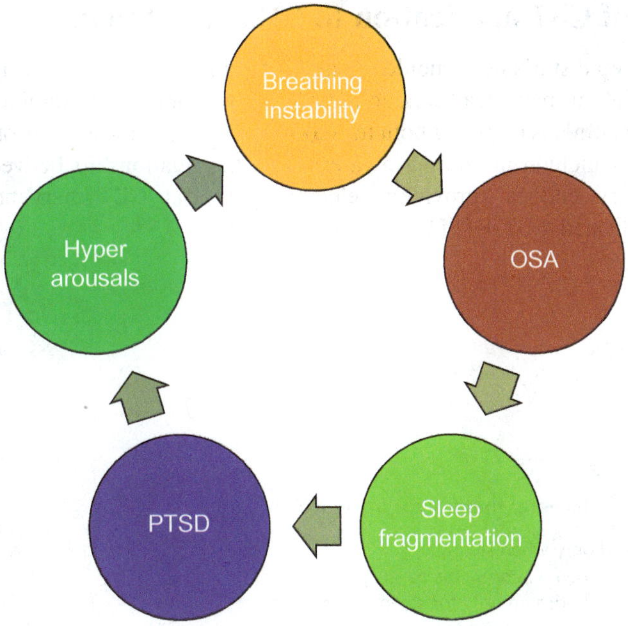

FIG. 1 The link between PTSD and OSA. PTSD, posttraumatic stress disorder; OSA, obstructive sleep apnea.

change. A CBT treatment to improve CPAP adherence has been successfully implemented in civilian population according to several RCTs (Aloia et al., 2001; Aloia et al., 2007). However, the efficacy of behavioral intervention on CPAP adherence has not been evaluated hitherto in PTSD patients with OSA.

To complicate matter further, it is a common observation for PTSD patients with OSA to have coexistent insomnia, a condition known as complex insomnia (Sweetman et al., 2017). Clinical investigations have indicated a lower CPAP adherence in patients with complex insomnia than OSA patients without insomnia (El-Solh, Kufel, & Adamo, 2018). Preliminary evidence from small trials in participants without PTSD has shown on average an increase of 1 h in CPAP utilization following CBT-I (Sweetman et al., 2019). Whether the same benefit can be observed in patients with PTSD remains to be seen. To our knowledge, there is no general consensus or evidence-based approach to the treatment of complex insomnia in Veterans with PTSD. The chronicity of insomnia and OSA in military personnel combined with the heightened risk of mental disorders and suicide suggest that sleep-specific treatments are urgently needed and may contribute to accelerated recovery and resilience in service members.

Barriers to effective treatment of PTSD with CBT

Access to effective CBT for PTSD remains a widespread clinical challenge. There is insufficient trained manpower to provide in-person CBT to the affected population. With the expansion of mobile coverage across rural areas, integrating evidence-based CBT protocols into patient-facing mobile applications have opened the door for previously unreachable segment of the population with PTSD to receive a proven therapy (Seyffert et al., 2016). Conversely, subsets of the affected PTSD population have been reported to demonstrate aversion to CBT (Doran, Pietrzak, Hoff, & Harpaz-Rotem, 2017; Zayfert et al., 2005). Whether this is due to stigmatization or perception of ineffectiveness of the mode of therapy remains a subject of debate (Dedert et al., 2020).

Poor patient adherence to therapy also constitutes a hindrance to achieving effective intervention for PTSD and related sleep dysregulation by available psychotherapy. Factors responsible for this have been highlighted by several studies in both the civilian and veteran cohorts (Doran, O'Shea, & Harpaz-Rotem, 2019). In addition, CBT dropout rates are reported to be high among PTSD sufferers (Alpert, Hayes, Barnes, & Sloan, 2020; Schottenbauer, Glass, Arnkoff, Tendick, & Gray, 2008). While risk factors for attrition during CBT are multifactorial, independent factors such as younger age, higher levels of youth and caregiver avoidance expressed during early sessions, as well as greater relationship difficulties between the patient and therapist are known to play a significant role in nonadherence and dropout from therapy (Schottenbauer et al., 2008; Yasinski et al., 2018).

Improving outcomes of CBT application to PTSD treatment

Treatment of PTSD-related sleep disturbances should be approached from a new perspective by moving away from considering sleep symptoms in isolation and instead conducting integrative studies that examine sequential or combined behavioral and/or pharmacological treatments targeting both the daytime and nighttime aspects of PTSD (Colvonen et al., 2018). Additional suggested measures include promoting awareness of the relationship between PTSD and sleep disorders, offering CBT training for primary healthcare providers, selecting appropriate CBT interventions to target specific attributes of PTSD, and establishing dedicated CBT-PTSD clinics and outreach.

Practice and procedures

Main components of cognitive-behavioral therapy for insomnia

Stimulus control	1) Going to bed only when sleepy 2) Getting out of bed when unable to sleep 3) Using the bed/bedroom only for sleep 4) Arising at the same time every morning 5) Avoiding naps
Sleep restriction	1) Calculate average time in bed (TIB) and total sleep time (TST) from sleep diaries over the last 2 weeks 2) Institute sleep time to equal baseline TIB + 30 min 3) Evaluate sleep diaries at each treatment session and adjust as follows: If average sleep efficiency (SE) > 90%, add 30 min to assigned TIB; If average SE < 85%, subtract 30 min from assigned TIB. 4) If SE is between 85% and 90%, do not adjust.

Insomnia severity index questionnaire

Please indicate the degree of severity for each question according to the following scale:
None (0), Mild (1), Moderate (2), Severe (3), Very severe (4)

a) How do you rate the severity of your difficulty in falling asleep?

b) How do you rate the severity of your difficulty in staying asleep?

c) How do you rate your problem waking up too early in the morning?

d) How do you rate how satisfied you are with your current sleep pattern?

e) How do you rate how your sleep problem interferes with your daily functioning, i.e. daytime fatigue, concentration, mood, ability to function at work, ability to complete tasks at home, etc.?

f) How do you rate the extent to which your sleep problem is noticeable to others?

g) How worried are you about your current sleep problem?

Guidelines for scoring:
Add scores for all seven items: Total score ranges 0–28
0–7 No clinically significant insomnia
8–14 Subclinical insomnia
15–21 Clinical insomnia (moderate)
22–28 Clinical insomnia (severe)

Mini-dictionary of terms

Posttraumatic stress disorder: a constellation of stress symptoms occurring as a result of traumatic injury or severe psychological shock, typically involving disturbance of sleep and constant vivid recall of the experience, with altered social and behavioral responses to others.

Obstructive sleep apnea: a sleep disorder characterized by repeated episodes of cessation of airflow despite breathing effort due to collapse of pharyngeal walls while sleeping followed by an arousal.

Continuous positive airway pressure: the term refers to a device that generates airflow in which a preset pressure is delivered to the nose and mouth via a mask with the intent of overcoming the upper airway collapse.

Nightmares: disturbing dreams that often occur during rapid eye movement sleep. The event can cause a sudden awakening and trigger various emotions ranging from anger, fear, and anxiety.

Insomnia: a sleep disorder characterized by difficulty falling asleep or maintaining sleep.

Cognitive-behavioral therapy: a form of psychotherapy aimed at altering the negative thoughts and attitudes toward self or social environment.

Stimulus control: a component of cognitive-behavioral therapy for insomnia whereby the presence of a stimulus increases the probability of a behavior.

Sleep restriction: a component of cognitive-behavioral therapy for insomnia whereby sleep deprivation is set to allow the body to reestablish proper sleeping dynamics and increase sleep efficiency.

Summary points

- Cognitive-behavioral therapy is an effective treatment of PTSD-associated sleep disorders.
- Although a scarce resource, CBT remains a valuable tool in the trained healthcare provider's armamentarium.
- Where available, CBT should be offered to every patient diagnosed with PTSD and its associated sleep symptoms.
- Efforts are needed to increase awareness of sustained, long-term benefit of CBT compared to short lasting effect of pharmacotherapy.
- Lack of ready access, poor adherence, and limited availability of trained manpower are recognized limitations to a satisfactory CBT response.
- Combination therapy with selected mood stabilizers or adjunct CBT components, where appropriate, will lead to improved outcomes from management of PTSD-associated sleep abnormalities and help reduce resource utilization due to these conditions.

References

Aloia, M. S., Di Dio, L., Ilniczky, N., Perlis, M. L., Greenblatt, D. W., & Giles, D. E. (2001). Improving compliance with nasal CPAP and vigilance in older adults with OAHS. *Sleep & Breathing, 5*(1), 13–21. https://doi.org/10.1007/s11325-001-0013-9.

Aloia, M. S., Smith, K., Arnedt, J. T., Millman, R. P., Stanchina, M., Carlisle, C., et al. (2007). Brief behavioral therapies reduce early positive airway pressure discontinuation rates in sleep apnea syndrome: Preliminary findings. *Behavioral Sleep Medicine, 5*(2), 89–104. https://doi.org/10.1080/15402000701190549.

Alpert, E., Hayes, A. M., Barnes, J. B., & Sloan, D. M. (2020). Predictors of dropout in cognitive processing therapy for PTSD: An examination of trauma narrative content. *Behavior Therapy, 51*(5), 774–788. https://doi.org/10.1016/j.beth.2019.11.003.

American Psychiatric Association. (2003). *Diagnostic and statistical manual of mental disorders (5)*. A. P. Association.

Bisson, J., & Andrew, M. (2007). Psychological treatment of post-traumatic stress disorder (PTSD). *Cochrane Database of Systematic Reviews, 3*, CD003388. https://doi.org/10.1002/14651858.CD003388.pub3.

Blanchard, E. B., Hickling, E. J., Devineni, T., Veazey, C. H., Galovski, T. E., Mundy, E., et al. (2003). A controlled evaluation of cognitive behavioural therapy for posttraumatic stress in motor vehicle accident survivors. *Behaviour Research and Therapy, 41*(1), 79–96. https://doi.org/10.1016/s0005-7967(01)00131-0.

Bradley, R., Greene, J., Russ, E., Dutra, L., & Westen, D. (2005). A multidimensional meta-analysis of psychotherapy for PTSD. *The American Journal of Psychiatry, 162*(2), 214–227. https://doi.org/162/2/214 [pii] https://doi.org/10.1176/appi.ajp.162.2.214.

Bryant, R. A., Creamer, M., O'Donnell, M., Silove, D., & McFarlane, A. C. (2010). Sleep disturbance immediately prior to trauma predicts subsequent psychiatric disorder. *Sleep, 33*(1), 69–74. https://www.ncbi.nlm.nih.gov/pubmed/20120622.

Bryant, R. A., Harvey, A. G., Dang, S. T., Sackville, T., & Basten, C. (1998). Treatment of acute stress disorder: A comparison of cognitive-behavioral therapy and supportive counseling. *Journal of Consulting and Clinical Psychology, 66*(5), 862–866. http://www.ncbi.nlm.nih.gov/pubmed/9803707.

Bryant, R. A., Kenny, L., Rawson, N., Cahill, C., Joscelyne, A., Garber, B., et al. (2019). Efficacy of exposure-based cognitive behaviour therapy for posttraumatic stress disorder in emergency service personnel: A randomised clinical trial. *Psychological Medicine, 49*(9), 1565–1573. https://doi.org/10.1017/S0033291718002234.

Bryant, R. A., Mastrodomenico, J., Felmingham, K. L., Hopwood, S., Kenny, L., Kandris, E., et al. (2008). Treatment of acute stress disorder: A randomized controlled trial. *Archives of General Psychiatry, 65*(6), 659–667. https://doi.org/10.1001/archpsyc.65.6.659.

Buysse, D. J., Germain, A., Moul, D. E., Franzen, P. L., Brar, L. K., Fletcher, M. E., et al. (2011). Efficacy of brief behavioral treatment for chronic insomnia in older adults. *Archives of Internal Medicine, 171*(10), 887–895. https://doi.org/10.1001/archinternmed.2010.535.

Casement, M. D., & Swanson, L. M. (2012). A meta-analysis of imagery rehearsal for post-trauma nightmares: Effects on nightmare frequency, sleep quality, and posttraumatic stress. *Clinical Psychology Review, 32*(6), 566–574. https://doi.org/10.1016/j.cpr.2012.06.002.

Cervena, K., Dauvilliers, Y., Espa, F., Touchon, J., Matousek, M., Billiard, M., et al. (2004). Effect of cognitive behavioural therapy for insomnia on sleep architecture and sleep EEG power spectra in psychophysiological insomnia. *Journal of Sleep Research, 13*(4), 385–393. https://doi.org/10.1111/j.1365-2869.2004.00431.x.

Clum, G. A., Nishith, P., & Resick, P. A. (2001). Trauma-related sleep disturbance and self-reported physical health symptoms in treatment-seeking female rape victims. *The Journal of Nervous and Mental Disease, 189*(9), 618–622. http://www.ncbi.nlm.nih.gov/pubmed/11580006.

Colvonen, P. J., Masino, T., Drummond, S. P., Myers, U. S., Angkaw, A. C., & Norman, S. B. (2015). Obstructive sleep apnea and posttraumatic stress disorder among OEF/OIF/OND veterans. *Journal of Clinical Sleep Medicine, 11*(5), 513–518. https://doi.org/10.5664/jcsm.4692.

Colvonen, P. J., Straus, L. D., Stepnowsky, C., McCarthy, M. J., Goldstein, L. A., & Norman, S. B. (2018). Recent advancements in treating sleep disorders in co-occurring PTSD. *Current Psychiatry Reports, 20*(7), 48. https://doi.org/10.1007/s11920-018-0916-9.

Coventry, P. A., Meader, N., Melton, H., Temple, M., Dale, H., Wright, K., et al. (2020). Psychological and pharmacological interventions for posttraumatic stress disorder and comorbid mental health problems following complex traumatic events: Systematic review and component network meta-analysis. *PLoS Medicine, 17*(8). https://doi.org/10.1371/journal.pmed.1003262, e1003262.

Davidson, J., Pearlstein, T., Londborg, P., Brady, K. T., Rothbaum, B., Bell, J., et al. (2001). Efficacy of sertraline in preventing relapse of posttraumatic stress disorder: Results of a 28-week double-blind, placebo-controlled study. *The American Journal of Psychiatry, 158*(12), 1974–1981. https://doi.org/10.1176/appi.ajp.158.12.1974.

Davis, J. L., Rhudy, J. L., Pruiksma, K. E., Byrd, P., Williams, A. E., McCabe, K. M., et al. (2011). Physiological predictors of response to exposure, relaxation, and rescripting therapy for chronic nightmares in a randomized clinical trial. *Journal of Clinical Sleep Medicine, 7*(6), 622–631. https://doi.org/10.5664/jcsm.1466.

Davis, J. L., & Wright, D. C. (2007). Randomized clinical trial for treatment of chronic nightmares in trauma-exposed adults. *Journal of Traumatic Stress, 20*(2), 123–133. https://doi.org/10.1002/jts.20199.

Davis, J. R., Byrd, P., Rhudy, J. L., & Wright, D. C. (2007). Characteristics of chronic nightmares in a trauma-exposed treatment seeking sample. *Dreaming, 17*, 187–198.

Dedert, E. A., LoSavio, S. T., Wells, S. Y., Steel, A. L., Reinhardt, K., Deming, C. A., et al. (2020). Clinical effectiveness study of a treatment to prepare for trauma-focused evidence-based psychotherapies at a veterans affairs specialty posttraumatic stress disorder clinic. *Psychological Services*. https://doi.org/10.1037/ser0000425.

Doran, J. M., O'Shea, M., & Harpaz-Rotem, I. (2019). In their own words: Clinician experiences and challenges in administering evidence-based treatments for PTSD in the veterans health administration. *The Psychiatric Quarterly, 90*(1), 11–27. https://doi.org/10.1007/s11126-018-9604-5.

Doran, J. M., Pietrzak, R. H., Hoff, R., & Harpaz-Rotem, I. (2017). Psychotherapy utilization and retention in a National Sample of veterans with PTSD. *Journal of Clinical Psychology, 73*(10), 1259–1279. https://doi.org/10.1002/jclp.22445.

Duffy, M., Gillespie, K., & Clark, D. M. (2007). Post-traumatic stress disorder in the context of terrorism and other civil conflict in Northern Ireland: Randomised controlled trial. *BMJ, 334*(7604), 1147. https://doi.org/10.1136/bmj.39021.846852.BE.

Edinger, J. D., Wohlgemuth, W. K., Radtke, R. A., Coffman, C. J., & Carney, C. E. (2007). Dose-response effects of cognitive-behavioral insomnia therapy: A randomized clinical trial. *Sleep, 30*(2), 203–212. https://doi.org/10.1093/sleep/30.2.203.

El-Solh, A., Kufel, T., & Adamo, D. (2018). Comorbid insomnia and sleep apnea in veterans with post traumatic stress disorder. *Sleep and Breathing, 22*(1), 23–31.

El-Solh, A. A., Ayyar, L., Akinnusi, M., Relia, S., & Akinnusi, O. (2010). Positive airway pressure adherence in veterans with posttraumatic stress disorder. *Sleep, 33*(11), 1495–1500. http://www.ncbi.nlm.nih.gov/pubmed/21102991.

El-Solh, A. A., Riaz, U., & Roberts, J. (2018). Sleep disorders in patients with posttraumatic stress disorder. *Chest, 154*(2), 427–439. https://doi.org/10.1016/j.chest.2018.04.007.

Foa, E. B., Rothbaum, B. O., Riggs, D. S., & Murdock, T. B. (1991). Treatment of posttraumatic stress disorder in rape victims: A comparison between cognitive-behavioral procedures and counseling. *Journal of Consulting and Clinical Psychology, 59*(5), 715–723. https://doi.org/10.1037//0022-006x.59.5.715.

Fonzo, G. A., Federchenco, V., & Lara, A. (2020). Predicting and managing treatment non-response in posttraumatic stress disorder. *Current Treatment Options in Psychiatry, 7*(2), 70–87. https://doi.org/10.1007/s40501-020-00203-1.

Forbes, D., Phelps, A., & McHugh, T. (2001). Treatment of combat-related nightmares using imagery rehearsal: A pilot study. *Journal of Traumatic Stress, 14*(2), 433–442. https://doi.org/10.1023/A:1011133422340.

Friedman, M. J., Resick, P. A., Bryant, R. A., & Brewin, C. R. (2011). Considering PTSD for DSM-5. *Depression and Anxiety, 28*(9), 750–769. https://doi.org/10.1002/da.20767.

Gilbert, K. S., Kark, S. M., Gehrman, P., & Bogdanova, Y. (2015). Sleep disturbances, TBI and PTSD: Implications for treatment and recovery. *Clinical Psychology Review, 40*, 195–212. https://doi.org/10.1016/j.cpr.2015.05.008.

Hansen, K., Hofling, V., Kroner-Borowik, T., Stangier, U., & Steil, R. (2013). Efficacy of psychological interventions aiming to reduce chronic nightmares: A meta-analysis. *Clinical Psychology Review, 33*(1), 146–155. https://doi.org/10.1016/j.cpr.2012.10.012.

Hansen, M., Hyland, P., Armour, C., Shevlin, M., & Elklit, A. (2015). Less is more? Assessing the validity of the ICD-11 model of PTSD across multiple trauma samples. *European Journal of Psychotraumatology, 6*, 28766. https://doi.org/10.3402/ejpt.v6.28766.

Harb, G. C., Cook, J., Gehrman, P., Gamble, G., & Ross, R. (2009). Post-traumatic stress disorder nightmares and sleep disturbance in Iraq war veteran: A feasible and promising treatment combination. *Journal of Agression, Maltreatment & Trauma, 15*(5), 516–531.

Harb, G. C., Cook, J. M., Phelps, A. J., Gehrman, P. R., Forbes, D., Localio, R., et al. (2019). Randomized controlled trial of imagery rehearsal for posttraumatic nightmares in combat veterans. *Journal of Clinical Sleep Medicine, 15*(5), 757–767. https://doi.org/10.5664/jcsm.7770.

Imel, Z. E., Laska, K., Jakupcak, M., & Simpson, T. L. (2013). Meta-analysis of dropout in treatments for posttraumatic stress disorder. *Journal of Consulting and Clinical Psychology, 81*(3), 394–404. https://doi.org/10.1037/a0031474.

Jaoude, P., Vermont, L. N., Porhomayon, J., & El-Solh, A. A. (2015). Sleep-disordered breathing in patients with post-traumatic stress disorder. *Annals of the American Thoracic Society, 12*(2), 259–268. https://doi.org/10.1513/AnnalsATS.201407-299FR.

Kline, A. C., Cooper, A. A., Rytwinksi, N. K., & Feeny, N. C. (2018). Long-term efficacy of psychotherapy for posttraumatic stress disorder: A meta-analysis of randomized controlled trials. *Clinical Psychology Review, 59*, 30–40. https://doi.org/10.1016/j.cpr.2017.10.009.

Kornor, H., Winje, D., Ekeberg, O., Weisaeth, L., Kirkehei, I., Johansen, K., et al. (2008). Early trauma-focused cognitive-behavioural therapy to prevent chronic post-traumatic stress disorder and related symptoms: A systematic review and meta-analysis. *BMC Psychiatry, 8*, 81. https://doi.org/10.1186/1471-244X-8-81.

Kubany, E. S., Hill, E. E., Owens, J. A., Iannce-Spencer, C., McCaig, M. A., Tremayne, K. J., et al. (2004). Cognitive trauma therapy for battered women with PTSD (CTT-BW). *Journal of Consulting and Clinical Psychology, 72*(1), 3–18. https://doi.org/10.1037/0022-006X.72.1.3.

Lebois, L. A. M., Seligowski, A. V., Wolff, J. D., Hill, S. B., & Ressler, K. J. (2019). Augmentation of extinction and inhibitory learning in anxiety and trauma-related disorders. *Annual Review of Clinical Psychology, 15*, 257–284. https://doi.org/10.1146/annurev-clinpsy-050718-095634.

Margolies, S. O., Rybarczyk, B., Vrana, S. R., Leszczyszyn, D. J., & Lynch, J. (2013). Efficacy of a cognitive-behavioral treatment for insomnia and nightmares in Afghanistan and Iraq veterans with PTSD. *Journal of Clinical Psychology, 69*(10), 1026–1042. https://doi.org/10.1002/jclp.21970.

Marks, I. (1978). Rehearsal relief of a nightmare. *The British Journal of Psychiatry, 133*, 461–465. https://www.ncbi.nlm.nih.gov/pubmed/31962.

Marks, I., Lovell, K., Noshirvani, H., Livanou, M., & Thrasher, S. (1998). Treatment of posttraumatic stress disorder by exposure and/or cognitive restructuring: A controlled study. *Archives of General Psychiatry, 55*(4), 317–325. https://doi.org/10.1001/archpsyc.55.4.317.

Marshall, R. D., Beebe, K. L., Oldham, M., & Zaninelli, R. (2001). Efficacy and safety of paroxetine treatment for chronic PTSD: A fixed-dose, placebo-controlled study. *The American Journal of Psychiatry, 158*(12), 1982–1988. https://doi.org/10.1176/appi.ajp.158.12.1982.

McDonagh, A., Friedman, M., McHugo, G., Ford, J., Sengupta, A., Mueser, K., et al. (2005). Randomized trial of cognitive-behavioral therapy for chronic posttraumatic stress disorder in adult female survivors of childhood sexual abuse. *Journal of Consulting and Clinical Psychology, 73*(3), 515–524. https://doi.org/10.1037/0022-006X.73.3.515.

Morin, C. B. D. (1993). *Insomnia: Psychological assesment and management*. Guilford Press.

Nacasch, N., Huppert, J. D., Su, Y. J., Kivity, Y., Dinshtein, Y., Yeh, R., et al. (2015). Are 60-minute prolonged exposure sessions with 20-minute imaginal exposure to traumatic memories sufficient to successfully treat PTSD? A randomized noninferiority clinical trial. *Behavior Therapy, 46*(3), 328–341. https://doi.org/10.1016/j.beth.2014.12.002.

O'Donnell, M. L., Alkemade, N., Nickerson, A., Creamer, M., McFarlane, A. C., Silove, D., et al. (2014). Impact of the diagnostic changes to posttraumatic stress disorder for DSM-5 and the proposed changes to ICD-11. *The British Journal of Psychiatry, 205*(3), 230–235. https://doi.org/10.1192/bjp.bp.113.135285.

Ohayon, M. M., & Shapiro, C. M. (2000). Sleep disturbances and psychiatric disorders associated with posttraumatic stress disorder in the general population. *Comprehensive Psychiatry, 41*(6), 469–478. https://doi.org/10.1053/comp.2000.16568.

Pitman, R. K., Rasmusson, A. M., Koenen, K. C., Shin, L. M., Orr, S. P., Gilbertson, M. W., et al. (2012). Biological studies of post-traumatic stress disorder. *Nature Reviews. Neuroscience, 13*(11), 769–787. https://doi.org/10.1038/nrn3339.

Pruiksma, K. E., Taylor, D. J., Wachen, J. S., Mintz, J., Young-McCaughan, S., Peterson, A. L., et al. (2016). Residual sleep disturbances following PTSD treatment in active duty military personnel. *Psychological Trauma, 8*(6), 697–701. https://doi.org/10.1037/tra0000150.

Raskind, M. A., Peterson, K., Williams, T., Hoff, D. J., Hart, K., Holmes, H., et al. (2013). A trial of prazosin for combat trauma PTSD with nightmares in active-duty soldiers returned from Iraq and Afghanistan. *The American Journal of Psychiatry, 170*(9), 1003–1010. https://doi.org/10.1176/appi.ajp.2013.12081133.

Rauch, S. A., Eftekhari, A., & Ruzek, J. I. (2012). Review of exposure therapy: A gold standard for PTSD treatment. *Journal of Rehabilitation Research and Development, 49*(5), 679–687. https://doi.org/10.1682/jrrd.2011.08.0152.

Reed, G. M., First, M. B., Kogan, C. S., Hyman, S. E., Gureje, O., Gaebel, W., et al. (2019). Innovations and changes in the ICD-11 classification of mental, behavioural and neurodevelopmental disorders. *World Psychiatry, 18*(1), 3–19. https://doi.org/10.1002/wps.20611.

Rothbaum, B., Hembree, E. A., Rauch, S. A. M., & Foa, E. B. (2007). *Prolonged exposure therapy for PTSD: Emotional processing of traumatic experiences*. Oxford University Press.

Schoenfeld, F. B., Deviva, J. C., & Manber, R. (2012). Treatment of sleep disturbances in posttraumatic stress disorder: A review. *Journal of Rehabilitation Research and Development, 49*(5), 729–752. https://doi.org/10.1682/jrrd.2011.09.0164.

Schottenbauer, M. A., Glass, C. R., Arnkoff, D. B., Tendick, V., & Gray, S. H. (2008). Nonresponse and dropout rates in outcome studies on PTSD: Review and methodological considerations. *Psychiatry, 71*(2), 134–168. https://doi.org/10.1521/psyc.2008.71.2.134.

Seyffert, M., Lagisetty, P., Landgraf, J., Chopra, V., Pfeiffer, P. N., Conte, M. L., et al. (2016). Internet-delivered cognitive behavioral therapy to treat insomnia: A systematic review and meta-analysis. *PLoS One, 11*(2). https://doi.org/10.1371/journal.pone.0149139, e0149139.

Shalev, A. Y., Ankri, Y., Gilad, M., Israeli-Shalev, Y., Adessky, R., Qian, M., et al. (2016). Long-term outcome of early interventions to prevent post-traumatic stress disorder. *The Journal of Clinical Psychiatry, 77*(5), e580–e587. https://doi.org/10.4088/JCP.15m09932.

Spoormaker, V. I., & Montgomery, P. (2008). Disturbed sleep in post-traumatic stress disorder: Secondary symptom or core feature? *Sleep Medicine Reviews, 12*(3), 169–184. https://doi.org/10.1016/j.smrv.2007.08.008.

Suliman, S., Seedat, S., Pingo, J., Sutherland, T., Zohar, J., & Stein, D. J. (2015). Escitalopram in the prevention of posttraumatic stress disorder: A pilot randomized controlled trial. *BMC Psychiatry, 15*, 24. https://doi.org/10.1186/s12888-015-0391-3.

Swanson, L. M., Favorite, T. K., Horin, E., & Arnedt, J. T. (2009). A combined group treatment for nightmares and insomnia in combat veterans: A pilot study. *Journal of Traumatic Stress, 22*(6), 639–642. https://doi.org/10.1002/jts.20468.

Sweetman, A., Lack, L., Catcheside, P. G., Antic, N. A., Smith, S., Chai-Coetzer, C. L., et al. (2019). Cognitive and behavioral therapy for insomnia increases the use of continuous positive airway pressure therapy in obstructive sleep apnea participants with comorbid insomnia: A randomized clinical trial. *Sleep, 42*(12). https://doi.org/10.1093/sleep/zsz178.

Sweetman, A. M., Lack, L. C., Catcheside, P. G., Antic, N. A., Chai-Coetzer, C. L., Smith, S. S., et al. (2017). Developing a successful treatment for co-morbid insomnia and sleep apnoea. *Sleep Medicine Reviews, 33*, 28–38. https://doi.org/10.1016/j.smrv.2016.04.004.

Talbot, L. S., Maguen, S., Metzler, T. J., Schmitz, M., McCaslin, S. E., Richards, A., et al. (2014). Cognitive behavioral therapy for insomnia in post-traumatic stress disorder: A randomized controlled trial. *Sleep, 37*(2), 327–341. https://doi.org/10.5665/sleep.3408.

Thompson, W. W., Gottesman, I. I., & Zalewski, C. (2006). Reconciling disparate prevalence rates of PTSD in large samples of US male Vietnam veterans and their controls. *BMC Psychiatry, 6*, 19. https://doi.org/10.1186/1471-244X-6-19.

Troxel, W. M., Germain, A., & Buysse, D. J. (2012). Clinical management of insomnia with brief behavioral treatment (BBTI). *Behavioral Sleep Medicine, 10*(4), 266–279. https://doi.org/10.1080/15402002.2011.607200.

Ulmer, C. S., Edinger, J. D., & Calhoun, P. S. (2011). A multi-component cognitive-behavioral intervention for sleep disturbance in veterans with PTSD: A pilot study. *Journal of Clinical Sleep Medicine, 7*(1), 57–68. https://www.ncbi.nlm.nih.gov/pubmed/21344046.

VA/DOD. (2017). *VA/DoD clinical practice guidelines for management of post-traumatic stress*. VA/DOD.

Williams, S. G., Collen, J., Orr, N., Holley, A. B., & Lettieri, C. J. (2015). Sleep disorders in combat-related PTSD. *Sleep & Breathing, 19*(1), 175–182. https://doi.org/10.1007/s11325-014-0984-y.

Wright, K. M., Britt, T. W., Bliese, P. D., Adler, A. B., Picchioni, D., & Moore, D. (2011). Insomnia as predictor versus outcome of PTSD and depression among Iraq combat veterans. *Journal of Clinical Psychology, 67*(12), 1240–1258. https://doi.org/10.1002/jclp.20845.

Yasinski, C., Hayes, A. M., Alpert, E., McCauley, T., Ready, C. B., Webb, C., et al. (2018). Treatment processes and demographic variables as predictors of dropout from trauma-focused cognitive behavioral therapy (TF-CBT) for youth. *Behaviour Research and Therapy, 107*, 10–18. https://doi.org/10.1016/j.brat.2018.05.008.

Yehuda, R. (1997). Sensitization of the hypothalamic-pituitary-adrenal axis in posttraumatic stress disorder. *Annals of the New York Academy of Sciences, 821*, 57–75. https://doi.org/10.1111/j.1749-6632.1997.tb48269.x.

Zayfert, C., Deviva, J. C., Becker, C. B., Pike, J. L., Gillock, K. L., & Hayes, S. A. (2005). Exposure utilization and completion of cognitive behavioral therapy for PTSD in a "real world" clinical practice. *Journal of Traumatic Stress, 18*(6), 637–645. https://doi.org/10.1002/jts.20072.

Zhang, Y., Weed, J. G., Ren, R., Tang, X., & Zhang, W. (2017). Prevalence of obstructive sleep apnea in patients with posttraumatic stress disorder and its impact on adherence to continuous positive airway pressure therapy: A meta-analysis. *Sleep Medicine, 36*, 125–132. https://doi.org/10.1016/j.sleep.2017.04.020.

Chapter 21

Psychosocial interventions for occupational stress and psychological disorders in humanitarian aid and disaster responders: A critical review

Cheryl Yunn Shee Foo[a], Helen Verdeli[a], and Alvin Kuowei Tay[b]

[a]Department of Counseling and Clinical Psychology, Teachers College, Columbia University of New York, New York, NY, United States [b]Faculty of Medicine, School of Psychiatry, University of New South Wales, Sydney, NSW, Australia

Abbreviations

COPSOQ	The Copenhagen Psychosocial Questionnaire
IASC	Inter-Agency Standing Committee
IAT	Integrative Adapt Therapy
ICD-11	The International Classification of Diseases 11th Revision
IPT	Interpersonal Psychotherapy
MBCT	Mindfulness-Based Cognitive Therapy
MHPSS	mental health and psychosocial support
NGO	nongovernmental organization
PFA	Psychological First Aid
PTE	potentially traumatic events
PTSD	Posttraumatic Stress Disorder
UN	United Nations
UNHCR	United Nations High Commissioner for Refugees
WHO	World Health Organization

Introduction

Mental health and psychosocial support (MHPSS) for humanitarian and disaster relief personnel gained prominence in the early 2000s following increased systematic threats and violence against humanitarian aid staff and organizations globally. In the last decade, mass emergencies—including armed conflict, civil unrest, terrorism, population displacement, natural hazards, pandemics—have occurred at alarming levels with significant sociopolitical, economic, and health consequences. Increasingly, occupational mental health is now regarded as a key priority by many international organizations, with accumulative research confirming the deleterious effects of various stressors on humanitarian workers (Brooks et al., 2015; Foo, Verdeli, & Tay, 2021). If left unaddressed, occupational stress and mental health problems among humanitarian staff can hurt the quality and sustainability of humanitarian aid work (UN, 2018).

Although several key policies and guidelines (e.g., Antares Foundation, 2012; UN, 2018; UNHCR, 2016) now underpin the approaches to MHPSS for humanitarian personnel, they vary in their approaches and have been inconsistently applied in practice across international organizations (Welton-Mitchell, 2013). This chapter reviews the existing literature and proposes a comprehensive and systematic framework for guiding the formulation and design of psychosocial interventions for humanitarian aid workers.

Occupational mental health of humanitarian aid and disaster responders

Humanitarian aid workers comprise national staff, international staff, and volunteers from various agencies, including the UN Agencies, Funds, and Programs (AFPs), International Committee of the Red Cross (ICRC) societies, and national and international Nongovernmental Organizations (NGOs). Humanitarian personnel is typically assigned to various roles in the field, including peacekeeping, security, operational support, human resource, medical services, and other specialized services, logistics, monitoring and evaluation, finances, law enforcement, auditing, and reporting. Guided by the mandates of respective organizations and the accountability to affected populations, humanitarian personnel are involved in all phases of humanitarian response and across different aid sectors (Fig. 1).

Much of the current literature on occupational mental health in emergency workers has mainly focused on military personnel and community first responders, such as police officers, firefighters, search and rescue personnel, and paramedics (Brooks et al., 2015; Brooks, Dunn, Amlôt, Greenberg, & Rubin, 2016). While these occupation groups overlap with humanitarian personnel in their core function of being on the frontline of crises, humanitarian personnel often face unique stressors and risks over an extended period due to the protracted and complex nature of humanitarian emergencies. For example, there have been numerous reports indicating the elevated rates of exposure to critical incidents while on duty (e.g., experiencing or witnessing threats and incidents of violence, witnessing the death of loved ones/friends/colleagues, etc.) (Stoddard, 2020). By being indirectly exposed to the trauma of the communities they serve, they may also experience secondary traumatic stress (Shah, Garland, & Katz, 2007). At the same time, workplace psychosocial hazards, such as high workload, job insecurity, lack of recognition for work by superiors, and difficult relationships with supervisors and colleagues, can also be unabating sources of stress for humanitarian workers (Foo et al., 2021).

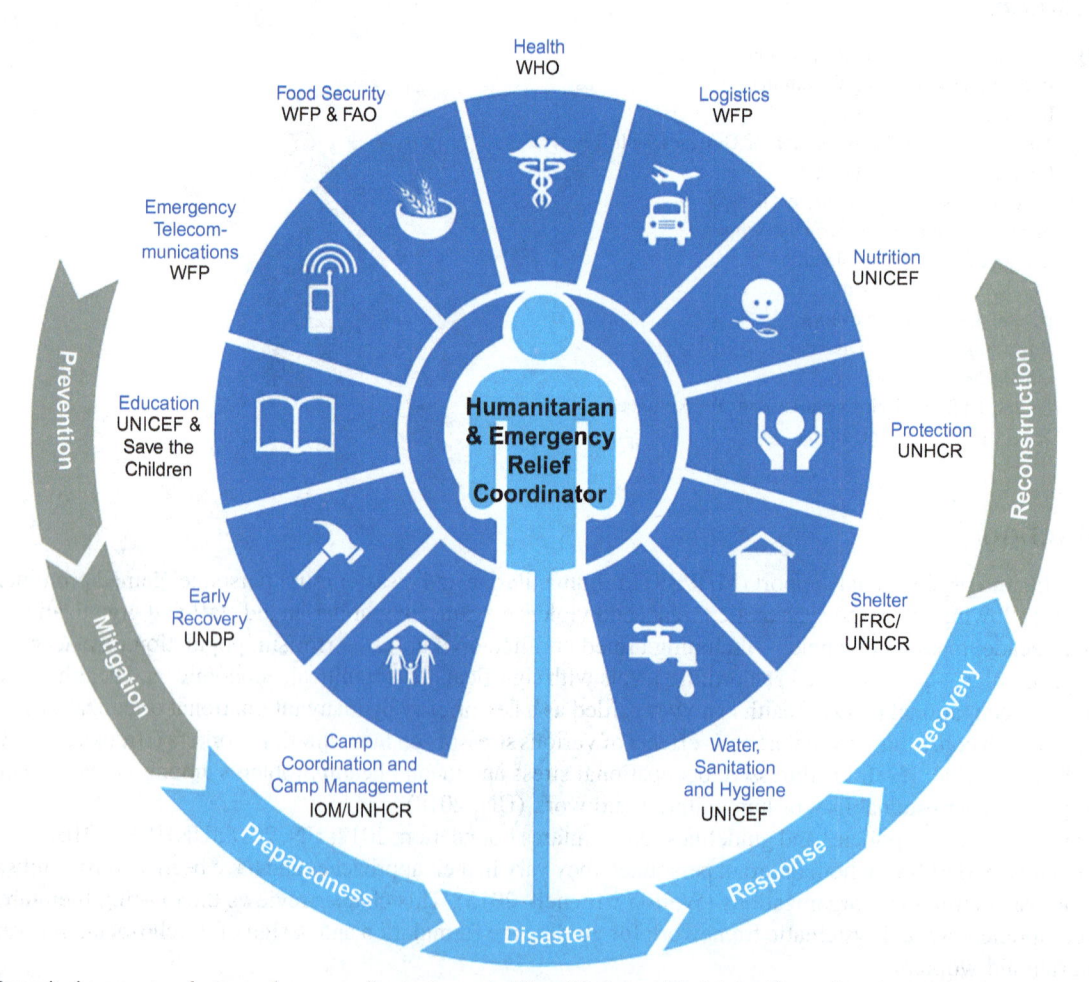

FIG. 1 Humanitarian response clusters and corresponding lead agencies. From "Reference Module for Cluster Coordination at Country Level," by Inter-Agency Standing Committee Sub-Working Group on the Cluster Approach and the Global Cluster Coordinators' Group, © 2015 United Nations. *(Reprinted with the permission of the United Nations.)*

In addition, specific demographic groups within the varied and diverse humanitarian workforce also experience unique challenges that can further contribute to mental health and psychosocial problems. For example, international staff is frequently re-deployed from one humanitarian emergency to another and faces challenges acculturating and maintaining social support (Antares Foundation, 2012). On the other hand, national staff and volunteers are often survivors of the humanitarian emergency exposed to the same trauma as their beneficiaries and face more financial concerns and job insecurities than their expatriate colleagues (Ager et al., 2012; Strohmeier, Scholte, & Ager, 2019). The increased vulnerability to potentially traumatic events (PTE), exacerbated by high levels of cumulative stress due to, inter alia, a lack of social support, interpersonal conflict, and chronic workplace stressors, place humanitarian aid workers at significant risk for occupational stress and mental health problems.

Occupational stress is a broad spectrum of maladaptive reactions to acute and chronic precipitants. Stress can be transitory, cumulative, or acute traumatic stress. Whereas most people have sufficient resources and skills to recover from transitory stress, cumulative stress is the most frequent form of stress encountered among humanitarian staff. For example, the prevalence of burnout, a result of unsuccessfully managed chronic workplace stress (ICD-11, WHO, 2019), is high among humanitarian workers. Over 45% of aid workers experienced an increased risk of burnout in at least one dimension of burnout (i.e., emotional exhaustion, depersonalization, or low personal accomplishment) (Eriksson et al., 2009).

Furthermore, unsurprisingly, chronic stress reactions and mental health sequelae are elevated in humanitarian workers, given the high exposure rate to PTE. Strohmeier and Scholte (2015) reviewed trauma-related mental health problems in national humanitarian staff. They found that the prevalence of posttraumatic stress disorder (PTSD), depression, anxiety, and hazardous alcohol consumption is high among this occupational group, ranging from 8% to 25%. These problems can lead to more serious accident and illness rates, absenteeism (UN, 2018), loss of efficiency and productivity, lower quality of service, lower work commitment and engagement, and high attrition rates (Welton-Mitchell, 2013).

Existing MHPSS guidelines and their limitations

While there is a substantial body of research on the prevalence, course, and risk factors of psychological and stress-related disorders, there is less evidence on effective MHPSS interventions for the prevention and recovery of psychological distress among humanitarian and disaster relief workers.

In an attempt to establish an industry standard, the Antares Foundation and the Centers for Disease Control and Prevention (CDC; Atlanta, Georgia, USA) used a consensus approach with NGOs to develop guidelines for the psychological protection of staff in humanitarian aid organizations (Antares Foundation, 2012). Their guideline addressed eight key domains of prevention and intervention that span the predeployment to postdeployment phases of humanitarian response (Fig. 2). More recently, the UN system described a five-year strategy to optimize the mental health and wellbeing of their staff, with priority actions cutting across four levels of intervention as outlined by the IASC (2007) (Fig. 3).

However, although guidelines for psychosocial staff care in the humanitarian sector are available, there is a paucity of empirically supported psychosocial interventions (Brooks, Dunn, Amlôt, Greenberg, & Rubin, 2018; Umeda, Chiba, Sasaki, Agustini, & Mashino, 2020). Indeed, MHPSS for staff is poorly implemented in aid organizations. For example, well-established organizations like UNHCR and Doctors without Borders [Medicins Sans Frontiers (MSF)] did not meet the compliance threshold on more than 50% of the indicators across practice standards based on the Antares framework (Welton-Mitchell, 2013). Furthermore, as noted in several qualitative studies, stigma and lack of confidentiality associated with help-seeking in this population pose a significant barrier to humanitarian aid workers accessing formal sources of support in their organizations (Brooks, Dunn, Amlôt, Rubin, & Greenberg, 2019; Skeoch, Stevens, & Taylor, 2017). There are also critiques that most of the MHPSS guidelines are reactive (i.e., management of adverse psychological effects after a critical incident) rather than proactive and preventive and may have iatrogenic effects (e.g., critical incident stress debriefing). Finally, following the more recent paradigm shift to resilience, more attention should also be given to how to promote positive adaptation following adversity (Luthar, 2006) as well as posttraumatic growth (i.e., the potential positive impact after experiencing a traumatic event) (Tedeschi & Calhoun, 1996) in humanitarian aid organizations.

Comprehensive and systematic MHPSS framework for psychological wellbeing of humanitarian aid personnel

Given these limitations, this chapter proposes a comprehensive and systematic framework for the promotion of wellbeing, prevention, and management of occupational stress-related mental health and psychosocial problems among staff in humanitarian aid organizations. The chapter summarizes prevention and intervention strategies into seven domains, cutting across staff, managerial, team, and organizational levels, and various deployment phases (Table 1).

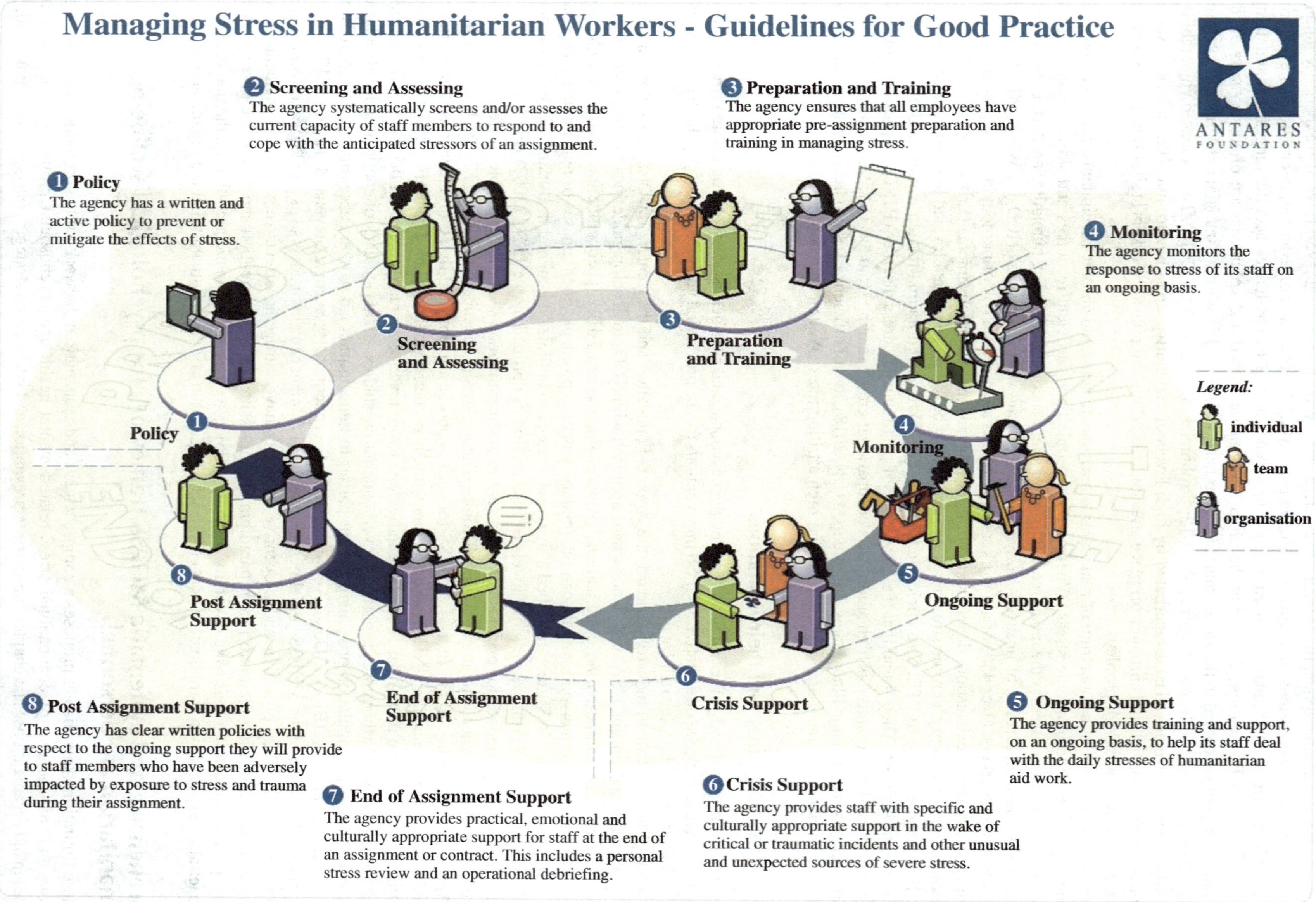
Managing Stress in Humanitarian Workers - Guidelines for Good Practice
ANTARES
FOUNDATION
1 Policy
The agency has a written and active policy to prevent or mitigate the effects of stress.
2 Screening and Assessing
The agency systematically screens and/or assesses the current capacity of staff members to respond to and cope with the anticipated stressors of an assignment.
3 Preparation and Training
The agency ensures that all employees have appropriate pre-assignment preparation and training in managing stress.
4 Monitoring
The agency monitors the response to stress of its staff on an ongoing basis.
Legend:
individual
team
organisation
1 Policy
2 Screening and Assessing
3 Preparation and Training
4 Monitoring
5 Ongoing Support
6 Crisis Support
7 End of Assignment Support
8 Post Assignment Support
8 Post Assignment Support
The agency has clear written policies with respect to the ongoing support they will provide to staff members who have been adversely impacted by exposure to stress and trauma during their assignment.
7 End of Assignment Support
The agency provides practical, emotional and culturally appropriate support for staff at the end of an assignment or contract. This includes a personal stress review and an operational debriefing.
6 Crisis Support
The agency provides staff with specific and culturally appropriate support in the wake of critical or traumatic incidents and other unusual and unexpected sources of severe stress.
5 Ongoing Support
The agency provides training and support, on an ongoing basis, to help its staff deal with the daily stresses of humanitarian aid work.

FIG. 2 Visual representation of Managing Stress for Humanitarian Workers—Guidelines for Good Practice by Antares Foundation, © 2012 Antares Foundation. *(Reprinted with the permission of Antares Foundation.)*

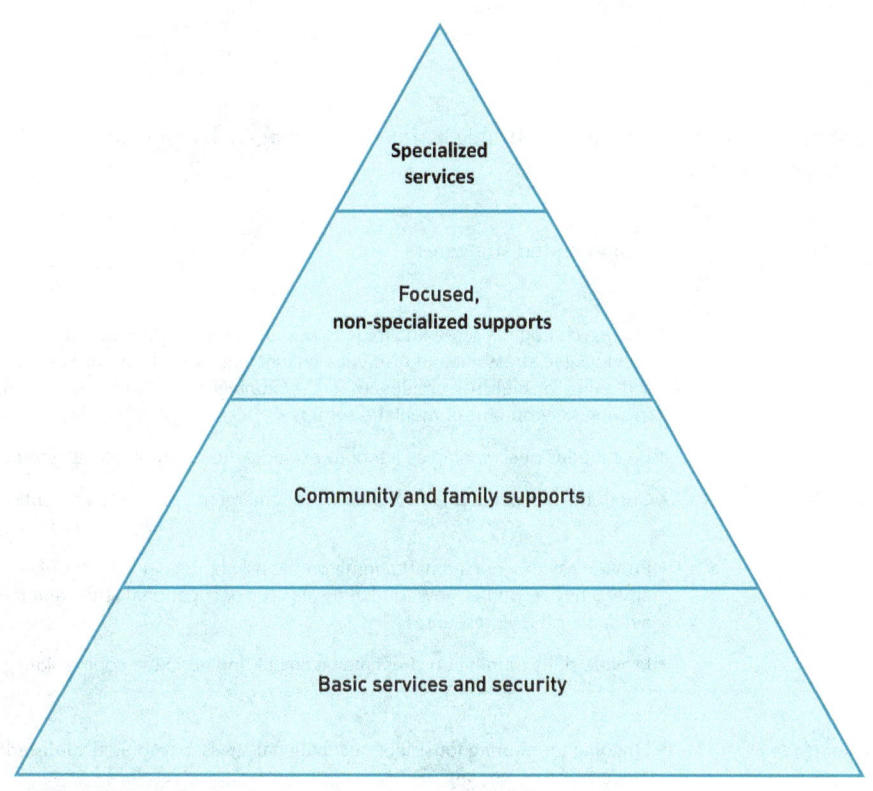

FIG. 3 IASC Intervention Pyramid. From "IASC Guidelines on Mental Health and Psychological Support in Emergency Settings," by IASC, © 2007 United Nations. *(Reprinted with the permission of the United Nations.)*

TABLE 1 Summary of prevention and intervention strategies for staff, managerial, team, and organizational levels at various deployment phases.

Domains of prevention and intervention	Recommended strategies
Organizational-Level Policy and Standards of Practice	
Prioritization and implementation of staff MHPSS	• Conduct comprehensive and participatory needs assessment with key stakeholders and staff on staff MHPSS needs and recommendations, with attention to specific demographic groups (e.g., working mothers, field staff, international and national staff)
	• Continued funding and lobbying with donors for MHPSS
	• Ongoing monitoring, evaluation, and accountability for implementation of staff MHPSS policies
Improve working conditions, quality of life, and benefits	• Provide better amenities and basic supplies for humane working conditions and quality of life (e.g., access to clean food and water; personal protective equipment; reliable transportation; safe accommodation and workspace; internet and phone service)
	• Enhance staff benefits and reward measures (e.g., health insurance, salary packages, time-off, professional development opportunities)
Enhance security and risk management	• Ensure equitable and available security training and resources for staff
	• Invest in staff with expertise in security risk management
	• Improve negation skills and outreach capacities
	• Stricter internal harassment policies, especially on sexual harassment
Reduce workplace psychosocial stressors	• Conduct psychosocial risk assessment and audit regularly to identify potential workplace stressors (e.g., COPSOQ)
	• Plan and institute corrective and preventive programs to ameliorate workplace stressors (see Table 2)

Continued

TABLE 1 Summary of prevention and intervention strategies for staff, managerial, team, and organizational levels at various deployment phases—cont'd

Domains of prevention and intervention	Recommended strategies
Predeployment	
Psychological screening and assessment	• Use psychological screening tools of risk factors (e.g., previous history of mental illness, previous life stressors) and resilience factors (e.g., social support) only as part of routine screening for high-risk employees (e.g., volunteer status, individual exposed to critical incidents, symptoms of mental disorders)
	• Be cautious of screening as a tool to exclude and discriminate against vulnerable groups
Psychological training and preparedness	• Continue to provide practical training for emergency response and critical incident response
	• Provide psychoeducational training on the psychological impact of humanitarian aid and emergency response, how to identify signs of occupational stress and mental disorders, and available MHPSS resources
	• Provide skills training on stress management and adaptive coping skills
Perideployment	
Ongoing MHPSS monitoring and support	• Ongoing monitoring for risk of occupational stress, psychological disorders, and job-related outcome
	• Continuing education on mental health issues
	• Readily available needs-based training
	• Ensure equitable access to staff MHPSS services for all staff
Manager-specific training	• Provide manager training for good management practices and leadership skills to reduce workplace psychosocial stressors
	• Provide psychoeducational training for managers on how to promote psychological wellbeing among the team, recognize warning signs of psychological stress among staff, and encourage self-care and help-seeking behaviors among staff
Social support and team cohesion	• Implement formal peer support programs for teams
	• Build capacity in mental health screening, providing psychoeducation and psychological first aid among staff or peer supporters
	• Ongoing supervision and informal check-ins for all levels of staff
	• Organize work-sponsored social activities
Crisis intervention and support	• Limit use of critical incident stress debriefing due to mixed evidence on its effectiveness
	• Implement stepped-care model of psychological interventions
	• Encourage use of psychological support services, ensuring confidentiality and lack of repercussions on job
	• Provide in-house or external crisis hotlines for psychological support
	• Provide readily available and time-limited psychological first aid and supportive counseling as first-line treatment for at-risk staff
	• Provide in-house evidence-based psychological therapies (e.g., CBT, IPT, IAT) (including via telehealth)
	• Provide access to external MHPSS referrals
Postdeployment	
Promote resilience-building and posttraumatic growth	• Continued emphasis on strengthening social support and adaptive coping skills
Rehabilitation and reintegration	• Continued access to confidential psychosocial support services
	• Nonstigmatizing reintegration policies after rehabilitation from mental illness or leave of absence

At the broader organizational level, *organizational policy and standards of practice* protecting physical and psychological safety are recommended, with attention paid to *promoting organizational wellbeing and reducing workplace psychosocial stressors*. We also examine the evidence base for *predeployment psychological screening* and *training and preparedness*. In addition, we consider the protective factors against psychological morbidity and review the evidence for *perideployment mental health monitoring and social support, crisis intervention, and psychological treatment*. Finally, we justify the need for *postdeployment rehabilitation and resilience-building*, including promoting resilience and posttraumatic growth.

Organizational policy and standards of practice

Working conditions, quality of life, and benefits

At the most basic level of MHPSS intervention (Fig. 3), humanitarian aid organizations must ensure optimal working conditions and quality of life for all staff. Especially for field staff who live and work in harsh environments, better amenities and basic supplies should be provided. These include clean drinking water, essential medicines, personal protective equipment, reliable transportation, accommodation and workspace, and liberal internet and phone service use policies.

In addition, according to the occupational health model of effort-reward imbalance (ERI), a lack of reciprocity between effort (i.e., working under pressure, heavy workload, and fast pace) and potential rewards (i.e., financial compensation, esteem reward, promotion aspects, and job security) will lead to higher levels of emotional distress, burnout, and negative health effects (Siegrist, 1996). In a survey with UNHCR staff, increased ERI scores were associated with a fourfold increase in risk for emotional exhaustion and a twofold increase in the depersonalization feature of burnout (UNHCR, 2016). A strong positive relationship between ERI and emotional exhaustion was also found among humanitarian aid workers in an international Geneva-based organization, even after adjusting for secondary traumatic stress and PTSD (Jachens, Houdmont, & Thomas, 2019). This evidence supports the need for organizations to reduce the imbalance between perceived effort and reward.

As found from a qualitative study with aid workers in South Sudan, examples of possible organizational measures that can be taken to improve ERI include: providing competitive benefit and salary packages, health insurance (including mental health care), skill enhancement and professional development, and career progression opportunities (Strohmeier et al., 2019).

Security and risk management

Given the security risks of working in high-risk environments, security and risk management policies must also be in place. Security training and resources (e.g., off-hours transportation, communication equipment, site security at home) must be equitably offered to all staff, including international and national, operational, and field staff. Furthermore, since most aid-worker attacks are committed by national-level nonstate armed groups (NSAGs), Stoddard (2020) recommended that organizations invest in negotiation skills and outreach capacities with NSAGs.

Reducing workplace psychosocial stressors

Workplace psychosocial hazards are the interactions between job content, work organization and management, and other environmental organizational contexts (Cox & Griffiths, 2005). In this occupational group, research has established the relationship between workplace psychosocial stressors and poor psychological health. In cross-sectional studies with local humanitarian aid workers from Sri Lanka and Northern Uganda, chronic stressors such as financial pressures, lack of recognition for work by supervisors, difficult relationship with management and colleagues, and disparity in treatment of international and national staff were related to higher levels of anxiety symptoms (Lopes Cardozo et al., 2012), and higher levels of emotional exhaustion (Ager et al., 2012). A qualitative study found that employees perceived an "emergency culture" in an UN-aligned organization, where the "constant feeling of crisis within the organization" and urgency to meet deadlines and humanitarian needs was a source of stress and accepted cultural norm that was reinforced in the workplace (Jachens, Houdmont, & Thomas, 2018). This finding was also corroborated by a qualitative study with international humanitarian aid workers by Young, Pakenham, and Norwood (2018), where organizational systems, structures, and demands were critical sources of stress.

Regular workplace psychosocial risk assessments or audits can help identify potential workplace stressors unique to the organizational context. For example, the Copenhagen Psychosocial Questionnaire (COPSOQ) (Burr et al., 2019) may be considered as an occupational risk assessment tool as it has been internationally validated. It has also been used to inform the UN Risk Assessment Framework for the psychosocial wellbeing of all security personnel worldwide (Tay, personal communication, May 19, 2020). Table 2 summarizes the psychosocial hazards commonly experienced by humanitarian staff, following the typology of the COPSOQ, and recommendations for intervention.

TABLE 2 Summary of common workplace psychosocial stressors and recommendations for interventions.

Dimension (based on COPSOQ)	Risk factors	Recommendations for interventions
Demands at work		
Workload and work pace	• Work overload • 'Emergency culture' where there may be pressure to respond immediately to crises and tasks • High pace and high levels of time pressure	• Organizational training aimed at promoting a better psychosocial safety climate • Redistribution of work to ensure reasonable workload and work pace • Increase staff hire and volunteer staff, especially during emergency response phase
Emotional demands	• Secondary exposure to traumatic experiences of aid recipients • Criticism of work by media or beneficiary community	• Staff wellbeing programs aimed at psychoeducation, stress management, coping strategies, and self-care • Increased advocacy and protection of humanitarian staff
Work organization and job contents		
Influence at work	• Low participation in decision making in job scope and work responsibilities • Lack of control over workload and pacing	• Implement open-door policy from managers • Better feedback mechanism with managers and HR for staff
Control over working time	• Inflexible work schedules • Unpredictable and long hours	• Implement rotating shifts or teams during the emergency response period to ensure healthy work distribution • Ensure that overtime work is compensated financially with additional paid time-off
Interpersonal relations and leadership		
Role clarity and role conflict	• Role ambiguity and role conflict due to multiple and competing responsibilities that may be out of job scope	• Managers to pay attention to the formation of the team and clear delegation of work roles and job scopes during project planning • Open-door policy from managers • Better feedback mechanism with managers and HR for staff
Recognition of effort at work	• Lack of recognition for work accomplishment from managers • Poor and inconsistent quality of supervision and management	• Manager-training on good management principles and skills on staff motivation, assessing and providing validation and feedback • Informal team social activities to show recognition and acknowledgment of team effort
Social support from colleagues and supervisors	• Poor relationship with supervisors • Interpersonal conflict with colleagues and supervisors	• Team-building activities and team retreats • Open-door policy from managers • Better feedback mechanism with managers and HR for staff • Manager-training on conflict resolution
Sense of community at work	• Weaker organizational identity and cohesion among smaller NGOs, compared to lead UN agencies and large international NGOs	• Team-building activities and team retreats • Informal team social activities to build a sense of community • Meaning creation and value addition by reinforcing the values and meaning of work
Work-individual interface		
Job insecurity	• Job insecurity due to short-term contracts and lack of long-term funding from donors • Large wage inequality between field staff and managerial staff • National staff may be paid less compared to international staff • National staff perceived fewer career progression opportunities	• Lobbying for continued funding and extension of contracts and mission duration • Fair wages for staff at all levels and nationalities • Professional development opportunities (e.g., training, student loans) for all staff

TABLE 2 Summary of common workplace psychosocial stressors and recommendations for interventions—cont'd

Dimension (based on COPSOQ)	Risk factors	Recommendations for interventions
Work-life conflict	• Difficult maintaining work-life balance • Conflicts between work and home demands, especially for working mothers	• More paid time-off and personal days that can be taken without repercussion • Staff wellbeing programs aimed at promoting a healthy work-life balance • Provision of recreational and exercise areas in work-space • Provision of child-friendly spaces at work with trained child-care volunteers for working mothers
Social capital		
Organizational justice	• Lack of trust in leadership • Lack of accountability, justice, and fairness in the organization in the handling of critical incidents, conflicts, and feedback from employees • National staff more frequently cite inequality in treatment between international and national staff	• Trust-building exercises and frequent sharing of information to ensure that all staff are informed • Conflict resolution training for managers • Open-door policy from managers • Better feedback mechanism with managers and HR for staff
Offensive behaviors at work		
Workplace violence and harassment	• Experience of gossip and slander; bullying; cyber-bullying; threats of violence; discrimination; and sexual harassment	• Staff training of workplace safety and ethics • Clear non-discrimination policies and protocols for reporting and taking action against offensive behaviors at work • Better feedback mechanism with managers, HR, Ombudsman for staff, with appropriate action taken against the perpetrator • Appropriate psychological support for staff who experienced workplace violence

After identifying risk factors, top-level management will have to play a leading role in planning and instituting corrective and preventive programs. Othman, Steel, Lawsin, and Wells (2018) provided a promising example of an integrated staff psychosocial support program for primary health care staff working with internally displaced Syrians in a conflict zone in Syria. They piloted a 6-month program of group sessions addressing workplace psychosocial challenges at the individual, team, organizational levels. The program was found to reduce role ambiguity and enhance personal relationships with colleagues and superiors (Othman et al., 2018). This example can be a blueprint for implementing psychosocial support programs targeted at identified workplace psychosocial risks in humanitarian aid organizations.

Predeployment

Key concerns about preemployment or predeployment psychological screening

Psychological screening has been carried out in several emergency services (Marshall, Milligan-Saville, Mitchell, Bryant, & Harvey, 2017) and recommended for humanitarian aid workers (Lopes Cardozo et al., 2012). The rationale is that preemployment or predeployment psychological screening can identify persons prone to psychological distress and mental health disorders. Those identified as being at higher risk may be advised against taking up physically and mentally intensive duties. They may also be closely monitored for their work performance and physical and mental health during their deployment. However, applying a screening procedure to identify these risk factors may not be practical or valuable. The evidence base for the risk factors included in the screening is equivocal, and the screening process can be imprecise (Opie, Brooks, Greenberg, & Rubin, 2020). For example, in a recent review, Opie et al. (2020) found that only demographic variables of volunteer status, previous history of mental illness, and previous life stressors emerged as reliable predictors for poor mental health among disaster relief workers.

Furthermore, there are also risks of excluding skilled and experienced staff who may have experienced prior challenges but show resilience. More importantly, psychological screening can reinforce stigma and discrimination against those identified as high risk and prevent disclosure and help-seeking (Brooks et al., 2015). Nevertheless, screening tools may be integrated as part of the routine monitoring of high-risk employees within the context of individual case management. For example, it may be useful for individuals who are recently exposed to critical incidents or those with suicidality or symptoms of severe mental disorders.

Psychoeducational training and preparedness

A significant predeployment protective factor is preparedness and training (Brooks et al., 2015). Predisaster training can prepare aid workers on practical aspects of disaster planning and humanitarian response and their duties and responsibilities. These can take the form of training workshops, computer-based training, or simulation exercises. However, as noted in one qualitative study, disaster relief staff reported that while they received practical training in case of emergencies, there is relatively less emphasis in trainings on the psychological impact of disasters and how to identify and take appropriate action to ameliorate distress (Brooks et al., 2019).

There is growing consensus that predeployment training should devote greater attention to psychoeducation about the common mental health concerns during humanitarian response and resources for further support and specialized care (Umeda et al., 2020). Especially since an avoidant coping style has been significantly associated with a higher risk of PTSD, anxiety, and depression among humanitarian aid workers (Eriksson, Vande Kemp, Gorsuch, Hoke, & Foy, 2001), predeployment training can also emphasize individual coping mechanisms and stress management during a humanitarian response.

Brooks' team's (2018) review found that aid workers reported increased confidence and self-efficacy in responding to disasters after receiving psychoeducational and skills training. A smaller set of studies demonstrated that predisaster training and preparation might translate to better coping in humanitarian response and psychological outcomes. For example, a dose-finding study of an online pandemic-preparation training course with hospital workers incorporating adaptive coping strategies found that the intervention led to pre- and postimprovements for those who initially did not utilize positive coping strategies (Maunder et al., 2010). This group of participants reported more problem-solving coping and social support seeking and were less likely to use avoidance-focused coping strategies after the course (Maunder et al., 2010). Finally, Okanoya et al. (2015) found that Japanese participants from disaster volunteer dispatch organizations who received predeparture psychoeducation on stress management were less likely to be at risk for PTSD than the nonintervention group postdeployment. These findings together show the potential applicability, effectiveness, and benefits of predeployment training and preparedness.

Peri-deployment mental health monitoring and support

Manager-specific training

Managerial support and prioritization of psychological support in teams and organizations can positively influence employee mental health (Dollard, Dormann, & Idris, 2019). Thematic analysis of interviews with aid workers in UN-partner organizations showed that managers were regarded as an "incredible source of support" in helping staff deal with work stress and overload (Jachens et al., 2018). Thus, to improve the quality of management and reduce workplace psychosocial stressors, specialized training can be provided to managers on good management principles and skills, such as project planning, time management, staff motivation, assessing, and providing validation and feedback.

Psychoeducational training can also be provided to managers to promote psychological wellbeing among the team, recognize warning signs of psychological stress, provide psychologically-minded supervision, and encourage self-care and help-seeking behaviors among staff. There is currently no systematic evidence supporting the effectiveness of manager training in humanitarian aid organizations, except for a mental health training program with Australia-based first responders (Milligan-Saville et al., 2017). Managers learned about common mental health concerns among staff and effective communication about mental health issues with their team. Results showed that managers in the intervention group experienced increased confidence and mental health literacy and improved attitudes and beliefs about the manager's role in recovery. Notably, there were also reduced rates of work-related sick leave in their employees, with effects retained 6 months postintervention. Compared with managers in the control group, managers in the intervention group applied their training and were more likely to have reached out and followed up with an employee suffering from stress or mental health problems. Although not in the humanitarian context, this study provides promising evidence that brief psychosocial and psychoeducational training for managers can positively impact employee wellbeing and the organization. Future development and assessment of the effectiveness of manager training programs are urgently needed.

Enhancing social support and team cohesion

Social support has been consistently confirmed as a protective factor against adverse mental health outcomes among humanitarian aid workers (Brooks et al., 2015; Lopes Cardozo et al., 2012). In organizations, perceived organizational support and team cohesion were also protective factors. Perceived organizational support was associated with lower perceived stress and greater mental wellbeing among humanitarian volunteers from the Sudanese Red Crescent Society, with evidence supporting reductions in feelings of helplessness and greater self-efficacy (Aldamann et al., 2019). Similarly, among humanitarian aid workers in South Sudan (Strohmeier, Scholte, & Ager, 2018) and national staff in Northern Uganda (Ager et al., 2012), higher team cohesion with co-workers was associated with better mental health outcomes and decreased burnout. The evidence for the protective effect of social support and team cohesion makes a strong case for strategies that enhance social support among staff.

Building social support and group cohesion into the procedural response to disasters by ensuring that staff remain together as a unit and support one another throughout the deployment. One such program is the Trauma Risk Management (TRiM) peer support system, which was first developed for the United Kingdom Armed Forces (Greenberg, Langston, & Jones, 2008) and used in other organizations with trauma-exposed personnel (Whybrow, Jones, & Greenberg, 2015). TRiM is a comprehensive peer-led approach that includes psychoeducational briefings, time for decompression, and follow-up after potentially traumatic events to detect colleagues with persistent psychological symptoms and encourage them to seek timely help. A review found that TRiM positively affected organizational functioning, such as reductions in sickness absence rates and disciplinary problems (Whybrow et al., 2015). UNHCR, for instance, is replicating a modified version of this program as an effective and cost-efficient approach to enhancing psychological support and reducing barriers to care for humanitarian aid workers via a formal peer support program (i.e., UNHCR Peer Support Personnel Network). Further research is needed to replicate and assess the acceptability and effectiveness of adapted peer support programs in humanitarian aid organizations.

Perideployment crisis intervention and psychological treatment

Mixed evidence for critical incident stress debriefing

As discussed, humanitarian work and settings put staff at a higher risk of exposure to critical incidents that are potentially traumatic and can overwhelm a person's coping capacity. Many organizations continue to practice critical incident stress debriefing, although the evidence for its effectiveness is mixed (Wessely, Rose, & Bisson, 2000). Critical incident stress debriefings are formal, structured procedures that usually occur 24–48 h following the disaster duty or critical incident and are carried out by mental health professionals (Mitchell, 1983). Debriefings can employ a psychoeducational approach or have therapeutic functions to prevent the development of posttraumatic stress response and symptoms. The psychoeducational debriefing provides staff with knowledge about common psychological and behavioral reactions following a critical incident, adaptive coping skills for self-management, and resources for appropriate mental health care.

Psychological debriefing may also comprise a component of emotional processing of feelings about the event and disclosure of signs or symptoms of a stress response in an individual or group setting. Several studies have found that debriefing intervention postdisaster reduced anxiety, depression, and PTSD symptoms (Tehrani, Walpole, Berriman, & Reilly, 2001). However, other studies have shown that debriefing can be unhelpful and even have negative consequences. For example, a controlled trial with emergency services personnel and disaster workers found that participants receiving group stress debriefing reported significantly higher psychological distress scores and showed less improvement in PTSD symptoms than those who were not debriefed after an earthquake in Australia (Wessely et al., 2000). The lack of standardization of debriefing protocols, including when it was carried out and for how many sessions, may contribute to the variable results found for debriefing effectiveness (Brooks et al., 2018). Given the mixed evidence and inconsistency in the literature on debriefing protocols and implementations, the National Institute for Health and Care Excellence (NICE, 2005), therefore, do not recommend the use of individual debriefing sessions focused on the trauma experience. Other psychosocial and psychological interventions may thus be more appropriate as perideployment crisis interventions.

Evidence-based psychotherapy interventions

Improving availability and access to quality, evidence-based psychological and psychosocial support services was the most frequent recommendation to improve organizational support and address mental health issues during deployment (UN, 2018). Psychotherapy interventions can include in-house and external, in-person, or virtual mental health services that protect employee confidentiality.

Psychosocial and psychological interventions can potentially be delivered in a stepped-care model, where more intensive and specialized treatments are reserved for people who do not benefit from first-line treatments. See Fig. 4 for recommended practice and procedures of a stepped-care model of crisis intervention and psychological support for those affected by critical incidents or potentially traumatic events during deployment.

Psychological First Aid (PFA; Pekevski, 2013) can be a first line of psychosocial support for employees who have experienced a potentially traumatic event or exhibit early signs of psychological distress, although more empirical studies evaluating its effectiveness are needed. Managers and staff can be easily trained in PFA to support their colleagues and aid in the early identification and intervention of psychological problems. Another example of an intervention that provided training and encouraged team cohesion and stress management is the "512 Psychological Intervention Model," delivered to military rescuers following an earthquake in China (Wu et al., 2012). The intervention group had significantly lower PTSD symptom scores and higher positive efficacy scores than the control group and the group that only received debriefing (Wu et al., 2012).

For aid workers who are already experiencing clinical symptoms of mental disorders, more intensive evidence-based psychological interventions, including cognitive-behavioral therapy (CBT), can be effective. In an intervention study with disaster workers who met the criteria for PTSD after the September 11th attacks in the United States, CBT was effective in decreasing PTSD scores compared to the treatment-as-usual group, where one-fifth had an increase in PTSD scores (Difede et al., 2007). Another randomized controlled trial tested the efficacy of a trauma-focused CBT for emergency personnel experiencing PTSD symptoms. They found that participants on the waitlist condition had smaller reductions in PTSD severity, depression scores, maladaptive appraisals about oneself and the world, and more minor improvements on psychological and social quality of life than the intervention condition at posttreatment and 6-month follow-up (Bryant et al., 2019). The treatment consisted of skills training to address comorbid problems commonly experienced by emergency responders and cognitive restructuring to restructure maladaptive cognitions. The study also showed that treatment with a brief 10-min in vivo imaginal exposure was as efficacious as treatment with 40-min prolonged in vivo imaginal exposure (Bryant et al., 2019). These results show that a shorter intervention with more cost savings can be equally effective.

Finally, a case study illustrated the delivery of a time-limited, group-based, Mindfulness-Based Cognitive Therapy (MBCT) targeted at burnout prevention and management for 15 humanitarian professions working in East Jerusalem and West Bank (Pigni, 2014). The course addressed burnout and its symptoms by enhancing more present and mindful awareness of their physical sensations, thoughts, and emotions as an instrument to notice burnout and their self-care needs. By suggesting that burnout stems from the organizational culture of overwork, Pigni (2014: 239) advocated for humanitarian organizations to foster "organizational mindfulness," where the work environment provides "time for reflection, learning, and care."

Other mental health and psychosocial interventions that have been effective for vulnerable persons in humanitarian contexts can also be adapted for humanitarian aid workers as they experience similar contextual stressors. For example, integrative adapt therapy (IAT; Tay & Silove, 2017) and its domains addressing safety and security, losses and separation from loved ones, access to justice, and existential meaning, may be appropriate for humanitarian staff working in postconflict settings. Interpersonal psychotherapy (IPT; Verdeli et al., 2003) can also be adapted for treating staff experiencing interpersonal stressors such as grief, role transitions, interpersonal conflicts, and social isolation.

These studies together suggest that evidence-based psychological treatments can be effective and applicable also to humanitarian aid workers. More studies are needed to test the effectiveness of various psychological interventions with this population group specifically and address other ongoing comorbid mental illnesses (e.g., depression, anxiety) other than traumatic stress. Dismantling designs can also advance our knowledge about the utility of CBT or other treatments to know which components and dosage are essential to optimally reducing clinical symptoms in this occupational group.

Postdeployment resilience-building and posttraumatic growth

Psychosocial support also needs to extend to postdeployment, when psychological distress and posttraumatic stress symptoms are likely to manifest (Lopes Cardozo et al., 2012). Resilience—most commonly understood as positive

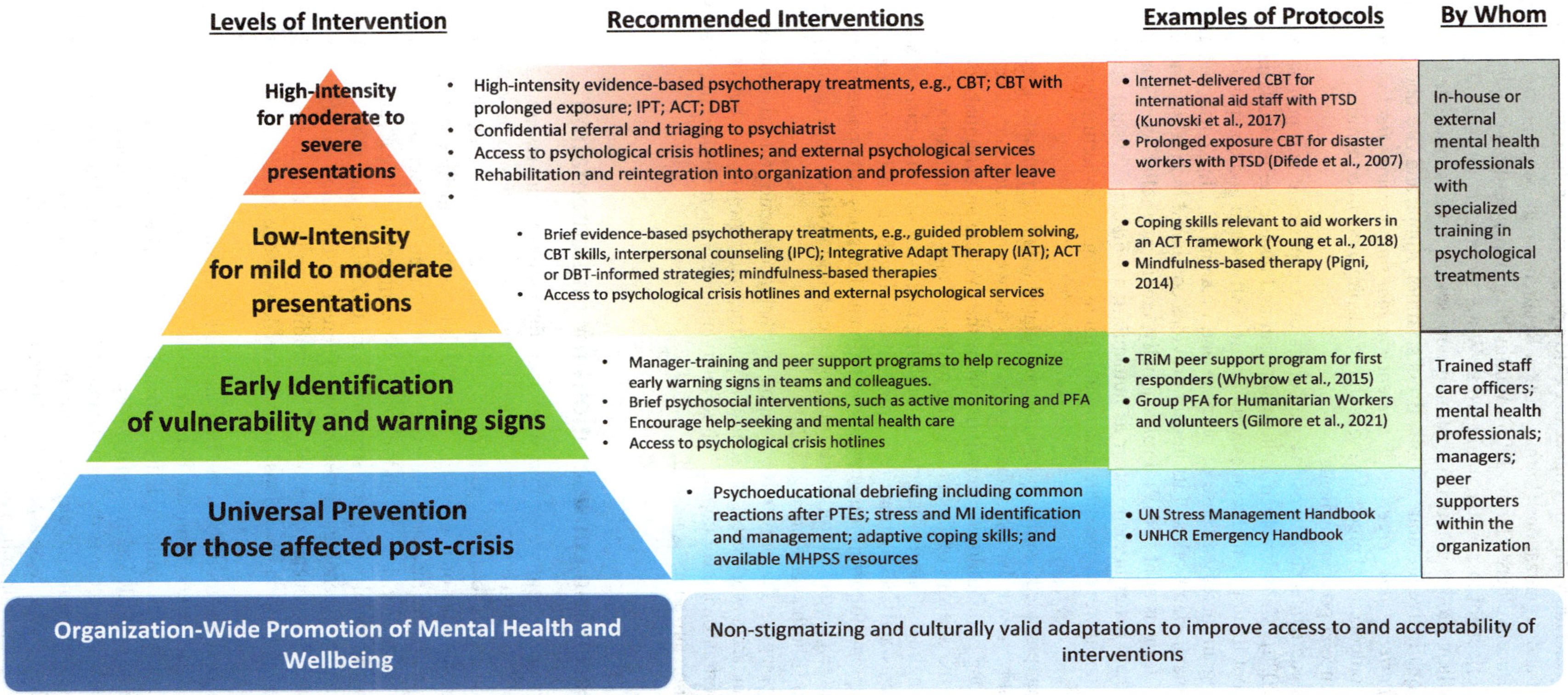

Levels of Intervention	Recommended Interventions	Examples of Protocols	By Whom
High-Intensity for moderate to severe presentations	• High-intensity evidence-based psychotherapy treatments, e.g., CBT; CBT with prolonged exposure; IPT; ACT; DBT • Confidential referral and triaging to psychiatrist • Access to psychological crisis hotlines; and external psychological services • Rehabilitation and reintegration into organization and profession after leave	• Internet-delivered CBT for international aid staff with PTSD (Kunovski et al., 2017) • Prolonged exposure CBT for disaster workers with PTSD (Difede et al., 2007)	In-house or external mental health professionals with specialized training in psychological treatments
Low-Intensity for mild to moderate presentations	• Brief evidence-based psychotherapy treatments, e.g., guided problem solving, CBT skills, interpersonal counseling (IPC); Integrative Adapt Therapy (IAT); ACT or DBT-informed strategies; mindfulness-based therapies • Access to psychological crisis hotlines and external psychological services	• Coping skills relevant to aid workers in an ACT framework (Young et al., 2018) • Mindfulness-based therapy (Pigni, 2014)	Trained staff care officers; mental health professionals; managers; peer supporters within the organization
Early Identification of vulnerability and warning signs	• Manager-training and peer support programs to help recognize early warning signs in teams and colleagues. • Brief psychosocial interventions, such as active monitoring and PFA • Encourage help-seeking and mental health care • Access to psychological crisis hotlines	• TRIM peer support program for first responders (Whybrow et al., 2015) • Group PFA for Humanitarian Workers and volunteers (Gilmore et al., 2021)	
Universal Prevention for those affected post-crisis	• Psychoeducational debriefing including common reactions after PTEs; stress and MI identification and management; adaptive coping skills; and available MHPSS resources	• UN Stress Management Handbook • UNHCR Emergency Handbook	
Organization-Wide Promotion of Mental Health and Wellbeing	Non-stigmatizing and culturally valid adaptations to improve access to and acceptability of interventions		

FIG. 4 Recommendations for a stepped-care model of crisis intervention and psychological support during deployment.

adaptation despite adversity and the ability to "bounce back" to a pretrauma baseline after crisis—is an essential capacity to build in humanitarian staff and organizations (Luthar, 2006). Resilience may be more of a norm than trauma following a disaster (Bonanno, Galea, Bucciarelli, & Vlahov, 2006). Within this group of resilient people, we can find protective factors against the mental health effects of humanitarian aid work. While there have not been moderation or mediation studies comparing at-risk and resilient groups in this occupation group, several factors associated with psychological resilience have been identified. In a qualitative literature review, Brooks, Amlôt, Rubin, and Greenberg (2020) identified three resilience factors: *training, experience, and perceived competence; social support;* and *effective coping strategies*. While these factors were addressed in components of predeployment and perideployment interventions, there is less research on optimal approaches to build resilience at postdeployment.

There has also been some evidence on positive outcomes on the personal and professional level following trauma exposure (Brooks et al., 2020). Many involved in humanitarian and disaster relief work have reported that feeling they have made a significant contribution, which comes with greater self-esteem and compassion, feeling more connected to the community, and having an increased sense of purpose and meaning in life (Zinsli & Smythe, 2009). The personal meaning derived from humanitarian work and its impact provides continued motivation and endurance for the hardships that come with the work (Jachens et al., 2019). Professionally, emergency relief work can also strengthen professional competency and prepare them better in their career as humanitarian workers (Bhushan & Kumar, 2012). In addition to considering protective factors in preventive strategies, organizations should also nurture posttraumatic growth and positive outcomes.

More recent research has provided insights into factors associated with posttraumatic growth to identify how to build it. For example, posttraumatic growth buffered the associations between inadequate disaster recovery and PTSD and depressive symptoms among civilian disaster survivors from the Wenchuan earthquake in China (Fu, Guo, Zhang, & Hall, 2021). In particular, the domains of "better relating to others" and "identification of new possibilities" carried this mediating effect, suggesting that strategies targeted at building interpersonal relationships and improving the ability to appraise social environments and seek options (e.g., problem-solving focused interventions) can be helpful to protect against psychological symptoms following a traumatic event. In another study with posttsunami relief volunteers in India, proactive coping strategies (e.g., goal setting and turning obstacles into positive experiences) were positively correlated with posttraumatic growth (Bhushan & Kumar, 2012).

Together, these findings suggest that preventive components that incorporate protective factors will impact building and ensuring a resilient humanitarian workforce. In addition, posttraumatic growth can be fostered by strategies aimed at strengthening social support and adaptive coping. More intervention research on the effectiveness of these strategies or programs in humanitarian organizations is needed.

Limitations of research and considerations for implementation

Despite the promising evidence, there are several significant limitations to the research base that warrant elaboration. Overall, research on the efficacy and effectiveness of psychosocial interventions to prevent and ameliorate occupational stress and mental health outcomes for this population is still limited due to the lack of randomized controlled trials or effectiveness-implementation studies on many of the suggested interventions in this chapter. The research also suffers from a lack of uniform and comprehensive measures that extend beyond conventional common mental disorders to include occupational stress-related conditions (e.g., burnout and substance use); impact on job (e.g., absenteeism, days out of role, leave of absence); indices of psychosocial workplace stressors (e.g., COPSOQ); and protective factors (e.g., organizational support, social support, coping, psychological flexibility). There is also a large variability and lack of standardization of intervention protocols and components. The lack of longitudinal studies also means that there is no evidence on the potential long-term effects of these interventions and whether knowledge and skills from predeployment training are retained and practiced on the job.

More importantly, less is understood about the translation of research to practice as there is a lack of effectiveness-implementation studies (Curran, Bauer, Mittman, Pyne, & Stetler, 2012). Implementation outcomes such as acceptability, adoption, appropriateness, costs, feasibility, fidelity, penetration, and sustainability of interventions and comprehensive MHPSS programs need to be evaluated at the consumer, provider, and organizational levels (Proctor et al., 2011). Monitoring and evaluation should also occur at the organizational level to ensure that organizations implement and meet the indicators of evidence-based recommendations and guidelines of staff support.

The following subsections elaborate on aspects that should be considered to ensure more successful implementation of evidence-based interventions into humanitarian organizations.

Reducing barriers to help-seeking and stigma

Previous research has illustrated that primary barriers to help-seeking include the availability, knowledge, and accessibility of a range of formal and informal MHPSS resources (UN, 2018). In addition, there are high attitudinal barriers to help-seeking, including self-stigma and external stigma toward acknowledging and disclosing a mental health problem (Cockcroft-McKay & Eiroa-Orosa, 2020). A thematic analysis of interviews with humanitarian aid trainees in Australia found that help-seeking is a "good idea that is not implemented" (Skeoch et al., 2017, p. 6) as the majority of the participants endorsed a perceived need for self-reliance. Attitudes of "martyrdom" are also common, where the organizational climate may implicitly encourage staff to sacrifice their wellbeing because of the greater needs of their beneficiaries and their work (Cockcroft-McKay & Eiroa-Orosa, 2020; Skeoch et al., 2017). This attitude may limit the staff's ability to acknowledge their psychological and emotional needs and reduce help-seeking behavior (Cockcroft-McKay & Eiroa-Orosa, 2020). The UN (2018) workplace mental health strategy for UN staff thus outlined plans for tackling stigma and lowering barriers to help-seeking, which includes: improving knowledge and quality of in-house resources; ensuring confidentiality and privacy of utilization of services; including a range of informal psychosocial support and external psychological services for staff to access; and regular promotion of mental wellness and communication of mental health among teams.

Applications and adaptations to vulnerable groups and cultural contexts

Most of the interventions reviewed focused on a general sample of disaster relief personnel or humanitarian workers and not on the particular vulnerable demographic groups, such as international workers, female aid workers, and field workers, who may have unique stressors and risk factors. For example, national staff is five times more likely to be victims of violence than international workers (Stoddard, 2020). National staff also report financial concerns as a chronic stressor as they are paid less than their expatriate counterparts and are often provided with less comprehensive employment benefits (Ager et al., 2012). Not surprisingly, they often cite injustice and tensions resulting from disparity of treatment between international and national staff as a frequent stressor (Ager et al., 2012). In contrast, international staff cite more struggles with acculturation and lack of social support (Antares Foundation, 2012). Finally, female aid workers are at a higher risk of burnout and mental illnesses than their male colleagues (Jachens et al., 2019; UNHCR, 2016). Especially for working mothers, this may be due to their greater involvement in the household which can create additional stress and challenges maintaining work-life balance. In addition, female aid workers also experience a high incidence of gender-based violence in the workplace (Stoddard, Harvey, Czwarno, & Brekenridge, 2019). This evidence for unique stressors for specific demographic groups among humanitarian aid workers supports a case for adapted and targeted interventions that address specific needs.

Furthermore, adaptations are also required for the intervention to be culturally valid and contextually sensitive. For example, De Fouchier and Kedia (2018) discussed integrating participants' conceptual model of mental health associated with spirit and witchcraft beliefs with explanations of stress reactions in an effective psychoeducational stress-management training for national humanitarian workers in the Central African Republic. The cultural adaptation of interventions will improve treatment acceptability and may improve engagement and responsiveness to treatment. Beyond the cultural context, different types of humanitarian organizations will also have various bureaucracies and systems and organizational cultures and norms, which will also be another contextual factor to consider. Hence, MHPSS interventions or programs need to also be adapted for specific cultural contexts and organizations.

Cost-effectiveness and funding

An important consideration and caveat is the limited funding and financial resources dedicated to staff welfare and psychosocial support in humanitarian aid organizations. Organizations must also ensure that staff support funding mechanisms are sustainable (e.g., factoring staff support into overhead costs of projects) and strengthening the role of donors with evidence-based lobbying (Strohmeier et al., 2019). In parallel, more cost-effectiveness analysis and research on briefer, scalable MHPSS interventions that can be task-shifted and delivered by nonspecialists (e.g., peer support programs) are needed.

Conclusions

It is our "duty of care" to the psychological safety of humanitarian and disaster relief staff who risk their lives in the face of humanitarian crises to serve and rebuild communities. While there is accumulating research on the prevalence of and risk factors influencing occupational stress and mental disorders among humanitarian aid and disaster relief personnel, less is known about empirically supported psychosocial interventions for this occupation group.

Notwithstanding the paucity of high quality randomized controlled trials and effectiveness-implementation studies of interventions with control groups, the summary of evidence provided in this chapter outlines a comprehensive and systematic framework for the MHPSS of humanitarian and disaster relief personnel that is not simply trauma-focused or reactive. Instead, the chapter proposed a broad spectrum of interventions (i.e., mental health promotion, prevention, early identification and intervention, treatment, and rehabilitation) informed by risk and protective factors from the psychosocial workplace environment unique to humanitarian and emergency relief work. These multilevel interventions are targeted at individual staff, managers, team, and organization-level policies and implemented at various deployment and emergency response phases.

At the organizational level, instituting *organizational MHPSS policy and standards of practice* are fundamental to protecting staff's physical and psychological safety, promoting organizational wellbeing, and systematically identifying and ameliorating workplace psychosocial stressors. By critically examining the evidence, we raised concerns about utilizing preemployment/predeployment psychological screening of staff, and critical incident stress debriefing, due to the possible iatrogenic effects for both interventions. Instead, following the promising evidence on protective and resilience factors and effective interventions, we suggested that *predeployment training* focus on psychoeducation and psychological preparedness (e.g., identification of warning signs of stress, stress management, adaptive coping skills), and that *perideployment mental health monitoring and support* focus on enhancing social support and team cohesion. *Manager-specific training* and *peer support programs* are cost-efficient and effective ways to build capacity within the organization to recognize mental health symptoms, provide supportive listening, and encourage help-seeking and mental health care among teams and colleagues. We recommended a stepped-care crisis intervention and psychological treatment model for at-risk staff exposed to potentially traumatic events or experience symptoms of burnout or mental disorders. In this model, psychoeducation and PFA can be the first line of treatment, and more intensive psychological treatments can be reserved for higher-risk staff (e.g., CBT; mindfulness-based therapies; ACT; IPT; IAT). Finally, more research on *building resilience and promoting posttraumatic growth* is recommended, such that organizations can help staff capitalize on resilience and motivational factors. Organizations must also consider implementation barriers and facilitators (e.g., stigma-related barriers to help-seeking, funding, and staff capacity) unique to the organization and cultural context, and consequently make appropriate adaptations to the interventions to ensure their successful implementation.

This evidence-informed MHPSS intervention framework can serve as a blueprint for implementing a comprehensive staff care program that promotes and protects the psychological health of humanitarian aid and disaster relief staff, hence ensuring the quality and sustainability of their essential work.

Mini-dictionary of terms

Burnout. An occupational syndrome recognized in the ICD-11 as the outcome of chronic workplace stress that has not been successfully managed. It is characterized by feelings of energy depletion or exhaustion, increased mental distance from one's job or feelings of negativism related to one's job, and reduced professional efficacy.

Copenhagen Psychosocial Questionnaire (COPSOQ). A comprehensive instrument for the psychosocial risk assessment of workplace stressors. It includes main dimensions of stressors from influential occupational stress theories, including Job-Strain, Demand-Control-Support, and Effort-Reward-Imbalance models and other aspects neglected in previous tools, including emotional demands and role clarity.

Implementation Science. A growing scientific field studying ways to promote the systematic uptake of evidence-based interventions into routine practice in real-world settings, to improve the quality and effectiveness of health services.

Mental health and psychosocial support. A broad term that refers to the need for diverse and complementary approaches that seek to promote psychosocial wellbeing and prevent, detect, mitigate, and ameliorate affected populations' mental health and psychosocial problems.

Psychological First Aid. An evidence-informed, first-line supportive intervention often used in the immediate aftermath of disasters for disaster-exposed populations. It aims to reduce the initial distress caused by potentially traumatic events and foster short- and long-term adaptive functioning and coping by providing psychoeducation, needs identification, enhancing adaptive coping, practical assistance, and connection with social supports and collaborative services.

Workplace psychosocial stressors. Refer to the properties and characteristics related to the work demands, organization, management and planning of work, work environment, work community, and interaction that strain employees.

Applications to other areas A critical area for future research is the translation of these findings into the development and evaluation of mental health resources and interventions in aid organizations. The stepped-care model of intervention developed for aid workers can also be adapted and tested in other high-stress, human-service occupations, such as emergency services, medicine, military personnel, nursing, clinical and social work, and teaching.

Key facts of burnout in humanitarian personnel

- There is a high rate of burnout among aid workers, comparable to that reported by US physicians.
- Cross-sectional studies found point prevalence rates of 24%–45% for emotional exhaustion, 9%–24% for depersonalization, and 10%–43% for low personal achievement, based on threshold scores on the Maslach Burnout Inventory.
- The caseness of burnout (i.e., scores above cut-offs on all three subscales) is surprisingly low, with rates of 1%–5%, suggesting that most aid workers can still stay resilient.
- Volunteer status, previous history of mental illness, and previous life stressors are the only reliable demographic variables consistently associated with poorer mental health in this population.
- Workplace psychosocial stressors are gaining attention as primary sources of stress with long-term effects, compared to acute impact of common aversities and potentially traumatic events
- Higher effort-reward-imbalance and lower perceived organizational support have been shown to increase the risk of burnout in aid workers.
- Resilience factors consistently found include training and psychological preparedness, social support, and adaptive coping skills.

Summary points

- This chapter proposed a comprehensive, systematic, multilevel framework for the MHPSS of humanitarian aid and disaster responders' occupational stress-related mental health problems.
- This chapter recommended a broad spectrum of interventions (i.e., mental health promotion, prevention, early identification and intervention, treatment, and rehabilitation) across staff, managerial, team, and organization levels at various deployment phases.
- Interventions are informed by mitigating risk factors (e.g., trauma exposure, workplace psychosocial stressors) and enhancing protective factors (e.g., social support, team cohesion) unique to this workforce.
- Preemployment screening for demographic risk variables and critical incident stress debriefing have mixed evidence for their effectiveness and are not recommended.
- Predeployment psychoeducational training (e.g., stress management and adaptive coping skills) can improve psychological preparedness and confidence.
- Manager training and peer support programs are cost-efficient and effective ways to build early identification and intervention capacity.
- A stepped-care psychological treatment model is recommended for postcrisis intervention, where psychoeducation and PFA are the first line of treatment, followed by more intensive psychological therapies for higher-risk staff.
- Consider reducing attitudinal barriers to help-seeking and making adaptations for specific demographic groups (e.g., female, international, national staff) to improve acceptability and use of interventions.
- More randomized controlled trials, longitudinal, and effectiveness-implementation studies are needed to understand the efficacy, long-term effects, and successful adoption and implementation of interventions in organizations' staff care policies and practices.

References

Ager, A., Pasha, E., Yu, G., Duke, T., Eriksson, C., & Lopes Cardozo, B. (2012). Stress, mental health, and burnout in national humanitarian aid workers in Gulu, Northern Uganda. *Journal of Traumatic Stress, 25*(6), 713–720.

Aldamann, K., Tamrakar, T., Dinesen, C., Wiedemann, N., Murphy, J., Hansen, M., et al. (2019). Caring for the mental health of humanitarian volunteers in traumatic contexts: The importance of organizational support. *European Journal of Psychotraumatology, 10*(1), 1694811.

Antares Foundation. (2012). *Managing stress in humanitarian workers: Guidelines for good practice*. Amsterdam: Antares Foundation.

Bhushan, B., & Kumar, J. S. (2012). A study of post-traumatic stress and growth in Tsunami relief volunteers. *Journal of Loss and Trauma, 17*, 113–124.

Bonanno, G. A., Galea, S., Bucciarelli, A., & Vlahov, D. (2006). Psychological resilience after disaster: New York City in the aftermath of the September 11th terrorist attack. *Psychological Science, 17*(3), 181–186.

Brooks, S. K., Dunn, R., Sage, C. A., Amlôt, R., Greenberg, N., & Rubin, G. J. (2015). *Risk and resilience factors affecting the psychological wellbeing of individuals deployed in humanitarian relief roles after a disaster*. BMC Psychology.

Brooks, S. K., Dunn, R., Amlôt, R., Greenberg, N., & Rubin, G. J. (2016). Social and occupational factors associated with psychological distress and disorder among disaster responders: A systematic review. *BMC Psychology, 4*, 18.

Brooks, S. K., Dunn, R., Amlôt, R., Greenberg, N., & Rubin, G. J. (2018). Training and post-disaster interventions for the psychological impacts on disaster-exposed employees: A systematic review. *Journal of Mental Health*, 1–25.

Brooks, S. K., Dunn, R., Amlôt, R., Rubin, G. J., & Greenberg, N. (2019). Protecting the psychological wellbeing of staff exposed to disaster or emergency at work: A qualitative study. *BMC Psychology, 7*, 78.

Brooks, S., Amlôt, R., Rubin, G. J., & Greenberg, N. (2020). Psychological resilience and post-traumatic growth in disaster-exposed organizations: Overview of the literature. *BMJ Military Health, 166*(1), 52–56.

Bryant, R. A., Kenny, L., Rawson, N., Cahill, C., Joscelyne, A., Garber, B., et al. (2019). Efficacy of exposure-based cognitive behavior therapy for post-traumatic stress disorder in emergency service personnel: A randomized clinical trial. *Psychological Medicine, 49*(9), 1565–1573.

Burr, H., Berthelsen, H., Moncada, S., Nübling, M., Dupret, E., Demiral, Y., et al. (2019). The third version of the Copenhagen Psychosocial Questionnaire. *Safety and Health at Work, 10*(4), 482–503.

Cockcroft-McKay, C., & Eiroa-Orosa, F. J. (2020). Barriers to accessing psychosocial support for humanitarian aid workers: A mixed methods enquiry. *Disasters, 45*(4), 762–796.

Cox, T., & Griffiths, A. (2005). The nature and measurement of work-related stress: Theory and practice. In J. R. Wilson, & N. Corlett (Eds.), *Evaluation of human work* (3rd ed.). Abingdon, UK: Routledge.

Curran, G. M., Bauer, M., Mittman, B., Pyne, J. M., & Stetler, C. (2012). Effectiveness-implementation hybrid designs: Combining elements of clinical effectiveness and implementation research to enhance public health impact. *Medical Care, 50*(3), 217–226.

De Fouchier, C., & Kedia, M. (2018). Trauma-related mental health problems and effectiveness of a stress management group in national humanitarian workers in the Central African Republic. *Intervention, 16*, 103–109.

Difede, J., Malta, L. S., Best, S., Henn-Haase, C., Metzler, T., Bryant, R., et al. (2007). A randomized controlled clinical treatment trial for world trade center attack-related PTSD in disaster workers. *The Journal of Nervous and Mental Disease, 195*(10), 861–865.

Dollard, M. F., Dormann, C., & Idris, M. A. (2019). *Psychosocial safety climate: A new work stress theory.* Switzerland: Springer Nature.

Eriksson, C. B., Vande Kemp, H., Gorsuch, R., Hoke, S., & Foy, D. W. (2001). Trauma exposure and PTSD symptoms in international relief and development personnel. *Journal of Traumatic Stress, 14*(1), 205–219.

Eriksson, C. B., Bjorck, J. P., Larson, L. C., Walling, S. M., Trice, G. A., Fawcett, J., et al. (2009). Social support, organizational support, and religious support in relation to burnout in expatriate humanitarian aid workers. *Mental Health, Religion and Culture, 12*(7), 671–686.

Foo, C. Y. S., Verdeli, H., Tay, A. K., & Brough, P. (2021). Humanizing work: Occupational mental health of humanitarian aid workers. In T. Wall, & C. B. E. Cary Cooper (Eds.), *The SAGE handbook of organizational wellbeing*. United Kingdom: SAGE Publications.

Fu, M., Guo, J., Zhang, Q., & Hall, B. J. (2021). Mediating role of post-traumatic growth in the relationship between inadequate disaster recovery and mental health outcomes: Long-term evidence from the Wenchuan earthquake. *European Journal of Psychotraumatology, 12*(1).

Greenberg, N., Langston, V., & Jones, N. (2008). Trauma risk management (TRiM) in the UK armed forces. *Journal of the Royal Army Medical Corps, 154*(2), 124–127.

Inter-Agency Standing Committee (IASC). (2007). *IASC guidelines on mental health and psychosocial support in emergency settings.* Geneva: IASC.

Jachens, L., Houdmont, J., & Thomas, R. (2018). Work-related stress in a humanitarian context: A qualitative investigation. *Disasters, 42*(4), 619–634.

Jachens, L., Houdmont, J., & Thomas, R. (2019). Effort-reward imbalance and burnout among humanitarian aid workers. *Disasters, 43*(1), 67–87.

Lopes Cardozo, B., Gotway Crawford, C., Eriksson, C., Zhu, J., Sabin, M., Ager, A., et al. (2012). Psychological distress, depression, anxiety, and burnout among international humanitarian aid workers: A longitudinal study. *PLoS One, 7*(9), e44948.

Luthar, S. S. (2006). Resilience in development: A synthesis of research across five decades. In D. Cicchetti, & D. J. Cohen (Eds.), *Developmental psychopathology: Risk, disorder, and adaptation* (pp. 740–795). New York: Wiley.

Marshall, R. E., Milligan-Saville, J. S., Mitchell, P. B., Bryant, R. A., & Harvey, S. B. (2017). A systematic review of the usefulness of pre-employment and pre-duty screening in predicting mental health outcomes amongst emergency workers. *Psychiatry Research, 253*, 129–137.

Maunder, R. G., Lancee, W. J., Mae, R., Vincent, L., Peladeau, N., Beduz, M. A., et al. (2010). Computer-assisted resilience training to prepare healthcare workers for pandemic influenza: A randomized trial of the optimal dose of training. *BMC Health Services Research, 10*, 72.

Milligan-Saville, J. S., Tan, L., Gayed, A., Barnes, C., Madan, I., Dobson, M., et al. (2017). Workplace mental health training for managers and its effect on sick leave in employees: A cluster randomized controlled trial. *The Lancet Psychiatry, 4*(11), 850–858.

Mitchell, J. T. (1983). When disaster strikes…the critical incident stress debriefing process. *JEMS: A Journal of Emergency Medical Services, 8*(1), 36–39.

National Institute for Health and Care Excellence (NICE). (2005). *Post-traumatic stress disorder (PTSD): The management of PTSD in adults and children in primary and secondary care.* http://www.nice.org.uk/guidance/cg26/chapter/guidance#the-treatment-of-ptsd.

Okanoya, J., Kimura, R., Mori, M., Nakamura, S., Somemura, H., Sasaki, N., et al. (2015). Psychoeducational intervention to prevent critical incident stress among disaster volunteers. *Kitasato Medical Journal, 45*, 62–68.

Opie, E., Brooks, S., Greenberg, N., & Rubin, G. J. (2020). The usefulness of pre-employment and pre-deployment psychological screening for disaster relief workers: A systematic review. *BMC Psychiatry, 20*(211), 1–13.

Othman, M., Steel, Z., Lawsin, C., & Wells, R. (2018). Addressing occupational stress among health staff in nongovernment controlled northern Syria: Supporting resilience in a dangerous workplace. *Torture: Quarterly Journal on Rehabilitation of Torture Victims and Prevention of Torture, 28*(3), 104–123.

Pekevski, J. (2013). First responders and psychological first aid. *Journal of Emergency Management (Weston, MA.), 11*(1), 39–48.

Pigni, A. (2014). Building resilience and preventing burnout among aid workers in Palestine: A personal account of mindfulness-based staff care. *Intervention: Journal of Mental Health and Psychosocial Support in Conflict Affected Areas, 12*(2), 231–239.

Proctor, E., Silmere, H., Raghavan, R., Hovmand, P., Aarons, G., Bunger, A., et al. (2011). Outcomes for implementation research: Conceptual distinctions, measurement challenges, and research agenda. *Administration and Policy in Mental Health, 38*(2), 65–76.

Shah, S. A., Garland, E., & Katz, C. (2007). Secondary traumatic stress: Prevalence in humanitarian aid workers in India. *Traumatology, 13*(1), 59–70.

Siegrist, J. (1996). Adverse health effects of high-effort/low-reward conditions. *Journal of Occupational Health Psychology, 1*(1), 27–41.

Skeoch, K., Stevens, G., & Taylor, M. (2017). Future role aspirations, achievement motivations and perceptions of personal help-seeking among humanitarian aid trainees. *Journal of International Humanitarian Action, 2*, 12.

Stoddard, A., Harvey, P., Czwarno, M., & Brekenridge, M. (2019). *Aid worker security report 2019: Addressing sexual violence and gender-based risk in humanitarian aid.* Humanitarian Outcomes. https://www.humanitarianoutcomes.org/sites/default/files/publications/awsr_2019_0.pdf.

Stoddard, A. (2020). *Necessary risks: Professional humanitarianism and violence against aid workers.* Palgrave Macmillan.

Strohmeier, H., & Scholte, W. F. (2015). Trauma-related mental health problems among national humanitarian staff: A systematic review of the literature. *European Journal of Psychotraumatology, 6*, 28541.

Strohmeier, H, Scholte, W. F., & Ager, A. (2018). Factors associated with common mental health problems of humanitarian workers in South Sudan. *PLoS ONE, 13*(10), e0205333.

Strohmeier, H., Scholte, W. F., & Ager, A. (2019). How to improve organisational staff support? Suggestions from humanitarian workers in South Sudan. *Intervention: Journal of Mental Health and Psychosocial Support in Conflict Affected Areas, 17*(1), 40–49.

Tay, A. K., & Silove, D. (2017). The ADAPT model: Bridging the gap between psychosocial and individual responses to mass violence and refugee trauma. *Epidemiology and Psychiatric Sciences, 26*(2), 142–145.

Tedeschi, R. G., & Calhoun, L. G. (1996). The posttraumatic growth inventory: Measuring the positive legacy of trauma. *Journal of Traumatic Stress, 9*(3), 455–471.

Tehrani, N., Walpole, O., Berriman, J., & Reilly, J. (2001). A special courage: Dealing with the Paddington rail crash. *Society of Occupational Medicine, 51*(2), 93–99.

World Health Organization (WHO). (2019). *International classification of diseases for mortality and morbidity statistics.* 11th revision https://icd.who.int/browse11/l-m/en.

Umeda, M., Chiba, R., Sasaki, M., Agustini, E. N., & Mashino, S. (2020). A literature review on psychosocial support for disaster responders: Qualitative synthesis with recommended actions for protecting and promoting the mental health of responders. *International Journal of Environmental Research and Public Health, 17*(6), 2011.

United Nations (UN). (2018). *A healthy workforce for a better world: United Nations system of mental health and wellbeing strategy.* Geneva: UN.

United Nations High Commissioner for Refugees (UNHCR). (2016). *Staff wellbeing and mental health in UNHCR.* Geneva: UNHCR.

Verdeli, H., Clougherty, K., Bolton, P., Speelman, L., Lincoln, N., Bass, J., et al. (2003). Adapting group interpersonal psychotherapy for a developing country: Experience in rural Uganda. *World Psychiatry: Official Journal of the World Psychiatric Association (WPA), 2*(2), 114–120.

Welton-Mitchell, C. E. (2013). *UNHCR's mental health and psychosocial support.* Geneva: UNHCR.

Wessely, S., Rose, S., & Bisson, J. (2000). Brief psychological interventions ("debriefing") for trauma-related symptoms and the prevention of post-traumatic stress disorder. *The Cochrane Database of Systematic Reviews*, (2), CD000560.

Whybrow, D., Jones, N., & Greenberg, N. (2015). Promoting organizational wellbeing a comprehensive review of trauma risk management. *Occupational Medicine (Oxford, England), 65*(4), 331–336.

Wu, S., Zhu, X., Zhang, Y., Liang, J., Liu, X., Yang, Y., et al. (2012). A new psychological intervention: "512 psychological intervention model" used for military rescuers in Wenchuan earthquake in China. *Social Psychiatry and Psychiatric Epidemiology, 47*, 1111–1119.

Young, T. K. H., Pakenham, K. I., & Norwood, M. F. (2018). Thematic analysis of aid workers' stressors and coping strategies: Work, psychological, lifestyle and social dimensions. *Journal of International Humanitarian Action, 3*(19), 1–16.

Zinsli, G., & Smythe, E. A. (2009). International humanitarian nursing work: Facing difference and embracing sameness. *Journal of Transcultural Nursing: Official Journal of the Transcultural Nursing Society, 20*(2), 234–241.

Further reading

Gilmore, B., Corey, J., Vallieres, F., Aldamman, K., Frawley, T., & Davidson, S. (2021). Group psychological first aid of humanitarian workers and volunteers. In *Wellcome 2020 Workplace Mental Health Commission.*

Kunovski, I., Donker, T., Driessen, E., Cuijpers, P., Andersson, G., & Sijbrandij, M. (2017). Internet-delivered cognitive behavioral therapy for posttraumatic stress disorder in international humanitarian aid workers: Study protocol. *Internet Interventions, 10*, 23–28.

Chapter 22

Social anxiety: Linking cognitive-behavioral therapy and strategies of third-generation therapies

Isabel C. Salazar[a], Stefan G. Hofmann[b], and Vicente E. Caballo[a]

[a]Department of Personality, Assessment and Psychological Treatment, Faculty of Psychology, University of Granada, Granada, Spain [b]Department of Clinical Psychology, Philipps-University Marburg, Germany, Marburg/Lahn, Germany

Abbreviations

ACT	acceptance and commitment therapy
APD	avoidant personality disorder
APA	American Psychiatric Association
BPD	borderline personality disorder
CBT	cognitive-behavioral therapy
CSISA	Clinical Semistructured Interview for Social Anxiety
DBT	dialectical behavior therapy
DSM-5	Diagnostic and Statistical Manual of mental disorders—5th edition
MBSR	mindfulness-based stress reduction
MISA	Multidimensional Intervention for Social Anxiety
REBT	rational emotive behavior therapy
SAD	social anxiety disorder
SAQ	Social Anxiety Questionnaire for adults
SST	social skills training

Introduction

Social anxiety disorder (SAD), also known as social phobia, is mainly characterized by an intense fear of being observed and negatively evaluated by others in social situations, whether performing in front of others or interacting with them. Anxiety (nervousness, overwhelm or jitters, as colloquially people call it) invariably appears in feared social encounters, and it is quite common for it to occur before the social event take place due to negative anticipations and cognitive biases of the individual. People with SAD also report the presence of a significant level of discomfort after the situation, as most continue with ruminations and negative evaluations of themselves and the situation once the social event has finished. Many people with SAD have deficits in their social behaviors or anxiety inhibits their social skills to behave appropriately, and therefore, they implement safety behaviors or escape from or avoid feared situations. This, generally, brings negative emotional consequences, such as disappointment, sadness, frustration, and anger (almost always directed toward himself), as well as impairment in his social, intimate, work, and/or home areas of functioning (Stein et al., 2017). Due to the importance of social contacts in our way of life, those who have SAD deal with the suffering of wanting to interact with others and find themselves unable to do so, and even more, their actions keep them away from reaching many of their life goals or to lead the life they would like.

☆ This study is part of the I+D+i project with reference RTI2018–093916-B-I00 funded by MCIN/AEI/10.13039/501100011033/ and FEDER "A way of doing Europe." Financial assistance from the Foundation for the Advancement of Behavioral Clinical Psychology (FUNVECA) is also acknowledged.

Social anxiety dimensions

Social situations triggering anxiety can greatly differ among people with SAD. Referring to them as individual situations does not help very much. For approximately 15 years, our group was investigating the most commonly feared social situations among the Spanish and Portuguese-speaking population, as well as with some samples from USA, and we were able to identify the multidimensional nature of social fears (Caballo et al., 2012, 2015; Caballo, Salazar, Arias, et al., 2010; Caballo, Salazar, Irurtia, et al., 2010). The five dimensions (Fig. 1) of social anxiety found were the following:

(1) *Interaction with strangers* refers to those situations in which social contact is established with little-known or unknown people, in which the skills to initiate and maintain conversations are necessary and in which it is feared to appear silly, uninteresting, give a negative impression of oneself, make them realize that you are nervous, or upsetting the other person. Examples of situations that could be included in this dimension are: "maintaining a conversation with someone I've just met" or "attending a social event where I know only one person."

(2) *Interaction with the opposite sex* includes situations in which the interaction is performed with people who physically attract us (no matter if they are same sex or not), situations requiring skills to initiate or respond to the approaches with these people and fearing to being exposed that we like her (and she don't like us), not being attractive, looking bored or stupid. Examples of situations that could be included in this dimension are: "initiating a conversation with someone of the opposite sex that I like" or "asking someone I find attractive to dance."

(3) *Assertive expression of annoyance, disgust, or displeasure* concern situations in which it is necessary to express points of view, preferences, or feelings of annoyance and for which assertive behaviors are necessary. The central fears are disapproval, anger, rejection, or abandonment by others. Examples of situations that could be included in this dimension are: "complaining to the waiter about my food" or "refusing when asked to do something I don't like doing."

(4) *Embarrassment or criticism* brings together situations in which we are the center of attention or in which our behavior could be criticized. The fears are of making a fool of yourself, appearing incompetent, and not being liked by others. Examples of situations that could be included in this dimension are: "talking to someone who isn't paying attention to what I am saying" or "being teased in public."

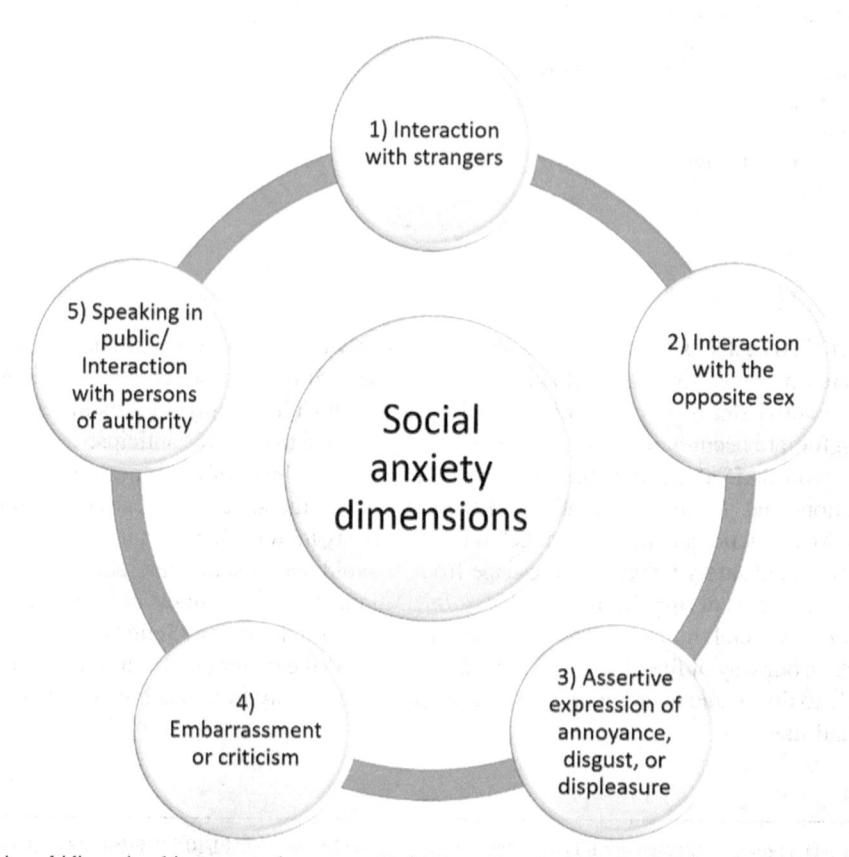

FIG. 1 Social anxiety is multidimensional in nature and a person with SAD may fear from one to the five facets or dimensions of it.

(5) *Speaking in public/Interaction with persons of authority* includes two types of situations, the first refers to when the person speaks in front of a relatively large group and the second would include contacts with authority figures (e.g., bosses, teachers, or people in a authority position). In either of these two types of situations we fear being wrong, appearing incompetent, nervous, uninteresting, and getting a negative impression of ourselves. Examples of situations that could be included in this dimension are: "being asked a question in class by the teacher or by a someone in authority at a meeting" or "while having dinner with colleagues, classmates or workmates, being asked to speak on behalf of the entire group."

Age of onset, associated factors and course

Adolescence, between 11 and 17 years of age, is the time of life most favorable to the onset of SAD, with 13–14 years of age being the most frequent ages of onset according to recent studies (e.g., Kessler, Petukhova, Sampson, Zaslavsky, & Wittchen, 2012; MacKenzie & Fowler, 2013; Sibrava et al., 2013; Stein, Lim, et al., 2017). However, it sometimes occurs in adulthood (e.g., 50% of South African cases occur by the age of 26) or before 11 years of age (e.g., Stein, Lim, et al., 2017).

The 12-month SAD *is associated with* earlier age of onset, female sex, being homemaker, being unmarried (never married, separated/divorced/widowed), lower educational status, and lower household income (Asher & Aderka, 2018; MacKenzie & Fowler, 2013; McLean, Asnaani, Litz, & Hofmann, 2011; Stein, Lim, et al., 2017). Lifetime risk of SAD is associated with earlier age of onset, female sex, being homemaker, being unmarried, low educational status, and low household income (Stein, Kawakami, de Girolamo, & Lépin, 2017; Stein, Lim, et al., 2017).

Its course is usually chronic if it does not receive psychological treatment and it seems to increase when the demand of social environment is higher. For example, there are people who, because of their work, must constantly be exposed to public performances and interactions with colleagues, bosses, and clients, but because they do not know how to cope with social situations, they endure them with great discomfort.

Prevalence and comorbidity

SAD affects the world's population unevenly. According to the latest WHO World Mental Health Surveys, there are important differences between countries (Kessler & Üstün, 2008; Stein, Lim, et al., 2017). The highest figures of SAD lifetime and in the last 12 months occur in high income countries (e.g., 12.1% and 7.1% in USA, 9.5% and 5.3% New Zealand, and 8.5% and 4.2% in Australia, respectively) and the lowest in low/lower-middle income countries (e.g., 0.2% and 0.2% in Nigeria; 0.5% and 0.4% in PRC China, and 0.8% and 0.7% in Iraq, respectively). SAD is a mental health problem that is among the first places of prevalence in each country. Taking as reference USA and Nigeria (countries with the highest and lowest prevalence), SAD ranks fourth and sixth among the 19 and 18 mental disorders evaluated, respectively (Gureje et al., 2008; Kessler et al., 2008).

Additionally, SAD is highly likely to occur *comorbidly* with another mental disorder, mainly with another anxiety disorder, a mood disorder, or a substance use disorder (MacKenzie & Fowler, 2013; Pignon et al., 2018; Quevedo et al., 2020; Schneier et al., 2010; Stein, Lim, et al., 2017).

Treatment

The most empirically supported nonpharmacological treatment for SAD is the cognitive-behavioral therapy (CBT) (e.g., Goldin et al., 2016; Kocovski, Fleming, Hawley, Huta, & Antony, 2013). More recently, the so-called "third wave behavior therapies," such as mindfulness-based stress reduction (MBSR), and acceptance and commitment therapy (ACT) have also been shown to be effective for social anxiety (e.g., Dalrymple & Herbert, 2007; Goldin et al., 2016; Goldin & Gross, 2010; Kocovski et al., 2013; Kocovski, Fleming, & Rector, 2009; Koszycki, Benger, Shlik, & Bradwejn, 2007; Liu et al., 2021; Norton, Abbott, Norberg, & Hunt, 2015; Shikatani, Antony, Kuo, & Cassin, 2014; Yadegari, Hashemiyan, & Abolmaali, 2014). These treatments are often used individually, but why not look for combined treatment approaches that could provide more benefits than individualized treatments for people with SAD?

The *Multidimensional Intervention for Social Anxiety* (MISA) program illustrates a blended approach to the treatment of SAD (Caballo, Salazar, & Garrido, 2018; Caballo, Salazar, Garrido, Irurtia, & Hofmann, 2018; Caballo, Salazar, & Hofmann, 2019). The MISA program combines mindfulness, experiential, acceptance and cognitive defusion exercises, plus education in values from ACT and MBSR, with social skill training, cognitive restructuring, exposure exercises (in session and in vivo) and psychoeducation from CBT. The main aim is that patients understand what are the factors involved in SAD, how these elements interact in maintaining their problem and, subsequently, learn to face social situations in a different way. Regarding this last aspect, the strategies derived from ACT and MBSR are very useful since patients are taught to stay focused on the present, detach themselves from dysfunctional thoughts, accept physical sensations and their associated emotions, and act according to their personal values, thus reducing avoidance or escape behaviors from social situations (Caballo et al., 2021).

Practice and procedures

Key elements of the MISA program application

The MISA program is composed by a therapist's guide (Caballo, Salazar, Garrido, Irurtia, & Hofmann, 2018) and a patient's workbook (Caballo, Salazar, & Garrido, 2018). The patients will use this workbook all along the program. It is designed to be applied to people with social anxiety, either SAD or avoidant personality disorder (APD). The inclusion criteria are quite simple: (a) be 18 years or older, (b) meet the SAD or APD criteria according to DSM-5 (APA, 2013), (c) score above the cut-off point in at least three dimensions of the SAQ, and (d) commitment to attend all sessions and doing homework. The exclusion criteria are also very simple: (a) not having SAD or APD as the main disorder, (b) having a serious comorbid disorder, such as a psychotic, bipolar, borderline personality disorder (BPD) or substance use disorder, and (c) not having enough time to attend regularly to the program. Admitted patients sign an informed consent before attending the program by which they agree not to miss two consecutive sessions or three alternate ones, in which case they would automatically be excluded from the program. Not doing homework can also be a reason for not continuing in the program.

On the other hand, the sessions are group sessions. The ideal size of the groups is 6–8 people, although initially groups of 10 people can be considered, since it is highly likely that a patient will be left out of the program for not meeting the conditions described above. It is convenient that there are people of both sexes, ideally 50% of each or 40% and 60% of both sexes.

The duration of the sessions is 2:30 h with a 15-min break in the middle of each session (approximately). These 15 min of "break" also constitute a space for the patients to socialize with each other and with the therapist. In Fig. 2 you can see a schedule of the development of the program, from its beginning to the follow-up sessions 1 year after the end of the program.

Of the 22 sessions that make up the MISA program, 6 are for assessment (2 at the beginning of the program, 2 at the end of the program, one 6 months after finishing the program and another at 12 months follow-up). There is also a booster session 3 months after the end of the program. The remaining 15 sessions constitute the core part of the treatment program. The first five treatment sessions are dedicated to teaching patients basic strategies, such as thinking about values, acceptance, mindfulness, detachment from thoughts (defusion), and cognitive restructuring, as well as behavior rehearsal and other elements of social skills training (SST). The next 10 sessions are devoted to patients learning to cope with the five dimensions of social anxiety using the strategies learned in the previous sessions. From the first treatment session, patients are accustomed to performing tasks at home, something that will be present throughout the entire MISA program.

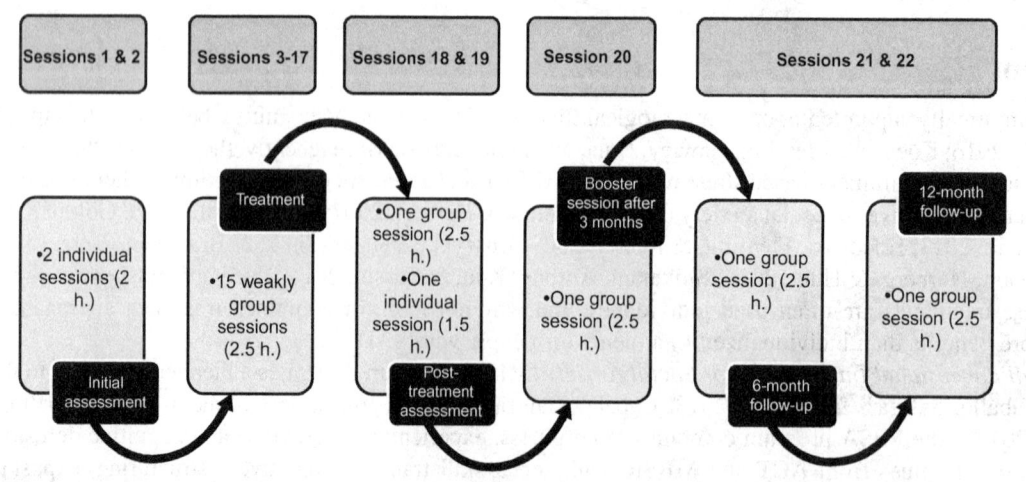

FIG. 2 General structure of the Multidimensional Intervention for Social Anxiety (MISA) program.

Instruments for assessing program outcome

In the first place, we use the *Clinical Semistructured Interview for Social Anxiety* (CSISA; Salazar & Caballo, 2017) that includes the diagnostic criteria of the DSM-5 (APA, 2013) but focused on the five dimensions that conform the program, as a pre/posttreatment measure to evaluate the results of it. It is the main evaluation instrument of the SAD (together with the SAQ). But we also include a battery of self-report measures that reflect well the progress (or lack thereof) of patients who have participated in the MISA program. These instruments are as follows:

(a) *Social Anxiety Questionnaire for Adults* (SAQ) (Caballo et al., 2012, 2015; Caballo, Salazar, Arias, et al., 2010; Caballo, Salazar, Irurtia, et al., 2010). Beside the semistructured interview, it is the self-report instrument on which the selection of patients is based. Developed by our research team, it has served as a source of information for the development of the MISA program. It evaluates five dimensions of social anxiety and constitutes a filter for the recruitment of patients, given that entering the treatment program they must score above the cut-off point in 3 of the 5 dimensions.

(b) *Penn State Worry Questionnaire* (PSWQ; Meyer, Miller, Metzger, & Borkovec, 1990). Assesses the general tendency to worry and consists of 16 items that are answered on a five-point Likert scale.

(c) *Beck Depression Inventory-II* (BDI-II; Beck, Steer, & Brown, 1996). The BDI-II measures the magnitude of depression through 21 items of four response alternatives that are ordered according to severity, from 0 to 3 points.

(d) *Alcohol Use Disorders Identification Test* (AUDIT; Babor, Higgins-Biddle, Saunders, & Monteiro, 2001). This measure attempts to identify people with a pattern of risk or harmful use of alcohol. The AUDIT consists of 10 questions in which the person must choose the option that best describes their current alcohol consumption.

(e) *Social Skills Questionnaire* (SOSAQ) (Caballo et al., 2017). Assesses the social skills of the participants. It is a new questionnaire that measures 10 dimensions of social skills. The results obtained with different Latin American countries are in the preparation stage for their publication.

(f) *Personal Sensibility Questionnaire* (PSQ; Caballo & Salazar, 2019]). It is a new questionnaire that assesses personal sensitivity considering the approaches of Aron (1999) and other authors who have worked on behavioral inhibition (e.g., Kagan, 1999). In addition, 7 items have been inserted at the end of the questionnaire that includes the diagnostic symptoms of APD according to the DSM-5 (APA, 2013).

(g) *Experiences Questionnaire* (EQ; Fresco et al., 2007). This measure consists of two subscales: (a) "Decentering subscale," which would assess the ability to distance oneself from one's own thoughts and feelings and (b) "Rumination subscale," that would indicate that the person turns his thoughts over and over again.

(h) *Multidimensional Perfectionism Scale* (MPS; Frost, Marten, Lahart, & Rosenblate, 1990). This scale assesses perfectionism through six subscales totaling 35 items.

Therapeutic strategies composing the MISA program

The procedures that we include in the MISA program, summarized in Fig. 3, are as follows:

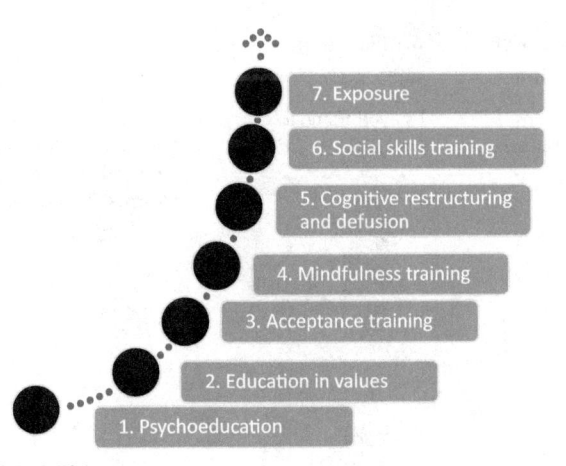

FIG. 3 General treatment components of the MISA program.

Psychoeducation

Psychoeducation is used to brief patients on a whole series of aspects related to SAD and relationships with other people. The therapist makes a clear, straightforward, and systematic presentation of the scientific aspects of the topic being addressed in the session, using the educational material in the patient's workbook and other sources (e.g., videos, pamphlets, testimonies, comic strips, etc.). At the same time, the aim is to actively help the patient understand and assimilate the information presented by using exercises, activities, discussions and, in general, any group strategy. Active listening is encouraged among patients at this stage, with them also asking questions out aloud, making comments or giving examples. As part of the psychoeducation strategy, Table 1 shows an example of how to identify feared social situations and some of their consequences.

Education in values

The patients are taught to recognize the values that guide their lives and act consistently with them. Values have been conceptualized in the ACT (e.g., Hayes & Smith, 2005; Hayes, Strosahl, & Wilson, 2012) and represent what people want their lives to be, what really matters to them, and the pillars that support their existence; in short, what gives their life meaning and purpose. Living a worthwhile life is acting according to what they value. Values ask *why* when it comes to an action, rather than just *what* the action is. Values are never completed.

Acceptance training

When we have to deal with things we cannot change, the best option is usually to accept them. Acceptance means coming to terms with the fact that we experience fear and anxiety, which means being ready to accept feelings, emotions, sensations, memories, and thoughts of anxiety for what they are, without trying to avoid or escape from those experiences and the circumstances that may have prompted them. To accept an emotion or feeling, we first have to detect its presence, and this is where acceptance and mindfulness overlap. What is more, acceptance is implicit in mindfulness.

Mindfulness training

Mindfulness training involves teaching patients to direct and focus their attention on what they are experiencing in the present, to live it fully, consciously, without judging and with acceptance. It involves learning to observe, describe, and actively take part in one's own life experiences without judging them and fully focusing on them. Indeed, we might say there are two types of mindfulness:

(a) One that favors concentration, *focusing one's attention* on a specific place or object.
(b) One that involves full awareness, without judgments, of internal and external experiences.

Both kinds of mindfulness feed off each other, and both are used in our program.

TABLE 1 Identification of feared social situations, the social anxiety dimension affected, avoidance or safety behaviors, and their effects and cost.

Feared social situations	Social anxiety dimension	Avoidance or safety behaviors	Cost of these behaviors
I was invited to go out with my classmates	Interactions with strangers	I stayed home to watch TV	- I felt lonely - I got angry with myself for being such a coward - I missed out on a good opportunity to get to know my classmate better, and of potentially making friends
1.			
2.			
3.			
...			
10.			

Cognitive restructuring and defusion

Two strategies are used in this section. The first one, cognitive restructuring, helps to reconsider thoughts (cognitions) that are maladaptive, identifying them and acknowledging their dysfunctionality. Cognitive distortions are questioned, and the way is opened for distancing or defusing from them. Rational emotive behavior therapy (REBT) could initially be used to identify and question dysfunctional thoughts (e.g., Lega, Caballo, & Ellis, 2002). This approach is used to help patients be more scientific, considering their automatic inferences and thoughts as hypotheses instead as facts, whereupon they then look for facts that may corroborate or refute those hypotheses. Patients are helped to identify their automatic thoughts and decide whether those thoughts contain distortions or are dysfunctional.

The second strategy therefore involves training patients to *defuse or detach from their dysfunctional thoughts*. It is clear that our thoughts have a major impact on our behavior and on how we feel, and we invest a great deal of time, effort and energy in responding to them. We may have spent (and still do) a large part of our lives struggling with thoughts that cause us distress. A different (and probably more effective) option for helping us in our problematic relationship with dysfunctional thoughts is to defuse from them. This strategy (or process, as some scholars affirm) encourages a patient to become an external observer of their own thoughts, looking in from the outside and not from them. Thoughts are then seen for what they are, a series of words and images, whereby they are taken into consideration in terms of their functionality instead of what they mean; in other words, the extent to which they are useful rather than whether they are true or not. When such thoughts are about oneself, this defusion may help to distinguish between the person having the thought (the real me) and the verbal categories applied to oneself though the thought (the other me).

Social skills training

SST allows acting out certain social situations within a controlled setting, such as group therapy, with a view to learning and rehearsing certain social behaviors that patients can then implement in their daily lives. Nevertheless, we should note that this roleplaying involves considering a large part of what we have already seen in the preceding sections, such as mindfulness, cognitive restructuring and defusing from thoughts. SST consists of several procedures (Caballo, 1997): (a) Instructions/coaching, (b) behavior rehearsal, (c) modeling or observational learning, (d) reinforcement by others/feedback, (e) exposure, and (f) homework. Behavior rehearsal, the key element of SST, includes the following steps:

(1) Role-play the problematic situation.
(2) Identify dysfunctional cognitions before and during the role-play.
(3) Identify the basic human assertive rights involved in the situation.
(4) Identify an appropriate target for the new patient's behavior.
(5) Suggestion of improvements by the group and the therapist.
(6) Role-play by the patient of the chosen response, performing it with defusion, acceptance, and mindfulness.
(7) Evaluation by the patient of the goals achieved.
(8) Positive feedback on the goals attained and suggestions for improvement by the group and the therapist.
(9) Covert practice of the new behavior.
(10) Role-play of the chosen response incorporating the improvements suggested by the therapist and the group, and applying defusion, acceptance, and mindfulness.
(11) Evaluation of the effectiveness of the behavior.
(12) Last instructions to the patient on the implementation of the rehearsed behavior to real life, emphasizing the use of defusion, acceptance, and mindfulness.

In those cases in which the patient is unable to incorporate the suggestions made by the therapist and the other group members in the new role-play of the situation, another group member may act out the behavior incorporating the suggestions ("*modeling*").

Exposure

Exposure is a key element of CBT and also of the MISA program. Patients must exposure themselves to social situations they have feared for a long time, but now incorporating what they are learning through the program. For instance, when exposing to an "interacting with strangers" situation in real life, patients use the social skills rehearsed at the group sessions regarding this dimension, but now including acceptance, defusion and mindfulness, and even values. Such a kind of exposure is much more improved from a simple exposure and the possibility of success is greatly enhanced. Furthermore, this will be the way people with social anxiety should cope with social situations in the future and as such, they should incorporate it in their life as a habit.

Homework

Homework assignments are tasks or exercises that patients have to carry out in their everyday lives, between each session, with a view to extrapolating into real life what they have learnt at the clinic. These assignments are an essential part of the MISA program, and its success depends largely on the tasks that patients carry out when they are not with the therapist. Each session in the program normally starts and ends with a discussion of homework, whose assignments are specifically designed to achieve the therapy's goals. As the program advances, a part of each session is used to prepare the patients for the next round of homework, with the tasks' difficulty gradually being increased as the treatment progresses. The following should be taken into account regarding homework tasks:

a. Some of them are designed and agreed with the patients, while many others are specified in the Patient's Workbook.
b. They should have detailed instructions, as in the case of those specified in the Patient's Workbook. Patients must be able to understand them.
c. Tasks should have a high probability of success, especially the initial ones. Successes may considerably increase patients' motivation and engagement regarding treatment.

Training in the dimensions of the MISA program

The MISA program devotes 10 sessions to work with the 5 specific dimensions conforming social anxiety. Table 2 shows a summary of the contents of the MISA program. Two sessions are devoted to each of these dimensions. In the training of each dimension, we use all the strategies composing the MISA program and that have been previously trained in patients before addressing each of the dimensions. We will now show some work with a specific social anxiety dimension to exemplify how we approach the different dimensions.

TABLE 2 Summary of the contents of MISA program.

Session number	Contents for each session
Pretreatment assessment (two individual sessions, 2h length every session)	
1 & 2	• Assessment of patients: interview and questionnaires • Definition of personal goals for therapy • Explaining the functioning of the group therapy
Treatment sessions (15 group sessions, 2 ½ hours length per session)	
3	• Constitution of the therapeutic group • Psychoeducation on SAD • Training in abdominal breathing • Training in progressive muscle attention • Homework assignments (this is always the last step of each session; we will not repeat it in the following sessions)
4	• Review of homework (this is always the first step for each session; we will not repeat it in the following sessions) • Identification of situational anxiety • Introduction to values for life • Introduction to dysfunctional cognitions in social anxiety • Mindful breathing
5	• Values for life and presenting our mind as a "storytelling machine" • Cognitive restructuring (1st part): ABC and maladaptive thoughts • Introduction to cognitive defusion • Mindful eating

6	• Cognitive restructuring (2nd part): The cognitive debate with defusion.
	  (a) Questioning of dysfunctional thoughts: The thoughts as "background noise"
	  (b) The cognitive debate with defusion
	• Practicing acceptance and cognitive defusion
	• Training in attention to bodily sensations through imagination (1st part)

TABLE 2 Summary of the contents of MISA program—cont'd

Session number	Contents for each session
7	• Review of values for life and defusion • Introduction to social skills • Components of social skills • Behavior styles: assertive, passive, and aggressive • Basic human assertive rights • Mindfulness training with the components of social behavior
8	*Training for Dimension 1: Interactions with strangers—1st part* • Dimension 1 (1st part). Response classes to be addressed: Starting, maintaining, and ending conversations • The multimodal self-monitoring form • Body mindfulness through imagination (2nd part)
9	*Training for Dimension 1: Interactions with strangers—2nd part* • Cognitive defusion for dimension 1 • Mindful observation of the outer world • Dimension 1 (2nd part). Response classes to be addressed: Making friends
10	*Training in Dimension 2: Interactions with the opposite sex—1st part* • Training in one-minute abdominal breathing • Defusing from thoughts in Interactions with the opposite sex • Mindfulness training: "mindful observation of the inner world" • Dimension 2 (1st part). Response classes to be addressed: Initiating interactions with people we find attractive, giving and receiving compliments, and expressing liking, love, and affection
11	*Training in Dimension 2: Interactions with the opposite sex—2nd part* • Cognitive defusion: "Repeat a word over and over again" • Acceptance of bodily sensations (1st part) • Dimension 2 (2nd part). Response classes to be addressed: Asking for a date
12	*Training for Dimension 3: Expressing annoyance, disgust, or displeasure—1st part* • Dimension 3 (1st part). Response classes to be addressed: Expression of negative feelings, asking someone to change their behavior • "Do not do what you say" and act in the opposite direction • Acceptance of bodily sensations (2nd part)
13	*Training for Dimension 3: Expressing annoyance, disgust, or displeasure—2nd part* • Cognitive defusion: "Get Jiminy Cricket out for a walk" • Acceptance of bodily sensations (3rd part) • Dimension 3 (2nd part). Response classes to be addressed: Making and refusing requests
14	*Training for Dimension 4: Criticism and embarrassment—1st part* • Review about the values for life and goals • Dimension 4 (1st part). Response classes to be addressed • Thinking in slow motion • Training in acceptance of embarrassing feelings

Continued

TABLE 2 Summary of the contents of MISA program—cont'd

Session number	Contents for each session
15	*Training for Dimension 4: Criticism and embarrassment—2nd part* • Dimension 4 (2nd part). Response classes to be addressed: Coping with criticism • Cognitive defusion: The big screen • Practice of acceptance of embarrassing feelings with cognitive defusion
16	*Training for Dimension 5: Speaking in public/Talking with people in authority—1st part* • Dimension 5 (1st part). Response classes to be addressed: Preparing a presentation, speaking in public • Training in cognitive defusion: Chain of thoughts and use of defusion strategies • Practice in values, mindfulness, and defusion all together
17	*Training for Dimension 5: Speaking in public/Talking with people in authority—2nd part* • Practice in values, mindfulness and defusion • Dimension 5 (2nd part). Response classes to be addressed: Expressing personal opinions
Posttreatment assessment (two sessions, the first in group, 2 ½ hours length and the second individual, 1 ½ hours length)	
18	• Group posttreatment assessment • Personal work planning for the next 3 months • Evaluation of the MISA program
19	• Posttreatment individual assessment (interview and questionnaires)
Booster session at 3 months (one groups session; 2 ½ hours length)	
20	• Review and reinforcement of the skills learned. • Programming the assessment session at 6 months
Follow-up (two group sessions: at 6 and 12 months; 1 ½ hours length)	
21	• Follow-up assessment at 6-months • Programming the assessment session at 12 months
22	• 12-month follow-up • End of the group therapy

Working with the "Criticism and embarrassment" dimension (1st part)

Sessions 14 and 15 are devoted to the training of the "Criticism and embarrassment" dimension (Caballo, Salazar, & Garrido, 2018; Caballo, Salazar, Garrido, Irurtia, & Hofmann, 2018). The situations included in this dimension are especially feared by people with social anxiety. In fact, the fear of negative evaluation is one of the central fears, perhaps the essence, of SAD. This fear usually includes concepts such as making a fool of himself, embarrassing oneself or being humiliated and that is why the idea of "wanting to please everyone" is addressed. Fear of criticism or embarrassment can be one of the big obstacles when it comes to meeting people, participating in activities, or having jobs that could be highly reinforcing for a person. If that were not enough, it seems there is the added fear of not being able to cope with the humiliation of rejection (Nelson-Jones, 1986). The fear of being judged by other people may be based in part on the implicit expectation that everyone should love and support us all the time. If this expectation is not fulfilled, we may develop that fear due to a tendency to consider rejection as a catastrophe and a reflection of our worth as individuals. The life values practiced by patients through the MISA program may constitute a sound base, regardless of other people's opinion, which affords us protection as we live our lives. To follow is a summary of sessions 14 and 15.

Summary of session 14
(1) Homework review.
(2) Patients review their life values and goals through an exercise which recover some of their values and goals, write in their workbook things they can do to achieve their goals, the obstacles they might encounter on the way, and how they can overcome those obstacles.
(3) Patients do an exercise designed to attract other people's attention. One by one, each patient stands in the middle of the group and tries to describe an object using only gestures. The other members of the group have to try to guess the object being described by the person in the middle. The aim is to experience being the focus of attention. The person going out into the middle of the group first has to prepare themselves through abdominal breathing and applying defusion from thoughts and mindfulness. Then therapist ask the patients to record the exercise in Table 3, reminding them to make a note of the situation, the initial emotion on the SUDS, dysfunctional anticipatory thoughts, the defusion strategy used, the application of mindfulness, the behavior finally used, and the final emotion on the SUDS.
(4) Psychoeducation regarding the topic of *dealing with criticism*. We teach the patients "negative assertion" and "negative inquiry" (Smith, 1977), accompanied with defusion and mindfulness, to be used when needed.
(5) We continue to work on defusion from thoughts, as it is a strategy that they are going to use both in the behavioral rehearsals and in the homework assignments. On this occasion, patients are going to choose several typical dysfunctional thoughts that appear when they are criticized. It may involve, for example, such thoughts as "if they are criticizing me, it's because they don't respect me," "when they criticize me, I feel they don't appreciate me," "if they are criticizing me, it's because I'm a failure," or "when they criticize me, I feel they are questioning my ability." They are going to slow down their thought by matching each word of each sentence with a part of the breath. Then it goes to the second thought and so with several habitual dysfunctional thoughts in situations in which they are criticized.
(6) Patients are trained in the acceptance of feelings of embarrassment through an exercise imagining some social situations eliciting these feelings.
(7) Some homework assignments are explained for patients to be done in real life until the next session.

TABLE 3 Worksheet for shame-attacking exercises.

Embarrassing situation	Initial emotion (0–100 SUD)	Initial dysfunctional thoughts	Defusion strategy	Mindfulness (center of attention)	Performed behavior (level of satisfaction: 1 = none– 10 = very high)	Final emotion(s) (0–100 SUD)
Go out into the middle of the group	Embarrassment (70) Anxiety (80)	I'm going to behave awkwardly I'm going to make a fool of myself Everyone will be looking at me People will realize how incompetent I am	There goes Jiminy Cricket ruining my life I'm not going to let Jiminy Cricket rule my life Jiminy Cricket can stay away while I'm in the middle of the group	I'm going to focus on what the other members of the group are doing I'm going to concentrate on the rest of the group's nonverbal communication	I've gone out into the middle of the group but it hasn't worried me I've done the task that the therapist indicated in a satisfactory manner = 8	Embarrassment (40) Anxiety (30)

Summary of session 15
(1) Homework assignments review.
(2) Behavioral rehearsal: challenging situations planned by the patients. Individual work is undertaken on the situation that each patient has chosen in relation to the dimension Criticism and embarrassment that they have brought in today as a homework assignment.
(3) Psychoeducation regarding the topic of *dealing with criticism* (2nd part). Here we teach patients the techniques of "separating subjects" and "disarming anger."
(4) A new exercise on "attacking the embarrassment of selling something" is practiced by patients. The aim is to prepare an advertisement to convince the audience to buy a product. The patients can use anything they like, from a slogan through to an advertising campaign, as they are the ones who will choose the format. They have a minute to do so. Each pair is given some envelopes to choose from. Each envelope contains the subject of an advertising campaign they have to arrange (a retirement home, a vacation hotel, a weekend trip, a vacuum-cleaner, a car, and marker pens). They will have to use more or less imagination depending on the subject. Always use defusion from thoughts and mindfulness when appearing before the rest of the group.
(5) Patients continue to work on defusion through the "big screen" exercise. They choose some of their previous dysfunctional thoughts and visualize them one-by-one on a big white screen they imagine in front of them. The therapist guides this exercise.
(6) Patients are trained in acceptance of embarrassment sensations with defusion. Through this exercise, patients are going to imagine and accept the sensations that arise in those situations in which they are embarrassed or make a fool of themselves, mindfully attending to the flow of sensations caused by the thoughts triggered by anticipating and evaluating threatening situations, then using defusion from thoughts and focusing our attention on the situation in which we are immersed.
(7) Finally, some homework assignments are prescribed to participants.

Key facts

- The MISA program is a comprehensive treatment program for SAD and APD (but not only) using traditional CBT and mindful and acceptance-based strategies.
- The MISA program uses a therapist's guide and a patient's workbook all along the treatment.
- The MISA program uses a group therapy format, although an individual therapy one is also possible.

Mini-dictionary of terms

- Acceptance: accepting feelings, emotions, sensations, memories, and thoughts for what they are.
- Cognitive restructuring: identifying dysfunctional thoughts and questioning them.
- Defusion: detaching from our thoughts, seeing them as an external observer.
- Exposure: exposing to the anxiety sources, mainly to feared social situations.
- Mindfulness: focusing our attention on what we are experiencing in the present.
- Social skills training: teaching of interpersonal strategies and skills to improve our competence in social situations.

Applications to other areas

The wide selection of therapeutic strategies in the MISA program makes it applicable to disorders other than SAD and APD. For example, a large part of the procedures that compose the MISA program could be used in the treatment of generalized anxiety disorder or in depressive disorders (Salazar et al., 2022). But where it is of special application is in the BPD, as we have done in the FUNVECA Clinical Psychology Center, Granada (Spain).

MISA program applied to BPD

Many of the activities proposed in the MISA program are useful in the treatment of people with BPD. There are two reasons for this. The first is that usually people with BPD, like people with SAD, experience excessive fear related to social situations. In these social interactions, they tend to feel under the scrutiny of others and then appear their fear of being inadequate, of not having or losing the others' approval, and eventually abandonment fears. The second reason is that people with BPD have sudden and intense ups and downs with most of the experienced emotions (sadness, anxiety, euphoria, anger, jealousy, guilt, shame, …) and, usually, their ways of reacting to their emotions tend to aggravate situations, cause more harm to themselves and their environment. Sometimes this end in impulsive acts and self-harm behaviors. All the above shows the difficulties in regulation that the person with BPD has and although anxiety is not the only emotion or the only problem to intervene, an adaptation of the MISA program can help the person with BPD.

Currently in our clinic, we are applying an individual treatment program for people with BPD based on the structure of the MISA program. This structure allows to harmoniously combine some techniques of traditional CBT with third-generation therapies in the different therapeutic sessions.

Example of a therapeutic session

One of the frequent complaints of these individuals is their overwhelming emotions and, for this reason, a therapeutic objective is usually to learn to experience emotions without rejecting them or increasing them to the point of losing control. The work begins by learning to observe the emotions to get to know them better, to know what they are like, when they occur and what keeps them going. *Psychoeducation* allows discussing with the patient about basic aspects of emotions (what they are, what they are for, how they express, and what are their relationship with thoughts and actions), and then the patient is asked to begin to attend to emotions (and everything that surrounds them) but in a different way from how he had been doing it. The challenge is to do it as a scientist or a child would, with curiosity and without judgment. It is recommended to use a self-monitoring format so that people begin to "take careful note" of what is happening.

Then, a *mindfulness* exercise (e.g., mindful eating) is carried out to introduce what this skill consists of and to begin working on the ability to focus on what is happening in the present moment, without judging or evaluating life experiences. As the patient is not used to detecting these judgments or evaluations, he is deliberately asked to be attentive to thoughts and sensations that appear while doing the exercise and given some examples ("this is not very pleasant," "I don't know if I'm doing the exercise right," "what nonsense am I doing") asking to patients to "note" that a thought or feeling has appeared and to refocus their attention on food and eating.

Judgments, ruminations, and other cognitive elements that arise during the previous exercise can be used by the therapist to introduce the concept of *cognitive defusion*, explaining the importance of differentiating the person who produces the thoughts from the thoughts themselves. Therapist and patient work on the creation of an "alter ego" to personify that part of the brain that insists on creating dysfunctional thoughts, causing discomfort, and taking you away from the way you want to conduct your life.

Then, we do an exercise in which we ask the patient to imagine her life without BPD and we guide him to visualize what and how he would do it, with whom and how he would relate. Again, he is asked to "take careful note" of what he seems to care about and then it is discussed how some of these things can be achieved if he identifies the *values* he needs to guide his actions.

The session ends with two other homework assignments: the daily practice of mindfulness with food and writing about the possible values that he recognizes that he has and those that he would require to have the life he imagined.

Summary points

- The MISA Program includes traditional CBT techniques together with ACT strategies.
- Length for the MISA program is 5 months (4th-month treatment plus 4th-week pre/postassessment sessions)
- The MISA program includes a therapist's guide and a patient's workbook. The patients will use this workbook all along the program.
- The MISA program revolves around five dimensions of social anxiety obtained through empirical research for more than 15 years.
- The MISA program is targeted for people with SAD or APD, but application to other disorders, such as BPD, is welcome.

References

American Psychiatric Association. (2013). *Diagnostic and statistical manual of mental disorders: DSM-5* (5th ed.). Arlington, VA: Author. https://doi.org/10.1176/appi.books.9780.

Aron, E. N. (1999). *The highly sensitive person*. London: HarperCollins.

Asher, M., & Aderka, I. M. (2018). Gender differences in social anxiety disorder. *Journal of Clinical Psychology, 74*(10), 1730–1741. https://doi.org/10.1002/jclp.22624.

Babor, T. F., Higgins-Biddle, J. C., Saunders, J. B., & Monteiro, M. G. (2001). *AUDIT. The alcohol use disorders identification test. Guidelines for use in primary care* (2nd ed.). Geneva, Switzerland: WHO.

Beck, A. T., Steer, R. A., & Brown, G. (1996). *Manual for the Beck depression inventory-II*. San Antonio, TX: Psychological Corporation.

Caballo, V. E., Salazar, I. C., Curtiss, J., Gómez, R. B., Rossitto, A. M., Coello, M. F., Hofmann, S. G., ... MISA Research Team. (2021). International application of the "Multidimensional Intervention for Social Anxiety" (MISA) program: I. Treatment effectiveness in patients with social anxiety. *Behavioral Psychology/Psicología Conductual, 29*(3), 517–547. https://doi.org/10.51668/bp.8321301n.

Caballo, V. E. (1997). *Manual de evaluación y entrenamiento de las habilidades sociales (2ª ed.) [Social skills assessment and training manual]* (2nd ed.). Madrid, Spain: Siglo XXI.

Caballo, V. E., & Salazar, I. C. (2019). *Personal Sensibility Questionnaire (PSQ)*. Unpublished manuscript Granada, Spain: FUNVECA Clinical Psychology Center.

Caballo, V. E., Arias, B., Salazar, I. C., Irurtia, M. J., Hofmann, S. G., & CISO-A Research Team. (2015). Psychometric properties of an innovative self-report measure: The social anxiety questionnaire for adults. *Psychological Assessment, 27*(3), 997–1012. https://doi.org/10.1037/a0038828.

Caballo, V. E., Salazar, I. C., & CISO-A Research Team Spain. (2017). Development and validation of a new social skills assessment instrument: The social skills questionnaire (CHASO). *Behavioral Psychology, 25*(1), 5–24.

Caballo, V. E., Salazar, I. C., & Garrido, L. (2018). *Programa de Intervención multidimensional para la ansiedad social (IMAS). Libro del paciente [Multidimensional Intervention for Social Anxiety (MISA) program. Patient's workbook]*. Madrid, Spain: Pirámide.

Caballo, V. E., Salazar, I. C., & Hofmann, S. G. (2019). A new multidimensional intervention for social anxiety: The MISA program. *Behavioral Psychology, 27*(1), 149–172.

Caballo, V. E., Salazar, I. C., Arias, B., Irurtia, M. J., Calderero, M., & CISO-A Research Team Spain. (2010). Validation of the "social anxiety questionnaire for adults" (SAQ-A30) with Spanish university students: Similarities and differences among degree subjects and regions. *Behavioral Psychology, 18*(1), 5–34.

Caballo, V. E., Salazar, I. C., Garrido, L., Irurtia, M. J., & Hofmann, S. G. (2018). *Programa de Intervención multidimensional para la ansiedad social (IMAS). Libro del terapeuta [Multidimensional Intervention for Social Anxiety (MISA) program. Therapist's guide]*. Madrid, Spain: Pirámide.

Caballo, V. E., Salazar, I. C., Irurtia, M. J., Arias, B., Hofmann, S. G., & CISO-A Research Team. (2012). The multidimensional nature and multicultural validity of a new measure of social anxiety: The social anxiety questionnaire for adults (SAQ-A30). *Behavior Therapy, 43*, 313–328. https://doi.org/10.1016/j.beth.2011.07.001.

Caballo, V. E., Salazar, I. C., Irurtia, M. J., Arias, B., Hofmann, S. G., & CISO-A Research Team. (2010). Measuring social anxiety in 11 countries: Development and validation of the social anxiety questionnaire for adults. *European Journal of Psychological Assessment, 26*(2), 95–107. https://doi.org/10.1027/1015-5759/a000014.

Dalrymple, K. L., & Herbert, J. D. (2007). Acceptance and commitment therapy for generalized social anxiety disorder. A pilot study. *Behavior Modification, 31*(5), 543–568. https://doi.org/10.1177/0145445507302037.

Fresco, D. M., Moore, M. T., van Dulmen, M. H. M., Segal, Z. V., Ma, S. H., Teasdale, J. D., & Williams, J. M. G. (2007). Initial psychometric properties of the experiences questionnaire: Validation of a self-report measure of decentering. *Behavior Therapy, 38*, 234–246. https://doi.org/10.1016/j.beth.2006.08.003.

Frost, R. O., Marten, P., Lahart, C., & Rosenblate, R. (1990). The dimensions of perfectionism. *Cognitive Therapy and Research, 14*(5), 449–468. https://doi.org/10.1007/BF01172967.

Goldin, P. R., Morrison, A., Jazaieri, H., Brozovich, F., Heimberg, R., & Gross, J. J. (2016). Group CBT versus MBSR for social anxiety disorder: A randomized controlled trial. *Journal of Consulting and Clinical Psychology, 84*(5), 427–437. https://doi.org/10.1037/ccp0000092.

Goldin, P., & Gross, J. (2010). Effects of mindfulness-based stress reduction (MBSR) on emotion regulation in social anxiety disorder. *Emotion, 10*(1), 83–91. https://doi.org/10.1037/a0018441.

Gureje, O., Adeyemi, O., Enyidah, N., Ekpo, M., Udofia, O., Uwakwe, R., & Wakil, A. (2008). Mental disorder among adult Nigerians: Risks, prevalence, and treatment. In R. C. Kessler, & T. B. Üstün (Eds.), *The WHO mental health surveys: Global perspectives on the epidemiology of mental disorders* (pp. 211–237). New York: Cambridge University.

Hayes, S. C., & Smith, S. (2005). *Get out of your mind & into your life: The new acceptance and commitment therapy*. Oakland, CA: New Harbinger.

Hayes, S. C., Strosahl, K. D., & Wilson, K. G. (2012). *Acceptance and commitment therapy* (2nd ed.). New York: Guilford.

Kagan, J. (1999). The concept of behavioral inhibition. In L. A. Schmidt, & J. Schulkin (Eds.), *Extreme fear, shyness, and social phobia* (pp. 3–13). New York: Oxford University Press.

Kessler, R. C., & Üstün, T. B. (Eds.). (2008). *The WHO mental health surveys: Global perspectives on the epidemiology of mental disorders*. New York: Cambridge University.

Kessler, R. C., Berglund, P. A., Chiu, W.-T., Demler, O., Glantz, M., Lane, M. C., ... Wells, K. B. (2008). The National Comorbidity Survey Replication (NCS-R): Cornerstone in improving mental health and mental health care in the United States. In R. C. Kessler, & T. B. Üstün (Eds.), *The WHO mental health surveys: Global perspectives on the epidemiology of mental disorders* (pp. 165–209). New York: Cambridge University.

Kessler, R. C., Petukhova, M., Sampson, N. A., Zaslavsky, A. M., & Wittchen, H.-U. (2012). Twelve-month and lifetime prevalence and lifetime morbid risk of anxiety and mood disorders in the United States. *International Journal of Methods in Psychiatric Research, 21*(3), 169–184. https://doi.org/10.1002/mpr.1359.

Kocovski, N. L., Fleming, J. E., & Rector, N. A. (2009). Mindfulness and acceptance-based group therapy for social anxiety disorder: An open trial. *Cognitive and Behavioral Practice, 16*(3), 276. https://doi.org/10.1016/j.cbpra.2008.12.004.

Kocovski, N. L., Fleming, J. E., Hawley, L. L., Huta, V., & Antony, M. M. (2013). Mindfulness and acceptance-based group therapy versus traditional cognitive behavioral group therapy for social anxiety disorder: A randomized controlled trial. *Behaviour Research and Therapy, 51*(12), 889–898. https://doi.org/10.1016/j.brat.2013.10.007.

Koszycki, D., Benger, M., Shlik, J., & Bradwejn, J. (2007). Randomized trial of a meditation-based stress reduction program and cognitive behavior therapy in generalized social anxiety disorder. *Behaviour Research and Therapy, 45*(10), 2518–2526. https://doi.org/10.1016/j.brat.2007.04.011.

Lega, L. I., Caballo, V. E., & Ellis, A. (2002). *Teoría y práctica de la terapia racional emotivo-conductual (2ª ed.) [Theory and practice of rational emotive-behavior therapy]* (2nd ed.). Madrid, Spain: Siglo XXI.

Liu, X., Yi, P., Ma, L., Liu, W., Deng, W., Yang, X., ... Li, X. (2021). Mindfulness-based interventions for social anxiety disorder: A systematic review and meta-analysis. *Psychiatry Research, 300*. https://doi.org/10.1016/j.psychres.2021.113935, 113935.

MacKenzie, M. B., & Fowler, K. F. (2013). Social anxiety disorder in the Canadian population: Exploring gender differences in sociodemographic profile. *Journal of Anxiety Disorders, 27*(4), 427–434. https://doi.org/10.1016/j.janxdis.2013.05.006.

McLean, C. P., Asnaani, A., Litz, B. T., & Hofmann, S. G. (2011). Gender differences in anxiety disorders: Prevalence, course of illness, comorbidity and burden of illness. *Journal of Psychiatric Research, 45*(8), 1027–1035. https://doi.org/10.1016/j.jpsychires.2011.03.006.

Meyer, T. J., Miller, M. L., Metzger, R. L., & Borkovec, T. D. (1990). Development and validation of the Penn State worry questionnaire. *Behaviour Research and Therapy, 28*, 487–495. https://doi.org/10.1016/0005-7967(90)90135-6.

Nelson-Jones, R. (1986). *Human relationship skills*. London: Holt, Rinehart and Winston.

Norton, A. R., Abbott, M. J., Norberg, M. M., & Hunt, C. (2015). A systematic review of mindfulness and acceptance-based treatments for social anxiety disorder. *Journal of Clinical Psychology, 71*(4), 283–301. https://doi.org/10.1002/jclp.22144.

Pignon, B., Amad, A., Pelissolo, A., Fovet, T., Thomas, P., Vaiva, G., ... Geoffroy, P. A. (2018). Increased prevalence of anxiety disorders in third-generation migrants in comparison to natives and to first-generation migrants. *Journal of Psychiatric Research, 102*, 38–43. https://doi.org/10.1016/j.jpsychires.2018.03.007.

Quevedo, L., de Mola, C., Pearson, R., Murray, J., Hartwig, F., Goncalves, H., ... Horta, B. L. (2020). Mental disorders, comorbidities, and suicidality at 30 years of age in a Brazilian birth cohort. *Comprehensive Psychiatry, 102*. https://doi.org/10.1016/j.comppsych.2020.152194.

Salazar, I. C., Caballo, V. E., Arias, V., Curtiss, J., Rossitto, A. M., Gómez, R. B., Hofmann, S. G., ... MISA Research Team. (2022). International application of the "Multidimensional Intervention for Social Anxiety" (MISA) program: II. Treatment effectiveness for social anxiety-related problems. *Behavioral Psychology/Psicología Conductual, 30*(1), 19–49. https://doi.org/10.51668/bp.8322102n.

Salazar, I. C., & Caballo, V. E. (2017). *Entrevista clínica semiestructurada para la ansiedad social (ECSAS) [Clinical structured Interview for Social Anxiety (CSISA)]*. Unpublished Manuscript Granada, Spain: FUNVECA Clinical Psychology Center.

Schneier, F. R., Foose, T. E., Hasin, D. S., Heimberg, R. G., Liu, S., Grant, B. F., & Blanco, C. (2010). Social anxiety disorder and alcohol use disorder comorbidity in the national epidemiologic survey on alcohol and related conditions. *Psychological Medicine, 40*(6), 977–988. https://doi.org/10.1017/S0033291709991231.

Shikatani, B., Antony, M. M., Kuo, J. R., & Cassin, S. E. (2014). The impact of cognitive restructuring and mindfulness strategies on postevent processing and affect in social anxiety disorder. *Journal of Anxiety Disorders, 28*(6), 570–579. https://doi.org/10.1016/j.janxdis.2014.05.012.

Sibrava, N. J., Beard, C., Bjornsson, A. S., Moitra, E., Weisberg, R. B., & Keller, M. B. (2013). Two-year course of generalized anxiety disorder, social anxiety disorder, and panic disorder in a longitudinal sample of African American adults. *Journal of Consulting and Clinical Psychology, 81*(6), 1052–1062. https://doi.org/10.1037/a0034382.

Smith, M. J. (1977). *Cuando digo NO me siento culpable [When I say no, I feel guilty]*. Barcelona, Spain: Grijalbo (Orig.: 1975).

Stein, D. J., Kawakami, N., de Girolamo, G., & Lépin, J.-P. (2017). Social anxiety disorder. In K. M. Scott, P. de Jonge, D. J. Stein, & R. C. Kessler (Eds.), *Mental disorders around the world. Facts and figures from the WHO World Mental Health Surveys* (pp. 120–133). New York: Cambridge University Press.

Stein, D. J., Lim, C. C. W., Roest, A. M., de Jonge, P., Aguilar-Gaxiola, S., Al-Hamzawi, A., ... WHO World Mental Health Survey Collaborators. (2017). The cross-national epidemiology of social anxiety disorder: Data from the world mental health survey initiative. *BMC Medicine, 15*, 143. https://doi.org/10.1186/s12916-017-0889-2.

Yadegari, L., Hashemiyan, K., & Abolmaali, K. (2014). Effect of acceptance and commitment therapy on young people with social anxiety. *International Journal of Scientific Research in Knowledge, 2*(8), 395–403. https://doi.org/10.12983/IJSRK-2014-P0395-0403.

Chapter 23

Implementing mindfulness-based cognitive therapy on dynamics of suicidal behavior: Understanding the efficacy and challenges

Debasruti Ghosh[a], Saurabh Raj[b], Tushar Singh[c], Sunil K. Verma[d], and Yogesh K. Arya[c]

[a]Department of Psychology, MDDM College (Babasaheb Bhimrao Ambedkar Bihar University), Muzaffarpur, Bihar, India, [b]Department of Psychology, Ramdayalu Singh College (Babasaheb Bhimrao Ambedkar Bihar University), Muzaffarpur, Bihar, India, [c]Department of Psychology, Faculty of Social Sciences, Banaras Hindu University, Varanasi, UP, India, [d]Department of Applied Psychology, Vivekananda College, University of Delhi, New Delhi, India

Abbreviations

ACT	acceptance and commitment therapy
CALM	coping with anxiety through living mindfully
CBT	cognitive-behavior therapy
DBT	dialectical behavior therapy
DSM-5	Diagnostic and Statistical Manual of Mental Disorders, 5th Edition
MBCT	mindfulness-based cognitive therapy
WHO	World Health Organization

The diverse aspects of suicidal behavior

The phenomenon of suicide is a grave concern, which is prevalent across all countries irrespective of age groups, class, creed, gender, and cultures. Suicidal behavior may consist of a set of self-inflicted potential injurious behaviors, initiated by an individual with a purpose to die, that may or may not result into death. In the year 2016, about 1.4% of deaths worldwide was estimated to have occurred by suicides and reported to be the 18th leading cause of death (WHO, 2019). In comparison to 2010, there has been reduction in death rate by suicide, due to increased mental health awareness. Nonetheless, it still remains a significant matter of concern as, 79% of suicides were from lower and middle income countries (WHO, 2019). These figures point out toward the complexities in prevention and treatment of suicidal behavior. There has been a long standing debate that whether suicidal behavior is an associated symptom or a separate nosological entity. Numerous evidences point out that the frequency of suicidal attempt is generally higher in people with a clinical condition, often as a symptom or consequential action of the underlying psychopathology such as depression, substance abuse, anxiety disorders, thought disorders, and physical health problems (Hawton, Sutton, Haw, Sinclair, & Harriss, 2005). Alternatively, another line of evidence suggests that suicidal ideation, thoughts and behavior can occur independently without the presence of core psychopathological features (Ahrens, Linden, Zäske, & Berzewski, 2000). Speculations in this direction gained momentum when it was explored that suicidal behavior was associated with the presence of more subtle forms of psychopathology such as social inhibition, poor family functioning, negative affectivity, emotion regulation difficulties and anxiety problems. More recently, DSM-5 mentioned about suicidal behavior disorder and the proposed criteria differentiates it from suicidal ideation and nonsuicidal self-injury and characterize it by the presence of suicidal attempt in past 2 years (American Psychiatric Association, 2013). In addition to the individual factors, certain environmental triggers can increase the risk for suicidality such as adjustment problems due to change of place, chronic illness, and sudden stressor like COVID-19 (Raj, Ghosh, Singh, Verma, & Arya, 2020). The WHO puts nonmedical factors such as social norms, developmental policies, economic and political systems under one umbrella term, i.e., *social determinants,* which can influence health outcomes. At individual level these causes manifest as job insecurity, work pressure, unemployment, poor family functioning, exposure to violence, failure in examination, and discrimination. These factors offer challenging circumstances and makes individual emotionally vulnerable by limiting their problem solving abilities, decision making process and compromise help seeking behavior that contribute to development of suicidal crisis.

The diversity of risk factors and causal pathways mentioned above increases the challenge for mental health professionals in planning efficient intervention strategies. The intervention methods need to be deliberated at two levels, i.e., as preventive approach (for vulnerable people and ideators) and as an adjunctive cure (for attempters). The prevention point of view may include handling stressors, awareness of situational triggers and dealing with thoughts and emotions that may lead to suicidal ideation. Whereas, as an adjunctive treatment, gaining better control over maladaptive thoughts that provoke suicide and develop coping mechanisms that may inhibit recurrence of suicidal behavior is required. The third wave of psychotherapeutic approaches that incorporate mindfulness techniques have gained immense popularity in both clinical and nonclinical settings. Among them, most popular are dialectical behavior therapy (DBT), acceptance and commitment therapy (ACT), and mindfulness-based cognitive therapy (MBCT). DBT is an evidence based approach for treating parasuicidal and self-harm behaviors in individuals diagnosed with borderline personality disorder (Linehan, Armstrong, Suarez, Allmon, & Heard, 1991). Nevertheless, DBT is an intensive and prolonged approach that not only integrates mindfulness techniques but includes other components too. ACT on the other hand has shown some efficacy in addressing suicidal ideation by reducing experiential avoidance in suicidal individuals. MBCT is a variant that combines cognitive behavioral strategies and mindfulness techniques to address various psychological issues and it can be a mode to address suicidal behavior issues as well. In addition to mindful awareness and acceptance, MBCT emphasizes on problem formulation model, goal identification, observation of thoughts and feelings, focusing on current experiences, treatment efficacy and evaluation, thereby retaining a similar structure as cognitive behavioral tradition (Teasdale et al., 2000). These elements may help in preventing occurrence and recurrence of suicidal behavior. However, suitability of an effective intervention in the context of suicide can only be determined by understanding the underlying mechanisms behind such acts.

Underpinnings of suicidal behavior from various perspectives

Suicidal nomenclature includes suicidal ideation, suicidal attempts, and suicidal cognition that are influenced by various psychological and social mechanisms. The cognitive theory of depression assumes that an individual's core belief, dysfunctional thinking patterns and negative automatic thoughts are the reasons behind their psychological condition. Grounded on these premises the cognitive model of suicidal behavior has identified a set of factors that activate a critical threshold where suicidal attempts most likely occur (Wenzel & Beck, 2008). For instance, a depressed individual (*predisposing vulnerability*) when experiences job loss (*situational trigger*) feels hopeless due to activation of dysfunctional thinking pattern, e.g., this is the worst" (*catastrophization*) and the individual might conclude that suicide is the way out to the current problems (*attention fixation*). In addition to these cognitive and emotional elements, several sub clinical psychopathological features or transdiagnostic processes possibly lie behind manifestation of suicidal behavior. The concept of suicidal mode which illustrates that activation of the *cognitive affective behavioral network* results into engaging in suicidal behavior mentions of such processes (Rudd, 2006). The network elaborates that emergence of suicidal cognition is a result of a synchronous process at different levels, i.e., experience of negative affect, increased physiological arousal and motivation to engage in suicidal behavior. Furthermore, it classifies three categories of core beliefs contribute to the cognitive component of suicidal mode and increases suicidal risk, i.e., helplessness, feeling unloved, and poor distress tolerance.

While working on clients with recurring depression Williams, Fennell, Barnhofer, Crane, and Silverton (2015) found that in the recurring episode one might not experience the same set of symptoms as they had in the last episode. However, suicidal ideation and tendency remained a relatively constant feature in the subsequent episodes with only a portion of depressive individuals actually acting on those ideations. In their attempt to explore on what activates the "suicidal mind," they found two possible mechanisms. The first one was *"cognitive reactivity"* that refers to the instant threshold where maladaptive thoughts are triggered by mere presence of a subtle nonpathological low mood (Ingram, Miranda, & Segal, 1998; Lau, Segal, & Williams, 2004; Segal, Teasdale, & Williams, 2004). Cognitive reactivity has the potential to reactivate the dysfunctional beliefs and create a condition for relapse of depressive episode. Many researches have indicated that depressive individuals with suicidal behavior have a high degree of cognitive reactivity (see Table 1). The second important mechanism is *"ruminative thinking"* which is a manner of responding to a stressful situation characterized by brooding and reflective pondering over the potential reasons and consequences of the symptoms (Treynor, Gonzalez, & Nolen-Hoeksema, 2003). It is not an active problem solving approach, rather it might lead to a fixated mode of feelings and

TABLE 1 Showing different underlying transdiagnostic processes behind suicidal behavior and indications on use of MBCT.

Transdiagnostic process	How does it contribute to suicidal crisis?	How MBCT can help?
Cognitive reactivity	In individuals with a history of suicidal depression the degree of cognitive reactivity is higher. Recurrent depressive patients with suicidal tendencies demonstrated higher cognitive reactivity in their recent episode in comparison to those who were not suicidal (Williams, 2008) Antypa, Van der Does, and Penninx (2010) found that cognitive reactivity may be the reason behind the relative stability of suicidal symptoms in depressive individuals. Further, they also found that distinct patterns of cognitive reactivity may be accountable for suicidal ideation and suicidal behavior	MBCT based exercises enable individuals to stabilize their attention. Focused attention will pave the way for such individuals to recognize their cognitive reactivity and decenter their negative patterns of thinking (Williams & Kabat-Zinn, 2011)
Attentional bias	Attentional bias in the context of suicide is fixation tendency on those negative thoughts that might increase the risk to get engaged in suicidal behavior It creates a block to effective problem solving strategy and an individual is unable to dis engage from the selective information processing and schemas related with it may be activated which increases vulnerability toward particular disorders (Beck, 2005)	Techniques of MBCT such as focusing on breathing, enhances awareness and makes people more attentive to present moment experience and in turn increases voluntary control of attention (Segal, Teasdale, Williams, & Gemar, 2002) Mindfulness helps in developing awareness on where one's attention is directed to. This enhanced awareness guides an individual in stabilizing their attention and making them capable of switching their attention between thoughts, bodily sensations and present stimuli (Bishop et al., 2004) MBCT also enhances executive attention which helps in reducing suicidal ideation (Chesin et al., 2016)
Ruminative thinking	Rumination is a common correlate of suicidal behavior and is regarded as a longitudinal predictor of suicidal thoughts (Smith, Alloy, & Abramson, 2006) In depressed individuals brooding tendency, i.e., dwelling on consequences and reflecting on exhausted problem solving abilities worsens the psychopathological experience and is related with suicidal attempts as well as suicidal ideation (Rogers & Joiner, 2017)	MBCT focuses on directing attention to present moment and staying in contact with reality that keeps a check on brooding and reflective tendencies involved in rumination Mindfulness increases the capacity to relate more compassionately to oneself and others, which might in turn reduce the tendency to become lost in negative ruminative thought patterns. (Gilbert & Tirch, 2009)
Hopelessness	Hopelessness depicts a solution-less state where an individual predominantly has negative expectations. It is one of the strongest indicators of current as well as eventual suicide	Mindfulness exercises which emphasize on focusing attention on daily life activities helps to build awareness about one's feelings, body sensations and thoughts as an automatic consequence of pleasant and unpleasant experiences. Gradually, suicidal individuals are able to identify the recurring nature of thoughts and feelings which help them gain access to their internal resources through which they improve their problem solving ability
Thought suppression	Thought suppression plays an vital role in mood disorders and suicide as it results in a paradoxical increase of unpleasant cognitions which are to be avoided including thoughts linked with suicide (Rosenthal, Cheavens, Lynch, & Follette, 2006)	MBCT discourages mental control strategies rather it emphasizes on acceptance and adaptive ways of managing feelings and thoughts. Acceptance yields better results in reducing aversiveness and intrusions in comparison to suppression (Marcks & Woods, 2005)
Thwarted belongingness	The interpersonal theory suggests indices of social isolation that are linked with suicide such as poor social support and living alone may indicate unmet belongingness needs. In order to avoid the negative emotions arising from this loneliness state an individual resorts to suicidal behavior	Mindfulness techniques focus on acceptance and attenuate emotional escaping tendencies arising from challenges to belongingness needs (Collins, Stebbing, Stritzke, & Page, 2017)

Continued

TABLE 1 Showing different underlying transdiagnostic processes behind suicidal behavior and indications on use of MBCT—cont'd

Transdiagnostic process	How does it contribute to suicidal crisis?	How MBCT can help?
Perceived burdensomeness	Suicidal behavior attributed to perceived burdensomeness occurs when one perceives that they are a burden on others, and that if they are gone then problems will solve. Examples: unemployment or functional impairment	Mindfulness techniques moderate perceived burdensomeness which can have a protective effect on suicidal ideation. MBCT can bring clarity in thoughts of burdensomeness, and may reduce distress by inducing problem solving when clear environmental clues of being a burden are absent (Collins, Best, Stritzke, & Page, 2016)
Poor emotional regulation	Inability to manage emotions has been strongly related to self-destructive behavior. Poor control over emotions result into impulsive actions which is often seen in adolescent suicidal behavior (Brown & Anderson, 1991)	MBCT enables decentering which has been hypothesized to be the core mechanism of change that produces greater affective stability. Mindfulness indirectly promotes adaptive emotion regulation strategies by improving the emotional differentiation capacity (Mandal, Arya, & Pandey, 2014)

people might not take action. Evidence suggests that depressive individuals engage in ruminative thinking and it has a relationship with both suicidal ideation and suicidal attempt (Surrence, Miranda, Marroquín, & Chan, 2009). Another thought related phenomenon in the context of suicide is "*thought suppression*" which is cognitive control mechanism involving a deliberate attempt to suppress negative thoughts and unpleasant cognitions in order to avoid them (Wegner, Schneider, Carter, & White, 1987). It may result in a "rebound effect," i.e., one might experience those suppressed thoughts more frequently (Abramowitz, Tolin, & Street, 2001).

The defeatist and impervious nature of thoughts of suicidal individuals, makes them wait for changes or someone who might pull them out of the situation instead of taking charge of the situation (Pollock & Williams, 2004). The cry for pain perspective explains how suicidal individuals feel that they are "caught up" feel helpless and become a prey of their own thoughts (Williams, 2014). Suicidal individuals have difficulty in decentering their thoughts, i.e., they have a tendency to dwell upon their thoughts, rather than seeing it from an independent perspective. This creates a gap between the thought and action relationships, followed by a poor control over situational determinants. In addition to this individuals with suicidal depression may have a compromised problem solving ability due to difficulty in recalling specific episodes or details from their autobiographical memory referred to as *overgeneral memory* (Evans, Williams, O'Loughlin, & Howells, 1992; Williams, Teasdale, Segal, & Soulsby, 2000).

In general, our day to day problem solving involves focusing on the mismatch, i.e., where we are and where we want to be, and accordingly we plan to act. In mindfulness paradigm, this is also referred to as the "doing mode" (Williams et al., 2015). However, when there are certain shifts in mood the focus on the mismatch is larger than the actual solution and that's when individual slips into the "driven doing mode" which impacts their problem solving abilities. Williams et al. (2015) explain the interaction of negative thoughts, mood and overgeneral memory results in a certain form of processing called as *discrepancy based processing* that leads to this driven doing mode. This form of processing has three components, i.e., negative views of self are perceived as reality, reaching to ideal self by a matching tendency to compensate this mismatch, and adopting rumination and avoidance to solve problems and address negative thoughts (Williams, 2008). Nonetheless, this mode is not healthy in addressing perceived discrepancy rather increases the negative thoughts and may be the reason behind emotion dysregulation in suicidal individuals (Hargus, Crane, Barnhofer, & Williams, 2010).

Another important phenomenon, i.e., *differential activation hypothesis* (Teasdale, 1988) refers to the point that during previous episodes of depression, low mood and other triggers gain potential to activate negative thinking in the recurrent depressive episodes. Williams et al. (2015) described analogical similarity between differential activation model of depression and Buddhist "*Sullatha suttha*" concept. It explains the difference between pain and suffering with a metaphor of two arrows whereby the first arrow is an inevitable emotional pain experience and second arrow is the suffering, i.e., the "thinking behind pain" reflective of the meaning one attaches with that pain. The meaning is generally drawn from one's past experiences, temperament and their affective states during that period. In case of depressive relapse, the previous episode could be the first arrow that caused emotional pain. And the suffering induced by the pain because of particular sets of learned beliefs and assumptions ("second arrow") would lead people to attach a negative meaning to this emotional pain, and then respond to this meaning by repetitively thinking about its causes and consequences. Adding to this Williams et al. (2015) proposed a differential activation profile of suicidal behavior and suggested the presence of a "third arrow." This begins when the negative thinking and sudden low mood, blocks the mind's access to positive events and triggers hopelessness which leads to a suicidal crises stage. Hence, a third element or third arrow which is a strong and captivating sense of *urgency* after the meaning is derived occurs. This sense of urgency is characterized by a sense of "having to do something" about the pain, or else it would be extremely distressing, and one might lose control of the situation and may have suicidal thoughts. The ideation to action framework explains that psychological pain and hopelessness generate passive suicidal ideation that becomes active when pain exceeds connectedness and when this state progresses and one has capacity then it turns into action (Klonsky, Saffer, & Bryan, 2018).

A number of social and cultural risk factors have been indicated in suicidal ideation and suicidal behavior. Social isolation, poor peer support, poor family functioning are the strongest risk factors for suicidal ideation, in contrast, peer support, social belongingness, marriage, and children are associated with reduced suicidal risk (Van Orden et al., 2010). As proposed by Joiner, Wingate, and Otamendi (2005) in the interpersonal theory of suicide, absence of reciprocal care and social isolation can challenge the belongingness needs of the individual making them feel less wanted leading to a *thwarted belongingness* state (Joiner et al., 2005). Additionally, factors such as old age, unemployment, chronic illness may instill a perception of being a burden to others and one might think that the world will be better without them, often referred to as a state of *perceived burdensomeness* (Joiner et al., 2005). The effects of above mentioned mechanisms in suicidal ideation and suicidal behavior have been illustrated in Table 1.

Why MBCT is relevant in the context of suicidal behavior?

MBCT can be one of the ways which can bring about a difference in the processes which build up suicidal thoughts (see Table 1). The concept of mindfulness has its roots in teachings of Buddhism but shares a conceptual similarity with existentialism, phenomenology, philosophy and humanistic perspectives (Brown, Ryan, & Creswell, 2007). Mindfulness has been defined as an awareness in which one purposively pays attention in the present moment in a nonjudgmental manner to unfold moment by moment experience (Kabat-Zinn, 2003). In Buddhist tradition mindfulness is regarded as a way to understand the origins and culmination of suffering. It is also regarded as a way to break the pattern of additional suffering and freeing oneself from pain and difficulty (Gunaratana, 2002). The two fundamental aspects of mindfulness are *awareness* and *attention*. Awareness involves directly experiencing the reality by consciously registering the activities of the mind and stimuli from the senses. Conversely, attention involves engaging or "taking notice of" a strong stimulus (Nyanaponika, 1994).

MBCT assumes that nonawareness and constant judgment of the thoughts are the root cause of such emotional disturbances. Williams and Swales (2004) assume that suicidal people experience an intolerable "psychic pain" generated from their attempts to change, reduce or fix their disturbing thoughts. Furthermore, they have highlighted four key themes of MBCT that emphasize the role of awareness in understanding thoughts that may be relevant in context of suicide. First, the reason behind this absence of awareness is that our minds are in "*automatic pilot*" mode, i.e., we always get indulged in analyzing about past or future activities and have the least awareness of what we are doing in the present moment. Second, in depressive individuals with suicidal tendencies, the state of *mindlessness* might trigger unwanted habit of indulging in negative thoughts, activate mechanisms like ruminative thinking, dwelling on past events or thoughts of what might happen in future often occupy their thoughts. Thirdly, the habitual pattern of these thoughts often leads to a self-perpetrated "*landslide effect*," i.e., worsening of mood, whereby awareness plays the most important role. Lastly, with MBCT one can be more aware of such thoughts, find alternate ways, and instead of being driven by unhelpful habits, one can freely choose on how to think and act. Williams et al. (2015) mentioned two indications showing MBCT can be a mode of psychotherapy for suicidal individuals. The first indication is the "solution blindness" situation in the part of the suicidal individual which is integrally related to the way they remember their past events. This generally happens due to the overgeneral and overabstract nature of memories. Studies have demonstrated the efficacy of MBCT reduces this overgeneral memory in previously depressed individuals (Williams et al., 2000). The second indication is that MBCT increases the present moment awareness and enhances attentiveness in the process of making such autobiographical memory more specific, thus reducing vulnerability. An example illustrating the mechanism of suicidal behavior explained using the above mentioned foundations has been represented in Fig. 1.

```
                    ┌─────────────────────────┐
                    │   'Automatic Pilot Mode'│
                    │ -Non Awareness to present│
                    │   moment activities     │
                    │ -Either dwells in past  │
                    │   or is apprehensive    │
                    │   about future          │
                    └─────────────────────────┘
```

```
   ┌──────────────────┐           ┌──────────────────┐
   │ Thoughts about   │           │ Thoughts about   │
   │ past             │           │ future           │
   │ 'I failed in the │           │ 'What will be the│
   │ last examination'│           │ result of my     │
   │                  │           │ upcoming         │
   │                  │           │ examination?'    │
   └──────────────────┘           └──────────────────┘
```

Habitual Negative Thinking Pattern
"I might fail in my upcoming examination'
"My performance will be the worst"
"People might ridicule me"

Thoughts shift away awareness of activities in the present moment and attention is fixed to this pattern (attentional bias).

Landslide effect
1. Worsening of mood.
2. Inability to decenter thoughts and a state of 'solution blindness'. (*I don't know how to overcome this*)

Maladaptive way/approach:

Outcome: Suicidal Behaviour

Adaptive way/approach (Use of intervention) MBCT
1. Increases awareness and attention to present moment experiences.

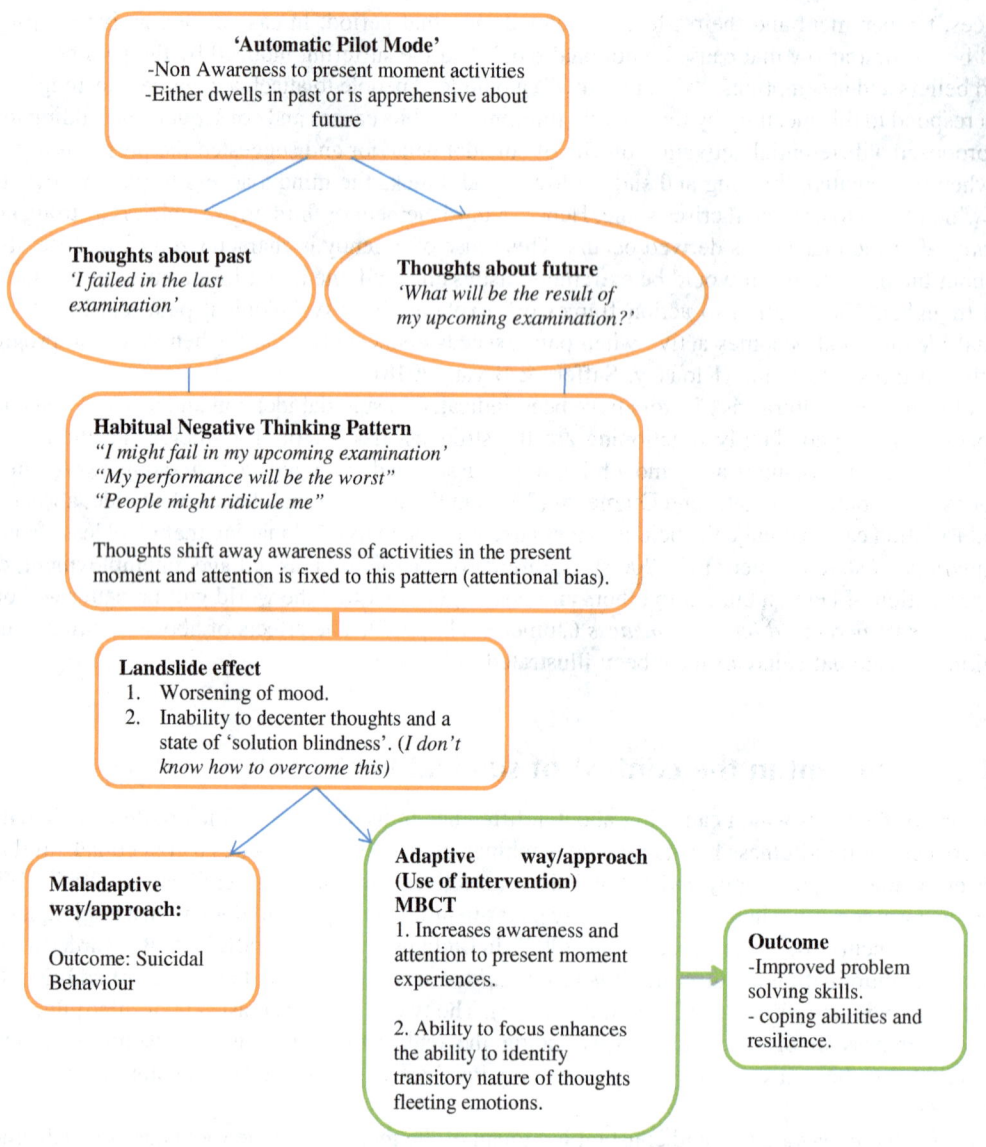

2. Ability to focus enhances the ability to identify transitory nature of thoughts fleeting emotions.

Outcome
-Improved problem solving skills.
- coping abilities and resilience.

FIG. 1 An example representing of theme behind use of MBCT in suicidal behavior. The figure represents an example based on themes for MBCT identified by Williams and Swales (2004) that the general reason behind any emotional pain and suffering is the automatic pilot mode. A person who has failed in examination might be preoccupied with thoughts about past or be apprehensive about future challenges and this tendency results into a habit of unhelpful negative thinking creating an attentional bias. The constant negative cognition has a landslide effect, i.e., a point where mood worsens and an individual might not be able to decenter negative thoughts leading to solution blindness situation (Williams et al., 2015). The overwhelming state makes an individual feel captivated and finding no way out the maladaptive outcome could be suicidal behavior. However, adaptive ways to deal with the problem could be through MBCT. Mindfulness based cognitive therapy can be implemented to enhance attention to present moment experience and to understand the transient state of emotional feelings. This ability builds coping strategies, problem solving and inculcates resilience.

The process and variations of MBCT

MBCT is an 8-week program which was developed by Teasdale et al. (2000) for patients with recurrent depression. It involves use of formal and informal meditative practices to enhance present moment awareness in the participants. While formal practices include mindfulness meditation, mindful eating, mindful walking and yoga; informal procedures may involve practicing mindfulness in daily life activities. The initial sessions involving mindfulness require participants to focus their attention to specific features like body sensation and breaths. In the subsequent sessions, participants attempt to extend their learning to a wider range of thoughts and emotions that they experience either internally and externally (Williams, Duggan, Crane, & Fennell, 2006). Cognitive techniques that are used in MBCT comprise of psycho-educating about basic symptomatology of depression, recognizing how suppressing and ruminating thoughts contribute to distressing

TABLE 2 Showing major techniques used in MBCT as illustrated by Williams et al. (2015) while working with clients with suicidal crisis.

Mindfulness techniques	How does it effect?	Cognitive techniques	How does it effect?
Body scan meditation and mindful eating exercises	These mindfulness based practices helps in building mindful awareness of one's thoughts and feelings and understanding the automatic pilot mode of mind Through body scan suicidal individuals learn the art of noticing, engaging and disengaging attention and recognize aversions	Cognitively oriented psycho-education	Participants learn about the symptoms of depression and vulnerability mechanisms leading to relapse
Stretching (Yoga) and scheduled breathing exercises	Individuals focus on accepting the thoughts and being in present moment. It involves understanding the wandering nature of mind, paying attention to it and bringing it back to the present activity and sensations. These practices also bring an awareness about the unhelpful and self-critical thoughts	Thoughts and feelings exercise	This is done by giving ambiguous scenarios through which the individual learns to reflect how body sensations, feelings, thoughts and behaviors are connected from a metacognitive viewpoint. This knowledge helps in gaining knowledge that thoughts are independent mental events
Meditation (sitting, seeing and hearing meditation)	These techniques enable individuals to identify aversive experiences. It involves an insight oriented approach on brooding and ruminative tendencies and recognizing how distractions occurs in one's thoughts. Gradually, acceptance to experiences and their impact on body is developed. Also, it helps to understand that change becomes spontaneous activity if one continues to accept these experiences rather than avoiding them	Identifying negative automatic thoughts	This exercise helps in understanding the relationship of thinking with emotional disturbances. It helps an individual to explore the connections between mood, thoughts and behavior and how a negative automatic thought comes with a slight change in mood. In combination with mindfulness these thoughts can be acknowledged and grounded
Mindful movement (mindful walking, stretching)	This requires the individuals to pay attention to their pleasant and unpleasant body sensations and notice aversions in the process	Recognizing signs of relapse	This is done through activities which may also include group based exercises which help in understanding the impact of mood on thoughts. Individuals are encouraged to observe and accept difficult thoughts from a decentering perspective and identify their relapse signatures
Mindfulness in everyday life (pleasant/unpleasant experiences, breathing spaces)	Practicing mindfulness in everyday activities helps in understanding the impact of mood on thoughts and promotes well-being		

experience, recognizing the origins of negative automatic thoughts and their impact on thinking and emotional processes (see Table 2). Gradually, with the combination of both the approaches the individuals learn to direct their attention to those experiences in their daily lives which they might have not noticed due to the "thought block." Though cognitive therapy techniques have a central role in MBCT, however, both differ in certain ways. First, while cognitive-behavior therapy (CBT) focuses on changing thoughts, MBCT emphasizes on acceptance. Second, MBCT offers to see the impact of negative affect on thoughts, body sensations independently, unlike CBT which focuses on avoidance and modification of content of thoughts. The MBCT practices adapted by researchers may vary, however a glimpse of the techniques used in the context of suicidal behavior as suggested by Williams et al. (2015) has been shown in Table 2.

MBCT can have good effect on suicidal clients as it does not involve changing the content of the thoughts, instead it focuses on enhancing awareness and modification of these processes (Williams & Swales, 2004). MBCT enables individuals to recognize the emotional responses and warning signals that arise from negative thoughts and makes them prepared to take relevant actions in the event of future depression and suicidal thoughts. It is not only effective in bringing about personal change but it also enhances metacognitive awareness in an individual. Metacognitive awareness is the capability to experience cognitions (thoughts, images, assumptions, attitudes, beliefs) as events in the mind, learned mental habits, rather than as a reflection of objective truth. Metacognitive awareness lets a person to delineate oneself from old patterns of thinking and feeling, and to step back and see the transitory nature of thoughts and feelings. These abilities help in developing coping and resilience against negative affect and thoughts and boost problem solving and decision making (Williams & Swales, 2004). Apart from this MBCT has proven quite effective in preventing relapse in depressives with suicidal tendencies. The literature indicates that MBCT is capable of bringing about changes by training the participants to get acquainted with their own specific relapse-signature and develop an action plan on their own, in case of a relapse (Williams et al., 2015).

Evidence of effectiveness of MBCT on suicidal ideation and other psychological dynamics related to suicide

The preceding sections have amply supplemented the view that the purpose behind a person resorting to suicidal behaviors is to escape from emotional pain and suffering. Mindfulness based interventions have proven effective in working with several aspects mentioned in the latent mechanisms. Based on the researches done in the context of suicide the efficacy of MBCT has been given below:

MBCT on suicidal ideation

Suicidal ideation can range from transient, fleeting thoughts, to a systematic and elaborate plan of attempting suicide. It has a significant association basic lifestyle stressors, anxiety, depression, and posttraumatic stress disorder. Forkmann et al. (2014) found that MBCT had a significant effect in reducing suicidal ideation in residual depression patients. In addition, effect of MBCT on suicidal ideation was independent of the impact it had on other symptoms of depression was also reported, which implies MBCT can work on suicidal ideation directly. Further studies found that MBCT has a superior effect in comparison to other treatment mode on suicidal items of Beck Depression Inventory in chronic depressive individuals (Forkmann, Brakemeier, Teismann, Schramm, & Michalak, 2016). The authors highlighted that one of the mechanism that could potentially enhance the impact of MBCT is that since it is a group based intervention, the feelings of social isolation are less and participants find peer fort which enhances their feelings of social belongingness and may be the reason for reduction in suicidal ideation. In adolescents with suicidal ideation MBCT improved self-awareness, enhanced understanding of faulty cognitions and reduced negative thoughts as result of which reduced suicidal ideation (Raj et al., 2019).

MBCT on cognitive mechanisms underlying suicide

Most research on cognitive mechanisms and executive functioning on suicidal depressive individuals report a positive impact of MBCT. Improved awareness through MBCT facilitates use of adaptive alternatives in depressive individuals with suicidal tendencies and helps them in modifying and breaking the vicious cycle of rumination and thought suppression (Hargus et al., 2010). Barnhofer et al. (2015) found that that MBCT skills that focus on decentering could bring a reduction in suicidal cognition and in turn reducing depressive symptoms. MBCT-S (Mindfulness based cognitive therapy for preventing suicidal behavior) enhanced executive attention and decreased ruminative thinking, hopelessness, and cognitive reactivity in high suicidal risk depressive patients (Chesin et al., 2016).

MBCT on associated psychological mechanisms related to suicide

Mindfulness and cognitive techniques such as identifying thoughts and feelings nonjudgmentally used in MBCT can lead to both lifestyle changes and cognitive changes that help in reducing suicidal thoughts and death anxiety in people with cancer (Nabipour, Rafiepour, & Haji Alizadeh, 2018). MBCT increases positive experiences and optimism leading to increased life satisfaction and life orientation in adolescents with suicidal ideation (Raj et al., 2019).

Implementing MBCT with suicidal clients: Challenges and way ahead

MBCT has an edge over other approaches since it is time bounded and can be implemented easily in groups, hence through this intervention a large number of people can be benefitted. Given these premises, techniques of MBCT can be used for suicide prevention in community level. The preceding sections adequately highlight the importance of MBCT in reducing depressive relapse and suicidal crisis. However, researches on the impact of MBCT specifically on suicidal risk factors are still in its formative years and limited data is available. Besides, in the context of suicidal behavior implementation of MBCT can be challenging due to different reasons. Firstly, attrition rates are a matter of concern reported in few of the studies (Raj, Ghosh, Verma, & Singh, 2020). Two factors lie behind the attrition or dropouts from the session, i.e., longer duration and associated negative emotion. MBCT involves an 8-week long training that requires patience and dedication from the participants. Many studies have pointed out that while working with suicidal clients one has to be careful about the negative emotions they might experience while focusing their attention. If it is not taken care of appropriately it can cause emotional uproar and distress in them and they might quit further sessions (Crane & Williams, 2010). Secondly, many researches back the view that MBCT cannot be used in acute suicidal crisis because of the overwhelming nature of the situation where suicidal client might not be in a state to start with mindfulness, limiting its use as an adjunctive approach rather than first line of treatment. Third, though MBCT can be imparted individually but it is primarily a group based intervention and sometimes, stigma associated with suicidal thoughts may prevent clients to openly share their emotions and thoughts in group. Williams et al. (2015) explain that the idea of sharing experiences with strangers can be anxiety provoking and intimidating for suicidal people. Fourthly, clinical competency of MBCT trainers in handling suicidal clients is extremely important. The competency characteristics of trainers include efficiency in the procedures as well as being empathetic toward the suicidal clients (Samra & Monk, 2007). This characteristic is an integral part of implementation fidelity as it determines the treatment efficacy and evaluation (Raj, Ghosh, Verma, & Singh, 2020). Though these challenges need precision, yet MBCT has proved to be effective in reducing suicidal behavior.

Future studies that aim to study on independent aspects of suicidal behavior are required in order so that impact of MBCT on suicidality can be clearly distinguished. Also, an objective assessment of trainer's abilities and competency in delivering mindfulness must be done, to ensure fidelity in treatment procedure. Further, we suggest that adaptable versions of MBCT to address suicidal behavior considering the diversity in its nomenclature and cultural specific practices must be designed. Culture based adaptation of MBCT would be beneficial in extending help and reaching out to maximum people with suicidal ideation. Suicide is a multifarious phenomenon occurring in all age groups and as per the reports of WHO, most suicides occur in 15–19 years old which is an alarming concern (WHO, 2019). MBCT-C is an adapted version of intervention for children with anxiety and depression, likewise adapted versions of MBCT for children and adolescent suicidal behavior can be designed. The world witnessed an unprecedented crisis in the form of COVID-19 and suicidal rates were higher in the beginning phases of the pandemic. In such a scenario web based or e-MBCT approaches for prevention from suicide could be a way ahead.

Mini-dictionary of terms

Suicidal ideation: It reflects thoughts and plan about suicide.
Overgeneral memory: Inability to recollect specific events from one's autobiographical memory.
Metacognitive awareness: It is the awareness of one's own thoughts and strategies.
Transdiagnostic process: A mechanism which may either pose as risk factor or maintaining factor of a disorder/condition.
Brooding: It is a passive and judgmental thinking mechanism, mainly dependent on one's mood.
Landslide effect: A condition characterized by worsening of mood which is self-perpetuated by unhelpful habits of thinking in depressive/suicidal people when triggered by certain incidents/events.
Automatic pilot: A state of mind whereby one acts without being consciously aware to their present moment experiences.
Psychoeducation: A process of spreading awareness and educating individuals about symptoms and presentation of a psychiatric condition.
Mindful eating: A mindfulness technique that involves focusing on sensations and experiences while having food.
Decentering: A strategy which involves seeing one's mental events (thoughts) objectively from an independent and non-judgmental perspective.

Key fact on suicide

1. According to WHO, every year approximately 8,00,000 people die by suicide. While the number of people attempting suicide is higher than the committed ones, but the number of people having suicidal ideation is highest.

Application to other areas

MBCT was initially developed for depressive individuals, but, it has widely been applied across other psychiatric disorders and general medical conditions as well. The ease of understanding the exercises and self-applicable nature of MBCT has made it a popular intervention in both face to face mode web based interfaces also. MBCT is impactful as an adjuvant treatment in the context panic disorder, generalized anxiety disorder, social phobia and anxiety in general clinical condition (Fumero, Peñate, Oyanadel, & Porter, 2020). MBCT components that emphasize on present moment focus have been linked to improved attention and mind wandering in adults with attention deficit hyperactivity disorder (Gu, Xu, & Zhu, 2018). The exercises used in MBCT helps in reducing cancer related fatigue by creating an insight to their inner and outer world boundaries leading to better recognition of the process that generate exhaustion (Bruggeman-Everts, 2019). The CALM pregnancy intervention which is an adapted MBCT approach has shown good short term and long term implication to ease perinatal anxiety and depression symptoms in pregnant women (Goodman et al., 2014). Not only this, recent evidences have identified that the ability to focus developed through MBCT enhances the use of adaptive coping strategies and self-regulation in nonclinical population (De la Fuente, Mañas, Franco, Cangas, & Soriano, 2018).

Summary points

- Researchers have considered suicidal behavior both as a consequence as well as independent nosological entity.
- Several transdiagnostic processes are accountable for development and maintenance of suicidal thoughts.
- MBCT is a variant of CBT that incorporates techniques of mindfulness and CBT.
- Mindfulness identifies that automatic pilot, a state of mindlessness and downward shift in mood (landslide effect) could result in maladaptive behavior (suicidal attempt).
- MBCT techniques enhances attention, builds awareness to current experiences which could be beneficial for suicidal individuals to recognize their maladaptive thoughts.
- Researchers have found that MBCT works on improving attention, metacognitive awareness, life satisfaction and decreases perceived burdensomeness in suicidal individuals.
- Apart from being an adjunctive approach, MBCT can be used as a suicide prevention approach in community level.

References

Abramowitz, J. S., Tolin, D. F., & Street, G. P. (2001). Paradoxical effects of thought suppression: A meta-analysis of controlled studies. *Clinical Psychology Review, 21*(5), 683–703.

Ahrens, B., Linden, M., Zäske, H., & Berzewski, H. (2000). Suicidal behavior—Symptom or disorder? *Comprehensive Psychiatry, 41*(2), 116–121.

American Psychiatric Association. (2013). *Diagnostic and statistical manual of mental disorders (DSM-5®)*. American Psychiatric Pub.

Antypa, N., Van der Does, A. W., & Penninx, B. W. (2010). Cognitive reactivity: Investigation of a potentially treatable marker of suicide risk in depression. *Journal of Affective Disorders, 122*(1–2), 46–52.

Barnhofer, T., Crane, C., Brennan, K., Duggan, D. S., Crane, R. S., Eames, C., ... Williams, J. M. G. (2015). Mindfulness-based cognitive therapy (MBCT) reduces the association between depressive symptoms and suicidal cognitions in patients with a history of suicidal depression. *Journal of Consulting and Clinical Psychology, 83*(6), 1013.

Beck, A. T. (2005). The current state of cognitive therapy: A 40-year retrospective. *Archives of General Psychiatry, 62*(9), 953–959.

Bishop, S. R., Lau, M., Shapiro, S., Carlson, L., Anderson, N. D., Carmody, J., ... Devins, G. (2004). Mindfulness: A proposed operational definition. *Clinical Psychology: Science and Practice, 11*(3), 230–241.

Brown, G. R., & Anderson, B. (1991). Psychiatric morbidity in adult inpatients with childhood histories of sexual and physical abuse. *American Journal of Psychiatry, 148*(1), 55–61.

Brown, K. W., Ryan, R. M., & Creswell, J. D. (2007). Addressing fundamental questions about mindfulness. *Psychological Inquiry, 18*(4), 272–281.

Bruggeman-Everts, F. Z. (2019). *Evaluation of two different web-based interventions for chronic cancer-related fatigue: online mindfulness-based cognitive therapy and ambulant activity feedback*. University of Twente. https://doi.org/10.3990/1.9789402813319. (Accessed 4 May 2022).

Chesin, M. S., Benjamin-Phillips, C. A., Keilp, J., Fertuck, E. A., Brodsky, B. S., & Stanley, B. (2016). Improvements in executive attention, rumination, cognitive reactivity, and mindfulness among high-suicide risk patients participating in adjunct mindfulness-based cognitive therapy: Preliminary findings. *The Journal of Alternative and Complementary Medicine, 22*(8), 642–649.

Collins, K. R., Best, I., Stritzke, W. G., & Page, A. C. (2016). Mindfulness and zest for life buffer the negative effects of experimentally-induced perceived burdensomeness and thwarted belongingness: Implications for theories of suicide. *Journal of Abnormal Psychology, 125*(5), 704.

Collins, K. R., Stebbing, C., Stritzke, W. G., & Page, A. C. (2017). A brief mindfulness intervention attenuates desire to escape following experimental induction of the interpersonal adversity implicated in suicide risk. *Mindfulness, 8*(4), 1096–1105.

Crane, C., & Williams, J. M. G. (2010). Factors associated with attrition from mindfulness-based cognitive therapy in patients with a history of suicidal depression. *Mindfulness, 1*(1), 10–20.

De la Fuente, J., Mañas, I., Franco, C., Cangas, A. J., & Soriano, E. (2018). Differential effect of level of self-regulation and mindfulness training on coping strategies used by university students. *International Journal of Environmental Research and Public Health, 15*(10), 2230.

Evans, J., Williams, J. M. G., O'Loughlin, S., & Howells, K. (1992). Autobiographical memory and problem-solving strategies of parasuicide patients. *Psychological Medicine, 22*(2), 399–405.

Forkmann, T., Brakemeier, E. L., Teismann, T., Schramm, E., & Michalak, J. (2016). The effects of mindfulness-based cognitive therapy and cognitive behavioral analysis system of psychotherapy added to treatment as usual on suicidal ideation in chronic depression: Results of a randomized-clinical trial. *Journal of Affective Disorders, 200*, 51–57.

Forkmann, T., Wichers, M., Geschwind, N., Peeters, F., van Os, J., Mainz, V., & Collip, D. (2014). Effects of mindfulness-based cognitive therapy on self-reported suicidal ideation: Results from a randomised controlled trial in patients with residual depressive symptoms. *Comprehensive Psychiatry, 55*(8), 1883–1890.

Fumero, A., Peñate, W., Oyanadel, C., & Porter, B. (2020). The effectiveness of mindfulness-based interventions on anxiety disorders. A systematic meta-review. *European Journal of Investigation in Health, Psychology and Education, 10*(3), 704–719.

Gilbert, P., & Tirch, D. (2009). Emotional memory, mindfulness and compassion. In *Clinical handbook of mindfulness* (pp. 99–110). New York, NY: Springer.

Goodman, J. H., Guarino, A., Chenausky, K., Klein, L., Prager, J., Petersen, R., ... Freeman, M. (2014). CALM pregnancy: Results of a pilot study of mindfulness-based cognitive therapy for perinatal anxiety. *Archives of Women's Mental Health, 17*(5), 373–387.

Gu, Y., Xu, G., & Zhu, Y. (2018). A randomized controlled trial of mindfulness-based cognitive therapy for college students with ADHD. *Journal of Attention Disorders, 22*(4), 388–399.

Gunaratana, B. (2002). *Buddhist concept of happiness*. Bhavana Society.

Hargus, E., Crane, C., Barnhofer, T., & Williams, J. M. G. (2010). Effects of mindfulness on meta-awareness and specificity of describing prodromal symptoms in suicidal depression. *Emotion, 10*(1), 34.

Hawton, K., Sutton, L., Haw, C., Sinclair, J., & Harriss, L. (2005). Suicide and attempted suicide in bipolar disorder: A systematic review of risk factors. *The Journal of Clinical Psychiatry, 66*(6), 693–704.

Ingram, R. E., Miranda, J., & Segal, Z. V. (1998). *Cognitive vulnerability to depression*. Guilford Press.

Joiner, T. E., Wingate, L. R., & Otamendi, A. (2005). An interpersonal addendum to the hopelessness theory of depression: Hopelessness as a stress and depression generator. *Journal of Social and Clinical Psychology, 24*(5), 649–664.

Kabat-Zinn, J. (2003). Mindfulness-based stress reduction (MBSR). *Constructivism in the Human Sciences, 8*(2), 73.

Klonsky, E. D., Saffer, B. Y., & Bryan, C. J. (2018). Ideation-to-action theories of suicide: A conceptual and empirical update. *Current Opinion in Psychology, 22*, 38–43.

Lau, M. A., Segal, Z. V., & Williams, J. M. G. (2004). Teasdale's differential activation hypothesis: Implications for mechanisms of depressive relapse and suicidal behaviour. *Behaviour Research and Therapy, 42*(9), 1001–1017.

Linehan, M. M., Armstrong, H. E., Suarez, A., Allmon, D., & Heard, H. L. (1991). Cognitive-behavioral treatment of chronically parasuicidal borderline patients. *Archives of General Psychiatry, 48*(12), 1060–1064.

Mandal, S. P., Arya, Y. K., & Pandey, R. (2014). Understanding emotion regulatory effect of mindfulness: Role of differentiation and range of emotional experiences. *Indian Journal of Positive Psychology, 5*(4), 356.

Marcks, B. A., & Woods, D. W. (2005). A comparison of thought suppression to an acceptance-based technique in the management of personal intrusive thoughts: A controlled evaluation. *Behaviour Research and Therapy, 43*(4), 433–445.

Nabipour, S., Rafiepour, A., & Haji Alizadeh, K. (2018). The effectiveness of mindfulness based cognitive therapy training on anxiety of death and thoughts of suicide of patients with cancer. *Zahedan Journal of Research in Medical Sciences, 20*(1).

Pollock, L. R., & Williams, J. M. G. (2004). Problem-solving in suicide attempters; problem-solving in suicide attempters; LR Pollock and JMG Williams. *Psychological Medicine, 34*(1), 163.

Raj, S., Ghosh, D., Singh, T., Verma, S. K., & Arya, Y. K. (2020). Theoretical mapping of suicidal risk factors during the COVID-19 pandemic: A mini-review. *Frontiers in Psychiatry, 11*.

Raj, S., Ghosh, D., Verma, S. K., & Singh, T. (2020). The mindfulness trajectories of addressing suicidal behaviour: A systematic review. *International Journal of Social Psychiatry, 67*(5), 507–519. https://doi.org/10.1177/0020764020960776.

Raj, S., Sachdeva, S. A., Jha, R., Sharad, S., Singh, T., Arya, Y. K., & Verma, S. K. (2019). Effectiveness of mindfulness based cognitive behavior therapy on life satisfaction, and life orientation of adolescents with depression and suicidal ideation. *Asian Journal of Psychiatry, 39*, 58–62.

Rogers, M. L., & Joiner, T. E. (2017). Rumination, suicidal ideation, and suicide attempts: A meta-analytic review. *Review of General Psychology, 21*(2), 132–142.

Rosenthal, M. Z., Cheavens, J. S., Lynch, T. R., & Follette, V. (2006). Thought suppression mediates the relationship between negative mood and PTSD in sexually assaulted women. *Journal of Traumatic Stress: Official Publication of the International Society for Traumatic Stress Studies, 19*(5), 741–745.

Rudd, M. D. (2006). Fluid vulnerability theory: A cognitive approach to understanding the process of acute and chronic suicide risk. In T. E. Ellis (Ed.), *Cognition and suicide: Theory, research, and therapy* (pp. 355–368). Washington, DC: American Psychological Association. https://doi.org/10.1037/11377-016.

Samra, J., & Monk, L. (2007). *Working with the client who is suicidal: A tool for adult mental health and addiction services*. Simon Fraser University. https://www.health.gov.bc.ca/library/publications/year/2007/MHA_WorkingWithSuicidalClient.pdf. (Accessed 4 May 2022).

Segal, Z. V., Teasdale, J. D., & Williams, J. M. G. (2004). Mindfulness-bases cognitive therapy: Theoretical rationale and empirical status. In S. C. Hayes, V. M. Follette, & M. M Linehan (Eds.), *Mindfulness and acceptance: Expending the cognitive behavioral tradition* (pp. 45–65). Guilford Press.

Segal, Z. V., Teasdale, J. D., Williams, J. M., & Gemar, M. C. (2002). The mindfulness-based cognitive therapy adherence scale: Inter-rater reliability, adherence to protocol and treatment distinctiveness. *Clinical Psychology & Psychotherapy*, 9(2), 131–138.

Smith, J. M., Alloy, L. B., & Abramson, L. Y. (2006). Cognitive vulnerability to depression, rumination, hopelessness, and suicidal ideation: Multiple pathways to self-injurious thinking. *Suicide and Life-threatening Behavior*, 36(4), 443–454.

Surrence, K., Miranda, R., Marroquín, B. M., & Chan, S. (2009). Brooding and reflective rumination among suicide attempters: Cognitive vulnerability to suicidal ideation. *Behaviour Research and Therapy*, 47(9), 803–808.

Teasdale, J. D. (1988). Cognitive vulnerability to persistent depression. *Cognition & Emotion*, 2(3), 247–274.

Teasdale, J. D., Segal, Z. V., Williams, J. M. G., Ridgeway, V. A., Soulsby, J. M., & Lau, M. A. (2000). Prevention of relapse/recurrence in major depression by mindfulness-based cognitive therapy. *Journal of Consulting and Clinical Psychology*, 68(4), 615.

Treynor, W., Gonzalez, R., & Nolen-Hoeksema, S. (2003). Rumination reconsidered: A psychometric analysis. *Cognitive Therapy and Research*, 27(3), 247–259.

Van Orden, K. A., Witte, T. K., Cukrowicz, K. C., Braithwaite, S. R., Selby, E. A., & Joiner, T. E., Jr. (2010). The interpersonal theory of suicide. *Psychological Review*, 117(2), 575.

Wegner, D. M., Schneider, D. J., Carter, S. R., & White, T. L. (1987). Paradoxical effects of thought suppression. *Journal of Personality and Social Psychology*, 53(1), 5.

Wenzel, A., & Beck, A. T. (2008). A cognitive model of suicidal behavior: Theory and treatment. *Applied and Preventive Psychology*, 12(4), 189–201.

Williams, J. M. G. (2008). Mindfulness, depression and modes of mind. *Cognitive Therapy and Research*, 32(6), 721–733.

Williams, J. M. G., & Kabat-Zinn, J. (2011). Mindfulness: Diverse perspectives on its meaning, origins, and multiple applications at the intersection of science and dharma. *Contemporary Buddhism*, 12(1), 1–18.

Williams, J. M. G., Duggan, D. S., Crane, C., & Fennell, M. J. (2006). Mindfulness-based cognitive therapy for prevention of recurrence of suicidal behavior. *Journal of Clinical Psychology*, 62(2), 201–210.

Williams, J. M. G., Fennell, M., Barnhofer, T., Crane, R., & Silverton, S. (2015). *Mindfulness and the transformation of despair: Working with people at risk of suicide*. Guilford Publications.

Williams, J. M. G., & Swales, M. (2004). The use of mindfulness-based approaches for suicidal patients. *Archives of Suicide Research*, 8(4), 315–329. https://doi.org/10.1080/13811110490476671.

Williams, J. M. G., Teasdale, J. D., Segal, Z. V., & Soulsby, J. (2000). Mindfulness-based cognitive therapy reduces overgeneral autobiographical memory in formerly depressed patients. *Journal of Abnormal Psychology*, 109(1), 150.

Williams, M. (2014). *Cry of pain: Understanding suicide and the suicidal mind*. Hachette UK.

World Health Organization. (2019, September). https://www.who.int/news-room/fact-sheets/detail/suicide.

Nyanaponika, T. (1994). The five mental hindrances and their conquest: Selected texts from the Pali Canon and the Commentaries. https://www.accesstoinsight.org/lib/authors/nyanaponika/wheel026.html (Accessed 04 May 2022).

Chapter 24

Cognitive-behavioral therapy for tobacco use disorder in smokers with depression: A critical review

Alba González-Roz[a], Sara Weidberg[a], and James MacKillop[b]
[a]Department of Psychology, University of Oviedo, Oviedo, Spain, [b]Department of Psychiatry and Behavioral Neurosciences, McMaster University & St. Joseph's Healthcare Hamilton, Hamilton, Canada

Introduction

Tobacco smoking is the leading cause of preventable death, estimated to cause more than 7 million annual deaths (World Health Organization [WHO], 2017). Each day over 4400 individuals from the United States (US), aged 12 or older, smoke their first cigarette (SAMHSA, 2017). The economic cost of smoking is substantial: a systematic review by Rezaei, Akbari Sari, Arab, Majzdadeh, and Mohammad Poorasl (2016) found that smoking-related diseases were responsible for 1.5%–6.8% of the national health system expenditures. According to the global trends provided by the WHO, tobacco use has decreased worldwide by approximately 60 million people from 2000 to 2018 (WHO, 2019). Nevertheless, it is estimated that 16% of world population will still smoke by 2030. The implementation of comprehensive and strict tobacco control policies over the last years among countries from the European Union (EU) has proven to be effective in reducing smoking prevalence rates (Feliu et al., 2019).

In contrast, smoking remains stable among smokers with psychiatric comorbidity, suggesting the need to improve tobacco control and systematically deliver effective treatments to this portion of the population (Steinberg, Williams, & Li, 2015). About a third of the cigarette smokers have also mental health problems (Brose, Brown, Robson, & McNeill, 2020). In particular, tobacco smoking has been linked to both major depressive disorder (depression) diagnosis and depressive symptoms (Dierker et al., 2015; Fluharty, Taylor, Grabski, & Munafò, 2017). Depression poses a challenge for health-care professionals, as it has been associated to more severe nicotine dependence (Bainter, Selya, & Oancea, 2020), and poorer cessation prognosis (Secades-Villa, González-Roz, García-Pérez, & Becoña, 2017). Unfortunately, smoking cessation treatments are not systematically offered to persons with cooccurrent mental health conditions, even when cumulative research has concluded a direct relationship between abstinence and improved mental health (Secades-Villa et al., 2015). In this chapter, we provide a critical review of cognitive-behavioral therapy (CBT) effectiveness for smoking cessation in persons with depression, one of the most widely studied treatments for this comorbidity. Both theoretical and practical knowledge on the relationship between smoking and depression is provided.

Explanatory models for the link between tobacco use and depression

Up to date, several hypotheses on the smoking and depression relationship have been proposed in the literature, but the directional nature of this relationship has not been disentangled (Fluharty et al., 2017). Although some studies have shown that smoking tobacco might cause depression (Boden, Fergusson, & Horwood, 2010), other longitudinal studies have also revealed that current or past depression can lead to cigarette smoking (McKenzie, Olsson, Jorm, Romaniuk, & Patton 2010). Four specific hypotheses have been consistently proposed:

a) *Causal models*: According to this model, one of the two conditions, either smoking or depression, leads to the other. The causal hypothesis has postulated that nicotine intake (or smoking cessation) precedes the presence of depression via the changes in the smokers' neurochemical circuits that increase susceptibility to environmental stressors, putting them at a greater risk to develop depression (Fowler, Logan, Wang, & Volkow, 2003). Nicotine withdrawal may also affect these neural factors (Rose, Behm, Ramsey, & Ritchie, 2001). Also, it encompasses the "self-medication hypothesis" which posits that individuals with depression start smoking to cope with distress. This is supported by evidence suggesting that nicotine subjectively enhances positive affect due to its antidepressant properties, as it increases the functioning of the dopaminergic reward system (see Cosci, Nardi, & Griez, 2014 for a review).

b) *Bidirectional models*: This hypothesis suggests that smoking or depression can increase vulnerability to the other. Longitudinal studies have supported this relationship (Leung Gartner, Hall, Lucke, & Dobson, 2012). The bidirectional hypothesis suggests an interaction effect, that is, while smoking initially alleviates depressive symptomatology, it ultimately exacerbates it over time (Munafò & Araya, 2010).

c) *Common factor models*: One possible explanatory mechanism underlying smoking and depression may be the common factor model. This posits that no causal relationship exists between both conditions, but instead a common genetic or environmental underlying factor (Tsuang, Francis, Minor, Thomas, & Stone, 2012; Tully, Iacono, & McGue, 2010).

d) *Transdiagnostic models:* These models propose a host of shared cognitive-affective factors accounting for both smoking and depression (e.g., negative affect, anhedonia, rumination, experiential avoidance, emotion regulation, or distress tolerance (Vujanovic et al., 2017). One of these transdiagnostic approaches that have been empirically tested suggests that depressed smokers are driven by three motivational mechanisms that increase the expected value of smoking, namely: high negative affect, low positive affect, and cognitive impairment (Leventhal & Zvolensky, 2015; Mathew, Hogarth, Leventhal, Cook, & Hitsman, 2017).

Clinical assessment: Diagnostic tools and screening measures for smoking and depression

Given the high depression prevalence in cigarette smokers, the assessment of people who smoke should rely on different diagnostic tools and screening measures targeting both nicotine dependence and depressive symptoms. A suggested list of screening tools for people who smoke and have depressive symptoms is presented in Table 3 (see the practice and procedures section). Depression screening tools and clinical interviews such as the Structural Clinical Interview of the DSM-5 (SCID-CV; First, Williams, Karg, & Spitzer, 2016) may help to understand the severity of the depression symptoms and assess for the presence of other cooccurrent disorders. The SCID-CV, like other diagnostic tools such as the Composite International Diagnostic Interview(Kessler & Üstün, 2004) can be administered in modules, facilitating the assessment of tobacco use disorder and depression.

The functional analysis of behavior, a powerful tool to understand the antecedents and consequences of the smoking behavior, is also an essential part of CBT. For this purpose, a self-report (diary) of tobacco use may help the client to get a sense on the main antecedents and consequences related to cigarette smoking. In doing this, it would be useful to ask the client to record the number of cigarettes smoked each day, the time of the day at which this behavior occurs, the pleasure (0–10) associated to each cigarette, and the situations or emotions linked to each cigarette.

Depression measures that could be used to monitoring changes in depression through treatment are the Beck Depression Inventory-II (BDI-II: Beck, Steer, & Brown, 1996), and the 9-item Patient Health Questionnaire (PHQ-9: Kroenke, Spitzer, & Williams, 2001). These are assessment tools that offer information on the level of depression severity. In addition to the above, it results essential to monitor depression symptoms through treatment. Self-reporting daily activities, the associated pleasure (e.g., from 0 to 10) and perceived relevance (e.g., from 0 to 10) is recommended. Other depression-related measures that can assist the therapist to evaluate the level of the client's activation and response-contingent positive reinforcement are the brief version of the Behavioral Activation for Depression Scale (BADS-SF: Manos, Kanter, & Luo, 2011), and the Environmental Reward Observation Scale (EROS: Armento & Hopko, 2007).

Empirical evidence on CBT for smokers with depression

CBT seeks to target clients' behaviors, emotions, and thoughts in a structured manner. Either solely (Martínez-Vispo et al., 2019), or combined with other pharmacological interventions (Aldi, Bertoli, Ferraro, Pezzuto, & Cosci, 2018), CBT is one of the most widely used smoking cessation therapies. The efficacy of CBT for cooccurrent smoking and depression has been supported in different systematic reviews and metaanalyses (Aldi et al., 2018; Gierisch, Bastian, Calhoun, McDuffie, & Williams 2012; Secades-Villa et al., 2017; van der Meer Willemsen, Smit, & Cuijpers 2013). Independent studies have also examined the clinical effectiveness of different CBT protocols in patients with history of depression diagnosis and subclinical depression. Effectiveness rates from these studies range between 13% and 57.58% (Bricker, Bush, Zbikowski, Mercer, & Heffner 2014; Haas, Muñoz, Humfleet, Reus, & Hall, 2004; Martínez-Vispo et al., 2019).

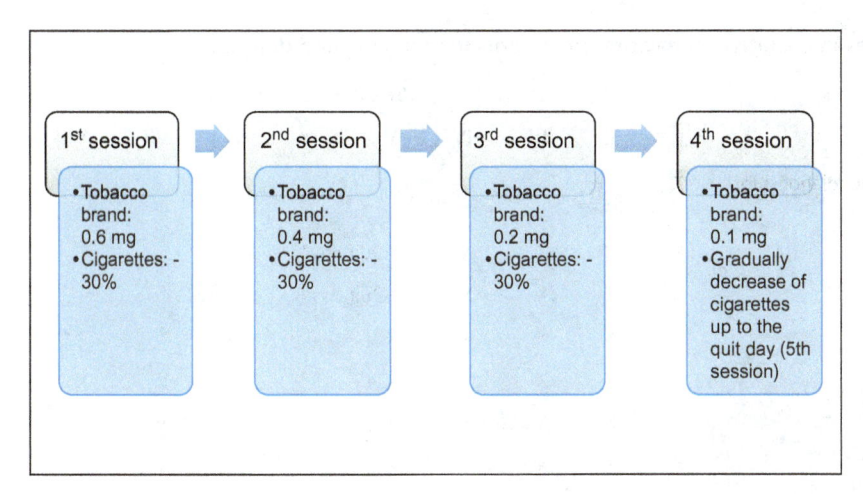

FIG. 1 Example of the nicotine fading procedure in a smoker using 30 cigarettes of a tobacco brand containing 0.8 mg of nicotine. This figure exemplifies how to implement the nicotine fading procedure in a cognitive-behavioral therapy protocol.

CBT can be delivered individually or in a group-based format, and usually during 8/12 sessions (Brown et al., 2001; Secades-Villa et al., 2019). Although different treatment protocols exist, the most effective ones integrate mood management components. Interventions usually integrate strategies addressing three sequential stages of change: (1) preparation for change (through motivational interviewing strategies), (2) initiating abstinence (through fading, quit date cessation initiation, refusal skills training, cognitive restructuring skills and mood management); (3) and maintenance of abstinence outcomes (through relapse prevention strategies). Several CBT protocols for smoking and depression also include nicotine fading to address the physical dependence on nicotine. This procedure (see Fig. 1 for an example) consists of gradually decreasing the number of cigarettes that can be smoked each day and progressively changing to lower nicotine content tobacco brands on a weekly basis.

Evidence of other behavioral combination treatments in smokers with depression

More recently, a number of studies have looked at the additive effects of other behavioral therapies on CBT abstinence outcomes (Vinci, 2020). More specifically, research has suggested that CBT can be even more effective if it is provided alongside other therapies (particularly behavioral activation [BA] and contingency management [CM]).

CM is a behavioral therapy that engineers the client environment and typically provides incentives in the form of vouchers for abstinence (Roll, McPherson, & McDonell, 2021). Several studies have provided superior abstinence rates when combined to CBT. For example, 100 treatment-seeking smokers with depression were randomly assigned to CBT+BA versus CBT+BA+CM (González-Roz, Weidberg, García-Pérez, Martínez-Loredo, & Secades-Villa, 2020; Secades-Villa et al., 2019). CM consisted of vouchers for tobacco abstinence, and therapies were delivered over 8 weeks in a group setting. One-year abstinence outcomes were superior in CBT+BA+CM (53.3%) as compared to CBT+BA only (23.3%). In another study by the same research group (Secades-Villa et al., 2015), 147 treatment-seeking smokers with subclinical depression were randomly allocated to either CBT ($n=74$), or CBT+CM ($n=73$), over 6 weeks of 1-h group-based sessions. Results indicated that the odd of being abstinent at the end of treatment was 5.25 times higher in CBT+CM than in CBT.

The benefits of adding BA to CBT protocols have also supported in a number of studies (Secades-Villa et al., 2017). A recent study by Martínez-Vispo et al. (2019), examined the effects of a smoking cessation CBT with components of BA versus a standard CBT, and a wait-list control group in a sample of 275 adult smokers. The therapy was delivered over 8 weekly 1-h face-to-face group sessions. Superior seven-day point prevalence abstinence rates were found for both active groups at the end of treatment. Abstinence rates at 12-months follow-up were 30% for the CBT+BA group, and 18% for CBT.

Core components and strategies in CBT for smoking and depression

General considerations

Based on the available evidence, a minimum of 1-h individual or 1.5-h group 8-week sessions should be considered (it is recommended to include a cotherapist if working in a group setting and to also consider the inclusion of up to four clients

TABLE 1 Core components and strategies in cognitive-behavioral treatments for smoking and depression.

Strategy	Session							
	I	II	III	IV	V	VI	VII	VIII
Treatment rationale and psychoeducation on the relationship between smoking and depression	X							
Nicotine fading	X	X	X	X	"Quit day"	[a]	[a]	[a]
Self-monitoring of cigarette smoking	X	X	X	X	"Quit day"			
Feedback of cigarette smoking	X	X	X	X	X	X	X	X
Cognitive restructuring	X	X	X	X	X	X	X	X
Stimulus control	X	X	X	X	X	X	X	X
Behavioral activation	X	X	X	X	X	X	X	X
Social skills assertiveness				X	X	X	X	X
Anger management				X	X	X	X	X
Weight concerns and exercise planning				X	X	X	X	X
Social support	X	X	X	X	X	X	X	X
Relapse prevention						X	X	X

This table provides a suggested sequencing of sessions and procedures.
[a] A 30%, 60%, and 90% decrease in nicotine can be followed during the first three sessions. If necessary, the nicotine fading procedure can be continued to facilitate abstinence in subsequent sessions.

per group). As part of the therapy, it is useful that the client self-monitors his/her cigarette smoking. This procedure requires the smoker to plot his/her daily cigarette consumption and facilitates to regulate clients' behavior. One essential part of CBT for people who smoke is providing feedback on their progress. Using a carbon monoxide (CO) "smokerlyzer" in each therapy session is strongly advised to enhance motivation and enhance clients' self-efficacy.

Following a CBT conceptual model, what follows is a detailed description on the main procedures and techniques that has shown effectiveness for smokers with depression. A suggested sequencing of procedures and techniques over an eight-week therapy is shown in Table 1.

Treatment rationale

This is an essential part in therapy that seeks to provide the patient evidence-based information on the CBT rationale and more specifically psychoeducation on the relationship between smoking and depression. The first CBT session seeks to establish a conceptual model that will guide the therapy while building a rapport. During this session, the therapist will introduce the CBT rationale, with particular focus on the A-B-C (antecedents-behaviors-consequences) chain (i.e., functional analysis of behavior) related to the smoking and depression relationship (see Fig. 2). Specifically, the therapist should raise the following themes:

- The therapy will target all the parts in the A (external and internal triggers)–B (emotion and distorted cognitions)–C (consequences) chain.
- As part of the therapy, they will be advised to adhere to different tasks to facilitate smoking abstinence that will not interfere with their daily life.
- During the therapy it is important to attend each therapy session, even if they feel depressed or not motivated to do so.
- Throughout the therapy, the person will be recommended to change her/his tobacco brands and gradually decrease the number of cigarettes. It is important that he/she adheres to each therapy guideline.

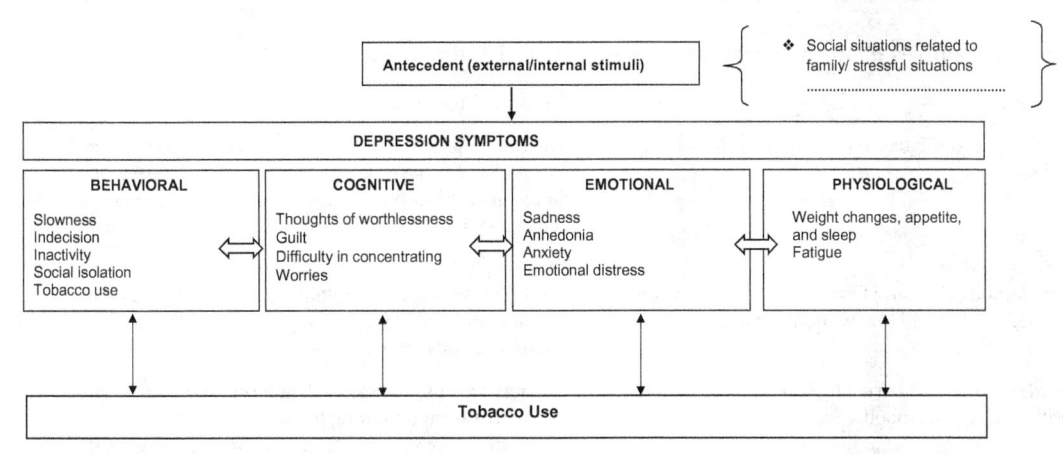

FIG. 2 Functional analysis on the smoking and depression association. This figure shows the A[antecedents]-B[behavior/s]-C[consequences] chain of the smoking and depression relationship.

Nicotine fading

This treatment strategy aims at gradually decreasing clients' dependence on nicotine typically over a 3- or 4-week period by weekly reductions of 30%, 60%, and 90% of the baseline level (Becoña & García, 1993). This may consist of fewer cigarettes and/or a set of weekly changes in cigarette brands to ones containing progressively less nicotine (Foxx & Axelroth, 1983). Fading reduces the nicotine withdrawal effects and allows the therapist to provide feedback on the clients' efforts to reduce dependence. This procedure is usually integrated in CBT protocols and is exemplified in Fig. 1. Support for its efficacy has been provided in different studies. Long-term follow-ups inform on abstinence rates that range between 33% (Foxx & Axelroth, 1983) and 46% (Brown, Lichtenstein, McIntyre, & Harrington-Kostur, 1984). After this gradual decrease, a quit day is set. If necessary, the nicotine fading procedure is continued to facilitate the client' abstinence.

Self-monitoring of cigarette consumption

Self-monitoring is an important part of treatments for smoking cessation that helps the client to reduce his/her consumption by increasing the awareness of both the number and the situations that are related to cigarette use (McFall, 1970). The therapist can ask the client to estimate the average number of cigarettes, the time at which he/she smokes each cigarette, and situations where smoking occurs. Self-monitored smoking data may also be used to identify particular targets throughout treatment (Korotitsch & Nelson-Gray, 1999). For example, the situations where the person uses cigarettes may help the therapist to train the client in stimulus control.

Feedback of cigarette smoking

Because of the relevance of providing immediate feedback of the client's progress, it is recommended to consider the inclusion of biochemical assessments in therapy. Apart from using self-registers to monitor the client's change, CO or cotinine urinalysis is suggested. The latter may represent relatively noninvasive procedures to inform on objective changes in nicotine use. Optimal carbon monoxide criteria to confirm 24-h smoking abstinence has been suggested to be ≤ 4 ppm (Perkins, Karelitz, & Jao, 2013). As regards to cotinine, a cut-off within the range of 3–15 ng/ML is suggested to confirm nonsmoking status (Benowitz et al., 2020).

Feedback on biochemical measures can facilitate the patients' education about the effects of smoking. Testing CO or cotinine levels can represent well the idea of *"seeing is believing"* (Goldstein, Gans, Ripley-Moffitt, Kotsen, & Bars, 2018), build the client's self-efficacy, and reinforce progressive reductions in cigarettes use. Apart from monitoring nicotine use, having documenting progress (while maintaining confidentiality) in the treatment facility may be of help. A white board can be used for plotting client's carbon monoxide or cotinine levels in each therapy session (an example is provided in figure under the practice and procedure section). It could also serve to exemplify the A-B-C chain of the smoking behavior. In the initial sessions, it can be a complement as an aid to foster motivation to change by focusing on the negative health-related consequences stemmed from cigarette smoking.

TABLE 2 Script examples on main worries and fears addressed at the first session.

Irrational cognition	Therapist intervention
"I am not sure if this is the best time to quit." "How about recovering from depression at first?"	"The best time to quit is probably now that you have decided to initiate a change. Studies have observed that smoking abstinence relates to an improvement in depression. Also, thanks to other people who have received this treatment, we know that quitting does not lead to worsened depression"
"Quitting smoking will be hard; I use cigarettes everywhere and every time." "What will I do without cigarettes?"	"Well, I understand that quitting may be hard. You have been doing the same for almost a life. Engaging in different positive and alternative activities to smoking may help"
"I think I need cigarettes, if I am not feeling good, a puff on a cigarette will enhance my mood"	"Cigarettes are usually smoked multiple times and are associated to almost each activity you do during the day. Reducing the dose of nicotine each week, will allow you to experience greater energy and release. You will also feel a boost in energy"

This table provides a script showing how the therapist can address worries and fears that usually arises in therapy.

Cognitive restructuring

Cognitive distortions or erroneous beliefs regarding smoking and depression are targeted through the therapy, especially during the first sessions. Several myths related to worsened depression and health-related impacts are also addressed. Example scripts for restructuring main worries and fears related to smoking cessation and depression are presented in Table 2. Typically, the latter are addressed during the first sessions, latter in treatment is recommended to pay attention to distorted cognitions related to perceived mental and physical negative health effects (i.e., *"abstinence will make me feel more anxious," "abstinence will worsen my physical health"*), and positive expectancies of cigarette smoking (i.e., *"smoking a cigarette will not lead me to smoke regularly again, it will probably help me to cope with stressful situations"*).

Stimulus control

Modifying the client's environment is an important part of CBT. It aims at reducing clients' exposure to smoking-related situations (Becoña & García, 1993). Smoking behavior is said to be under stimulus control when the presence of a given stimulus (or stimuli) changes the likelihood of its occurrence (Shiffman, Dunbar, & Ferguson, 2015). The aim of training in stimulus control is to gradually break the learned association between specific situational cues and smoking behavior, by encouraging the client to choose 2–3 situations where he/she will no longer smoke on a weekly basis, starting from the easiest ones. Smoking behavior is gradually restricted to specific situations from sessions 2–4, with the expectation that the client will only smoke in few situations by last sessions or the week before the quit day (Secades-Villa et al., 2019). In addition, the clients are trained in other self-control strategies: not to smoke the final part of the cigarette, refuse smoking offers, taking gradually fewer deep puffs, or putting the cigarette only in the mouth to smoke.

Behavioral activation

BA aims to enhance the client's contact with more valued environments through systematic efforts to increase nonsmoking rewarding experiences of positive activities, which may reduce negative affect and improve positive affect (Lejuez, Hopko, Acierno, Daughters, & Pagoto, 2011). This is important as both negative and positive affect represent the most frequent relapse situations (Rodríguez-Cano, López-Durán, Martínez-Vispo, & Becoña, 2021). The client is highly encouraged to develop a plan to engage in nonsubstance using behaviors. A menu of resources and positive activities, such as physical exercise is provided to encourage the client to substitute his/her smoking behavior. BA focus on the functional association between depression and smoking, and consists of self-monitoring of daily activities, its importance and enjoyment, the identification of life areas and values that guide the generation of meaningful positive activities, engagement in 2–3 weekly activation goals, and social support through behavioral contracts.

Social skills assertiveness

This module offers the smoker the opportunity to actively work on social skills that may prevent him/her from relapsing, namely assertive communication to refuse offers of cigarettes. The training includes psychoeducation on the interpersonal communication styles (e.g., assertive, aggressive, passive), modeling of effective styles of communication (i.e., how the voice and message needs to accompany the physical gesture), and group role-playing exercises to consolidate acquired skills (Brown et al., 2001).

Anger management

Anger management seeks to train patients in the following skills: to gain awareness of anger and aggressiveness-related situations that increase smoking craving (Yalcin, Unal, Pirdal, & Karahan, 2014), thinking before acting under such situations (e.g., postponing important decisions to make them out of anger situations, trying to preserve the calm when someone gets angry), and to increase the use of relaxation for the control of anger in daily life stressful situations.

Weight concerns and exercise planning

Clients are trained on weight control to prevent them from weight increases while quitting smoking. This module has relevance for facilitating abstinence, particularly among females. Research has suggested that psychological distress relates to using cigarettes to suppress appetite, prevent weight gain, and manage withdrawal symptoms associated to quitting smoking (Burr et al., 2020). For this reason, CBT protocols typically include cognitive restructuring of misconceptions on the relationship between smoking and weight (e.g., smoking abstinence will necessarily lead to gaining weight), and training in how to control their eating topography (e.g., eating slowly, leaving the cutlery between bites, controlling overeating, cooking and/or going shopping while not being hungry, etc.) (Martínez-Vispo et al., 2019; Secades-Villa et al., 2019). To prevent weight gain, a plan to practice physical activity regularly with a minimum frequency of two periods of 30 min per week is also recommended. Importantly, research has shown that incorporating weight management components substantially increases abstinence in weight-concerned females (Perkins et al., 2001) while increasing self-efficacy for weight control, diet quality, and weight loss (Sallit, Ciccazzo, & Dixon, 2009).

Social support

Emotional, instrumental, and informational support is intrinsic to group settings. In this line, enlisting social support resources while recognizing that sharing with others this decision might be difficult is an important part of CBT for smoking cessation, with arguably relapse prevention effects. Bolstering peer support is important at the intra-treatment group and the community level (Westmaas, Bontemps-Jones, & Bauer 2010). In doing this, it is also relevant to train the client in how he/she could manage negative support (e.g., discouragement) from others.

Examining social support can be effectively addressed if the therapist and the client collaboratively develop a social support contract indicating which persons may help the client to engage in nonsmoking behaviors. In this regard, it will be important to ask the client regarding the smoking status of his/her social network. Just telling them that the client is now working on smoking cessation, can pose an effective way to handle smoking offers and cravings while sharing time with friends/significant others who are smokers.

Relapse prevention

As part of relapse prevention, clients are trained in identifying and problem solving high-risk situations. This is a CBT strategy for enhancing success and also self-efficacy toward preventing smoking relapse (Marlatt & Gordon, 1980). It involves defining life problems in terms of the A-B-C chain, evaluating the pros and cons of each raised solution, and practice and re-assessment of the chosen solution. In addition to the above, clients are encouraged to choose a high-risk relapse situation, identify any alternative solution (without judging their potential efficacy), evaluating the pros and cons of each behavior and choose the one to be implemented (Livingstone-Banks et al., 2019). An individualized plan of action is also encouraged to outline specific coping responses that may be implemented in case a relapse occurs.

Conclusions

Due to the high prevalence of depression in the cigarette smoking population, it is essential to screen for depression in tobacco use cessation services. Relative to people with no cooccurring mental health disorders, people who smoke and have depression show higher nicotine dependence levels and acute withdrawal symptoms during cessation attempts. Importantly, even when effective therapies are delivered, studies consistently show low abstinence rates in smokers with depression. Of the psychological therapies, CBT is among the most effective to date for smokers with depressive symptoms and incorporating BA and CM leads to improved cessation rates. Irrespective of depression severity, CBT promotes superior abstinence rates than nonintervention or less intensive therapies (e.g., self-help). Based on the reviewed evidence, group therapy delivered over 6 to 8 weeks stands as the most common implementation procedure. Continued monitoring and additional support in the form of reminders of effective relapse prevention strategies (e.g., at 1, 2, 3, 6, and 12 months since quitting) is recommended over a 1-year period.

Unfortunately, smokers who have depression or other mental-health disorders are often not systematically offered smoking cessation treatments. Several barriers, such as misperceptions on the relationship between smoking cessation and worsening in depression, and perception that clients are not motivated to quit are related to low provision of effective smoking cessation therapies. On the contrary, there exists mounting evidence on the association between smoking abstinence and reduced depression and anxiety symptomatology (Secades-Villa et al., 2017; Taylor et al., 2021). Furthermore, compared to persistent smoking, quality of life has been found to improve substantially as well (Taylor et al., 2014).

Practice and procedures

See Table 3 and Fig. 3.

Mini-dictionary of terms

- **Behavioral activation:** a structured intervention (although flexible and with an idiographic nature) that helps clients to increase the number and type of positive activities they are engaged with.
- **Cognitive restructuring:** structured and collaborative therapeutic approach in which clients are trained in how to identify, evaluate, and modify the distorted thoughts, and beliefs that are considered responsible for their emotional distress.
- **Cooccurring condition:** a person who presents more than one mental health disorder at a given time (e.g., smoking, anxiety, and depression).

TABLE 3 Suggested assessment tools for smokers with depression.

Depression

- Beck Depression Inventory-II (BDI-II: Beck, Steer, & Brown, 1996)
- The 9-item Patient Health Questionnaire (PHQ-9: Kroenke, Spitzer, & Williams, 2001)
- The Structured Clinical Interview for DSM-5-[depression module] (SCID-5: First, Williams, Karg, & Spitzer, 2016)
- The Behavioral Activation for Depression Scale—short form (BADS-SF: Manos, Kanter, & Luo, 2011)
- The Environmental Reward Observation Scale (Armento & Hopko, 2007)
- Self-monitoring of activity and mood (including, day/time, activity, level of engagement from 0 to 10, feelings after and before from 0 to 10)

Smoking

- Clinical interview comprising items to assess number of cigarettes, tobacco brands, and previous quit attempts
- Self-registers of cigarette smoking (e.g., number of cigarettes, time of the day at which smoking occurs, perceived pleasure associated to each cigarette from 0 to 10, situations where smoking occurs)
- Carbon monoxide and cotinine monitoring
- Stages of change—University of Rhode Island Change Assessment Scale (URICA: McConnaughy, Prochaska, & Velicer, 1983)
- The Fagerstrom Test for Cigarette Dependence (FTCD: Heatherton, Kozlowski, Frecker, & Fagerstrom, 1991)
- The Structured Clinical Interview for DSM-5—The Tobacco Use Disorder module (SCID-5: First, Williams, Karg, & Spitzer, 2016)

This table provides several assessment and diagnostic tools for smokers with depression.

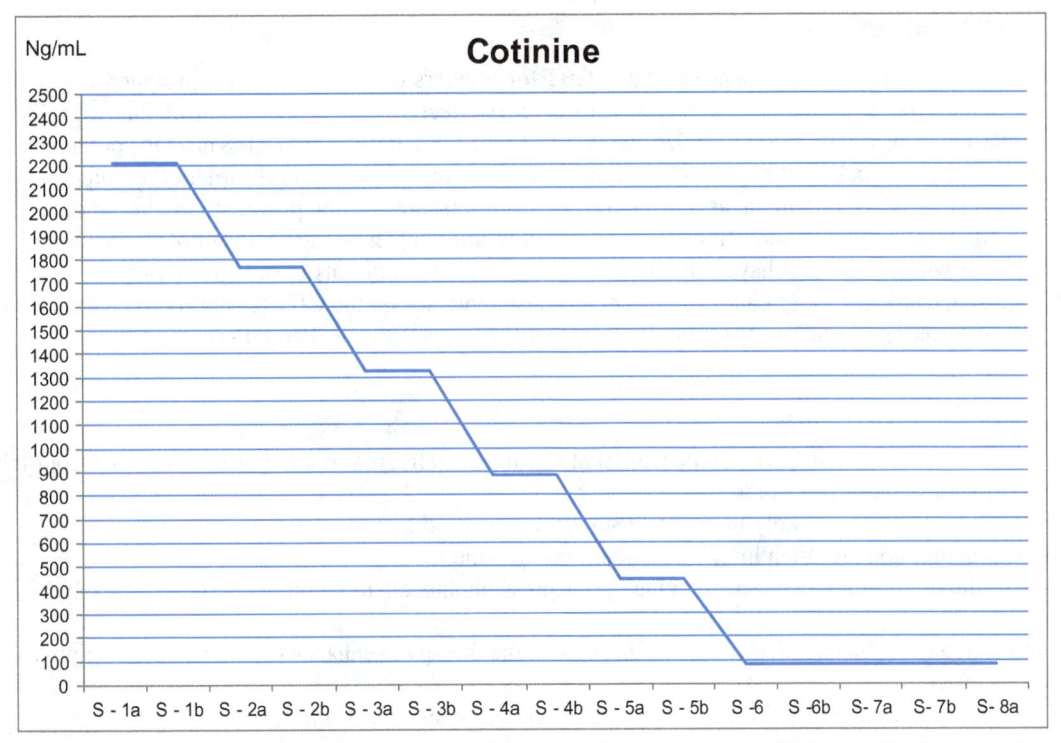

FIG. 3 Sample sheet for registering cotinine levels and providing feedback to the client on decreases in nicotine use. This figure provides an example on how to register urine cotinine levels. Both therapy (as denoted by the letter "a") and control sessions (as denoted by the letter "b") refer to occasions when clients attend the clinic to provide biochemical samples only. *Note.* ng/mL = nanograms per milliliter. Letters "a" and "b" denote therapy and control visits, respectively.

- **Functional analysis of behavior:** key step in CBT that refers to the identification of relevant, controllable, and causal functional relations that apply to a particular behavior.
- **Major depressive disorder:** clinical diagnosis defined by five or more DSM-5 symptoms that is characterized by depressed mood and/or loss of interest continuously, over at least a 2-week period.
- **Nicotine fading:** this strategy seeks to gradually reduce the nicotine intake by changes in cigarettes and tobacco brands.
- **Stimulus control:** CBT strategy that aims to train the client in how to overcome the situations conditioned to smoking.
- **Subclinical depression:** threshold depression that occurs when a person presents some symptoms without meeting the full criteria for depression diagnosis according to the diagnostic systems.
- **Tobacco use disorder:** clinical diagnosis characterized by using cigarettes over a longer period than was intended and unsuccessful efforts to control tobacco use, among others.
- **Relapse prevention:** a cognitive-behavioral strategy focused on training clients with effective coping skills.

Key facts

Key facts of cognitive-behavioral therapy for smokers and depression

- The assessment of people who smoke and have depression relies on different screening tools, although none of them can substitute the clinical interview.
- CBT is among the most effective psychosocial therapies for promoting abstinence in people who smoke and have depression.
- It is recommended to implement a minimum of six-to-eight therapy sessions.
- Combining CM and/or BA with CBT, produces additive effects on abstinence outcomes.
- To prevent relapses, it is recommended to schedule follow-up visits (e.g., 1, 3, 6, and 12 months) after treatment termination.

Applications to other areas

In this chapter we have reviewed the evidence-based of CBT for smokers with depression and provided an overview of key procedures and strategies. This treatment has shown to be cost-effective in the population of interest (González-Roz, Weidberg, García-Pérez, Martínez-Loredo, & Secades-Villa, 2020). CBT-based approaches have the potential to be implemented in cases of minor/subclinical depression and patients with other cooccurrent disorders (i.e., bipolar or persistent depression disorder). This is important as mental health disorders are highly prevalent in the smoking population (Conway et al., 2017). The intervention components reviewed in this chapter are also susceptible to be used in treatments delivered to people who smoke and have other cooccurrent physical conditions (e.g., cardiovascular diseases, cancer) (Reavell, Hopkinson, Clarkesmith, & Lane, 2018). There is also evidence on the CBT feasibility to be delivered remotely as it is the case of smoking cessation apps (García-Pazo, Fornés-Vives, Sesé, & Pérez-Pareja, 2020).

Summary points

- This chapter outlines essential CBT components that should be delivered to any person with comorbid tobacco use disorder and major depressive disorder.
- Cigarette smoking is twice as likely in cigarette smokers compared to nonusers.
- Smokers with depression are often highly dependent on nicotine.
- The nature of the relationship between smoking and depression appears to be bidirectional, with each contributing to the other.
- It is recommended using biochemical measures throughout the therapy to enhance the clients' motivation to change and reinforce self-efficacy.
- Smoking cessation treatment should be systematically offered to smokers with depression.

References

Aldi, G. A., Bertoli, G., Ferraro, F., Pezzuto, A., & Cosci, F. (2018). Effectiveness of pharmacological or psychological interventions for smoking cessation in smokers with major depression or depressive symptoms: A systematic review of the literature. *Substance Abuse*, *39*, 289–306.

Armento, M. E. A., & Hopko, D. R. (2007). The Environmental Reward Observation Scale (EROS): Development, validity, and reliability. *Behavior Therapy*, *38*, 107–119.

Bainter, T., Selya, A. S., & Oancea, S. C. (2020). A key indicator of nicotine dependence is associated with greater depression symptoms, after accounting for smoking behavior. *PLoS One*, *15*, e0233656.

Beck, A. T., Steer, R. A., & Brown, G. K. (1996). *Manual for the beck depression inventory-II*. San Antonio, TX: Psychological Corporation.

Becoña, E., & García, M. P. (1993). Nicotine fading and smokeholding methods to smoking cessation. *Psychological Reports*, *73*, 779–786.

Benowitz, N. L., Bernert, J. T., Foulds, J., Hecht, S. S., Jacob, P., Jarvis, M. J., ... Piper, M. E. (2020). Biochemical verification of tobacco use and abstinence: 2019 update. *Nicotine and Tobacco Research*, *7*, 1086–1097.

Boden, J. M., Fergusson, D. M., & Horwood, L. J. (2010). Cigarette smoking and depression: Tests of causal linkages using a longitudinal birth cohort. *British Journal of Psychiatry*, *196*, 440–446.

Bricker, J. B., Bush, T., Zbikowski, S. M., Mercer, L. D., & Heffner, J. L. (2014). Randomized trial of telephone-delivered acceptance and commitment therapy versus cognitive behavioral therapy for smoking cessation: A pilot study. *Nicotine & Tobacco Research*, *16*, 1446–1454.

Brose, L. S., Brown, J., Robson, D., & McNeill, A. (2020). Mental health, smoking, harm reduction and quit attempts—A population survey in England. *BMC Public Health*, *20*, 201237.

Brown, R. A., Kahler, C. W., Niaura, R., Abrams, D. B., Sales, S. D., Ramsey, S. E., ... Miller, I. W. (2001). Cognitive-behavioral treatment for depression in smoking cessation. *Journal of Consulting and Clinical Psychology*, *69*, 471–480.

Brown, R. A., Lichtenstein, E., McIntyre, K. O., & Harrington-Kostur, J. (1984). Effects of nicotine fading and relapse prevention on smoking cessation. *Journal of Consulting and Clinical Psychology*, *52*, 307–308.

Burr, E. K., O'Keeffe, B., Kibbey, M. M., Coniglio, K. A., Leyro, T. M., & Farris, S. G. (2020). Distress intolerance in relation to reliance on cigarettes for weight, shape, and appetite control. *International Journal of Behavioral Medicine*, *27*, 247–254.

Conway, K. P., Green, V. R., Kasza, K. A., Silveira, M. L., Borek, N., Kimmel, H. L., ... Compton, W. M. (2017). Co-occurrence of tobacco product use, substance use, and mental health problems among adults: Findings from wave 1 (2013–2014) of the Population Assessment of Tobacco and Health (PATH) study. *Drug and Alcohol Dependence*, *177*, 104–111.

Cosci, F., Nardi, A., & Griez, E. (2014). Nicotine effects on human affective functions: A systematic review of the literature on a controversial issue. *CNS & Neurological Disorders - Drug Targets*, *13*, 981–991.

Dierker, L., Rose, J., Selya, A., Piasecki, T. M., Hedeker, D., & Mermelstein, R. (2015). Depression and nicotine dependence from adolescence to young adulthood. *Addictive Behaviors*, *41*, 124–128.

Feliu, A., Filippidis, F. T., Joossens, L., Fong, G. T., Vardavas, C. I., Baena, A., ... Fernández, E. (2019). Impact of tobacco control policies on smoking prevalence and quit ratios in 27 European Union countries from 2006 to 2014. *Tobacco Control*, *28*, 101–109.

First, M. B., Williams, J. B. W., Karg, R. S., & Spitzer, R. L. (2016). *User's guide for the SCID-5-CV structured clinical interview for DSM-5® disorders: Clinical version*. American Psychiatric Publishing, Inc.

Fluharty, M., Taylor, A. E., Grabski, M., & Munafò, M. R. (2017). The association of cigarette smoking with depression and anxiety: A systematic review. *Nicotine & Tobacco Research, 1*, 3–13.

Fowler, J. S., Logan, J., Wang, G. J., & Volkow, N. D. (2003). Monoamine oxidase and cigarette smoking. *Neurotoxicology, 24*, 75–82.

Foxx, R. M., & Axelroth, E. (1983). Nicotine fading, self-monitoring and cigarette fading to produce cigarette abstinence or controlled smoking. *Behaviour Research and Therapy, 21*, 17–27.

García-Pazo, P., Fornés-Vives, J., Sesé, A., & Pérez-Pareja, F. J. (2020). Apps for smoking cessation through cognitive behavioural therapy. A review. *Adicciones, 33*, 359–368.

Gierisch, J. M., Bastian, L. A., Calhoun, P. S., McDuffie, J. R., & Williams, J. W. (2012). Smoking cessation interventions for patients with depression: A systematic review and meta-analysis. *Journal of General Internal Medicine, 3*, 351–360.

Goldstein, A. O., Gans, S. P., Ripley-Moffitt, C., Kotsen, C., & Bars, M. (2018). Use of expired air carbon monoxide testing in clinical tobacco treatment settings. *Chest, 2*, 554–562.

González-Roz, A., Weidberg, S., García-Pérez, Á., Martínez-Loredo, V., & Secades-Villa, R. (2020). One-year efficacy and incremental cost-effectiveness of contingency management for cigarette smokers with depression. *Nicotine & Tobacco Research, 23*, 320–326.

Haas, A. L., Muñoz, R. F., Humfleet, G. L., Reus, V. I., & Hall, S. M. (2004). Influences of mood, depression history, and treatment modality on outcomes in smoking cessation. *Journal of Consulting and Clinical Psychology, 72*, 563–570.

Heatherton, T. F., Kozlowski, L. T., Frecker, R. C., & Fagerstrom, K.-. O. (1991). The Fagerström test for nicotine dependence: A revision of the Fagerstrom tolerance questionnaire. *British Journal of Addiction, 86*, 1119–1127.

Kessler, R. C., & Üstün, T. B. (2004). The World Mental Health (WMH) Survey Initiative Version of the World Health Organization (WHO) Composite International Diagnostic Interview (CIDI). *International Journal of Methods in Psychiatric Research, 13*, 93–121.

Korotitsch, W. J., & Nelson-Gray, R. O. (1999). An overview of self-monitoring research in assessment and treatment. *Psychological Assessment, 4*, 415–425.

Kroenke, K., Spitzer, R. L., & Williams, J. B. (2001). The PHQ-9: Validity of a brief depression severity measure. *Journal of General Internal Medicine, 16*, 606–613.

Lejuez, C. W., Hopko, D. R., Acierno, R., Daughters, S. B., & Pagoto, S. L. (2011). Ten year revision of the brief behavioral activation treatment for depression: Revised treatment manual. *Behavior Modification, 35*, 111–161.

Leung, J., Gartner, C., Hall, W., Lucke, J., & Dobson, A. (2012). A longitudinal study of the bi-directional relationship between tobacco smoking and psychological distress in a community sample of young Australian women. *Psychological Medicine, 42*, 1273–1282.

Leventhal, A. M., & Zvolensky, M. J. (2015). Anxiety, depression, and cigarette smoking: A transdiagnostic vulnerability framework to understanding emotion-smoking comorbidity. *Psychological Bulletin, 141*, 176–212.

Livingstone-Banks, J., Norris, E., Hartmann-Boyce, J., West, R., Jarvis, M., & Hajek, P. (2019). Relapse prevention interventions for smoking cessation. *The Cochrane Database of Systematic Reviews, 2*, CD003999.

Manos, R. C., Kanter, J. W., & Luo, W. (2011). The behavioral activation for depression scale-short form: Development and validation. *Behavior Therapy, 42*, 726–739.

Marlatt, G. A., & Gordon, J. R. (1980). Determinants of relapse: Implications for the maintenance of behavior change. In P. O. Davidson, & S. M. Davidson (Eds.), *Behavioral medicine: Changing health lifestyles*. New York: Brunner/Mazel.

Martínez-Vispo, C., Rodríguez-Cano, R., López-Durán, A., Senra, C., Fernández del Río, E., & Becoña, E. (2019). Cognitive-behavioral treatment with behavioral activation for smoking cessation: Randomized controlled trial. *PLoS One, 4*, e0214252.

Mathew, A. R., Hogarth, L., Leventhal, A. M., Cook, J. W., & Hitsman, B. (2017). Cigarette smoking and depression comorbidity: Systematic review and proposed theoretical model. *Addiction, 112*, 401–412.

McConnaughy, E. A., Prochaska, J. O., & Velicer, W. F. (1983). Stages of change in psychotherapy: Measurement and sample profiles. *Psychotherapy: Theory, Research and Practice, 20*, 368–375.

McFall, R. M. (1970). Effects of self-monitoring on normal smoking behavior. *Journal of Consulting and Clinical Psychology, 35*, 135–142.

McKenzie, M., Olsson, C. A., Jorm, A. F., Romaniuk, H., & Patton, G. C. (2010). Association of adolescent symptoms of depression and anxiety with daily smoking and nicotine dependence in young adulthood: Findings from a 10-year longitudinal study. *Addiction, 9*, 1652–1659.

Munafò, M. R., & Araya, R. (2010). Cigarette smoking and depression: A question of causation. *British Journal of Psychiatry, 6*, 425–426.

Perkins, K. A., Marcus, M. D., Levine, M. D., D'Amico, D., Miller, A., Broge, M., … Shiffman, S. (2001). Cognitive-behavioral therapy to reduce weight concerns improves smoking cessation outcome in weight-concerned women. *Journal of Consulting and Clinical Psychology, 69*, 604–613.

Perkins, K. A., Karelitz, J. L., & Jao, N. C. (2013). Optimal carbon monoxide criteria to confirm 24-hr smoking abstinence. *Nicotine & Tobacco Research, 15*, 978–982.

Reavell, J., Hopkinson, M., Clarkesmith, D., & Lane, D. A. (2018). Effectiveness of cognitive behavioral therapy for depression and anxiety in patients with cardiovascular disease: A systematic review and meta-analysis. *Psychosomatic Medicine, 80*, 742–753.

Rezaei, S., Akbari Sari, A., Arab, M., Majdzadeh, R., & Mohammad Poorasl, A. (2016). Economic burden of smoking: A systematic review of direct and indirect costs. *Medical Journal of the Islamic Republic of Iran, 30*, 397.

Rodríguez-Cano, R., López-Durán, A., Martínez-Vispo, C., & Becoña, E. (2021). Causes of smoking relapse in the 12 months after smoking cessation treatment: Affective and cigarette dependence-related factors. *Addictive Behaviors*, 106903.

Roll, J. M., McPherson, S. M., & McDonell, M. G. (2021). Contingency management as a behavioral approach in addiction treatment. In *Textbook of addiction treatment* (pp. 417–432). Springer International Publishing.

Rose, J. E., Behm, F. M., Ramsey, C., & Ritchie, J. C., Jr. (2001). Platelet monoamine oxidase, smoking cessation, and tobacco withdrawal symptoms. *Nicotine & Tobacco Research, 3*, 383–390.

Sallit, J., Ciccazzo, M., & Dixon, Z. (2009). A cognitive-behavioral weight control program improves eating and smoking behaviors in weight-concerned female smokers. *Journal of the American Dietetic Association, 109*, 1398–1405.

Secades-Villa, R., González-Roz, A., García-Pérez, Á., & Becoña, E. (2017). Psychological, pharmacological, and combined smoking cessation interventions for smokers with current depression: A systematic review and meta-analysis. *PLoS One, 12*, e0188849.

Secades-Villa, R., González-Roz, A., Vallejo-Seco, G., Weidberg, S., García-Pérez, Á., & Alonso-Pérez, F. (2019). Additive effectiveness of contingency management on cognitive behavioral treatment for smokers with depression: Six-month abstinence and depression outcomes. *Drug and Alcohol Dependence, 204*, 107495.

Secades-Villa, R., Vallejo-Seco, G., García-Rodríguez, O., López-Núñez, C., Weidberg, S., & González-Roz, A. (2015). Contingency management for cigarette smokers with depressive symptoms. *Experimental and Clinical Psychopharmacology, 23*, 351–360.

Shiffman, S., Dunbar, M. S., & Ferguson, S. G. (2015). Stimulus control in intermittent and daily smokers. *Psychology of Addictive Behaviors, 29*, 847–855.

Steinberg, M. L., Williams, J. M., & Li, Y. (2015). Poor mental health and reduced decline in smoking prevalence. *American Journal of Preventive Medicine, 49*, 362–369.

Substance Abuse and Mental Health Services Administration (SAMHSA). (2017). *2017 National survey on drug use and health detailed tables*. Retrieved from: https://www.samhsa.gov/data/sites/default/files/cbhsq-reports/NSDUHDetailedTabs2017/NSDUHDetailedTabs2017.pdf.

Taylor, G. M., Lindson, N., Farley, A., Leinberger-Jabari, A., Sawyer, K., Te Water Naudé, R., … Aveyard, P. (2021). Smoking cessation for improving mental health. *The Cochrane Database of Systematic Reviews, 3*, CD013522.

Taylor, G., McNeill, A., Girling, A., Farley, A., Lindson-Hawley, N., & Aveyard, P. (2014). Change in mental health after smoking cessation: Systematic review and meta-analysis. *BMJ, 348*, g1151.

Tsuang, M. T., Francis, T., Minor, K., Thomas, A., & Stone, W. S. (2012). Genetics of smoking and depression. *Human Genetics, 131*, 905–915.

Tully, E. C., Iacono, W. G., & McGue, M. (2010). Changes in genetic and environmental influences on the development of nicotine dependence and major depressive disorder from middle adolescence to early adulthood. *Development and Psychopathology, 22*, 831–848.

van der Meer, R. M., Willemsen, M. C., Smit, F., & Cuijpers, P. (2013). Smoking cessation interventions for smokers with current or past depression. *Cochrane Database of Systematic Reviews, 8*.

Vinci, C. (2020). Cognitive behavioral and mindfulness-based interventions for smoking cessation: A review of the recent literature. *Current Oncology Reports, 22*, 58.

Vujanovic, A. A., Meyer, T. D., Heads, A. M., Stotts, A. L., Villarreal, Y. R., & Schmitz, J. M. (2017). Cognitive-behavioral therapies for depression and substance use disorders: An overview of traditional, third-wave, and transdiagnostic approaches. *American Journal of Drug and Alcohol Abuse, 43*, 402–415.

Westmaas, J. L., Bontemps-Jones, J., & Bauer, J. E. (2010). Social support in smoking cessation: Reconciling theory and evidence. *Nicotine & Tobacco Research, 12*, 695–707.

World Health Organization. (2017). *WHO report on the global tobacco epidemic, 2017: Monitoring tobacco use and prevention policies*. Switzerland: World Health Organization.

World Health Organization. (2019). *WHO global report on trends in prevalence of tobacco use 2000–2025* (3rd ed.). Geneva: World Health Organization.

Yalcin, B. M., Unal, M., Pirdal, H., & Karahan, T. F. (2014). Effects of an anger management and stress control program on smoking cessation: A randomized controlled trial. *Journal of the American Board of Family Medicine, 27*, 645–660.

Section C

International aspects

Chapter 25

Psychopathophysiology and compassion-based cognitive-behavior group therapy for patients with coronary artery disease

Chia-Ying Weng[a], Tin-Kwang Lin[b,c], and Bo-Cheng Hsu[a]

[a]Department of Psychology, National Chung Cheng University, Chiayi, Taiwan, [b]Department of Internal Medicine, School of Medicine, Tzu Chi University, Hualien, Taiwan, [c]Department of Internal Medicine, Dalin Tzu Chi Hospital, Buddhist Tzu Chi Medical Foundation, Chiayi, Taiwan

Abbreviations

ACT	acceptance and commitment therapy
ANS	autonomic nerve system
BDI	beck depression inventory
BP	blood pressure
BVA	blood volume amplitude
BVP	blood volume pulse
CAD	coronary artery disease
CBGT	cognitive-behavioral group therapy
CBT	cognitive-behavioral therapy
CHI-SF	Chinese hostility inventory-short form
CI	confidence interval
HR	hazard ratio
HPA	hypothalamic-pituitary-adrenal
HF	high frequency
HRV	heart rate variability
LF	low frequency
MI	myocardial infarction
OR	odd ratio
PAR	population attributable risks
PSS	perceived stress scale
PSP	psychophysiological stress profile
RR	risk ratio
RAAS	renin-angiotensin-aldosterone system
RCT	randomized controlled trials
STAI	state-trait anxiety inventory
WLC	wait-list control

Introduction

Psychological factors and coronary artery disease

Cardiovascular disease has always been the leading cause of morbidity and mortality worldwide, and it has great impact on human health and health care costs (Virani et al., 2021). Coronary artery disease (CAD) that manifests clinically as myocardial infarction and ischemic cardiomyopathy ranks as the most prevalent among all cardiovascular diseases (Roth et al., 2017). Traditional risk factors (age, sex, race, hypertension, hyperlipidemia, diabetes, and smoking status) only partially explain the occurrence of and prognosis for CAD, and psychosocial risk factors have not received much attention until recently. In the INTERHEART study, Yusuf et al. (2004) investigated the relation of psychosocial factors to risk of myocardial infarction in 24,767 people from 52 countries. The critical result showed that presence of psychosocial factors (odds ratio [OR] 2.67 [99% CI 2.20–3.22]) is associated with increased risk of acute myocardial infarction, with the population attributable risks (PAR) exceeding 32.5%.

Moreover, psychosocial factors play a decisive role for the occurrence and prognosis of CAD, as important as traditional risk factors. Clinicians not only require psychosocial knowledge but also professional judgment and courage to intervene psychosocial distress appropriately. Here, we will look at some important psychosocial factors that have been identified as help to predict morbidity and prognosis in CAD population.

Hostility and anger

Hostility is a personality trait described as an individual's negative orientation toward interpersonal transactions. Based on the Chinese culture that insists on circumspect outward expression of emotions, Lin and Weng (2002) defined hostility as a multidimensional construct, hostile individuals adopt negative appraisals or cognitive processes in their daily lives to display hostility behavior overtly (expressive) or suppress their hostility (suppressive). Among these multidimensional constructs, a metaanalysis study showed that anger (hostile affect) and hostile cognition are significantly associated with increased CAD events in the healthy population (combined hazard ratio [HR]: 1.19 [95% CI 1.05–1.35]), and with prognosis in the CAD population (HR: 1.24 [95% CI 1.08–1.42]) (Chida & Steptoe, 2009). Furthermore, expressive and suppressive hostility behaviors are also related to CAD course by alterations of autonomic dysfunctions during different stages of anger task (Lin, Weng, Lin, & Lin, 2015). The evidence documenting these associations showed that hostility has been considered as an important psychological risk factor for coronary artery disease (CAD) and is related to CAD course and mortality.

Anxiety

Anxiety is an experienced sense of alarm in generalized response to an unknown threat or internal conflict that includes pronounced apprehension of *future* uncertainties. In terms of the risk models, Roest, Martens, de Jonge, and Denollet (2010) have suggested that anxious persons were at risk of CAD (HR: 1.26 [95% CI 1.15–1.38]) and cardiac mortality (HR: 1.48 [95% CI 1.14–1.92]), independent of demographic variables, biological risk factors, and health behaviors. The long-term effect of anxiety, especially trait anxiety, could also help in risk stratification and identification of CAD patients at risk of mortality (OR 1.07 [99% CI 1.01–1.15]) and rehospitalization (OR 1.06 [99% CI 1.01–1.13]) after cardiac surgery (Székely et al., 2007) as well as new cardiac events after myocardial infarction (OR 1.71 [99% CI 1.31–2.23]) (Roest et al., 2010). Most researches have suggested that anxiety symptom is an independent risk factor for the morbidity and mortality of CAD in initially healthy or CAD populations.

Depression

Depression, including feeling sad, down or miserable most of the time, or losing interest in usual activities, is the main psychosocial risk factor for CAD (Serrano Jr, Setani, Sakamoto, Andrei, & Fraguas, 2011). A metaanalysis of 11 studies found that subjects suffering depression had an overall relative risk of 1.64 for developing CAD compared to nondepressed subjects (Katon et al., 2004). Also, when other risk factors (i.e., smoking) for CAD were controlled, depression independently predicted fatal and nonfatal CAD (Srinivasan, 2011). Moreover, the risk models in patients with CAD appears to be proportionate to the severity of the depression. More severe depression in CAD patients confers a relative risk between 1.5 and 2.5 for cardiac morbidity and mortality (Lett et al., 2004). The moderate to severe levels of depression symptoms during admission for MI is more closely linked to cardiac mortality (Lespérance, Frasure-Smith, Talajic, & Bourassa, 2002). There is now convincing evidence linking depression and morbidity or mortality among CAD population.

Stress

There is an enormous amount of researches on psychological stress and cardiovascular disease (Dimsdale, 2008). Types of stress, such as work stress and marital stress have consistently been associated with increased CAD risk (Eaker, Sullivan, Kelly-Hayes, D'Agostino Sr, & Benjamin, 2007; Kivimäki et al., 2006). Perceived stress, the general perception that environmental demands exceed perceived capacity (RR 1.27 [95% CI 1.12–1.45]), is also consistently associated with incident CAD (Richardson et al., 2012). High levels of perceived stress are a potent risk factor for the development and progression of CAD and cardiac mortality (Rosengren et al., 2004). Furthermore, a person with negative traits or vulnerability to emotional disturbances could suffer from stresses. The stress-induced effect also plays the interactive role of morbidity or mortality in patients with CAD (Dimsdale, 2008).

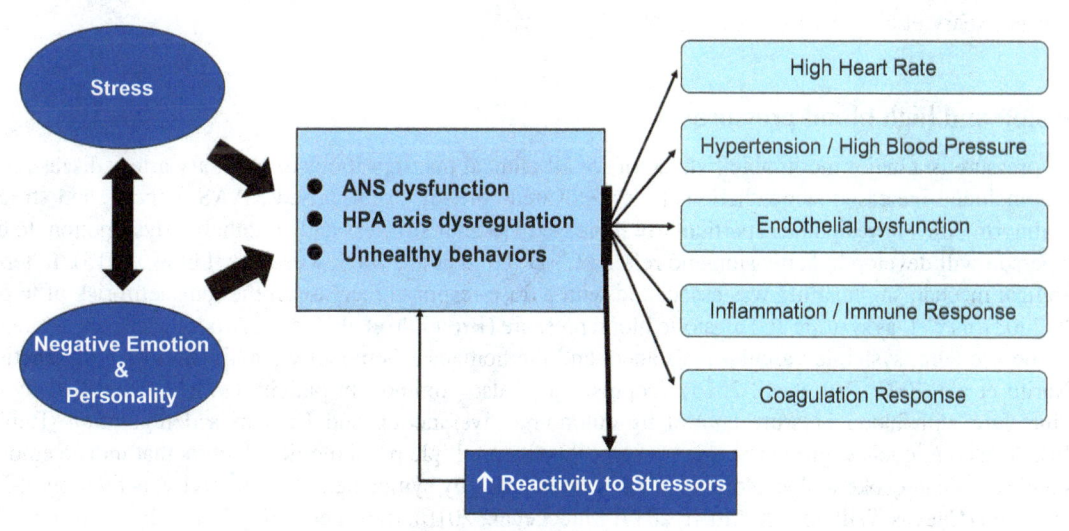

Psychopathological mechanisms of coronary artery disease

Coronary artery disease and its pathological and clinical consequences are resulted from the complex interplay between several underlying processes, such as traditional risk factors, physiological modifiers, and psychosocial factors, that continuously emerge the development of atherosclerosis (Dimsdale, 2008).

An extensive body of human and animal research reveals that psychological factors could affect cardiovascular functions, possibly via autonomic dysfunctions (including excessive sympathetic activation or parasympathetic withdrawal) and the hypothalamic-pituitary-adrenal (HPA) axis, to the development and progression of coronary artery atherosclerosis as well as hypertension and even endothelial dysfunction. Evidence also evaluates that psychological factors frequently contributes to inflammations, immune responses, and platelet activation, leading to exacerbating atherosclerosis (Rozanski, Blumenthal, Davidson, Saab, & Kubzansky, 2005; Rozanski, Blumenthal, & Kaplan, 1999). Furthermore, psychopathological mechanisms underlying the coronary artery disease can separate into altered behavioral pathways, whereby psychological factors result in a higher rate of unhealthy behaviors, such as smoking, physical inactivity, and poor diet (Rozanski et al., 1999, 2005) (see Fig. 1). It can be seen that an understanding of these psychopathological mechanisms will enable clinicians and psychologists to establish the potential role of psychological factors in CAD and identify more appropriate treatments or interventions.

In this section, we review various important psychopathological mechanisms linking psychological factors to coronary artery atherosclerosis, including autonomic dysfunctions, HPA axis, high blood pressure, endothelial dysfunctions, and altered unhealthy behaviors and lifestyle.

Autonomic dysfunctions

The autonomic nerve system (ANS) regulates cardiac function and maintain homeostasis to control over the heart, heart rate, and contractile force. The ANS dysfunction, such as sympathetic hyperactivity or parasympathetic hypoactivity, also plays an important role in atherosclerosis and CAD (Brosschot & Thayer, 1998; Rozanski et al., 2005). Sympathetic hyperactivity, which predominates in the response to psychological factors, or the fight-flight response, is a key feature of CAD to increase heart rate, blood pressure, cardiac output, inflammation and immune response, and endothelial dysfunctions. Parasympathetic hypoactivity via vagal withdrawal, assessed by high frequency (HF) component of heart rate variability (HRV), is associated with increased myocardial infarction, atherosclerosis, heart failure, hypertension, and sudden cardiac death. Low HRV, reflecting ANS dysfunction, was predictive of cardiovascular risk factors and poor prognoses among healthy adults and CAD population (Hillebrand et al., 2013; Thayer, Yamamoto, & Brosschot, 2010).

FIG. 1 Psychopathological mechanisms of coronary artery disease.

There is convincing evidence linking psychological factors and ANS dysfunction, promoting atherosclerosis and CAD development. Compared with low hostility individuals, those with high hostility manifest greater cardiovascular response and sympathetic hyperactivity to experimental harassment tasks or interpersonal challenges (al'Absi, Bongard, & Lovallo, 2000; Powch & Houston, 1996), as well as lower parasympathetic activation (parasympathetic hypoactivity) during the recovery stage (Lin et al., 2015). There is a positive relationship between hostility and sympathetic activity [low frequency (LF)/high frequency ratio, LF/HF ratio] during the resting baseline in healthy adults (Sloan et al., 2001) and during the neutral and anger stage in CAD population (Lin et al., 2015). In the long run, hostility in individuals may cause the recurrent psychopathological mechanism of ANS dysfunction, which promotes atherosclerosis (Angerer et al., 2000) and CAD development (Brydon et al., 2010; Lin et al., 2015; Sloan et al., 2001). Furthermore, a review showed that anxiety and chronic stresses have been almost characterized by sympathetic hyperactivity and parasympathetic hypoactivity together with higher resting heart rate and blood pressure (Kreibig, 2010). Reports on anxiety or its disorder indicate diminished HRV, increased LF and LF/HF (sympathetic hyperactivity), and decreased HF (parasympathetic hypoactivity) in healthy and CAD individuals (Kreibig, 2010), reflecting the obvious psychopathological mechanism of ANS dysfunction. Stresses-induced ANS dysfunction also predisposes to coronary events by contributing to platelet activation, endothelial dysfunction, and coronary vasoconstriction, accelerating atherosclerosis and triggering acute myocardial infraction (Rozanski et al., 1999, 2005). Finally, depressed patients show higher resting heart rate than nondepressed patients and exhibit ANS dysfunction, including reduced HRV and baroreflex function, raised total body sympathetic activity, and increased plasma norepinephrine (Carney, Freedland, & Veith, 2005).

Hypothalamic-pituitary-adrenal (HPA) axis

The hypothalamic-pituitary-adrenal (HPA) axis comprises an integration of central nervous system and peripheral tissues to regulate adrenal hormones, especially glucocorticoids, which help to preserve or restore homeostasis in stress conditions (Burford, Webster, & Cruz-Topete, 2017). Dysregulation of HPA axis and glucocorticoids is linked to a variety of physiopathologies including hypercortisolism, hypertension and subsequent vascular damage, and even cardiac dysfunction (such as coronary artery calcification) (Burford et al., 2017).

Emotional disturbance and chronic stress can have a prolonged impact on the HPA axis and cortisol release. Evidence relating dysregulation of cortisol rhythms to hostility is limited. Several studies show that hostility is associated with hypercortisolemia (Ranjit et al., 2009; Sjögren, Leanderson, & Kristenson, 2006), but may have limited the ability to detect cortisol alterations of the diurnal rhythm. Chronic stimulation of the HPA axis by anxiety and depression results in blunted HPA axis and hypercortisolemia (Nandam, Brazel, Zhou, & Jhaveri, 2020; Zorn et al., 2017). An abundance of empirical studies have supported that chronic stress triggers an allostatic shift in the normal circadian rhythm of cortisol release as well as in stress-induced cortisol levels. This allostatic condition makes the HPA axis more sensitive, resulting in higher cortisol exposure or greater cortisol burden following each stressful episode (Hamer, Endrighi, Venuraju, Lahiri, & Steptoe, 2012; Iob & Steptoe, 2019). Suffered from psychological factors, hypercortisolemia may occur an increase in cardiac and peripheral fatty acids, coronary calcifications, and noncalcified coronary plaque volumes, contributing to the progression of atherosclerosis (Neary et al., 2013).

Hypertension and high blood pressure

High blood pressure is a major modifiable risk factor for all clinical manifestations of coronary artery disease (CAD), the mechanisms included increased sympathetic activity, elevated cortisol levels, raised RAAS activity, and structural and functional abnormalities in vesiculates, particularly increased vascular stiffness and endothelial dysfunction, to determine whether a person will develop hypertension and related CAD (Gold et al., 2005; Rosendorff et al., 2015). In term of psychopathological mechanism, hostility was associated with a dose-response increase in the long-term risk of hypertension (Yan et al., 2003), as well as systolic and diastolic blood pressure (Brondolo et al., 2003). Anxiety and stress-related factors increase blood pressure, systemic vascular resistance, and the homeostasis model via activation of sympathetic nervous system (Narita et al., 2007; Pan et al., 2015). Depression is also common in patients with uncontrolled hypertension and may interfere with blood pressure control by autonomic dysfunction and HPA axis deregulation (Rubio-Guerra et al., 2013). As a consequence, growing evidence indicates that multiple psychological factors that induce emotional disturbances and stress can evoke a physiological response meditated by sympathetic hyperactivity, activation of HPA axis, and inflammation (Cuevas, Williams, & Albert, 2017; Mucci et al., 2016). Repeated activation of these psychopathological mechanisms can result in failing to return to resting blood pressure (BP) levels (Cuevas et al., 2017).

Endothelial dysfunctions

Endothelial dysfunction in response to specific pathophysiological stimuli (i.e., hypercholesterolemia and other dyslipidemias, diabetes, obesity, hypertension, aging) has important local manifestations within lesion-susceptible regions of arteries, contributing to the pathobiology of atherosclerosis and CAD (Gimbrone Jr & García-Cardeña, 2016). More recent data indicate that psychological factors are associated with a heighted incidence of endothelial dysfunction among healthy and CAD population. Hostility is positively associated with endothelial dysfunction, as indicated by less vasodilation of the brachial artery and flow-mediated dilation in healthy adults (Lin et al., 2008; Schott, Kamarck, Matthews, Brockwell, & Sutton-Tyrrell, 2009). Anxiety and depression are also strongly related to excessive microvasculature dysfunctions, including endothelial dysfunction (Do, Dowd, Ranjit, House, & Kaplan, 2010; Munk et al., 2012) and vascular stiffness (Logan, Barksdale, Carlson, Carlson, & Rowsey, 2012; Seldenrijk et al., 2013). Chronic stress can induce or precede the onset of emotional disturbances to promote endothelial dysfunction via activation of ANS and HPA axis (Sher et al., 2020). Above all, an abundance of empirical studies has supported the view that psychological factors, interacted with ANS dysfunction and HPA axis dysregulation, lead to excessive endothelial dysfunction and even the accelerating atherosclerosis and CAD symptoms (Rozanski et al., 1999, 2005; Sher et al., 2020).

Altered unhealthy behaviors and lifestyle

One of the pathways mediating the indirect effects of psychological factors on the pathogenesis and expression of coronary artery disease is unhealthy behavioral and lifestyle factors, such as smoking (Kassel, Stroud, & Paronis, 2003), physical inactivity (Dimsdale, 2008), unhealthy diet (Lopresti, Hood, & Drummond, 2013), and alcohol consumption (Airagnes et al., 2017). Psychological factors (i.e., depression, anxiety, and stress) are significantly associated with nonadherence to medical orientations (Crawshaw, Auyeung, Norton, & Weinman, 2016; Sundbom & Bingefors, 2013), increasing poor progression and prognoses of myocardial infarction. Furthermore, patients suffering from emotional disturbance and stress are more likely to have maladaptive coping strategies (Littleton, Horsley, John, & Nelson, 2007). These behaviors and lifestyle factors contribute to the lethal properties of psychological factors in coronary artery disease.

Psychotherapy and coronary artery disease

Systematic review of randomized controlled trials (RCT) indicated that cognitive-behavioral therapy significantly reduced perceived stress, anxiety, and depression (Magán et al., 2020; Sommaruga, 2016), as well as hostility and cardiac death (Whalley, Thompson, & Taylor, 2014) among patients with CAD.

In Taiwan, Weng and Lin developed the face to face (Weng et al., 2016) and internet-based (Lin et al., 2018) versions of cognitive-behavioral group therapy (CBGT) protocol aimed to reduce hostility and anxiety for patients with CAD, which includes eight 2-h sessions with the following components: (1) Increasing motivations by viewing psychophysiological stress profile; (2) Acknowledging the difficulty to change; (3) Controlling impulsive behavior by lowering physical arousal; (4) Reinforcing compassion behavior; (5) Cardiovascular health CBT strategies including cognitive and behavioral change, and emotion regulation; (6) Increasing social support by facilitating reciprocal sharing in group. The results indicated that patients with CAD in the CBGT group experience reduced anxiety, hostility, and respiration rate, together with improved vasomotor function indexed by blood volume amplitude (BVA) compared with the wait-list control (WLC) group.

The purpose of this case study is to demonstrate the practice procedures of the CBGT group, including psychological and physical measurement materials, personal case report provided to patients, group dynamic following the Compassion theory (Neff, 2003), and patients' experiences shared in session.

Practice and procedures

(1) Increasing motivations by viewing psychophysiological stress profile (Session 1).

Patients participated in psychotherapy program for their health concerns, not only mental health but also physical health conditions, such as heart rate, blood pressure and vasomotor function, these are directly related to disease progression of atherosclerosis. The first and essential task is showing patients their individual physical responses under emotional stress.

Psychological measurements (PSS, BDI, STAI, CHI-SF) and psychophysiological stress profile (PSP) of anger recall task (Lin et al., 2015) were examined preintervention with the help of the biofeedback equipment (BioGraph Infiniti version 5.0.3, Thought Technology Ltd., Montreal, Quebec, Canada). Test results were provided to patients at first session. Fig. 2 is

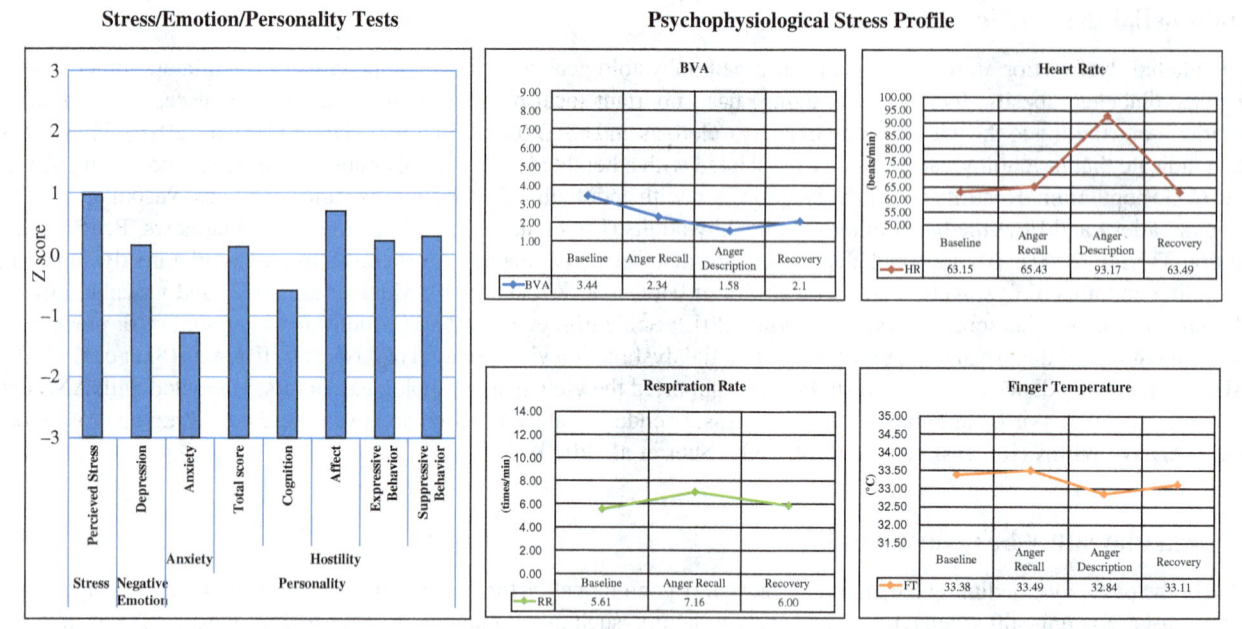

FIG. 2 Test results of psychological measurements and PSP_ Pre-intervention.

the preintervention test results of a 53-year-old male CAD patient. The PSP showed personal autonomic nervous system responses including increased heart rate, increased blood pressure, and decreased finger BVA during anger recall stage.

Fig. 3 showed the vessel activities of participants with different hostility behavior pattern. Blood volume pulse (BVP) signal indicates the relative changes in the perfusion of the blood through the finger of the participant and the effect may be influenced by autonomic related vasoconstriction caused by emotional disturbance. Fig. 3A showed the vessel activity of participant A without hostility behavior pattern (low expressive and low suppressive hostility score in CHI-SF scale). In contrast, Fig. 3B showed the vessel of participant B with an expressive hostility pattern (high expressive and low suppressive hostility score in CHI-SF scale) constricted in anger recall stage, then moderately delayed in dilatation during recovery stage that came after anger recall stage. Fig. 3C showed the vessel of participant C with a suppressive hostility pattern (low expressive and high suppressive hostility score in CHI-SF scale) constricted in anger recall stage; but still could not dilate to baseline level in the sit silently recovery stage. As participants knew that CAD and heart attack are caused by atherosclerosis and plaque rupture with thrombus, they were appalled when witted their own vessels constrict during anger recall and cannot dilate even though they just sit silently in the recovery stage. .

The second task is translating their individual cardiovascular responses under emotional disturbance to the psychopathophysiology of CAD. The psychoeducation contents include (1) how chronic psychosocial distress affects the pathophysiological mechanism and prognosis of CAD through the autonomic nervous system and hypothalamus-pituitary-adrenal axis. The pathways include enhanced heart rate, blood pressure, inflammation, and coagulation response. These mechanisms will further accelerate atherosclerosis; (2) how acute violent heart rate, blood pressure fluctuations, and extreme coronary artery vasoconstriction under extraordinary emotional reactions cause atheroma plaque rupture of coronary arteries to format thrombus, which may obstruct coronary arteries immediately. Furthermore, the stenotic vessel is worsened as it constricted under emotional disturbances.

The importance of the psychoeducation is to improve participants' self-awareness about their emotional responses, which related directly to their disease control, and further making patients realized it's their responsibility to improve their health conditions. As a patient said after reviewing his PSP, "I think I learned that—the heart and vessels are affected by emotions. The most important thing is to create mood within mind." The level of motivation to change determines how diligent and the frequency they practice CBT skills and thus the strength of newly learned adaptive behaviors, which is vital for the effects of CBT.

(2) Acknowledging the difficulty to change (Sessions 1–2)

Although patients want to maintain emotional stability, their frustrated experiences make them less confident in taking actions. As a patient participated in CBT group said in the first meeting, "This program won't work, I spend my whole life trying to cure my foul temper, but it just won't work!" After explaining behavior strength is determined by frequency

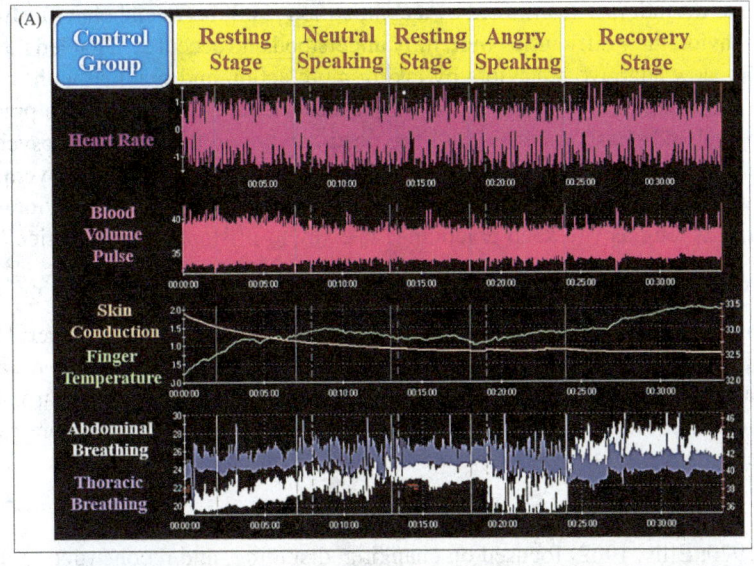

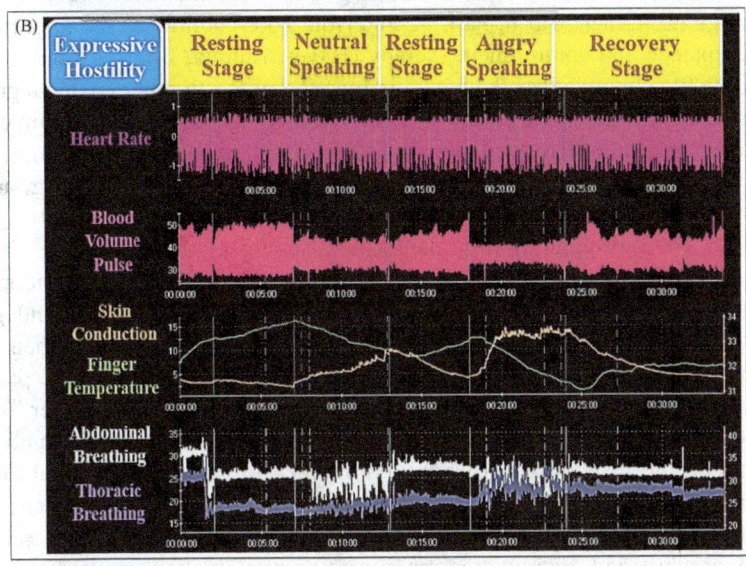

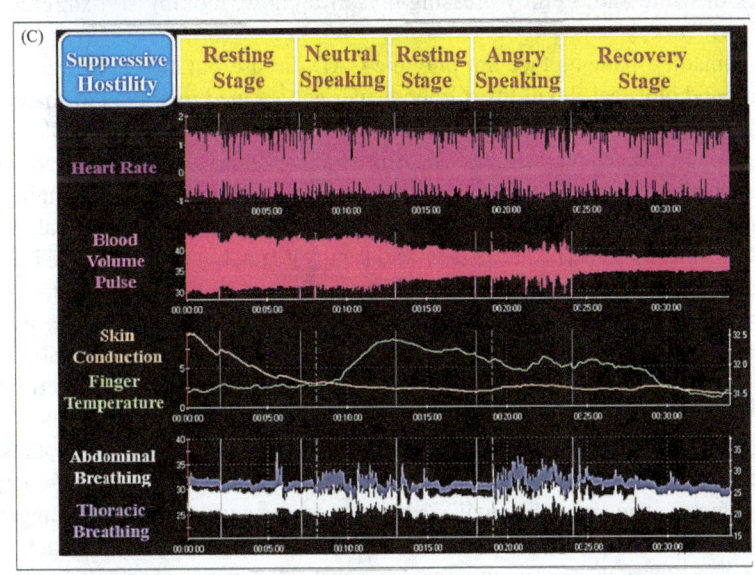

FIG. 3 PSP of expressive vs. suppressive hostility pattern during anger recall task. PSP patterns of (A) control group, (B) expressive hostility, and (C) suppressive hostility during anger recall task. *(With permission of Professor I-Mei Lin.)*

of practice; and behavior manifestation is determined by a competition result between behavior strength of old foul temper and that of new adapted behavior, their difficulties were fully understood and accepted. That led them to commit to increase the behavior strength of the new adaptive behaviors by taking more action and practice with the help of clinical health psychology (group leader) and other group members. Group leader encouraged members to practice and work together to improve both psychological and physical conditions. The postintervention test results of psychological measurements and PSP were provided to every member to exhibit psychological and cardiovascular improvements of their work.

Besides, the tolerance of failure was increased through accepting it as the outcome of behavioral competition. "It's okay if you don't do well at first, as long as you keep practicing. The key to it is one word, practice!"

(3) Controlling impulsive behavior by lowering physical arousal (Sessions 1–2)

In addition to diaphragmatic breath training, the portable biofeedback equipment (StressEraser; Helicor, Inc., New York, USA) was provided to build the connections between relaxation practice and breathe conditions, and further help patients concentrate while they practice at home. Relaxation training is adopted to lower the physical arousal level and prolong the reaction time of a tantrum, which often appears in a high arousal state. That gives them more capacity to re-decide and choose a new adaptive behavior before they reapply foul temper.

(4) Reinforcing compassion behavior (Session 3)

Traditional CBT (Beck, 1970; Ellis, 1962) focused on changing, disputing, and reconstructing the maladaptive thought contents. Positive psychology (Seligman, Steen, Park, & Peterson, 2005) addresses the importance of positive as well as negative emotions. Acceptance and commitment therapy (Hayes, Luoma, Bond, Masuda, & Lillis, 2006; Powers, Vörding, & Emmelkamp, 2009) acknowledges the adaptive function of the thought contents grounded in patients' life experiences. Group leader of ACT address the psychological flexibility to change or maintain certain behavior/thought to serve key personal value of patients as they perceive the present inner and environment status fully. Hence, increasing awareness is the key issue, which can be accomplished by relaxation, breathing, and many kinds of meditation training.

Compassion theory (Neff, 2003) originating from the Eastern Buddhist culture addresses the importance of the following three concepts: kindness, common humanity, and mindfulness. This CBT program adopted Compassion theory to focus not only on releasing negative emotions but also building kind, noncritical, and genuine relationship to oneself. For kindness, the group leader created a warm and noncritical circumstance, then demonstrated an open, understanding, and accepting attitude toward distress in group session. Furthermore, patients' self-compassion behaviors were increased through behavior activation strategy entitled "One happy offer." Members of the group are encouraged to do one happy thing, as an offering, to themselves. As understanding and acceptance for oneself increased through focusing on the inside, patients would know how to care, treat, protect, and nourish themselves. The self-compassion behaviors are reinforced by self-reinforcement and social reinforcement, and that will substitute harmful and unhealthy impulsive behaviors. A patient who is a CEO of a large company said, "I used to have only baggy clothes, all in black or white and. I enjoy dressing myself up now, looking stylish, greeting people happily and accepting their compliments."

Besides, for common humanity, patients face not only the CAD disease but also aging and death threatens, that constrained their psychological and physical flexibility. Group leader encouraged members to share their own effective cognition or behavior change experiences with other patients, those with the same disease and similar age. In the supporting and sharing atmosphere of the group, patients realized disease distress experience is a common matter instead of personal misfortune, their self-indulgence or self-isolation gradually dissipated. The happy self-challenging experiences shared by fellow members, acts as compassion coming from others, encouraged patients' psychological flexibility and the passion to live. "I thought having CAD means that all I can do is to wait for my death, but everyone here is ill, some of them are even more severe than me, I realize there are a lot more things I can do."

Fig. 4 shows the Compassion based CBT model of this study. The vertical axis indicates the psychological flexibility of mind to behave in the surrounding world. Group leaders did not need to give advice, but only explain the health consequence of daily repetitive behavior, and principles of behavior strength and behavioral competition, then respect patients' autonomy and decisions. As they explore more and have deeper understanding of themselves, patients would have more "One happy offer," which increase the flexibility and permit themselves to contact inner needs, thus make better choices for their health. The horizontal axis indicates the compassion toward oneself and others. The self-compassion behaviors came from the active creation, which are in line with their sincere needs, instead of rule/order/suggestions from outside. As self-compassion and self-acceptance increased, and anxiety reduced. As the group supports each other and works together to be healthier, compassion for others increased, and hostility substituted and reduced.

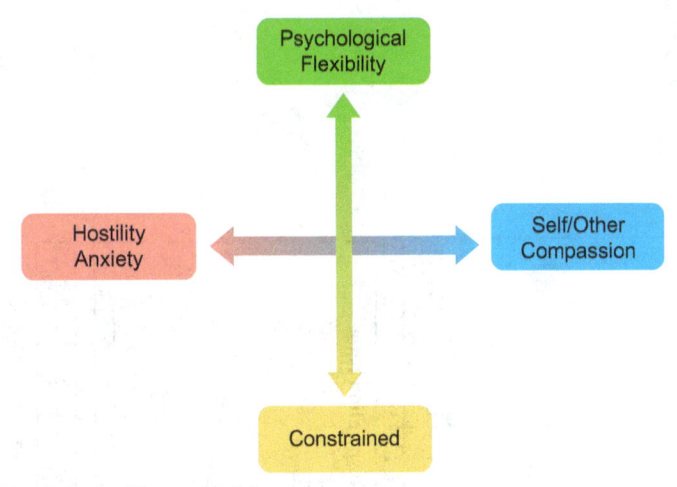

FIG. 4 Compassion-Based Cognitive-Behavior Therapy Model.

(5) Cardiovascular health CBT strategies (Sessions 4–8)

As patients' psychological flexibility and the passion to live were raised, they would try any effective strategies including CBT to improve their cardiovascular health conditions. The followings are CBT strategies provided in program and patients' feedback.

I was the king of being nervous. As I get nervous, my heart beats rapidly, the practices really help alleviate it effectively. I used to have problem controlling my blood pressure, now it drops from 140–150 to 110–120, and my medication usage is reduced after consulted doctor.

Cognitive change: Providing psychoeducation on how cognitive thoughts impact emotional and behavioral responses, and instructing CBT strategies to foster cognitive flexibility and thought change (shifting perspective).

I undergone cardiac catheterization when I was young, and I'm always worried about heart diseases. Whenever my heart feels uncomfortable, it worries me. Sometimes I can shift perspective, quiet down and come to think that I have done everything that I could, sometimes I can calm myself down.

Master our own emotion: Providing psychoeducation on the effect of repetitive emotional impulses on the associated neural circuits, instructing self-monitoring strategies for emotional reactions and enhancing strength of self-compassion behavior.

I was about to throw a tantrum at my husband, then I remembered that I should be kind to myself, don't get angry and hurt my heart and vessels. So I it as a chance to practice shifting perspective, and at the same time I lowered my voice, and I felt less angry.

Behavior change: Reinforcing behavior changes initiated by participants.

For my own health, I slow down my actions and take more rest, instead of having tight schedule like I used to. Now I have more time to be with my family and meet new friends.

(6) Increasing social support by facilitating reciprocal sharing in group, including effective strategies of cognitive and behavioral change, and emotion regulation

I'm glad to meet friends with CAD here, we're sharing effective ways to relax, be happy and shifting perspective, we're working together to make everyone healthier.

The effect of this group intervention program to reduce hostility and anxiety was supported, as well as improve vasomotor and ANS regulation for patients with CAD (Lin et al., 2018; Weng et al., 2016). Fig. 5 showed the postintervention (with 6 months follow-up) tests results of a 53 years old male CAD patient. The hostility score examined postintervention was lower than that examined preintervention, and went still lower at the 6 month follow-up. The perceived stress, anxiety and depression scores examined at postintervention were lower than those examined preintervention, but slightly rebound at the

Psychophysiological Stress Profiles

BVA

	B	AR	AD	R
Pre	3.44	2.34	1.58	2.10
Post	5.82	4.71	3.16	3.83
Follow	8.30	6.97	5.12	5.92

Heart Rate (beats/min)

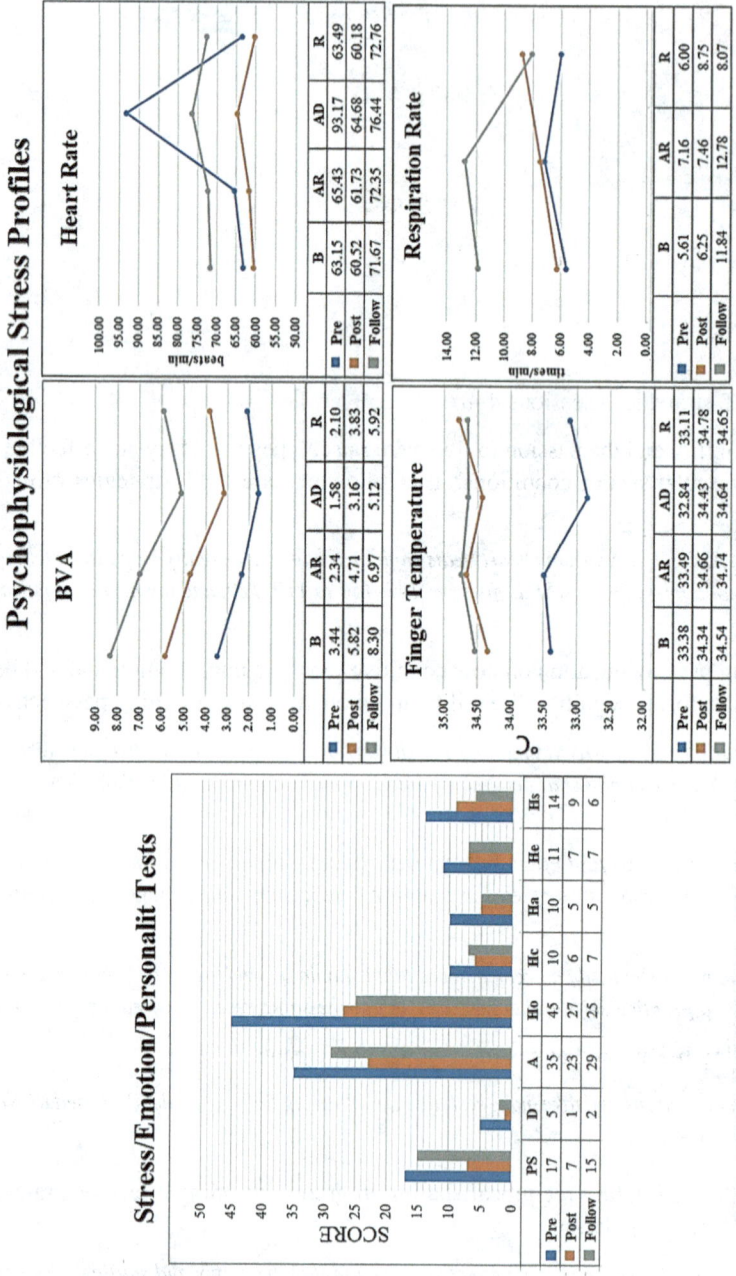

	B	AR	AD	R
Pre	63.15	65.43	93.17	63.49
Post	60.52	61.73	64.68	60.18
Follow	71.67	72.35	76.44	72.76

Finger Temperature (°C)

	B	AR	AD	R
Pre	33.38	33.49	32.84	33.11
Post	34.34	34.66	34.43	34.78
Follow	34.54	34.74	34.64	34.65

Respiration Rate (times/min)

	B	AR	AD	R
Pre	5.61	7.16	—	6.00
Post	6.25	7.46	—	8.75
Follow	11.84	12.78	—	8.07

Stress/Emotion/Personalit Tests

	PS	D	A	Ho	Hc	Ha	He	Hs
Pre	17	5	35	45	10	10	11	14
Post	7	1	23	27	6	5	7	9
Follow	15	2	29	25	7	5	7	6

Abbreviations of psychological factors: PS, perceived stress; D, depression; A, anxiety; Ho, total hostility; Hc, cognitive hostility; Ha, affective hostility; He, expressive hostility; Hs, suppressive hostility

Abbreviations of PSP stages: B, baseline; AR, anger recall; AD, anger description; R, recovery

FIG. 5. Test results of psychological measurements and PSP_ Pre- vs. postintervention and 6-month follow-up.

6-month follow-up. With regard of PSP, the increased BVA after intervention indicated better vasodilation and ANS regulation. Moderated heart rate fluctuation under anger recall after intervention indicated the better emotion regulation and ANS balance.

Mini-dictionary of terms

Autonomic dysfunctions. The condition in which the autonomic nervous system (ANS) does not work properly, damage, or dysregulation.

Coronary artery disease. The most common type of *heart disease* that the reduction of blood flow to the heart due to atherosclerosis in the arteries of the heart.

Heart rate variability. The physiological phenomenon (measure) of the variation in the time interval between adjacent heartbeats.

Hostility. Individuals adopt negative appraisals or cognitive processes in their daily lives to display hostility behavior overtly (expressive) or suppress their hostility (suppressive).

Hypothalamic-pituitary-adrenal (HPA) axis. The interactive neuroendocrine unit between the hypothalamus, pituitary gland, and adrenal glands.

Psychopathological mechanisms. The study of the mental disorders or psychosocial distresses which affect or interact to physiopathological (direct) and behavioral (indirect) pathways to illness conditions.

Psychophysiological stress profile. The records of a baseline for physiological readings to the stress condition or stressors in order to guide treatment efforts and gauge treatment progress.

Self-compassion behaviors. The behaviors that extend compassion to one's self in instances of perceived inadequacy, failure, or general suffering.

Unhealthy behaviors. The patterned behaviors that is harmful to physical or psychological health.

Key facts

- Coronary artery disease (CAD) ranks as the most prevalent among all cardiovascular diseases, which has been the leading cause of morbidity and mortality worldwide.
- Psychosocial factors, as important as traditional risk factors, play a decisive and important role for the occurrence and prognosis of CAD.
- Psychosocial factors contribute 32.5% of acute myocardial infarction incidence. Among all the modifiable psychological and biological risk factors, smoking has the highest odds ratio of 2.87.
- Psychosocial risk factor was a close second, odds ratio of 2.67, which is higher than that of diabetes, 2.04; and hypertension, 1.91.
- The understanding of these psychopathological mechanisms (i.e., autonomic dysfunctions, HPA axis, high blood pressure and heart rate, inflammation, coagulation response, endothelial dysfunctions, and unhealthy behaviors and lifestyle) will enable clinicians and psychologists to establish the potential role of psychological factors in CAD and identify more appropriate treatments or interventions.
- Psychosocial intervention and psychotherapy significantly reduced perceived stress, anxiety, depression, hostility and cardiac death among patients with CAD.
- High level effectiveness of the group psychotherapy, including face to face, internet-based, and mobile-based interventions, could have a significant improvement in psychosocial and cardiovascular health, as well as quality of life for patients with CAD.

Applications to other areas

In this chapter, we have reviewed psychosocial risk factors and psychopathological mechanisms, as well as evidence-based cognitive-behavioral therapy (CBT) for coronary artery disease (CAD). Metaanalyses of longitudinal studies validated psychosocial factors play an important role for the occurrence and prognosis of CAD. Hostility and anxiety personality trait, depression, and stress affect cardiovascular functions, these psychosocial risk factors impact the progression of coronary artery atherosclerosis through autonomic nervous system, HPA axis, and unhealthy behaviors. The mechanisms include enhanced heart rate, blood pressure, inflammation, coagulation response, and endothelial dysfunction. CBT could significantly reduce psychological risk factors and prognosis related to CAD. Weng et al. (2016) and Lin et al. (2018) developed the Compassion-based cognitive-behavioral group therapy (CBGT) protocol for CAD patients. The effect of this protocol to reduce hostility and anxiety was supported, as well as improve vasomotor and ANS regulation for patients with CAD. This CBGT program increases motivations by providing psychoeducation of CAD psychopathology, along with personal psychophysiological stress profile illustrating enhanced heart rate, blood pressure, and vasoconstriction under emotional disturbance. Therefore, patients are highly motivated to attend psychotherapy session and adhere to medical advices, thus improve their psychological and even cardiovascular health. In addition, our CBGT protocol for CAD population has established to extend to other intervention guideline.

In the cardiology, there are many types of cardiovascular diseases, such as hypertension, arrhythmia, and even heart failure, some of which are preventable. Future psychotherapy, based on this CBT protocol, might even reveal new directions to these cardiovascular diseases. Furthermore, according to psychopathological mechanisms, this CBT protocol could be revised to apply on lifestyle modification and related psychosomatic diseases, such as metabolic syndromes, psycho-neuroimmunology, or psycho-gastroenterology.

This CBT protocol has been transferred into internet-based and mobile-based intervention for primary care and disease control for CAD and heart failure patients. Besides, internet-based and mobile-based health education and psychoeducation, has adopted app-assisted relaxation training and cognitive-behavioral therapy strategies, including cognitive flexibility and behavioral activation, to reduce the severity of psychosocial distresses, and further prevent disease recurrence.

Summary points

- Metaanalyses of longitudinal studies validated psychosocial factors play an important role for the occurrence and prognosis of coronary artery disease (CAD).
- Hostility and anxiety personality trait, depression, and stress affect cardiovascular functions, these psychosocial risk factors impact the progression of coronary artery atherosclerosis through autonomic nervous system, HPA axis, and unhealthy behaviors. The mechanisms include enhanced heart rate, blood pressure, inflammation, coagulation response, and endothelial dysfunction.
- The effect of this Compassion-based CBGT to reduce hostility and anxiety was supported, as well as improve vasomotor and ANS regulation for patients with CAD.
- Motivation is increased by providing psychoeducation of CAD psychopathology, along with personal psychophysiological stress profile illustrating enhanced heart rate, blood pressure, and vasoconstriction under emotional disturbance.
- This cognitive-behavioral group therapy program increases patients' motivations to attend psychotherapy session and adhere to medical advices, thus improves psychological and even cardiovascular health.

References

Airagnes, G., Lemogne, C., Gueguen, A., Hoertel, N., Goldberg, M., Limosin, F., & Zins, M. (2017). Hostility predicts alcohol consumption over a 21-year follow-up in the Gazel cohort. *Drug and Alcohol Dependence, 177*, 112–123.

al'Absi, M., Bongard, S., & Lovallo, W. R. (2000). Adrenocorticotropin responses to interpersonal stress: Effects of overt anger expression style and defensiveness. *International Journal of Psychophysiology, 37*(3), 257–265.

Angerer, P., Siebert, U., Kothny, W., Mühlbauer, D., Mudra, H., & von Schacky, C. (2000). Impact of social support, cynical hostility and anger expression on progression of coronary atherosclerosis. *Journal of the American College of Cardiology, 36*(6), 1781–1788.

Beck, A. T. (1970). Cognitive therapy: Nature and relation to behavior therapy. *Behavior Therapy, 1*(2), 184–200.

Brondolo, E., Rieppi, R., Erickson, S. A., Bagiella, E., Shapiro, P. A., McKinley, P., & Sloan, R. P. (2003). Hostility, interpersonal interactions, and ambulatory blood pressure. *Psychosomatic Medicine, 65*(6), 1003–1011.

Brosschot, J. F., & Thayer, J. F. (1998). Anger inhibition, cardiovascular recovery, and vagal function: A model of the link between hostility and cardiovascular disease. *Annals of Behavioral Medicine, 20*(4), 326–332.

Brydon, L., Strike, P. C., Bhattacharyya, M. R., Whitehead, D. L., McEwan, J., Zachary, I., & Steptoe, A. (2010). Hostility and physiological responses to laboratory stress in acute coronary syndrome patients. *Journal of Psychosomatic Research, 68*(2), 109–116.

Burford, N. G., Webster, N. A., & Cruz-Topete, D. (2017). Hypothalamic-pituitary-adrenal axis modulation of glucocorticoids in the cardiovascular system. *International Journal of Molecular Sciences, 18*(10), 2150.

Carney, R. M., Freedland, K. E., & Veith, R. C. (2005). Depression, the autonomic nervous system, and coronary heart disease. *Psychosomatic Medicine, 67*, S29–S33.

Chida, Y., & Steptoe, A. (2009). The association of anger and hostility with future coronary heart disease: A meta-analytic review of prospective evidence. *Journal of the American College of Cardiology, 53*(11), 936–946.

Crawshaw, J., Auyeung, V., Norton, S., & Weinman, J. (2016). Identifying psychosocial predictors of medication non-adherence following acute coronary syndrome: A systematic review and meta-analysis. *Journal of Psychosomatic Research, 90*, 10–32.

Cuevas, A. G., Williams, D. R., & Albert, M. A. (2017). Psychosocial factors and hypertension: A review of the literature. *Cardiology Clinics, 35*(2), 223–230.

Dimsdale, J. E. (2008). Psychological stress and cardiovascular disease. *Journal of the American College of Cardiology, 51*(13), 1237–1246.

Do, D. P., Dowd, J. B., Ranjit, N., House, J. S., & Kaplan, G. A. (2010). Hopelessness, depression, and early markers of endothelial dysfunction in US adults. *Psychosomatic Medicine*, *72*(7), 613.

Eaker, E. D., Sullivan, L. M., Kelly-Hayes, M., D'Agostino, R. B., Sr., & Benjamin, E. J. (2007). Marital status, marital strain, and risk of coronary heart disease or total mortality: The Framingham offspring study. *Psychosomatic Medicine*, *69*(6), 509–513.

Ellis, A. (1962). *Reason and emotion in psychotherapy*.

Gimbrone, M. A., Jr., & García-Cardeña, G. (2016). Endothelial cell dysfunction and the pathobiology of atherosclerosis. *Circulation Research*, *118*(4), 620–636.

Gold, S. M., Dziobek, I., Rogers, K., Bayoumy, A., McHugh, P. F., & Convit, A. (2005). Hypertension and hypothalamo-pituitary-adrenal axis hyperactivity affect frontal lobe integrity. *The Journal of Clinical Endocrinology & Metabolism*, *90*(6), 3262–3267.

Hamer, M., Endrighi, R., Venuraju, S. M., Lahiri, A., & Steptoe, A. (2012). Cortisol responses to mental stress and the progression of coronary artery calcification in healthy men and women. *PLoS One*, *7*(2), e31356.

Hayes, S. C., Luoma, J. B., Bond, F. W., Masuda, A., & Lillis, J. (2006). Acceptance and commitment therapy: Model, processes and outcomes. *Behaviour Research and Therapy*, *44*(1), 1–25.

Hillebrand, S., Gast, K. B., de Mutsert, R., Swenne, C. A., Jukema, J. W., Middeldorp, S., ... Dekkers, O. M. (2013). Heart rate variability and first cardiovascular event in populations without known cardiovascular disease: meta-analysis and dose–response meta-regression. *Europace*, *15*(5), 742–749.

Iob, E., & Steptoe, A. (2019). Cardiovascular disease and hair cortisol: A novel biomarker of chronic stress. *Current Cardiology Reports*, *21*(10), 1–11.

Kassel, J. D., Stroud, L. R., & Paronis, C. A. (2003). Smoking, stress, and negative affect: Correlation, causation, and context across stages of smoking. *Psychological Bulletin*, *129*(2), 270.

Katon, W. J., Lin, E. H., Russo, J., Von Korff, M., Ciechanowski, P., Simon, G., ... Young, B. (2004). Cardiac risk factors in patients with diabetes mellitus and major depression. *Journal of General Internal Medicine*, *19*(12), 1192–1199.

Kivimäki, M., Virtanen, M., Elovainio, M., Kouvonen, A., Väänänen, A., & Vahtera, J. (2006). Work stress in the etiology of coronary heart disease—A meta-analysis. *Scandinavian Journal of Work, Environment & Health*, 431–442.

Kreibig, S. D. (2010). Autonomic nervous system activity in emotion: A review. *Biological Psychology*, *84*(3), 394–421.

Lespérance, F., Frasure-Smith, N., Talajic, M., & Bourassa, M. G. (2002). Five-year risk of cardiac mortality in relation to initial severity and one-year changes in depression symptoms after myocardial infarction. *Circulation*, *105*(9), 1049–1053.

Lett, H. S., Blumenthal, J. A., Babyak, M. A., Sherwood, A., Strauman, T., Robins, C., & Newman, M. F. (2004). Depression as a risk factor for coronary artery disease: Evidence, mechanisms, and treatment. *Psychosomatic Medicine*, *66*(3), 305–315.

Lin, I. M., & Weng, C. Y. (2002). Relationship between hostility pattern and psychophysiological disorders: Cases of coronary artery disease and headache. *Chinese Journal of Psychology*, *44*(2), 211–226.

Lin, I. M., Weng, C. Y., Lin, T. K., & Lin, C. L. (2015). The relationship between expressive/suppressive hostility behavior and cardiac autonomic activations in patients with coronary artery disease. *Acta Cardiologica Sinica*, *31*(4), 308.

Lin, T. K., Weng, C. Y., Wang, W. C., Chen, C. C., Lin, I. M., & Lin, C. L. (2008). Hostility trait and vascular dilatory functions in healthy Taiwanese. *Journal of Behavioral Medicine*, *31*(6), 517–524.

Lin, T. K., Yu, P. T., Lin, L. Y., Liu, P. Y., Li, Y. D., Hsu, C. T., ... Weng, C. Y. (2018). A pilot-study to assess the feasibility and acceptability of an Internet-based cognitive-behavior group therapy using video conference for patients with coronary artery heart disease. *PLoS One*, *13*(11), e0207931.

Littleton, H., Horsley, S., John, S., & Nelson, D. V. (2007). Trauma coping strategies and psychological distress: A meta-analysis. *Journal of Traumatic Stress: Official Publication of the International Society for Traumatic Stress Studies*, *20*(6), 977–988.

Logan, J. G., Barksdale, D. J., Carlson, J., Carlson, B. W., & Rowsey, P. J. (2012). Psychological stress and arterial stiffness in Korean Americans. *Journal of Psychosomatic Research*, *73*(1), 53–58.

Lopresti, A. L., Hood, S. D., & Drummond, P. D. (2013). A review of lifestyle factors that contribute to important pathways associated with major depression: Diet, sleep and exercise. *Journal of Affective Disorders*, *148*(1), 12–27.

Magán, I., Casado, L., Jurado-Barba, R., Barnum, H., Redondo, M. M., Hernandez, A. V., & Bueno, H. (2020). Efficacy of psychological interventions on psychological outcomes in coronary artery disease: Systematic review and meta-analysis. *Psychological Medicine*, 1–15.

Mucci, N., Giorgi, G., De Pasquale Ceratti, S., Fiz-Pérez, J., Mucci, F., & Arcangeli, G. (2016). Anxiety, stress-related factors, and blood pressure in young adults. *Frontiers in Psychology*, *7*, 1682.

Munk, P. S., Isaksen, K., Brønnick, K., Kurz, M. W., Butt, N., & Larsen, A. I. (2012). Symptoms of anxiety and depression after percutaneous coronary intervention are associated with decreased heart rate variability, impaired endothelial function and increased inflammation. *International Journal of Cardiology*, *158*(1), 173–176.

Nandam, L. S., Brazel, M., Zhou, M., & Jhaveri, D. J. (2020). Cortisol and major depressive disorder—Translating findings from humans to animal models and back. *Frontiers in Psychiatry*, *10*, 974.

Narita, K., Murata, T., Hamada, T., Takahashi, T., Omori, M., Suganuma, N., & Wada, Y. (2007). Interactions among higher trait anxiety, sympathetic activity, and endothelial function in the elderly. *Journal of Psychiatric Research*, *41*(5), 418–427.

Neary, N. M., Booker, O. J., Abel, B. S., Matta, J. R., Muldoon, N., Sinaii, N., ... Gharib, A. M. (2013). Hypercortisolism is associated with increased coronary arterial atherosclerosis: Analysis of noninvasive coronary angiography using multidetector computerized tomography. *The Journal of Clinical Endocrinology & Metabolism*, *98*(5), 2045–2052.

Neff, K. (2003). Self-compassion: An alternative conceptualization of a healthy attitude toward oneself. *Self and Identity*, *2*(2), 85–101.

Pan, Y., Cai, W., Cheng, Q., Dong, W., An, T., & Yan, J. (2015). Association between anxiety and hypertension: A systematic review and meta-analysis of epidemiological studies. *Neuropsychiatric Disease and Treatment*, *11*, 1121.

Powch, I. G., & Houston, B. K. (1996). Hostility, anger-in, and cardiovascular reactivity in White women. *Health Psychology*, *15*(3), 200.

Powers, M. B., Vörding, M. B. Z. V. S., & Emmelkamp, P. M. (2009). Acceptance and commitment therapy: A meta-analytic review. *Psychotherapy and Psychosomatics, 78*(2), 73–80.

Ranjit, N., Diez-Roux, A. V., Sanchez, B., Seeman, T., Shea, S., Shrager, S., & Watson, K. (2009). Association of salivary cortisol circadian pattern with cynical hostility: The multi-ethnic study of atherosclerosis. *Psychosomatic Medicine, 71*(7), 748.

Richardson, S., Shaffer, J. A., Falzon, L., Krupka, D., Davidson, K. W., & Edmondson, D. (2012). Meta-analysis of perceived stress and its association with incident coronary heart disease. *The American Journal of Cardiology, 110*(12), 1711–1716.

Roest, A. M., Martens, E. J., de Jonge, P., & Denollet, J. (2010). Anxiety and risk of incident coronary heart disease: A meta-analysis. *Journal of the American College of Cardiology, 56*(1), 38–46.

Rosendorff, C., Lackland, D. T., Allison, M., Aronow, W. S., Black, H. R., Blumenthal, R. S., ... White, W. B. (2015). Treatment of hypertension in patients with coronary artery disease: A scientific statement from the American Heart Association, American College of Cardiology, and American Society of Hypertension. *Circulation, 131*(19), e435–e470.

Rosengren, A., Hawken, S., Ôunpuu, S., Sliwa, K., Zubaid, M., Almahmeed, W. A., & INTERHEART Investigators. (2004). Association of psychosocial risk factors with risk of acute myocardial infarction in 11 119 cases and 13 648 controls from 52 countries (the INTERHEART study): case-control study. *The Lancet, 364*(9438), 953–962.

Roth, G. A., Johnson, C., Abajobir, A., Abd-Allah, F., Abera, S. F., Abyu, G., ... Ukwaja, K. N. (2017). Global, regional, and national burden of cardiovascular diseases for 10 causes, 1990 to 2015. *Journal of the American College of Cardiology, 70*(1), 1–25.

Rozanski, A., Blumenthal, J. A., Davidson, K. W., Saab, P. G., & Kubzansky, L. (2005). The epidemiology, pathophysiology, and management of psychosocial risk factors in cardiac practice: The emerging field of behavioral cardiology. *Journal of the American College of Cardiology, 45*(5), 637–651.

Rozanski, A., Blumenthal, J. A., & Kaplan, J. (1999). Impact of psychological factors on the pathogenesis of cardiovascular disease and implications for therapy. *Circulation, 99*(16), 2192–2217.

Rubio-Guerra, A. F., Rodriguez-Lopez, L., Vargas-Ayala, G., Huerta-Ramirez, S., Serna, D. C., & Lozano-Nuevo, J. J. (2013). Depression increases the risk for uncontrolled hypertension. *Experimental & Clinical Cardiology, 18*(1), 10.

Schott, L. L., Kamarck, T. W., Matthews, K. A., Brockwell, S. E., & Sutton-Tyrrell, K. (2009). Is brachial artery flow-mediated dilation associated with negative affect? *International Journal of Behavioral Medicine, 16*(3), 241–247.

Seldenrijk, A., van Hout, H. P., van Marwijk, H. W., de Groot, E., Gort, J., Rustemeijer, C., ... Penninx, B. W. (2013). Sensitivity to depression or anxiety and subclinical cardiovascular disease. *Journal of Affective Disorders, 146*(1), 126–131.

Seligman, M. E., Steen, T. A., Park, N., & Peterson, C. (2005). Positive psychology progress: Empirical validation of interventions. *American Psychologist, 60*(5), 410.

Serrano, C. V., Jr., Setani, K. T., Sakamoto, E., Andrei, A. M., & Fraguas, R. (2011). Association between depression and development of coronary artery disease: Pathophysiologic and diagnostic implications. *Vascular Health and Risk Management, 7*, 159.

Sher, L. D., Geddie, H., Olivier, L., Cairns, M., Truter, N., Beselaar, L., & Essop, M. F. (2020). Chronic stress and endothelial dysfunction: Mechanisms, experimental challenges, and the way ahead. *American Journal of Physiology-Heart and Circulatory Physiology, 319*(2), H488–H506.

Sjögren, E., Leanderson, P., & Kristenson, M. (2006). Diurnal saliva cortisol levels and relations to psychosocial factors in a population sample of middle-aged Swedish men and women. *International Journal of Behavioral Medicine, 13*(3), 193–200.

Sloan, R. P., Bagiella, E., Shapiro, P. A., Kuhl, J. P., Chernikhova, D., Berg, J., & Myers, M. M. (2001). Hostility, gender, and cardiac autonomic control. *Psychosomatic Medicine, 63*(3), 434–440.

Sommaruga, M. (2016). Cognitive and behavioral psychotherapy in coronary artery disease. In *Psychotherapy for ischemic heart disease* (pp. 159–172). Cham: Springer.

Srinivasan, K. (2011). "Blues" ain't good for the heart. *Indian Journal of Psychiatry, 53*(3), 192.

Sundbom, L. T., & Bingefors, K. (2013). The influence of symptoms of anxiety and depression on medication nonadherence and its causes: A population based survey of prescription drug users in Sweden. *Patient Preference and Adherence, 7*, 805.

Székely, A., Balog, P., Benkö, E., Breuer, T., Székely, J., Kertai, M. D., ... Thayer, J. F. (2007). Anxiety predicts mortality and morbidity after coronary artery and valve surgery—A 4-year follow-up study. *Psychosomatic Medicine, 69*(7), 625–631.

Thayer, J. F., Yamamoto, S. S., & Brosschot, J. F. (2010). The relationship of autonomic imbalance, heart rate variability and cardiovascular disease risk factors. *International Journal of Cardiology, 141*(2), 122–131.

Virani, S. S., Alonso, A., Aparicio, H. J., Benjamin, E. J., Bittencourt, M. S., Callaway, C. W., ... American Heart Association Council on Epidemiology and Prevention Statistics Committee and Stroke Statistics Subcommittee. (2021). Heart disease and stroke statistics—2021 update: A report from the American Heart Association. *Circulation, 143*(8), e254–e743.

Weng, C. Y., Lin, C. L., Lin, T. K., Chen, C. W., Li, Y. D., Hsu, C. T., & Pai, S. A. (2016). Cardiac rehabilitation program for patients with coronary heart disease: A cognitive-behavior group therapy approach. *Chinese Journal of Psychology, 58*, 143–167.

Whalley, B., Thompson, D. R., & Taylor, R. S. (2014). Psychological interventions for coronary heart disease: Cochrane systematic review and meta-analysis. *International Journal of Behavioral Medicine, 21*(1), 109–121.

Yan, L. L., Liu, K., Matthews, K. A., Daviglus, M. L., Ferguson, T. F., & Kiefe, C. I. (2003). Psychosocial factors and risk of hypertension: The Coronary Artery Risk Development in Young Adults (CARDIA) study. *JAMA, 290*(16), 2138–2148.

Yusuf, S., Hawken, S., Ôunpuu, S., Dans, T., Avezum, A., Lanas, F., ... INTERHEART Study Investigators. (2004). Effect of potentially modifiable risk factors associated with myocardial infarction in 52 countries (the INTERHEART study): Case-control study. *The Lancet, 364*(9438), 937–952.

Zorn, J. V., Schür, R. R., Boks, M. P., Kahn, R. S., Joëls, M., & Vinkers, C. H. (2017). Cortisol stress reactivity across psychiatric disorders: A systematic review and meta-analysis. *Psychoneuroendocrinology, 77*, 25–36.

Chapter 26

Application of mindfulness-based cognitive therapy and health qigong-based cognitive therapy among Chinese people with mood disorders

Sunny Ho-Wan Chan[a] and Charlie Lau[b]

[a]School of Health and Social Wellbeing, University of the West of England, Bristol, United Kingdom [b]Department of Rehabilitation Sciences, The Hong Kong Polytechnic University, Hung Hom, Hong Kong

Abbreviations

CBT	cognitive behavior therapy
MBI	mind-body interventions
MM	mindfulness meditation
HQ	health qigong
MBCT	mindfulness-based cognitive therapy
HQCT	health qigong-based cognitive therapy
WC	waitlist control

Introduction

Prevalence of depression and anxiety

Mood disorders such as depression and anxiety are prevalent globally (Kessler, Berglund, Demler, Jin, & Walters, 2005; Phillips et al., 2009). Within the Chinese population, the first territory-wide epidemiological survey in Hong Kong revealed that the prevalence of common mental disorders in adults was 13.3%, with mixed anxiety and depressive disorder being the most frequent diagnoses (Lam et al., 2015).

Limitation of traditional cognitive behavioral therapy

Considering the potential side effects and dependence on pharmaceutical treatment, individuals may apply psychotherapy instead of medication. While previous research has established that cognitive behavioral therapy (CBT) is an effective psychosocial treatment in depression and anxiety management (Twomey, O'Reilly, & Byrne, 2015; Zhang et al., 2019), some reviews demonstrated a steady decline of effect size of CBT in alleviating depression and anxiety since its rise four decades ago (Cuijpers et al., 2011; Johnsen & Friborg, 2015; Lynch, Laws, & McKenna, 2010). It is possible that the approach of CBT was over mechanistic and unable to address the concerns of the "whole" individual (Gaudiano, 2008). Therefore, CBT per se might not be sufficient in alleviating mood disorders.

Alternative forms of therapy: Mindfulness-based cognitive therapy

In view of the limitations of traditional treatments, much focus shifted to alternative forms of therapy. Recently, researchers have shown an increased interest in mind-body interventions (MBIs) depression and anxiety (Hoch et al., 2012; Kinser, Elswick, & Kornstein, 2014). For example, mindfulness meditation (MM) is regarded as a contemplative approach to foster nonjudgmental moment-to-moment awareness and experiences (Kabat-Zinn, 1990). Studies have reported its effectiveness in treating anxiety and depression, particularly its potential relevance to psychiatric comorbidity (Khusid & Vythilingam, 2016; Ren, Zhang, & Jiang, 2018; Wielgosz, Goldberg, Kral, Dunne, & Davidson, 2019). Integrating MM with CBT was put forward to promote awareness of internal experiences and acceptance which is crucial for subsequent cognitive restructuring (Radkovsky, McArdle, Bockting, & Berking, 2014). In particular, the integration of MM and CBT engenders a fabrication of mindfulness-based cognitive therapy (MBCT; Segal, Williams, & Teasdale, 2013). Metaanalyses have indicated that MBCT is an effective approach for mental health and well-being (Querstret, Morison, Dickinson, Cropley, & John, 2020) as well as depressive or anxiety symptoms (Goldberg et al., 2019; Thomas, Chur-Hansen, & Turner, 2020).

Alternative forms of therapy: Health qigong cognitive therapy

However, MM only reflects the static form of MBIs where mediative body movements are scarcely involved. There is another long-established dynamic form of MBI namely health qigong (HQ; Chan & Tsang, 2019) which centers on body movement coupled with introspective focus, an awareness of breathing and the natural force or energy in the body (Jahnke, Larkey, Rogers, Etnier, & Lin, 2010). Not only have previous studies found that HQ such as Baduanjin exercises could alleviate depression or anxiety in individuals with physical or mental illness (Wang et al., 2014; Zou et al., 2018), comparable therapeutic effects of HQ and CBT on depressive and anxiety symptoms have also been demonstrated (Wang et al., 2013). Combining HQ and CBT could promote the application of traditional Chinese meditative movement as one of the major behavioral strategies in combating mood symptoms (Chow & Tsang, 2007). While a study integrating HQ and CBT revealed beneficial effects on self-perceived personal well-being of community-dwelling elderly (Liu & Tsui, 2014), the application on individuals with mood disorders has not been investigated.

Study objective

Collectively, while MM and HQ are both considered as MBIs and share the common characteristics of a focus on breathing, they reflect two distinct approaches with different mechanisms and efficacy. MM rests on mind-based practice such as cultivation of nonjudgmental acceptance, whereas HQ centers on body-based movement practice. Although integrating MM into CBT has been shown to be effective in improving mood symptoms, the effectiveness of combining HQ and CBT remains speculative. The overarching objective of the study was to examine and compare the relative therapeutic effects of MBCT and HQCT in managing depression and anxiety with WC among Chinese adults with mood disorders.

Candidates for the interventions

Inclusion criteria for the interventions are as follows:

- Diagnosed with depression or anxiety disorder
- Age 18–70
- Regular psychiatric follow-up
- No active suicidal ideation or self-harm behaviors
- Primary education level of above and able to communicate in Cantonese
- No previous experience with cognitive therapy, mindfulness-based intervention, or HQ

The only exclusion criteria is comorbid diagnoses of schizophrenia, schizoaffective disorder, substance misuse, organic brain syndrome, or intellectual disabilities.

Reduced mood symptoms

The study compared the treatment outcome of MBCT and HQCT with the WC group in treating depression and anxiety among Chinese adults. Corroborating previous findings revealing the beneficial effects of MM and HQ on depression and anxiety (Wielgosz et al., 2019; Zhou et al., 2016), significant reductions of mood symptoms were found following both the MBCT and HQCT. In contrast to earlier findings which showed small-to-moderate efficacy for symptoms reduction of depression and anxiety (Yin & Dishman, 2014) or lack of reliable effects at follow-up assessments (Ren et al., 2018), the present study found a sustained moderate effect for both MBCT and HQCT during follow-up assessments, implying the enhanced effect of integration of CBT with different forms of MBI.

Improved health status

Along with the reduction in mood symptoms, both MBCT and HQCT groups demonstrated improved health status as compared with WC. Specifically, MBCT tended to promote mental health status whereas HQCT is more conducive to physical health outcomes. In the MBCT group, individual growth models further indicated that the improved mental health status led to mood symptom reduction. As such, the essence of mindfulness awareness and cognitive restructuring of thoughts in MBCT should be the crucial component that contributes to the mental health outcomes (Gu, Strauss, Bond, & Cavanagh, 2015). Thus, in line with prior studies (Godfrin & van Heeringen, 2010; Kaviani, Hatami, & Javaheri, 2012), the current results highlighted the significant impact of MBCT on mental health such that alterations in mental health can be regarded as essential element in reducing mood symptoms in MBCT.

HQCT & physical health

On the other hand, the role of physical health in HQCT was underlined, as evinced by the individual growth model. While previous studies focus on the effect of mood symptoms on physical health (Shen, Fan, Lim, & Tay, 2019; van Milligen, Lamers, de Hoop, Smit, & Penninx, 2011), the current findings adopted an alternative perspective in which physical health was viewed as an predictor of mood changes. In line with previous results (Jing et al., 2018; Zou et al., 2017), HQ by itself, as well as the combined effect with CBT, could foster positive outcomes in physical health which in turn resulted in attenuation of mood symptoms.

HQCT led to more benefits than MBCT

Additionally, despite the small effect size, participants in the HQCT group reported slightly more reduction of mood symptoms during posttreatment and follow-up assessments than those in the MBCT group, suggesting movement-based MBI intervention brought about more benefits than mind-based MBI in a Chinese context. The prevailing emphasis on physical health in HQCT and the primary focus on mental health in MBCT may elucidate their applicability in Chinese population. Since HQ is considered as mind–body exercise or meditative movement within traditional Chinese medicine treatment modalities (Abbott & Lavretsky, 2013; Payne & Crane-Godreau, 2013), individuals with a Chinese cultural background perceive such exercises as a health-preserving activity which is an important index of health status (Liu, Speed, & Beaver, 2015). Furthermore, in comparison with conventional exercise, HQ is a form of exercise that makes use of movement in concert with moving vital energy (or qi) throughout the entire body. Hence, it is possible that HQ leads to additional neurophysiological or biological effects on mood symptoms (Chan & Tsang, 2019).

Somatization tendency in Chinese culture

The somatization tendency as endorsed in Chinese culture may also unfold the greater effect of HQCT on mood change. Intrinsically, strong "holism" ideation is present in Chinese culture such that an individual is usually perceived "holistically" as a psychosomatic process (Jullien, 2007), hence there is no dichotomy between mind and body as both are considered as the same construct (Lewis, 2006). As a result, such holistic conception may precipitate individuals to associate physical and mental health readily under the influence of Chinese culture. This holistic conception lends support to the association between physical and mental health under the influence of Chinese culture. Therefore, movement-based mediation such as HQCT should be more effective than static mediation such as MBCT in ameliorating depression and anxiety (Chen et al., 2012).

Conclusion

Taken together, the study revealed the beneficial effects of HQCT and MBCT over WC in precipitating mood changes. Specifically, the emphasis on physical health of HQCT enhanced its acceptability and effectiveness in alleviating mood disorders in a Chinese population.

Practice and procedures

MBCT

MBCT is a 8-week group program, each weekly session lasts for about 2h. It incorporates mindful breathing, sitting meditation, body scan technique, 3-min breathing space, mindful movement and other mindfulness activities relevant to ordinary daily activities (Segal et al., 2013). The following MBCT outline is referenced from Segal et al., 2013 (Table 1).

TABLE 1 Content of the 8-week MBCT group program.

Session 1	Rationale: Mindfulness begins when we are aware of our tendency to be an automatic pilot and devoted to learning how to step out of it and become mindful of each moment. Practices involve moving attention around the body deliberately	
	Practice	Exercises
	• Eating meditation • Body scan • Routine activity (home practice) • Noticing (homework which continues throughout)	• N/A
Session 2	Rationale: The chatter of the mind which tends to control our reactions to everyday events starts to manifest more clearly after further focus on the body including emotion, body sensations, behavior, and thoughts	
	Practice	Exercises
	• Body scan • Breath • Routine activity (home practice)	• Thoughts and feelings • Focus on pleasant experiences (home practice)
Session 3	Rationale: The mind is often occupied and scattered. Awareness of the breath paves the way for being more focused on the present moment. Mindful movement improves awareness of the body	
	Practice	Exercises
	• Seeing/hearing • Sitting with breath and body; responding to painful sensations or mindful movements (lying down) (home practice) • Breathing space (home practice) • Standing stretches and sitting with breath and body (home practice)	• Focus on unpleasant experiences (home practice)
Session 4	Rationale: The mind is the most scattered when we try to hold on to some things and keep away from others. Mindfulness presents a way to focus on the present by providing another and wider perspective to relate differently to experience (including thoughts and feelings)	
	Practice	Exercises
	• Sitting with breath, body, sounds, thoughts, open awareness (home practice) • Breathing space: regular and when distressed or disturbed (home practice) • Mindful walking	• Defining the territory of depression
Session 5	Rationale: Relating differently involves bringing acceptance and nonjudgment to experience. Such an attitude of acceptance is a crucial aspect of taking care of oneself and seeing clearly what (if any) needs to be changed	
	Practice	Exercises
	• Sitting with breath, body and difficulty (home practice) • Expanded breathing space (first step to attending to the body) (home practice; whenever there are unpleasant feelings)	• N/A
Session 6	Rationale: Negative feelings and the associated thoughts impact our ability to relate to experience. Our thoughts are merely thoughts and that we can choose whether to engage with them, even those thoughts which appear they are not just thoughts. Thoughts often appear out of context and mood. Recognizing the recurrence of the same patterns of thought can help us to step back from our thoughts, without necessarily questioning them or seeking alternatives	
	Practice	Exercises
	• Sitting with breath, body, thoughts and feelings, open awareness (or sitting with breath, body, difficulty) • Breathing space (first step to a wider view of thoughts) • Selection of practices (home practices)	• Moods, thoughts and alternative viewpoints

TABLE 1 Content of the 8-week MBCT group program—cont'd

Session 7	Rationale: Specific steps can be taken when depression threatens. Take a breathing space first, and then decide what action (if any) to take. Each person has his or her own unique relapse signs. Participants in the program can help each other to plan how best to cope with those signs
	Practice • Sitting with breath, body, sounds, thoughts, and choiceless awareness (noticing and responding to difficulty) • Breathing space (first step to wise action) • Selecting forms of practice you will be able to continue with (homework)	**Exercises** • Activity and mood (nourishing vs. depleting activities) • Identifying relapse signatures
Session 8	Rationale: Regular mindfulness practice helps maintain balance in life. Good intentions can be strengthened by linking the practice with reasons for taking care of oneself
	Practice • Body scan • Closing sitting (wishing one another well)	**Exercises** • Personal reflections on the program • What are the things in your life that you most value, that the practice could help you with? • Preparing for the future (what to do when I notice early warning signs)

HQCT

HQCT is a 8-week program integrating CBT and HQ technique. Baduanjin was adopted as the HQ practice since it has been standardized and considered as less complicated and cognitively or physically demanding (Ho et al., 2012). CBT components were adapted from a Changeways core program which is a psychoeducational group therapy for mood disorders (Paterson, Alden, & Koch, 2008) (Table 2).

TABLE 2 Contents of the 8-week HQCT group program.

Session	HQ	CT

1	Prop up the sky with both hands to regulate the triple warmer	• Introduction and goal setting
		   ○ Introduction to the group
		   ○ The triangle: Thoughts, emotions, and behaviors
		   ○ Setting attainable goals

2	Draw a bow on both sides like shooting a vulture	• Home practice review
		 • Stress, depression and lifestyle
		   ○ The nature of stress
		   ○ The nature of depression

3	Raise single arm to regulate spleen and stomach	• Home practice review • Stress, depression and lifestyle (cont.) ○ The sustaining style

4	Look back to treat five strains and seven impairments	• Home Practice Review
		 • The role of social life
		   ○ Social network
		   ○ Creating a support team

5	Sway head and buttocks to expel heart-fire	• Home practice review
		 • The role of social life (cont.)
		   ○ Social balancing
		   ○ Assertiveness

Continued

TABLE 2 Contents of the 8-week HQCT group program—cont'd

Session	HQ	CT
		• Thinking about thinking ○ Introduction

6	Pull toes with both hands to reinforce kidney and waist	• Home practice review
		• Think about thinking (cont.)
		    ○ Handling changes in mood
		    ○ Styles of distorted thinking

7	Clench fists and look with eyes wide open to build up strength and stamina	• Home practice review • Think about thinking (cont.) ○ Overcoming negative thinking
8	Rise and fall on tiptoes to dispel all disease	• Home practice review • Preventing future difficulties

Mini-dictionary of terms

- **Mind-body interventions.** It focuses on the relationships among the brain, mind, body, and behavior, and their effect on health and disease and encompasses a wide variety of therapies such as meditation, yoga, biofeedback and tai chi.
- **Mindfulness meditation.** MM involves the intentional self-regulation of attention to present moment experience, combined with a nonjudgmental and accepting attitude.
- **Health qigong.** "Qi" literally refers to vital energy within the body. Free flow of "Qi" within the system is crucial for good health whereas blockage would lead to illness. "Gong" refers to training and practice.
- **Mindfulness-based cognitive therapy.** It couples cognitive behavioral techniques with mindfulness techniques so as to help individuals to better understand and manage their emotions and thoughts and in turn alleviate distress.
- **Health qigong-based cognitive therapy.** It combines cognitive techniques with HQ techniques.
- **Baduanjin.** One of the various forms of qigong. It means "8 pieces of silken brocade" in Mandarin and consists of 8 movements that are performed in a smooth and graceful manner, hence the name. The movements, coupled with breathing and meditation, exercise the mind and body for healing.
- **Mindful breathing.** It is to teach participants to have a continuous feeling and observation on their breath while focusing on and returning to their breath without judgment if their mind is having any intrusive thought.
- **Body scan.** It invites participants to pay attention to the sensation that they feel the persistent change in different regions of the body nonjudgmentally.
- **Sitting meditation**: It requires participants to pay attention to their breath, sound from the environment, body sensation, their thoughts and emotions with an alert state and erect sitting posture.
- **Mindful movement.** It needs participants to build mindful awareness on different parts of the body when doing different stretching postures. It can be carried out in sitting, lying, or standing postures.
- **3-min breathing space.** It is composed of three steps, which are awareness, gathering and expanding. Through this practice, it is believed that participants can better utilize mindfulness skills to tackle emotion challenges in daily life.

Key facts

Key facts of health qigong cognitive therapy and mindfulness-based cognitive therapy

- In China, it is estimated that over 100 million people practice HQ.
- Both mindfulness-based cognitive therapy (MBCT) and health qigong-based cognitive therapy (HQCT) had significantly reduced anxiety, depression, and perceived stress, and significantly increased sleep quality and self-efficacy.
- HQCT is better in producing more improvements in physical health than MBCT.
- MBCT is superior in making greater improvements in overall mental health than HQCT.
- It cannot be overlooked when applying MBCT or HQCT in a population with emphasis on Chinese culture.

Applications to other areas

In this chapter, we reviewed the application of MBCT and HQCT among Chinese people with mood disorders. The study showed that HQCT was more effective in alleviating mood disorders among Chinese. HQCT can be potentially applied to improve sleep disturbance in the general population. Sleep disturbance which comprises disorders including insomnia, hypersomnia and disturbed circadian rhythm (Cormier, 1990) is a major public health issue because of its high prevalence in both general public (Leger et al., 2011; Wong & Fielding, 2011) and individuals with depression and anxiety (Hombali et al., 2019). Not only is sleep disturbance often comorbid with depression and anxiety (Smith, Huang, & Manber, 2005), it is also considered as a core symptom of depression and anxiety (American Psychiatric Association, 2013). Longitudinal studies revealed that sleep disturbance is an independent risk factor for subsequent depression and anxiety (Buysse et al., 2008; Chang, Ford, Mead, CooperPatrick, & Klag, 1997; Jaussent et al., 2011). Moreover, studies also have also revealed a bidirectional relationship between sleep disturbance and the two disorders (Alvaro, Roberts, & Harris, 2013; Fang, Tu, Sheng, & Shao, 2019; Jansson-Frojmark & Lindblom, 2008; Roberts, Kamruzzaman, & Tholen, 2009). Considering the complex interplay between sleep disturbance, depression and anxiety and the positive effects of HQCT on ameliorating mood disorders, HQCT targeting sleep may be a potential early intervention for mood disorders.

Summary points

- MBCT and HQCT both resulted in greater improvements in depressive and anxiety symptoms, physical and mental health status, perceived stress, sleep quality, and self-efficacy.
- In comparison, the HQCT group experienced more mood symptoms reduction than the mindfulness based cognitive therapy group.
- HQCT is more conducive to physical health status.
- MBCT has more positive mental health outcomes.
- The prevailing emphasis on physical health in HQCT contributes to its more favorable acceptability and effectiveness than MBCT among Chinese adults with mood disorders.

References

Abbott, R., & Lavretsky, H. (2013). Tai Chi and Qigong for the treatment and prevention of mental disorders. *Psychiatric Clinics of North America, 36*, 109–119.

Alvaro, P. K., Roberts, R. M., & Harris, J. K. (2013). A systematic review assessing bidirectionality between sleep disturbances, anxiety, and depression. *Sleep, 36*(7), 1059–1068. https://doi.org/10.5665/sleep.2810.

American Psychiatric Association. (2013). *Diagnostic and statistical manual of mental disorders* (5th ed.). Washington, DC: American Psychiatric Association.

Buysse, D. J., Angst, J., Gamma, A., Ajdacic, V., Eich, D., & Rossler, W. (2008). Prevalence, course, and comorbidity of insomnia and depression in young adults. *Sleep, 31*(4), 473. https://doi.org/10.1093/sleep/31.4.473.

Chan, S. H. W., & Tsang, H. W. H. (2019). The beneficial effects of Qigong on elderly depression. In S. Y. Yau, & K. F. So (Eds.), *Vol. 147. International review of neurobiology* (pp. 155–188). US: Academic Press.

Chang, P. P., Ford, D. E., Mead, L. A., CooperPatrick, L., & Klag, M. J. (1997). Insomnia in young men and subsequent depression—The Johns Hopkins Precursors Study. *American Journal of Epidemiology, 146*(2), 105–114.

Chen, K. W., Berger, C. C., Manheimer, E., Forde, D., Magidson, J., Dachman, L., & Lejuez, C. W. (2012). Meditative therapies for reducing anxiety: A systematic review and meta-analysis of randomized controlled trials. *Depress & Anxiety, 29*, 545–562.

Chow, Y. W. Y., & Tsang, H. W. H. (2007). Biopsychosocial effects of Qigong as a mindful exercise for people with anxiety disorders: A speculative review. *The Journal of Alternative and Complementary Medicine, 13*, 831–839.

Cormier, R. E. (1990). Sleep disturbances. In H. K. Walker, W. D. Hall, & J. W. Hurst (Eds.), *Clinical methods: The history, physical, and laboratory examinations* (3rd ed.). Boston: Butterworths (Chapter 77).

Cuijpers, P., Clignet, F., van Meijel, B., van Straten, A., Li, J., & Andersson, G. (2011). Psychological treatment of depression in inpatients: A systematic review and meta-analysis. *Clinical Psychology Review, 31*, 353–360.

Fang, H., Tu, S., Sheng, J. F., & Shao, A. W. (2019). Depression in sleep disturbance: A review on a bidirectional relationship, mechanisms and treatment. *Journal of Cellular and Molecular Medicine, 23*(4), 2324–2332. https://doi.org/10.1111/jcmm.14170.

Gaudiano, B. A. (2008). Cognitive-behavioural therapies: Achievements and challenges. *Evidence-Based Mental Health, 11*, 5–7.

Godfrin, K. A., & van Heeringen, C. (2010). The effects of mindfulness-based cognitive therapy on recurrence of depressive episodes, mental health and quality of life: A randomized controlled study. *Behaviour Research and Therapy, 48*, 738–746.

Goldberg, S. B., Tucker, R. P., Greene, P. A., Davidson, R. J., Kearney, D. J., & Simpson, T. L. (2019). Mindfulness-based cognitive therapy for the treatment of current depressive symptoms: A meta-analysis. *Cognitive Behaviour Therapy, 48*, 445–462.

Gu, J., Strauss, C., Bond, R., & Cavanagh, K. (2015). How do mindfulness-based cognitive therapy and mindfulness-based stress reduction improve mental health and wellbeing? A systematic review and meta-analysis of mediation studies. *Clinical Psychology Review*, 37, 1–12.

Ho, R. T. H., Au Yeung, F. S. W., Lo, P. H. Y., Law, K. Y., Wong, K. O. K., Cheung, I. K. M., & Ng, S. M. (2012). Tai-Chi for residential patients with schizophrenia on movement coordination, negative symptoms, and functioning: A pilot randomized controlled trial. *Evidence-based Complementary and Alternative Medicine*, 2012, 10. https://doi.org/10.1155/2012/923925.

Hoch, D. B., Watson, A. J., Linton, D. A., Bello, H. E., Senelly, M., Milik, M. T., ... Kvedar, J. C. (2012). The feasibility and impact of delivering a mind-body intervention in a virtual world. *PLoS One*, 7(3), e33843.

Hombali, A., Seow, E., Yuan, Q., Chang, S. H. S., Satghare, P., Kumar, S., ... Subramaniam, M. (2019). Prevalence and correlates of sleep disorder symptoms in psychiatric disorders. *Psychiatry Research*, 279, 116–122. https://doi.org/10.1016/j.psychres.2018.07.009.

Jahnke, R., Larkey, L., Rogers, C., Etnier, J., & Lin, F. (2010). A comprehensive review of health benefits of Qigong and Tai Chi. *American Journal of Health Promotion*, 24(6), e1–e25.

Jansson-Frojmark, M., & Lindblom, K. (2008). A bidirectional relationship between anxiety and depression, and insomnia? A prospective study in the general population. *Journal of Psychosomatic Research*, 64(4), 443–449. https://doi.org/10.1016/j.jpsychores.2007.10.016.

Jaussent, I., Bouyer, J., Ancelin, M. L., Akbaraly, T., Peres, K., Ritchie, K., ... Dauvilliers, Y. (2011). Insomnia and daytime sleepiness are risk factors for depressive symptoms in the elderly. *Sleep*, 34(8), 1103–1110. https://doi.org/10.5665/sleep.1170.

Jing, L., Jin, Y., Zhang, X., Wang, F., Song, Y., & Xing, F. (2018). The effect of Baduanjin qigong combined with CBT on physical fitness and psychological health of elderly housebound. *Medicine*, 97, e13654.

Johnsen, T. J., & Friborg, O. (2015). The effects of cognitive behavioral therapy as an anti-depressive treatment is falling: A meta-analysis. *Psychological Bulletin*, 141, 747–768.

Jullien, F. (2007). *Vital nourishment: Departing from happiness*. New York, NY: Zone Books.

Kabat-Zinn, J. (1990). *Full catastrophe living: Using the wisdom of your body and mind to face stress, pain, and illness*. New York: Delta Books.

Kaviani, H., Hatami, N., & Javaheri, F. (2012). The impact of mindfulness-based cognitive therapy (MBCT) on mental health and quality of life in a sub-clinically depressed population. *Archives of Psychiatry and Psychotherapy*, 14, 21–28.

Kessler, R. C., Berglund, P., Demler, O., Jin, R., & Walters, E. E. (2005). Lifetime prevalence and age-of-onset distributions of DSM-IV disorders in the National Comorbidity Survey Replication. *Archives of General Psychiatry*, 62, 593–602.

Khusid, M. A., & Vythilingam, M. (2016). The emerging role of mindfulness meditation as effective self-management strategy. Part 1. Clinical implications for depression, post-traumatic stress disorder, and anxiety. *Military Medicine*, 181, 961–968.

Kinser, P. A., Elswick, R. K., & Kornstein, S. (2014). Potential long-term effects of a mind-body intervention for women with major depressive disorder: Sustained mental health improvements with a pilot yoga intervention. *Archives of Psychiatric Nursing*, 28, 377–383.

Lam, L. C. W., Wong, C. S. M., Wang, M. J., Chan, W. C., Chen, E. Y. H., Ng, R. M. K., ... Bebbington, P. (2015). Prevalence, psychosocial correlates and service utilization of depressive and anxiety disorders in Hong Kong: The Hong Kong mental morbidity survey (HKMMS). *Social Psychiatry and Psychiatric Epidemiology*, 50, 1379–1388.

Leger, D., du Roscoat, E., Bayon, V., Guignard, R., Paquereau, J., & Beck, F. (2011). Short sleep in young adults: Insomnia or sleep debt? Prevalence and clinical description of short sleep in a representative sample of 1004 young adults from France. *Sleep Medicine*, 12(5), 454–462. https://doi.org/10.1016/j.sleep.2010.12.012.

Lewis, M. E. (2006). *The construction of space in early China*. Albany, NY: State University of New York Press.

Liu, Y. W. J., & Tsui, C. M. (2014). A randomized trial comparing tai chi with and without cognitive-behavioral intervention (CBI) to reduce fear of falling in community-dwelling elderly people. *Archives of Gerontology and Geriatrics*, 59, 317–325.

Liu, Z., Speed, S., & Beaver, K. (2015). Perceptions and attitudes towards exercise among Chinese elders—the implications of culturally based self-management strategies for effective health-related help seeking and person-centred care. *Health Expectations*, 18, 262–272.

Lynch, D., Laws, K. R., & McKenna, P. J. (2010). Cognitive behavioural therapy for major psychiatric disorder: Does it really work? A meta-analytical review of well-controlled trials. *Psychological Medicine*, 40, 9–24.

Paterson, R. J., Alden, L. E., & Koch, W. J. (2008). *The Core progra A cognitive behavioural guide to depression*. Toronto: Caversham Booksellers.

Payne, P., & Crane-Godreau, M. A. (2013). Meditative movement for depression and anxiety. *Frontiers in Psychiatry*, 4, 71.

Phillips, M. R., Zhang, J., Shi, Q., Song, Z., Ding, Z., Pang, S., ... Wang, Z. (2009). Prevalence, treatment, and associated disability of mental disorders in four provinces in China during 2001–05: An epidemiological survey. *Lancet*, 373, 2041–2053.

Querstret, D., Morison, L., Dickinson, S., Cropley, M., & John, M. (2020). Mindfulness-based stress reduction and mindfulness-based cognitive therapy for psychological health and well-being in nonclinical samples: A systematic review and meta-analysis. *International Journal of Stress Management*, 27, 394–411. https://doi.org/10.1037/str0000165.

Radkovsky, A., McArdle, J. J., Bockting, C. L. H., & Berking, M. (2014). Successful emotion regulation skills application predicts subsequent reduction of symptom severity during treatment of major depressive disorder. *Journal of Consulting and Clinical Psychology*, 82, 248–262.

Ren, Z., Zhang, Y., & Jiang, G. (2018). Effectiveness of mindfulness meditation in intervention for anxiety: A meta-analysis. *Acta Psychologica Sinica*, 50, 283–305.

Roberts, K., Kamruzzaman, P., & Tholen, J. (2009). Young people's education to work transitions and inter-generational social mobility in post-soviet Central Asia. *Young*, 17(1), 59–80. https://doi.org/10.1177/110330880801700105.

Segal, Z. V., Williams, J. M. G., & Teasdale, J. D. (2013). *Mindfulness-based cognitive therapy for depression* (2nd ed.). New York, NY: Guilford.

Shen, B. J., Fan, Y., Lim, K. S. C., & Tay, H. Y. (2019). Depression, anxiety, perceived stress, and their changes predict greater decline in physical health functioning over 12 months among patients with coronary heart disease. *International Journal of Behavioral Medicine*, 26, 352–364.

Smith, M. T., Huang, M. I., & Manber, R. (2005). Cognitive behavior therapy for chronic insomnia occurring within the context of medical and psychiatric disorders. *Clinical Psychology Review, 25*(5), 559–611. https://doi.org/10.1016/j.cpr.2005.04.004.

Thomas, R., Chur-Hansen, A., & Turner, M. (2020). A systematic review of studies on the use of mindfulness-based cognitive therapy for the treatment of anxiety and depression in older people. *Mindfulness, 11*, 1599–1609. https://doi.org/10.1007/s12671-020-01336-3.

Twomey, C., O'Reilly, G., & Byrne, M. (2015). Effectiveness of cognitive behavioural therapy for anxiety and depression in primary care: A meta-analysis. *Family Practice, 32*, 3–15.

van Milligen, B. A., Lamers, F., de Hoop, G. T., Smit, J. H., & Penninx, B. W. J. H. (2011). Objective physical functioning in patients with depressive and/or anxiety disorders. *Journal of Affective Disorders, 131*, 193–199.

Wang, C. W., Chan, C. H. Y., Ho, R. T. H., Chan, J. S. M., Ng, S. M., & Chan, C. L. W. (2014). Managing stress and anxiety through qigong exercise in healthy adults: A systematic review and meta-analysis of randomized controlled trials. *BMC Complementary and Alternative Medicine, 14*, 8.

Wang, C. W., Chan, C. L. W., Ho, R. T. H., Tsang, H. W. H., Chan, C. H., & Ng, S. M. (2013). The effect of qigong on depressive and anxiety symptoms: A systematic review and meta-analysis of randomized controlled trials. *Evidence-Based Complementary and Alternative Medicine, 2013*, 716094.

Wielgosz, J., Goldberg, S. B., Kral, T. R. A., Dunne, J. D., & Davidson, R. J. (2019). Mindfulness meditation and psychopathology. *Annual Review of Clinical Psychology, 15*, 285–316.

Wong, W. S., & Fielding, R. (2011). Prevalence of insomnia among Chinese adults in Hong Kong: A population-based study. *Journal of Sleep Research, 20*(1), 117–126. https://doi.org/10.1111/j.1365-2869.2010.00822.x.

Yin, J., & Dishman, R. K. (2014). The effect of tai chi and Qigong practice on depression and anxiety symptoms: A systematic review and meta-regression analysis of randomized controlled trials. *Mental Health and Physical Activity, 7*, 135–146.

Zhang, A., Bornheimer, L. A., Weaver, A., Franklin, C., Hai, A. H., Guz, S., & Shen, L. (2019). Cognitive behavioral therapy for primary care depression and anxiety: A secondary meta-analytic review using robust variance estimation in meta-regression. *Journal of Behavioral Medicine, 42*, 1117–1141.

Zhou, X., Peng, Y., Zhu, X., Yao, S., Dere, J., Chentsova-Dutton, Y. E., & Ryder, A. G. (2016). From culture to sympto Testing a structural model of "Chinese somatization". *Transcultural Psychiatry, 53*, 3–23.

Zou, L., SasaKi, J. E., Wang, H., Xiao, Z., Fang, Q., & Zhang, M. (2017). A systematic review and meta-analysis Baduanjin qigong for health benefits: Randomized controlled trials. *Evidence-Based Complementary & Alternative Medicine*, 1–17.

Zou, L., Yeung, A., Quan, X., Hui, S. S. C., Hu, X., Chan, J. S. M., … Wang, H. (2018). Mindfulness-based Baduanjin exercise for depression and anxiety in people with physical or mental illnesses: A systematic review and meta-analysis. *International Journal of Environmental Research and Public Health, 15*(2), 321.

Chapter 27

Bipolar disorder in Japan and cognitive-behavioral therapy

Yasuhiro Kimura[a], Sayo Hamatani[b,c], and Kazuki Matsumoto[b,d]

[a]*Department of Welfare Psychology, Faculty of Welfare, Fukushima College, Fukushima, Japan* [b]*Research Center for Child Mental Development, Chiba University, Chiba, Japan* [c]*Research Center for Child Mental Development, University of Fukui, Fukui, Japan* [d]*Division of Clinical Psychology, Kagoshima University Hospital, Kagoshima, Japan*

Abbreviations

APA	American Psychiatric Association
BDI-II	Beck Depression Inventory-II
CBT	cognitive-behavioral therapy
FFT	family-focused therapy
HAMD	Hamilton Depression Rating Scale
IPSRT	interpersonal and social rhythm therapy
JSMD	Japanese Society of Mood Disorders
RCT	randomized controlled trial
Rework	return to work
TCA	tricyclic antidepressants
WMHJ	World Mental Health Japan Survey
YMRS	Young Mania Rating Scale

Introduction

Bipolar disorder is a mental disorder described by the American Psychiatric Association's Diagnostic and Statistical Manual of Mental Disorders, 5th edition (DSM-5) as a group of brain disorders that cause extreme fluctuations in a person's mood, energy, and functional capacity. It is broadly divided into three categories: bipolar I disorder, bipolar II disorder, and cyclothymic disorder (APA, 2013). Bipolar I disorder is characterized by manic episodes. Bipolar II disorder consists of episodes of depression and hypomania. In cyclothymic disorder, hypomanic symptoms that do not satisfy hypomanic episodes and depressive symptoms that do not satisfy depressive episodes last for more than 2 years. People living with bipolar disorder experience periods of great excitement, overactivity, delusions, and euphoria—collectively known as mania—and periods of sadness and hopelessness are collectively known as depression.

The reported recurrence rate of bipolar disorder is about 90%, and nearly 50% of patients experience a remission relapse within the first 2.5 years (Bowden & Krishnan, 2004). A 5-year follow-up study of patients who had had manic episodes in remission reported that 73% had a recurrence, of which two-thirds had multiple recurrences (Gitlin, Swendsen, Heller, & Hammen, 1995).

Patients with bipolar disorder who are taking medication may also improve their symptoms with psychotherapy. In particular, high-intensity forms of psychotherapy, such as cognitive-behavioral therapy (CBT), interpersonal social rhythm therapy (IPSRT), and family-focused therapy (FFT), have demonstrated positive effects (Miklowitz et al., 2007). Therefore, Japanese guidelines for the treatment of bipolar disorder are beginning to recognize that psychotherapy may play an adjunct role to pharmacotherapy in treating the disorder (So, 2017).

Effectiveness of CBT for bipolar disorder

Treating bipolar disorder with both pharmacotherapy and CBT reduces the rate of subsequent recurrence (44% in the pharmacotherapy and CBT group; 75% in the pharmacotherapy-only group) (Lam et al., 2003). The effectiveness of CBT as an adjunct to pharmacotherapy for preventing relapse in bipolar disorder has been examined in a systematic review that included randomized controlled trials (RCTs) and a metaanalysis of 19 RCTs. CBT proved to significantly decrease bipolar symptoms (pooled odd rate = 0.506; 95% CI = 0.278 to −0.921), including depressive symptoms ($g = -0.494$; 95% CI = −0.963 to −0.026) and mania severity ($g = -0.581$; 95% CI = −1.127 to −0.035); psychosocial functioning ultimately improved ($g = 0.457$; 95% CI = 0.106–0.809) (Chiang et al., 2017). A 12-month follow-up after intervention also showed a significantly lower recurrence rate in the CBT and drug therapy group than in the pharmacotherapy-only group (mood episode: $n = 20$ in normal care group, $n = 12$ in CBT added group, $\chi2 = 7.64$, $P < 0.006$, estimated HR = 0.38, 95% CI 0.18–0.78) (Jones et al., 2015). Regarding the recurrence of manic episodes, the difference was even more pronounced ($n = 10$ in the normal care group, $n = 3$ in the CBT additional group, $\chi2 = 6.77$, $P < 0.009$, estimated HR = 0.38, 95% CI 0.19–0.79).

Epidemiology of bipolar disorder in Japan

The lifetime prevalence of bipolar disorder in general population is approximately 1.0% (Bebbington & Ramana, 1995; Pini et al., 2005). According to a World Health Organization survey, 45 million people worldwide suffer from bipolar disorder (GBD 2017 Disease and Injury Incidence and Prevalence Collaborators, 2018). Overall, bipolar disorder appears to be almost evenly distributed among genders and ethnic groups (Rowland & Marwah, 2018).

Two large-scale studies in Japan suggested that the combined prevalence of bipolar I disorder and bipolar II disorder is approximately 0%–0.4% (Kawakami, 2007, 2016) (Table 1). Although the differences between the figures for Japan and those of the above studies may be ascribed to cultural and social factors, previous research has been limited and it is not clear what causes these differences. Anxiety disorders are well known as a comorbidity of bipolar disorder, and it has been reported also in Japan that the comorbidity rate of bipolar disorder and anxiety disorder is high (Inoue, Kimura, Inagaki, & Shirakawa, 2020).

Social status of bipolar disorder in Japan

Generally, there is little public awareness of what constitutes bipolar disorder. According to an Internet survey of the general population, 72.8% of respondents said they did not recognize the term "bipolar disorder" (Katagiri & Nakane, 2013). One reason for this may be because media and national statistics tend to refer to it as "manic-depressive illness" (Katagiri & Nakane, 2013).

A manic state tends to be perceived as a personality or mood problem more than a depressive state, with only about 50% of those with knowledge of bipolar disorder and 30% of those without knowledge of bipolar disorder recognizing the need

TABLE 1 Prevalence of bipolar disorder in Japan.

Survey	Male	Female	Total
WMHJ 1			
Bipolar I disorder	0.1%	0.06%	0.08%
Bipolar II disorder	0.11%	0.16%	0.13%
WMHJ 2			
Bipolar I disorder	0.6%	0.3%	0.4%
Bipolar II disorder	0.0%	0.1%	0.0%

This table shows the lifetime prevalence of bipolar I disorder and bipolar II disorder in Japan.
Data from Kawakami, N. (2007). *Kokoro no kenko ni tsuite no ekigaku cyosa ni kansuru kenkyu* [Report of epidemiologic study of mental health of Japan]. Health labour science research grant, 2004–2006 research on psychiatric and neurological diseases and mental health and Kawakami, N. (2016). *Seishin shikkan no yubyo ritsu to ni kansuru daikibo ekigaku cyosa kenkyu* [Large-scale epidemiologic study on the prevalence of mental disorders: Report of world mental health Japan 2nd survey]. Health labour science research grant, Comprehensive research on disability health and welfare. Retrieved from: http://wmhj2.jp/WMHJ2-2016R.pdf.

for treatment (Katagiri & Nakane, 2013). In another study of patients with bipolar disorder ($N=457$), approximately 40% had delays in reporting to psychiatrists because of a lack of knowledge of bipolar disorder (Watanabe, Harada, Inoue, Tanji, & Kikuchi, 2016).

In the survey by Watanabe et al. (2016), 44% of the participants were college graduates, but 48% were working, and 44% of those working had an annual income of one million yen or less. This was about one quarter of the average annual income in Japan the year the survey was taken (National Tax Agency Japan, 2014). Of those respondents who had been married at least once, 34% were divorced (Watanabe et al., 2016). The bipolar disorder patients who could not be diagnosed early and received appropriate treatment ($N=392$) often experience social problems, such as an inability to work or study (65%), conflict with family (46%), discharge from work or dropping out of school (46%), financial difficulties (43%), and interpersonal conflict with nonfamily members (42%) (Watanabe et al., 2016). Although most participants accept the disease as part of their life, they admit that they have difficulty controlling their symptoms (Watanabe et al., 2016).

Medical care for patients with bipolar disorder in Japan

In Japan, there are guidelines for the treatment of bipolar disorder (Japanese Society of Mood Disorders, 2020). The guidelines recommend treatment for each phase of bipolar disorder (Table 2). When it comes to treating bipolar disorder, pharmacotherapy is generally the first option (Yathma et al., 2018) and lithium is one of the major pharmacological agents (Volkman, Bschor, & Köhler, 2020). Lithium has remained the gold standard in the treatment of bipolar disorder since the discovery of its antimanic properties in 1949 (Cade, 2000). A metaanalysis performed by Geddes, Burgess, Hawton, Jamison, and Goodwin (2004) shows that lithium reduces manic episodes significantly more than a placebo. In this study, the recurrence rate was 14% for lithium, compared to 24% for the placebo. In the treatment of bipolar disorder, the prevention of recurrence is as important as or better than the suppression of symptoms.

However, the recommended treatment of these guidelines are not always consistent with the treatment environment. Under Japan's universal health insurance system, medical practices for which medical fees are paid are prescribed by the government. Described in the guidelines may not be covered depending on the phase of illness. For example, lithium is the first choice for the acute treatment of mania as well as for maintaining the suppression of bipolar symptoms, but insurance covers only acute treatment; for maintenance treatment, only lamotrigine is covered.

TABLE 2 The recommended psychosocial treatment in Japan.

Treatment	Most recommended	Next recommended	Not Recommended
Acute treatment			
Manic episode	Carbamazepine(O) Olanzapine(O) Aripiprazole(O) Quetiapine Risperidone	No applicable treatment	
Depressive episode	Lamotrigine Olanzapine(O) Lithium Lurasid	No applicable treatment	
Maintenance treatment			
	Psychoeducation	CBT IPSRT FFT	Psychosocial treatment alone without pharmacotherapy

This table shows recommended psychosocial treatment for bipolar disorder in Japan.
Data from Japanese Society of Mood Disorders. (2020). *Nihon utsu byo gakkai chiryo gaidorain: Sokyoku sei syogai 2020* [*Treatment guideline by Japanese Society of Mood Disorders: Bipolar disorder 2020*]. Retrieved from: https://www.secretariat.ne.jp/jsmd/iinkai/katsudou/data/guideline_sokyoku2020.pdf.

Psychosocial treatment for bipolar disorder in Japan

As shown in Table 2, four forms of psychotherapy are currently recommended, but psychosocial treatment for bipolar disorder is not covered by insurance, and the recommended high-intensity treatment is rarely provided due to cost and lack of specialists. Generally, therapeutic psychotherapy for bipolar disorder in Japan are conducted using some of the techniques based on studies that confirm that such programs had beneficial effects in other countries because it is difficult to carry out intensive psychotherapy in Japan's treatment environment. Therefore, empirical knowledge about psychosocial approaches to bipolar disorder is limited to case reports, and only a few case reports exist. This section describes three types of psychosocial approaches other than CBT: psychoeducation programs, rework programs, and family support.

Psychoeducation program

Psychoeducation is divided into an individual method and a group method. Psychoeducation is often incorporated into daily medical examinations and other programs. Notably, psychoeducation is listed as the most recommended treatment for bipolar disorder because it is easier to implement than other psychosocial treatments (Table 2). On the other hand, few reports have examined the effect of a program comprising only psychoeducation as other psychosocial interventions are often provided under the name of psychoeducation programs. Saito-Tanji, Tsujimoto, Taketani, Yamamoto, and Ono (2016) developed a text-based "simple individual psychoeducation" protocol tailored to Japan's treatment environment and reported its efficacy. Simple individual psychoeducation is a program in which a therapist and a patient alternately read the document "Living with bipolar" (JSMD, 2021) while waiting for a doctor to examine the patient. Saito-Tanji et al. (2016) reported that the program improved understanding of bipolar disorder and awareness of mood states, allowing patients to better communicate with family members and medical staff and greatly improve their quality of life. The group psychoeducation are often implemented as part of inpatient treatment, day care programs, and rework programs (described below), these programs have mixed patients and other interventions, and the effects of group psychoeducation on bipolar disorder patients alone have not been reported.

Rework program

A rework program is a psychosocial treatment used in Japan. While it is not included in official treatment guidelines, it is offered at more than 220 medical facilities (Igarashi, 2018). Rework is a rehabilitation program for workers who are on leave due to depression; its goal is to prevent them from taking another leave of absence (Igarashi, 2018). Although there have been no reports of structured IPSRT for bipolar disorder in Japan, rework programs for bipolar disorder often use techniques designed to stabilize social rhythms (Okuyama & Akiyama, 2011). Fig. 1 shows an example of a rework program for bipolar disorder (Okuyama & Akiyama, 2011). Okuyama (2012) conducted a rework program for bipolar disorder focusing on psychological education and stabilization of social rhythm and reported a case in which the patients actually returned to work. On the other hand, treatment of bipolar disorder varies among institutions (Tokura & Ozaki, 2012), partly because rework programs have focused on depression. It is also pointed out that program content is not uniform across facilities, whether it is group psychotherapy, IPSRT, CBT, occupational therapy, or return-to-work training, and that the effectiveness of these rework program has not been sufficiently proven (Kato, 2019).

FIG. 1 Rework program for bipolar II disorder. This figure shows example of rework program for bipolar II disorder. (Data from Okuyama, S., & Akiyama, T. (2011). Sokyoku sei syogai no fuku-syoku ni saishite: Sokyoku II gata wo cyushin ni [Re-Work in bipolar disorders: mainly bipolar II disorder]. Japanese Journal of Clinical Psychiatry, 40, 349–360.)

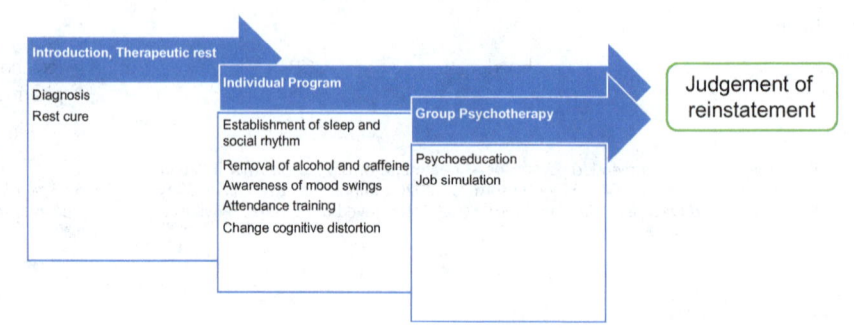

Introduction, Therapeutic rest
- Diagnosis
- Rest cure

Individual Program
- Establishment of sleep and social rhythm
- Removal of alcohol and caffeine
- Awareness of mood swings
- Attendance training
- Change cognitive distortion

Group Psychotherapy
- Psychoeducation
- Job simulation

Judgement of reinstatement

Family support

There have been few reports of family support as part of formal treatment for bipolar disorder. Sakai and Akiyama (2012) conducted a survey of medical facilities ($N=43$) that implement rework programs and found that only 18.6% of them implement programs that involve families. In addition, although approximately 90% of the medical facilities that do not implement such support stated that they felt the need to provide family support, it was suggested that such support was not implemented due to staff and cost issues. Mino et al. (2009) divided 15 hospitalized patients with bipolar disorder into a family psychological education group and a control group and compared the risk of recurrence and medical costs during the 9 months after discharge. Although there was no difference in the risk of recurrence, they reported that the medical costs of the psychological education group were one third of the costs of the control group. The results suggest an economic impact of family support, although there is room for improvement in the program.

As noted above, sufficient evidence does not yet exist for any psychosocial treatment for bipolar disorder in Japan. The psychosocial treatment of bipolar disorder in Japan involves a few common issues, namely: this lack of research evidence and issues related to the treatment environment, such as cost and providers. Although these issues are difficult to resolve in the short term, Japanese practice reports suggest that some intervention techniques may also be useful in Japan.

CBT for bipolar disorder in Japan

We have already discussed the problems related to psychosocial treatment in Japan; CBT involves similar issues. Unlike other psychosocial treatments, CBT is covered by insurance for mood disorders, but this is mainly for depression and must be done according to a manual developed in Japan. Thus, CBT for bipolar disorder is also not fully utilized. Currently, there are also a small number of practice reports of CBT for bipolar disorder.

Kimura, Hamatani, Matsumoto, and Shimizu (2020) reported a case series of CBT for three patients with bipolar II disorder during a depressive episode, which consisted of six modules: psychological education, case conceptualization and treatment goal setting, activity and mood monitoring, acquisition of prodromal coping behaviors, establishment of sleep and routines, and correction of nonfunctional beliefs. Ogai (2019) reported a case study of CBT based on psychoeducation, mood monitoring, establishing routines and sleep, and coping skills to deal with prodrome for a student with bipolar II disorder as student counseling. In addition, in Ogai's case, coping behaviors included defusion exercises and value clarification based on Acceptance and Commitment Therapy. MIyazaki (2016) reported a case of behavioral activation for a patient with a major depressive episode of bipolar II disorder who was hospitalized.

The reports to date have been for bipolar II disorder, and there are no reports of CBT for bipolar I disorder. This is also true for other psychosocial treatments, and it may be that bipolar II disorder is more likely to lead to psychosocial treatment in the Japanese treatment environment. In any case, the above reports were based on chance encounters between clients and therapist—this situation is not common in Japan. Surprisingly, the overall rate of implementation of CBT was 37.9% in

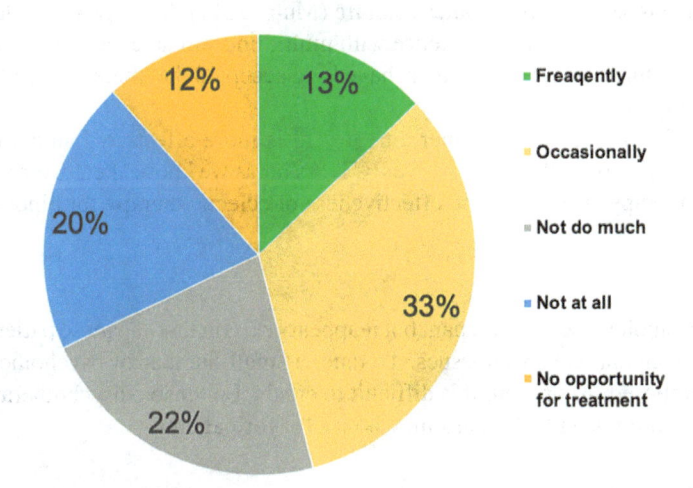

FIG. 2 Implementation rate of CBT for bipolar disorder in Japanese psychiatric clinics. This figure shows the percentage of psychiatric clinics in Japan that provide CBT for bipolar disorder. *(Data from Takahashi, F., Takegawa, S. Okumura. Y., & Suzuki, S. (2018).* Nihon no seishinka shinryojyo ni okeru ninchi kodo ryoho no teikyo taisei ni kansuru jittai cyosa *[Actual condition survey on the implementation of cognitive behavioral therapy at psychiatric clinics in Japan]. Retrieved from: http://ftakalab.jp/wordpress/wp-content/uploads/2011/08/japancbtclinic_report.pdf.)*

psychiatric clinics in Japan ($N = 1019$), and only half of these clinics provided CBT for bipolar disorder (Fig. 2; Takahashi, Takegawa, Okumura, & Suzuki, 2018). It seems that there are not many opportunities to offer CBT for bipolar disorder even in facilities that implement CBT.

Relationship between CBT and universal health insurance in Japan

A universal health insurance system has been established in Japan, which provides all citizens with medical insurance. Naturally, insurance-covered treatments are prioritized in psychiatric care. The three most common factors that have been identified in previous surveys as preventing the widespread use of high-intensity psychotherapy, including CBT, in Japanese medical institutions were: lack of time to provide psychotherapy, unprofitable service, and lack of staff who can provide psychotherapy (Fujisawa, 2005; Takahashi et al., 2018). These factors may be related to the fact that psychologists have not been nationally licensed in Japan for a long time. Takahashi et al.'s (2018) survey of psychiatric clinics shows that more than half of psychiatric clinics have turned down requests for CBT from patients. Another patient survey ($N = 434$) also reported that the experience rate with individual CBT was 5.5% (Haraguchi, Shimizu, Nakazato, Kobori, & Iyo, 2015), indicating that CBT may be difficult to access.

In a survey of psychiatric outpatients in a Japanese prefecture, the most common form of psychotherapy received within the past 3 years was brief psychotherapy of less than 20 min (54.0%); when continuous psychotherapy of more than 50 min was conducted, group psychotherapy was the most common form (49.5%) (Haraguchi et al., 2015), and may refer primarily to brief psychotherapy given by psychiatrists at the time of examination or to day care group programs. In other words, the main form of psychotherapy in Japan is short or group therapy, and individual psychotherapy for more than 50 min seems to be rare. Although this survey was conducted in only one region in Japan, regional difference is likely to be small because the medical service fee system is uniform throughout the country. As noted above, although the implementation rate of CBT in Japan is low, 35.2% (number of subjects: 118/335) of patients mentioned that they would like to receive individual CBT as a psychotherapy in the future (Haraguchi et al., 2015). Therefore, in Japan, patients would prefer to access CBT therapies, but expanding its use remains difficult.

Eclectic therapies in Japan

Another aspect is that in Japan, specific forms of psychotherapy are very rarely used alone and appear to be usually combined with other psychotherapies. In a survey by Takahashi et al. (2018), about 80% of psychiatric clinics that reported offering CBT combined it with different therapies (20.0% for single use). The percentages of other psychotherapy combinations were as follows: psychoanalyticical therapy, 80.3% (19.7% for single use); interpersonal psycotherapy, 82.7% (0.05% for single use); person centered therapy, 82.7% (17.3% for single use); and family therapy, 95.9% (4.1% for single use) (Takahashi et al., 2018). This survey suggested that the eclectic therapies were most common therapy in Japan. This may be due to cultural influences: it has been pointed out that clinical psychology in Japan has a unique structure in which counseling and psychotherapy are mixed; this was not established by chance because of the import of western culture, but instead as a result of unconscious selection by Japanese culture (Muto, 2011). In other words, Japanese people tend to keep things ambiguous; they prefer flexibility and coexistence with multidimensional elements and do not decide on one single solution (Minami, 1983); for example, residents of Japan have long accepted that there are "eight million Gods (Yaoyorozu no kami)" in the Japanese religion Shinto.

Of course, it is not possible to determine whether a treatment is more effective than a control without conducting a clinical trial that includes a strictly controlled prospective RCT. As far as we know, there were no RCTs that have evaluated the effectiveness of eclectic therapy. Therefore, the effectiveness of eclectic therapy for bipolar disorder remains unclear.

Conclusion

There is a lack of research on bipolar disorder in Japan, but it appears certain that Japanese patients with bipolar disorder are struggling with disease management and social issues. To date, a small number of psychotherapies for bipolar disorder, including CBT, have been proposed. At present, it is difficult to conduct intensive psychotherapy in Japan in terms of cost and staff, but providing some promising techniques may be useful for patients.

Practice and procedures

The basic technique of each stage of CBT for bipolar disorder is shown below. In Japan, CBT is so prevalent in a stylized set of ways that novices tend to misunderstand that all the techniques need to be applied. In our experience, however, patients

TABLE 3 Basic techniques of CBT for bipolar disorder.

Assessment
Psychoeducation
Case formulation
Therapeutic goal setting
Self-monitoring
Establishment of daily routine and sleep habits
Behavioral experiment
Development of coping skill to prodrome
Summarize his/her CBT treatment

This table shows basic techniques of CBT for bipolar disorder easy to use in Japan.

benefit from techniques that are easy to incorporate, even if they are not perfect. What is easy to do in Japan is to implement psychoeducation as a first step and introduce other techniques according to the particular situations of patients and their therapists (So, 2017). This section describes the basic CBT techniques for bipolar disorder that are likely to be particularly useful in Japan (Table 3). Much of the techniques are based on the modules of Lam et al. (2010), but the description is based on materials and other resources that are readily available in Japan.

Assessment

Efficient CBT requires a comprehensive assessment of the patient during the intake interview before the intervention. With sufficient information, cognitive-behavioral therapists will be able to accurately formulate cases and develop treatment plans. In CBT for bipolar disorder, important information includes family history, life history, medical history, social function, medication content, social resources, stress coping strategies, prodromal symptoms and corresponding behavior patterns, and circadian rhythm disorders. A life chart can be used to summarize the course of previous disorders and to help patients understand patterns of improvement and deterioration in their condition. A life chart is available to anyone on the JSMD website. When asking a patient to write a life chart, it is advisable not only to hand it to the patient, but also to write it with the patient. Patients with a long history often forget about past events and are more likely to recall when the therapist asks questions. The assessment is not limited to the first session. Data from the course of treatment can be added to reinforce the assessment. Japanese versions of the following tools have been developed as assessment tools for bipolar disorder: the Young Mania Rating Scale (YMRS; Young, Biggs, Ziegler, & Meyer, 1978), the Hamilton Depression Rating Scale (HAMD; Hamilton, 1960), and the Beck Depression Inventory-II (BDI-II; Beck, Steer, & Brown, 1996).

Psychoeducation

Patients can understand their mental disorder by receiving a general explanation of bipolar disorder. It is usually difficult to remember what you were told about your disorder because you are exposed to a lot of information. Therefore, it is often recommended to use pamphlets in psychoeducation (JSMD, 2021; Lam et al., 2010). The therapist can read the same pamphlet as the patient and mark the relevant parts, which will be useful for later case conceptualization. The patient does not have to read or understand it all in one session; they can take it home and read it. It has also been mentioned that reading aloud can be beneficial (Saito-Tanji et al., 2016), but the important thing is not to just get the reading done and then move on, but to communicate with the patient based on the text. The therapist asks the patient whether there are any unclear points, or the patient asks the therapist about points that are unclear to them. Therapists should not try to persuade patients of claims if they are not convinced; instead, they should use repeated reminders during and after psychotherapy. A document entitled *"Living with Bipolar Disorder"* (JSMD, 2021) has been published on the JSMD website and is available to anyone. This psychoeducational pamphlet describes the symptoms, causes, what the patient should be careful of, the effects and side effects of therapeutic drugs, psychotherapy (ISPRT), requests for family members, how to find a specialist in bipolar disorder, and research.

Case formulation

Case formulation is a framework for the therapist to understand under what circumstances the patient's problems are maintained and sustained by the patient's reasonable and improper coping. If the client knows in advance that the main problem is bipolar disorder, the patient can be understood relatively easily by following the cognitive-behavioral model. It is recommended that the therapist continue to formulate the case from the first contact with the patient to the end (Beck, 2011). Cognitive-behavioral therapists can tailor treatment to be more effective by envisioning a provisional case formulation and corresponding treatment model and modifying the cognitive-behavioral model if necessary (Persons, 2008; Tarrier, 2006).

Therapeutic goal setting

Setting treatment goals is a crucial part of CBT for bipolar disorder (Lam et al., 2010). In the case of people with bipolar disorder, some want to maintain the prodrome of a manic state, while others want to reexperience a manic episode. In that case, it's a good idea to remind them of the catastrophic results in the medium to long term, as a substitute for feeling the best in the short term. Treatment goals for most patients with bipolar disorder include reducing depressive or mania symptoms, repairing family and interpersonal functions, preventing recurrence, dealing with reinstatement and financial problems, and using support services. Treatment goals are voluntarily set for the patient's own values and long-term adaptation. A constructive dialogue between the therapist and the patient leads to an agreement on treatment goals.

It is important to set specific and realistic treatment goals. Our recent study found that the strength of consensus between treatment goals and treatment challenges predicts a good prognosis (Matsumoto et al., 2019).

Self-monitoring

Self-monitoring is a fundamental technique in CBT and has implication for both assessment and intervention. Assessment involves collecting information about changes in lifestyle and mood to help case formulation and manage prodrome. The purpose of intervention is to improve self-control by the habituation of observation behavior. Self-monitoring often focuses on mood swings and activities, sleep, and eating. It is well known that disorderly life events, sleep, daily routines, and circadian rhythms can cause manic episodes (Healy & Williams, 1989). Bipolar disorder is a highly biologically driven disorder but controlling patients' moods and lifestyles even slightly can improve their quality of life. Patients are asked to keep records of their experiences using a monitoring sheet, and this should be continued throughout treatment. Recording these details can be cumbersome and is often interrupted. Therapists need to fully inform and psychologically support at the beginning of monitoring because patients will be adding new tasks to their lives.

Establishment of daily routine and sleep habits

This behavioral approach to maintaining daily routines and sleep habits have played a central role in the management of bipolar disorder (APA, 2002; JSMD, 2020). People with bipolar disorder easily change their sleep times and sleep time zones as their moods swing. Therefore, in CBT therapy for bipolar disorder, it is important to secure the minimum sleep time and not to change the sleep time zone after confirming the patient's current sleep pattern. As for food, some people spare time to eat during manic episodes—the amount of caffeine may increase, and some people have no appetite and avoid eating during depressive episodes. Unstable diets like this will eventually have a negative impact on your overall lifestyle. A good dietary goal would be to be able to eat regularly three times a day. It is also important to plan appropriate activities. If patients act according to their moods, they may do nothing when depressed and continue to ruminate on depressive thoughts; meanwhile, they may ignore daily routines and indulge in fun during manic episodes. Patients with bipolar disorder can experience a stable mood at both extremes by performing prescheduled activities rather than following their moods. The sleep-wake schedule record sheet is available on the JSMD website for anyone to access.

Behavioral experiment

Some patients may be reluctant to change their usual behavior. It is relatively natural to worry that new actions will unintentionally exacerbate symptoms. Behavioral experiments are useful as a way to discuss the validity of a patient's ideas (Beck, 2011). The patient and the therapist can experimentally observe whether new behaviors change mood and yield unusual results. As an example of a behavioral experiment, one patient planned to "make tea in a 'Japanese tea ceremony' manner as a calming behavior." Interestingly, she chose not to make tea because it calmed her down: she interpreted teatime as the opposite of pleasure and happiness. However, the results of her behavioral experiment showed her that she had developed an increased ability to observe the experience of being calm and relaxed and that on days she practiced this behavior she wasted less money and time going out.

Development of coping skills at prodrome

Known prodrome in patients with bipolar disorder include decreased sleep cravings, increased goal-oriented activity, loss of interest, and anxiety (Lam & Wong, 1997). Cognitive-behavioral therapists work with patients to compile prodromal patterns in detail and devise coping strategies. A comprehensive review can be made by discussing with the patient and writing out the prodrome corresponding to the areas of mood, cognition, and behavior. Classification of prodromes in chronological order, early, middle, and late, also prepares the patient for choosing the coping behavior. Means of identifying prodrome include open-ended questions and the use of the previously described life charts and monitoring sheets. Patient values will be the basis of the activities and priorities to be undertaken. The therapist does not direct coping strategies.

Cognitive-behavioral therapists should note that the prodrome of a manic episode is different from that of a depressed episode. When the former prodrome is noticed, patients are encouraged to choose to avoid irritation, engage in calming activities, and prioritize and perform them one by one. When a patient becomes aware of the prodrome of a depressive episode, the patient is encouraged to perform activities such as reducing time to worry alone or increasing time for hobbies. Prodrome and coping in cases reported by authors are shown in Table 4 (Kimura et al., 2020). In case of recurrence, you can create a coping card that describes how to respond appropriately to the prodrome and have the patient use it later in the treatment.

Summarize his/her CBT treatment

Summarizing of past efforts in preparation for future difficulties is able to a memorandum of understanding for patients (Lam et al., 2010). In the final session, we will review the experience from the patient's efforts and the cognitive-behavioral model. Even after CBT is complete, patients must maintain their daily routine, manage their sleep habits, become aware of their prodromal symptoms, and actively work to prevent recurrence. To that end, therapists can encourage them to remember how they changed and what they learned. The therapist and the patient should reconsider any self-management and sleep issues and the dangers of seeking stimuli such as caffeine, tobacco, cannabis, alcohol, and sex.

Some patients may feel anxious as sessions come to a close; anxiety may be lessened by reducing the frequency of sessions from once a week to twice a month as the end approaches. Patients who are worried about reducing the number of sessions should discuss the advantages and disadvantages of reducing the number of sessions and be aware of the advantages of reducing the number of sessions (Beck, 2011). The following examples was a clinical practice by the third author. KM: "*Your treatment will end next month, so we will meet only four more times, including today. Would you like do a session every two weeks from next?*" The Patient: "*OK, yes. But I'm worried. Do you have any idea why?*" KM: "*It's natural to be anxious. But, if you can do it alone for two weeks, you'll be confident.*" The patient: "*That's right. If I can't handle it myself, I can go to the hospital, right?*" KM: "*Of course.*"

TABLE 4 Example of prodrome and coping in Japanese patient.

	Prodrome	Coping
Hypomanic	Decreased sleep time Increased activity Talkativeness Extravagance	Taking a break Calling a family member Consulting a doctor
Depression	Think about self-responsibility Feel worthlessness Loss of interest	Meeting friends or family Watching favorite pictures

This table shows example of prodrome and coping by Japanese bipolar disorder patient.
Data from Kimura, Y., Hamatani, S., Matsumoto, K., & Shimizu, E. (2020). Cognitive behavioral therapy for three patients with bipolar II disorder during depressive episodes. *Case Reports in Psychiatry*, 3892094. https://doi.org/10.1155/2020/3892024.

Mini-dictionary of terms

- **Euthymic state:** Period of no symptoms.
- **Maintenance treatment:** Treatment intended to prevent relapse and loss of quality of life during remission.
- **Manic-depressive illness:** Previous term for bipolar disorder.
- **Prodrome:** Minute changes or symptoms before the manic or depressive phase of bipolar disorder.
- **Rework program:** Program to help employees on leave for depression to return to work.
- **Universal health insurance system.** A system in which all citizens must join public insurance in Japan.

Key facts

Key facts of suicide in Japan

- Since 1998, the suicide rate has been 30,000 per year, but since 2010 it has been decreasing, and in 2017 it was 20,465.
- Japan's suicide rate is the highest of all seven industrialized countries (G7).
- Over the past 20 years, men have consistently accounted for about 70% of suicides, with men in their 40s and 60s accounting for about 40% of suicides.
- About 60% of suicides are by unemployed people.
- The Basic Act on Suicide Prevention was enacted in 2006.
- In 2008, mental health support centers were established in each prefecture.
- In 2015, a stress check system became mandatory for companies with more than 50 employees.
- For the past 10 years, "concerns and effects of depression including bipolar disorder" has statistically been the most frequent cause of and factor in suicides.

Applications to other areas

This chapter reviews the potential of CBT for the long-term management of bipolar disorder in Japan. Though the effectiveness of the program in Japan has not been sufficiently examined, its level of practicability and applicability is promising. Also, it was indicated that in Japan, it is difficult to carry out intensive psychotherapy. However, combined CBT and pharmacotherapy can also be applied to chronic physical disorders in which drug therapy is essential, barring cost, and staff problems. In fact, self-administered education and behavioral approaches have already been recommended by the Japan Diabetes Care Guidelines (Araki et al., 2020). Bipolar disorder and diabetes have a cognitive-behavioral commonality in that activities that are comforting only in the short term lead to destabilization; bipolar disorder is sometimes described as "mental diabetes" (So, 2020).

Summary points

- The prevalence of bipolar disorder in Japan is estimated to be about 0%–0.4%.
- Japanese bipolar disorder patients have various social challenges.
- The Japanese guidelines recommend four psychosocial interventions: psychoeducation, CBT, IPSRT, and FFT.
- The recommended treatment in guidelines and treatment settings are inconsistent, and intensive psychotherapy is rarely offered in Japan.
- Although CBT is widely known in Japan as a treatment for depression, it has not spread to the treatment of bipolar disorder due to practical problems.
- It may be more appropriate to offer a few promising technique than to provide a perfect program of practice at this stage.

References

American Psychiatric Association. (2002). *Practice guideline for the treatment of patients with bipolar disorder* (2nd ed.). Retrieved from: https://psychiatryonline.org/pb/assets/raw/sitewide/practice_guidelines/guidelines/bipolar.pdf.

American Psychiatric Association. (2013). *Diagnostic and statistical manual of mental disorders* (5th ed.). Arlington, TX: American Psychiatric Publishing.

Araki, E., Goto, A., Kondo, T., Noda, M., Noto, H., Origasa, H., … Yoshioka, N. (2020). Japanese clinical practice guideline for diabetes 2019. *Diabetology International, 11*, 165–223. https://doi.org/10.1007/s13340-020-00439-5.

Bebbington, P., & Ramana, R. (1995). The epidemiology of bipolar affective disorder. *Social Psychiatry and Psychiatric Epidemiology, 30*, 279–292. https://doi.org/10.1007/BF00805795.

Beck, A. T., Steer, R. A., & Brown, G. K. (1996). *Manual for the beck depression inventory-II*. San Antonio, TX: Psychological Corporation.

Beck, J. S. (2011). *Cognitive behavior therapy: Basics and beyond* (2nd ed.). New York, NY: Guilford Press.

Bowden, C. L., & Krishnan, A. (2004). Pharmacotherapy for bipolar depression: An economic assessment. *Expert Opinion on Pharmacotherpy, 5*, 1101–1107. https://doi.org/10.1517/14656566.5.5.1101.

Cade, J. F. (2000). Lithium salts in the treatment of psychotic excitement. 1949. *Bulletin of the World Health Organization Supplement, 78*, 518–520.

Chiang, K.-J., Tsai, J.-C., Liu, D., Lin, C.-H., Chiu, H.-L., & Chou, K.-R. (2017). Efficacy of cognitive-behavioral therapy in patients with bipolar disorder: A meta-analysis of randomized controlled trials. *PLoS One, 12*(5). https://doi.org/10.1371/journal.pone.0176849, e0176849.

Fujisawa, D. (2005). *Honpo ni okeru seishin ryoho no jissi jyokyo ni kansuru kenkyu [Research on the status of psychotherapy in Japan] : Research on methods and effectiveness of psychotherapy (H16-kokoro-014)*. Health labour sciences research grant, research on psychiatric and neurological diseases and mental health. Retrived from MHLW grants system https://mhlw-grants.niph.go.jp/system/files/2005/057131/200500784A/200500784A0002.pdf.

GBD 2017 Disease and Injury Incidence and Prevalence Collaborators. (2018). Global, regional, and national incidence, prevalence, and years lived with disability for 354 diseases and injuries for 195 countries and territories, 1990–2017: A systematic analysis for the global burden of disease study 2017. *The Lancet, 392*, 1789-1858. https://doi.org/10.1016/S0140-6736(18)32279-7.

Geddes, J. R., Burgess, S., Hawton, K., Jamison, K., & Goodwin, G. M. (2004). Long-term lithium therapy for bipolar disorder: Systematic review and meta-analysis of randomized controlled trials. *The American Journal of Psychiatry, 161*, 217–222. https://doi.org/10.1176/appi.ajp.161.2.217.

Gitlin, M. J., Swendsen, J., Heller, T. L., & Hammen, C. (1995). Relapse and impairment in bipolar disorder. *The American Journal of Psychiatry, 152*, 1635–1640. https://doi.org/10.1176/ajp.152.11.1635.

Hamilton, M. (1960). Development of rating scale for primary depressive illness. *British Journal of Clinical Psychology, 6*, 278–296.

Haraguchi, T., Shimizu, E., Nakazato, M., Kobori, O., & Iyo, M. (2015). Chiba ken ni okeru ninchi kodo ryoho no jittai: Sitsumon shi cyosa [Current status of cognitive behavioral therapy in Chiba prefecture: A questionaire survey]. *Chiba Igaku, 91*, 177–184. Retrieved from: https://core.ac.uk/download/pdf/97064926.pdf.

Healy, D., & Williams, J. M. (1989). Moods, misattributions and mania: An interaction of biological and psychological factors in the pathogenesis of mania. *Psychiatric Developments, 7*, 49–70.

Igarashi, Y. (2018). Riwāku proguramu no genjyo to kadai [Current status and challenges of rework programs]. *The Japanese Journal of Labour Studies, 695*(6), 62–70. Retrieved from: https://www.jil.go.jp/institute/zassi/backnumber/2018/06/pdf/062-070.pdf.

Inoue, T., Kimura, T., Inagaki, Y., & Shirakawa, O. (2020). Prevalence of comorbid anxiety disorders and their associated factors in patients with bipolar disorder or major depressive disorder. *Neuropsychiatric Disease and Treatment, 16*, 1695–1704. https://doi.org/10.2147/NDT.S246294.

Japanese Society of Mood Disorders. (2020). *Nihon utsu byo gakkai chiryo gaidorain: Sokyoku sei syogai 2020 [Treatment guideline by Japanese Society of Mood Disorders: Bipolar disorder 2020]*. Retrieved from: https://www.secretariat.ne.jp/jsmd/iinkai/katsudou/data/guideline_sokyoku2020.pdf.

Japanese Society of Mood Disorders. (2021). *Sokyoku sei syogai to tsukiau tame ni [Living with bipolar disorder]*. Retrieved from: https://www.secretariat.ne.jp/jsmd/gakkai/shiryo/data/bd_kaisetsu_ver10-20210324.pdf.

Jones, S. H., Smith, G., Mulligan, L. D., Lobban, F., Law, H., Dunn, G., ... Morrison, A. P. (2015). Recovery-focused cognitive-behavioural therapy for recent-onset bipolar disorder: Randomized controlled pilot trial. *The British Journal of Psychiatry, 206*, 58–66. https://doi.org/10.1192/bjp.bp.113.141259.

Katagiri, H., & Nakane, H. (2013). Sokyoku sei syogai ni kansuru intānetto ni yoru ninchi do cyosa [An internet-based survey on awareness of bipolar disorder]. *Seishin Igaku, 55*, 383–390.

Kato, T. (2019). *Sokyoku sei syogai [Bipolar disorder]* (3rd ed.). Tokyo, Japan: Igaku-Syoin.

Kawakami, N. (2007). Kokoro no kenko ni tsuite no ekigaku cyosa ni kansuru kenkyu [Report of epidemiologic study of mental health of Japan]. In *Health labour science research grant, 2004–2006 research on psychiatric and neurological diseases and mental health*.

Kawakami, N. (2016). Seishin shikkan no yubyo ritsu to ni kansuru daikibo ekigaku cyosa kenkyu [Large-scale epidemiologic study on the prevalence of mental disorders: Report of world mental health Japan 2nd survey]. In *Health labour science research grant, comprehensive research on disability health and welfare*. Retrieved from: http://wmhj2.jp/WMHJ2-2016R.pdf.

Kimura, Y., Hamatani, S., Matsumoto, K., & Shimizu, E. (2020). Cognitive behavioral therapy for three patients with bipolar II disorder during depressive episodes. *Case Reports in Psychiatry*. https://doi.org/10.1155/2020/3892024, 3892094.

Lam, D. H., Jones, S. H., & Hayward, P. (2010). *Cognitive therapy for bipolar disorder: A therapist's guide to concepts, methods and practice*. Chichester, UK: Wiley-Blackwell.

Lam, D. H., Watkins, E. R., Hayward, P., Bright, J., Wright, K., Kerr, N., ... Sham, P. (2003). A randomized controlled study of cognitive therapy for relapse prevention for bipolar affective disorder: Outcome of the first year. *Archives of General Psychiatry, 60*, 145–152. https://doi.org/10.1001/archpsyc.60.2.145.

Lam, D. H., & Wong, G. (1997). Prodromes, coping strategies, insight and social functioning in bipolar affective disorders. *Psychological Medicine, 27*, 1091–1100. https://doi.org/10.1017/s0033291797005540.

Matsumoto, K., Yoshida, T., Hamatani, S., Sutoh, C., Hirano, Y., & Shimizu, E. (2019). Prognosis prediction using therapeutic agreement of video conference-delivered cognitive behavioral therapy: Retrospective secondary analysis of a single-SRM pilot trial. *JMIR Mental Health, 6*(11). https://doi.org/10.2196/15747, e15747.

Miklowitz, D. J., Otto, M. W., Frank, E., Reilly-Harrington, N. A., Wisniewski, S. R., Kogan, J. N., ... Sachs, G. S. (2007). Psychosocial treatments for bipolar depression: A 1-year randomized trial from the systematic treatment enhancement program. *Archives of General Psychiatry, 64*, 419–426. https://doi.org/10.1001/archpsyc.64.4.419.

Minami, H. (1983). *Nihon teki jiga [Japanese ego]*. Tokyo, Japan: Iwanami Shoten.

Mino, Y., Shimodera, S., Fukuzawa, Y., Morokuma, I., Fujita, H., Yonekura, Y., & He, L. (2009). Nihon ni okeru sokyoku sei syougai no kazoku shinri kyoiku no iryohi he no eikyo [A medical cost analysis of family psychoeducation for bipolar disorder in Japan]. *The Journal of Social Problems, 58*, 13–17. https://doi.org/10.24729/00003124.

MIyazaki, T. (2016). Sokyoku II gata syogai no dai utsubyo episodo ni taishite kodo kasseika ga choko shi fukusyoku ni itatta 1 rei [Remarkable Efficacy of behavioral Actication for Major Depressive Episode in Bipolar II Disorder: A Case Study]. *Japanese Journal of Behavior Therapy, 42*(2), 267–277. https://doi.org/10.24468/jjbt.42.2_267.

Muto, T. (2011). Nihon teki na amarini nihon teki na: Nihon bunka ni kansuru kodo bunseki gaku teki ichi shiron no ichi tenkai [Japanese, all too Japanese: What "a Japanese behavior analyst analyzes his own culture" suggest to us]. *Japanese Journal of Behavior Analysis, 26*, 9–12.

National Tax Agency Japan. (2014). *2013 private sector salary survey: Survey results report*. Retrieved from: https://www.nta.go.jp/publication/statistics/kokuzeicho/minkan2013/pdf/001.pdf.

Ogai, H. (2019). Sokyoku II gata syogai to shindan sare kesseki wo kurikaeshita jyosidaigakusei ni taisuru ninchi kodo ryoho no kokoromi [Cognitive behavioral therapy attempted for a female university student diagnosed as bipolar II disorder with repeated absences]. *Japanese Journal of Counseling Science, 51*, 189–196.

Okuyama, S. (2012). Sokyoku sei syogai no riwāku puroguramu: Sokyoku sei syogai ni rikan shite chiryocyu ni kyumu ni itatta kinrosya heno fukusyoku ni saishite no seishin ka rihabiritēsyon puroguramu. [Re-work program for bipolar disorders: Psychiatric rehabilitation program designed to help patients with bipolar disorders to return to their workplace]. *Job Stress Research, 19*, 227–234.

Okuyama, S., & Akiyama, T. (2011). Sokyoku sei syogai no fukusyoku ni saishite: Sokyoku II gata wo cyushin ni [Re-work in bipolar disorders: Mainly bipolar II disorder]. *Japanese Journal of Clinical Psychiatry, 40*, 349–360.

Persons, J. B. (2008). *The case formulation approach to cognitive-behavior therapy*. New York, NY: Guilford Press.

Pini, S., de Queiroz, V., Pagnin, D., Pezawas, L., Angst, J., Cassano, G. B., & Wittchen, H.-U. (2005). Prevalence and burden of bipolar disorders in European countries. *European Neuropsychopharmacology, 15*, 425–434. https://doi.org/10.1016/j.euroneuro.2005.04.011.

Rowland, T. A., & Marwah, S. (2018). Epidemiology and risk factor for bipolar disorder. *Therapeutic Advances in Psychopharmacology, 8*, 251–269. https://doi.org/10.1177/2045125318769235.

Saito-Tanji, Y., Tsujimoto, E., Taketani, R., Yamamoto, A., & Ono, H. (2016). Effectiveness of simple individual psychoeducation for bipolar II disorder. *Case Report in Psychiatry, 6062801*. https://doi.org/10.1155/2016/6062801.

Sakai, Y., & Akiyama, T. (2012). Utsubyo no riwāku puroguramu no genjyo to kadai [The present and future possibility of return-to-work assist programs patient with depressive disorders]. *Job Stress Research, 19*, 217–225.

So, M. (2017). Sokyoku sei syogai II gata ni okeru seishin ryoho [Psychotherapies for bipolar disorder type II]. *Japanese Journal of Clinical Psychiatry, 46*, 271–278.

So, M. (2020). Taijin kankei syakai rizumu ryoho: Iki kata no sutairu wo kaenai kagiri sokyoku sei syogai ha saihatsu shi tuzukeru [Interpersonal and social rhythm therapy (IPSRT) for bipolar disorders]. *Japanese Journal of Clinical Medicine, 78*, 1716–1724.

Takahashi, F., Takegawa, S., Okumura, Y., & Suzuki, S. (2018). *Nihon no seishinka shinryojyo ni okeru ninchi kodo ryoho no teikyo taisei ni kansuru jittai cyosa [Actual condition survey on the implementation of cognitive behavioral therapy at psychiatric clinics in Japan]*. Retrieved from: http://ftakalab.jp/wordpress/wp-content/uploads/2011/08/japancbtclinic_report.pdf.

Tarrier, N. (Ed.). (2006). *Case formulation in cognitive behaviour therapy: The treatment of challenging and complex cases*. Hove, UK: Routledge.

Tokura, T., & Ozaki, N. (2012). Ri wāku puroguramu ni okeru sokyoku sei syogai no atsukai [How to treat the patients with bipolar disorder in the re-work program]. *Japanese Journal of Clinical Psychiatry, 41*, 1535–1542.

Volkman, C., Bschor, T., & Köhler, S. (2020). Lithium treatment over the lifespan in bipolar disorders. *Frontiers in Psychiatry, 11*, 377. https://doi.org/10.3389/fpsyt.2020.00377.

Watanabe, K., Harada, E., Inoue, T., Tanji, Y., & Kikuchi, T. (2016). Perceptions and impact of bipolar disorder in Japan: Results of an internet survey. *Neuropsychiatric Disease and Treatment, 12*, 2981–2987. https://doi.org/10.2147/NDT.S113602.

Yathma, L. N., Kennedy, S. H., Parikh, S. V., Schaffer, A., Bond, D. J., Frey, B. N., ... Berk, M. (2018). Canadian Network for Mood and Anxiety Treatments (CANMAT) and International Society for Bipolar Disorder (ISBD) 2018 guidelines for the management of patients with bipolar disorder. *Bipolar Disorder, 20*, 97–170. https://doi.org/10.1111/bdi.12609.

Young, R. C., Biggs, J. T., Ziegler, V. E., & Meyer, D. A. (1978). A rating scale for mania. *The British Journal of Psychiatry, 133*, 429–435.

Chapter 28

Cognitive-behavioral therapy for anxiety disorders in Italian mental health services

Laura Giusti, Silvia Mammarella, Anna Salza, and Rita Roncone

Department of Life, Health and Environmental Sciences, University of L'Aquila, L'Aquila, Italy

Abbreviations

ADs anxiety disorders
CBT cognitive-behavioral therapy
MHSs mental health services
PD   panic disorder
SP   social phobia

Introduction

Anxiety disorders (ADs) are the most common mental health problem, characterized by impaired personal and social functioning and low quality of life (Alonso et al., 2004; Carta et al., 2015; Preti et al., 2021).

Based on the first Italian epidemiological study on the prevalence of mental disorders, part of the European project European Study on the Epidemiology of Mental Disorders (ESEMeD) (Alonso et al., 2004), 11.1% of Italian people reported a lifetime history of any AD, and a 12-month prevalence of 5.1%, lower than in any of the other European countries surveyed, with women being more likely to have anxiety disorders (de Girolamo et al., 2006).

A more recent Italian nationwide community survey showed a relatively low generalized anxiety disorder (GAD) frequency in the community (2.3% lifetime prevalence) (Preti, Demontis, et al., 2021), and an overall lifetime prevalence of agoraphobia of 1.5%, with a higher prevalence among women (2.0%) than men (0.9%), often comorbid with other psychiatric disorders (Preti, Piras, et al., 2021). Agoraphobia showed an attributable burden in health-related quality of life comparable to that observed for chronic mental disorders, such as major depression, posttraumatic stress disorder, or obsessive–compulsive disorder (Preti, Piras, et al., 2021). The severe impact of ADs on the lives of individuals with a relevant quality of life impairment has promoted the development of several therapeutic approaches.

Among the psychotherapeutic approaches, international guidelines suggest that cognitive-behavioral therapy (CBT), associated with pharmacological treatment, represents the preferred evidence-based treatment for ADs (National Institute for Health and Care Excellence, 2011). Significant and positive long-term outcomes for youth and adults were found (Kaczkurkin & Foa, 2015). CBT is a directive, problem-oriented, structured, time-limited therapeutic approach that emphasizes the importance of a collaborative therapeutic relationship (Beck, 1995). According to CBT principles, it is not the context or circumstances that make a person suffer emotionally, but one's perceptions, beliefs, and assumptions about the situation. CBT's general purpose is to guide persons to modify the content of problematic cognitions and thereby break the cycle of maladaptive thoughts, emotions, and behavior. A recent exhaustive systematic review and meta-analysis assessed the long-term outcomes after CBT (compared with usual care, relaxation, psychoeducation, pill placebo, supportive treatment, or waiting list) for ADs, posttraumatic stress disorder (PTSD), and obsessive–compulsive disorder (OCD) (van Dis et al., 2020). The paper suggested that, compared with control conditions, CBT was generally associated with lower anxiety symptoms within 12 months after treatment completion, although few studies examined longer-term outcomes (van Dis et al., 2020).

The Italian scenario and CBT treatments

The Mental Health Law 180 in 1978 changed the history of Italian psychiatry, moving from a hospital-based system of mental healthcare to a community-based one. Since then, the Italian mental health context has been focused on the community treatment of psychotic disorders, prioritizing severe mental illness.

The "Italian way" of "doing community psychiatry" had, and still has, many positive effects, not always documented in the scientific literature, along with some reasons for concerns (Amaddeo & Barbui, 2018).

For the theme of this work, among the difficulties of Italian MHSSs, we must include the development and implementation of innovative ways of treatment. In most cases, the organization of services has hardly changed over the last 40 years, and very few evidence-based interventions, the majority of them based on the cognitive-behavioral paradigm, have been implemented in the routine of clinical practice. In Italy, one of the main problems in implementing effective psychosocial intervention could be attributed to a strong aversion to behavioral-cognitive methods, which are considered too "simple" and "unsophisticated," unable to capture the "complexity of mental illness," far from deep-rooted cultural models (Casacchia & Roncone, 1999). This is likely one of the main reasons why the Italian MHSS community has been so slow and reluctant to adopt CBT-based interventions.

Concerning the integrated CBT treatment of psychotic disorders with the involvement of relatives of the affected person, between 1991 and 1997, 32 training courses were conducted for clinicians and mental health staff in Italian MHSSs (Falloon et al., 1999). The results of 6-year training courses and consultation designed to develop a widespread application of CBT evidence-based strategies throughout Italy, as reported by Falloon et al., highlighted that the lack of training for mental health professionals was not the only problem to be challenged. Limited structural resources, community stigma, and health administrators more focused on service delivery than outcome seemed to be obstacles to the introduction of innovative clinical strategies into routine clinical services. Currently, despite guideline recommendations for psychosis, CBT evidence-based family interventions are not yet available in most Italian MHSSs, unlike in many countries (Casacchia & Roncone, 2014).

CBT anxiety in mental health service in Italy: The first steps

A vital training

Despite the epidemiological prevalence and the severe impact on the quality of life of persons affected by ADs (Mendlowicz & Stein, 2000), within the Italian MHSS in the beginning, CBT interventions for ADs management encountered the same suspicion as family CBT for psychosis. The psychoanalytic orientation of most therapists and the lack of CBT specialization schools could be among the causes of this reluctance (Fava, 1997).

To our knowledge, the introduction of CBT for AD treatment in the Italian MHSS can be traced back to 1997 with the Milano training course at the Regional Training Institute of the Lombardy Region (Istituto Regionale per la Formazione della Lombardia, IREF). The Australian Andrews Gavin, Director of the Clinical Research Unit for Anxiety Disorders (CRUFAD) of the University of New South Wales, Sidney, was the principal teacher of the course. The training opportunity was promoted and coordinated by Pierluigi Morosini (1941–2008), an epidemiologist and psychiatrist working at the Laboratory of Epidemiology and Biostatistics of the Rome National Health Institute. He gave a tremendous and influential impulse to implement integrated, evidence-based CBT interventions, promoting the philosophy of simple although reliable psychiatric outcome evaluation, with a particular interest in developing a proper scientific evaluation movement in health services. His conviction was that responding effectively to users results from a balance between staff training and professionalism and the availability of resources and organizational capacity (Morosini, Casacchia, & Roncone, 2000).

Beginning with the 1997 Milano training course, some MHSSs in Northern Italy (Milan, Bergamo, Genoa) and Central Italy (L'Aquila, Arezzo, Grosseto) started the application of CBT in the treatment of ADs. The original work titled *The Treatment of Anxiety Disorders: Clinician Guides and Patients Manuals, Second Edition*, including the manual for clinicians and the manual for users, by Andrews et al. (2003b) was translated into Italian (Andrews et al., 2003a).

Many experiences were "informally" clinically conducted by the trainees in their Italian MHSS following the enthusiasm promoted by the course, although very few of them have been reported in the international scientific literature (Table 1).

TABLE 1 "First generation" Italian CBT published studies on anxiety disorders (ADs).

Authors	Sample size	Diagnosis	Protocol	Assessment	Outcome
Leveni, Mazzoleni, and Piacentini (1999)	22 (15 women; 7 men)	Panic Disorder with agoraphobia	Replication of the original Australian protocol (Andrews et al., 2003b) Three-week treatment: 5 days of full-time therapy; 9 days of application of strategy in everyday life; 5 days of full-time active treatment. 6-month follow-up. At the entry in the study 13 subjects (69%) were taking psychotropic drugs.	**Mark Sheehan Phobia Scale, MSPS** (Marks & Sheehan, 1983); **Short-form Health Survey, SF-36** (Ware & Sherbourne, 1992); **Panic Attacks Anticipatory Anxiety Scale, PAAAS** (Sheehan, 1983a); **State–Trait Anxiety Inventory, STAI-X1, STAI-X2** (Sanavio, Bertolotti, Michielin, Vidotto, & Zotti, 1986).	20 subjects (91%) completed the treatment; 2 subjects (9%) dropped out. At the end of the treatment and 6-month follow-up: - all investigated variables improved - statistically significant reduction in the number of panic attacks - only 7 subjects (35%) were still taking drugs
Leveni, Piacentini, and Campana (2002)	11 (6 women; 5 men)	Social Phobia	Replication of the original Australian protocol (Andrews et al., 2003b) Three-week treatment: 5 days of full-time therapy; 9 days of application of strategy in everyday life; 5 days of full-time active treatment; 6-month follow-up. Subjects were not taking psychotropic drugs.	**Liebowitz Social Phobia Scale**, LSPS (Liebowitz, 1987); **Short-form Health Survey**, SF-36 (Ware & Sherbourne, 1992); **Clinical Global Impression**, CGI (Guy, 1976)	Any drop-out. At the end of the treatment: 7/11 much improved—improved in all investigated variables At 6-month follow-up: 11/11 much improved—improved in all investigated variables

Continued

TABLE 1 "First generation" Italian CBT published studies on anxiety disorders (ADs)—cont'd

Authors	Sample size	Diagnosis	Protocol	Assessment	Outcome
Mirabella et al. (2009)	70 (58 women; 12 men)	Panic Disorder	Adaptation to an outpatient setting of the original protocol (Andrews et al., 2003) 10 weekly 2-h group session treatment Prepost study Spontaneous organization of self-help groups	**Panic Attacks Anticipatory Anxiety Scale, PAAAS** (Sheehan, 1983a); **Hamilton Anxiety Scale, HAM-A** (Hamilton, 1959); **Sheehan Patient Rated Anxiety Scale, SPRAS** (Sheehan, 1983b); **Mark Sheehan Phobia Scale, MSPS** (Marks & Sheehan, 1983); **Short-form Health Survey, SF-36** (Ware & Sherbourne, 1992).	52 subjects (74%) completed the treatment; 18 subjects (26%) dropped out At the end of the treatment: marked improvements in somatic anxiety, psychological anxiety, the extent of fear and avoidance phobia, self-esteem regarding phobias, specific symptoms, anticipatory anxiety, the number of expected panic attacks, and the dosage of pharmacological treatment
Mastrocinque, De Wet, and Fagiolini (2013)	96 (70 women; 26 men)	Panic Disorder without agoraphobia	Adaptation to an outpatient setting of the original protocol (Andrews et al., 2003b) 10 weekly 2-h group session treatments Prepost study	**Clinical Outcome Routine Evaluation—Outcome Measure, CORE-OM** (Barkham et al., 2001; Evans, Connell, Barkham, Marshall, & Mellor-Clark, 2003; Palmieri et al., 2009) **Symptom Check List, SCL 90; Hamilton Anxiety Scale, HAM-A** (Hamilton, 1959).	60 subjects (62.5%) completed the treatment; 36 subjects (37.5%) dropped out At the end of the treatment: almost all scores of the clinical scales used improved significantly. Only 42% of subjects initially above the clinical cut-off on the CORE-OM showed clinically significant changes, and 20% were below the clinical cut-off after treatment.

The table shows the first CBT group interventions on effectiveness in the treatment of ADs.

The first effectiveness studies

The first published study gained from the "Australian training" was conducted in the MHSS, the Psychosocial Center, of a village of around 9000 inhabitants in Lombardy, Zogno (Bergamo, Italy). Leveni et al. applied a group CBT treatment for **panic disorder (PD) with agoraphobia** (Leveni et al., 1999). The treatment involved 22 subjects, mean age 39 years, some treated with antidepressants, others with benzodiazepines, and others without psychopharmacological support. The adopted three-week treatment replicated the protocol proposed at the Clinical Research Unit for Anxiety Disorders (CRUFAD) in Sydney (Andrews et al., 1994). The treatment had an intensive structure over 3 weeks: 5 days of full-time therapy, 9 days of application of the teachings in the context of the everyday life of the users, and an additional 5 days of full-time active treatment at the center. The short-term and 6-month follow-up results of such an intensive CBT group treatment showed a statistically significant improvement in the various dimensions of anxiety symptoms with a decrease in panic attacks frequency. In particular, the findings showed a perceived significant global improvement, and a substantial reduction in fears and avoidance behaviors. The authors underlined that these results represented a valuable new starting point for Italian psychiatry, because they demonstrated the possibility of introducing an evidence-based psychotherapeutic strategy at low cost, with minimal investment, and with an optimization of resources, within a day-hospital clinical setting (Leveni et al., 1999).

Subsequently, in the same center, the same CBT protocol was adopted by the Leveni's research group (Leveni et al., 2002) to treat **social phobia (SP)**. Once again, the aim was to generalize the feasibility of CBT treatment protocols to other ADs in Italian public health care settings. From 1999 to 2001, 11 subjects with SF that matched the DSM-IV diagnostic criteria were recruited for treatment, and none of them was treated with psychotropic drugs. Nine subjects underwent group treatment, while two of them underwent individual treatment. They were investigated for their general conditions, symptoms, and life satisfaction at the end of the CBT treatment and after 6 months: the results obtained showed a statistically significant improvement in most of the areas investigated by the end of the treatment. Moreover, the benefits persisted and increased over time, reaching the highest statistical significance at 6 months follow-up. The innovative contribution of this study was that, in addition to the clinical variables, importance was given to the user's perception of improvement, displaying their appreciation of the benefits. The authors confirmed the applicability of CBT treatment for SP in an outpatient clinical setting (Leveni et al., 2002).

With the support of the Italian National Institute of Health, the therapists of Zogno developed additional specific competencies in CBT interventions, and they applied a CBT treatment for depression in the postnatal period to be routinely implemented in the clinical MHSS practice (Piacentini et al., 2011).

The Tuscany replication studies

In Tuscany, a study involving 70 subjects diagnosed with **PD** aimed to assess the effectiveness of CBT group therapy in the local Mental Health Department of Pisa (Mirabella et al., 2009). The study was conducted for 2 years, organizing groups of 5–6 users. Compared to the original Australian protocol, the manualized treatment (Andrews et al., 2003a) was adapted to an outpatient clinical format. Instead of an "intensive" 3-week period, it became a 10-week treatment with 2-h weekly treatments. The methodological design was based on a prepost study and assessments were obtained after 2 ½ months by comparing the results before and after the intervention. Marked improvements were observed in the follow-up, particularly for somatic anxiety, psychological anxiety, the extent of fear and avoidance phobia, self-esteem regarding phobias, specific symptoms, anticipatory anxiety, and the number of expected minor or major panic attacks. The authors commented that the uncontrolled design of the study did not exclude other active factors in the intervention; however, the results showed that group CBT could improve patients' general condition, regardless of whether the users were taking psychotropic drugs. The authors reported that this type of therapy needs only a small medical team and a limited number of instruments, making it suitable for public service because the cost-effectiveness is favorable. Moreover, the authors reported that this experience gave rise to self-help groups attended mainly by those who needed some form of aggregation and those who still had difficulty managing anxious symptoms (the latter situation was more frequent in those with a personality structure more rigid). The self-help group met with an average frequency of about 8–10 people meeting once a week in a room made available by the Municipality of Pontedera, a little town close to Pisa (Mirabella et al., 2009).

In the Mental Health Department of Pisa, Andrea Fagiolini replicated the study of Mirabella et al., involving 96 subjects with **PD with and without agoraphobia** (Mastrocinque et al., 2013). The study challenged three main criticisms about the transferability of research interventions in the clinical setting concerning the following issues: (1) users: treatments carried out in research settings will not work in clinical practice because subjects in clinical settings have more severe and complex psychopathological conditions; (2) clinicians: specialists in services outside the research setting often do not have access to training; (3) protocols: interventions and therapy manuals are less used in clinical practice. The study included "real-world" subjects aged between 20 and 75 years, and comorbidity with other psychiatric disorders was not considered an exclusion criterion. The study used a battery of psychometric instruments to assess a wide range of psychological variables, subjective well-being, and psychosocial functioning. The results showed that 15% of the subjects presented a complete recovery from PD at the end of treatment. Despite experiencing decreased panic symptoms, the rest of the sample could not be considered fully recovered, although the clinicians could report many benefits. The subsample of subjects with more severe depressive symptoms showed a marked improvement. Almost all the scores of the symptom scales used were reduced in the posttest, particularly somatic anxiety and psychic anxiety. Based on the results obtained, the authors concluded that their intervention could produce significant and clinically relevant improvements in most symptoms that characterize PD, including quality of life (Mastrocinque et al., 2013). Although the treatment successfully brought the treated sample up to the level of the general population, the PD-affected subjects still deviated significantly from the average normal population. The authors expressed their conceivable prediction that the users' symptoms would have continued to improve after treatment, bringing the average of the treated group closer to the population average (Mastrocinque et al., 2013).

The Italian "second generation" of CBT intervention for ADs

CBT interventions of AD management have seen progressive dissemination in Italian MHSS, mainly conducted by the "psychiatric rehabilitation technician," a specific mental-health professional with academic and professional profile skills, trained to properly conduct psychosocial interventions (Roncone, Ussorio, Salza, & Casacchia, 2016). Globally, our mental health professionals show a good global psychosocial orientation toward recovery and a growing interest in evidence-based practice (Giusti et al., 2019).

We can define a "second generation" CBT intervention for ADs, treatments that "enriched" the original Australian protocol, as the intervention we developed for young users at the TRIP- Psychosocial Rehabilitation Treatment, Early Interventions University Unit, at the San Salvatore Hospital, in L'Aquila (Giusti et al., 2018). We used a combined CBT intervention based on the protocol developed by Andrews et al. (Andrews et al., 2003a) together with two MetaCognitive Training (MCT) modules (Moritz & Woodward, 2010) selected from previous work conducted by the same research group in which a revised and adapted MCT version was developed for subjects with psychosis (Ussorio et al., 2016).

The two MetaCognitive Training (MCT) modules (modules 2 and 7, respectively) included in the anxiety CBT protocol aim to modify "the interpersonal sensitivity," as a precursor of paranoid ideation. The protocol also considered the cognitive "jumping to conclusions" bias (JTC) often displayed by young people with ADs (Giusti et al., 2018). The study's rationale considered the mechanism underlying the anxious experience that appears to have many similarities with the mechanism underlying the paranoid ideation, characteristic of psychotic illness, also widespread in nonclinical populations (Freeman et al., 2005). Freeman et al. proposed a threat anticipation model in which paranoid ideation seems to arise from an interaction of vulnerability, emotional processes, and reasoning biases (Freeman, 2007). Freeman's paranoia hierarchy includes five levels (Fig. 1). The first level of paranoia ideation was identified as "social evaluative concerns, such as fears of rejection, feeling of vulnerability, thoughts that the world is potentially dangerous." Correlation studies suggest a significant association between anxiety and paranoid ideation among clinical and nonclinical populations (Freeman, Pugh, & Garety, 2008). Some studies (Lincoln, Lange, Burau, Exner, & Moritz, 2010) suggest that anxiety may increase paranoid ideation. Such an increase may be moderated by the level of individual vulnerability and mediated by the jumping to conclusions (JTC) reasoning bias.

Psychopathology, social functioning, metacognition, and the JTC bias were investigated in 60 subjects, randomly assigned to the enriched combined CBT group plus treatment-as-usual (TAU) ($n=35$) or to a wait-list group ($n=25$) receiving only TAU. Each group was divided into two subgroups based on the score of the SCL-90 subscale paranoid ideation (high paranoid ideation, HP; low paranoid ideation, LP). The enriched combined CBT group received a weekly session of a CBT for 3 months (Giusti et al., 2018). The 3-month CBT intervention included eight group sessions focusing on six specific content areas: Session 1 focused on psychoeducation on anxiety disorders. Sessions 2 and 3 focused on learning self-control strategies to manage anxiety, controlled breathing, progressive muscle relaxation, and isometric relaxation. Session 4 focused on gradual exposure to achieve personal goals. Sessions 5 and 6 were based on cognitive restructuring. Sessions 7 and 8 focused on structured problem-solving methods. Sessions 9 and 10 focused on MCT modules 2 and 7, respectively, presented through slides. At the end of the intervention, greater effectiveness in improving anxious symptoms, paranoid ideation, interpersonal sensitivity, and interpersonal relationship was reported in the enriched combined CBT with a statistically significant reduction of the JTC bias compared to the TAU group. More significant benefits were displayed by the subgroup showing high paranoid ideation (Giusti et al., 2018).

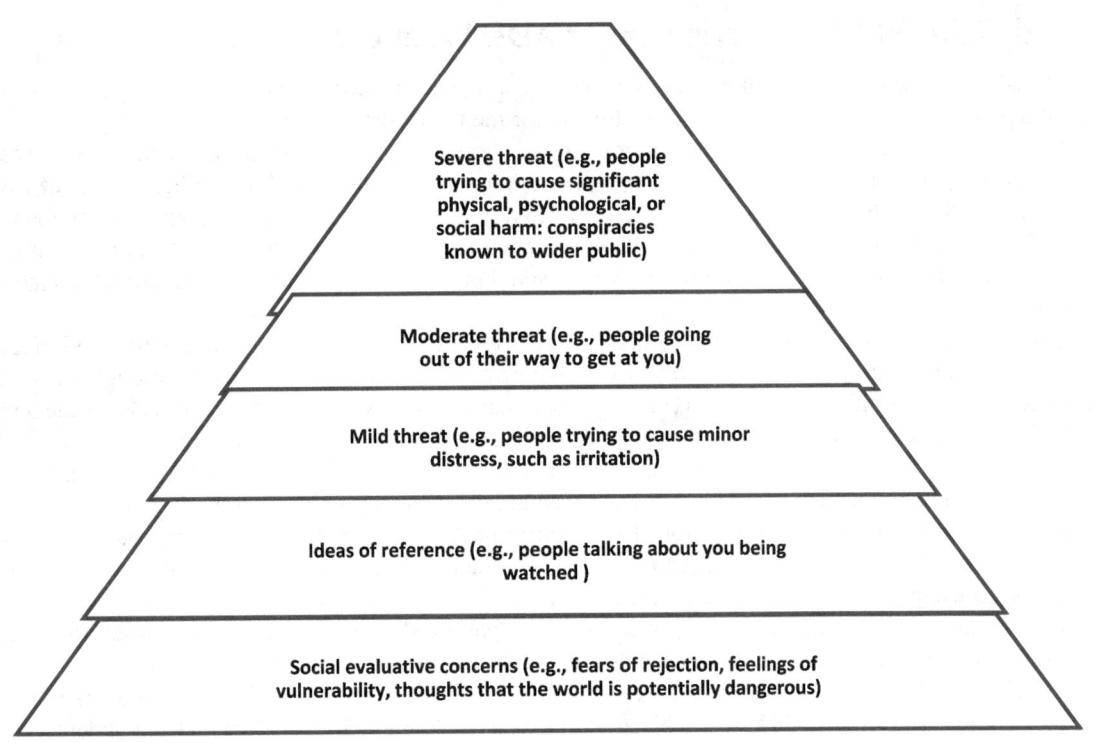

FIG. 1 The paranoia hierarchy. This figure shows the five levels identified by Freeman (2007). His paranoia hierarchy includes five levels in ascending order as (1) social anxiety or interpersonal worry theme; (2) ideas of reference build upon these sensitivities; (3–4) persecutory thoughts closely associated with the attributions of significance; as the severity of the threatened harm increases, the less common the thought; and (5) suspiciousness involving severe harm and organizations and conspiracy is at the top of the hierarchy that the world is potentially dangerous. Based on this model, severe paranoia may build upon common emotional concerns. *(From Freeman, D. (2007). Suspicious minds: The psychology of persecutory delusions.* Clinical Psychology Review, *27(4), 425–457. No permission requested Elsevier.)*

This experience provided encouraging clinical and cognitive evidence of the effectiveness of an integrated rehabilitation strategy designed for young people with ADs to help them manage anxiety and develop a more functional cognitive pattern. Furthermore, it suggested the need to assess the presence of paranoid ideation and JTC bias to complement CBT intervention on anxiety with specific training modules (Giusti et al., 2018).

Postdisaster distress transdiagnostic CBT

The earthquake that on April 6, 2009 struck L'Aquila, the capital city of Abruzzo, Italy, killing 309 residents, injuring over 2500, leaving 28,000 homeless and 66,000 displaced, causing destruction and severe damage, challenged the research group of the L'Aquila University (Casacchia, Pollice, & Roncone, 2012). They applied CBT intervention to individuals who experienced postdisaster distress in transdiagnostic approaches.

A postdisaster CBT treatment training, "Cognitive-Behavioral Therapy for Distress Following Natural Disaster," was held in February 2011 at the University of Aquila, conducted by Dr. Jennifer Gottlieb, Research Assistant Professor in the Department of Psychiatry of the Dartmouth School of Medicine, United States. Our research group adapted and manualized this CBT intervention focused on the psychological distress following natural disasters and major traumas (Hamblen, Roncone, Giusti, & Casacchia, 2018a, 2018b).

Thirty-nine young users received a CBT intervention (Bianchini et al., 2013), based on 12 weekly sessions, for individuals who experienced trauma-related symptoms, particularly associated with anxiety or mood disorders. At the end of the treatment compared to a control population, the young people (with an average age of 27 years) undergoing therapy, primarily women, showed a significant improvement in perceived well-being and the dimension of avoidance, improving coping with earthquake-related situations. Cognitive restructuring (CR) changed dysfunctional thoughts into more functional ones. Some factors, such as religious attitude, played the role of mediators of improvement, especially in women, contributing to a positive outcome. A good improvement was also displayed by young people exposed to the trauma of the earthquake and associated medium-long term earthquake consequences, such as displacement and the loss of social relations (Bianchini et al., 2013).

Online and digitized CBT intervention for ADs: What experiences in Italy?

In the last decade, the increasing rate of access to technology, which overcomes the barriers of place and time, has highlighted the potential of technology to ensure treatments for mental disorders within people's everyday life.

The sustainability of "telepsychiatric" intervention in the current Italian mental health system was discussed by Favaretto & Zanalda (Favaretto & Zanalda, 2018), as reported in the official journal of the Italian Society of Psychiatry. In 2018, the Italian MHSSs did not seem ready to deliver telepsychiatry services, neither from an organizational point of view nor a cultural one. Besides, despite the well-known advantages of videoconferencing for team reunions or service comparisons, no data suggested wider use of this modality, even though the mental health services are dislocated all over the territory.

The Department of Mental Health of Treviso and Torino were two Italian pilot centers for the CBT treatment of depression in the European project *MASTERMIND* (MAnagement of mental health diSorders Through advancEd technology and seRvices—telehealth for the MIND) (Favaretto & Zanalda, 2018; Mol et al., 2016). The Department of Mental Health of Torino is also participating in the international study *ImpleMentAll*. Funded by the European Union's Horizon 2020 program, the project intends to study and improve methods for implementing evidence-based internet-based CBT (iCBT) services for common mental disorders in routine mental health care (Buhrmann et al., 2020).

With the exception of the previously mentioned research projects, digitalized CBT interventions developed and/or implemented in our country have been scarce. To offer a more attractive way to be involved in the treatment to a young population prone to using smartphones, personal computers, and internet sites, the research group of the University of L'Aquila, Italy developed an experimental therapist-assisted computerized CBT (TacCBT) intervention for young adults affected by ADs (Salza, Giusti, Ussorio, Casacchia, & Roncone, 2020). The study was conducted at the San Salvatore Hospital, in the TRIP—Psychosocial Rehabilitation Treatment, Early Interventions in Mental Health, University Unit, at the University of L'Aquila. The study included 50 subjects suffering from ADs (70% of the sample were undergoing psychopharmacological treatment) consecutively referred in 8 months (September 2018 to April 2019). This study aimed to evaluate in young adults with AD the feasibility of the proposed "prototype" of therapist-assisted computerized CBT (TacCBT) and the effectiveness of two different interventions—group CBT and TacCBT—in an "enriched" format for managing anxiety and modifying reasoning biases compared to a control group (Table 2). Both "enriched" CBT programs used in this protocol demonstrated efficacy in young adults with AD and cognitive biases (Giusti et al., 2018). Anxiety, cognitive flexibility, and social functioning represented the outcome variables. Participants were allocated to three treatment conditions (Andrews et al., 2003a): (1) "person-to-person" CBT group and drug treatment (CBT; $n=25$), i.e., the groups included 5–6 users conducted by a Psychiatric Rehabilitation Technician and a clinical psychologist for 3 months (12 weekly sessions, each lasting 90 min); (2) TacCBT group and drug treatment (TacCBT; $n=13$) developed through an internet platform, Moodle, a "virtual clinic," and users could access the virtual clinic only when they were in the service; and (3) treatment as usual group (TAU; $n=12$): subjects received drug treatment and bimonthly clinical consultation with simple CBT strategies. After the intervention, all groups showed significant improvement in anxiety symptoms. Both CBT groups showed an improvement in cognitive flexibility compared to TAU and a reduction in their overconfidence in reasoning. Preliminary results show the benefits of the CBT program and highlight its advantages (Salza et al., 2020). Namely, in both treatment conditions, participants with different diagnoses (PD, GAD, SAD) showed similar rates of variable changes. At the end of the study, all three groups showed a significant improvement in symptoms, whereas both CBT groups improved social functioning and cognitive flexibility compared to the TAU group. In particular, the TacCBT Group showed more significant improvement in the "Participation" domain than the other groups. In contrast, the CBT group showed more significant improvement in the "Getting along" domain. Regarding cognitive flexibility, both CBT groups showed substantial improvement in the "Self-certainty" domain as measured by the BCIS instrument. Therefore these preliminary results seemed to encourage the future application of the "prototype" TacCBT program, with clinically verified feasibility in treating anxiety in young adults who are unwilling or unable to follow face-to-face CBT therapy (Salza et al., 2020).

The impulse given by COVID-19 pandemic to digitalized CBT for ADs

The first major trial of telepsychiatry on a national scale took place in Italy during the pandemic. During the first phase of the Covid-19 epidemic, the Italian mental health system reported a substantial change in the standard operation model, with only urgent psychiatric interventions, compulsory treatments, and consultations for imprisoned people continuing unchanged (Carpiniello et al., 2020). All other activities were reduced to some extent, with a drastic reduction in levels of care. Remote contacts with users were set up in about 75% of cases. Thus the emergency COVID-19 pandemic context

TABLE 2 Session contents of the "Enriched" CBT for anxiety management and reasoning bias modification training (Salza et al., 2020).

	Session content	Homework assignment
Sessions 1–2 Orient the patient to CBT/psychoeducation	**Orient the patient to CBT** • Psychoeducation about the common signs and symptoms of anxiety disorders • Set initial treatment plan/goals	1. Read the user's manual section on anxiety disorders 2. Monitor the achievement of established weekly goals
Sessions 3–4 Anxiety management strategies	**Acquire specific relaxation skills** • Explain the rationale for relaxation strategies • Deep breathing • Muscle relaxation	1. Read the user's manual section on specific relaxation skills 2. Daily diary of deep breathing exercises 3. Daily diary of muscle relaxation exercises
Sessions 5–8 Cognitive therapy/thinking strategies	**Introducing the cognitive model** • Explain the rationale for examining thinking patterns • Review the relationship between thoughts, feelings, and behavior • Explain the ABC model (activating event, beliefs, emotional and behavioral consequences) • Identifying maladaptive thoughts and beliefs • Focus on 'jumping to conclusions' bias • Bias against disconfirmatory evidence, BADE • Suggest or generate alternative, more functional thoughts/beliefs • Challenge of self-injurious thoughts and feelings through Cognitive Restructuring form	1. Read the user's manual section on specific problematic thinking styles 2. Daily diary of unpleasant situations 3. Daily diary of maladaptive thoughts and beliefs 4. Practice with the cognitive restructuring module
Sessions 9–11 Structured problem solving	**Introduce rationale and when to problem-solve** • Explain the steps to effective structured problem-solving and practice	1. Read the user's manual section on structured problem-solving 2. Daily schedule of applied problem-solving for practical problems
Session 12 Relapse prevention	**Prepare a relapse prevention plan** • Strategies for encouraging generalization and maintenance	

has promoted a substantial acceleration of the process of incorporating CBT approaches within websites and web-application apps to help people self-manage their difficulties and provide help and psychoeducation for anxiety symptoms (Giusti et al., 2020).

Our research group used a protected digital platform to deliver psychological interventions to students and young people help-seeking from the Counseling and Consultation Service for Students, SACS, of the University of L'Aquila (Italy). The project proposed a free online emotional support service and was promoted through various channels (e-mails, WhatsApp, Facebook, and university institutional site). This digital intervention combined online therapy modalities (synchronous and asynchronous, automatic and interpersonal, narrative, and suggestions of cognitive-behavioral strategies for anxiety). The platform also included the following: (1) digital narrative diary available to the user to tell their story, through a guided tour of narrative stimuli on cognitive, emotional, and behavioral states; (2) messaging and video counseling sessions, based on a shared calendar; and (3) structured CBT Program for Anxiety Management (Salza et al., 2020). When the students displayed high levels of anxiety, they were invited to access the structured online CBT intervention for anxiety. Regarding the CBT Program for Anxiety Management, the data are going to be published. The digital platform (psydit.com) was also used to analyze the emotional and cognitive experiences and the psychopathological symptoms of University students seeking help during the COVID-19 lockdown, focusing on narrative dimensions analyses reported on the digital diary (Giusti et al., 2020).

Conclusions

The experiences of CBT interventions for AD treatments in Italian MHSs demonstrated good feasibility and a favorable cost-effectiveness ratio since they need only small teams and a limited number of instruments, making it suitable for public service. The spread of CBT interventions for ADs in Italian public services is still limited, considering obstacles such as the scarce availability of therapists. Despite the growing culture of evidence-based interventions and the Italian community psychiatry context, characterized by a diffuse recovery-centered psychosocial approach (Giusti et al., 2019), these interventions show a slow, progressive, although insufficient, implementation in the clinical practice.

The lesson we learned from the COVID-19 pandemic will encourage new strategies based on the following central concerns. First, the most severe consequences on mental health will be manifested in the near future, with an increase of common mental disorders, including ADs, hand in hand with the evolution of the severe economic crisis (Carpiniello et al., 2020). Second, the most effective means for mental health services to deliver interventions under the associated circumstances relies on telepsychiatry. A wider diffusion of CBT group "face-to-face" intervention for ADs and the development of online digital intervention appear as the most significant challenge for every country, as for Italian community psychiatry.

Practice and procedures

The CBT interventions for treatments of ADs in this chapter are described in manuals (both for clinicians and for users) translated from English into Italian (Andrews et al., 2003a).

Mini-dictionary of terms

The Italian Mental Health Law 180 in 1978: Basaglia Law or Law 180 is the Italian Mental Health Act of 1978, which signified an extensive reform of the psychiatric system in Italy, contained directives for the closing down of all psychiatric hospitals, and led to their gradual replacement with a whole range of community-based services, including settings for acute in-patient care in general hospitals with a limited number of beds (no more than 14–16) for short admissions.

Italian National Health System: The Italian National Health Service is based on universality of healthcare, solidarity of financing through general taxation, and equitable access to services. The Italian Health Service makes the right to health accessible to all citizens, without discrimination based on income, gender, or age. These are the three guiding principles of our national health service, to achieve uniform levels of care throughout the territory, equitable access to services for all citizens, and fiscal solidarity as the fundamental way of financing the health system. This means that all services included in the core benefit package (LEA) must be equally accessible in all Italian regions. https://www.salute.gov.it

Key facts

- Cognitive-behavioral treatments are not diffusely popular among Italian mental health professionals.
- The scientific literature on CBT intervention for anxiety disorders in Italian mental health community psychiatry is scarce, despite a slow progressive implementation in clinical practice.
- Cognitive-behavioral group treatment for anxiety disorders has shown good feasibility and a favorable cost-effectiveness ratio since only small teams and a limited number of instruments are needed, making it suitable for public service.
- The second generation of "enriched" CBT intervention for anxiety disorders adds on new therapeutic modules and comparisons with different treatment administration modalities (as for therapist-assisted computerized CBT, TacCBT), showing positive outcomes.
- After the COVID-19 pandemic, a growing interest is rising concerning the potential need for treatment in numerous distressed people and the online application of structured CBT intervention for anxiety disorders.

Applications to other areas

In this chapter on CBT intervention in the Italian mental health services, we underlined that the application in public services is suitable. In analogy, structured, manualized CBT interventions can be conducted in mental health after relatively short training of professionals. The benefits for users seem substantially assessed, considering as the primary endpoint health-related quality of life. The group format was the more frequently investigated and the less time consumable for professionals. In the treatment of posttraumatic distress, the practical application of several strategies used in the CBT treatment of ADs, widely used in natural disasters in Italy, could be seen.

Summary points

- Anxiety disorders are common and disabling. A lifetime history of any AD at 11.1% and a 12-month prevalence of 5.1% were reported in epidemiological studies in the Italian population.
- The Mental Health Law 180 in 1978 established a favorable scenario for community psychiatry, but implementing effective psychosocial CBT intervention was considered far from deep-rooted cultural models.
- Good training on CBT intervention for anxiety disorders for Italian mental health professionals gave a vital impulse to the implementation in clinical practice.
- Effectiveness in clinical practice was demonstrated out of the research academic settings.

References

Alonso, J., Angermeyer, M. C., Bernert, S., Bruffaerts, R., Brugha, T. S., Bryson, H., et al. (2004). Prevalence of mental disorders in Europe: Results from the European Study of the Epidemiology of Mental Disorders (ESEMeD) project. *Acta Psychiatrica Scandinavica. Supplementum*, (420), 21–27.

Amaddeo, F., & Barbui, C. (2018). Celebrating the 40th anniversary of the Italian mental health reform. *Epidemiology and Psychiatric Sciences*, 27(4), 311–313.

Andrews, G., Creamer, M., Crino, R., Hunt, C., Lampe, L., & Page, A. (1994). *The treatment of anxiety disorders. Clinician guides and patients manuals*. Cambridge, UK: Cambridge University Press.

Andrews, G., Creamer, M., Crino, R., Hunt, C., Lampe, L., & Page, A. (2003a). *Trattamento dei Disturbi d'Ansia. Guide per il clinico e Manuali per chi soffre del Disturbo* (Traduzione italiana a cura di Morosini P, Leveni D, Piacentini D, Aloja, E.). Torino: Centro Scientifico Editore.

Andrews, G., Creamer, M., Crino, R., Hunt, C., Lampe, L., & Page, A. (2003b). *The treatment of anxiety disorders. Clinician guides and patients manuals*. Cambridge, UK: Cambridge University Press.

Barkham, M., Margison, F., Leach, C., Lucock, M., Mellor-Clark, J., Evans, C., … McGrath, G. (2001). Service profiling and outcomes benchmarking using the CORE-OM: Toward practice-based evidence in the psychological therapies. Clinical Outcomes in Routine Evaluation-Outcome Measures. *Journal of Consulting and Clinical Psychology*, 69, 184–196.

Beck, E. (1995). *Cognitive behavior therapy: Basics and beyond*. New York: Guilford Press.

Bianchini, V., Roncone, R., Tomassini, A., Necozione, S., Cifone, M. G., Casacchia, M., et al. (2013). Cognitive behavioral therapy for young people after l'aquila earthquake. *Clinical Practice and Epidemiology in Mental Health*, 9, 238–242.

Buhrmann, L., Schuurmans, J., Ruwaard, J., Fleuren, M., Etzelmuller, A., Piera-Jimenez, J., et al. (2020). Tailored implementation of internet-based cognitive behavioural therapy in the multinational context of the ImpleMentAll project: A study protocol for a stepped wedge cluster randomized trial. *Trials*, 21(1), 893.

Carpiniello, B., Tusconi, M., Zanalda, E., Di Sciascio, G., Di Giannantonio, M., & Executive Committee of The Italian Society of, P. (2020). Psychiatry during the Covid-19 pandemic: A survey on mental health departments in Italy. *BMC Psychiatry*, 20(1), 593.

Carta, M. G., Moro, M. F., Aguglia, E., Balestrieri, M., Caraci, F., Dell'Osso, L., et al. (2015). The attributable burden of panic disorder in the impairment of quality of life in a national survey in Italy. *International Journal of Social Psychiatry*, 61(7), 693–699.

Casacchia, M., Pollice, R., & Roncone, R. (2012). The narrative epidemiology of L'Aquila 2009 earthquake. *Epidemiology and Psychiatric Sciences*, 21(1), 13–21.

Casacchia, M., & Roncone, R. (1999). Family psychoeducational treatment in schizophrenia: Love for foreigners or application of evidence-based interventions? *Epidemiologia e Psichiatria Sociale*, 8(3), 183–189.

Casacchia, M., & Roncone, R. (2014). Italian families and family interventions. *The Journal of Nervous and Mental Disease*, 202(6), 487–497.

de Girolamo, G., Polidori, G., Morosini, P., Scarpino, V., Reda, V., Serra, G., et al. (2006). Prevalence of common mental disorders in Italy: Results from the European Study of the Epidemiology of Mental Disorders (ESEMeD). *Social Psychiatry and Psychiatric Epidemiology*, 41(11), 853–861.

Evans, C., Connell, J., Barkham, M., Marshall, C., & Mellor-Clark, J. (2003). Practice-based evidence: Benchmarking NHS primary care counselling services at national and local levels. *Clinical Psychology & Psychotherapy*, 10, 374–388.

Falloon, I. R. H., Casacchia, M., Lussetti, M., Magliano, L., Morosini, P., Piani, F., et al. (1999). The development of cognitive-behavioral therapies within Italian mental health services. *International Journal of Mental Health*, 28(3), 60–67.

Fava, G. A. (1997). Psychotherapy research: Why is it neglected in Italy? *Epidemiologia e Psichiatria Sociale*, 6(2), 81–83.

Favaretto, G., & Zanalda, E. (2018). Telepsychiatry in Italy, from premises to experiences. *Evidence-Based Psychiatric Care*, 4, 25–32.

Freeman, D. (2007). Suspicious minds: The psychology of persecutory delusions. *Clinical Psychology Review*, 27(4), 425–457.

Freeman, D., Garety, P. A., Bebbington, P., Slater, M., Kuipers, E., Fowler, D., et al. (2005). The psychology of persecutory ideation II: A virtual reality experimental study. *The Journal of Nervous and Mental Disease*, 193(5), 309–315.

Freeman, D., Pugh, K., & Garety, P. (2008). Jumping to conclusions and paranoid ideation in the general population. *Schizophrenia Research*, 102(1–3), 254–260.

Giusti, L., Salza, A., Mammarella, S., Bianco, D., Ussorio, D., Casacchia, M., et al. (2020). #Everything will be fine. Duration of home confinement and "all-or-nothing" cognitive thinking style as predictors of traumatic distress in young university students on a digital platform during the COVID-19 Italian lockdown. *Frontiers in Psychiatry, 11*.

Giusti, L., Ussorio, D., Salza, A., Malavolta, M., Aggio, A., Bianchini, V., et al. (2018). Preliminary study of effects on paranoia ideation and jumping to conclusions in the context of group treatment of anxiety disorders in young people. *Early Intervention in Psychiatry, 12*(6), 1072–1080.

Giusti, L., Ussorio, D., Salza, A., Malavolta, M., Aggio, A., Bianchini, V., et al. (2019). Italian investigation on mental health workers' attitudes regarding personal recovery from mental illness. *Community Mental Health Journal, 55*(4), 680–685.

Guy, W. (1976). *ECDEU assessment manual for psychophannacology*. DHEW Publication No. (ADM) 76–338.

Hamblen, J. L., Roncone, R., Giusti, L., & Casacchia, M. (2018a). La sofferenza psicologica da disastri naturali e traumi importanti. In *Trattamento cognitivo-comportamentale. Manuale per gli operatori*. Roma: Il Pensiero Scientifico Editore.

Hamblen, J. L., Roncone, R., Giusti, L., & Casacchia, M. (2018b). La sofferenza psicologica da disastri naturali e traumi importanti. In *Trattamento cognitivo-comportamentale. Quaderno di lavoro per l'utente*. Roma: Il Pensiero Scientifico Editore.

Hamilton, M. (1959). The assessment of anxiety states by rating. *British Journal of Medical Psychology, 32*, 50.

Kaczkurkin, A. N., & Foa, E. B. (2015). Cognitive-behavioral therapy for anxiety disorders: An update on the empirical evidence. *Dialogues in Clinical Neuroscience, 17*(3), 337–346.

Leveni, D., Mazzoleni, D., & Piacentini, D. (1999). Cognitive-behavioral group treatment of panic attacks disorder: A description of the results obtained in a public mental health service. *Epidemiologia e Psichiatria Sociale, 8*(4), 270–275.

Leveni, D., Piacentini, D., & Campana, A. (2002). Effectiveness of cognitive-behavioral treatment in social phobia: A description of the results obtained in a public mental health service. *Epidemiologia e Psichiatria Sociale, 11*(2), 127–133.

Liebowitz, M. R. (1987). Social phobia. *Modern Problems on Pharmacopsychiatry, 22*, 141.

Lincoln, T. M., Lange, J., Burau, J., Exner, C., & Moritz, S. (2010). The effect of state anxiety on paranoid ideation and jumping to conclusions. An experimental investigation. *Schizophrenia Bulletin, 36*(6), 1140–1148.

Marks, I. M., & Sheehan, D. V. (1983). *The anxiety disease*. New York: Scribner's.

Mastrocinque, C., De Wet, D., & Fagiolini, A. (2013). Group cognitive-behavioral therapy for panic disorder with and without agoraphobia: An effectiveness study. *Rivista di Psichiatria, 48*(3), 240–251.

Mendlowicz, M. V., & Stein, M. B. (2000). Quality of life in individuals with anxiety disorders. *The American Journal of Psychiatry, 157*(5), 669–682.

Mirabella, F., Mastrocinque, C., Giunti, F., Galardini, P., Meduri, C., & Morosini, P. (2009). Valutazione di efficacia di un intervento di psicoterapia cognitivo-comportamentale di gruppo per il disturbo di panico. *Psichiatria di Comunità, 8*, 156–164.

Mol, M., Dozeman, E., van Schaik, D. J., Vis, C. P., Riper, H., & Smit, J. H. (2016). The therapist's role in the implementation of internet-based cognitive behavioural therapy for patients with depression: Study protocol. *BMC Psychiatry, 16*(1), 338.

Moritz, S., & Woodward, T. S. (2010). *Metacognition study group. Metacognitive training for schizophrenia (MCT). Manual* (4th ed.). Hamburg: VanHam Campus. The MCT manual and modules can be obtained cost-free in over 15 languages at www.uke.de/mkt.

Morosini, P., Casacchia, M., & Roncone, R. (2000). *Qualità dei servizi di salute mentale. Manuale questionario per la formazione organizzativa, l'autovalutazione e l'accreditamento tra pari*. Roma: Il Pensiero Scientifico.

National Institute for Health and Care Excellence, N. (2011). *Generalised anxiety disorder and panic disorder in adults: Management*. London: National Institute for Health and Clinical Excellence. Last update 2019. Retrieved from https://www.nice.org.uk/guidance/cg113.

Palmieri, G., Evans, C., Hansen, V., Brancaleoni, G., Ferrari, S., Porcelli, P., … Rigatelli, M. (2009). Validation of the Italian version of the Clinical Outcomes in Routine Evaluation Outcome Measure (CORE-OM). *Clinical Psychology & Psychotherapy, 16*, 444–449.

Piacentini, D., Mirabella, F., Leveni, D., Primerano, G., Cattaneo, M., Biffi, G., et al. (2011). Effectiveness of a manualized cognitive-behavioural intervention for postnatal depression. *Rivista di Psichiatria, 46*(3), 187–194.

Preti, A., Demontis, R., Cossu, G., Kalcev, G., Cabras, F., Moro, M. F., et al. (2021). The lifetime prevalence and impact of generalized anxiety disorders in an epidemiologic Italian National Survey carried out by clinicians by means of semi-structured interviews. *BMC Psychiatry, 21*(1).

Preti, A., Piras, M., Cossu, G., Pintus, E., Pintus, M., Kalcev, G., et al. (2021). The burden of agoraphobia in worsening quality of life in a community survey in Italy. *Psychiatry Investigation, 18*(4), 277–283.

Roncone, R., Ussorio, D., Salza, A., & Casacchia, M. (2016). Psychiatric rehabilitation in Italy: Cinderella no more—The contribution of psychiatric rehabilitation technicians. *International Journal of Mental Health, 45*(1), 24–31.

Salza, A., Giusti, L., Ussorio, D., Casacchia, M., & Roncone, R. (2020). Cognitive behavioral therapy (CBT) anxiety management and reasoning bias modification in young adults with anxiety disorders: A real-world study of a therapist-assisted computerized (TACCBT) program Vs. "person-to-person" group CBT. *Internet Interventions-The Application of Information Technology in Mental and Behavioural Health, 19*.

Sanavio, E., Bertolotti, G., Michielin, P., Vidotto, G., & Zotti, A. M. (1986). *CBA 2.0 Scale Primarie*. Firenze: O.S.

Sheehan, D. V. (1983a). *Sheehan anxiety and panic attack scales*. Kalamazoo, MI: Upjohn.

Sheehan, D. V. (1983b). *The anxiety disease*. New York: Scribner's.

Ussorio, D., Giusti, L., Wittekind, C. E., Bianchini, V., Malavolta, M., Pollice, R., et al. (2016). Metacognitive training for young subjects (MCT young version) in the early stages of psychosis: Is the duration of untreated psychosis a limiting factor? *Psychology and Psychotherapy, 89*(1), 50–65.

van Dis, E. A. M., van Veen, S. C., Hagenaars, M. A., Batelaan, N. M., Bockting, C. L. H., van den Heuvel, R. M., et al. (2020). Long-term outcomes of cognitive behavioral therapy for anxiety-related disorders: A systematic review and meta-analysis. *JAMA Psychiatry, 77*(3), 265–273.

Ware, J. E., Jr., & Sherbourne, C. D. (1992). The MOS 36-item short-form health survey (SF-36). I. Conceptual framework and item selection. *Medical Care, 30*(6), 473–483.

Chapter 29

Mood and anxiety disorders in Japan and cognitive-behavioral therapy

Naoki Yoshinaga and Hiroki Tanoue
School of Nursing, Faculty of Medicine, University of Miyazaki, Miyazaki, Japan

Abbreviations

CBT   Cognitive-Behavioral Therapy
IAPT   Improving Access to Psychological Therapies
JABA   The Japanese Association for Behavior Analysis
JACT   The Japanese Association for Cognitive Therapy
JABCT The Japanese Association of Behavioral and Cognitive Therapies

Introduction

Mood and anxiety disorders are common and lead to a significant economic burden both in Japan and worldwide. Cognitive-behavioral therapy (CBT) is a global-standard empirically-supported psychological treatment for a wide variety of mental disorders, and people with mental disorders (including mood and anxiety disorders) often prefer psychological treatment rather than pharmacological treatment (McHugh, Whitton, Peckham, Welge, & Otto, 2013). However, in Japan's mental health services, pharmacotherapy has historically been more common, and patients' access to psychological treatment has been severely limited. For example, a nationwide survey reported that only 28% of all mental healthcare institutions in Japan were capable of providing any form of psychotherapy in a satisfactory manner (Fujisawa et al., 2006). Therefore, disseminating empirically supported psychological treatment (i.e., CBT) across Japan is an important issue not only for individual patients but also for society.

Mood and anxiety disorders in Japan

Mood and anxiety disorders are prevalent; they tend to recur, causing patients to miss work and/or perform poorly at work, and also lead to a considerable social economic loss around the world (e.g., Thornicroft et al., 2017). In Japan, the lifetime prevalence of mood and anxiety disorders is 7.0% and 4.2%, respectively (Ishikawa, Kawakami, & Kessler, 2016; Kawakami, Tachimori, & Takeshima, 2016). A previous study demonstrated that the total cost (including direct and indirect costs) of mood and anxiety disorders in Japan reached 2.0 trillion JPY (18.2 billion USD [exchange rate: 1 USD = 110 JPY]) and 2.4 trillion JPY (21.8 billion USD), respectively (Keio University, 2011). Although mood and anxiety disorders in Japan have been reported to be less prevalent than they are in Western countries, they have become more common since the 2000s (Nishi, Ishikawa, & Kawakami, 2019). Along with this, the Japanese government has attempted to raise citizens' general awareness of mental health with advertisements on television and in newspapers (Ishikawa et al., 2018). The 12-month prevalence of mental health service use in Japan has also increased from the 2000s to the 2010s (it has increased about 1.2–1.6 times) (Kawakami et al., 2016; Nishi et al., 2019).

Development of academic societies for CBT in Japan

In Japan, CBT began in the late 1950s with, for example, the application of learning theory on behavioral problems. There are three major CBT-related academic societies—The Japanese Association of Behavioral and Cognitive Therapies (JABCT: formally established as The Japanese Association for Behavior Therapy in 1976), The Japanese Association for Behavior Analysis (JABA: formally established in 1983), and The Japanese Association for Cognitive Therapy (JACT: formally established in 2001). As of April 2021, JABCT, JABA, and JACT now have over 2300, 1000, and 1700 official members, respectively. Along with this progress, awareness of CBT has widely spread, not only among healthcare professionals and academics but also to the general public.

In 2004, the Japanese Association for Behavior Therapy (renamed as JABCT), JABA, and JACT jointly held the World Congress of Behavioral and Cognitive Therapies in Kobe, Japan. Over 1400 international delegates attended the conference, which featured over 500 presentations. From that time, interest in CBT has continued to expand among mental health professionals and academics in the mental health field (Ishikawa et al., 2016).

Research on CBT for mood and anxiety disorders in Japan

In order to capture a current picture of high-quality research on CBT for mood and anxiety disorders in Japan, we conducted a brief literature search using the MEDLINE (via PubMed) on March 9th, 2021. The following search formula was used: ((("Anxiety Disorders" [Mesh]) OR ("Mood Disorders" [Mesh])) AND ("Cognitive Behavioral Therapy" [Mesh])) AND ("Japan" [Mesh]). A total of 42 articles were identified; of these, three articles were randomized clinical trials of CBT targeting mood and anxiety disorders (studies of preventive intervention and targeting subthreshold disorders were excluded). In addition to the online database search, we performed a hand search by scanning the reference lists of those articles identified by the database search. Experts in the field were also consulted to identify any other key research. This hand search identified three additional articles. Table 1 summarizes the characteristics of identified studies evaluating CBT for mood and anxiety disorders in Japan; there are four articles on mood disorders, and two articles on anxiety disorders. Details of the identified studies are described below.

Randomized clinical trials for mood disorders ($n=4$)

Nakagawa et al. (2017) conducted an assessor-blinded randomized controlled trial for Japanese patients with medication-resistant major depressive disorder. A total of 80 patients were randomly allocated to either the CBT combined with usual care group ($n=40$) or the usual care alone group ($n=40$). CBT was conducted in 16 weekly 50-min individual sessions. Four psychiatrists, one clinical psychologist, and one psychiatric nurse provided CBT based on the standardized CBT manual for depression. All the study therapists completed a 2-day intensive workshop before the study, and thereafter, they attended 2-h ongoing group supervision sessions once every 2 weeks. The results demonstrated that the CBT plus usual care group showed significant improvements in depressive symptoms (GRID-Hamilton Rating Scale [blinded assessor-rated]) compared to the usual care alone group at postintervention (week 16), and the treatment effect was maintained for at least 12 months.

Nakao et al. (2018) evaluated the efficacy of internet-based CBT blended with individual face-to-face sessions (blended CBT) for Japanese patients with medication-resistant major depressive disorder through an assessor-blinded, waiting-list controlled, randomized trial. Forty patients were randomized to either the blended CBT or the waiting-list groups at a ratio of 1:1. Blended CBT was provided as 12 weeks of internet-based CBT blended with 12 weekly, 45-min face-to-face sessions. The internet-based program included five core components: psychoeducation, assessment, and problem clarification; behavioral activation; cognitive restructuring; problem solving; and relapse prevention. Two clinical psychologists, one psychiatrist, and one psychiatric nurse, who had received a 2-day intensive CBT workshop and 1-h group supervision every week, provided the face-to-face sessions. The main role of therapists in the face-to-face sessions were reviewing the internet-based program material, evaluating/discussing patient-specific problems, practicing CBT skills, and setting homework with each patient through the internet-based program. At posttreatment assessment (week 12), patients in the blended CBT group showed significant improvements in depressive symptoms (GRID-Hamilton Rating Scale [blinded assessor-rated]) when compared to those in the waiting-list group. The short-term treatment effects in the blended CBT group were maintained until the final 3-month follow-up assessment (week 24). In addition, blended CBT successfully reduced patient-therapist contact time by about two-thirds compared to the standard CBT protocol.

Mantani et al. (2017) evaluated the efficacy of a self-help smartphone CBT app as an adjunctive therapy among patients with medication-resistant major depressive disorder through an assessor-blinded randomized controlled trial. A total of 164 patients were randomly allocated to either the smartphone CBT combined with antidepressant switch group (either to escitalopram [5–10 mg/d] or to sertraline [25–100 mg/d]) or the antidepressant switch only group ($n=81$ and 83, respectively). The self-help smartphone CBT app (called *Kokoro*-app [*kokoro* means "mind" in Japanese]) included eight sessions on self-monitoring, behavioral activation, cognitive restructuring, and relapse prevention. At the end of each session, the central office sent a personalized text message to patients to congratulate them on their progress. Results demonstrated that adding smartphone CBT to antidepressant switch was more effective in improving depressive symptoms (Patient Health Questionnaire-9 [assessor-rated]) than treatment by antidepressant switch alone during the short-term

TABLE 1 Summary of characteristics of randomized clinical trials evaluating CBT for mood and anxiety disorders in Japan.

Category	Study	Target population	Intervention/control	N	Primary disorder-specific outcomes
Mood disorders	Nakagawa et al. (2017)	Adults (aged 20–65) with MDD who were resistant to antidepressants	Individual CBT for MDD (50 min, weekly, total 16 sessions [provider: MD, RN, CP]) + UC	81	PHQ-9 (assessor-rated), BDI-II (self-reported)
			UC	83	
	Nakao et al. (2018)	Adults (aged 20–65) with MDD who were resistant to antidepressants	Individual CBT with web-based program for MDD (45 min, weekly, total 12 sessions [provider: MD, RN, CP]) + UC	40	GRID-HAMD (assessor-rated), BDI-II (self-reported), QIDS (self-reported)
			UC	40	
	Mantani et al. (2017)	Adults (aged 25–59) with MDD who were resistant to antidepressants	Smartphone CBT for MDD (approx. 20 min, weekly, total 8 sessions [provider: N/A]) + antidepressant switch	20	GRID-HAMD (assessor-rated), BDI-II (self-reported), QIDS (self-reported)
			Antidepressant switch	20	
	Watanabe et al. (2011)	Adults (aged 20–70) with MDD who were resistant to antidepressants and had chronic comorbid insomnia	Individual CBT for insomnia (50 min, weekly, total 4 sessions [provider: MD, RN]) + UC	21	GRID-HAMD (assessor-rated), ISI for insomnia (self-reported; primary focus of CBT)
			UC	21	
Anxiety disorders	Yoshinaga et al. (2016)	Adults (aged 18–65) with SAD who were resistant to antidepressants	Individual CBT for SAD (50–90 min, weekly, total 16 sessions [provider: MD, RN, CP, PSW]) + UC	26	LSAS (assessor-rated)
			UC	25	
	Ishikawa et al. (2019)	Children and adolescents (aged 8–15) with anxiety disorders	Group CBT for anxiety (60–90 min, weekly, total 8 sessions [provider: CP])	21	SCAS (self-reported)
			WLC	21	

Abbreviations: BDI-II, beck depression inventory-second edition; CBT, cognitive-behavioral therapy; CP, clinical psychologist; GRID-HAMD, GRID-Hamilton depression rating scale; ISI, insomnia severity index; LSAS, Liebowitz social anxiety scale; MD, medical doctor (psychiatrist); MDD, major depressive disorder; PHQ-9, patient health questionnaire-9 item; PSW, psychiatric social worker; QIDS, quick inventory of depressive symptomatology; RN, registered nurse (psychiatric nurse); SAD, social anxiety disorder; SCAS, Spence children's anxiety scale; UC, usual care; MLC, wait-list control.

period (up to week 9). The treatment effects of the smartphone CBT app were maintained until follow-up assessment (week 17).

Watanabe et al. (2011) conducted an assessor-blinded randomized controlled trial for major depressive disorder, but primarily focused on comorbid chronic insomnia. Thirty-seven patients were randomly assigned to either brief CBT for insomnia (4 weekly 1-h individual sessions) in addition to usual care, or to usual care alone. CBT was provided by five psychiatrists and one psychiatric nurse who had completed a 2-day intensive training program and received ongoing supervision once a month. The CBT plus usual care group showed significant improvements in all sleep parameters as well as depressive symptoms (GRID-Hamilton Rating Scale [blinded assessor-rated]) at both the posttreatment (week 4) and 1-month follow-up (week 8) assessments. The follow-up analyses further demonstrated that adding CBT for insomnia to usual care could produce substantive benefits in some aspects of quality of life, and is highly likely to be cost-effective (e.g., Watanabe et al., 2015).

In order to synthesize the immediate postintervention outcomes of CBT for major depressive disorder, we performed a brief metaanalysis based on these four trials. The depression indicator (continuous variable) was identified as the main

Study or Subgroup	Intervention (CBT) Mean	SD	Total	Control Mean	SD	Total	Weight	Std. Mean Difference IV, Random, 95% CI
Mantani et al. (2017)	19.3	10.4	81	23.3	7.8	83	44.0%	-0.43 [-0.74, -0.12]
Nakagawa et al. (2017)	15	11.8	40	17.9	12	40	27.3%	-0.24 [-0.68, 0.20]
Nakao et al. (2018)	18.5	12.3	20	20.8	9.1	20	15.6%	-0.21 [-0.83, 0.41]
Watanabe et al. (2011)	9.9	7.6	20	17.8	7.8	17	13.0%	-1.00 [-1.70, -0.31]
Total (95% CI)			161			160	100.0%	-0.42 [-0.69, -0.15]

Heterogeneity: Tau² = 0.02; Chi² = 3.84, df = 3 (P = 0.28); I² = 22%
Test for overall effect: Z = 3.10 (P = 0.002)

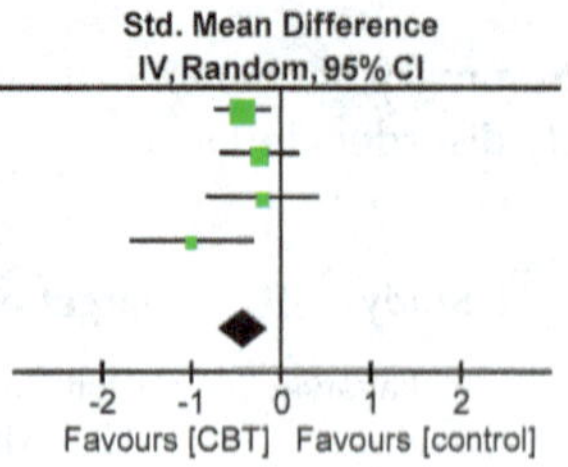

Favours [CBT] Favours [control]

FIG. 1 Forest plot of randomized clinical trials for major depressive disorder in the metaanalysis of immediate effects of CBT on depressive symptoms.

outcome in this analysis. The severity of depression was measured by the beck depression inventory-second edition (Beck, Steer, & Brown, 1996). As an alternative, the GRID-Hamilton Rating Scale was used (Williams et al., 2008). The standardized Hedges' mean difference was used to estimate the effect sizes with Review Manager Version 5.4.1 (Cochrane Collaboration, Software Update, Oxford, UK). The results showed that CBT had a significant immediate effect on depressive symptoms with a g value of -0.42 (95% CI $[-0.69, -0.15]$, $P < .001$; Fig. 1). There is no statistically significant heterogeneity between studies (Chi$^2 = 3.84$; df $= 3$; $P = 0.28$; I$^2 = 22\%$; Fig. 1).

Randomized clinical trials for anxiety disorders ($n = 2$)

Yoshinaga et al. (2016) examined the efficacy of CBT as an adjunct to usual care compared with usual care alone in 42 patients with social anxiety disorder who remained symptomatic following antidepressant treatment through an assessor-blinded randomized controlled trial. CBT was provided over 16 weekly, 50- to 90-min individual sessions. Four clinical psychologists, one psychiatrist, one psychiatric nurse, and one psychiatric social worker delivered CBT based on the standardized manual for social anxiety disorder. All the study therapists attended a two-day intensive CBT workshop before the study, and thereafter, they also participated in/received weekly peer-group supervision as well as individual supervision with a senior supervisor. Over 16 weeks, CBT was effective as an adjunct to usual care in reducing the severity of social anxiety symptoms (Liebowitz Social Anxiety Scale [blinded assessor-rated]). Significant differences were also found in favor of the CBT group in improving depressive symptoms and functional disability. Their follow-up study (Yoshinaga et al., 2019) also confirmed that the effects of adding CBT to usual care were well-maintained until the end of the 1-year follow-up after completion of treatment.

Ishikawa et al. (2019) conducted a randomized controlled trial evaluating the efficacy of a group CBT program with a waiting-list control for children and adolescents aged 8–15 with anxiety disorders. Fifty-one children and adolescents were randomly allocated to either a group-CBT program or a waiting-list control condition. Two clinical psychologists delivered the group CBT program, which consisted of 8 weekly, 60- to 120-min group sessions. In order to increase suitability and acceptability of the Western-originated group CBT program for Japanese children/adolescents, several cultural adaptations and modifications of both context and content were made. Based on diagnostic-free proportions in the group-CBT and waiting-list control conditions, a significant improvement was found in the group-CBT at posttreatment (50% in the group-CBT condition and 12% in the waiting-list condition). Significant improvements in other parent-reported anxiety and child-reported depression were also found in the group-CBT at posttreatment, and these improvements were maintained for 6 months.

Clinical practice under the health insurance scheme in Japan

Japan uses a universal healthcare system, and patients pay only part of their total medical fees (10%–30%) according to patients' age and socioeconomic status. Patients in Japan generally do not have a consistent "general practitioner" or "family doctor," and there is no gate-keeping to advanced medical treatment; thus, everyone in Japan can freely choose their medical institution and specialists depending on their symptoms (e.g., if someone feels a certain mental problem, he/she can directly access a psychiatric clinic or a psychiatrist).

Based on findings from the above-mentioned studies and other related preliminary studies (Ono et al., 2011), in 2010, individual CBT for adult outpatients with mood disorders provided by a medical doctor was initially added to Japan's health insurance scheme. This is the first time that CBT has been recognized as an officially accepted structured psychotherapy for outpatients by the national health insurance.

However, the provision of CBT in Japan did not increase after its initial coverage by national health insurance. More specifically, the number of patients who received CBT decreased by 1.8% from FY2010 to FY2015 (10,746 in FY2010 and

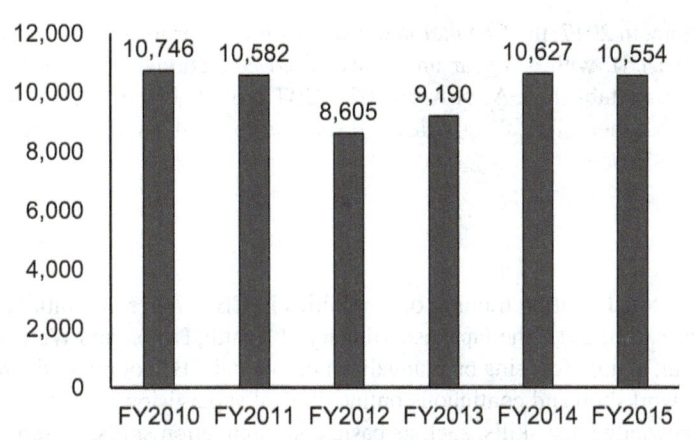

FIG. 2 The number of outpatients with mood disorders who received CBT in Japan from FY2010 to FY2015. *(Adapted from Hayashi, Y., Yoshinaga, N., Yonezawa, Y., Tanoue, H., Arimura, Y., Yoshimura, K., Yanagita, T., Aoishi, K., & Ishida, Y. (2018). Dissemination of cognitive behavioral therapy for mood disorder under the national health insurance scheme in Japan: A descriptive study using the National Database of Health Insurance Claims of Japan with special focus on Japan's southwest region. Asian Pacific Journal of Health Economics and Policy, 1, 2–10.)*

10,554 in FY2015; Fig. 2) (Hayashi et al., 2018). There was also a large regional variation in CBT utilization (Fig. 3) (Hayashi et al., 2020). There were some practical obstacles for mental health professionals in providing CBT under the health insurance scheme (Ono, 2017). For example, under the health insurance scheme pre-FY2016, the CBT provider had to be a psychiatrist, and could target only mood disorders. As for CBT providers (i.e., psychiatrists), they could only bill for individual CBT if each session lasted more than 30 min. In Japan, psychiatrists see seven outpatients per hour on average in routine clinical practice; so, these requirements made it quite difficult for Japanese psychiatrists to deliver CBT due to its low profitability and time-consuming nature (Hayashi et al., 2020). Furthermore, CBT for depression in Japan was mainly provided by clinical psychologists in routine settings (Sato & Tanno, 2012), but was provided outside the health insurance system (e.g., private counseling). This was mainly because the license for clinical psychologists was not a nationally-recognized license (i.e., it was a private-sector license), so CBT administered by clinical psychologists was not covered by the national health insurance system.

In order to disseminate CBT across Japan, from FY2016, registered nurses were added as CBT providers, and target disorders were expanded to include anxiety disorders (panic disorder and social anxiety disorder) under the national health

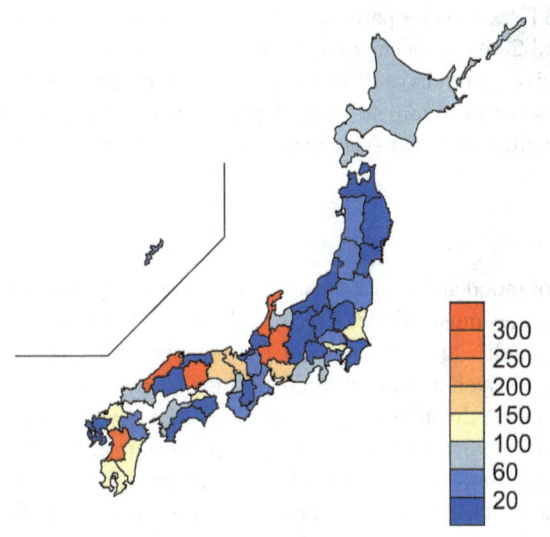

FIG. 3 Geographical distribution of standardized claim ratios for the number of outpatients with mood disorders who received CBT in Japan from FY2010 to FY2015. The color bar shows the degree of standardized claim ratio. Standardized claim ratio of 100 indicates the national mean. *(Adapted from Hayashi, Y., Yoshinaga, N., Sasaki, Y., Tanoue, H., Yoshimura, K., Kadowaki, Y., Arimura, Y., Yanagita, T., & Ishida, Y. (2020). How was cognitive behavioural therapy for mood disorder implemented in Japan? A retrospective observational study using the nationwide claims database from FY2010 to FY2015. BMJ Open, 10(5), e033365 with permission from BMJ Publishing Group Limited.)*

insurance scheme. Furthermore, in 2017, the first national qualification for professional psychologists (called *Konin-Shinrishi* [Professional Psychologists]), with a 6-year university-based educational program (4 years of undergraduate and 2 years of Master's course), was established. As of April 2021, CBT provided by professional psychologists is not a health insurance-covered treatment, but their clinical activities will soon be covered; so, it is expected that this will also improve patients' access to CBT across Japan.

CBT training in Japan

In Japan, there are several postqualification training opportunities in CBT. After the initial health insurance coverage of CBT for mood disorders, since April 2011, the Japanese Ministry of Health, Labor, and Welfare has started to organize and sponsor a CBT training system, mainly focusing on manualized individual CBT for major depressive disorder. The training consists of a two-day onsite workshop and continuous online clinical supervision. In the two-day onsite workshop, practitioners learn basic CBT knowledge and skills, such as basic communication skills, therapy structure, case conceptualization, and cognitive/behavioral techniques (e.g., behavioral activation, cognitive restructuring, problem-solving, and assertiveness training). As for continuous online clinical supervision, practitioners (supervisees) receive clinical supervision for individual cases before/after each session using video-conferencing technology (Skype). Supervisees need to submit case descriptions at the initial stage of the therapy, and after each session, they also submit a video recording of therapy sessions as well as a self-reflection sheet to their supervisors. Since FY2016, this training system has gradually expanded its target disorders to include anxiety disorders (panic and social anxiety disorders) that are also covered by the national health insurance. As of April 2021, under this training system, practitioners can receive supervision for up to two individual cases.

Outside the government-funded CBT training system, academic societies (e.g., JABCT and JACT) also provide workshops at their respective annual conferences. Generally, these are half-day or full-day workshops without ongoing clinical supervision. Of course, such a one-off workshop is insufficient to develop adequate skills in providing CBT, but it encourages and allows opportunities for an increase in the dissemination of basic knowledge of CBT among interdisciplinary practitioners.

Furthermore, several institutions provide CBT training opportunities. For example, The National Center for Cognitive Behavior Therapy and Research in Tokyo delivers a series of one-off CBT workshops for a wide variety of mental disorders. In another example, Chiba University set up a 2-year systematic CBT training course in FY2010, which is the first postqualification course for CBT in Japan (Kobori et al., 2014). This training course, Improving Access to Psychological Therapies in Chiba (Chiba-IAPT), was inspired and influenced by the IAPT in the UK. The full course of training has more than 400h including clinical supervision. Chiba-IAPT also provides placement at Chiba University Hospital, where trainees can provide individual CBT sessions for patients with mental disorders. The process and outcomes of Chiba-IAPT have been evaluated, and the initial 2-year data showed that trainee-delivered individual CBT for social anxiety disorder and other mental disorders (obsessive compulsive disorder and bulimia nervosa) led to significant reductions in symptom severity (e.g., primary disorder-specific symptom severity, depressive symptoms, and general anxiety symptoms) (Kobori et al., 2014). However, such an institution-based systematic CBT training system/course is rarely available in Japan.

Future direction

Especially since the 2000s, CBT for mood and anxiety disorders has gradually become acknowledged, accepted, and practiced in Japan; however, many obstacles must still be overcome. First, CBT is covered by Japan's health insurance only for the highly structured, face-to-face individual treatment of outpatients with mood disorder, panic disorder, and social anxiety disorder when provided by a psychiatrist or psychiatric nurse. So we need to expand the range of CBT providers (e.g., professional psychologists), target disorders (e.g., generalized anxiety disorder), treatment settings (e.g., inpatient, community, home-visit care), and delivery methods (e.g., group format, online treatment). Second, patient access to CBT is severely limited, mainly because there are only a few competent CBT therapists. To solve this problem, the Japanese government needs to invest in the further development of a systematic CBT training system. More reasonable medical fees are also required to motivate practitioners and mental healthcare facilities to use CBT under the national health insurance scheme. In addition to governmental support, both pre and postqualification CBT training for various mental health professionals are required. As Gunter and Whittal (2010) proposed, practitioners and academics need to continue accumulating real-world clinical data of CBT outcomes, and then advocate and appeal for the required funding and organizational support necessary for the wide-scale dissemination of CBT for mood and anxiety disorders across Japan.

Practice and procedures

All of the CBT protocols as well as the assessment tools that appeared in this chapter are largely based on Western-developed protocols/assessment tools. So, there would be no serious issue in using the practices and procedures of the above-mentioned protocols/assessment tools in other countries. However, readers need to careful about professional licenses (e.g., clinical psychologist in Japan is not a national license), CBT training systems/courses, and health insurance systems which vary across countries.

Mini-dictionary of terms

- *Mood disorders*. All types of depressive and bipolar disorders that affect mood, such as major depressive disorder and bipolar I/II disorders.
- *Anxiety disorders*. All types of anxiety disorders that share features of excessive fear/anxiety and related behavioral disturbances, such as social anxiety disorder, panic disorder, agoraphobia, generalized anxiety disorder, specific phobias, separation anxiety disorder, and selective mutism.
- *Japan*. Developed island country in East Asia, located in the northwest Pacific Ocean, which is divided into 47 prefectures. The total population in 2021 is over 126 million, and the country is known as the world's most super-aged society.

Key facts

Key facts of mood and anxiety disorders in Japan and CBT

- According to the World Mental Health Japan Survey (2013–15) (Ishikawa, Kawakami, & Kessler, 2016; Kawakami et al., 2016), the lifetime prevalence of mood and anxiety disorders in Japan is 7.0% and 4.2%, respectively. In terms of disability-adjusted life years among hundreds of diseases, injuries, and risk factors in Japan, depressive disorders ranked 20th, anxiety disorders ranked 31st, and self-injury ranked 12th (as of 2019) (Institution for Health Metrics and Evaluation, 2019).
- In FY2010, Japan's health care insurance scheme initially covered individual CBT for outpatients with mood disorders provided by a psychiatrist. As of FY2015, the maximum fee for CBT is 5000 JPY per session (approx. 45.5 USD).
- In the FY2016 revision of the national health care insurance scheme for CBT, psychiatric nurses were added as CBT providers, and target disorders were expanded to include social anxiety disorder, panic disorder, and other related disorders. As of FY2021, the maximum fee per session is 4800 JPY (approx. 43.6 USD) for CBT provided by a psychiatrist, and 3500 JPY (approx. 31.8 USD) for CBT provided by a psychiatric nurse.

Applications to other areas

As briefly stated before, all of the Japanese versions of CBT protocols are largely based on Western-developed versions. However, some have made cultural adaptations to their protocols/manuals, in addition to literal language translation procedures. More specifically, the Japanese version of the CBT protocol/manual for depression is based on Beck's original treatment manual (Beck, 1979); the Japanese version of the CBT protocol/manual for social anxiety disorder is based on the cognitive model of social anxiety proposed by Clark and Wells (1995); and the Japanese version of the group CBT program is based on standardized group programs in Western countries (Barrett, 1998; Kendall, 1994) (Tables 2–4 present core program components addressed in Japanese versions of the CBT protocol/manual for each disorder). During the development and standardization process for Japanese versions, several minor cultural adaptations in context and content have been made to increase treatment compliance as well as outcomes. More detailed cultural modifications and considerations in therapy can be found elsewhere (e.g., Fujisawa et al., 2010; Ishikawa et al., 2019; Yoshinaga, Kobori, Iyo, & Shimizu, 2013). Although approaches to making cultural adaptations to CBT may vary (Rathod, Phiri, & Naeem, 2019), several studies have indicated that Western-originated CBT, without major modifications, can be an acceptable and effective treatment for a culturally diverse population with mood and anxiety disorders (Griner & Smith, 2006; Hernandez, Waller, & Hardy, 2020; Jankowska, 2019). In fact, there has been no clear evidence that treatment responses are different between cultures and/or ethnic groups. Thus, Western-originated CBT for mood and anxiety disorders can be used among culturally diverse patients, and is also effective when working with native Japanese patients living outside Japan.

TABLE 2 Program components of the Japanese version of the CBT manual for depression.

Session	Agenda	Contents/techniques
1–2	Understanding problems; Psychoeducation; Motivate the patient; Socialize the patient	Assessment; Psychoeducation
3–4	Case conceptualization; Goal setting; Activating the patient	Setting treatment goals; Activity scheduling
5–6	Identifying mood and automatic thoughts	Dysfunctional thought record (triple column)
7–12	Testing automatic thoughts (Optional—dissolving interpersonal conflicts/problem solving)	Dysfunctional thought record (seven columns) (Optional—assertive training/problem solving)
13–14	Identifying schemas	Dysfunctional thought record; Discussion on schemas
15–16	Termination; Relapse prevention	Review of the therapy; Relapse prevention; Preparation for booster sessions

Treatment comprises 16 weekly 50-min individual sessions.

TABLE 3 Program components of the Japanese version of the CBT manual for social anxiety disorder.

Session	Agenda [Adolescent version]	Contents/techniques
1	Identifying problems and setting goals	Assessment; Psychoeducation; Developing an individualized version of the cognitive model
2	Discovering the unhelpful effects of safety behaviors and self-focus	Role play-based behavioral experiments, with and without safety behaviors
3	Updating distorted negative self-images	Video feedback
4	Shifting attention from an internal to an external focus	Systematic attention training
5–11	Discovering that the things one fears are less likely to happen than originally thought	Behavioral experiments
12	Modifying anticipatory worry and postevent processing	Establishing unhelpfulness of processes and then testing out inhibiting them through distraction
13	Discovering other people's view of the feared outcome	Opinion survey
14	Dealing with socially traumatic memories	Rescripting early memories linked to negative self-images in social situations
15	Dealing with the remaining assumptions	Schema work
16	Relapse prevention and ending treatment	Developing my therapy summary

Treatment comprises 16 weekly, 50- to 90-min individual sessions.

TABLE 4 Program components of the Japanese Anxiety Children/Adolescents Cognitive-Behavior Therapy program.

Session	Agenda [Adolescent version]	Contents/techniques
1	Let's think about your problem!!	Psychoeducation
2	Be aware of your feelings!	Psychoeducation
3	Be aware of your own thoughts! [Try to think flexibly!]	Cognitive restructuring
4	Try to think flexibly! [Be aware of irrational thoughts!]	Cognitive restructuring
5	Try to think even more flexibly!	Cognitive restructuring
6	Build your anxiety staircase!	Anxiety hierarchy
7	Walk up your staircase!	In vivo exposure; Relaxation
8	Review and party!	Homework exposure

Treatment comprises 8 weekly, 60- to 90-min group sessions.

Summary points

- Mood and anxiety disorders are prevalent mental disorders in Japan, and are associated with functional disability, and place a substantial economic burden on society.
- The use of CBT in Japanese clinical practice began in the late 1950s, and several academic societies have made continuous efforts to spread awareness and knowledge of CBT among mental healthcare professionals as well as the general public.
- There have been several randomized clinical trials conducted in Japan, demonstrating the efficacy of individual CBT (face-to-face or blended) and a self-help smartphone CBT app for adults with major depressive disorder and social anxiety disorder, and the efficacy of group CBT for children/adolescents with anxiety disorders.
- As of FY2021, among mood and anxiety disorders, Japan's health insurance covers CBT under the following conditions: (a) it targets outpatients with mood disorder, social anxiety disorder, and panic disorder; (b) it provides treatment in an individual face-to-face session (with each session lasting more than 30 min); and (c) psychiatrists or psychiatric nurses offer the treatment.
- There are several postqualification CBT training opportunities in Japan, such as the government-funded training system, workshops provided by academic societies, and other institution-based workshops and/or training courses.
- Japanese practitioners and academics need to continue accumulating real-world evidence of CBT outcomes for mood and anxiety disorders, and then advocate and appeal for the required funding and organizational support necessary for its wide-scale dissemination across Japan.

References

Barrett, P. M. (1998). Evaluation of cognitive-behavioral group treatments for childhood anxiety disorders. *Journal of Clinical Child Psychology, 27*(4), 459–468. https://doi.org/10.1207/s15374424jccp2704_10.

Beck, A. T. (1979). *Cognitive therapy of depression.* Guilford press.

Beck, A. T., Steer, R. A., & Brown, G. K. (1996). *Beck depression inventory manual (2nd ed.).* San Antonio, TX: The Psychological Corporation.

Clark, D. M., & Wells, A. (1995). A cognitive model of social phobia. In R. G. L. Heimberg, R. Michael, D. A. Hope, & F. R. Schneier (Eds.), *Social phobia: Diagnosis, assessment, and treatment* (pp. 69–93). New York: Guilford Press.

Fujisawa, D., Nakagawa, A., Sado, M., Kikuchi, T., Sakamoto, S., Yamauchi, K., et al. (2006). *Current status and dissemination of psychotherapies in Japan.* Japanese Ministry of Health, Labor and Welfare.

Fujisawa, D., Nakagawa, A., Tajima, M., Sado, M., Kikuchi, T., Hanaoka, M., et al. (2010). Cognitive behavioral therapy for depression among adults in Japanese clinical settings: A single-group study. *BMC Research Notes, 3,* 160. https://doi.org/10.1186/1756-0500-3-160.

Griner, D., & Smith, T. B. (2006). Culturally adapted mental health intervention: A meta-analytic review. *Psychotherapy (Chicago, IL), 43*(4), 531–548. https://doi.org/10.1037/0033-3204.43.4.531.

Gunter, R. W., & Whittal, M. L. (2010). Dissemination of cognitive-behavioral treatments for anxiety disorders: Overcoming barriers and improving patient access. *Clinical Psychology Review, 30*(2), 194–202. https://doi.org/10.1016/j.cpr.2009.11.001.

Hayashi, Y., Yoshinaga, N., Sasaki, Y., Tanoue, H., Yoshimura, K., Kadowaki, Y., et al. (2020). How was cognitive behavioural therapy for mood disorder implemented in Japan? A retrospective observational study using the nationwide claims database from FY2010 to FY2015. *BMJ Open, 10*(5). https://doi.org/10.1136/bmjopen-2019-033365, e033365.

Hayashi, Y., Yoshinaga, N., Yonezawa, Y., Tanoue, H., Arimura, Y., Yoshimura, K., et al. (2018). Dissemination of cognitive behavioral therapy for mood disorder under the national health insurance scheme in Japan: A descriptive study using the National Database of health insurance claims of Japan with special focus on Japan's southwest region. *Asian Pacific Journal of Health Economics and Policy, 1,* 2–10. https://doi.org/10.6011/apj.2018.02.

Hernandez, M. E., Waller, G., & Hardy, G. (2020). Cultural adaptations of cognitive behavioural therapy for Latin American patients: Unexpected findings from a systematic review. *The Cognitive Behaviour Therapist, 13.* https://doi.org/10.1017/S1754470X20000574, e57.

Ishikawa, H., Kawakami, N., & Kessler, R. C. (2016). Lifetime and 12-month prevalence, severity and unmet need for treatment of common mental disorders in Japan: Results from the final dataset of world mental health Japan survey. *Epidemiology and Psychiatric Sciences, 25*(3), 217–229. https://doi.org/10.1017/s2045796015000566.

Ishikawa, H., Tachimori, H., Takeshima, T., Umeda, M., Miyamoto, K., Shimoda, H., et al. (2018). Prevalence, treatment, and the correlates of common mental disorders in the mid 2010's in Japan: The results of the world mental health Japan 2nd survey. *Journal of Affective Disorders, 241,* 554–562. https://doi.org/10.1016/j.jad.2018.08.050.

Ishikawa, S. I., Hida, N., Kishida, K., Ueda, Y., Nakanishi, Y., & Kaneyama, Y. (2016). The empirical review of academic activities regarding cognitive behavioral therapies for children and adolescents in Japan: Before and after the world congress of behavioral and cognitive therapies in 2004, Kobe. *The Japanese Journal of Cognitive Psychology, 9,* 34–43.

Ishikawa, S. I., Kikuta, K., Sakai, M., Mitamura, T., Motomura, N., & Hudson, J. L. (2019). A randomized controlled trial of a bidirectional cultural adaptation of cognitive behavior therapy for children and adolescents with anxiety disorders. *Behaviour Research and Therapy, 120.* https://doi.org/10.1016/j.brat.2019.103432, 103432.

Jankowska, M. (2019). Cultural modifications of cognitive behavioural treatment of social anxiety among culturally diverse clients: A systematic literature review. *The Cognitive Behaviour Therapist, 12*, e7. https://doi.org/10.1017/S1754470X18000211.

Kawakami, N., Tachimori, H., & Takeshima, T. (2016). *Report of the World Mental Health Japan Survey 2nd (2013–2015)*. Retrieved April 1, 2021 from: wmhj2.jp/WMHJ2-2016R.pdf.

Keio University. (2011). *Estimation of social cost of mental disorders*. Retrieved April 1, 2021 from: https://www.mhlw.go.jp/bunya/shougaihoken/cyousajigyou/dl/seikabutsu30-2.pdf.

Kendall, P. C. (1994). Treating anxiety disorders in children: Results of a randomized clinical trial. *Journal of Consulting and Clinical Psychology, 62*(1), 100–110. https://doi.org/10.1037//0022-006x.62.1.100.

Kobori, O., Nakazato, M., Yoshinaga, N., Shiraishi, T., Takaoka, K., Nakagawa, A., et al. (2014). Transporting cognitive behavioral therapy (CBT) and the improving access to psychological therapies (IAPT) project to Japan: Preliminary observations and service evaluation in Chiba. *The Journal of Mental Health Training, Education and Practice, 9*(3), 155–166. https://doi.org/10.1108/JMHTEP-10-2013-0033.

Mantani, A., Kato, T., Furukawa, T. A., Horikoshi, M., Imai, H., Hiroe, T., et al. (2017). Smartphone cognitive behavioral therapy as an adjunct to pharmacotherapy for refractory depression: Randomized controlled trial. *Journal of Medical Internet Research, 19*(11). https://doi.org/10.2196/jmir.8602, e373.

McHugh, R. K., Whitton, S. W., Peckham, A. D., Welge, J. A., & Otto, M. W. (2013). Patient preference for psychological vs pharmacologic treatment of psychiatric disorders: A meta-analytic review. *Journal of Clinical Psychiatry, 74*(6), 595–602. https://doi.org/10.4088/JCP.12r07757.

Nakagawa, A., Mitsuda, D., Sado, M., Abe, T., Fujisawa, D., Kikuchi, T., et al. (2017). Effectiveness of supplementary cognitive-behavioral therapy for pharmacotherapy-resistant depression: A randomized controlled trial. *Journal of Clinical Psychiatry, 78*(8), 1126–1135. https://doi.org/10.4088/JCP.15m10511.

Nakao, S., Nakagawa, A., Oguchi, Y., Mitsuda, D., Kato, N., Nakagawa, Y., et al. (2018). Web-based cognitive behavioral therapy blended with face-to-face sessions for major depression: Randomized controlled trial. *Journal of Medical Internet Research, 20*(9). https://doi.org/10.2196/10743, e10743.

Nishi, D., Ishikawa, H., & Kawakami, N. (2019). Prevalence of mental disorders and mental health service use in Japan. *Psychiatry and Clinical Neurosciences, 73*(8), 458–465. https://doi.org/10.1111/pcn.12894.

Ono, Y. (2017). Cognitive behavior therapy in Japan. *Journal of Japanese Association of Psychiatric Hospitals, 36*, 98–102.

Ono, Y., Furukawa, T. A., Shimizu, E., Okamoto, Y., Nakagawa, A., Fujisawa, D., et al. (2011). Current status of research on cognitive therapy/cognitive behavior therapy in Japan. *Psychiatry and Clinical Neurosciences, 65*(2), 121–129. https://doi.org/10.1111/j.1440-1819.2010.02182.x.

Rathod, S., Phiri, P., & Naeem, F. (2019). An evidence-based framework to culturally adapt cognitive behaviour therapy. *The Cognitive Behaviour Therapist, 12*. https://doi.org/10.1017/S1754470X18000247, e10.

Sato, H., & Tanno, Y. (2012). The effect of cognitive behavioral therapy for depression delivered by Japanese psychologists: A systematic review. *Japanese Journal of Behavior Therapy, 38*(3), 157–167. https://doi.org/10.24468/jjbt.38.3_157.

Thornicroft, G., Chatterji, S., Evans-Lacko, S., Gruber, M., Sampson, N., Aguilar-Gaxiola, S., et al. (2017). Undertreatment of people with major depressive disorder in 21 countries. *British Journal of Psychiatry, 210*(2), 119–124. https://doi.org/10.1192/bjp.bp.116.188078.

Watanabe, N., Furukawa, T. A., Shimodera, S., Katsuki, F., Fujita, H., Sasaki, M., et al. (2015). Cost-effectiveness of cognitive behavioral therapy for insomnia comorbid with depression: Analysis of a randomized controlled trial. *Psychiatry and Clinical Neurosciences, 69*(6), 335–343. https://doi.org/10.1111/pcn.12237.

Watanabe, N., Furukawa, T. A., Shimodera, S., Morokuma, I., Katsuki, F., Fujita, H., et al. (2011). Brief behavioral therapy for refractory insomnia in residual depression: An assessor-blind, randomized controlled trial. *Journal of Clinical Psychiatry, 72*(12), 1651–1658. https://doi.org/10.4088/JCP.10m06130gry.

Williams, J. B., Kobak, K. A., Bech, P., Engelhardt, N., Evans, K., Lipsitz, J., et al. (2008). The GRID-HAMD: Standardization of the Hamilton depression rating scale. *International Clinical Psychopharmacology, 23*(3), 120–129. https://doi.org/10.1097/YIC.0b013e3282f948f5.

Yoshinaga, N., Kobori, O., Iyo, M., & Shimizu, E. (2013). Cognitive behaviour therapy using the Clark & amp; Wells model: A case study of a Japanese social anxiety disorder patient. *The Cognitive Behaviour Therapist, 6*. https://doi.org/10.1017/S1754470X13000081, e3.

Yoshinaga, N., Kubota, K., Yoshimura, K., Takanashi, R., Ishida, Y., Iyo, M., et al. (2019). Long-term effectiveness of cognitive therapy for refractory social anxiety disorder: One-year follow-up of a randomized controlled trial. *Psychotherapy and Psychosomatics, 88*(4), 244–246. https://doi.org/10.1159/000500108.

Yoshinaga, N., Matsuki, S., Niitsu, T., Sato, Y., Tanaka, M., Ibuki, H., et al. (2016). Cognitive behavioral therapy for patients with social anxiety disorder who remain symptomatic following antidepressant treatment: A randomized, assessor-blinded, controlled trial. *Psychotherapy and Psychosomatics, 85*(4), 208–217. https://doi.org/10.1159/000444221.

Institution for Health Metrics and Evaluation. (2019). Global burden of disease calculator. Retrieved April 1, 2021 from: https://vizhub.healthdata.org/gbd-compare/.

Chapter 30

Cognitive behavioral therapy for posttraumatic stress disorder in Pakistan

Anwar Khan
Department of Psychology and Management Sciences, Khushal Khan Khattak University, Karak, Khyber Pakhtunkhwa, Pakistan

Abbreviations

CBT cognitive behavioral therapy
PTSD posttraumatic stress disorder
TF-CBT trauma-focused CBT

Introduction

Posttraumatic stress disorder (PTSD) is psychological disorder, which is triggered by experiencing or witnessing traumatic events. PTSD is one of the most prevalent mental health problems around the world. Recent epidemiological studies have reported relatively high global prevalence rates of PTSD among different populations, for example it was 24% among health workers (Li, Scherer, Felix, & Kuper, 2021) and 23% among general public (Yuan et al., 2021). PSTD is also prevalent in Pakistan, and Shah et al. (2020) reported 48.61% prevalence rate of PTSD in Pakistan. It is primarily because Pakistani society is full of torture, violence, deprivation, and conflicts. Moreover, the ongoing war on terrorism has exposed the Pakistani people to an unending series of traumatic experiences. Mental Health Practitioners in Pakistan treat PTSD mostly with psychotherapy mostly because Affective Disorders like PTSD can be successfully treated with psychotherapeutic interventions as compared to pharmacotherapeutic. Keeping in mind the importance of research on the symptomology and treatment of PTSD in Pakistan, the current study has reviewed different aspects related to the symptomology and treatment of PTSD in Pakistan by using a non-systematic narrative review technique.

Posttraumatic stress disorder: A brief overview and history

Posttraumatic stress disorder (PTSD) is caused by witnessing or experiencing traumatic events. It has a broad range of symptoms including distressed recollection of traumatic events (Jarero, Artigas, Uribe, & Miranda, 2014); hyper-vigilance, problems in concentration (Hardy et al., 2020); numbing of body sensation (Kuester et al., 2017); and feelings of detachment (McNally, 2016). PTSD also has certain comorbid symptoms, like, depression & anxiety (Price & van Stolk-Cooke, 2015); paranoid disorder (de Bont et al., 2016); insomnia (Raboni, Alonso, Tufik, & Suchecki, 2014); substance use & addiction disorder (Najavits, 2014); and panic disorder (Katz et al., 2017).

PTSD can be clinically diagnosed through two criteria, one is the Diagnostic and Statistical Manual of Mental Disorders, Fifth Edition (DSM-5) criteria and other is the International Classification of Diseases 11th (ICD-11) guidelines for PTSD. According to the DSM-5, PTSD can be diagnosed in a person if he or she has experienced or witnessed traumatic event, like serious injury or death (Criterion: A); reexperiencing intrusive distressing memories, flashbacks & recurrent distressing dreams (Criterion: B); avoidance of internal & external memories or thoughts of trauma (Criterion: C); alterations in cognitions & mood with inability to experience positive emotions (Criterion: D); and irritability, anger, self-destruction, & exaggerated startle response (Criterion: E). Whereas the recent diagnostic guidelines of ICD-11 have adopted a more narrow focus of PTSD by including few symptoms like reexperiencing of the painful event(s); avoidance of painful thoughts; and feelings of constant threat, which is reflected by various types of arousal. The ICD-11 has also introduced the complex PTSD, which has additional symptoms like, conflicts in self-identity (negative self-concept), emotional reactivity with violent outbursts, and serious difficulties in personal relationships.

Both PTSD and complex PTSD can be emerge after experiencing or witnessing different kinds of traumatic events, like, sexual abuse (Priebe et al., 2018); war, torture & genocide (Ibrahim & Hassan, 2017); violence from close relationship, rape, imprisonment or taken hostage, natural catastrophe & road accidents (Lassemo, Sandanger, Nygård, & Sørgaard, 2017); suffering from serious illness (Lewis et al., 2019); and sociodemographic problems, like, lower socioeconomic status, minority race or lower ethnic status & financial losses (Verbitsky, Dopfel, & Zhang, 2020).

The modern history of PTSD can be traced back to the early 20th century, when it was first identified among the soldiers during the World War: I, however at that time it was termed as Shell Shock (Leese, 2002). During Korean and Vietnam wars, similar symptoms were experienced by the soldiers, and it was termed as Combat or Battle Exhaustion and Combat Neurosis (Jones & Wessely, 2005). During 1970s, PTSD was studied among the civilian population. Two famous psychologists, that is., Ann Wolbert Burgess and Lynda Lytle Holmstrom formed a term "Rape Trauma Syndrome" after studying the rape victims (Burgess & Holmstrom, 1974). During late 1980s, the PTSD was finally included in the Diagnostic and Statistical Manual of Mental Disorders-III and in this way it was recognized as psychiatric illness (Regier, Kuhl, & Kupfer, 2013). More recently, PTSD has been included in the International Classification of Diseases 11th Revision (Jowett, Karatzias, Shevlin, & Albert, 2020).

Cognitive behavioral therapy: As a first line treatment choice

Cognitive Behavioral Therapy (CBT) is a psychosocial and evidence based psychotherapeutic approach that combine both psychological and social factors for the treatment of common mental health problems (Beck, 2020). CBT is a multidimensional approach that focuses on changing the cognitive distortions through improving emotional regulations, boosting positive behaviors and developing individual coping skills (Beck & Carlson, 2012). Recently CBT is added with a more transdiagnostic-based approach that covers both the individual and environmental aspects in treating the mental health problems (Dobson & Dozois, 2019). Moreover, the positive psychology aspects with solution focused brief therapy approach have also been added into the CBT (Bannink & Geschwind, 2021).

CBT as a psychotherapeutic approach has relatively rigorous criteria of evidence-based framework, that is why, CBT has been used as a reference treatment due to its overall efficacy in the treatment of different mental health problems (David, Cristea, & Hofmann, 2018). With the passage of time, the CBT became not only a therapy, but also an "umbrella term" that covered all cognitive based psychotherapies (Kazantzis et al., 2019). These therapies include Cognitive Therapy, Dialectical Behavior Therapy, Rational Emotive Behavior Therapy, and Cognitive Processing Therapy (Ruggiero, Spada, Caselli, & Sassaroli, 2019). This blend of psychotherapies constituted the "Third Wave CBT," which is expected to be more efficacious in terms of its applicability in the treatment of various mental health problems (Hayes & Hofmann, 2017).

As earlier stated CBT focuses on changing the cognitive distortions with more effective & real thoughts. This is done through improving emotional regulations and reducing of the emotional distress (Beck, 2020). In this whole process, the psychotherapy related skills of the psychologist matters a lot. Moreover, the individual coping skills of the patients also have a role, since the more a patient is self-efficacious and resilient the less he/she will be affected by the cognitive and emotional distortions.

The whole CBT is typically implemented in the following six major phases (Turk & Flor, 2013):

1. Psychological Assessment or Diagnosis;
2. Reconceptualization;
3. Skill Acquisition;
4. Skill Consolidation;
5. Generalization & Maintenance;
6. Posttreatment Assessment Follow-up.

The above mentioned are general phases of CBT, whereas there are different protocols of CBT, which are designed according to the nature of mental illness, like, for the treatment of PTSD, Mueser et al. (2008) and Cohen, Mannarino, Kliethermes, and Murray (2012) have proposed a trauma-focused CBT protocol, which has following eight phases:

1. Introduction
2. Crisis Plan Review;
3. Psycho-education, Part one;
4. Breathing Retraining;
5. Psycho-education, Part two;

6. Cognitive Restructuring, Part one;
7. Cognitive Restructuring, Part two;
8. Generalization Training and Termination.

Pakistan aspects of posttraumatic stress disorder

Pakistan is situated in the southern region of Asia and it is the fifth populous country in the world (Qureshi, 2020). Pakistan has heterogeneous demographics with many ethnic groups and languages, although the official languages are Urdu and English (Karim, Saeed, & Akber, 2019). Due to large population, and lack of proper infrastructure & governance, Pakistan has been facing multiple challenges, including terrorism, poverty, illiteracy, corruption, unfair distribution of resources and other similar like problems (Nawaz, Khan, Batool, & Rasool, 2021). These challenges have badly hampered the routine lives, mental health and well-being of the common persons in Pakistan. For such reasons, the status of mental health in Pakistan has remained a topic of debate since the last few years (Javed, Khan, Nasar, & Rasheed, 2020). Mental health facilities are provided by the Government of Pakistan, moreover, the private sector is also engaged in providing mental health care to the common person, however, such facilities are very scarce. There are only 400 psychiatrists in Pakistan, which are dealing around millions of patients (Siddiqi, 2020).

Due to the prevalent socioeconomic situation in Pakistan, the mental health status of a common person in Pakistan is not that much good. This situation has been further exacerbated by the lack of mental health facilities. People in Pakistan are suffering from different mental health problems, for example, anxiety & depressive disorders (Farooq, Khan, Zaheer, & Shafique, 2019); schizophrenia (Hussain et al., 2019); obsessive compulsive disorder (Asghar et al., 2020), and psychosomatic disorders (Qayyum et al., 2021). Among all these mental health problems, the prevalence rate of PTSD is high, which is evident from the results of recent research studies, like for example, Shah et al. (2020) reported 48.61% prevalence rate of PTSD in Pakistan. This indicates that PTSD is one of the common health issues in Pakistan (Junaid, Haar, & Brougham, 2020). It is primarily because Pakistani society is full of acts and events related to torture, violence, deprivation, and conflicts. Moreover, the ongoing war on terrorism has exposed the Pakistani people to an unending series of traumatic experiences. The review of existing literature reveals interesting facts about the prevalence of PTSD in Pakistan. Details show that there were various sources of PTSD, including community violence, terrorism, natural disaster, marital conflicts and domestic violence, etc., as shown in Table 1. Such diverse sources of PTSD denotes that the prevalence of PTSD in Pakistan is because of existing of wide variety causes, ranging from the individual causes, like for example, sexual abuse, to

TABLE 1 Sources of PTSD in Pakistan.

Sources of PTSD	Reference
COVID-19 related crisis	Ashraf et al. (2020)
Burns Wound	Bibi, Kalim, and Khalid (2018)
Parents Separation	Ishfaq and Kamal (2020)
Community Violence	Khan et al. (2016)
Work-related trauma	Shah et al. (2020)
Physical and Emotional Abuse	Ishfaq and Kamal (2020)
War related Internal Displacement	Ali, Farooq, Bhatti, and Kuroiwa (2012)
Natural Disaster (Earthquake)	Ahmad and Hussain (2020)
Terrorism & Terrorist Attacks	Junaid et al. (2020)
Marital Conflict and Bereavement	Agha (2020)
Witnessing Violence	Ishfaq and Kamal (2020)
Domestic Violence	Latif et al. (2021)
Physical Disability and Organ Loss	Babar and Dildar (2021)
Sexual Abuse (Child)	Ashraf, Niazi, Masood, and Malik (2019)
Serious Accident	Ishfaq and Kamal (2020)

TABLE 2 Prevalence of PTSD in Pakistan.

Type of affected population	Gender (Age)	Reference
Incarcerated Convicts	Male (19 to 75 years)	Ishfaq and Kamal (2020)
Burn Survivors	Female (15 to 25 years)	Bano and Naz (2020)
Retired Soldiers (Veterans)	Male (45 to 75 years)	Shehzad and Ahsan (2020)
Terrorism Affected Population	Male & Female (25 to 55 years)	Zarak et al. (2020)
Victims of COVID-19	Male & Female (18 to 53 years)	Ashraf et al. (2020)
Family Conflict Victims	Male & Female (Note given)	Agha (2020)
Natural Disaster Victims	Male & Female (18 to 45 years)	Ali et al. (2012)
Child Abuse Victims	Male & Female (11 to 18 years)	Ashraf et al. (2019)
University Students	Male & Female (18 to 23 years)	Khan et al. (2016)
Journalists	Male & Female (26 to 45 years)	Shah et al. (2020)
University Employees	Male & Female (18 to 34 years)	Zaman and Munib (2020)
Emergency Medical Services Staff	Male & Female (25 to 46 years)	Kerai et al. (2017)
Internally Displaced Population	Mothers (20 to 35 years)	Rashid et al. (2020)

the organizational causes, like for example, work-related trauma, and finally, to the community level, like for example, community violence. Such diverse sources also denote that the social environment in Pakistan is getting full of different traumatic events, which ultimately creates such diverse sources of PTSD.

Furthermore, the prevalence of PTSD was among different population, including imprisoned convicts, military veterans, natural disaster victims details and internally displaced population, etc., as shown in Table 2. It means that the PTSD is a very common and wide spread mental illness, which occurs among different population. Further details showed that the symptomology of PTSD varied across the populations, and most of the common symptoms of PTSD included reexperiencing of traumatic thought, avoidance, hyperarousal, amnesia, diminished interest in life, detachment, emotional numbing, irritability, difficulty concentrating, and nervousness. The main symptoms of PTSD also existed with certain comorbid symptoms, including depression, anxiety, problematic substance abuse, and psychosomatic problems, as shown in Table 3.

The PTSD had badly affected the routine working & social lives of Pakistani people and it had also affected the economic life of common persons in Pakistan. Results reported from the published studies show that PTSD had badly affected the working lives by seriously hampering the performance and overall work productivity of common persons in Pakistan. For example Khan et al. (2016) found that the PTSD had affected the academic performance and overall campus life of the University students in Pakistan.

Similarly, Kerai et al. (2017) found that the PTSD had affected the working performance and overall work output of the Emergency Medical Service Personnel in Pakistan. PTSD also place huge medical and other costs on its victims and during the year 2016 the economic costs of different mental illnesses (including PTSD and its comorbidities) was around Rs: 250,483/- million (Malik & Khan, 2016). Finally, the social costs of PTSD include poor marital & family life (Naz & Malik, 2018), stigma and negative beliefs (Shah et al., 2020), lack of acceptance in community (Shah, Ahmad, & Hallahan, 2019) and poor social support from the family, colleagues and friends (Ashfaq et al., 2018). The poor social support is mostly because of negative beliefs and stigma associated with the mental illness.

Cognitive behavioral therapy in Pakistan

As earlier stated that the mental health of a common person in Pakistan is not that much good, mostly because of lack of awareness about mental health illness and due to lack of mental health facilities. Overall, there are only 400 psychiatrists in Pakistan (Siddiqi, 2020) and 478 consultant clinical psychologists (WHO, 2019), which are dealing around millions of patient in Pakistan. The mental healthcare professionals are trying their best to provide good quality mental health treatment facilities to the citizens of Pakistan. The available treatment facilities are either pharmacotherapeutic or psychotherapeutic,

TABLE 3 Sources of PTSD in Pakistan.

Main symptoms of PTSD	Comorbid symptoms of PTSD	Reference
Reexperiencing, Avoidance, Hyper-arousal, Numbing, Arousal, Self-persecution	Nil	Bibi et al. (2018)
Reexperiencing, Avoidance, Hyper-arousal, Helplessness, Intense Fear	Nil	Khan et al. (2016)
Reexperiencing, Avoidance, Hyper-arousal, Amnesia, diminished Interest in Life, Detachment, Emotional Numbing, Trouble Sleeping, Irritability, Difficulty Concentrating, Nervousness	Nil	Rashid et al. (2020)
Dissociation, Repetitive Thoughts, Depersonalization	Anxiety, Substance Abuse, Psychosomatic Disorder, Depression, Mania, Psychosis	Khan et al. (2016)
Reexperiencing, Avoidance, Hyper-arousal, Helplessness, Hopelessness, Self-blame, Preoccupation with Danger, LOSS of Belongings	Anxiety, Depression, Social Dysfunction, Somatic problems	Chung, Jalal, and Khan (2014)
Intrusion, Avoidance, Hyper-arousal	Anxiety, Depression, Insomnia, Problematic Substance Abuse	Kerai et al. (2017)

however, mental healthcare professionals sometimes prefer psychotherapeutic interventions, especially in a situation where the pharmacotherapeutic interventions are not that much effective or costly (Lee et al., 2016). The cost factor is very important in the underdeveloped eastern countries, including Pakistan, because, most of the patients in Pakistan have less buying power, therefore they are sometimes not able to continue expensive multinational brand medications.

There are different psychotherapeutic interventions, including Cognitive Behavioral Therapy, Client-centered Therapy, Eye Movement Desensitization and Reprocessing. Among these all psychotherapeutic choices, the Cognitive Behavioral Therapy is most widely used, mainly due to its efficacy and overall simplicity of use (David et al., 2018). Presently, the underdeveloped eastern countries, including Pakistan are working on making CBT available to the local patients for the treatment of different mental health problems, therefore, it is expected that in near future CBT will be one of the most commonly adopted psychotherapeutic interventions in Pakistan (Masud, 2019). The Government of Pakistan with the help of Higher Education Commission and Pakistan Medical and Dental Counsel are introducing Clinical Psychology related programs and diplomas both at undergraduate and postgraduate levels in the different universities and colleges in Pakistan. Moreover, Pakistani Psychologists have established their own societies, like for example, Pakistan Psychological Society and Pakistan Association of Clinical Psychologists. All such efforts are expected to promote the field of Clinical Psychology to a greater extent (Naeem, Gobbi, Ayub, & Kingdon, 2017).

Despite these efforts, there are certain challenges, which might hamper all such developments. These challenges include, first, the unavailability of any accreditation or licensing body for the registration of Clinical Psychologists in Pakistan. Second, lack of established Psychology departments in the teaching hospitals of Pakistan (there are only Psychiatry departments). Third, there are no locally developed protocols for the psychotherapeutic interventions, like CBT or EMDR, which sometimes create serious problems during the process of treatment because some of the aspects of the psychotherapeutic interventions could not be easily culturally adopted. Fourth, the buying power of patients in Pakistan is low, so sometimes it might be difficult for the patients to pay the fees of 12 or 14 sessions, which might prevent the popularity of CBT or other similar interventions in Pakistan. Finally, there is a relatively low level of Mental Health Literacy in Pakistan and most of the people believe in supernatural origins of mental illnesses (Khan & Anwar, 2018).

CBT as an evidence based psychotherapeutic interventions is used for the treatment of different mental health issues in Pakistan and the existing review of literature documents various mental health issues that have been treated by using CBT, details given in Table 4.

Moreover, CBT has been used as a first line treatment among different populations, like for example, women with menstrual problems (Babar & Ahmad, 2021), females who had experienced and witnessed trauma (Bukhari & Afzal, 2020), school children with social anxiety (Amin et al., 2020), females who had suffered from domestic violence (Latif et al., 2021), healthcare workers affected by COVID-19 crisis (Naeem et al., 2020), disabled children (Nasir, 2020) and mothers with depressive symptoms (Rahman et al., 2008).

Since the basic protocols of CBT were developed in western countries, therefore, to be effective in the non-western cultures various adaptations and changes need to be made in the originally developed protocols of CBT. However, due

TABLE 4 Symptoms treated with CBT in Pakistan.

Main symptoms	Comorbid symptoms	Reference
PTSD	Nil	Bukhari and Afzal (2020)
Erectile Dysfunction	Anxiety, Depression,	Khan, Amjad, and Rowland (2017)
Depression, Anxiety, Somatic Complaints	Nil	Naeem, Gul, et al. (2015)
Conversion Disorder	Nil	Choudhry, Munawar, Khaiyom, and Lian (2020)
Learning Related Disabilities	Nil	Nasir (2020)
Social Anxiety	Nil	Amin, Iqbal, Naeem, and Irfan (2020)
Body Image Dissatisfaction	Nil	Babar and Ahmad (2021)
Maternal Depression	Anxiety	Rahman et al. (2008)
COVID-19 Associated Fear	Depression	Naeem, Irfan, and Javed (2020)
Domestic Violence, Depression, Anxiety	Nil	Latif et al. (2021)

to lack of locally adaptive CBT protocols, most of the psychotherapists in Asian countries (including Pakistan) apply the western CBT protocols, which may undermine the efficacy of CBT (Naeem et al., 2021). Realizing this challenge, researchers in Pakistan have done different efforts to develop some culturally adaptive protocols of CBT, details are given in Table 5.

Cognitive behavioral therapy for posttraumatic stress disorder in Pakistan

The prevalence of PTSD is high in Pakistan and it was reported 48.61% during year 2020 (Shah et al., 2020). The high occurrence rate of PTSD in Pakistan indicates that PTSD is one of the common health issues in Pakistan (Junaid et al., 2020). It is primarily because Pakistani society is full of torture, violence, deprivation, and conflicts. Moreover, the ongoing war on terrorism has exposed the Pakistani people to an unending series of traumatic experiences. The available treatment facilities are either pharmacotherapeutic or psychotherapeutic. PTSD affected patients in Pakistan are mostly treated with psychotherapy (Khan et al., 2016). It is because Affective Disorders like PTSD can be more successfully treated with psychotherapeutic intervention as compared to pharmacotherapeutic (Stewart & Wrobel, 2010). The review of existing literature on the application of CBT for the treatment of PTSD in Pakistan revealed that most of the studies have adopted the Trauma-Focused CBT protocols for the treatment of symptoms PTSD like, Reexperiencing, Avoidance, Hyper-arousal, Changes in Cognitions, Negative Beliefs, Feeling of Detachment, and Difficulty in Falling Sleep. Moreover, most of the victims of PTSD were females and children, who had suffered from domestic violence and marital conflicts (Bukhari & Afzal, 2020; Latif et al., 2021).

Psychologists and Mental Health Professionals in Pakistan face different challenges and issues while implementing CBT and other similar evidence based psychotherapies. First, the basic protocols of CBT were developed in western

TABLE 5 Culturally adaptive CBT protocols in Pakistan.

Type of CBT protocol	Symptoms treated	Reference
Culturally Adaptive Trauma-Focused CBT	PTSD	Latif et al. (2021)
Culturally Adapted CBT-based Guided Self-help	Social Anxiety	Amin et al. (2020)
Culturally adapted Psycho-education intervention	Bipolar Disorder	Husain et al. (2017)
Brief Culturally Adapted CBT	Depression	Naeem, Gul, et al. (2015)
Culturally adapted CBT-P	Psychosis	Naeem, Saeed, et al. (2015)
Culturally Adaptive CBT-OCD	Obsessive Compulsive Disorder	Aslam, Irfan, and Naeem (2015)

countries; therefore, locally adaptive CBT protocols need to be developed in Pakistan (Naeem et al., 2011). Second, people in Pakistan have lack awareness about their mental health and overall there is low level of Mental Health Literacy due to which treatment compliance is low and dropout rate is high. Due to low Mental Health Literacy, people have negative beliefs about mental illnesses, and they attribute mental illnesses to supernatural powers. People prefer to visit faith healer and magicians for treating their mental illnesses (Khan & Anwar, 2018). Third, there is an overall shortage of Mental Health Professionals, and millions of patients are treated by only 400 psychiatrists (Siddiqi, 2020) and 478 consultant clinical psychologists (WHO, 2019) in Pakistan, therefore shortage of Mental Health Professionals in Pakistan is one of major setback in the success of mental health sector (Masud, 2019). Fourth, Mental Health Professionals, particularly, psychologists face different hurdles in therapy, like, most of the patients are poor and they cannot afford treatment expenses, especially of 12 to 14 sessions of CBT. Due to the less paying power of patients, most of the patients visit government hospitals, which obviously means that the psychotherapists have to deal a higher than usual number of patients. Such workload may undermine the performance of psychotherapists (Naeem et al., 2017). Finally, patients do not show compline in taking homework, which is one of the important aspects of CBT.

Conclusion

PSTD is the most prevalent mental health problem in Pakistan, primarily because Pakistani society is full of torture, violence, deprivation, and conflicts. Moreover, the ongoing war on terrorism has exposed the Pakistani people to an unending series of traumatic experiences. The available treatment facilities are either pharmacotherapeutic or psychotherapeutic. However, Mental Health Practitioners in Pakistan treat PTSD mostly with psychotherapy (Khan et al., 2016). It is because Affective Disorders like PTSD can be more successfully treated with psychotherapeutic intervention as compared to pharmacotherapeutic (Stewart & Wrobel, 2010). Due to its wider efficacy, CBT is one of the first line treatment choices in Pakistan for PTSD and other commonly occurring Affective Disorders. The Government of Pakistan should work proactively for the development of different evidence based psychotherapies, including CBT in Pakistan.

Practice and procedure

The current study has adopted a non-systematic narrative review methodology. The non-systematic narrative review is a less structured method of collection information about the topic of interest since it allow researchers to collect information with freedom (Ferrari, 2015). The non-systematic narrative review was conducted in three steps. In the first step, the key terms and key headings were decided, which were searched in the major online databases, like Google Scholar, PubMed and ScienceDirect. In the next step, all relevant studies and reports were downloaded and then reviewed for the collection of related information. In the final step, the major sections of review were decided according to the collected information, and write up was completed in such a way that in the beginning of review the introductory sections were given that were followed by sections containing specific information. Whereas at the end of review the conclusion was given.

Mini-dictionary of terms

Posttraumatic Stress Disorder (PTSD): It is a mental health disorder that occurs among the people who have either experienced or witnessed traumatic events, such as accidents, natural disasters, death and sexual violence.

Cognitive Behavioral Therapy (CBT): It is an evidence-based psychotherapeutic intervention that combines the psychological and social factors for the treatment of common mental health problems. Recently CBT is added with a more trans-diagnostic-based approach that covers both the individual and environmental aspects in treating the mental health problems.

Trauma Focused Cognitive Behavioral Therapy (TF-CBT): It is one of the types of CBT that is used for addressing the treatment needs of patients suffering from PTSD.

Key facts

1. PTSD was first discovered during the First World War and later observed in the Second World War, Korean War, and Vietnam War.
2. PTSD is most commonly occurring mental health problem with global prevalence rate of 24% among health workers and 23% rate among general public.
3. PTSD is also very commonly occurring mental health problem in Pakistan with prevalence rate of 48.6%.

4. CBT is a first line treatment choice for PTSD because recently CBT is added with a more trans-diagnostic-based approach that covers both the individual and environmental aspects in treatment.
5. Mental health of a common person in Pakistan is not that much good, mostly because of lack of awareness about mental health illness and due to lack of mental health facilities.
6. Pakistan has only 400 psychiatrists and 478 consultant clinical psychologists who are dealing around millions of patient in Pakistan.
7. Presently, Pakistani researchers are working on making CBT available to the local patients for the treatment of different mental health problems, therefore, it is expected that in near future CBT will be one of the most commonly adopted psychotherapeutic interventions in Pakistan.

Applications to other areas

This review has provided a bird eye view of the key information about the symptomology and treatment aspects of PTSD in Pakistan. Findings obtained from this review will be significant to the researchers in the field of Clinical Psychology and to the mental health practitioners, since this review can provide to them valuable insights regarding the Pakistan aspects of PTSD and CBT. Findings of this review will be also significant to the administrators of mental health care institutions, who can understand the problems and challenges faced by the psychologists in dealing patients and in implementing the evidence based psychotherapeutic interventions in Pakistan. Finally, findings of this review will be important to the policy makers in the healthcare sector of Pakistan, who also can understand challenges faced by the healthcare practitioners in Pakistan. In this way, the administrators can provide a conducive working environment to the healthcare practitioners in Pakistan and the policymakers can work on formulating good policies for the development of healthcare sector in Pakistan.

Summary points

1. PTSD is one the most commonly occurring mental health problem around the world.
2. PTSD is more common among the victims of social and domestic violence.
3. PSTD is the most prevalent mental health problem in Pakistan, primarily because of the torture, violence, deprivation, terrorism and conflicts.
4. PTSD can be more successfully treated with psychotherapeutic interventions as compared to pharmacotherapeutic.
5. CBT has relatively rigorous criteria of evidence-based framework, that is why, CBT has been used as a reference treatment in the treatment of different mental health problems.
6. CBT is expected to be one of the most commonly adopted psychotherapeutic interventions in Pakistan

Funding

This study has been funded by the EMDR Research Foundation, USA.

References

Agha, S. A. U. (2020). Bereavement, post-traumatic growth, and the role of cognitive processes: Study of bereaved parents and spouses in Baluchistan, Pakistan. *Pakistan Journal of Psychological Research*, 125–139.

Ahmad, N., & Hussain, S. (2020). Internal displacement: Relationship of mental health and education of children in swat, Pakistan. *Pakistan Journal of Medical Sciences*, *36*(5), 909.

Ali, M., Farooq, N., Bhatti, M. A., & Kuroiwa, C. (2012). Assessment of prevalence and determinants of posttraumatic stress disorder in survivors of earthquake in Pakistan using Davidson Trauma Scale. *Journal of Affective Disorders*, *136*(3), 238–243.

Amin, R., Iqbal, A., Naeem, F., & Irfan, M. (2020). Effectiveness of a culturally adapted cognitive behavioural therapy-based guided self-help (CACBT-GSH) intervention to reduce social anxiety and enhance self-esteem in adolescents: A randomized controlled trial from Pakistan. *Behavioural and Cognitive Psychotherapy*, *48*(5), 503–514. https://www.cambridge.org/core/article/effectiveness-of-a-culturally-adapted-cognitive-behavioural-therapybased-guided-selfhelp-cacbtgsh-intervention-to-reduce-social-anxiety-and-enhance-selfesteem-in-adolescents-a-randomized-controlled-trial-from-pakista.

Asghar, M. A., et al. (2020). Relationship of obsessive-compulsive disorders with religion and psychosocial attitude among local medical college students of Karachi: An epidemiological study. *Journal of the Pakistan Medical Association*, *70*(1563).

Ashfaq, A., et al. (2018). Exploring symptoms of post-traumatic stress disorders and perceived social support among patients with burn injury. *Cureus*, *10*(5), e266 a9. https://pubmed.ncbi.nlm.nih.gov/30042920.

Ashraf, F., Niazi, F., Masood, A., & Malik, S. (2019). Gender comparisons and prevalence of child abuse and post-traumatic stress disorder symptoms in adolescents. *Journal of the Pakistan Medical Association, 69*(3), 320–324.

Ashraf, et al. (2020). The burden of quarantine on mental health amidst Covid-19 pandemic: A cross sectional study. *Pakistan Armed Forces Medical Journal, 70*(2), S584–S589.

Aslam, M., Irfan, M., & Naeem, F. (2015). Brief culturally adapted cognitive behaviour therapy for obsessive compulsive disorder: A pilot study. *Pakistan Journal of Medical Sciences, 31*(4), 874.

Babar, I., & Dildar, S. (2021). Emotion reactivity a key factor in post-traumatic stress disorder symptoms severity among amputees. *Journal of the Pakistan Medical Association*, 1–10.

Babar, S., & Ahmad, K. (2021). Efficacy of cognitive Behvioral therapy on body dissatisfaction of females with PCOS. *Bahria Journal of Professional Psychology, 20*(1), 40–51.

Bannink, F., & Geschwind, N. (2021). *Positive CBT: Individual and group treatment protocols for positive cognitive behavioral therapy*. Hogrefe Publishing. https://books.google.com.pk/books?id=E50JzgEACAAJ.

Bano, Z., & Naz, I. (2020). Post-traumatic stress disorder, cognitive function and adjustment problems in women burn survivors: A multi-center study. *Journal of the Pakistan Medical Association*, 1–16.

Beck, J., & Carlson, J. (2012). *Cognitive therapy*. American Psychological Association.

Beck, J. S. (2020). Cognitive behavior therapy. In *Basics and beyond* (3rd ed.). Guilford Publications. https://books.google.com.pk/books?id=yb_nDwAAQBAJ.

Bibi, A., Kalim, S., & Khalid, M. A. (2018). Post-traumatic stress disorder and resilience among adult burn patients in Pakistan: A cross-sectional study. *Burns & Trauma, 6*.

Bukhari, S. R., & Afzal, F. (2020). Treating post traumatic stress disorder with cognitive behavior therapy: A case study. *Pakistan Journal of Medical Research, 59*(3), 120–123.

Burgess, A. W., & Holmstrom, L. L. (1974). Rape trauma syndrome. *American Journal of Psychiatry, 131*(9), 981–986.

Choudhry, F. R., Munawar, K., Khaiyom, J. H. A., & Lian, T. C. (2020). Cognitive behavior therapy with a migrant Pakistani in Malaysia: A single case study of conversion disorder. *Jurnal Psikologi Malaysia, 33*(2).

Chung, M. C., Jalal, S., & Khan, N. U. (2014). Posttraumatic stress disorder and psychiatric comorbidity following the 2010 flood in Pakistan: Exposure characteristics, cognitive distortions, and emotional suppression. *Psychiatry, 77*(3), 289–304. https://www.tandfonline.com/doi/abs/10.1521/psyc.2014.77.3.289.

Cohen, J. A., Mannarino, A. P., Kliethermes, M., & Murray, L. A. (2012). Trauma-focused CBT for youth with complex trauma. *Child Abuse & Neglect, 36*(6), 528–541.

David, D., Cristea, I., & Hofmann, S. G. (2018). Why cognitive behavioral therapy is the current gold standard of psychotherapy. *Frontiers in Psychiatry, 9*, 4. https://pubmed.ncbi.nlm.nih.gov/29434552.

de Bont, P. A. J. M., et al. (2016). Prolonged exposure and EMDR for PTSD v. a PTSD waiting-list condition: Effects on symptoms of psychosis, depression and social functioning in patients with chronic psychotic disorders. *Psychological Medicine, 46*(11), 2411–2421.

Dobson, K. S., & Dozois, D. J. A. (2019). *Handbook of cognitive-behavioral therapies* (4th ed.). Guilford Publications. https://books.google.com.pk/books?id=3T2EDwAAQBAJ.

Farooq, S., Khan, T., Zaheer, S., & Shafique, K. (2019). Prevalence of anxiety and depressive symptoms and their association with multimorbidity and demographic factors: A community-based, cross-sectional survey in Karachi, Pakistan. *BMJ Open, 9*(11), e029315.

Ferrari, R. (2015). Writing narrative style literature reviews. *Medical Writing, 24*(4), 230–235.

Hardy, A., et al. (2020). A network analysis of post-traumatic stress and psychosis symptoms. *Psychological Medicine*, 1–8.

Hayes, S. C., & Hofmann, S. G. (2017). The third wave of cognitive behavioral therapy and the rise of process-based care. *World Psychiatry: Official Journal of the World Psychiatric Association (WPA), 16*(3), 245–246. https://pubmed.ncbi.nlm.nih.gov/28941087.

Husain, M. I., et al. (2017). Pilot study of a culturally adapted psychoeducation (CaPE) intervention for bipolar disorder in Pakistan. *International Journal of Bipolar Disorders, 5*(1), 1–9.

Hussain, S., et al. (2019). Illness perceptions in patients of schizophrenia: A preliminary investigation from Lahore, Pakistan. *Pakistan Journal of Medical Sciences, 33*(4), 829–834. https://pubmed.ncbi.nlm.nih.gov/29067048.

Ibrahim, H., & Hassan, C. Q. (2017). Post-traumatic stress disorder symptoms resulting from torture and other traumatic events among Syrian Kurdish refugees in Kurdistan region, Iraq. *Frontiers in Psychology, 8*, 241.

Ishfaq, N., & Kamal, A. (2020). Explaining the predictive relationship between early life trauma and comorbid psychiatric symptoms among convicts in Pakistan. *Journal of Police and Criminal Psychology*. https://doi.org/10.1007/s11896-020-09408-9.

Jarero, I., Artigas, L., Uribe, S., & Miranda, A. (2014). EMDR therapy humanitarian trauma recovery interventions in Latin America and the Caribbean. *Journal of EMDR Practice and Research, 8*(4), 260–268.

Javed, A., Khan, M. N. S., Nasar, A., & Rasheed, A. (2020). Mental healthcare in Pakistan. *Taiwanese Journal of Psychiatry, 34*(1), 6.

Jones, E., & Wessely, S. (2005). *Shell shock to PTSD: Military psychiatry from 1900 to the Gulf War*. Taylor & Francis. https://books.google.com.pk/books?id=5xt5AgAAQBAJ.

Jowett, S., Karatzias, T., Shevlin, M., & Albert, I. (2020). Differentiating symptom profiles of ICD-11 PTSD, complex PTSD, and borderline personality disorder: A latent class analysis in a multiply traumatized sample. *Personality Disorders: Theory, Research, and Treatment, 11*(1), 36.

Junaid, F. A., Haar, J., & Brougham, D. (2020). Post-traumatic stress, job stressors, psychological capital and job outcomes: A study of Pakistan employees living under ongoing terrorism. *Labour & Industry: A Journal of the Social and Economic Relations of Work*, 1–23.

Karim, A. S., Saeed, S., & Akber, N. (2019). Ethnic diversity and political development in Pakistan. *The Government-Annual Research Journal of Political Science*, 7(7).

Katz, A. C., et al. (2017). *Effect of comorbid post-traumatic stress disorder and panic disorder on defensive responding.*

Kazantzis, N., et al. (2019). The processes of cognitive behavioral therapy: A review of meta-analyses. *Cognitive Therapy and Research*, 42(4), 349–357.

Kerai, S. M., et al. (2017). Post-traumatic stress disorder and its predictors in emergency medical service personnel: A cross-sectional study from Karachi, Pakistan. *BMC Emergency Medicine*, 17(1), 26. https://pubmed.ncbi.nlm.nih.gov/28851280.

Khan, A., & Anwar, M. (2018). Dynamics of mental health literacy among the academic staff: A developing country perspective. *Global Educational Studies Review*, 3(1).

Khan, A. A., et al. (2016). Prevalence of post-traumatic stress disorder due to community violence among university students in the World's Most dangerous megacity: A cross-sectional study from Pakistan. *Journal of Interpersonal Violence*, 31(13), 2302–2315.

Khan, S., Amjad, A., & Rowland, D. (2017). Cognitive behavioral therapy as an adjunct treatment for Pakistani men with ED. *International Journal of Impotence Research*, 29(5), 202–206.

Kuester, A., et al. (2017). Comparison of DSM-5 and proposed ICD-11 criteria for PTSD with DSM-IV and ICD-10: Changes in PTSD prevalence in military personnel. *European Journal of Psychotraumatology*, 8(1), 1386988.

Lassemo, E., Sandanger, I., Nygård, J. F., & Sørgaard, K. W. (2017). The epidemiology of post-traumatic stress disorder in Norway: Trauma characteristics and pre-existing psychiatric disorders. *Social Psychiatry and Psychiatric Epidemiology*, 52(1), 11–19.

Latif, M., et al. (2021). Culturally adapted trauma-focused CBT-based guided self-help (CatCBT GSH) for female victims of domestic violence in Pakistan: Feasibility randomized controlled trial. *Behavioural and Cognitive Psychotherapy*, 49(1), 50–61.

Lee, D., et al. (2016). Psychotherapy versus pharmacotherapy for posttraumatic stress disorder: Systemic review and meta-analyses to determine first-line treatments. *Depression and Anxiety*, 33(9), 792–806.

Leese, P. (2002). *Shell shock: Traumatic neurosis and the British soldiers of the first world war.* Palgrave Macmillan UK.

Lewis, S. J., et al. (2019). The epidemiology of trauma and post-traumatic stress disorder in a representative cohort of young people in England and Wales. *The Lancet Psychiatry*, 6(3), 247–256.

Li, Y., Scherer, N., Felix, L., & Kuper, H. (2021). Prevalence of depression, anxiety and post-traumatic stress disorder in health care workers during the COVID-19 pandemic: A systematic review and meta-analysis. *PLoS One*, 16(3), e0246454.

Malik, M. A., & Khan, M. M. (2016). Economic burden of mental illnesses in Pakistan. *Journal of Mental Health Policy and Economics*, 19(3), 155.

Masud, S. (2019). Evaluation of psychotherapy practice in Pakistan—A qualitative exploration. *Journal of Research & Reviews in Social Sciences Pakistan*, 2(1), 290–302.

McNally, R. J. (2016). Can network analysis transform psychopathology? *Behaviour Research and Therapy*, 86, 95–104.

Mueser, K. T., et al. (2008). A randomized controlled trial of cognitive-behavioral treatment for posttraumatic stress disorder in severe mental illness. *Journal of Consulting and Clinical Psychology*, 76(2), 259.

Naeem, F., Gobbi, M., Ayub, M., & Kingdon, D. (2017). Psychologists experience of cognitive behaviour therapy in a developing country: A qualitative study from Pakistan. *International Journal of Mental Health Systems*, 4(1), 1–9.

Naeem, F., Gul, M., et al. (2015). Brief culturally adapted CBT (CaCBT) for depression: A randomized controlled trial from Pakistan. *Journal of Affective Disorders*, 177, 101–107.

Naeem, F., Irfan, M., & Javed, A. (2020). Coping with COVID-19: Urgent need for building resilience through cognitive behaviour therapy. *Khyber Medical University Journal*, 12(1), 1–3.

Naeem, F., Saeed, S., et al. (2015). Brief culturally adapted CBT for psychosis (CaCBTp): A randomized controlled trial from a low income country. *Schizophrenia Research*, 164(1–3), 143–148.

Naeem, F., et al. (2011). Preliminary evaluation of culturally sensitive CBT for depression in Pakistan: Findings from developing culturally-sensitive CBT project (DCCP). *Behavioural and Cognitive Psychotherapy*, 39(2), 165–173.

Naeem, F., et al. (2021). Transcultural adaptation of cognitive behavioral therapy (CBT) in Asia. *Asia-Pacific Psychiatry*, 13(1), e12442.

Najavits, L. M. (2014). *PTSD and substance abuse.* Treatment Innovations LLC Newton Center MA.

Nasir, M. S. I. (2020). Sensory integration versus cognitive behavioral therapy on behavioral issues in learning-disabled children. *Pakistan Journal of Rehabilitation*, 9(2), 11–17.

Nawaz, D., Khan, S. A., Batool, S., & Rasool, A. (2021). Governance issues and deep-rooted corruption in Pakistan. *International Journal of Modern Agriculture*, 10(2), 2699–2706.

Naz, S., & Malik, N. I. (2018). Domestic violence and psychological well-being of survivor women in Punjab, Pakistan. *Journal of Psychology & Clinical Psychiatry*, 9(2), 184–189.

Price, M., & van Stolk-Cooke, K. (2015). Examination of the interrelations between the factors of PTSD, major depression, and generalized anxiety disorder in a heterogeneous trauma-exposed sample using DSM 5 criteria. *Journal of Affective Disorders*, 186, 149–155.

Priebe, K., et al. (2018). Defining the index trauma in post-traumatic stress disorder patients with multiple trauma exposure: Impact on severity scores and treatment effects of using worst single incident versus multiple traumatic events. *European Journal of Psychotraumatology*, 9(1), 1486124.

Qayyum, W., et al. (2021). Endoscopic findings in patients with refractory dyspepsia at a tertiary care hospital in Peshawar, KPK Province, Pakistan. *The Professional Medical Journal*, 28(04), 585–591.

Qureshi, Z. (2020). *Pakistan surpasses Brazil to become World's 5th most populous country.* https://gulfnews.com/world/asia/pakistan-surpasses-brazil-to-become-worlds-5th-most-populous-country-1.72557051.

Raboni, M. R., Alonso, F. F. D., Tufik, S., & Suchecki, D. (2014). Improvement of mood and sleep alterations in posttraumatic stress disorder patients by eye movement desensitization and reprocessing. *Frontiers in Behavioral Neuroscience*, 8, 209.

Rahman, A., et al. (2008). Cognitive behaviour therapy-based intervention by community health workers for mothers with depression and their infants in rural Pakistan: A cluster-randomised controlled trial. *The Lancet, 372*(9642), 902–909.

Rashid, H. U., et al. (2020). Post-traumatic stress disorder and association with low birth weight in displaced population following conflict in Malakand division, Pakistan: A case control study. *BMC Pregnancy and Childbirth, 20*(1), 1–8.

Regier, D. A., Kuhl, E. A., & Kupfer, D. J. (2013). The DSM-5: Classification and criteria changes. *World Psychiatry: Official Journal of the World Psychiatric Association (WPA), 12*(2), 92–98.

Ruggiero, G. M., Spada, M. M., Caselli, G., & Sassaroli, S. (2019). A historical and theoretical review of cognitive behavioral therapies: From structural self-knowledge to functional processes. *Journal of Rational-Emotive & Cognitive-Behavior Therapy, 36*(4), 378–403. https://doi.org/10.1007/s10942-018-0292-8.

Shah, M. K., Ahmad, I., & Hallahan, B. (2019). Impact of conventional beliefs and social stigma on attitude towards access to mental health services in Pakistan. *Community Mental Health Journal, 55*(3), 527–533.

Shah, S. F. A., et al. (2020). Trauma exposure and post-traumatic stress disorder among regional journalists in Pakistan. *Journalism*. https://doi.org/10.1177/1464884920965783, 1464884920965783.

Shehzad, G., & Ahsan, S. (2020). PTSD symptomatology and social anxiety among retired army officers: Mediating role of internalized shame. *Pakistan Journal of Psychological Research*, 559–575.

Siddiqi, K. (2020). *Mental health challenges*. https://tribune.com.pk/story/2267953/mental-health-challenges.

Stewart, C. L., & Wrobel, T. A. (2010). Evaluation of the efficacy of pharmacotherapy and psychotherapy in treatment of combat-related post-traumatic stress disorder: A meta-analytic review of outcome studies. *Military Medicine, 174*(5), 460–469.

Turk, D., & Flor, H. (2013). The cognitive-behavioral approach to pain management. In S. McMahon, & M. Koltzenburg (Eds.), *Textbook of pain e-book* (pp. 592–602). Elsevier Health Sciences.

Verbitsky, A., Dopfel, D., & Zhang, N. (2020). Rodent models of post-traumatic stress disorder: Behavioral assessment. *Translational Psychiatry, 10*(1), 1–28.

WHO. (2019). *WHO report on mental health system in Pakistan*. https://www.who.int/mental_health/pakistan_who_aims_report.pdf.

Yuan, K., et al. (2021). Prevalence of posttraumatic stress disorder after infectious disease pandemics in the twenty-first century, including COVID-19: A meta-analysis and systematic review. *Molecular Psychiatry*, 1–17.

Zaman, N. I., & Munib, P. M. (2020). Post traumatic stress disorder and resilience: An exploratory study among survivors of bacha Khan University Charsadda, Pakistan. *FWU Journal of Social Sciences, 14*(2).

Zarak, M. S., et al. (2020). Assessment of psychological status (PTSD and depression) among the terrorism affected Hazara Community in Quetta, Pakistan. *Assessment, 12*.

Chapter 31

Schizophrenia in Japan and cognitive behavioral therapy

Hiroki Tanoue and Naoki Yoshinaga
School of Nursing, Faculty of Medicine, University of Miyazaki, Miyazaki, Japan

Abbreviations

CBT   Cognitive Behavioral Therapy
MCT   Metacognitive Training
JABC Japanese Association of Behavioral and Cognitive Therapies
JACT Japanese Association for Cognitive Therapy
MHLW Ministry of Health, Labour and Welfare
TAU   treatment as usual

Introduction

Schizophrenia causes a significant economic burden in Japan and globally (Crown et al., 2001). Pharmacotherapy, especially antipsychotic medication, is the primary treatment strategy for schizophrenia. However, 20%–30% of patients with schizophrenia are resistant to antipsychotics (Elkis & Buckley, 2016). Furthermore, patients are often dissatisfied with antipsychotics (Lieberman et al., 2005). In Japan, treatment of schizophrenia has relied on hospitalization and pharmacotherapy. It has become clear that combining pharmacotherapy and psychosocial treatment for schizophrenia has a synergistic effect (Morrison et al., 2018). CBT for schizophrenia is an empirically supported psychosocial intervention. It is essential to popularize CBT in Japan and provide it to various medical and welfare services in order to solve problems facing schizophrenia care in Japan.

Schizophrenia and its current status in Japan

In Japan, the lifetime prevalence of schizophrenia is estimated to be approximately 1%, which is similar to that in the rest of the world (Nakane, Ohta, & Radford, 1992). The number of patients with schizophrenia in Japan, calculated from the life time prevalence, is estimated to be between 700,000 and 800,000 (Ministry of Health, Labour and Welfare of Japan, 2017). Schizophrenia is associated with high relapse rates and follows a chronic course leading to long-term hospitalization and impaired social functioning. Additionally, cognitive deficits in schizophrenia are found in multiple domains, impairing daily functioning and instrumental daily living activities (Bowie & Harvey, 2005). These difficulties may be partially responsible for the chronicity of the condition and institutionalization of patients, limiting access to full-time employment, residential independence, and causing poor social outcomes (Amado et al., 2016; Harvey & Penn, 2010). This in turn leads to high health care costs and increased morbidity costs. Notably, in developed countries around the world, approximately 2%–3% of all health care resources are devoted to the treatment and care of schizophrenia, resulting in considerable social and economic losses (Keio University, 2011). As in other countries, the social burden arising from schizophrenia is enormous in Japan. Previous studies have shown that the total cost of schizophrenia in Japan, including direct and indirect costs, amounts to about 2.8 trillion JPY (25.5 billion USD [exchange rate: 1 USD = 110 JPY]), and the absolute amount of nonwork costs is large compared to other disorders (Sado et al., 2013).

The challenge of long-term hospitalization for schizophrenia in Japan

Because Japan has traditionally relied on hospitalization for the treatment of schizophrenia, the number of psychiatric beds in Japan is considerably higher than in other countries. A total of 261 psychiatric beds per 100,000 population is approximately three times the average in Organization for Economic Co-operation and Development (OECD) countries (Fig. 1) (OECD Health Statistics, 2018). Because of this, Japan has lagged in community life support systems, and long-term hospitalization has become the norm. Until around 2000, people with schizophrenia accounted for approximately two-thirds of the 330,000 psychiatric hospital beds in Japan. Moreover, schizophrenia accounted for more than half of patients hospitalized for more than 1 year (Ministry of Health, Labour and Welfare of Japan, 2017).

To solve these problems, the government articulated a vision for the future called the "Vision for the Reform of Mental Health and Welfare Policies" in 2004 and initiated policies to promote the basic concept of "shifting from a focus on

FIG. 1 Psychiatric beds per 100,000 population, 2018. *(Source: OECD Health Statistics, https://data.oecd.org/healtheqt/hospital-beds.htm.)*

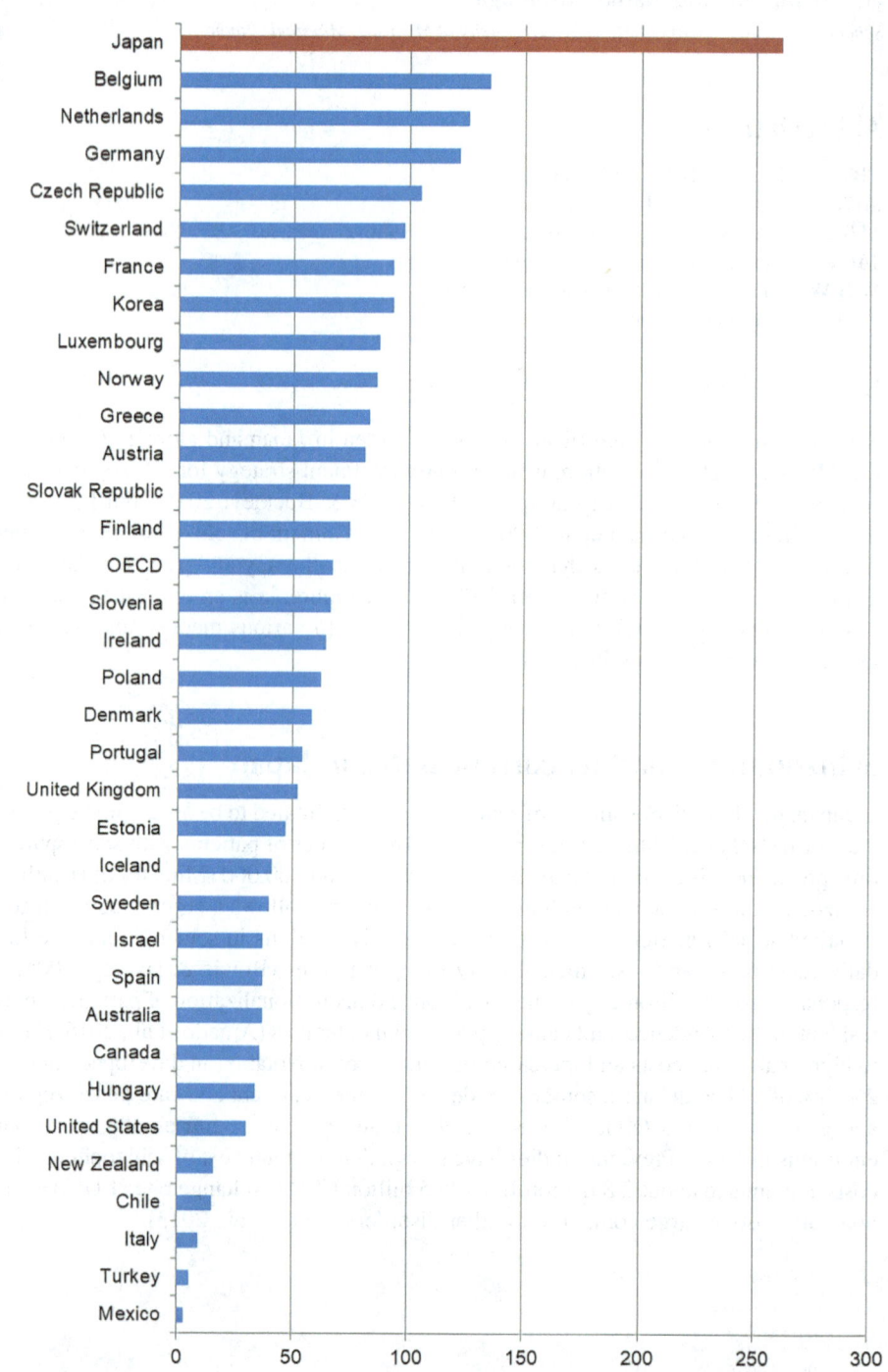

inpatient care to community life" (MHLW, 2004). These policies include deepening and improving public understanding of schizophrenia, correcting discrimination, reforming psychiatric care, and strengthening community life support. A survey conducted as part of these measures revealed approximately 70,000 patients who could be discharged from hospitals (socially hospitalized) (Matsubara, 2008). Further, measures have been taken to strengthen cooperation among medical and welfare services, promote the discharge of long-term hospitalized patients, prevent new long-term hospitalization (early intervention, early treatment, early discharge), and support for continuing community life and development. After implementing hospital discharge facilitation projects and regional transition support projects, regional consultation support efforts under Act on the Comprehensive Support for the Daily and Social Life of Persons with Disabilities enacted after that were carried out in various places, and regional supporters (e.g., psychiatric social workers (PSWs)) and psychiatric hospitals cooperated and integrated their services. Such efforts promote the discharge of schizophrenic patients into the community and provide them with a secure place for living. The number of long-term hospitalized psychiatric patients who are discharged from hospitals and shift to community life is steadily increasing. However, more than 10 years after the "Reform Vision" was clearly stated, 180,000 patients remain in hospital for more than 1 year. About 70% of those hospitalized for more than 5 years are moved from one facility to another until they die (MHLW, 2014). Even in recent years, 10% of new hospitalizations have resulted in long-term hospitalization of 1 year or more (MHLW, 2014). Therefore, it is imperative that those involved in mental healthcare and welfare understand this reality and that all supporters involved with patients with schizophrenia and other mental illnesses work together to support their transition into the community by making full use of all resources and measures.

Where CBT is provided to people with schizophrenia

In recent years, community-based services for the mentally ill in Japan have been enhanced in parallel with shortening hospital stays. In addition to outpatient services, medical system services include daycare (which is similar to partial hospitalization in Western countries), psychiatric home-care nursing, and employment support. CBT and related techniques are provided in inpatient and other settings. Healthcare professionals, such as psychiatric nurses, occupational therapists, and social workers, are assigned, and work together to provide psychosocial interventions and community life support for patients with schizophrenia. In Japanese psychiatric outpatient departments, psychiatrists cannot devote enough time to individual patients (they see an average of seven patients in 1h) (Hayashi et al., 2020). Hence, psychiatric nurses and clinical psychologists are the main psychotherapy providers.

The public medical services that are most commonly used by patients with schizophrenia, especially in Japan, are psychiatric daycare and psychiatric home-care nursing. Psychiatric daycare is a one-day psychiatric rehabilitation service, similar to a partial hospitalization in Western countries, and is covered by public medical insurance. In psychiatric daycare, people experiencing difficulties due to mental illness go to the center at a fixed time and engage in various activities such as arts and crafts, and exercise(Iwasaki, Hirosawa, & Nakamura, 2006). These activities aim to prevent the recurrence of mental illness and hospitalization, and the benefits of these activities in Japanese settings have been shown (Miyaji et al., 2008). More than 8,000,000 people use daycare services in Japan every year, and more than half of them are schizophrenic patients. Psychiatric daycare programs are often open to participation regardless of diagnosis (Kasahara et al., 1995). Daycare programs also include interactions and activities among daycare users, which helps them practice interpersonal relationships. Many programs also provide schooling for children/adolescents, and give support to those seeking employment; this serves as a stepping stone for reintegration into society. There are many preliminary reports of psychiatric daycare programs that have offered group CBT (Yoshinaga et al., 2015). On the other hand, home-care nursing is a service where psychiatric nurses and other rehabilitation staff visit the patient's home to help maintain daily life and acquire/expand life skills. The care provided by visiting psychiatric nurses is diverse. It includes monitoring symptoms, stabilizing and improving symptoms, support for taking medication and going to the hospitals, and assistance with interpersonal/family relationships. CBT techniques are also utilized in home-visit psychiatric nursing services. Thus, the places where CBT is provided to people with schizophrenia are not limited to inpatient and outpatient settings, but have expanded to community-based services.

Spread of CBT for schizophrenia in Japan

In 2002, the 12th World Psychiatric Association (WPA) was held in Japan, and the WPA Yokohama Declaration was issued at this meeting. Simultaneously, the Japanese diagnostic name and the concept of schizophrenia was reviewed and modified in order to change public perception of this condition in Japan (Desapriya & Nobutada, 2002). In the Medical Care and Supervision Act (MHLW, 2003), which came into effect in 2003, CBT was incorporated as an essential treatment program in inpatient facilities, based on the British system and its implementation (Kikuchi, 2019). This Japanese law aims to provide appropriate medical care and promote social reintegration for people who have committed severe harm while in a state of insanity or deprivation of mind—a state in which criminal responsibility cannot be imposed because of mental disorders, such as an inability to distinguish between right and wrong. Since about 80% of the people covered by this law were schizophrenic patients, CBT for psychosis (CBTp) became the focus of attention among psychiatrists and clinical psychologists (Ishigaki, 2019).

Furthermore, the World Congress on Behavioral and Cognitive Therapies was held in Japan in 2004, which led to the spread of CBT in Japan (Ishikawa et al., 2016). Subsequently, a group of experts on CBTp called the "CBTp Network," which was established in 2011 by experts in Japan, has been leading the way in CBTp's dissemination and application in clinical practice (Ishigaki, 2019). Additionally, the approach collectively called CBTp includes various CBT techniques (Thomas, 2015). New programs such as metacognitive training (MCT), social cognition and interaction training (SCIT), and cognitive remediation therapy, which are derivatives of CBTp, have been introduced at conference workshops, and several preliminary reports in Japan have been published (Iwata et al., 2017; Tsukagoshi, Tawara, Matsuoka, Ubukata, & Naya, 2016). Increasingly, MCT has been implemented for inpatients and daycare patients with schizophrenia because of its ease of use. The "MCT-J Network (http://mct-j.jpn.org/)," which aims to spread MCT and improve the quality of practice in Japan, has about 1500 members. The MCT-J Network includes a variety of professionals such as psychiatrists, psychiatric nurses, clinical psychologists, occupational therapists, and social workers. A Japanese research group recently conducted a multicenter randomized controlled trial in a psychiatric daycare setting that also demonstrated the efficacy of MCT on positive symptoms and overall functioning in people with schizophrenia (Ishikawa et al., 2020). Other so-called third-generation CBTs, such as acceptance and commitment therapy, mindfulness, compassion-focused therapy, and metacognitive therapy, are also gradually gaining recognition among professionals. Workshops are held annually by the Japanese Association for Behavior Therapy and Cognitive Therapy (JABCT) and the Japanese Association for Cognitive Therapy (JACT). The concept of recovery is now being taken into account in treatment outcomes for schizophrenia. Accordingly, CBT providers are striving to incorporate the concept of recovery into their perspectives.

Research on CBT for schizophrenia in Japan

To determine the current status of high-quality research on CBT for schizophrenia in Japan, we conducted a simple literature search using MEDLINE (via PubMed) on March 15, 2021, using the following search formula: (("Schizophrenia" [Mesh]) or ("Psychosis") and ("Psychotherapy" [Mesh])) and ("Japan" [Mesh]). In addition to the online database search, a manual search was conducted by scanning the articles' reference lists identified in the database search. We also consulted experts in the field to see if there were any other significant studies. Three additional papers were found in this manual search. Our search established that there are no randomized controlled trials of CBTp in Japan. However, there were two reports on group therapy for schizophrenia; one study employed MCT (one randomized controlled trial), and the other study employed group CBT aiming to restore self-esteem in people with schizophrenia and other disorders. Furthermore, one report targeted people with "at-risk" status of psychosis and examined the feasibility of individual CBTp in this population. Table 1 summarizes the characteristics of identified studies evaluating CBT and related intervention for schizophrenia in Japan. Details of the identified studies are described below:

Research on group CBT for schizophrenia ($n = 3$)

Ishikawa et al. (2020) conducted an assessor-blinded, multicenter, randomized controlled trial in Japanese patients with schizophrenia. A total of 50 patients were randomly assigned to either MCT plus treatment as usual (MCT+TAU) ($n=24$) or TAU alone ($n=26$). MCT was given once a week in 50-min group sessions. Four psychiatric nurses, a clinical psychologist, an occupational therapist, and a PSW performed MCT based on the latest and extended versions of the standardized MCT manual. The results showed that the MCT+TAU group showed significant improvement in positive symptoms (Positive and Negative Syndrome Scale [PNASS: blind assessor rating]) and general assessment of functioning (GAF) after the intervention (week 10) compared to the TAU alone group. The treatment effect was later maintained for at least 1 month.

Tanoue et al. (2020) evaluated the clinical effectiveness of MCT as a transdiagnostic program in the context of a routine daycare setting in Japan involving individuals with various mental disorders. The study employed a prospective, multicenter, single-group, pre-post design in which 34 participants (schizophrenia=22, nonschizophrenia=12) underwent MCT. Participants diagnosed with various psychiatric disorders received MCT once a week for 10- to 50-min group sessions. Significant improvements in quality of life (EQ-5D-5L) and global functioning (GAF) were evident after the intervention, and further improvements were observed 4 weeks after completion.

TABLE 1 Summary of characteristics of clinical trials evaluating CBT for schizophrenia in Japan.

	Study/type	Target population	Intervention/control	N	Primary disorder-specific outcomes
Group CBT	Ishikawa et al. (2020)	Adults (aged 20–65) with Schizophrenia	Group MCT (50 min, weekly, total 10 sessions [provider: RN, CP, OTR, PSW]) + TAU	24	PANSS (assessor-rated), GAF (assessor-rated), CBQp (self-reported), BDI-II (self-reported), BCIS (self-reported)
			WLC	26	
	Tanoue et al. (2020)	Adults (aged 20–) with psychiatric disorder	Group MCT (50 min, weekly, total 10 sessions [provider: RN, CP, OTR, PSW]) + TAU	34 (Schizophrenia = 22)	EQ-5D-5L (self-reported), GAF (assessor-rated), BDI-II (self-reported), BCIS (self-reported) PANSS (assessor-rated), CBQp (self-reported)
	Kunikata, Yoshinaga, and Nakajima (2016)	Adults (aged 20–65) with psychiatric disorder	Group CBT for recovery of self-esteem (50 min, weekly, total 12 sessions [provider: RN]) + TAU	41 (Schizophrenia = 25)	RSES (assessor-rated), BDI-II (self-reported), POMS (self-reported), CBS (self-reported), SUBI (self-reported), BPRS (assessor-rated)
			TAU	21 (Schizophrenia = 13)	
Individual CBT	Matsumoto et al. (2019)	Adults (aged 20–65) with at-risk metal state	Individual CBT (50 min, weekly, total 12 sessions [provider: MD, CP]) + TAU	14	PANSS (assessor-rated)

Abbreviations: BCIS, beck cognitive insight scale; BDI-II, beck depression inventory-second edition; BPRS, brief psychiatric rating scale; CBQp, cognitive biases questionnaire for psychosis; CBS, cognitive bias scale; CBT, cognitive behavioral therapy; CP, clinical psychologist; EQ-5D-5L, EuroQOL 5 dimensions 5-level; GAF, global assessment of functioning; POMS, profile of mood states; PSW, psychiatric social worker; RN, registered nurse (psychiatric nurse); RSES, Rosenberg self-esteem scale; SUBI, subjective well-being inventory; TAU, treatment as usual; WLC, waiting-list control.

Kunikata et al. (2016) examined the feasibility of group CBT for recovery of self-esteem delivered to community-dwelling individuals with mental illness in a nonrandomized controlled trial. The group CBT program consisted of 12 sessions, and was a blended program that included general CBT techniques and acceptance commitment therapy elements. The study did not limit the participants' clinical diagnosis or life circumstances/history, and 25 of the 41 participants in the intervention group had schizophrenia. Although diagnosis-specific results were not presented, outcome measures such as self-esteem and subjective well-being improved predominantly after the intervention compared to the control group. Based on within-group trends and between-group differences in self-esteem, the group CBT program has a relatively long-term impact on self-esteem recovery.

Research on individual CBTp for at-risk status of schizophrenia (*n* = 1)

Matsumoto et al. (2019) examined the feasibility of individual CBTp for individuals with at-risk mental states in Japan through a prospective, multicenter, single-group, pre-post design. Fourteen at-risk participants received individual CBTp over 6 months and were followed-up for 6 months. Results showed that 13 individuals completed the CBTp intervention and assessment; the PANSS mean total score improved from 60.2% to 46.0% after the intervention. The effect was maintained for 6 months. The effect was also maintained during the follow-up.

Clinical practice in Japan

In Japan, in addition to the universal health insurance system, there is a public support system for the mentally ill called the "Medical Care System for Independence Support." This is a publicly funded medical system that reduces out-of-pocket medical expenses for medical treatment to cure or alleviate mental illness. Most people with schizophrenia use this additional support system (MHLW, 2006), which makes it easier for them to receive necessary medical care. However, CBT for schizophrenia patients is not yet covered by the Japanese health insurance system. Thus, CBT has only been partially implemented in inpatient, outpatient, daycare, and home-care nursing. Whether or not CBT is provided is up to individual institutions/providers.

The symptoms of schizophrenia are often comorbid with depression and anxiety disorders, so there is more than one treatment goal for schizophrenia. Therefore, the approach collectively referred to as "CBTp," consists of a wide variety of CBT packages (Thomas, 2015). Only therapists who are able to use a wide range of CBT techniques to address the various treatment goals can practice CBTp, and training of therapists is time-consuming and expensive. Because the symptoms of schizophrenia are highly individualized, it is difficult to develop a manual. These issues are also common in other countries. In Japan, there are reports of excellent CBTp practice and several textbooks have been published. Therapists who provide care for schizophrenia are expected to adapt to highly individualized psychosocial interventions for schizophrenia by using practice reports and textbooks.

Future directions

Especially in the last decade or so, CBT for schizophrenia has been gradually recognized, accepted, and practiced in Japan. However, it has been pointed out that there are many difficulties in introducing CBTp to schizophrenic patients in Japan (similar to the problems faced in other countries). It is expected that CBTp will be further developed and practiced in Japan to improve the quality of life of patients with schizophrenia in line with further improvements to the national psychiatric service system. Qualification standards, development and improvement of therapist skills, and standardized programs are needed. Practitioners and researchers need to continue to accumulate actual clinical data on CBT outcomes and advocate for the necessary funding and organizational support to widely disseminate CBT for schizophrenia throughout Japan. Moreover, Japanese institutions should continue to secure staff who can provide a high quality of CBT. We recommend that both group format, which is cost-effective and reproducible, and individual format, which allows for case formulation and a coordinated approach, be developed and offered to meet patients' needs.

Practice and procedure

The CBT protocols and assessment tools for schizophrenia used in Japan are based on protocols and assessment tools developed in Europe and the United States. For example, MCT for psychosis/schizophrenia, originally developed in Germany, has been translated into more than 30 languages (see www.uke.de/mkt for the full list of languages); therefore, it is recommended to use the version in one's own native language. Additionally, the qualifications of mental healthcare professionals and health insurance systems vary from country to country.

Mini-dictionary of terms

- *Schizophrenia*. A disorder with symptoms characterized by distortions in thought, perception, emotion, language, sense of self, and behavior in relation to others.
- *Cognitive behavioral therapy for psychosis (CBTp)*. Includes a variety of symptom-based CBT programs for psychosis and/or schizophrenia.
- *Metacognitive training (MCT) for psychosis*. Standardized and manualized group training for individuals with schizophrenia, targeting common cognitive errors and problem-solving biases.

Key facts

Key facts of CBT for schizophrenia in Japan

- In terms of disability-adjusted life years among hundreds of diseases, injuries, and risk factors in Japan, schizophrenia ranked 10th (as of 2011) (Institution for Health Metrics and Evaluation, 2019).
- Under the Japanese medical insurance system, individual and group CBT for inpatients and outpatients with schizophrenia are not covered.

TABLE 2 Content and target domain of the MCT group modules.

Module number	Title	Target domain
1	Attribution: blaming and taking credit	Mono causal inferences
2	Jumping to conclusions I	Jumping to conclusions/liberal acceptance
3	Changing beliefs	Bias against disconfirmatory evidence
4	To empathize I	Theory of mind first order
5	Memory	Over-confidence in errors
6	To empathize II	Theory of mind second order/need for closure
7	Jumping to conclusions II	Jumping to conclusions/ liberal acceptance
8	Mood	Depressive cognitive patterns
9 (additional module I)	Self esteem	Low self-esteem
10 (additional module II)	Dealing with prejudices (Stigma)	Dealing with prejudices

Applications to other areas

As mentioned earlier, CBTp includes a wide variety of CBT techniques. New approaches are being developed, and MCT is one of them. An MCT program is based on the theoretical foundations of the cognitive-behavioral model of schizophrenia. MCT has been translated and published in over 30 languages. The program is comprised of eight modules and two additional modules targeting common cognitive errors and problem-solving biases in schizophrenia (Table 2) (Moritz et al., 2014; Moritz, Veckenstedt, Bohn, Köther, & Woodward, 2013). But, most of the cognitive errors and problem-solving biases addressed in MCT are also common in other mental disorders (e.g., monocausal attributions, jumping to conclusions, inflexibility, problems in social cognition, overconfidence with memory errors, depressive thought patterns, low self-esteem, etc.). Hence, an MCT program is not limited to people with schizophrenia (i.e., individuals with a broad range of disorders can attend MCT group sessions). In addition, many other versions of MCTs for different disorders are being developed, and their introduction is being promoted around the world.

Summary points

- Schizophrenia is a common mental disorder in Japan and globally. It is associated with functional disability and social and economic burdens.
- In Japan, the treatment of schizophrenia has traditionally relied on hospitalization and pharmacotherapy. Many long-term hospitalized patients cannot be discharged.
- CBT has been used in clinical practice in Japan since the late 1950s, but it was not until the 2000s that it was used for schizophrenia. Several societies and professional associations have made continuous efforts to spread awareness and knowledge of CBT among mental healthcare professionals and the general public.
- Randomized clinical trials of psychosocial interventions for schizophrenia conducted in Japan have demonstrated the efficacy of group MCT on positive symptoms as well as improvement of quality of life.
- There is no health insurance coverage of CBT for schizophrenia. CBT for schizophrenia has been practiced in various settings, including inpatient, outpatient, daycare, and home-care nursing, but the number is still small.
- Practitioners and researchers in Japan need to expand the practice and training of CBT for schizophrenia, accumulate evidence, and build a foundation for providing CBT.

References

Amado, I., Brénugat-Herné, L., Orriols, E., Desombre, C., Dos Santos, M., Prost, Z., et al. (2016). A serious game to improve cognitive functions in schizophrenia: A pilot study. *Frontiers in Psychiatry*, 7, 64. https://doi.org/10.3389/fpsyt.2016.00064.

Bowie, C. R., & Harvey, P. D. (2005). Cognition in schizophrenia: Impairments, determinants, and functional importance. *The Psychiatric Clinics of North America*, 28(3), 613–626. https://doi.org/10.1016/j.psc.2005.05.004.

Crown, W. H., Neslusan, C., Russo, P. A., Holzer, S., Ozminkowski, R., & Croghan, T. (2001). Hospitalization and total medical costs for privately insured persons with schizophrenia. *Administration and Policy in Mental Health*, 28(5), 335–351. https://doi.org/10.1023/a:1011139215761.

Desapriya, E. B., & Nobutada, I. (2002). Stigma of mental illness in Japan. *Lancet (London, England)*, 359(9320), 1866. https://doi.org/10.1016/s0140-6736(02)08698-1.

Elkis, H., & Buckley, P. F. (2016). Treatment-resistant schizophrenia. *The Psychiatric Clinics of North America*, 39(2), 239–265. https://doi.org/10.1016/j.psc.2016.01.006.

Harvey, P. D., & Penn, D. (2010). Social cognition: The key factor predicting social outcome in people with schizophrenia? *Psychiatry (Edgmont (Pa: Township))*, 7(2), 41–44.

Hayashi, Y., Yoshinaga, N., Sasaki, Y., Tanoue, H., Yoshimura, K., Kadowaki, Y., et al. (2020). How was cognitive behavioural therapy for mood disorder implemented in Japan? A retrospective observational study using the nationwide claims database from FY2010 to FY2015. *BMJ Open*, 10(5). https://doi.org/10.1136/bmjopen-2019-033365, e033365.

Ishigaki, T. (2019). History and development of CBTp in Japan. *Japanese Journal of Clinical Psychology*, 19(2), 129–132.

Ishikawa, R., Ishigaki, T., Shimada, T., Tanoue, H., Yoshinaga, N., Oribe, N., et al. (2020). The efficacy of extended metacognitive training for psychosis: A randomized controlled trial. *Schizophrenia Research*, 215, 399–407. https://doi.org/10.1016/j.schres.2019.08.006.

Ishikawa, S. I., Hida, N., Kishida, K., Ueda, Y., Nakanishi, Y., & Kaneyama, Y. (2016). The empirical review of academic activities regarding cognitive behavioral therapies for children and adolescents in Japan: Before and after the World Congress of Behavioral and Cognitive herapies in 2004, Kobe. *The Japanese Journal of Cognitive Psychology*, 9, 34–43.

Iwasaki, K., Hirosawa, M., & Nakamura, K. (2006). A re-examination of day care programs for people with mental disorders. *Journal of Health and Sports Science Juntendo University*, 10, 9–20.

Iwata, K., Matsuda, Y., Sato, S., Furukawa, S., Watanabe, Y., Hatsuse, N., et al. (2017). Efficacy of cognitive rehabilitation using computer software with individuals living with schizophrenia: A randomized controlled trial in Japan. *Psychiatric Rehabilitation Journal*, 40(1), 4–11. https://doi.org/10.1037/prj0000232.

Kasahara, Y., Kato, H., Tsukahara, T., Takano, Y., Yakajima, S., & Miura, S. (1995). The demographic and clinical characteristics of long stay patients in the psychiatric day care. *Japan Journal of Clincal Psychiatry*, 24, 577–583.

Kunikata, H., Yoshinaga, N., & Nakajima, K. (2016). Effect of cognitive behavioral group therapy for recovery of self-esteem on community-living individuals with mental illness: Non-randomized controlled trial. *Psychiatry and Clinical Neurosciences*, 70(10), 457–468. https://doi.org/10.1111/pcn.12418.

Keio University. (2011). Estimation of social cost of mental disorders. Retrieved April 1, 2021 from: https://www.mhlw.go.jp/bunya/shougaihoken/cyousajigyou/dl/seikabutsu30-2.pdf.

Kikuchi, A. (2019). CBTp global standard in United Kingdom. *Japanese Journal of Clinical Psychology*, 19(2), 133–138.

Lieberman, J. A., Stroup, T. S., McEvoy, J. P., Swartz, M. S., Rosenheck, R. A., Perkins, D. O., et al. (2005). Effectiveness of antipsychotic drugs in patients with chronic schizophrenia. *The New England Journal of Medicine*, 353(12), 1209–1223. https://doi.org/10.1056/NEJMoa051688.

Matsubara, S. (2008). *Comprehensive research on understanding and optimizing the quality of divine medical care shared research; Survey on utilization of mental hospital beds* (pp. 9–51). Journal of Japanese Association of Psychiatric Hospitals.

Matsumoto, K., Ohmuro, N., Tsujino, N., Nishiyama, S., Abe, K., Hamaie, Y., et al. (2019). Open-label study of cognitive behavioural therapy for individuals with at-risk mental state: Feasibility in the Japanese clinical setting. *Early Intervention in Psychiatry*, 13(1), 137–141. https://doi.org/10.1111/eip.12541.

Ministry of Health, Labour and Welfare of Japan. (2003). Services and supports for persons with disabilities act. Retrieved April 1, 2021 from: http://www.mhlw.go.jp/topics/2005/02/dl/tp0214-1b.pdf.

Ministry of Health, Labour and Welfare of Japan. (2004). About the "vision for reforming mental health and medical welfare". Retrieved April 1, 2021 from: https://www.mhlw.go.jp/topics/2004/09/dl/tp0902-1a.pdf.

Ministry of Health, Labour and Welfare of Japan. (2006). System of medical payment for services and supports for persons with disabilities. Retrieved April 1, 2021 from: https://www.mhlw.go.jp/english/wp/wp-hw4/dl/health_care_and_welfare_measures_for_people_with_physical_disabillities/2011071904.pdf.

Ministry of Health, Labour and Welfare of Japan. (2014). Patient survey in 2014. Retrieved April 1, 2021 from: https://www.mhlw.go.jp/toukei/saikin/hw/kanja/14/dl/03.pdf.

Ministry of Health, Labour and Welfare of Japan. (2017). Patient survey in 2017. Retrieved April 1, 2021 from: https://www.mhlw.go.jp/toukei/saikin/hw/kanja/17/dl/05.pdf.

Miyaji, S., Yamamoto, K., Morita, N., Tsubouchi, Y., Hoshino, S., Yamamoto, H., et al. (2008). The relationship between patient characteristics and psychiatric day care outcomes in schizophrenic patients. *Psychiatry and Clinical Neurosciences*, 62(3), 293–300. https://doi.org/10.1111/j.1440-1819.2008.01796.x.

Moritz, S., Veckenstedt, R., Bohn, F., Köther, U., & Woodward, T. S. (2013). Metacognitive training in schizophrenia: Theoretical rationale and administration. In D. L. Roberts, & D. L. Penn (Eds.), *Social cognition in schizophrenia. From evidence to treatment* (pp. 358–383). New York, NY: Oxford University Press.

Moritz, S., Andreou, C., Schneider, B. C., Wittekind, C. E., Menon, M., Balzan, R. P., et al. (2014). Sowing the seeds of doubt: A narrative review on metacognitive training in schizophrenia. *Clinical Psychology Review*, 34, 358–366.

Morrison, A. P., Law, H., Carter, L., Sellers, R., Emsley, R., Pyle, M., et al. (2018). Antipsychotic drugs versus cognitive behavioural therapy versus a combination of both in people with psychosis: A randomised controlled pilot and feasibility study. *The Lancet. Psychiatry*, 5(5), 411–423. https://doi.org/10.1016/S2215-0366(18)30096-8.

Nakane, Y., Ohta, Y., & Radford, M. H. (1992). Epidemiological studies of schizophrenia in Japan. *Schizophrenia Bulletin, 18*(1), 75–84. https://doi.org/10.1093/schbul/18.1.75.

OECD Health statistics. (2018). Retrieved April 1, 2021 from: https://data.oecd.org/healtheqt/hospital-beds.htm.

Sado, M., Inagaki, A., Koreki, A., Knapp, M., Kissane, L. A., Mimura, M., et al. (2013). The cost of schizophrenia in Japan. *Neuropsychiatric Disease and Treatment, 9*, 787–798. https://doi.org/10.2147/NDT.S41632.

Tanoue, H., Yoshinaga, N., Hayashi, Y., Ishikawa, R., Ishigaki, T., & Ishida, Y. (2020). Clinical effectiveness of metacognitive training as a transdiagnostic program in routine clinical settings: A prospective, multicenter, single-group study. *Japan Journal of Nursing Science*, e12389. Advance online publication https://doi.org/10.1111/jjns.12389.

Thomas, N. (2015). What's really wrong with cognitive behavioral therapy for psychosis? *Frontiers in Psychology, 6*, 323. https://doi.org/10.3389/fpsyg.2015.00323.

Tsukagoshi, C., Tawara, A., Matsuoka, K., Ubukata, S., & Naya, A. (2016). A group treatment using "SCIT" for individuals with social cognitive impairments following acquired brain injury—A preliminary study. *Higher Brain Function Research, 36*(3), 450–458.

Yoshinaga, N., Nosaki, A., Hayashi, Y., Tanoue, H., Shimizu, E., Kunikata, H., et al. (2015). Cognitive behavioral therapy in psychiatric nursing in Japan. *Nursing Research and Practice, 2015*, 529107. https://doi.org/10.1155/2015/529107.

Institution for Health Metrics and Evaluation. (2019). Global burden of disease calculator. Retrieved from: https://vizhub.healthdata.org/gbd-compare/. (Accessed 1 April 2021).

Chapter 32

Tinnitus and psychological and cognitive behavioral therapies in Japan

Sho Kanzaki, Mami Tazoe, Chinatsu Kataoka, and Tomomi Kimizuka
Department of Otolaryngology Head and Neck Surgery, Keio University School of Medicine, Tokyo, Japan

Abbreviations

CBT cognitive behavioral therapy
TCQ Tinnitus Cognitions Questionnaire
THI tinnitus handicap inventory
TRT tinnitus retraining therapy

Introduction

Tinnitus is defined as an auditory perception without external sound. Currently, there is no treatment for tinnitus. Particularly, severe chronic tinnitus usually has psychological effects, such as depression and anxiety disorder.

We have shown that patients with severe tinnitus have single nucleotide polymorphism, which is also associated with bipolar depression (Watabe et al., 2020). Chronic tinnitus is difficult to treat, but the purpose of treatment is alleviating distress but not resolving tinnitus. Cognitive behavioral therapy (CBT) has been shown to be effective and have long-term effects (Beukes, Andersson, Allen, Manchaiah, & Baguley, 2018; Fuller et al., 2020).

CBT and tinnitus treatment

CBT is one of the psychotherapy for tinnitus. Systematic review of seven papers revealed good validity of this treatment. Four metaanalyses of CBT for tinnitus have shown that CBT is both effective and associated with a high level of evidence. Hesser et al. found that the effects of CBT were maintained over time (Hesser et al., 2012), and Martinez Devesa, Waddell, Perera, and Theodoulou (2007) and Grewal, Spielmann, Jones, and Hussain (2014) observed significant improvements in both QOL and depression scores following CBT. Interestingly, the loudness of tinnitus did improve significantly following therapy. CBT is also recommended by tinnitus guidelines generated in the United States, Germany, the Netherlands, and Sweden (Idrizbegovic et al., 2011; German S3 Tinnitus Guideline, 2015; Tunkel et al., 2014) (Table 1).

The Tinnitus Clinical Practice Guidelines were first published in Japan in May 2019. The guideline is based on evidence and recommends CBT, for which strong evidence has been accumulated. However, in Japan, while CBT is used for the treatment of conditions such as depression, it is still rarely used for tinnitus.

Evidence level and recommended strength

The Tinnitus Guideline has shown that CBT is one of the best evidenced-based therapies (Ogawa et al., 2020) (Table 2). The *Guideline also does not recommend drug treatment.* Tinnitus retraining therapy (TRT) consists of two types, that is, wearing a hearing aid for tinnitus with deafness and listening to music or wearing sound generators. These therapies should partially mask tinnitus and divert attention to outside sounds.

We usually evaluate patient's tinnitus distress but not sound level because we cannot resolve tinnitus itself.

TABLE 1 Recommendation of tinnitus treatment.

1A: Strong evidence and strongly recommended
  ① TRT (wearing a hearing aid for tinnitus with deafness)
  ② CBT

1B: Medium evidence, highly recommended
  ① Educational counseling

2C: Weak evidence, weakly suggested, or conditionally recommended
  ① Drug therapy
  ② TRT (music and sound generators)
  ③ Cochlear implant
  ④ Repeated transcranial magnetic stimulation

2D: Uncertain evidence and weakly suggested or not recommended
  ① Acupuncture
  ② Laser treatment

TABLE 2 Cognitive behavior therapy.

Advantages	Disadvantages
• It is more cost-effective that conventional treatment (Kallio et al., 2008)	• Treatment by experts is necessary (Currently, there are few facilities that administer this therapy in Japan)
• High level of evidence	• Time required for treatment
• No side effects (Olze, 2015)	

CBT for tinnitus in Japan

CBT shows grade "1A" in the tinnitus guidelines but is not covered by reimbursement in Japan (Ogawa et al., 2020). Moreover, CBT is indicated for depression, anxiety disorder, and panic disorder and performed by psychiatrists. However, CBT for tinnitus has not been performed in otolaryngology because there are no psychiatrists in the otolaryngology clinic. We have psychotherapists in the otolaryngology clinic. Patients with tinnitus are treated in the otolaryngology clinic.

Patients usually do not understand the reason that they should be treated in the psychiatry clinic. CBT captures the human mind from the four aspects of emotion, behavior, cognition, and body, and these factors are related to each other. On the premise that the patient's symptoms and problem behaviors occur with the interaction between the individual and environment, a vicious cycle will occur. Moreover, CBT uses various behavioral techniques and cognition to promote changes in the patient's behavior and cognition. Therapeutic approaches aimed to reduce symptoms and distress using conventional techniques (Hesser, Weise, Westin, & Andersson, 2011). Currently, CBT is available from the first to third generations, but CBT for tinnitus has evidence and is recommended by Japanese (Ogawa et al., 2020) and tinnitus guidelines in Germany, United States, Netherlands, and Sweden (Fuller et al., 2020). These evidences are mainly second-generation CBT, but recently, third-generation CBT based on a mindfulness approach, which aims to take a bird's eye view of thinking and be free from thinking, is becoming widespread. In European countries and United States, the effectiveness of third-generation CBT for chronic tinnitus has already been reported (Fuller et al., 2020).

Practice and procedures

We initially performed educational counseling, including tinnitus mechanism, and how to change thinking on tinnitus. If educational counseling is an inadequate treatment, we recommend TRT or CBT. If patients prefer CBT, we perform psychological tests and initiate CBT as a protocol (Table 3).

We have a small CBT group, including seven members at maximum. Patients have 7 or 8 sessions in the CBT course. In the first two sessions, psychotherapists instruct patients to concentrate on stress responses individually, not on tinnitus. In the third session, medical doctors explain educational counseling and tinnitus treatment.

Basically, we will perform the 2nd generation of CBT to intervene focusing on both behavioral and cognitive aspects. When anxiety and tension caused by tinnitus interfere with daily life, systematic sensitization method is used as a method to increase anxiety and tension in small steps by using relaxation method. For hyperacusis, we also use exposure methods that allow us to get used to the sound in small steps. In addition, theories such as applied behavior analysis may be used to promote self-understanding of patients to understand avoidance behavior due to hyperacusis and tinnitus. In addition, as a third-generation CBT, mindfulness techniques may be used as an intervention in the state of mind, such as how tinnitus is currently perceived.

TABLE 3 Protocol of group CBT for tinnitus.

		Protocol		
Visit	What we do	Contents	Examination HW (homework) others	
1	Introduction	• Inspections (before and after the session) • Self-introduction: patient/certified public psychologist • Group description • Explanation of future schedule • Q & A	• POMS (before and after the session) • THI, TCQ, BDI II, STAI	Psychotherapist
2	Psychoeducation ① Know your stress monitoring Externalization and mindfulness	• Inspections (before and after the session) • What is stress? • Stressor and stress response • Capture of the stress response with the basic model of CBT • Understanding of the characteristics of stressors and stress reactions (cognitive, behavioral, physical/physiological, emotional) in daily life from the basic model of CBT • Implementation of stress checklist and feedback/review • Filling out and review of the CBT basic model sheet • Group discussion • Q & A • HW	• POMS (before and after the session) • Stress checklist • Tool: CBT basic model sheet • HW: "Write down daily stressors and stress reactions and apply them to the basic model of CBT"	Psychotherapist

Continued

TABLE 3 Protocol of group CBT for tinnitus—cont'd

Visit	What we do	Protocol Contents	Examination HW (homework) others	
3-1	Psychoeducation ② Know your stress response Two assessments	• Inspections (before and after the session) • Looking back on HW • Dealing with stress • Knowing the characteristics of your own stress coping • Implementation of TAC24 and feedback/review • Filling out the CBT basic model sheet • Understanding of the characteristics of stressors and stress reactions (cognitive, behavioral, physical/physiological, emotional) in daily life from the basic model of CBT • Group discussion • Q & A • HW	• POMS (before and after the session) • Tri-Axial Coping Scale-2 (TAC24) (stress coping and features) • HW: "Write down your daily stressor and stress response and apply it to the basic model of CBT. Moreover, write about the stress coping you did at that time"	Psychotherapist
3-2	Education of tinnitus	• Mechanism of onset of tinnitus • Treatment of tinnitus • Tinnitus and stress • Q & A		Medical doctor
4	Know your stress response 3 Cognition (automatic thinking)	• Conducting inspections (before and after the session) • Looking back on HW • Understanding the characteristics of stressors and stress reactions (cognitive, behavioral, physical/physiological, emotional) in daily life from the basic model of CBT • What is cognition (automatic thinking)? • Test to know the habit of thinking Feedback/reflection • Considering your own cognition (automatic thinking) (automatic thinking) and stress response • Group discussion • Q & A • HW	• POMS (before and after the session) • Automatic Thinking Test: A test to know the habit of thinking • HW: "Write down the daily stressor and stress response and apply it to the basic model of CBT. Furthermore, write down the stress coping that was performed at that time and examine the effect by scaling"	Psychotherapist
5	Consider a wide range of ideas ① Cognitive reconstruction method	• Inspections (before and after the session) • Looking back on HW • Examination of the relationship between stress response and cognition (automatic thinking) • Considering cognition (automatic thinking) • Examination of changes in stress response due to new cognition (automatic thinking) • Group discussion • Q & A	• POMS (before and after the session) • HW: "Write down daily stressors and stress responses and apply them to the basic model of CBT. Examine changes in stress responses due to new cognition (automatic thinking)"	Psychotherapist

TABLE 3 Protocol of group CBT for tinnitus—cont'd

Visit	What we do	Protocol Contents	Examination HW (homework) others	
6	Consider a wide range of ideas ① Cognitive reconstruction method	• Understanding tinnitus and stress response • Considering tinnitus and cognition (automatic thinking) • Considering tinnitus recognition (automatic thinking) • Considering change in stress response (cognition, behavior, physical/physiological, emotion) due to new cognition (automatic thinking) • Group discussion • Q & A • HW	• λ HW: "Write down the stress response due to tinnitus and apply it to the basic model of CBT. Examine (scaling) the change in stress response due to new cognition (automatic thinking)." Then, examine the change in stress response due to it is also examined (scaling)"	Psychotherapist
7	Considering a wide range of ideas ③ About tinnitus Cognitive reconstruction method	• Inspections (before and after the session) • Looking back on HW • Understanding tinnitus and stress response • Considering tinnitus and cognition (automatic thinking) • Considering tinnitus recognition (automatic thinking) • Considering change in stress response (cognition, behavior, physical/physiological, emotion) due to new cognition (automatic thinking) • Group discussion • Q & A • HW	• POMS (before and after the session) • HW: "Write down the stress response due to tinnitus and describe the basic model of CBT." Moreover, "describe the measures taken at that time and examine (scaling) the change in the stress response due to it." "New cognition (automatic thinking) and change (scaling) in the stress response will be examined"	Psychotherapist
8	Summary	• Inspections (before and after the session) • Looking back on HW • Overall summary • Implementation of inspection • Group discussion • Q & A • Implementation of inspection • Feedback of test results (changes from the first time) • Looking back • Group discussion • Q & A	• POMS (before and after the session) • THI, TCQ, BDI II, and STAI	Psychotherapist
Follow-up (FU)	One month after final CBT	• Psychoeducation and remotivation as appropriate • Implementation of inspection • Feedback of test results (changes from the first time) • Group discussion • Q & A • Psychoeducation and remotivation as appropriate	• THI, TCQ, BDI II, and STAI	Psychotherapist

Continued

TABLE 3 Protocol of group CBT for tinnitus—cont'd

		Protocol		
Visit	What we do	Contents	Examination HW (homework) others	
FU	Three months after final CBT	• Implementation of inspection • Feedback of test results (changes from the first time) • Looking back • Group discussion • Psychoeducation and remotivation as appropriate	• THI, TCQ, BDI II, and STAI	Psychotherapist
FU	Six months after final CBT	• Implementation of inspection • Feedback of test results (changes from the first time) • Looking back • Group discussion • Psychoeducation and remotivation as appropriate	• THI, TCQ, BDI II, and STAI	Psychotherapist
FU	One year after initial treatment	• Implementation of inspection • Feedback of test results (changes from the first time) • Looking back • Group discussion • Psychoeducation and remotivation as appropriate	• THI, TCQ, BDI II, and STAI	Psychotherapist

The number of patients in the therapy group is 5–7. CBT has 7–8 sessions. HW, homework; Psychological tests: POMS, Profile of Mood States 2nd edition; BDI II, Beck Depression Inventory-Second Edition; STAI, State-Trait Anxiety Inventory.

Mini-dictionary of terms

Examples of mini-dictionary of terms

We have shown the protocol of CBT for patients with tinnitus in Table 3.

Key facts

Purpose of treatment for tinnitus is to reduce attention to tinnitus

Physicians explained to patients with tinnitus that "do not worry about tinnitus," and many patients will have almost no problems with it. However, if it is not possible, they will continue to have chronic tinnitus.

From a clinical psychological point of view, "do not worry about tinnitus" is a negative instruction. The procedure misleads the patient to "pay attention to tinnitus, then remove attention from tinnitus." This treatment cannot completely divert the attention from tinnitus.

We do not use the word "tinnitus" when we explain to patients to "direct attention to another thing other than tinnitus."

A better instruction to "find something that can be concentrated" can be considered, but it is difficult to always concentrate on something in daily life, and its effectiveness is low.

Practically, we recommend patients to perform activities, such as reading a newspaper, talking with people, cleaning, and taking a walk, to easily divert the attention from tinnitus.

However, patients do not completely pay attention to something during the abovementioned activities. Recently, CBT combined with TRT have been used for a relatively short period. In the meantime, a program has been developed to improve the difficulty of treating tinnitus (Grewal et al., 2014).

Applications to other areas

In this chapter on tinnitus, we reviewed the group CBT protocol on tinnitus. Several studies have shown the outcome measures with the CBT (Ogawa et al., 2020). In a Cochrane study in 2010 (Martinez-Devesa, Perera, Theodoulou, & Waddell, 2010), significant improvements in depression scores and quality of life were found. They also found that CBT had a positive effect on the management of tinnitus. However, they did not find evidence of a significant change in the subjective loudness of tinnitus. Andersson demonstrated a moderate to strong effect on tinnitus annoyance in a metaanalytic review (Andersson & Lyttkens, 1999). They also found some effects on tinnitus loudness. However, CBT had less effect on negative affect and sleep problems.

Summary points

- Many difficulties with chronic tinnitus are caused by the inability to divert attention from tinnitus because severe chronic tinnitus is associated with depression and anxiety disorder.
- When patients with tinnitus feel anxious, attention is directed, and anxiety causes a stress response, which is conditioned on tinnitus. The sound itself can be threatening. However, no stress response occurs when attention is removed from tinnitus.
- Before treatment, we perform educational counseling (understanding tinnitus mechanism and treatment), which is one of the steps in CBT.
- Our protocol of group CBT has 7–8 sessions, including counseling and homework.
- In a metaanalysis, the effectiveness of CBT for tinnitus has been shown.

References

Andersson, G., & Lyttkens, L. (1999). A meta-analytic review of psychological treatments for tinnitus. *British Journal of Audiology, 33*(4), 201–210. https://doi.org/10.3109/03005369909090101.
Beukes, E. W., Andersson, G., Allen, P. M., Manchaiah, V., & Baguley, D. M. (2018). Effectiveness of guided internet-based cognitive behavioral therapy vs face-to-face clinical care for treatment of tinnitus: A randomized clinical trial. *JAMA Otolaryngology—Head & Neck Surgery, 144*(12), 1126–1133. https://doi.org/10.1001/jamaoto.2018.2238.
Fuller, T., Cima, R., Langguth, B., Mazurek, B., Vlaeyen, J. W., & Hoare, D. J. (2020). Cognitive behavioural therapy for tinnitus. *Cochrane Database of Systematic Reviews, 1*, CD012614. https://doi.org/10.1002/14651858.CD012614.pub2.
German S3 Tinnitus Guideline. (2015). *017/064: Chronic tinnitus [AWMF]-Register Nr.* 017/064 Klasse2015.
Grewal, R., Spielmann, P. M., Jones, S. E., & Hussain, S. S. (2014). Clinical efficacy of tinnitus retraining therapy and cognitive behavioural therapy in the treatment of subjective tinnitus: A systematic review. *Journal of Laryngology and Otology, 128*(12), 1028–1033. https://doi.org/10.1017/S0022215114002849.
Hesser, H., Gustafsson, T., Lunden, C., Henrikson, O., Fattahi, K., Johnsson, E., ... Andersson, G. (2012). A randomized controlled trial of Internet-delivered cognitive behavior therapy and acceptance and commitment therapy in the treatment of tinnitus. *Journal of Consulting and Clinical Psychology, 80*(4), 649–661. https://doi.org/10.1037/a0027021.
Hesser, H., Weise, C., Westin, V. Z., & Andersson, G. (2011). A systematic review and meta-analysis of randomized controlled trials of cognitive-behavioral therapy for tinnitus distress. *Clinical Psychology Review, 31*(4), 545–553. https://doi.org/10.1016/j.cpr.2010.12.006.
Idrizbegovic, E., Kjerulf, E., & Team for Diagnostics Hearing Habilitation, & Adults., C. a. Y. a. H. R. f. (2011). *Tinnitus Care Program [Tinnitus Vårdprogram].* Stockholm: Karolinska Institute.
Kallio, H., Niskanen, M. L., Havia, M., Neuvonen, P. J., Rosenberg, P. H., & Kentala, E. (2008). I.V. ropivacaine compared with lidocaine for the treatment of tinnitus. *British Journal of Anaesthesia, 101*(2), 261–265. https://doi.org/10.1093/bja/aen137.
Martinez Devesa, P., Waddell, A., Perera, R., & Theodoulou, M. (2007). Cognitive behavioural therapy for tinnitus. *Cochrane Database of Systematic Reviews,* (1), CD005233. https://doi.org/10.1002/14651858.CD005233.pub2.
Martinez-Devesa, P., Perera, R., Theodoulou, M., & Waddell, A. (2010). Cognitive behavioural therapy for tinnitus. *Cochrane Database of Systematic Reviews,* (9), CD005233. https://doi.org/10.1002/14651858.CD005233.pub3.
Ogawa, K., Sato, H., Takahashi, M., Wada, T., Naito, Y., Kawase, T., ... Kanzaki, S. (2020). Clinical practice guidelines for diagnosis and treatment of chronic tinnitus in Japan. *Auris, Nasus, Larynx, 47*(1), 1–6. https://doi.org/10.1016/j.anl.2019.09.007.
Olze, H. (2015). Cochlear implants and tinnitus. *HNO, 63*(4), 291–297. https://doi.org/10.1007/s00106-014-2975-5.
Tunkel, D. E., Bauer, C. A., Sun, G. H., Rosenfeld, R. M., Chandrasekhar, S. S., Cunningham, E. R., Jr., ... Whamond, E. J. (2014). Clinical practice guideline: tinnitus executive summary. *Otolaryngology—Head and Neck Surgery, 151*(4), 533–541. https://doi.org/10.1177/0194599814547475.
Watabe, T., Kanzaki, S., Sato, N., Matsunaga, T., Muramatsu, M., & Ogawa, K. (2020). Single nucleotide polymorphisms in tinnitus patients exhibiting severe distress. *Scientific Reports, 10*(1), 13023. https://doi.org/10.1038/s41598-020-69467-0.

Chapter 33

Cognitive-behavioral interventions for mental health conditions among women in sub-Saharan Africa

Huynh-Nhu Le[a], Kantoniony M. Rabemananjara[a], and Deepika Goyal[b]
[a]Department of Psychological and Brain Sciences, George Washington University, Washington, DC, United States [b]The Valley Foundation School of Nursing, San José State University, San Jose, CA, United States

Abbreviations

ART antiviral therapy
CBT cognitive-behavioral therapy
CHW community health worker
CHV community health volunteer
CPT cognitive processing therapy
EPDS Edinburgh Postnatal Depression Scale
GBV gender-based violence
HIV+ human immunodeficiency virus positive
IPV intimate partner violence
PD perinatal depression
PMTCT prevention of mother-to-child transmission
PM+ problem management plus
PST problem-solving therapy
PTSD posttraumatic stress disorder
RCT randomized controlled trial
WHO World Health Organization

Introduction

Common mental health conditions, such as depression and anxiety, are the third leading global burden of disease worldwide for women between the ages of 14 and 44 years (Mayosi et al., 2009). Women are at higher risk for depression and anxiety than men in low and middle-income countries (LMICs), including countries in sub-Saharan Africa (Honikman, van Heyningen, Field, Baron, & Tomlinson, 2012; World Health Organization, 2020), often due to a range of interpersonal, psychosocial, and social risk factors experienced across this vast and diverse region. In this chapter, we focus on the mental health conditions of women in sub-Saharan Africa for several reasons. First, women in this region represent slightly over 50% of the continent's human resources (WHO, 2012). Second, Africa's population is rapidly increasing and expected to double from 1.2 billion to 2.4 billion by 2050. This growth is due mostly to the increasing number of women of reproductive age, expected to reach almost 1 billion by 2050 (You, Hug, & Anthony, 2015). Third, rates of maternal mortality and HIV/AIDS disproportionately affect women in Africa (Lathrop, Jamieson, & Danel, 2014). Fourth, women are at high risk for mental health conditions, such as depression, anxiety, trauma-related disorders, and alcohol abuse (Gibbs, Jewkes, Willan, & Washington, 2018; Moultrie & Kleintjes, 2006). Fifth, health and mental health conditions are intrinsically interrelated; for example, the rates of gender-based violence (GBV), such as intimate partner violence (IPV), are higher among women living with HIV (Mitchell, Wight, Van Heerden, & Rochat, 2016). Thus, gender inequity, poverty among women, limited economic capacity, sexual and GBV including female genital mutilation contribute to women's vulnerability to mental health conditions (Ramjee & Daniels, 2013; WHO, 2012).

In turn, these mental health conditions are associated with deleterious effects on maternal and child physical and psychological health and well-being (Gelaye, Rondon, Araya, & Williams, 2016; WHO, 2019).

This chapter reviews cognitive-behavioral therapy (CBT) interventions to address mental health conditions that disproportionately affect women in sub-Saharan Africa, focusing on three areas: (1) HIV and comorbid mental health conditions, (2) GBV and other traumatic experiences, and (3) perinatal depression (PD). CBT interventions are widely used to effectively treat and prevent many mental health conditions around the globe (Kuo, 2019). Despite the high prevalence and the harmful consequences associated with mental disorders (Gelaye et al., 2016), many women in sub-Saharan Africa, especially those who live in resource-restricted settings (e.g., rural areas), are left untreated largely due to a shortage or lack of mental health professionals (Marangu, Sands, Rolley, Ndetei, & Mansouri, 2014). To address this mental health treatment gap, the WHO has recommended incorporating *task-shifting*, where nonspecialist mental health workers (e.g., nurses, community health workers) are trained to deliver mental health interventions (Ginneken et al., 2011). Task-shifting allows women to access psychological services in settings where they would typically receive healthcare or community services (e.g., primary health facilities). In these settings, CBT has been used as psychotherapeutic interventions that are delivered by a variety of nonspecialist providers in sub-Saharan Africa and other LMICs (Chowdhary et al., 2013; Patel, Chowdhary, Rahman, & Verdeli, 2011).

In this chapter, we review interventions that are CBT-based, defined as psychological interventions that comprise at least one CBT component: behavioral activation, cognitive reframing/restructuring/problem-solving, relaxation training, and/or exposure therapy. This chapter describes intervention studies that include trials with quantitative results describing mental health outcomes or protocol descriptions for studies currently taking place and do not yet have results. Articles in this chapter focus on women's health and mental health related to HIV, trauma, and PD in sub-Saharan Africa. Studies with ≥80% of women in the sample were also included to maximize the number of interventions evaluated in this region.

HIV and comorbid mental health conditions

Sub-Saharan Africa is a region in the world that is most affected by HIV. Girls and women are disproportionately affected accounting for 59% of all new HIV infections (UNAIDS, 2020). Women living with HIV are more likely to experience depression, anxiety, trauma, and IPV (Dos Santos & Wolvaardt, 2016; Mitchell et al., 2016; UNAIDS, 2020). These mental health conditions can significantly decrease engagement in HIV testing and compromise antiviral therapy (ART) adherence (Tao, Vermund, & Qian, 2018). The disparate impact of HIV on women is associated with multiple factors, ranging from biological, socioeconomic, to systemic vulnerabilities (Ramjee & Daniels, 2013).

HIV comorbidity and CBT

There are five studies that have used CBT to address HIV and comorbid mental health conditions focusing on women (see Table 1). The first study evaluated the *Ziphamandla* (Empower Yourself) intervention in 14 individuals in South Africa (93% female) living with HIV and meeting major depression criteria (Andersen et al., 2018). *Ziphamandla* was adapted from a CBT intervention for ART and depression that was effective in improving depression and ART adherence among HIV individuals in the United States (Safren et al., 2016). The *Ziphamandla* program includes 6 to 8 individual sessions, with 5 modules addressing problem-solving, behavioral activation, and relaxation. Delivered by nurses, participants found *Ziphamandla* acceptable and feasible. Results supported depressive symptom improvement but not in ART-adherence post-intervention and at 3 months. Limitations included low fidelity to intervention delivery and that the intervention was not designed specifically for women.

Futterman et al. (2010) conducted a pilot study evaluating the *Mamekhaya* ("respect for women" in the Xhosa language) program, an 8-session culturally adapted CBT group aimed to improve adherence to the Prevention of Mother-to-Child Transmission (PMTCT) among HIV+ pregnant women in South Africa recruited at two clinics. The program was delivered by HIV+ peer mentor mothers who provided psychosocial support and psychoeducation covering four broad topics: HIV adherence; psychological well-being (e.g., depression, substance use, stigma); partnering and prevention transmission methods; and parenting, including attachment and immunization. Sessions included role plays, didactic, paired and group discussions, music, meditation, breathing exercises, and goal-setting activities. The intervention site received Mamekhaya in addition to PMTCT and the control site received only PMTCT. Both groups reported high levels of adherence to core practices of PMTCT with greater improvement in HIV knowledge, establishing social support, improved coping, and deceased depressive symptoms in the intervention group versus the control group. A limitation included the low follow-up rate at 6 months postpartum (44%).

To address HIV care-seeking, adherence behaviors, and clinical outcomes among HIV+ women with sexual trauma histories, Sikkema et al. (2018) conducted a pilot RCT evaluating a culturally adapted trauma-focused intervention,

TABLE 1 CBT-based interventions addressing HIV+ and mental health comorbidity.

Mental health condition	Author/year	Sample/design	CBT intervention	Main intervention components	Facilitator(s)	Country	Results/outcomes
Depression	Andersen et al. (2018)	13 of 14 (92.9%) female HIV+	Ziphamndla (Empower yourself)	6–8 individual sessions. Focus areas: problem-solving, behavioral activation, and relaxation.	Nurses	South Africa	Decreased self-report and clinician-report in depressive symptoms from baseline to 3-month follow-up treatment. No significant improvement in ART adherence following treatment.
Depression, anxiety	Chibanda et al. (2016)	573 (86.4% women); 41.7% HIV positive. Cluster RCT	Friendship Bench	6-sessions, individual PST. Focus on identifying and resolving problems, increasing coping and control over one's life. Plus optional 6-sessions peer support program.	Lay health workers	Zimbabwe	Intervention group had decreased depressive symptoms, anxiety, and disability. Improved health-related quality of life compared to usual care, no differences in results with HIV status.
Depression	Futterman et al. (2010)	160 pregnant females, HIV+; non-RCT	Mamekhaya	Combined mothers2mothers peer-mentoring program with 8-sessions of culturally adapted CBT group. Focus areas: HIV adherence, prevention transmission methods, well-being, and parenting.	Peer mentor mothers, HIV+	South Africa	HIV knowledge and social support increased, decreased depressive symptoms between intervention and control groups. No significant improvement in PMCTC-related actions or transition risk behaviors.
Sexual trauma	Sikkema et al. (2018)	64 HIV+ females with sexual abuse histories; pilot RCT	Improving AIDS Care After Trauma (ImpACT)	7 sessions: 4 individual + 3 group - trauma-focused, culturally adapted coping model. Goals: increase awareness of personal values, understand impact of sexual trauma, encourage adaptive strategies for coping to promote long-term HIV engagement using CBT and coping framework.	Lay providers and community care workers	South Africa	Significant decrease in avoidance, arousal symptoms, and PTSD symptoms; increased ART adherence at 3 months in intervention versus control group.
Stigma, depression	Tshabalala and Visser (2011)	20 HIV+ women experiencing stigma; RCT	CBT	8 individual sessions exploring role of HIV in one's life, feelings of powerlessness, guilt, anger, negative self-evaluation, low self-worth, and stigma impact. Cognitive restructuring, alternative coping skills, and behavioral activation.	Clinical psychologist	South Africa	Intervention group had significant decrease in depression, lower internalized stigma, improved ways of coping, higher self-esteem than control group. No group difference in enacted stigma.

Improving AIDS CARE after Trauma (ImpACT) in South Africa. ImpACT includes 7 sessions aimed to reduce traumatic stress and increase HIV care engagement by teaching women coping strategies with HIV and trauma. CBT components include psychoeducation to address stress from having HIV and sexual trauma. The intervention was delivered by a lay provider with supervision from a clinical psychologist. Women in the Standard of Care (SoC) group received 3 sessions of adherence counseling, including education on HIV and ART readiness and engagement to care, required for patients prior to initiating ART in public HIV clinics and is delivered by clinic counselors. Following the ImpACT intervention, participants received the SoC sessions. High rates of sexual abuse were reported, ranging from 18.9% in childhood to 76.6% in adulthood, with 41% reporting current sexual abuse. IPV rates were also high; 84.4% experienced in one's lifetime and 23.4% currently. Results indicate that at both 3 and 6 months follow up, intervention participants reported significant decreases in overall PTSD symptom severity scores and all PTSD subscales (avoidance, hyperarousal). ART adherence was higher in the intervention group compared to the control group at 3 but not 6 months follow-up.

Tshabalala and Visser (2011) developed a CBT intervention for Black South African women living with HIV to address issues of internalized and enacted stigma to improve their self-worth and enhance health-seeking behaviors and outcomes. Although stigma is not a mental health condition, stigma adversely impacts help seeking for HIV care and depression ((Parker & Aggleton, 2003). The intervention was adapted to meet the needs of HIV+ women based on focus groups with clinical psychologists and patient stakeholders. The resulting intervention includes 8 individual sessions, delivered by a male clinical psychologist with CBT expertise, and addressed 5 common themes typically experienced by women living with HIV+ and taught ways to challenge feelings of powerlessness and limited self-worth, negative self-evaluation, destructive behavior, stigma, and uncertainty regarding the future using positive reframing strategies, assertiveness training, and goal setting. Results indicate that women in the intervention group reported improved coping (less negative and more positive coping), decreased internalized stigma and depression, and increased self-esteem at postintervention. There were no differences in enacted stigma. Women in the intervention condition also qualitatively reported improved ways of coping with their HIV and "rediscovered their meaning in life" (p. 23). One limitation of this study was the lack of a follow-up evaluation period to assess whether these positive outcomes are enduring.

The Friendship Bench project, based on problem-solving therapy (PST) to address common mental health disorders (depression and anxiety), has been widely accepted in Zimbabwe, with over 7000 participating in this intervention over a five-year period (Chibanda et al., 2011). Over 80% of this program's users live with HIV (Chibanda et al., 2017). In a cluster RCT, Chibanda et al. (2016) evaluated the effectiveness of the Friendship Bench; trained female lay health workers provided six sessions of PST, teaching participants to identify and resolve problems through active coping skills and behavioral activation strategies. Following the individual sessions, participants can attend a 6-session peer group support. Participants identify and work on problems with a feasible action plan, and text messages, phone calls, or both aim to reinforce PST to implement an action plan. Participants included majority women (86.4%); approximately 42% were living with HIV+ who screened positive for depression and anxiety on a locally validated Shona Symptom Questionnaire (Patel, Simunyu, Gwanzura, Lewis, & Mann, 1997). Compared to the control group (usual care plus information, education and support on common mental disorders), the intervention group had fewer symptoms of depression and anxiety, a lower risk of developing depression symptoms, and improved quality of life at six months follow-up. A limitation was that the intervention was not designed specifically for women, and it is unclear if results are generalizable beyond the 6-month period.

Trauma and gender-based violence against women

This section reviews interventions focusing specifically on trauma-related stressors stemming from GBV, including IPV or sexual violence against women (see Table 2). A public health issue worldwide, GBV contributes to increased mortality and morbidity of women with deleterious effects on women's health and their children (Firoz et al., 2013; WHO, 2013). GBV is more prevalent in developing countries with higher rates of poverty, food insecurity, and low education, as in the sub-Saharan Africa region (Gibbs et al., 2018). Recent research indicates that the prevalence of all forms of IPV against women in this region is among the highest in the world at 36% (Muluneh, Stulz, Francis, & Agho, 2020). Moreover, IPV rates are higher among women who have experienced child abuse (Machisa, Christofides, & Jewkes, 2017). Additionally, GBV is more prevalent among women living with HIV/AIDS (Ramjee & Daniels, 2013) and in high-conflict settings (e.g., postwar; Greene et al., 2019). Women who experience GBV are at higher risk for developing depression, anxiety, and PTSD (Gibbs et al., 2018; Tsai, Tomlinson, Comulada, & Rotheram-Borus, 2016). Women during their childbearing years, for instance, are more likely to develop postpartum depression (PPD) and PTSD, which were also associated with a greater risk of subsequent revictimization leading to a vicious cycle of trauma that not only affect their mental well-being but also their offspring (Tsai et al., 2016). Another type of trauma that is commonly experienced by women in sub-Saharan Africa

TABLE 2 CBT-based interventions addressing trauma and mental health comorbidity.

Mental health condition	Author/year	Sample/design	CBT intervention	Main intervention components	Facilitator(s)	Country	Results/outcomes
PTSD, depression, anxiety	Bass et al. (2013)	Female sexual violence survivors with high levels of PTSD, depression, and anxiety: 157 provided therapy and 248 with individual support	Cognitive Processing Therapy	CPT including 1 individual session and 11 group sessions. Includes cognitive components of CPT.	Psychosocial assistants (community-based paraprofessionals supervised by psychosocial staff and clinical experts.	Democratic Republic of Congo	Depression and anxiety improved for both groups, but were significantly improved in the therapy group and similar patterns found in PTSD improvement.
Gender-based violence	Greene et al. (2019)	Congolese refugee female IPV survivors	Cognitive Processing Therapy	8-sessions of CPT and advocacy counseling.	Lay refugee facilitators	Tanzania	Protocol and intervention development and cohort data indicating high acceptability and attendance for the brief CPT intervention.
Gender-based violence	Bryant et al. (2017); Sijbrandij et al. (2016)	Women who experienced gender-based violence: 209 were assigned to PM+ and 212 to EUC.	Problem Management Plus (PM+)	5 weekly 90-min sessions. Focus areas: (1) stress management; (2) problem solving; (3) behavioral activation; (4) skills to strengthen social support.	Female CHWs	Kenya	Small effect sizes in the reduction of posttraumatic stress and functional impairment between PM+ participants and control group.
Psychological distress from birth surgery/trauma	Watt et al. (2015); Watt et al. (2017)		Mental health intervention for obstetric fistula patients	6, 60-min individual counseling sessions. Two before surgery, four after surgery.	Nurses	Tanzania	There were significant improvements in mental health outcomes over time, with no evidence of differences by condition.

during the reproductive years is associated from surgery to correct an obstetric fistula (Oluwasola & Bello, 2020). Affecting 1.57 per 1000 women in sub-Saharan Africa, an obstetric fistula is a painful and an often paralyzing childbirth injury that has been highly associated with depression, anxiety, trauma, self-esteem and stigma (Oluwasola & Bello, 2020; Watt et al., 2017).

Trauma and GBV against women and CBT

There are four studies that have used CBT-based interventions to treat trauma-related mental health conditions. Three studies address trauma stemming from GBV and one study addresses obstetric fistula. The first study (Bass et al., 2013) was a RCT of cognitive processing therapy (CPT), a protocol-based CBT for trauma symptoms, adapted and tested among Congolese survivors of sexual violence with high levels of PTSD, depression, and anxiety. CPT is a type of CBT that is generally delivered over 12 sessions and helps participants learn how to challenge unhelpful thoughts related to trauma (Resick & Schnicke, 1992). A total of 405 women in the Democratic Republic of Congo were randomly assigned to CPT ($n=248$ in 8 villages) or individual support ($n=157$ in 7 villages). The intervention condition consisted only of the cognitive parts of CPT (without a trauma narrative) and included one individual session and 11 group sessions. In the comparison condition, women received support services, such as psychosocial, economic, medical support and/or legal referrals. The intervention was led by trained psychological assistants with ongoing supervision. While both conditions showed improvements during treatment, the intervention group had greater improvements than those in the comparison group, effectively reducing PTSD symptoms and combined depression and anxiety symptoms at the end of treatment and at 6 months after treatment. This adaptation was made for low literacy participants who may experience ongoing violence. One limitation mentioned by the authors is that the measures used to identify clinical cases of PTSD and combined depression and anxiety were not validated for this sample.

The second study (Greene et al., 2019) was conducted in a refugee camp in Tanzania and described the development of an integrated intervention including CPT and advocacy counseling for Congolese refugee women who were IPV survivors. This formative research included four phases of the intervention development. The treatment, *Nguvu* (meaning strength in Swahili), consisted of 8 sessions including one individual session of advocacy counseling aimed at improving autonomy and empowerment of women affected by IPV, 6 group sessions of CPT, and a final group session of advocacy counseling. The intervention was delivered by trained local refugee incentive workers. To test the delivery, acceptability, and feasibility of the intervention, a treatment cohort mixed-methods study included 60 participants who experienced IPV and had elevated psychological distress. Results revealed that participants attended two-thirds of the sessions. One limitation with this study was that the key adaptations of the intervention were mainly drawn from the feedback of the facilitators, with limited feedback from participants.

The third study (Bryant et al., 2017) uses Problem Management Plus (PM+), an intervention originally developed by the WHO as a CBT-based individualized program to treat common mental disorders and that can be delivered by nonspecialist counselors. Sijbrandij et al. (2016) developed a study protocol for a RCT specifically for women affected by gender-based violence and urban adversity in Kenya. In the RCT (Bryant et al., 2017), 209 women were assigned to PM+ and 212 women to Enhanced Usual Care (EUC). The PM+ sessions were delivered by trained and supervised community health workers (CHWs) in home visits. PM+ includes five sessions that cover motivational interviewing, psychoeducation, stress management; CBT components included problem-solving and behavioral activation and strengthening social supports and relapse prevention education. Those in the comparison group were referred to primary healthcare centers and were seen by nurses who provided counseling but did not follow the PM+ manual. Results indicated that both conditions significantly decreased psychological distress overtime. However, relative to EUC, the PM+ condition moderately reduced psychological distress and PTSD posttreatment and from baseline to 3- month follow-up. Another key finding from this study was that women who experienced GBV can be screened and detected for distress and impairment. One limitation from this study included no long-term follow-up assessments to determine sustainability of intervention effects.

The fourth study by Watt et al. (2015, 2017) addresses psychological distress resulting from obstetric fistula surgery in Tanzania. Watt et al. (2015) developed a CBT and coping psychological intervention which includes six individual sessions delivered by a community health nurse: recounting the fistula story, creating a new story about the fistula, loss, grief, and shame, specific strategies for coping, social relationships, and planning for the future. In 2017, Watt et al. conducted a pilot RCT to determine acceptability of the intervention for patients with obstetric fistulas ($n=60$) in two conditions (30 per group) and moderate to high levels of depression, anxiety, PTSD, self-esteem. The goal of the intervention was to help patients reframe their fistula experience and develop coping skills to manage the physical and psychosocial impact. Results found high intervention fidelity, high attendance, and high satisfaction post treatment. Moreover, in both conditions, there were significant improvements in all mental health outcomes from baseline to post treatment and from baseline to 3-month follow-up, but the difference between the conditions in these changes was small. The small sample size is a limitation in this study, which was underpowered to achieve significant effects.

Perinatal depression

A growing area of research in sub-Saharan Africa focuses on perinatal depression (PD), a common disorder among childbearing women. PD is defined as the onset of depressive symptoms during the pregnancy (antenatal depression) or first year after the birth of an infant (postpartum depression, PPD) (American College of Obstetricians and Gynecologists, 2018). PD is multifactorial and contributing issues for women living in sub-Saharan Africa include HIV, IPV, and cultural factors (Gelaye et al., 2016; Sowa, Cholera, Pence, & Gaynes, 2015; Wittkowski, Gardner, Bunton, & Edge, 2014). Antenatal depression rates in LMICs range between 10% and 35% (Kaaya et al., 2016) and are associated with poor child growth (Sikander et al., 2019; Surkan, Patel, & Rahman, 2016). PPD rates range between 6.9% and 50% (Atuhaire, Brennaman, Cumber, Rukundo, & Nambozi, 2020; Pellowski et al., 2019) and contribute to poor maternal child bonding, poor infant cognitive development, and lower breastfeeding rates (Rahman et al., 2016).

Perinatal depression and CBT

CBT is well documented as an effective treatment and prevention for PD (Nyatsanza, Schneider, Davies, & Lund, 2016; Sockol, 2015). Four studies have used CBT-based interventions to prevent PD among women who are at high risk for PD in sub-Saharan Africa (see Table 3). The first, a pilot feasibility study (Green et al., 2020), assessed the CBT-based *Healthy Moms* intervention with 47 perinatal women recruited from two public hospitals in Kenya. Modeled after the *Thinking Healthy* program (WHO, 2015), *Healthy Moms* provides an evidence-based task-sharing approach to prevent PD by utilizing CHWs to conduct home visits using CBT techniques. *Healthy Moms* was adapted for automated delivery via a mobile phone and includes three prenatal and 12 postpartum sessions. Sessions are delivered via an app, Zuri, with journaling in between sessions. Women rated their mood via SMS text messaging upon enrollment and then every 3 days thereafter. Ten-month postpartum results indicated the intervention was feasible and acceptable to participants. Although not conclusive, Zuri was associated with improved mood. Mobile phone and SMS access were required to participate, thereby limiting generalizability.

The second study was also an RCT to prevent PD among women receiving care in two South African antenatal clinics (Lund et al., 2020). Pregnant women with Edinburgh Postnatal Depression Scale (EPDS; Cox, Holden, & Sagovsky, 1987) scores ≥ 13 were randomized into an intervention ($n=184$) or control group ($n=200$). Intervention group participants received six CBT-based sessions covering six areas: (1) psychoeducation; (2) problem solving; (3) behavioral activation; (4) healthy thinking; (5) relaxation training; and (6) birth preparation. Sessions were facilitated by trained CHWs and findings revealed no difference in depression scores between the groups.

Using a longitudinal cluster RCT design, the third study evaluated the effectiveness of a CBT-based home visiting intervention delivered by CHW to address perinatal and infant health risks (Rotheram-Borus et al., 2014). Neighborhoods were randomized into intervention ($n=12$; 644 women) or control ($n=12$; 594 women) groups. Women at risk of developing PPD (EPDS scores ≥ 13) were recruited in pregnancy and followed through 18 months postpartum. Women in the control group received usual care. Women in the intervention received the *Philani Plus* mentor mother CHW home visiting program that consisted of four antenatal visits and four visits during the first 2 months postpartum, and then monthly up to 18 months postpartum. Visits lasted between 20 and 60 min. Randomization of entire neighborhoods and supervision of CHWs maintained intervention fidelity. Findings indicated no significant differences in depression scores. Using trained CHWs for home visits was feasible; however, a longer follow-up period was needed.

To determine if PPD could be prevented with longer follow-up, in the last study, Tomlinson et al. (2016) followed the same women as in Rotheram-Borus et al. (2014) and analyzed 36-month outcome data. Intervention group participants continued to receive home visits from trained CHWs with depressive symptoms assessed at each visit. Results indicated women in the intervention group had significantly lower depressive symptoms versus women receiving standard care. The authors discussed that the positive outcomes may be related to the caliber of CHWs who delivered the intervention as they were paid a stipend and only worked part-time.

Summary of CBT interventions across mental health conditions

This chapter reviewed 12 studies that used CBT-based interventions to address common mental health issues, including depression and anxiety (specifically trauma-related), among women in sub-Saharan Africa. The majority of these studies

TABLE 3 CBT-based interventions addressing perinatal depression.

Author/year	Sample/design	CBT intervention	Main intervention components	Facilitator(s)	Country	Results/outcomes
Green et al. (2020)	Pregnant and postpartum women ($N=47$); pilot feasibility study	Healthy Moms Program	CBT-based Thinking Healthy program	Artificial intelligence (Zuri) SMS texts delivered by mobile phone	Kenya	Healthy Moms delivered via Zuri was feasible.
Lund et al. (2020)	RCT of pregnant women scoring ≥ 13 on EPDS, intervention ($n=184$) and control ($n=200$) group	Thinking Healthy Program	6 psycho-education sessions, addressing problem solving, behavioral activation, healthy thinking, relaxation training, birth preparation	CHW	South Africa	No significant difference in depression between intervention and control at 3 months postpartum.
Rotheram-Borus et al. (2014)	24 RCT matched neighborhoods of perinatal women, intervention (12 neighborhoods, 644 women) or control (12 neighborhoods, 594 women)	Pilani Plus mother mentoring program	Pilani Plus program, 4 home visits during pregnancy monthly postpartum up to 18 months postpartum.	Trained mentor mothers	South Africa	No significant differences in depression scores between intervention and control groups
Tomlinson et al. (2016)	See Rotheram-Borus et al. (2014)	See Rotheram-Borus et al. (2014)	Same as Rotheram-Borus et al. (2014) above with additional support and monitoring visits up to 36 months postpartum.	See Rotheram-Borus et al. (2014)	South Africa	Significantly lower depression scores at 36 months in intervention versus control group

were conducted in the Southern region (South Africa, $n=6$; Zimbabwe $n=1$) or Eastern region (Kenya, $n=2$; Tanzania, $n=2$), and one in the Democratic Republic of Congo (Bass et al., 2013). There was no one standard model of "CBT" used; some studies focused exclusively on behavioral interventions (e.g., Andersen et al., 2018) or purely cognitive techniques (e.g., CPT; Bass et al., 2013; Greene et al., 2019). These interventions all differed across studies; however, a commonality was that these interventions were culturally and contextually adapted to meet the needs of African women using existing evidence-based CBT interventions (e.g., Thinking Healthy Program; Green et al., 2020). Additionally, studies differed in: (a) the research design, ranging from the pilot feasibility testing (e.g., Green et al., 2020) to clustered RCTs (e.g., Chibanda et al., 2016; Rotheram-Borus et al., 2014); (b) follow-up periods, ranging from immediately postintervention (e.g., Bass et al., 2013) to 36 months (Tomlinson et al., 2016); (c) settings, with most interventions integrated into settings where women are already receiving care (e.g., HIV clinics; Futterman et al., 2010); and (d) modality: individual, group, combined, or online. Except for Tshabalala and Visser (2011), most studies utilized nonmental health providers who were trained and supervised, utilizing task-shifting approaches to address the mental health gap (Ginneken et al., 2011). And, most interventions were developed to meet women's needs, except for the HIV-related interventions (Andersen et al., 2018; Chibanda et al., 2016). Furthermore, the CBT studies here are mostly treatment based, suggesting that more prevention interventions are warranted. Finally, except for Green et al. (2020) who used a texting platform to deliver their intervention and Chibanda et al. (2016) who included texts as an optional component, the remaining studies focus on utilizing in-person resources. Overall, these studies are promising in addressing the mental health of African women.

Future directions

This review suggests that CBT-based interventions are acceptable, feasible, and overall effective in reducing depression and anxiety among adult women and PD among pregnant and postpartum women in sub-Saharan Africa. These interventions are brief, structured, and can be effectively delivered by trained nonmental health providers, such as nurses and CHWs. These task-shifting approaches have been particularly beneficial for regions of the world where there is a documented shortage of mental health providers. Moreover, these interventions can be delivered within settings in which women are likely to be already be seen (e.g., HIV clinics). Although the focus of this review has been on adult-aged women, women at other periods of the developmental spectrum, including children, adolescents, and elderly are also at risk for mental health conditions (WHO, 2012, 2019). Culturally adapted CBT-based interventions for these populations are also needed.

Given the small number of studies found in this review, there is room for more methodologically rigorous trials to take place across a larger number of countries that make up the sub-Saharan region. There is the need to scale up the small number of studies and determine the cost effectiveness of these larger studies to assess the feasibility of integration at the health systems level, particularly in resource-restricted areas and high-conflict settings. Additionally, there is a need to identify which components of CBT are most effective for subgroups of women. For example, Tshabalala and Visser (2011) reviewed their therapy transcripts and process notes to identify the most and least effective CBT techniques used in their intervention to address HIV-related stigma. The most effective techniques focused on positive cognitive reframing and coping strategies, whereas the least effective were the "advanced" cognitive techniques, including identifying negative thought patterns and Socratic questioning. This study suggests that including mixed methods research can be used to identify which techniques work best for specific populations. Finally, more research is needed to examine the psychometric properties of screening tools and measures used to assess mental health conditions in sub-Saharan Africa to incorporate idioms of distress and symptom presentations in different cultural settings, which may have implications for intervention development and evaluation. Furthermore, the ongoing COVID-19 pandemic has negatively impacted women's mental health in sub-Saharan Africa (Rafaeli & Hutchinson, 2020; Semo & Frissa, 2020), suggesting an avenue of research for testing additional CBT interventions. Overall, CBT interventions can be part of the response aimed to improve the lives of women and their families in sub-Saharan Africa.

Practice and procedures

The main intervention components are described in Tables 1–3 and in the chapter.

Mini-dictionary of terms

We defined selected terms in the chapter.

Key facts

- Mental health conditions, such as depression and anxiety, are the third leading global burden of disease worldwide for women.
- Girls and women are disproportionately affected by HIV and account for 59% of all new HIV infections.
- Task-shifting approaches are feasible to address common mental health conditions using CBT-based interventions.

Applications to other areas

CBT-based interventions have the potential to be applied to three different areas. First, cognitive-behavioral therapy has been effectively used to treat more severe mental illnesses, such as schizophrenia and chronic treatment-resistant mood disorders (Thase, Kingdon, & Turkington, 2014). Although the first line of treatment for severe mental disorders are psychotropic medications, CBT can be used as adjunctive nonpharmacological treatments to reduce symptoms, improve the quality of the response to medication, or increase the amount of recovery time after achieving a treatment response, particularly among individuals (Thase et al., 2014). CBT can also be used to help families understand the impact of these disorders, especially in regions of the world where stigma related to mental health is high. This research, however, has been limited in sub-Saharan Africa. Second, CBT-based interventions can address the mental health conditions that may differ in presentation and needs across the developmental lifespan, including children, adolescents, young adults, and older populations (WHO, 2012). Third, given the increase of internet access and utilization of mobile phones worldwide, CBT-based interventions can be used as psychoeducational and prevention tools to reach more people online via apps or tele-health (Muñoz et al., 2018). More evidence is needed to determine whether it is feasible to use other types of technology such internet-based CBT, and tele-mental health in sub-Saharan Africa.

Summary points

- Women in sub-Saharan Africa are at high risk for poor mental health partly due to macro-level social determinants, such as poverty, gender-based violence, and living with HIV/AIDS.
- Culturally and contextually adapted CBT is feasible, acceptable, and effective as an intervention to treat depression, perinatal depression, anxiety, trauma related stressors, and associated with living HIV among women in sub-Saharan Africa.
- CBT is structured, time-limited, and manualized—aspects that can be adapted across different populations (e.g., people with low literacy) and contexts (e.g., rural and urban settings).
- Nonspecialist providers can be trained and supervised to deliver CBT interventions with fidelity.
- More research is needed to identify which components of CBT are most effective for women in sub-Saharan Africa.

References

American College of Obstetricians and Gynecologists. (2018). *Committee opinion number 757: Screening for perinatal depression.* https://www.acog.org/clinical/clinical-guidance/committee-opinion/articles/2018/11/screening-for-perinatal-depression.

Andersen, L. S., Magidson, J. F., O'Cleirigh, C., Remmert, J. E., Kagee, A., Leaver, M., et al. (2018). A pilot study of a nurse-delivered cognitive behavioral therapy intervention (Ziphamandla) for adherence and depression in HIV in South Africa. *Journal of Health Psychology, 23*(6), 776–787. https://doi.org/10.1177/1359105316643375.

Atuhaire, C., Brennaman, L., Cumber, S. N., Rukundo, G. Z., & Nambozi, G. (2020). The magnitude of postpartum depression among mothers in Africa: A literature review. *Pan African Medical Journal, 37*, 89. https://doi.org/10.11604/pamj.2020.37.89.23572.

Bass, J. K., Annan, J., McIvor Murray, S., Kaysen, D., Griffiths, S., Cetinoglu, T., et al. (2013). Controlled trial of psychotherapy for Congolese survivors of sexual violence. *New England Journal of Medicine, 368*(23), 2182–2191. https://doi.org/10.1056/NEJMoa1211853.

Bryant, R. A., Schafer, A., Dawson, K. S., Anjuri, D., Mulili, C., Ndogoni, L., et al. (2017). Effectiveness of a brief behavioural intervention on psychological distress among women with a history of gender-based violence in urban Kenya: A randomised clinical trial. *PLoS Medicine, 14*(8). https://doi.org/10.1371/journal.pmed.1002371, e1002371.

Chibanda, D., Cowan, F., Verhey, R., Machando, D., Abas, M., & Lund, C. (2017). Lay health workers' experience of delivering a problem solving therapy intervention for common mental disorders among people living with HIV: A qualitative study from Zimbabwe. *Community Mental Health Journal, 53*(2), 143–153. https://doi.org/10.1007/s10597-016-0018-2.

Chibanda, D., Mesu, P., Kajawu, L., Cowan, F., Araya, R., & Abas, M. A. (2011). Problem-solving therapy for depression and common mental disorders in Zimbabwe: Piloting a task-shifting primary mental health care intervention in a population with a high prevalence of people living with HIV. *BMC Public Health, 11*, 828. https://doi.org/10.1186/1471-2458-11-828.

Chibanda, D., Weiss, H. A., Verhey, R., Simms, V., Munjoma, R., Rusakaniko, S., et al. (2016). Effect of a primary care-based psychological intervention on symptoms of common mental disorders in Zimbabwe: A randomized clinical trial. *JAMA, 316*(24), 2618–2626. https://doi.org/10.1001/jama.2016.19102.

Chowdhary, N. M. D., Sikander, S. P., Atif, N. M. S., Singh, N. M. P. H., Ahmad, I. M. B. A. M. S. M. P., Fuhr, D. C. D., et al. (2013). The content and delivery of psychological interventions for perinatal depression by non-specialist health workers in low and middle income countries: A systematic review. *Best Practice and Research: Clinical Obstetrics & Gynaecology, 28*(1), 113–133. https://doi.org/10.1016/j.bpobgyn.2013.08.013.

Cox, J. L., Holden, J. M., & Sagovsky, R. (1987). Detection of postnatal depression: Development of the 10-item Edinburgh Postnatal Depression Scale. *British Journal of Psychiatry, 150*, 782–786.

Dos Santos, M., & Wolvaardt, G. (2016). Integrated intervention for mental health co-morbidity in HIV-positive individuals: A public health assessment. *African Journal of AIDS Research, 15*(4), 325–331. https://doi.org/10.2989/16085906.2016.1229683.

Firoz, T., Chou, D., von Dadelszen, P., Agrawal, P., Vanderkruik, R., Tunçalp, O., et al. (2013). Measuring maternal health: Focus on maternal morbidity. *Bulletin of the World Health Organization, 91*(10), 794–796. https://doi.org/10.2471/BLT.13.117564.

Futterman, D., Shea, J., Besser, M., Stafford, S., Desmond, K., Comulada, W. S., et al. (2010). Mamekhaya: A pilot study combining a cognitive-behavioral intervention and mentor mothers with PMTCT services in South Africa. *AIDS Care, 22*(9), 1093–1100. https://doi.org/10.1080/09540121003600352.

Gelaye, B., Rondon, M. B., Araya, R., & Williams, M. A. (2016). Epidemiology of maternal depression, risk factors, and child outcomes in low-income and middle-income countries. *Lancet Psychiatry, 3*(10), 973–982. https://doi.org/10.1016/S2215-0366(16)30284-X.

Gibbs, A., Jewkes, R., Willan, S., & Washington, L. (2018). Associations between poverty, mental health and substance use, gender power, and intimate partner violence amongst young (18-30) women and men in urban informal settlements in South Africa: A cross-sectional study and structural equation model. *PLoS One, 13*(10). https://doi.org/10.1371/journal.pone.0204956, e0204956.

Ginneken, N., Tharyan, P., Lewin, S., Rao, G., Romeo, R., & Patel, V. (2011). Non-specialist health worker interventions for mental health care in low- and middle-income countries. *The Cochrane Database of Systematic Reviews, 2011*. https://doi.org/10.1002/14651858.CD009149.

Green, E. P., Lai, Y., Pearson, N., Rajasekharan, S., Rauws, M., Joerin, A., et al. (2020). Expanding access to perinatal depression treatment in Kenya through automated psychological support: Development and usability study. *JMIR Formative Research, 4*(10). https://doi.org/10.2196/17895, e17895.

Greene, M. C., Rees, S., Likindikoki, S., Bonz, A. G., Joscelyne, A., Kaysen, D., et al. (2019). Developing an integrated intervention to address intimate partner violence and psychological distress in Congolese refugee women in Tanzania. *Conflict and Health, 13*(1), 38. https://doi.org/10.1186/s13031-019-0222-0.

Honikman, S., van Heyningen, T., Field, S., Baron, E., & Tomlinson, M. (2012). Stepped care for maternal mental health: A case study of the perinatal mental health project in South Africa. *PLoS Medicine, 9*(5). https://doi.org/10.1371/journal.pmed.1001222, e1001222.

Kaaya, S., Garcia, M. E., Li, N., Lienert, J., Twayigize, W., Spiegelman, D., et al. (2016). Association of maternal depression and infant nutritional status among women living with HIV in Tanzania. *Maternal & Child Nutrition, 12*(3), 603–613. https://doi.org/10.1111/mcn.12154.

Kuo, C. (2019). Cognitive behavioral therapy around the globe. In D. Stein, J. Bass, & S. Hofmann (Eds.), *Global mental health and psychotherapy* (1st ed., pp. 87–126). Academic Press.

Lathrop, E., Jamieson, D. J., & Danel, I. (2014). HIV and maternal mortality. *International Journal of Gynecology & Obstetrics, 127*(2), 213–215. https://doi.org/10.1016/j.ijgo.2014.05.024.

Lund, C., Schneider, M., Garman, E. C., Davies, T., Munodawafa, M., Honikman, S., ... Susser, E. (2020). Task-sharing of psychological treatment for antenatal depression in Khayelitsha, South Africa: Effects on antenatal and postnatal outcomes in an individual randomised controlled trial. *Behaviour Research and Therapy, 130*, 103466. https://doi.org/10.1016/j.brat.2019.103466.

Machisa, M. T., Christofides, N., & Jewkes, R. (2017). Mental ill health in structural pathways to women's experiences of intimate partner violence. *PLoS One, 12*(4). https://doi.org/10.1371/journal.pone.0175240, e0175240.

Marangu, E., Sands, N., Rolley, J., Ndetei, D., & Mansouri, F. (2014). Mental healthcare in Kenya: Exploring optimal conditions for capacity building. *African Journal of Primary Health & Family Medicine, 6*(1), E1–E5. https://doi.org/10.4102/phcfm.v6i1.682.

Mayosi, B. M., Flisher, A. J., Lalloo, U. G., Sitas, F., Tollman, S. M., & Bradshaw, D. (2009). The burden of non-communicable diseases in South Africa. *The Lancet, 374*(9693), 934–947. https://doi.org/10.1016/S0140-6736(09)61087-4.

Mitchell, J., Wight, M., Van Heerden, A., & Rochat, T. J. (2016). Intimate partner violence, HIV, and mental health: A triple epidemic of global proportions. *International Review of Psychiatry, 28*(5), 452–463. https://doi.org/10.1080/09540261.2016.1217829.

Moultrie, A., & Kleintjes, S. (2006). Women's mental health in South Africa: Women's health. *South African Health Review, 2006*(1), 347–366. https://doi.org/10.10520/EJC35453.

Muluneh, M. D., Stulz, V., Francis, L., & Agho, K. (2020). Gender based violence against women in sub-Saharan Africa: A systematic review and meta-analysis of cross-sectional studies. *International Journal of Environmental Research Public Health, 17*(3). https://doi.org/10.3390/ijerph17030903.

Muñoz, R. F., Chavira, D. A., Himle, J. A., Koerner, K., Muroff, J., Reynolds, J., et al. (2018). Digital apothecaries: A vision for making health care interventions accessible worldwide. *mHealth, 4*, 18. https://doi.org/10.21037/mhealth.2018.05.04.

Nyatsanza, M., Schneider, M., Davies, T., & Lund, C. (2016). Filling the treatment gap: Developing a task sharing counselling intervention for perinatal depression in Khayelitsha, South Africa. *BMC Psychiatry, 16*(167), 164. https://doi.org/10.1186/s12888-016-0873-y.

Oluwasola, T. A. O., & Bello, O. (2020). Clinical and psychosocial outcomes of obstetrics fistulae in Sub-Saharan Africa: A review of literature. *Factors Affecting Postnatal Quality of Life, Vol. 9*, 8–16. https://doi.org/10.4103/2278-960X.1945139.

Parker, R., & Aggleton, P. (2003). HIV and AIDS-related stigma and discrimination: A conceptual framework and implications for action. *Social Science and Medicine, 57*(1), 13–24. https://doi.org/10.1016/S0277-9536(02)00304-0.

Patel, V., Chowdhary, N., Rahman, A., & Verdeli, H. (2011). Improving access to psychological treatments: Lessons from developing countries. *Behaviour Research and Therapy, 49*(9), 523–528. https://doi.org/10.1016/j.brat.2011.06.012.

Patel, V., Simunyu, E., Gwanzura, F., Lewis, G., & Mann, A. (1997). The shona symptom questionnaire: The development of an indigenous measure of common mental disorders in Harare. *Acta Psychiatrica Scandinavica, 95*(6), 469–475. https://doi.org/10.1111/j.1600-0447.1997.tb10134.x.

Pellowski, J. A., Bengtson, A. M., Barnett, W., DiClemente, K., Koen, N., Zar, H. J., et al. (2019). Perinatal depression among mothers in a South African birth cohort study: Trajectories from pregnancy to 18 months postpartum. *Journal of Affective Disorders, 259*, 279–287. https://doi.org/10.1016/j.jad.2019.08.052.

Rafaeli, T., & Hutchinson, G. (2020). *The secondary impacts of COVID-19 on women and girls in sub-Saharan Africa*. https://opendocs.ids.ac.uk/opendocs/handle/20.500.12413/15408.

Rahman, A., Hafeez, A., Bilal, R., Sikander, S., Malik, A., Minhas, F., et al. (2016). The impact of perinatal depression on exclusive breastfeeding: A cohort study. *Maternal Child Nutrition, 12*(3), 452–462. https://doi.org/10.1111/mcn.12170.

Ramjee, G., & Daniels, B. (2013). Women and HIV in sub-Saharan Africa. *AIDS Research and Therapy, 10*(1), 30. https://doi.org/10.1186/1742-6405-10-30.

Resick, P. A., & Schnicke, M. K. (1992). Cognitive processing therapy for sexual assault victims. *Journal of Consulting and Clinical Psychology, 60*(5), 748–756. https://doi.org/10.1037//0022-006x.60.5.748.

Rotheram-Borus, M. J., Tomlinson, M., le Roux, I. M., Harwood, J. M., Comulada, S., O'Connor, M. J., et al. (2014). A cluster randomised controlled effectiveness trial evaluating perinatal home visiting among South African mothers/infants. *PLoS One, 9*(10). https://doi.org/10.1371/journal.pone.0105934, e105934.

Safren, S. A., Bedoya, C. A., O'Cleirigh, C., Biello, K. B., Pinkston, M. M., Stein, M. D., et al. (2016). Cognitive behavioural therapy for adherence and depression in patients with HIV: A three-arm randomised controlled trial. *The Lancet. HIV, 3*(11), e529–e538. https://doi.org/10.1016/S2352-3018 (16)30053-4.

Semo, B. W., & Frissa, S. M. (2020). The mental health impact of the COVID-19 pandemic: Implications for sub-Saharan Africa. *Psychology Research and Behavior Management, 13*, 713–720. https://doi.org/10.2147/PRBM.S264286.

Sijbrandij, M., Bryant, R. A., Schafer, A., Dawson, K. S., Anjuri, D., Ndogoni, L., et al. (2016). Problem management plus (PM+) in the treatment of common mental disorders in women affected by gender-based violence and urban adversity in Kenya: Study protocol for a randomized controlled trial. *International Journal of Mental Health Systems, 10*(1), 44. https://doi.org/10.1186/s13033-016-0075-5.

Sikander, S., Ahmad, I., Bates, L. M., Gallis, J., Hagaman, A., O'Donnell, K., et al. (2019). Cohort profile: Perinatal depression and child socioemotional development: The bachpan cohort study from rural Pakistan. *BMJ Open, 9*(5). https://doi.org/10.1136/bmjopen-2018-025644, e025644.

Sikkema, K. J., Mulawa, M. I., Robertson, C., Watt, M. H., Ciya, N., Stein, D. J., et al. (2018). Improving AIDS care after trauma (ImpACT): Pilot outcomes of a coping intervention among HIV-infected women with sexual trauma in South Africa. *AIDS and Behavior, 22*(3), 1039–1052. https://doi.org/10.1007/s10461-017-2013-1.

Sockol, L. E. (2015). A systematic review of the efficacy of cognitive behavioral therapy for treating and preventing perinatal depression. *Journal of Affective Disorders, 177*, 7–21. https://doi.org/10.1016/j.jad.2015.01.052.

Sowa, N. A., Cholera, R., Pence, B. W., & Gaynes, B. N. (2015). Perinatal depression in HIV-infected African women: A systematic review. *Journal of Clinical Psychiatry, 76*(10), 1385–1396. https://doi.org/10.4088/JCP.14r09186.

Surkan, P. J., Patel, S. A., & Rahman, A. (2016). Preventing infant and child morbidity and mortality due to maternal depression. *Best Practice & Research Clinical Obstetrics & Gynaecology, 36*, 156–168. https://doi.org/10.1016/j.bpobgyn.2016.05.007.

Tao, J., Vermund, S. H., & Qian, H. Z. (2018). Association between depression and antiretroviral therapy use among people living with HIV: A meta-analysis. *AIDS and Behavior, 22*(5), 1542–1550. https://doi.org/10.1007/s10461-017-1776-8.

Thase, M. E., Kingdon, D., & Turkington, D. (2014). The promise of cognitive behavior therapy for treatment of several mental disorders: A review of recent developments. *World Psychiatry, 13*, 244–250. https://doi.org/10.1002/wps.20149.

Tomlinson, M., Rotheram-Borus, M. J., le Roux, I. M., Youssef, M., Nelson, S. H., Scheffler, A., et al. (2016). Thirty-six-month outcomes of a generalist paraprofessional perinatal home visiting intervention in South Africa on maternal health and child health and development. *Prevention Science, 17*(8), 937–948. https://doi.org/10.1007/s11121-016-0676-x.

Tsai, A. C., Tomlinson, M., Comulada, W. S., & Rotheram-Borus, M. J. (2016). Intimate partner violence and depression symptom severity among South African women during pregnancy and postpartum: Population-based prospective cohort study. *PLoS Medicine, 13*(1). https://doi.org/10.1371/journal.pmed.1001943, e1001943.

Tshabalala, J., & Visser, M. (2011). Developing a cognitive behavioural therapy model to assist women to deal with HIV and stigma. *South Africa Journal of Psychology, 41*(1), 17–28. https://doi.org/10.1177/008124631104100103.

UNAIDS. (2020). *Global HIV & AIDS statistics—2020 fact sheet*. https://www.unaids.org/sites/default/files/media_asset/UNAIDS_FactSheet_en.pdf.

Watt, M. H., Mosha, M. V., Platt, A. C., Sikkema, K. J., Wilson, S. M., Turner, E. L., et al. (2017). A nurse-delivered mental health intervention for obstetric fistula patients in Tanzania: Results of a pilot randomized controlled trial. *Pilot and Feasibility Studies, 3*(1), 35. https://doi.org/10.1186/s40814-017-0178-z.

Watt, M. H., Wilson, S. M., Sikkema, K. J., Velloza, J., Mosha, M. V., Masenga, G. G., et al. (2015). Development of an intervention to improve mental health for obstetric fistula patients in Tanzania. *Evaluation and Program Planning, 50*, 1–9. https://doi.org/10.1016/j.evalprogplan.2015.01.007.

Wittkowski, A., Gardner, P. L., Bunton, P., & Edge, D. (2014). Culturally determined risk factors for postnatal depression in Sub-Saharan Africa: A mixed method systematic review. *Journal of Affective Disorders, 163*, 115–124. https://doi.org/10.1016/j.jad.2013.12.028.

World Health Organization. (2012). *Addressing the challenge of women's health in Africa: Report of the commission on women's health in the African region*. https://www.afro.who.int/sites/default/files/2017-06/report-of-the-commission-on-womens-health-in-the-african-region- -full-who_acreport-comp%20%281%29.pdf.

World Health Organization. (2013). *Global and regional estimates of violence against women: Prevalence and health effects of intimate partner violence and non-partner sexual violence*. https://apps.who.int/iris/bitstream/handle/10665/85239/9789241564625_eng.pdf?sequence=1.

World Health Organization. (2015). *Thinking healthy: A manual for psychological management of perinatal depression*. https://www.who.int/mental_health/maternal-child/thinking_healthy/en/.

World Health Organization. (2019). *The WHO special initiative for mental health (2019–2023)*. Universal Health Coverage for Mental Health. https://apps.who.int/iris/bitstream/handle/10665/310981/WHO-MSD-19.1-eng.pdf?ua=1.

World Health Organization. (2020). *Depression*. https://www.who.int/news-room/fact-sheets/detail/depression.

You, D., Hug, L., & Anthony, D. (2015). UNICEF report generation 2030 Africa calls upon investing in and empowering girls and young women. *Reproductive Health, 12*(1), 18. https://doi.org/10.1186/s12978-015-0007-x.

Section D

Case studies

Chapter 34

Application of online cognitive-behavioral therapy for insomnia among individuals with epilepsy

Zainab Alimoradi[a], Mark D. Griffiths[b], and Amir H. Pakpour[c]

[a]Social Determinants of Health Research Center, Research Institute for Prevention of Non-Communicable Diseases, Qazvin University of Medical Sciences, Qazvin, Iran [b]International Gaming Research Unit, Psychology Department, Nottingham Trent University, Nottingham, United Kingdom [c]Department of Nursing, School of Health and Welfare, Jönköping University, Jönköping, Sweden

Abbreviations

CBT   cognitive-behavioral therapy
CBT-I   cognitive-behavioral therapy for insomnia

Introduction

Epilepsy is a disease with abnormal brain activity, seizures, and/or unusual behaviors and feelings, and sometimes includes loss of consciousness that requires ongoing treatment (Hamedi-Shahraki et al., 2019; Kearney, Byrne, Cavalleri, & Delanty, 2019). Approximately 50 million individuals worldwide are affected by epilepsy, which results in various neurological, cognitive, psychological, and/or social consequences (Lin & Pakpour, 2017; Lin, Updegraff, & Pakpour, 2016; Pakpour et al., 2015). There are relatively few studies on insomnia among individuals with epilepsy, but insomnia appears to be common among individuals with epilepsy (Latreille, Louis, & Pavlova, 2018). A recent systematic review of recurrent insomnia was reported in approximately half of patients with epilepsy (Macêdo, de Oliveira, Foldvary-Schaefer, & da Mota Gomes, 2017). However, the prevalence varies depending on the detection or inclusion criteria. The prevalence of insomnia among adults with epilepsy varies from 36% to 74%, and the symptoms of moderate to severe insomnia are observed among 15% to 51% of patients (Macêdo et al., 2017).

According to the pathophysiological model of insomnia, several specific factors may explain the high frequency of insomnia among individuals with epilepsy. New diagnoses of epilepsy, particularly the consequences of changes at work, school, and social role changes, and anxiety and insecurity about seizure control, can contribute to insomnia by creating stress and anxiety. Many other contributing factors can also occur. For example, a daily nap, after a seizure, may lead to the next problem in sleeping or staying asleep the next night (Latreille et al., 2018). Adequate sleep is necessary to maintain the health of the body, fight pathogens, and play an important role in optimal cognitive function (Pakpour, Griffiths, Ohayon, Broström, & Lin, 2020). In contrast, sleep deprivation is associated with negative consequences including worsening of symptoms of other underlying diseases, endocrine dysfunction, and central nervous system consequences (Latreille et al., 2018).

Despite the high prevalence and complications of insomnia, only a small number of individuals are treated for sleep disorders. The first stage of insomnia treatment usually involves self-medication with herbal or dietary supplements, or excessive use of hypnotics (Bonnar et al., 2015; Leichsenring, Hiller, Weissberg, & Leibing, 2006). Prescribing a sleeping pill is usually the first medical recommendation, often effective in the short-term, but their long-term use is associated with risks of side effects and dependence. In addition, the effect of drugs is reduced in the long run (Bonnar et al., 2015). Non-pharmacological methods to solve the problem of insomnia include lifestyle changes, touch therapy, yoga, acupuncture, light therapy, relaxation, cognitive-behavioral therapy, and self-help (Carlbring & Andersson, 2006; Cunnington, Junge, & Fernando, 2013; Williams, Roth, Vatthauer, & McCrae, 2013).

Cognitive-behavioral therapy

Cognitive-behavioral therapy (CBT) can be used to treat a wide range of individuals with mental health problems. CBT is based on the idea of how individuals think (cognition), how individuals feel (emotions), and how individuals interact with each other (behavior). In particular, it can be said that individuals' thoughts, feelings determine their behaviors. Therefore, negative and imaginative thoughts can cause individuals discomfort and problems. When individuals suffer from emotional distress, they react to different situations, which in turn has a negative impact on their actions. CBT aims to educate individuals because they create negative interpretations in their minds and follow behavioral patterns that lead to their perceptual thinking. CBT helps individuals develop alternative ways of thinking and behaving that aims to reduce their mental disorders. CBT is actually an umbrella term (meaning it covers many cases) for many different therapies that share some common elements (Ahorsu et al., 2020; Hofmann, 2011; Majd et al., 2020). CBT is based on a series of general assumptions includes the following:

- The cognitive approach believes that the disorder is caused by misdiagnosis of individuals as well as the world of others and the individual's own world. This misconception may be due to cognitive deficiencies (lack of planning) or cognitive distortions (incorrect information processing).
- These cognitions distort the way individuals look at different things due to irrational thinking.
- Individuals communicate with the world through their mental representation of it. If mental imagery is incorrect or reasoning is inadequate, individuals' emotions and behaviors may also be disrupted (Simmons & Griffiths, 2017).

The cognitive therapist teaches clients how to identify distorted cognitions through an evaluation process. Clients learn to distinguish between their thoughts and reality. They learn the effect that cognition has on their emotions, and therapists teach them how to observe their thoughts and monitor their existence (Milne & Reiser, 2017).

Part of the treatment process involves assigning specific tasks to clients (for example, writing in a diary). The therapist assigns tasks to the client to help them challenge their unreasonable beliefs. The idea is that the client identifies their useless beliefs and realizes that they are wrong. As a result, their beliefs begin to change (Gordon, 2012).

In general, cognitive-behavioral therapy (CBT) is a type of psychotherapy that focuses on the effect of an individual's beliefs, thoughts, and attitudes on their feelings and behaviors. The purpose of cognitive-behavioral therapy is to teach an individual how to actively face and overcome various problems or events during their life (Harris, 2013).

Cognitive-behavioral therapy for insomnia

Cognitive-behavioral therapy for insomnia (CBT-I) is one of the therapies in sleep science and uses the general principles of cognitive-behavioral therapy and is designed to eliminate the symptoms of insomnia (Pietrzak, Johnson, Goldstein, Malley, & Southwick, 2009). CBT-I is an effective psychotherapy for insomnia that has been shown to be even more effective than medication in the long run (Arnedt et al., 2013). In this method, the principles of CBT are presented to the patient as a set of scientific methods whose effectiveness has been tested (Bee, Lovell, Lidbetter, Easton, & Gask, 2010). These principles include behavioral modification, such as stimulus control, sleep hygiene, sleep restriction, relaxation training, and cognitive therapy. The aim is to adapt sleep habits through cognitive strategies to improve sleep thoughts, feelings, and expectations (Arnedt et al., 2013; Bee et al., 2010; Insel & Cuthbert, 2009). Most CBT-I treatments contain "exercise" (Pietrzak et al., 2009).

The method of performing CBT-I is determined in individual or group sessions, face-to-face or virtually based on the therapist's priority, patient progress, preference, and patient access to various facilities (Chan, Torous, Hinton, & Yellowlees, 2014). One of the obstacles to face-to-face treatment is that patients with insomnia, as a psychological disorder, often do not know where to seek treatment. Most patients report their sleep concerns and problems to a healthcare provider, and less than 50% of individuals who suffer from insomnia symptoms seek help (Torous, Friedman, & Keshavan, 2014). Patients with insomnia often express their illness in the context of other health concerns and seek to address the problems caused by insomnia that have arisen for them (Rooksby, Elouafkaoui, Humphris, Clarkson, & Freeman, 2015).

There are several barriers to getting CBT by most patients, including access, cost, time constraints, and social stigma. The problem of access to CBT services can be attributed to the lack of trained therapists and mental health professionals (Hedman et al., 2011; Savard, Ivers, Savard, & Morin, 2014). The spatial dimension, especially in rural areas, can also impede access to treatment. In addition to cost and time constraints, individuals living in rural areas have face-to-face negative beliefs about mental health and psychotherapy that prevent them from receiving CBT services (Freedman, Dayan, Kimelman, Weissman, & Eitan, 2015). These barriers and beliefs, along with economic and geographical barriers, can make access to CBT difficult (Mozer, Franklin, & Rose, 2008). Due to the various problems that can hinder access to

CBT treatment, various strategies, including the use of technology to increase access to psychiatric and psychological services, have been proposed to maximize the availability and effectiveness of mental health services (Yuen et al., 2013).

Use of technology in the treatment of insomnia

The use of technology to provide CBT includes methods such as telephone, internet and mobile applications, due to convenience. Privacy and increase trust in treatment reduces the symptoms of mental illness and increases individuals' desire for it (Blom et al., 2015). The internet is an emerging platform for mental health services worldwide (Alimoradi et al., 2019; Alimoradi, Lin, Imani, Griffiths, & Pakpour, 2019; Babson, Ramo, Baldini, Vandrey, & Bonn-Miller, 2015). The internet serves as a potential solution to geographic and transportation challenges and to facilitate the provision of mental health services to patients in their own homes. Internet-based CBT or CBTI has been used for insomnia with promising results (Lancee, van den Bout, van Straten, & Spoormaker, 2012; Ritterband et al., 2012, 2009; Vincent & Lewycky, 2009). If the content of CBT-I-based technology, which is disseminated via the internet, e-mail, or mobile phone, is of good quality and has credible evidence-based experiences, it will have positive therapeutic effects and initial patient acceptance (Ben-Zeev, Davis, Kaiser, Krzsos, & Drake, 2013; Ruggiero, Peach, & Gaultney, 2019).

Despite these definitions, there are some disadvantages to this technology, including the fact that it is not possible to get good feedback from the patient, which can reduce the patient's ability to learn strategies to relieve symptoms (Sivertsen, Vedaa, & Nordgreen, 2013). Patients' access to mobile devices is still poor, but the portability of mobile phones and low-consumption needs and the lack of need for basic electricity communication by mobile phones and the establishment of fast and cheap internet through it are promising (Andersen & Svensson, 2013). Patients of all ages can use mobile apps that help their health. Even patients aged over 60 years, who tend to use smartphones less than younger people, express more interest in using smartphones and have mobile applications (Andersen & Svensson, 2013). Therefore, in general, CBT-I is a good treatment option for patients with insomnia, including patients with epilepsy who experience insomnia.

Procedure

Three patients with epilepsy were referred for routine examination and psychological counseling. The characteristics of these individuals based on initial assessments were as follows:

- Case 1: A 39-year-old married female housewife with a high school education and focal epilepsy.
- Case 2: A 41-year-old married woman, employed, with a college education and focal epilepsy.
- Case 3: A 33-year-old single man with a university degree and generalized epilepsy.

These individuals were asked to check for other possible psychological problems such as anxiety and depression, insomnia, sleep quality. In the initial study, the level of anxiety and depression, insomnia, sleep quality, sleep hygiene behavior, and sleep characteristics were determined based on actinography and quality of life specific to patients with epilepsy. Given that these individuals were suffering from insomnia, they were jointly invited to participate in an internet-based cognitive-behavioral insomnia therapy program.

CBT-I program content

The CBT-I program is a six-week program that is presented to individuals using the Self Help Sleep Health (SHSH) application designed for the same purpose. The SHSH application is based on the theory of transient self-regulation (TST) (Hall & Fong, 2007, 2010, 2013) and cognitive-behavioral therapy for insomnia (CBT-I) (Williams et al., 2013) designed to help individuals improve their sleep (Majd et al., 2020) Several behavior change techniques, including habit formation, behavior self-monitoring, problem solving, habit reversal, and action planning, are integrated into the program (Ahorsu et al., 2020). The training section is presented in six weekly sessions, which are described in Table 1.

Outcome assessment was performed 6 months after the end of the intervention. Table 2 shows the values obtained in the initial evaluation and follow-up evaluation. Follow-up evaluation findings showed that using this application-based CBT-I intervention, the participants' condition improved in terms of all the outcomes examined.

Application to other factors

Other epilepsy-related consequences including anxiety, depression, sexual dysfunction, and adherence to treatment can be treated using online CBT interventions.

TABLE 1 Educational content based on cognitive-behavioral therapy.

Week	Content
1	Participants are informed about the importance of sleep, the amount of sleep required, the main causes of insomnia, short-term and long-term complications and consequences of insufficient sleep and insomnia treatments (e.g., CBT-I and drug use), the prevalence of insomnia in Iran. In addition, participants are asked to list the potential cost of insufficient sleep and the potential benefits of good sleep. At the end of the first week, participants are encouraged to record sleep information (bedtime, wake-up time) using a notebook (located on the front page of the app for added convenience and availability).
2	This week emphasizes healthy sleeping habits. Participants are trained to consider negative habits that can disrupt sleep, such as smoking at night, drinking tea, coffee or alcohol, the bedroom environment (such as a dark, quiet, cool room) and worrying. Participants are encouraged to practice these sleep habits by reading a list of tips. For example, "listening" to their body, eating healthy foods, drinking safely, and engaging in relaxing exercises before bed. Participants are asked to write down new habits that could be accepted as an activity for this week. Participants are also asked to list the habits that affect their sleep and that they are trying to change as the next exercise. At the end of the week, relaxing exercises are presented in text and audio, along with music for the exercises. Participants are also asked to do these exercises every night before going to bed in comfortable places (for example, in bed) and to set their phones to silent.
3	Relaxing exercises (continued). Participants are asked to imagine a place where they feel comfortable (for example, a pleasant place) every night before bed. In addition, participants are taught how to connect the bedroom with sleep alone. This is done by asking participants to limit the amount of time spent in bed. To know if this exercise is done correctly, participants are asked to calculate their waking time during the night using a sleep performance chart. Participants are asked to rate their sleep performance using their sleep charts and/or sleep notes. At the end of each week, they are able to see their sleep performance on a sleep chart (for example, to see if their sleep increased by limiting the amount of time they spent in bed).
4	At the beginning of Week 4, participants are asked to check their sleep patterns using a sleep chart to ensure that they are using the third week of exercise. The goal of Week 4 is to change misconceptions about sleep. A list of common misconceptions about sleep experienced by individuals with insomnia is provided with the correct information. Differences between thoughts and feelings are examined. In addition, participants are asked to try to sleep and write down situations, thoughts, feelings, and behaviors on a board to help them identify unpleasant thoughts.
5	At the beginning of Week 5, participants are asked to look at their sleep chart and check their sleep efficiency. In addition, participants are asked to plan for sleep hygiene behaviors, when and where to sleep. They are asked to write this information down in a chart, and they are encouraged to identify barriers they may encounter and find ways to overcome them.
6	Participants are helped to learn how to connect the bed with sleep rather than waking. They are reminded not to stay too awake in bed. They are reminded of all the aforementioned techniques (e.g., sleep restriction), relaxation, sleep training, cognitive reconstruction, action planning, and coping planning. During Week 6, the goal is a paradox in which participants are asked to sleep an hour less than normal to show them that sleep deprivation did not necessarily affect daily activities. At the end of Week 6, participants are asked to read their notebooks daily.

Key facts

- Epilepsy, like many common medical disorders, can cause insomnia.
- Several factors may explain insomnia among individuals with epilepsy including new diagnoses of epilepsy, anxiety, stress, and insecurity about seizure control, changes at work, in education, and social role changes.
- Cognitive-behavioral therapy for insomnia (CBT-I) is one of the therapies in sleep science. If the content of CBT-I-based technology (which is disseminated via the internet, email or smartphone), is of good quality and has credible evidence-based experiences, it will have positive therapeutic effects and initial patient acceptance.

Summary points

- Cognitive-behavioral therapy can be an effective way to manage insomnia among individuals with epilepsy.
- Evaluation of patients with epilepsy for insomnia should be considered by healthcare providers.
- Providing cognitive-behavioral therapy for insomnia (CBT-I) based on internet platforms such as mobile applications can increase access and content to patients
- CBT-I can reduce the need for physical presence to facilitate the implementation of interventions to improve the sleep status of patients.

TABLE 2 Outcome assessment at baseline and six-month follow-up.

Case No	QOLIE-31		Sleep hygiene behavior		Anxiety		Depression		Insomnia		PSQI		SoL (mins)		WAS (mins)		TST (mins)		SE (%)	
	Before	After	Before	After	Before	After	Before	After	Before	After	Before	After	Before	After	Before	After	Before	After	Before	After
1	55.60	68.11	6.41	15.74	10	1	5	4	20	6	16	4	28.40	23.91	42.94	40.22	380.81	396.24	70.83	76.35
2	52.93	62.81	7.25	14.68	8	2	11	5	22	8	14	3	28.39	24.01	42.11	38.41	382.10	398.73	71.19	77.51
3	53.07	66.22	6.96	15.12	6	3	5	2	19	8	13	3	25.12	24.23	42.68	37.34	384.5	395.2	71.64	76.94

PSQI, Pittsburgh Sleep Quality Index; *SOL*, sleep onset latency; *TST*, total sleep time; *WASO*, wake after sleep; *SE*, sleep efficiency.

References

Ahorsu, D. K., Lin, C.-Y., Imani, V., Carlbring, P., Nygårdh, A., Broström, A., ... Pakpour, A. H. (2020). Testing an app-based intervention to improve insomnia in patients with epilepsy: A randomized controlled trial. *Epilepsy & Behavior, 112*, 107371.

Alimoradi, Z., Lin, C.-Y., Broström, A., Bülow, P. H., Bajalan, Z., Griffiths, M. D., ... Pakpour, A. H. (2019). Internet addiction and sleep problems: A systematic review and meta-analysis. *Sleep Medicine Reviews, 47*, 51–61.

Alimoradi, Z., Lin, C.-Y., Imani, V., Griffiths, M. D., & Pakpour, A. H. (2019). Social media addiction and sexual dysfunction among Iranian women: The mediating role of intimacy and social support. *Journal of Behavioral Addictions, 8*(2), 318–325.

Andersen, A. J. W., & Svensson, T. (2013). Internet-based mental health services in Norway and Sweden: Characteristics and consequences. *Administration and Policy in Mental Health and Mental Health Services Research, 40*(2), 145–153.

Arnedt, J. T., Cuddihy, L., Swanson, L. M., Pickett, S., Aikens, J., & Chervin, R. D. (2013). Randomized controlled trial of telephone-delivered cognitive behavioral therapy for chronic insomnia. *Sleep, 36*(3), 353–362.

Babson, K. A., Ramo, D. E., Baldini, L., Vandrey, R., & Bonn-Miller, M. O. (2015). Mobile app-delivered cognitive behavioral therapy for insomnia: Feasibility and initial efficacy among veterans with cannabis use disorders. *JMIR Research Protocols, 4*(3), e87.

Bee, P. E., Lovell, K., Lidbetter, N., Easton, K., & Gask, L. (2010). You can't get anything perfect: "User perspectives on the delivery of cognitive behavioural therapy by telephone". *Social Science & Medicine, 71*(7), 1308–1315.

Ben-Zeev, D., Davis, K. E., Kaiser, S., Krzsos, I., & Drake, R. E. (2013). Mobile technologies among people with serious mental illness: Opportunities for future services. *Administration and Policy in Mental Health and Mental Health Services Research, 40*(4), 340–343.

Blom, K., Jernelöv, S., Kraepelien, M., Bergdahl, M. O., Jungmarker, K., Ankartjärn, L., ... Kaldo, V. (2015). Internet treatment addressing either insomnia or depression, for patients with both diagnoses: A randomized trial. *Sleep, 38*(2), 267–277.

Bonnar, D., Gradisar, M., Moseley, L., Coughlin, A.-M., Cain, N., & Short, M. A. (2015). Evaluation of novel school-based interventions for adolescent sleep problems: Does parental involvement and bright light improve outcomes? *Sleep Health, 1*(1), 66–74.

Carlbring, P., & Andersson, G. (2006). Internet and psychological treatment. How well can they be combined? *Computers in Human Behavior, 22*(3), 545–553.

Chan, S. R., Torous, J., Hinton, L., & Yellowlees, P. (2014). Mobile tele-mental health: Increasing applications and a move to hybrid models of care. *Healthcare, 2*(2), 220–233.

Cunnington, D., Junge, M. F., & Fernando, A. T. (2013). Insomnia: Prevalence, consequences and effective treatment. *Medical Journal of Australia, 199*, S36–S40.

Freedman, S. A., Dayan, E., Kimelman, Y. B., Weissman, H., & Eitan, R. (2015). Early intervention for preventing posttraumatic stress disorder: An internet-based virtual reality treatment. *European Journal of Psychotraumatology, 6*(1), 25608.

Gordon, P. K. (2012). Ten steps to cognitive behavioural supervision. *The Cognitive Behaviour Therapist, 5*(4), 71–82.

Hall, P. A., & Fong, G. T. (2007). Temporal self-regulation theory: A model for individual health behavior. *Health Psychology Review, 1*(1), 6–52.

Hall, P. A., & Fong, G. T. (2010). Temporal self-regulation theory: Looking forward. *Health Psychology Review, 4*(2), 83–92.

Hall, P. A., & Fong, G. T. (2013). Temporal self-regulation theory: Integrating biological, psychological, and ecological determinants of health behavior performance. In P. Hall (Ed.), *Social neuroscience and public health* (pp. 35–53). Springer.

Hamedi-Shahraki, S., Eshraghian, M.-R., Yekaninejad, M.-S., Nikoobakht, M., Rasekhi, A., Chen, H., & Pakpour, A. (2019). Health-related quality of life and medication adherence in elderly patients with epilepsy. *Neurologia i Neurochirurgia Polska, 53*(2), 123–130.

Harris, S. (2013). *Cognitive behavioural therapy: Basics and beyond* (2nd ed.). The Guilford Press.

Hedman, E., Andersson, G., Ljótsson, B., Andersson, E., Rück, C., Mörtberg, E., & Lindefors, N. (2011). Internet-based cognitive behavior therapy vs. cognitive behavioral group therapy for social anxiety disorder: A randomized controlled non-inferiority trial. *PLoS One, 6*(3), e18001.

Hofmann, S. G. (2011). *An introduction to modern CBT: Psychological solutions to mental health problems*. John Wiley & Sons.

Insel, T. R., & Cuthbert, B. N. (2009). Endophenotypes: Bridging genomic complexity and disorder heterogeneity. *Biological Psychiatry, 66*(11), 988–989.

Kearney, H., Byrne, S., Cavalleri, G. L., & Delanty, N. (2019). Tackling epilepsy with high-definition precision medicine: A review. *JAMA Neurology, 76*(9), 1109–1116.

Lancee, J., van den Bout, J., van Straten, A., & Spoormaker, V. I. (2012). Internet-delivered or mailed self-help treatment for insomnia? A randomized waiting-list controlled trial. *Behaviour Research and Therapy, 50*(1), 22–29.

Latreille, V., Louis, E. K. S., & Pavlova, M. (2018). Co-morbid sleep disorders and epilepsy: A narrative review and case examples. *Epilepsy Research, 145*, 185–197.

Leichsenring, F., Hiller, W., Weissberg, M., & Leibing, E. (2006). Cognitive-behavioral therapy and psychodynamic psychotherapy: Techniques, efficacy, and indications. *American Journal of Psychotherapy, 60*(3), 233–259.

Lin, C.-Y., & Pakpour, A. H. (2017). Using Hospital Anxiety And Depression Scale (HADS) on patients with epilepsy: Confirmatory factor analysis and Rasch models. *Seizure, 45*, 42–46.

Lin, C.-Y., Updegraff, J. A., & Pakpour, A. H. (2016). The relationship between the theory of planned behavior and medication adherence in patients with epilepsy. *Epilepsy & Behavior, 61*, 231–236.

Macêdo, P. J. O. M., de Oliveira, P. S., Foldvary-Schaefer, N., & da Mota Gomes, M. (2017). Insomnia in people with epilepsy: A review of insomnia prevalence, risk factors and associations with epilepsy-related factors. *Epilepsy Research, 135*, 158–167.

Majd, N. R., Broström, A., Ulander, M., Lin, C.-Y., Griffiths, M. D., Imani, V., ... Pakpour, A. H. (2020). Efficacy of a theory-based cognitive behavioral technique app-based intervention for patients with insomnia: Randomized controlled trial. *Journal of Medical Internet Research, 22*(4), e15841.

Milne, D. L., & Reiser, R. P. (2017). *A manual for evidence-based CBT supervision*. John Wiley & Sons.

Mozer, E., Franklin, B., & Rose, J. (2008). Psychotherapeutic intervention by telephone. *Clinical Interventions in Aging, 3*(2), 391–396.

Pakpour, A. H., Gholami, M., Esmaeili, R., Naghibi, S. A., Updegraff, J. A., Molloy, G. J., & Dombrowski, S. U. (2015). A randomized controlled multimodal behavioral intervention trial for improving antiepileptic drug adherence. *Epilepsy & Behavior, 52,* 133–142.

Pakpour, A. H., Griffiths, M. D., Ohayon, M. M., Broström, A., & Lin, C.-Y. (2020). A good sleep: The role of factors in psychosocial health. *Frontiers in Neuroscience, 14,* 520.

Pietrzak, R. H., Johnson, D. C., Goldstein, M. B., Malley, J. C., & Southwick, S. M. (2009). Perceived stigma and barriers to mental health care utilization among OEF-OIF veterans. *Psychiatric Services, 60*(8), 1118–1122.

Ritterband, L. M., Bailey, E. T., Thorndike, F. P., Lord, H. R., Farrell-Carnahan, L., & Baum, L. D. (2012). Initial evaluation of an internet intervention to improve the sleep of cancer survivors with insomnia. *Psycho-Oncology, 21*(7), 695–705.

Ritterband, L. M., Thorndike, F. P., Gonder-Frederick, L. A., Magee, J. C., Bailey, E. T., Saylor, D. K., & Morin, C. M. (2009). Efficacy of an internet-based behavioral intervention for adults with insomnia. *Archives of General Psychiatry, 66*(7), 692–698.

Rooksby, M., Elouafkaoui, P., Humphris, G., Clarkson, J., & Freeman, R. (2015). Internet-assisted delivery of cognitive behavioural therapy (CBT) for childhood anxiety: Systematic review and meta-analysis. *Journal of Anxiety Disorders, 29,* 83–92.

Ruggiero, A. R., Peach, H. D., & Gaultney, J. F. (2019). Association of sleep attitudes with sleep hygiene, duration, and quality: A survey exploration of the moderating effect of age, gender, race, and perceived socioeconomic status. *Health Psychology and Behavioral Medicine, 7*(1), 19–44.

Savard, J., Ivers, H., Savard, M.-H., & Morin, C. M. (2014). Is a video-based cognitive behavioral therapy for insomnia as efficacious as a professionally administered treatment in breast cancer? Results of a randomized controlled trial. *Sleep, 37*(8), 1305–1314.

Simmons, J., & Griffiths, R. (2017). *CBT for beginners* (3rd ed.). London: UK: Sage.

Sivertsen, B., Vedaa, Ø., & Nordgreen, T. (2013). *The future of insomnia treatment—The challenge of implementation.* Oxford University Press.

Torous, J., Friedman, R., & Keshavan, M. (2014). Smartphone ownership and interest in mobile applications to monitor symptoms of mental health conditions. *JMIR mHealth and uHealth, 2*(1), e2994.

Vincent, N., & Lewycky, S. (2009). Logging on for better sleep: RCT of the effectiveness of online treatment for insomnia. *Sleep, 32*(6), 807–815.

Williams, J., Roth, A., Vatthauer, K., & McCrae, C. S. (2013). Cognitive behavioral treatment of insomnia. *Chest, 143*(2), 554–565.

Yuen, E. K., Herbert, J. D., Forman, E. M., Goetter, E. M., Comer, R., & Bradley, J.-C. (2013). Treatment of social anxiety disorder using online virtual environments in second life. *Behavior Therapy, 44*(1), 51–61.

Chapter 35

CASE STUDY: Borderline personality disorder and cognitive behavioral therapy in an adult

Jaiganesh Selvapandiyan
Department of Psychiatry, All India Institute of Medical Sciences, Vijayawada, Andhra Pradesh, India

Abbreviations

BPD borderline personality disorder
CBT cognitive behavior therapy

Introduction

Borderline personality disorder (BPD) is estimated to affect around 1% of the population (Coid, Yang, Tyrer, Roberts, & Ullrich, 2006). The affected individuals face problems in managing their emotions, most often leading to self-harm behaviors and suicidal attempts (Blasco-Fontecilla et al., 2009; Conklin, Bradley, & Westen, 2006; Paris, 2002; Reisch, Ebner-Priemer, Tschacher, Bohus, & Linehan, 2008). Repeated self-harm attempts create havoc in their lives by disrupting interpersonal relationships and occupational functioning (Hill et al., 2008; Sansone, Butler, Dakroub, & Pole, 2006). Also, health care costs would increase because of repeated presentations in emergency services department (Bender et al., 2001). They may also experience stress-induced dissociative or psychotic spells which add to the already existing burden of emotional dysregulation (Glaser, Van Os, Thewissen, & Myin-Germeys, 2010; Stiglmayr et al., 2008). Considering the complexity involved in treating these individuals, effective and evidence-based treatment protocols have to be put into practice for optimal remission of the illness.

Psychotherapy and borderline personality disorder

It is a common perception among clinicians that BPD patients are difficult to treat and many even avoid making a diagnosis of BPD and continue focusing on a comorbid Axis I disorder (especially depressive disorder), which is thought to be more amenable for treatment (Paris, 2005). However, unless the Axis II disorder is effectively treated, the clinician cannot expect successful recovery of his/her patient from comorbid depression (Shea et al., 1990). Contrary to the long-held belief that these individuals are resistant to treatment and the disorder would remain stable for most of their lives, epidemiological evidence has emerged stating that the remission rates are higher than previously thought (Zanarini et al., 2008; Zanarini, Frankenburg, Jager-Hyman, Reich, & Fitzmaurice, 2008). Two major factors would influence the resistance of a BPD patient to a psychotherapeutic intervention. One is the severity of the illness, and the other is the treatment module chosen for treating that individual. In clinical practice, therapies such as dialectical behavior therapy (Linehan, 1993), which are based on cognitive behavioral paradigm, are recommended as the treatment of choice for BPD. Evidence base supporting the efficacy of cognitive behavioral therapy (CBT) in BPD has expanded many folds in recent years (Bohus et al., 2004; Davidson et al., 2006; Davidson, Tyrer, Norrie, Palmer, & Tyrer, 2010; Goodman et al., 2016; Koons et al., 2001; Linehan et al., 1999, 2006; Linehan, Armstrong, Suarez, Allmon, & Heard, 1991; Verheul et al., 2003). With this background information, we shall discuss the application of CBT in an individual suffering from BPD.

Application of cognitive behavior therapy in borderline personality disorder

Mrs. X, 34 years old, homemaker, is brought for consultation with a history of frequent conflict with her spouse, which are mostly accompanied by self-harm behaviors such as hitting herself, slashing her wrists, contemplating suicide and heavy screaming spells. Her relationship with other family members was also dysfunctional and she got into frequent fights with them for trivial reasons. She also faced significant interpersonal conflicts in her workplace. These problematic behavioral responses were present for the past 10 years. On evaluation, she reported a sense of feeling emptiness and insecurity during difficult times. After two sessions of insight facilitation, she reported experiencing overwhelming emotional states and her inability to handle them. She could also recognize that her behavioral responses were out of proportion to the presenting problem and she requested for therapeutic interventions to get out these persistent and problematic emotional states. Based on the symptom profile, a formal diagnosis of BPD was made based on ICD-10 DCR. Zanarini rating scale for BPD was administered and her score was 26.

She was taken up for CBT and the following methods were implemented to train her in CBT techniques. Total number of sessions planned were 32. Each session would last for an hour and was scheduled to happen every week. She was introduced to the cognitive way of human life by presenting examples from her life with embedded cognitions, behaviors and emotional phenomena. Cognitive formulation of her problems revealed multiple distorted cognitions related to her interpersonal functioning, perception of future, appraisal of self and interpretation of the world. Most of the cognitive distortions (superficial automatic thoughts) she experienced during crisis periods could be grouped into one among the following cognitive varieties. Arbitrary inferences, maladaptive should statements and maladaptive comparisons were the common distortions that were noted when she was dealing with stressful interpersonal situations.

Examples

Arbitrary Inferences (without obvious evidence, the patient believed her thoughts to be the truth):
"In order to take revenge for my past mistakes, my spouse has done this act."
"Her behavior indicates that she is trying to avoid me."
"My relatives will not interact with me like before, after this incident."

Maladaptive should statements (harboring this should statement acts as the engine room for interpersonal conflicts):
"My spouse should do the activities as I instruct him."
"He should obey my instructions."

Maladaptive comparisons (based on single instances):
"My sister is enjoying a happy life whereas I am unable to lead a happy life."
"Everyone around me is enjoying whereas I am suffering."

At the peak of emotional crisis, the overgeneralization type of cognitive distortions predominated. Example: "For my whole life I will suffer like this." The personalization type of error thoughts were noted when she anticipated feared consequences for her loved ones because of her problem filled life. Example: "I am the only person to be blamed if my father suffers a heart attack as a result of he worrying about my life." In the later stages of therapy, after gaining much insight into her cognitive processing, she often engaged in "selective use of out of context facts" type of distorted thoughts. Example: "If I had not handled that event in that manner, I would not have had the need to endure the current problem in this manner."

To teach her about the methods to manage her cognitive distortions, both didactic methods and Socratic questioning methods were used. Every cognition mentioned above was put into therapeutic testing. Cognitive experiments such as collecting evidence for and against the cognitions, invoking alternative explanations, asking for the worst and best outcomes were practiced. Finally, the false representation of the cognitions was unearthed and her involvement in dysfunctional cognitive processing was regulated. As a homework practice, the patient was instructed to generalize the techniques learned to other spontaneous distorted cognitions that would arise between two therapy sessions. This she practiced using either the paper-pen method (thought records) or mind-based mentalization method. Also, during intense emotional crisis, she was trained to do chain analysis of the back-to-back cognitive intrusions and to break free from the cyclical thought processes.

As the sessions progressed, the cognitive conceptualization expanded to include the intermediate beliefs, which were the underlying reasons for the rise of automatic cognitive errors. Most of her intermediate beliefs (assumptions, rules, and attitudes) concerning her major life events and significant others were dysfunctional. These maladaptive intermediate cognitions decided the way she processed information about her spouse and the significant others in her life. Most of her maladaptive assumptions covered the areas of subjugation and entitlement.

Examples

"Whatever the situation be, things should happen the way I want."
"Others should serve me even if I have not done any good to them."

These compensatory beliefs have a deeper rooting in the underlying core beliefs, which were mostly negative in content. Her entrenched core beliefs covered the areas of loneliness, dependence, mistrust and insecurity. With regular practice involving superficial automatic thoughts, the intermediate beliefs and core beliefs became amenable for therapeutic manipulation. The same strategies, which were used for superficial cognitions (didactic learning and Socratic questioning), were applied to intermediate and core beliefs. If these core beliefs are not addressed, there is a greater chance of relapse and recurrence of symptoms once the therapy ends.

Even though the behavioral work done with the patient is presented here separately, it was actually combined with the cognitive strategies during the implementation of cognitive techniques. Her behavioral repertoires, apart from self-harm behaviors, included an array of dysfunctional coping strategies such as frequent engagement in social media, binging on movies, stress-induced eating, and compulsive shopping. These behaviors, by providing her a temporary relief from emotional distress, managed to stick on closely to her perceived identity. In order to practice the recommended cognitive techniques, she was motivated to abandon these maladaptive behaviors and engage in cognition-oriented focus. She was clearly informed that if she continued to engage in dysfunctional behaviors, she wouldn't find any mental space to practice the enlightening cognitive techniques. As the patient was inquisitively practicing the therapy, she could assimilate the techniques easily and the final score on Zanarini rating scale dropped down to 5. During the follow up period of 1 year, her score on Zanarini scale was 4 ± 2.

Treatment setting

The treatment procedure described above has been done on an outpatient setting. Research evidence suggests that BPD patients falling into mild to moderate spectrum can be managed through day-treatment programs. In-patient care is especially recommended only during crisis periods and the majority of treatment sessions can be done on outpatient basis, which would help the patient to deal with real-life problems (Livesley, 2003).

Conclusion

The therapeutic approach described in this case gave more emphasis on therapeutic alliance and problem solving within the cognitive behavioral framework. CBT, by specifically targeting the dysfunctional cognitive patterns inherent to BPD patients, helps patients to achieve optimal remission.

Summary points

- Pessimism in treating BPD patients using psychotherapy has to be revisited in the light of current evidence base for CBT.
- Cognitive behavioral techniques serve as effective emotion regulation tools in BPD patients.
- Patients who fall within mild to moderate spectrum of illness severity show significant response with CBT.
- Holistic CBT, addressing all the three levels of cognitions (from superficial automatic thoughts to deeper core beliefs), provides superior outcome.

References

Bender, D. S., Dolan, R. T., Skodol, A. E., Sanislow, C. A., Dyck, I. R., McGlashan, T. H., et al. (2001). Treatment utilization by patients with personality disorders. *The American Journal of Psychiatry, 158*, 295–302.

Blasco-Fontecilla, H., Baca-Garcia, E., Dervic, K., Perez-Rodriguez, M. M., Saiz-Gonzalez, M. D., Saiz-Ruiz, J., et al. (2009). Severity of personality disorders and suicide attempt. *Acta Psychiatrica Scandinavica, 119*, 149–155.

Bohus, M., Haaf, B., Simms, T., Limberger, M. F., Schmahl, C., Unckel, C., et al. (2004). Effectiveness of inpatient dialectical behavioral therapy for borderline personality disorder: A controlled trial. *Behaviour Research and Therapy, 42*, 487–499.

Coid, J., Yang, M., Tyrer, P., Roberts, A., & Ullrich, S. (2006). Prevalence and correlates of personality disorder in Great Britain. *The British Journal of Psychiatry, 188*, 423–431.

Conklin, C. Z., Bradley, R., & Westen, D. (2006). Affect regulation in borderline personality disorder. *The Journal of Nervous and Mental Disease, 194*, 69–77.

Davidson, K., Norrie, J., Tyrer, P., Gumley, A., Tata, P., Murray, H., et al. (2006). The effectiveness of cognitive behavior therapy for borderline personality disorder: Results from the borderline personality disorder study of cognitive therapy (BOSCOT) trial. *Journal of Personality Disorders, 20*, 450–465.

Davidson, K. M., Tyrer, P., Norrie, J., Palmer, S. J., & Tyrer, H. (2010). Cognitive therapy v. usual treatment for borderline personality disorder: Prospective 6-year follow-up. *The British Journal of Psychiatry, 197*, 456–462.

Glaser, J. P., Van Os, J., Thewissen, V., & Myin-Germeys, I. (2010). Psychotic reactivity in borderline personality disorder. *Acta Psychiatrica Scandinavica, 121*(2), 125–134.

Goodman, M., Banthin, D., Blair, N. J., Mascitelli, K. A., Wilsnack, J., Chen, J., et al. (2016). A randomized trial of dialectical behavior therapy in high-risk suicidal veterans. *The Journal of Clinical Psychiatry, 77*, e1591–e1600.

Hill, J., Pilkonis, P., Morse, J., Feske, U., Reynolds, S., Hope, H., et al. (2008). Social domain dysfunction and disorganization in borderline personality disorder. *Psychological Medicine, 38*, 135–146.

Koons, C. R., Robins, C. J., Tweed, J. L., Lynch, T. R., Gonzalez, A. M., Morse, J. Q., et al. (2001). Efficacy of dialectical behavior therapy in women veterans with borderline personality disorder. *Behavior Therapy, 32*, 371–390.

Linehan, M. M. (1993). *Dialectical behavioral therapy of borderline personality disorder*. New York: Guilford.

Linehan, M. M., Armstrong, H. E., Suarez, A., Allmon, D., & Heard, H. L. (1991). Cognitive-behavioral treatment of chronically parasuicidal borderline patients. *Archives of General Psychiatry, 48*, 1060–1064.

Linehan, M. M., Comtois, K. A., Murray, A. M., Brown, M. Z., Gallop, R. J., Heard, H. L., et al. (2006). Two-year randomized controlled trial and follow-up of dialectical behavior therapy vs therapy by experts for suicidal behaviors and borderline personality disorder. *Archives of General Psychiatry, 63*, 757–766.

Linehan, M. M., Schmidt, H., Dimeff, L. A., Craft, J. C., Kanter, J., & Comtois, K. A. (1999). Dialectical behavior therapy for patients with borderline personality disorder and drug-dependence. *The American Journal on Addictions, 8*, 279–292.

Livesley, W. J. (2003). *The practical management of personality disorder*. New York: Guilford.

Paris, J. (2002). Chronic suicidality among patients with borderline personality disorder. *Psychiatric Services, 53*, 738–742.

Paris, J. (2005). The diagnosis of borderline personality disorder: Problematic but better than the alternatives. *Annals of Clinical Psychiatry, 17*, 41–46.

Reisch, T., Ebner-Priemer, U. W., Tschacher, W., Bohus, M., & Linehan, M. M. (2008). Sequences of emotions in patients with borderline personality disorder. *Acta Psychiatrica Scandinavica, 118*, 42–48.

Sansone, R. A., Butler, M., Dakroub, H., & Pole, M. (2006). Borderline personality symptomatology and employment disability: A survey among outpatients in an internal medicine clinic. *Primary Care Companion to the Journal of Clinical Psychiatry, 8*, 153–157.

Shea, M. T., Pilkonis, P. A., Beckham, E., Collins, J. F., Elkin, I., Sotsky, S. M., et al. (1990). Personality disorders and treatment outcome in the NIMH treatment of depression collaborative research program. *The American Journal of Psychiatry, 147*, 711–718.

Stiglmayr, C. E., Ebner-Priemer, U. W., Bretz, J., Behm, R., Mohse, M., Lammers, C. H., et al. (2008). Dissociative symptoms are positively related to stress in borderline personality disorder. *Acta Psychiatrica Scandinavica, 117*, 139–147.

Verheul, R., Van Den Bosch, L. M., Koeter, M. W., De Ridder, M. A., Stijnen, T., & Van Den Brink, W. (2003). Dialectical behaviour therapy for women with borderline personality disorder: 12-month, randomised clinical trial in The Netherlands. *The British Journal of Psychiatry, 182*, 135–140.

Zanarini, M. C., Frankenburg, F. R., Jager-Hyman, S., Reich, D. B., & Fitzmaurice, G. (2008). The course of dissociation for patients with borderline personality disorder and axis II comparison subjects: A 10-year follow-up study. *Acta Psychiatrica Scandinavica, 118*, 291–296.

Zanarini, M. C., Frankenburg, F. R., Reich, D. B., Fitzmaurice, G., Weinberg, I., & Gunderson, J. G. (2008). The 10-year course of physically self-destructive acts reported by borderline patients and axis II comparison subjects. *Acta Psychiatrica Scandinavica, 117*, 177–184.

Chapter 36

CASE STUDY: Cognitive behavioral therapy for an adult smoker receiving substance use treatment

Alba González-Roz, Gema Aonso-Diego, and Roberto Secades-Villa

Addictive Behaviors Research Group, Department of Psychology, Faculty of Psychology, University of Oviedo, Oviedo, Principality of Asturias, Spain

Abbreviations

BDI-II beck depression inventory, II
CBT cognitive behavioral therapy
CO carbon monoxide
FTCD Fagerström test for cigarette dependence
ng/mL nanograms milliliter
ppm parts per million

Introduction: Prevalence rates of cigarette smoking in persons with substance use disorder and its associated consequences

Recent estimates from Europe and the United States indicate that cigarette smoking ranges between 48.7% and 68.9% in substance use treatment settings (Guydish et al., 2020; Hayhurst et al., 2021). Epidemiological research studies have also revealed that mental health disorders are overrepresented among people who smoke. Furthermore, cigarette smoking is about twice as prevalent in individuals with any mental health disorder compared to those with no comorbid conditions (Parker & Villanti, 2021).

Cannabis and alcohol are the most frequently used substances among cigarette smokers (Schauer, Berg, Kegler, Donovan, & Windle, 2015). Synergistic and compensatory health effects of concurrent use of tobacco and other drugs have also been reported in the literature (Rabin & George, 2015). Importantly, smoking is well recognized as the main cause of mortality in people who use substances, representing up to 49% of the total deaths (Callaghan, Gatley, Sykes, & Taylor, 2018).

The relevance of providing smoking cessation support while in substance use treatment is well supported by independent clinical studies, as well as a recent metaanalysis, evidencing a relationship between smoking cessation and abstinence from alcohol and illicit drugs (Secades-Villa, Aonso-Diego, García-Pérez, & González-Roz, 2020).

To date, the most consistent evidence supports, inter alia, the use of cognitive behavioral therapy (CBT) for facilitating smoking cessation in people who are in substance use treatment and recovery. Smoking cessation has a positive effect on substance use outcomes (McKelvey, Thrul, & Ramo, 2017). Dual substance interventions seem feasible (Walsh, McNeill, Purssell, & Duaso, 2020) and are often delivered in 5 to 6 weekly group therapy sessions (Becker, Haug, Kraemer, & Schaub, 2015), which integrate motivation interviewing, training in self-control strategies, and relapse prevention, among other aspects.

This chapter seeks to describe the clinical report of an individual (client P.A.) who received CBT for smoking cessation.

Case study

Case formulation

P.A. was a 43-year-old male, who was single and lived with his mother in Oviedo (Spain). He worked full time, and his hobbies were playing instruments, especially the guitar and drums, and participating in competitions with remote-controlled cars.

He requested smoking cessation treatment because he wanted to start a "new life," completely eliminating substance use, including tobacco. The client reported using 14 hand-rolled cigarettes per day. This was his first quit attempt. At the time of the baseline assessment, he had been receiving outpatient substance use treatment for cannabis and cocaine consumption for approximately 4 months. He started using cannabis at 14, and cocaine at 30. His average number of days of abstinence from cocaine was 160 days, and from cannabis it was 130 days.

The intervention that P.A. received was an empirically validated CBT protocol for tobacco use (Secades-Villa, García-Rodríguez, López-Núñez, Alonso-Pérez, & Fernández-Hermida, 2014) that had been tailored to this patient in order to specifically address the use of both tobacco and other substances.

Clinical assessment

P.A. attended a single baseline assessment. Two measures were used: a clinical interview and self-reported questionnaires. What follows is a description of the main variables that were collected. All the assessments were also repeated at post-treatment (eighth session) and in the follow-ups (at 1, 2, 3, 6, and 12 months).

P.A. had been smoking regularly for 20 years, since he was 23 years old. He indicated that he started using tobacco at the same time as cannabis, because he used the two substances in combination. He had always feared returning to drug consumption. When P.A. started treatment for substance use, he successfully quit cocaine and cannabis but continued to consume tobacco, "for fear of using cannabis again," even when the flavor and smell of tobacco were unpleasant for him.

A Fagerström Test for Cigarette Dependence (FTCD: Heatherton, Kozlowski, Frecker, & Fagerström, 1991) score of 4 indicated that P.A. had moderate nicotine dependence. He reported that his first cigarette of the day was the most pleasant. In the mornings, his tobacco consumption began no later than 30 to 60 min after waking up.

Exposure to tobacco use was biochemically assessed through carbon monoxide (CO) in expired air and cotinine urinalysis. At the baseline assessment, CO levels were 15 ppm (ppm), and urine cotinine concentrations were 2263.1 nanograms per milliliter (ng/mL). These assessments occurred at each session and sought to provide feedback on decreasing CO and cotinine as well as to enhance motivation to change. Additionally, P.A. was asked to record his daily cigarette consumption. In particular, the therapist asked him to indicate the following aspects: (1) number and time of cigarettes smoked each day; (2) pleasure (from 0 to 10); (3) situations or events where cigarette smoking occurred.

P.A. informed of no lifetime history of mental health disorders. The Beck Depression Inventory-II (BDI-II: Beck, Steer, & Brown, 1996) suggested nonclinical depressive symptomatology (total score of 7/63). Symptoms with the highest scores were sense of guilt, low self-confidence, high self-criticism, impaired decision making, loss of energy, and changes in sleep.

Cognitive behavioral therapy: Treatment implementation

General considerations

The CBT consisted of 8 weeks of therapy sessions, and 7 mid-week sessions were additionally scheduled to collect biochemical samples (CO and cotinine), which took approximately 30 min. The mid-week sessions sought to reinforce the client's progress and self-efficacy. At the beginning of the sessions, P.A. provided a CO and urine sample and was informed immediately on his CO and cotinine levels. Reductions (see Figs. 1 and 2) were plotted on a board and served to discuss barriers and difficulties throughout the treatment.

Given the scope of this chapter, what follows is a session-by-session description of the main CBT components. The CBT treatment protocol is presented in Table 1. However, the reader should consider that a nicotine fading component was implemented in this intervention as well (see for further details Foxx & Axelroth, 1983). The purpose was to gradually reduce the nicotine intake by 20% each week. At each therapy session, the therapist advised the client on the number of daily cigarettes and which tobacco brands he could smoke. Other recommendations were as follows: (1) to inhale shallower puffs, (2) to reject cigarette offers from his friends, (3) to smoke a third less of the cigarette each week, and (4) only to bring the cigarette to the mouth to smoke. Each week, the therapist also requested the patient to delay the first cigarette—in the morning, after meals, and after drinking coffee—by at least 15 min.

FIG. 1 In-treatment reductions in carbon monoxide levels. *Note*: The letters A and B refer to the therapy and mid-week sessions, respectively. *ppm*, parts per million; *BL*, baseline.

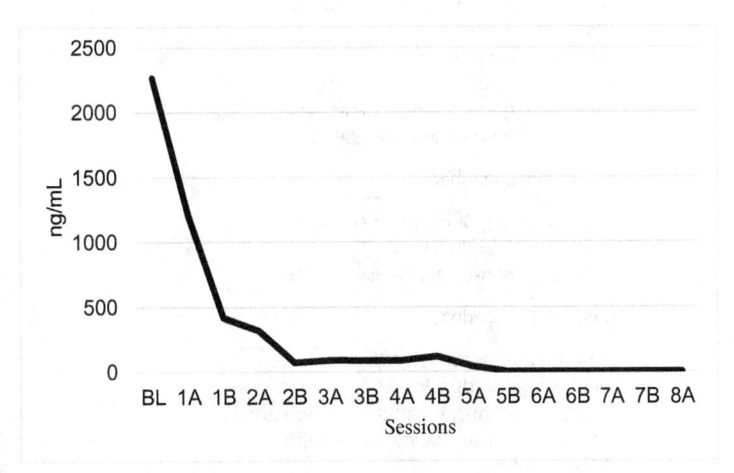

FIG. 2 In-treatment reductions in urine cotinine levels. *Note*: The letters A and B refer to the therapy and mid-week sessions, respectively. *ng/mL*, nanograms per milliliter; *BL*, baseline.

Sessions 1 and 2

In the first session, P.A. self-reported an average of 6.6 cigarettes per day. The situations in which he smoked the most were on breaks—from work and from therapy at the substance use treatment center—and at home (watching TV, and after lunch and dinner).

P.A. was ambivalent regarding his decision to quit smoking. He reported that using both cannabis and tobacco was a way of dealing with distress, especially when facing stressful situations. Using a motivational interviewing therapeutic style, psychoeducation on the relationship between tobacco and substance consumption, mental health, and substance use outcomes was utilized as an effective way to reduce positive expectancies related to substance use. What follows is an example of how the therapist worked collaboratively with the client on enhancing motivation to maintain cannabis abstinence.

TABLE 1 Sequence and main procedures used in therapy.

Sessions	CBT component
1A	Self-monitoring of cigarette smoking Biochemical feedback (CO and urine cotinine) Reasons to quit smoking Psychoeducation on the relationship between tobacco and substance use Functional analysis of smoking behavior Reduction of nicotine intake by 20%
1B	Biochemical feedback
2A	Self-monitoring of cigarette smoking Biochemical feedback Stimulus control Alternative nonsubstance use activities Management of stress and anger Diaphragmatic breathing Reduction of nicotine intake by 20%
2B	Biochemical feedback
3A	Self-monitoring of cigarette smoking Biochemical feedback Weight control strategies Stimulus control Alternative nonsubstance use activities Reduction of nicotine intake by 20%
3B	Biochemical feedback
4A	Self-monitoring of cigarette smoking Biochemical feedback Problem solving skills Stimulus control Alternative nonsubstance use activities Reduction in nicotine intake by 20%
4B	Biochemical feedback
5A	Self-monitoring of cigarette smoking Biochemical feedback Alternative nonsubstance use activities
5B	Biochemical feedback
6A	Tobacco abstinence Biochemical feedback Strategies to manage withdrawal symptoms Alternative nonsubstance use activities Problem solving skills
6B	Biochemical feedback
7A	Tobacco abstinence Biochemical feedback Managing smoking withdrawal symptoms Alternative nonsubstance use activities Relapse prevention
7B	Biochemical feedback
8A	Tobacco abstinence Biochemical feedback Alternative nonsubstance use activities Relapse prevention strategies

Note: The letters A and B refer to the therapy and mid-week sessions, respectively.

Client: "I am sure that I will never use cocaine again, and now I am abstinent, I am certain that I have quit forever." "But I cannot assure you that I will never use cannabis again."
Therapist: "Why do you think you will return to smoking joints again in the future?"
Client: "Because I like cannabis, its effect, its flavour..."
Therapist: "I understand that stopping using cannabis may be difficult for you at this point." However, what people do not often consider is that cannabis may increase the likelihood of smoking tobacco again, since cannabis and tobacco share several aspects, such as the route of administration, the gesture, the smoking... "
Client: "You are right." "I started smoking tobacco at the same time I started smoking joints, and it is something that I have never liked."

Once the basic concepts of tobacco use had been addressed, P.A. was asked to provide his main reasons for quitting smoking. He indicated the following motives: to improve his health and his aesthetic perception (i.e., to avoid the tobacco smell, yellow fingers, and black teeth). Based on the smoking self-reports, the therapist worked collaboratively with P.A. on identifying the antecedents (i.e., people, situations, times of the day), and consequences of tobacco use.

Stimulus control training occurred in session 2. The client was provided a list of situations where smoking typically occurs (e.g., waiting, talking on the phone) and was subsequently told to choose two situations where he would not smoke from then on. P.A. indicated that not smoking while in the car or when playing instruments could be feasible. The client was also trained in several strategies to deal with withdrawal symptoms: he was encouraged to drink water, chew gum, eat fruit and nuts, practice diaphragmatic breathing, or engage in pleasant activities, among other things.

At the end of the second week, P.A. self-reported an average of 5.15 cigarettes per day. The client agreed with the therapist to switch to cigarettes with lower nicotine content (0.3 mg of nicotine) and to smoke a maximum of 3 cigarettes each day. To enhance self-control, he was advised to record his cigarette consumption.

Sessions 3 and 4

P.A. reported that he had no difficulties in meeting the guidelines to reduce his tobacco use. At that time, he self-reported using 3.42 cigarettes per day. His CO was also significantly decreased after the two first sessions.

The client informed that he had not smoked in any of the chosen situations. He actively decided to work on several new situations and extend the time elapsed after eating, waiting (e.g., for the bus, a date), working, and smoking. Postcessation weight gain concerns were also addressed. The therapist discussed with P.A. the relationship between smoking and weight gain and introduced the following topics:

Therapist: "A common belief is that smokers who quit tobacco gain a lot of weight. However, you should note that the consequences of gaining a few kilograms are considerably less harmful to your health than continuing to smoke." "How many kilograms do you think a person gains on average?"
Client: "I know people who gained more than 10 kilograms, but I do not know if that is normal."
Therapist: "Well, that can happen, but there are also people who quit smoking and lose weight." "On average, quitters gain about 4.5 kg. Your weight depends on your eating and exercise habits." "Exercising a minimum of two hours per week, and some basic guidelines about food may help" ...

At the fourth session, P.A. informed of using two cigarettes per day. To help him cope with the identification of temptations and impulses to smoke, the metaphor of the wave was introduced:

Therapist: "The urge to smoke is like a huge wave." "Like you, many people want to get rid of the craving. However, you should be aware that it does not last long; it increases and decreases rapidly, just like a wave" ..." Have you ever focused on how the urge to smoke fluctuates?"
Client: "I see...Actually, I've never paid attention to my smoking urges. I only know they hurt" ...
Therapist: "Well, it is true, sometimes it can be difficult, especially at the beginning." "You should pay attention to how your cravings fluctuate during the day. You will notice that, as you increasingly delay your first cigarette in the morning, the urges to smoke end up disappearing."
Therapist: "It is a matter of engaging in non-substance-using activities. Whenever you feel cravings, you should engage in positive and alternative behaviors to smoking (e.g., going for a walk, eating a piece of fruit) ..."

The problem-solving technique was also introduced in the context of relapse prevention, with the aim of practicing new and more effective ways to cope with stressful situations and more broadly solving problems without using substances. P.A. used tobacco to handle anger, and he informed of several family conflicts due to his angry outbursts while reducing tobacco use. The problem-solving protocol included: (1) defining life problems in terms of A[trigger]-B[behavior]-C[consequence (s)] chains; (2) brainstorming solutions to problems; (3) identifying pros and cons of each of the solutions raised, and selecting the one with the best potential to promote positive (desirable) outcomes; (4) practicing the chosen solution in treatment sessions; (5) envisaging a plan to implement the solution effectively in vivo; (6) monitoring and revising the plan during the next therapy sessions.

P.A. chose a future situation that was to occur the next week, "a remote-control car race." This situation was associated with smoking for P.A. because tobacco helped him to concentrate better. P.A. would have to face this situation for the first time since being in treatment. To effectively manage the situation, he worked on each of the problem-solving steps and selected different alternatives to using cigarettes: going for a walk or asking for help from trusted people.

As P.A. was smoking a low number of cigarettes each day (between 1 and 2) he was asked to quit tobacco by the next session. Several recommendations were given to the client: (1) to remember his reasons for quitting, and the perceived benefits associated with smoking cessation; (2) to remove all smoking-related cues (e.g., ashtrays, lighters); (3) to set a quit day. P.A. chose Sunday and several strategies were also practiced during this session: diaphragmatic breathing, problem solving, and coping strategies for withdrawal symptoms.

Session 5 (quit day)

At the fifth session, P.A. had been abstinent for the previous 43 h. His CO level was 1 ppm. In the fifth session, the difference between lapse and relapse was discussed with the client. The former refers to a well-delimited consumption (e.g., a puff), while relapse refers to returning to baseline consumption (e.g., 20 cigarettes per day). This differentiation is important since it makes it easier to return to abstinence if the patient takes a puff or smokes the occasional cigarette. P.A. worried about a number of potential negative consequences during his first weeks of smoking abstinence. He started noticing several physical sensations related to smoking cessation: an increased taste of nicotine in the mouth, dry mouth, and constipation. P.A. was told that these were positive signs which meant that his body was eliminating the toxins from tobacco. He was told that these signs were only temporary. Cognitive restructuring was conducted to address distorted cognitions on the effects of smoking abstinence:

- Does quitting smoking make your health worse?
- Am I at risk of relapsing to other substance use?
- Will I be more nervous/anxious/angry or irritable?
- Will I lose concentration?

Sessions 6–7

P.A. had not smoked since he quit. However, he was still worried about experiencing urges to smoke but claimed he was noticing a gradual decrease since the first session. He also informed of having difficulty in managing his cravings when his mother smoked close to him. The therapist reminded him the metaphor of the wave and encouraged him to maintain abstinence so that the urge to smoke became progressively less intense. During the sixth session, the therapist worked collaboratively with the client on a weekly exercise plan. In the session, P.A. scheduled at least two days when he could walk for at least 30 min. He was told to reward himself after each practice by preparing a different dinner and sharing with his mother how good he felt after doing some sport.

A relapse prevention plan was developed during session seven, following a two-step procedure (see Fig. 3). Firstly, high-risk situations were defined as any situation that could potentially trigger a lapse or relapse. To this end, P.A. wrote down a list of situations (both internal and external) that could pose a threat to smoking abstinence (e.g., negative moods, interpersonal conflicts). The action plan sought to provide P.A. with effective strategies to deal with his smoking cravings.

Session 8

The eighth session was the last therapy session. The therapist and P.A. worked collaboratively to review the action plan. The following themes were raised:

Triggers: *"You have now identified several places and things that have the potential to trigger a relapse."* *"You will probably have to use your action plan in the following weeks."* *"It is also important to let you know that you won't be able to list every single potentially triggering situation. However, you now know how to act in risky situations."*

- Craving and negative symptoms: *"You now know how to cope with the urges to smoke."* *"Sometimes it is common to experience craving, but you will also notice that the urgency to smoke decreases in intensity."*
- Lifestyle changes: *"As you have experienced over these past weeks, quitting smoking is easier than you thought at first."* *"As time goes by, it becomes easier for you to maintain abstinence."*

> **What can I do in high-risk situations?**
>
> **My action plan**
>
> To develop your action plan, you will have to consider the following:
>
> (1) Be specific and realistic: write down what you want to do (must be achievable and precise). For example: do physical activity with a friend on Tuesdays and Thursdays from 6.00 p.m. to 7.00 pm.
> (2) Be action-oriented: plan what you will do without anticipating the emotions or consequences that may result from it. For example: Have a good time in the countryside with your family next Saturday morning.
> (3) Make it limited in time: this facilitates the motivation to take action while Increasing the feeling of well-being or self-efficacy.
>
> **Please indicate any potentially triggering situations. How will you deal with them?**
>
> _____
> _____
> _____

FIG. 3 Relapse prevention plan work sheet.

The posttreatment assessment also occurred during this session. P.A. self-reported 22 days of continuous abstinence. CO levels were 1 ppm, and urinalyses yielded 0 ng/mL in the cotinine analyses. Substance use tests also confirmed that P.A. was abstinent from cocaine and cannabis. He was certain that he would never smoke again and said that quitting tobacco was the best thing he had ever done. His depressive symptomatology was also significantly reduced, as he scored 0 on the BDI-II. Sustained abstinence from both tobacco and other substances was observed at one year.

Applications to other areas

We described the application of CBT for smoking cessation in a person undergoing substance use treatment. The CBT protocol we presented includes motivational interviewing, nicotine fading, stimulus control, and strategies to deal with withdrawal symptoms. This intervention protocol has been evaluated in a Randomized Control Trial and showed acceptable smoking cessation rates (20%) in substance users at the posttreatment (Aonso-Diego, González-Roz, Krotter, García-Pérez, & Secades-Villa, 2021).

This CBT protocol has the potential to be implemented in cases presenting other co-occurring disorders (i.e., depression, anxiety) (Secades-Villa et al., 2019). In this protocol, activation is targeted throughout treatment, meaning the therapist encourage the patient to engage in positive and reinforcing activities. This is clinically relevant given the high prevalence of other mental health disorders in substance users (Hindocha, Brose, Walsh, & Cheeseman, 2021) and the impact that they may have on tobacco and substance use recovery (Bartoli, Carreta, & Carrà, 2021; Charney, Palacios-Boix, Negrete, Dobkin, & Gill, 2005; Wai, Shulman, Nunes, Hasin, & Weiss, 2021).

Further, CBT can be easily implemented in existing substance use treatments, as part of broader interventions without jeopardizing abstinence from other drug use outcomes. Recent research has evidenced that CBT can be combined with other intervention approaches, such as behavioral activation, contingency management, and pharmacotherapy, suggesting it can be even more effective if it is delivered as combined with these former therapies (Secades-Villa et al., 2019). The use of e-health interventions is promising as it can make interventions more broadly available. Currently, there exists evidence supporting mobile-based CBT is both feasible and effective for substance users (Shams et al., 2021).

Key facts

- Cigarette smoking is highly prevalent (48.7%–68.9%) in substance users.
- Relative to the general population, substance users show higher nicotine dependence.

- Cigarette smoking increases poor substance use treatment outcomes.
- A large majority of smokers in substance use treatment experience anxiety and fear of relapsing back to drug use.
- Smoking abstinence may improve patients' chances for sustained substance use abstinence.
- CBT is amongst the most effective therapies for smoking cessation.

Summary points

- CBT facilitates sustained smoking abstinence in individuals in early-stage substance use recovery.
- Nicotine fading is a feasible procedure that can be implemented within CBT interventions for smoking cessation.
- CBT based approaches for smoking cessation can be used to prevent substance use relapse.
- Smoking cessation did not jeopardize abstinence from other substances.
- Providing effective smoking cessation treatments should be a priority for individuals in substance use treatment or in recovery.

References

Aonso-Diego, G., González-Roz, A., Krotter, A., García-Pérez, A., & Secades-Villa, R. (2021). Contingency management for smoking cessation among individuals with substance use disorders: In-treatment and post-treatment effects. *Addictive Behaviors, 119*, 106920.

Bartoli, F., Carreta, D., & Carrà, G. (2021). Comorbid anxiety and alcohol or substance use disorders: An overview. In N. el-Guebaly, G. Carrà, M. Galanter, & A. M. Baldacchino (Eds.), *Textbook of Addiction Treatment* (pp. 1315–1325). Springer. https://doi.org/10.1007/978-3-030-36391-8_91.

Beck, A. T., Steer, R. A., & Brown, G. K. (1996). *Manual for the Beck depression inventory–II*. San Antonio, TX: Psychological Corporation.

Becker, J., Haug, S., Kraemer, T., & Schaub, M. P. (2015). Feasibility of a group cessation program for co-smokers of cannabis and tobacco. *Drug and Alcohol Review, 34*, 418–426.

Callaghan, R. C., Gatley, J. M., Sykes, J., & Taylor, L. (2018). The prominence of smoking-related mortality among individuals with alcohol- or drug-use disorders. *Drug and Alcohol Review, 37*, 97–105.

Charney, D. A., Palacios-Boix, J., Negrete, J. C., Dobkin, P. L., & Gill, K. J. (2005). Association between concurrent depression and anxiety and six-month outcome of addiction treatment. *Psychiatric Services, 56*(8), 927–933.

Foxx, R. M., & Axelroth, E. (1983). Nicotine fading, self-monitoring and cigarette fading to produce cigarette abstinence or controlled smoking. *Behaviour Research and Therapy, 21*, 17–27.

Guydish, J., Kapiteni, K., Le, T., Campbell, B., Pinsker, E., & Delucchi, K. (2020). Tobacco use and tobacco services in California substance use treatment programs. *Drug and Alcohol Dependence, 214*, 108173.

Hayhurst, K. P., Jones, A., Cairns, D., Jahr, S., Williams, E., Eastwood, B., et al. (2021). Tobacco smoking rates in a national cohort of people with substance use disorder receiving treatment. *European Addiction Research, 27*, 151–155.

Heatherton, T., Kozlowski, L., Frecker, R., & Fagerström, K. (1991). The Fagerström test for nicotine dependence: A revision of the Fagerström tolerance questionnaire. *British Journal of Addiction, 86*, 1119–1127.

Hindocha, C., Brose, L. S., Walsh, H., & Cheeseman, H. (2021). Cannabis use and co-use in tobacco smokers and non-smokers: Prevalence and associations with mental health in a cross-sectional, nationally representative sample of adults in Great Britain, 2020. *Addiction, 116*(8), 2209–2219.

McKelvey, K., Thrul, J., & Ramo, D. (2017). Impact of quitting smoking and smoking cessation treatment on substance use outcomes: An updated and narrative review. *Addictive Behaviors, 65*, 161–170.

Parker, M. A., & Villanti, A. C. (2021). Relationship between comorbid drug use disorders, affective disorders, and current smoking. *Substance Use & Misuse, 56*, 93–100.

Rabin, R. A., & George, T. P. (2015). A review of co-morbid tobacco and cannabis use disorders: Possible mechanisms to explain high rates of co-use. *The American Journal on Addictions, 24*, 105–116.

Schauer, G. L., Berg, C. J., Kegler, M. C., Donovan, D. M., & Windle, M. (2015). Assessing the overlap between tobacco and marijuana: Trends in patterns of co-use of tobacco and marijuana in adults from 2003-2012. *Addictive Behaviors, 49*, 26–32.

Secades-Villa, R., Aonso-Diego, G., García-Pérez, Á., & González-Roz, A. (2020). Effectiveness of contingency management for smoking cessation in substance users: A systematic review and meta-analysis. *Journal of Consulting and Clinical Psychology, 88*, 951–964.

Secades-Villa, R., García-Rodríguez, O., López-Núñez, C., Alonso-Pérez, F., & Fernández-Hermida, J. R. (2014). Contingency management for smoking cessation among treatment-seeking patients in a community setting. *Drug and Alcohol Dependence, 140*, 63–68.

Secades-Villa, R., González-Roz, A., Vallejo-Seco, G., Weidberg, S., García-Pérez, Á., & Alonso-Pérez, F. (2019). Additive effectiveness of contingency management on cognitive behavioural treatment for smokers with depression: Six-month abstinence and depression outcomes. *Drug and Alcohol Dependence, 204*, 107495.

Shams, F., Wong, J., Nikoo, M., Outadi, A., Moazen-Zadeh, E., Kamel, M. M., ... Krausz, R. M. (2021). Understanding eHealth cognitive behavioral therapy targeting substance use: Realist review. *Journal of Medical Internet Research, 23*(1), e20557.

Wai, J. M., Shulman, M., Nunes, E. V., Hasin, D. S., & Weiss, R. D. (2021). Co-occurring mood and substance use disorders. In N. el-Guebaly, G. Carrà, M. Galanter, & A. M. Baldacchino (Eds.), *Textbook of Addiction Treatment* (pp. 1297–1313). Springer. https://doi.org/10.1007/978-3-030-36391-8_91.

Walsh, H., McNeill, A., Purssell, E., & Duaso, M. (2020). A systematic review and Bayesian meta-analysis of interventions which target or assess co-use of tobacco and cannabis in single- or multi-substance interventions. *Addiction*, *115*, 1800–1814.

Chapter 37

CASE STUDY: Cultural diversity and cognitive-behavioral therapy

Esteban V. Cardemil, Sarah J. Hartman, and José R. Rosario
Frances L. Hiatt School of Psychology, Clark University, Worcester, MA, United States

Abbreviations

CBT   cognitive-behavioral therapy
CES   Compañeros En Salud
CHW   community health worker
FCSP   Family Coping Skills Program
NHANES   National Health and Nutrition Examination Survey

Introduction

As has been documented in the other chapters in this book, research on cognitive-behavioral interventions has grown tremendously over the past 50 years (Waltman & Sokol, 2017). This research has directly contributed to the identification and dissemination of numerous effective interventions for a variety of disorders. However, mental health services research has found that many individuals in need do not receive adequate treatment (Kessler, Merikangas, & Wang, 2010). For example, data from the National Health and Nutrition Examination Survey (NHANES) indicate that over 70% of individuals with self-reported depressive symptoms did not receive either pharmacotherapy or psychotherapy, including almost 50% of individuals with severe levels of symptoms (Wittayanukorn, Qian, & Hansen, 2014).

Moreover, research has consistently shown that individuals from racial and ethnic minority groups are especially unlikely to receive mental health services (Galvan & Gudiño, 2021; Kearney, Draper, & Barón, A., 2005; Kim, Park, La, Chang, & Zane, 2016; Kohn-Wood & Hooper, 2014). Further, there is evidence that when individuals from these groups do seek out mental health services, they are less likely to receive an adequate dosage of treatment (Fortuna, Alegria, & Gao, 2010; Simpson, Krishnan, Kunik, & Ruiz, 2007), and more likely to report negative treatment experiences (Bartholomew, Pérez-Rojas, Bledman, Joy, & Robbins, 2021; Imel et al., 2011; Ortega et al., 2007). Concerningly, recent research suggests that some of these disparities are not improving, and in some cases may be worsening (Blanco et al., 2007; Cook, McGuire, & Miranda, 2007; Lagomasino, Stockdale, & Miranda, 2011).

These disparities in mental healthcare have numerous and interrelated causes, some of which are long-term societal issues, including inadequate funding of mental health services, insufficient insurance policies, and limitations in training/education (Cardemil, Nelson, & Keefe, 2015; Cook et al., 2019; Miranda, Snowden, & Legha, 2020). Nevertheless, it is also clear that service providers can play an important role in contributing to, or ameliorating, these disparities. For example, research has consistently found that therapist cultural competence is associated with client satisfaction, engagement, and treatment outcome (Soto, Smith, Griner, Domenech Rodríguez, & Bernal, 2018).

Cognitive-behavioral therapy (CBT) has been at the forefront of research examining the generalizability of interventions to diverse populations, including the development and evaluation of cultural adaptations of standard interventions for particular populations. In addition, the growing field of implementation science has advanced our understanding of how systems of care can be adapted to increase the uptake of particular interventions. However, there remain important gaps in both of these research literatures that contribute to the continuing disparities in mental healthcare. In this chapter, we briefly review relevant literature on cultural adaptations and implementation science, identify gaps that are relevant to disparities, and then present recommendations for integrating these approaches in targeted ways that have potential to reduce disparities.

Definitions of disparities

One of the challenges facing efforts to address disparities in mental healthcare is the lack of agreement regarding what constitutes a disparity (Cardemil et al., 2015). Most definitions focus on differences between groups in the receipt or quality of healthcare (Smedley, Stith, & Nelson, 2003). However, there is inconsistency in the extent to which factors outside of the healthcare system should be taken into consideration when assessing the existence and extent of disparities. For example, the Institute of Medicine has defined disparities as "differences in the quality of healthcare that are not due to access-related factors or clinical needs, preferences, and appropriateness of intervention" (Smedley et al., 2003). This definition of disparities does not include group preferences in clinical care, which could plausibly include preferences that result from negative experiences with the healthcare system. For example, some racial and ethnic groups have negative historical experiences with the medical system that lead to mistrust of medical providers (Hammond, 2010; Oakley, López-Cevallos, & Harvey, 2019). A group preference that is shaped by negative historical experiences would seem critical to include in definitions of disparities, given that utilization of healthcare and negative experiences with the healthcare system are interconnected, particularly in the area of mental health (Jackson, 2006).

In addition, it is important to distinguish among interrelated terms, such as *access*, *utilization*, and *engagement* (see Table 1). Although similar, these concepts have important differences among them that have substantive implications for efforts to address disparities. We define access as those factors that influence the availability of services for individuals and communities. Barriers to access include structural factors, like limited availability of community health centers and number of bilingual providers. In contrast, when referring to utilization, we generally reference individual acts of seeking out services; barriers to utilization thus include both logistical and psychological factors. Some examples of psychological factors include negative attitudes toward mental health services, stigma, and particular cultural values. Finally, the term engagement refers to ongoing interaction with health services; this is especially relevant to mental health services given the generally chronic nature of mental disorders. Barriers to engagement can be the same ones that also affect access and utilization, but they can also include factors related to the patient-provider interactions.

Taken together, we define disparities in mental healthcare as group differences in access, utilization, or engagement in particular services that are not the result of clinical need. We now briefly review the literature on cultural adaptations of cognitive-behavioral interventions, focusing on how these adaptations have addressed disparities.

Brief review of cultural adaptation literature

Over the past 25 years, there has been a growing recognition that standard interventions might not be adequately situated to best meet the needs of the changing demographics in the United States (Cardemil, 2010). In particular, more and more research has documented sociocultural influences on prevalence of mental disorders, as well as group differences in treatment acceptability, engagement, and outcome (Acevedo-Polakovich et al., 2017; Rodriguez-Seijas, Eaton, & Pachankis, 2019; Yu, Pope, & Perez, 2019). This theoretical and empirical literature set the foundation for researchers to develop and evaluate adaptations of standard interventions for different social groups (Cardemil, 2010; Domenech Rodríguez & Bernal, 2012; Hwang, 2012). Most of the initial waves of adapted interventions were focused on racial and ethnic minority groups in the United States, hence the name "cultural adaptations."

Several meta-analyses have reviewed this body of research, and the growing consensus is that cultural adaptations have generally positive effects for the participants (e.g., Benish, Quintana, & Wampold, 2011; Griner & Smith, 2006;

TABLE 1 Different elements of Disparities.

Term	Definition	Example barriers
Access	Availability of mental health services	Structural factors, including limited availability of community health centers and number of bilingual providers
Utilization	Individual acts of seeking mental health services	Logistical barriers, such as work limitations on time off, childcare needs, transportation issues Psychological barriers, including negative attitudes toward mental health services, stigma, and particular cultural values
Engagement	Ongoing individual interactions with mental health services	Same barriers as above, but also include interactions with providers (e.g., cultural competence, therapeutic alliance)

Soto et al., 2018). These meta-analyses have also begun to explore what particular adaptations might be most associated with positive effects. For example, Griner and Smith (2006) found larger effect sizes in those studies that had high numbers of Latinx participants, and especially in those studies with larger numbers of less acculturated Latinxs. Similarly, Soto et al. (2018), in their meta-analysis of 99 different studies, found that language adaptations, culturally consonant metaphors, and culturally relevant techniques were associated with larger effect sizes.

More recent cultural adaptation research has attempted to make sociopolitical factors more central. For example, Robinson et al. (2016) adapted a stress reduction intervention to focus on culturally relevant factors like experiences of violence, racism, and suicide-related stigma relevant to many Black youth living in the United States. Preliminary findings suggested that these adaptations were associated with positive effects. Similarly, Stewart, Orengo-Aguayo, Wallace, Metzger, and Rheingold (2019) adapted trauma-focused CBT for Black youth to include racial socialization and considerations of discrimination; the authors noted increased engagement and patient acceptability of the adapted intervention.

Despite the growing consensus that culturally adapted interventions can yield positive effects across a number of different disorders and with a number of different populations, several important questions remain. In particular, there is currently no consensus regarding the best process for making cultural adaptations to existing treatments (Domenech Rodríguez & Bernal, 2012). As a result, one issue that has emerged in the cultural adaptation literature has been the overfocus on the intervention itself, and less attention to the larger context in which these interventions might be delivered. This is likely due to the fact that cultural adaptation research is grounded in the literature on psychosocial treatment development and evaluation, which has primarily emphasized decontextualized individual change. Thus, it is unsurprising that societal context and systems of care have generally not been emphasized in cultural adaptation research. This limitation is an important one, since inattention to systems of care can make it more difficult for interventions to address disparities in the mental healthcare system, thus undermining the primary goal of cultural adaptations. Implementation science, however, through its focus on larger context and systems issues, has the potential to address this limitation. We now briefly review some of this relevant research.

Brief review of implementation science literature

Implementation science is a branch of health services research focused on the development and evaluation of strategies to increase the use and effectiveness of evidence-based practices, including mental health interventions (Peters, Tran, & Adam, 2013; Vroom & Massey, 2021). Various implementation science frameworks exist across psychology, medicine, and related fields (Tabak, Khoong, Chambers, & Brownson, 2013), with differing terminology and concepts (Chambers, 2014). Regardless of the variation in the models, however, implementation science researchers often explore outcomes such as access, acceptability/utilization, dose, reach, and effectiveness (Craig et al., 2008; Moore et al., 2015; Peters et al., 2013). By focusing on this range of outcomes, implementation researchers aim to disentangle which aspects of intervention delivery might contribute to positive results and which may need to be altered, oftentimes with the goal of decreasing disparities in care (Cabassa & Baumann, 2013).

Mental health services are increasingly turning to implementation science to improve service delivery, particularly in community mental health and other settings that prioritize reducing disparities in care (Dixon & Patel, 2020; Shelton et al., 2020). Because implementation science focuses on contextual factors (Peters et al., 2013), it is more readily able to attend to the various cultural and structural contexts that contribute to mental health disparities. Across the globe, researchers and practitioners have demonstrated an increase in access, engagement, utilization, and/or service quality among marginalized groups once effective implementation strategies are identified (e.g., Abas et al., 2018; Hutchison, Karpov, Deegan, MacDonald-Wilson, & Schuster, 2015; McKay, Sensoy Bahar, & Ssewamala, 2020; Miguel-Esponda, Bohm-Levine, Rodríguez-Cuevas, Cohen, & Kakuma, 2020). For example, in one study of the Ugandan national mental health services for children, McKay et al. (2020) found that involving community members in the research process helped to increase engagement in mental health services. With this inclusive and collaborative method, materials were more adequately adapted and community members felt more connected to and in control of their care.

While there is overlap between cultural adaptation research and implementation science, the latter tends to focus on implementation at the systemic level, while cultural adaptation research often attends to provider- and client-level adaptations (Cabassa & Baumann, 2013). The drawback to implementation science is that it rarely delves into the how and why of behavior change, as psychologists do in cultural adaptation research. Implementation scientists tend to focus on how to integrate and deliver intervention manuals, but they spend less time building and adapting the manuals themselves. At times, they assume that interventions shown to be effective in clinical trials will be effective in all settings, as long as the delivery is adapted, disregarding the need to adapt the intervention material or include community stakeholders. Additionally, despite being focused on implementation, at times implementation science strategies are not translated into practice once the research study ends (Baumann, Cabassa, & Wiltsey Stirman, 2017). As with the limitations of cultural adaptation efforts, these limitations contribute to the continued disparities that exist in mental healthcare.

Cultural adaptations and implementation science: Working together to reduce disparities

Given that their relative strengths and limitations are complementary, public health approaches that integrate implementation science and cultural adaptation have significant potential to address disparities (see Table 2). In particular, as we have noted, the field of cultural adaptation has not paid sufficient attention to the systems of care and care delivery that have important influences on whether and how a new intervention may be accessed and utilized by individuals from a particular community. In contrast, while the field of implementation science has emphasized the systems of care, it has paid less attention to how particular interventions might best be adapted to respond to local cultural and contextual factors. Thus, bringing the strengths of each of these perspectives offers the global mental health community an opportunity to directly address and potentially reduce disparities in mental health.

In this section, we elaborate on this idea by providing examples from two community-based interventions in which we have been involved. The first was a cognitive-behavioral depression preventive intervention that was developed for financially challenged Latina mothers (Cardemil, 2010; Cardemil, Kim, Pinedo, & Miller, 2005). Named the Family Coping Skills Program (FCSP), this intervention integrated group and family sessions to bolster the resiliency and coping skills of participants. The group sessions offered participants psychoeducation around a variety of different cognitive-behavioral emotion regulation skills in a supportive environment with other Latina mothers. The two family sessions, which were grounded in a theoretical model that emphasizes the interrelatedness of family members across a variety of domains (Miller, Ryan, Keitner, Bishop, & Epstein, 2000; Ryan, Epstein, Keitner, Miller, & Bishop, 2005), consisted of the participant and one other adult family member who participated in child rearing (e.g., partner, parent, friend). These sessions provided participants with the opportunity to discuss a range of parenting and relationship issues with a family member, as well as receive emotional and logistical support.

The second intervention is a CBT-informed psychoeducation course that was developed by Partners In Health Mexico/ *Compañeros En Salud (CES)* for primary care patients with depressive and anxious disorders in impoverished communities in Chiapas, Mexico (Hartman, Miguel-Esponda, Watson, Cardemil, & Rodríguez-Cuevas, 2018; Rodríguez-Cuevas, Hartman, Aguerrebere, & Palazuelos, 2020). The CES team adapted the course, *El Curso de los Triángulos* or The Triangles Course, from Lewinsohn, Steinmetz, Antonuccio, and Teri's (1984) *Coping with Depression Course* to the context of rural Chiapas. The Triangles Course consists of nine group sessions, including one family session, with CBT and psychoeducational topics such as the symptoms and causes of depression, cognitive restructuring, and behavioral activation exercises. Individuals who attend the course learn about mental health, the connections among thoughts, feelings, and behaviors, and coping strategies. Moreover, participants connect with others from their community with similar experiences.

Both the FCSP and the Triangles Course were novel CBT interventions designed for particular contexts (i.e., low-income Latina mothers in the United States and primary care patients in Chiapas, Mexico). As such, both research teams had to consider issues relevant to both the literature on cultural adaptation and implementation science. Specifically, these issues included attending to culture and context in (1) the creation and adaptation of the content and focus of the intervention itself, (2) the delivery of the intervention, and (3) the training of the interventionists (see Table 3). We now describe each in turn.

TABLE 2 Strengths and limitations of cultural adaptation work and implementation science.

	Strengths	Limitations
Cultural Adaptation	Theoretically driven attention to modifications	Less focus on systems and context
Implementation Science	Systemic focus	Less focus on actual content of intervention

TABLE 3 Culturally attuned content, delivery, and interventionists for two CBT-informed interventions.

	Family coping skills program	Triangles course
Content	Material available in Spanish and English Culturally relevant examples, role-plays, and stories Opportunities for participants to share stories of family immigration, difficulties navigating systems of care, and experiences with discrimination	Manual translated into Spanish Adaptations included culturally relevant vignettes, adaptation of content to a more visual format, simplification of exercises
Delivery	Warm and personable interactions Informal style (e.g., provided refreshments, facilitators engaged in more self-disclosure) Use of more formal Spanish when speaking with participants Flexible scheduling and provision of bus passes or taxi vouchers Onsite childcare	Community health workers delivered the course to increase attendance Provision of snacks for more relaxed atmosphere and socialization Course took place in clinic so participants wouldn't have to travel far to participate Young children allowed to accompany parents to course
Interventionists	Culturally attuned clinicians who held similar identities (e.g., Latinx) Supervision and training around engaging with participants in a culturally sensitive manner Supervision and training on issues of power and marginalization	Interventionists switched from physicians to community health workers (CHWs) in order to increase course attendance and cultural resonance CHWs came from the same community as the participants

Attending to culture and context through adaptations in intervention content

Interventions must be relevant to the population in which they will be used, and so researchers who want to use standard interventions in novel cultural contexts need to consider what adaptations to the curriculum of interventions might make the intervention more relevant. Many cognitive-behavioral interventions bring the theory to life through use of examples, role-play situations, and stories in both therapist and client manuals. A common approach to cultural adaptations is to use topics and themes that are more relevant to the lived experiences of the population of interest. In addition, it is important to make use of culturally relevant metaphors and stories in the delivery of particular concepts.

In the FCSP, we incorporated culturally-relevant material throughout the curriculum. In introducing cognitive skills like recognizing and challenging negative and unrealistic thoughts, we used relevant examples of stressful situations in the lives of the participants. Moreover, we created opportunities for the women to share stories of family immigration and adaptation, difficulties navigating systems of care, and experiences with prejudice and discrimination. Sharing these stories allowed the participants to connect with each other around their common experiences, and also engage in helpful problem-solving. In the family sessions, we helped participants explore how they navigated the stress of immigration, their financial limitations, and worries about raising their children in a different culture.

Similarly, with the Triangles Course, *Compañeros En Salud* adapted Lewinsohn et al.'s (1984) original manual so it was culturally relevant to patients. Beyond translating the original content into Spanish, the research team incorporated culturally relevant vignettes, adapted the text to a more visual format given the high level of illiteracy in the region, and simplified the information and exercises to make them accessible to participants from diverse educational backgrounds. One vignette, for example, highlighted the social determinants of mental health that are common in the participants' context; the woman in the story faced economic oppression and childhood trauma that contributed to her depression, and her symptoms manifested in ways common in this context (e.g. aches and pains, fatigue, irritability). Participants could identify with these culturally-relevant vignettes, and often corresponding role-plays, and many participants reported that these stories were one of the most memorable pieces of the course. The CES team also adapted the manual so it could be useful for participants with various levels of education, given the lack of access to education in these isolated rural areas. The team changed "automatic negative thoughts" to "skewed thoughts" ("*pensamientos chuecos*"), for example, and simplified the corresponding exercises. Such changes were necessary to make evidence-based interventions culturally resonant in new contexts.

Both the FCSP and the Triangles Course took standard cognitive-behavioral techniques and reworked them to make them relevant to their particular populations. Stories and sayings were created that were culturally congruent, allowing for the participants to see the relevance of the techniques in their own personal lives. Without these adaptations to the content, it is unlikely that the participants would have found the interventions as appealing as they did.

Attending to disparities through adaptations to the delivery of the intervention

The initial wave of cultural adaptation research focused on adaptations to how therapists or intervention leaders could deliver the intervention in ways that would make the intervention more appealing to participants. Much of this work was in recognition of the fact that psychotherapy is traditionally delivered in a formal manner within a medical setting. As such, cultural adaptations often focused on efforts to deliver the program in more communal and egalitarian ways (e.g., more informal and relaxed style, provision of food during therapy sessions).

In our work with the FCSP, we recognized that the cultural values of *personalismo* and *respeto* were important guides for how many Latinxs approach interpersonal interactions (Edwards & Cardemil, 2015). Thus, we worked to welcome participants through warm and personable interactions, provided food to support an informal style, and engaged in more interventionist self-disclosure than is typical in psychotherapy. However, because many Latinxs engage in professional interactions in more formal ways, we also used more formal Spanish when speaking with our participants (i.e., Usted vs. Tú) and showed deference to the expertise that each participant had in their own lives. We also made considerable effort to flexibly schedule assessment and intervention sessions, offering bus passes or taxi vouchers to all participants, and providing onsite childcare for those participants who needed it. These efforts were made in recognition of the limited resources and supports that the participants had in their lives. We understood that it would be difficult for the women to attend the FCSP without these supports.

In the case of the Triangles Course, the CES team also developed and continually adapted the delivery of the course to make it more accessible to participants. The course took place in private spaces in the primary care clinics so most participants did not have to travel far to participate. Moreover, participants' young children were welcomed into the course and often played while their parents (usually mothers) participated in the sessions. The facilitator also brought snacks to each session for a more relaxed atmosphere and to encourage socialization. Nevertheless, despite these adaptations, attendance was low and so in response, community health workers (CHWs) were trained to deliver the course. The CES Mental Health team trained the CHWs in therapeutic skills, such as active listening, and in the delivery of the course. The CHWs were able to implement all of the previously mentioned adaptations, as well as hold the course more consistently and remind participants of the next session more frequently than the clinic supervisors who previously facilitated the course.

Again, both the FCSP and the Triangles Course needed to be delivered in innovative ways, as standard mental health delivery practices had already been shown to be limited. In the case of the FCSP, adaptations to delivery were made in recognition of both cultural norms around interpersonal interactions and the very real life constraints experienced by the participants. Similar adaptations were made initially by the CES team, but then because attendance was low, CHWs were used to deliver the intervention. It is likely that as members of the community in Chiapas, the CHWs were able to overcome some of the concerns that participants may have had with engaging in the Triangles Course. The flexibility demonstrated in adapting to the needs of the community is a model for intervention development and dissemination in other contexts and communities.

Attending to disparities with culturally sensitive interventionists

In addition to attending to issues of culture in the content and delivery of the intervention, it is critical that the interventionists are also skilled at working with the population of interest. Considerable attention has been given to training and supporting the development of cultural competency (Clauss-Ehlers, Chiriboga, Hunter, Roysircar, & Tummala-Narra, 2019; Patallo, 2019). Different definitions of cultural competence exist, but in general, there exist a consensus that interventionists should have and understanding and self-awareness of one's own worldview and biases, knowledge of the cultural group with which one is working, and the skills to work with that individuals from that cultural group (Sue & Sue, 2013). More recently, scholars have expanded this conceptualization to recognize the influence of power, privilege, and oppression (Hays, 2016).

In our work with the FCSP, most of the interventionists self-identified as Latinx and bicultural and bilingual (Spanish and English). Moreover, all interventionists, irrespective of identity, had extensive experience working with individuals from Latinx backgrounds and felt comfortable engaging participants in discussion relevant to their lives. In addition, we developed a specific training program for all interventionists that was complemented by ongoing supervision. The training program focused on teaching the interventionists how to deliver the FCSP with a high degree of fidelity and included review of sessions guides, as well as watching and discussing recordings of the previous groups. Understanding how to engage with the participants in sensitive and connecting ways was also a critical part of the training. In addition to discussing issues related to Latinx culture (e.g., norms and values), considerable attention was devoted to issues of power and marginalization, given the differences in social status between the interventionists and the participants. This focus was particularly important for the interventionists who identified as Latinx, as it was clear that there were important differences between them and the participants, despite the fact that they shared a similar ethnic background.

With the Triangles Course, the providers changed in order to adapt to the program needs. As mentioned above, at first the providers were the clinic supervisors. These physicians were all Mexican, like the participants, but most often came from urban, more economically-advantaged areas of Mexico compared to the participants. When the team decided to train CHWs, the idea was to increase not just course attendance but also cultural resonance. Prior to this training, CHWs made home visits to patients with chronic diseases, and CES noted the positive impact on patients from having someone from their community and context accompany them. For the Triangles Course, for example, participants reported strong therapeutic alliance with the CHWs.

Both of the interventions demonstrate how important it is for the interventionists to be well-trained in both the content of the intervention and in working with the community. Both interventions used interventionists who were connected to the communities by virtue of ethnic background and, in the case of the Triangles Course, actual members of the community. The advantages to interventionists from the community are apparent; however, there are also challenges to consider. For example, in the Triangles Course, there was some participant concern about the CHWs possibly spreading rumors given their connection to others in the community. This concern was not as relevant to the participants in the FCSP, since the interventionists were not from the actual community.

Concluding thoughts

In this chapter, we have argued for the importance of attending to disparities in mental health care that disproportionately affect individuals from racial and ethnic minority backgrounds. We posit that approaches that integrate the strengths of the cultural adaptation literature with the strengths of the implementation science literature hold significant potential to ameliorate disparities. This integrated approach would require attention to cultural and contextual factors in the content and focus of the intervention, approaches to delivery of the intervention, and to the identification and training of interventionists. Grounded in two community-based interventions addressing depression, we provided examples of how this could be done.

These examples also raise areas for future research. In particular, the field continues to struggle with how to evaluate the effectiveness of particular adaptations. Critical to addressing this issue is careful assessment of particular adaptations, whether they are located in the content, delivery, or training of interventionists. For example, interventions that create space for participants to bring children may seem well-received, but there have been no efforts to quantify this effect and investigate its association with outcome of the intervention. Relatedly, efforts to assess therapist cultural competency have not achieved consensus regarding who is the primary evaluator of this competence (i.e., patient, therapist, or observer).

Another area in need of additional research is in the growing recognition of the complexity of intersectional identities and structural forces. In this chapter, although we have focused primarily on Latinx ethnicity, we have also recognized the importance of social class and gender. However, in this work, we have not attended to the nuances that arise with respect to other identities, including race and sexual and gender diversity. Indeed, recent scholarship has highlighted the dearth of attention to the diverse experiences that exist within particular cultural groups, including Latinxs (e.g., Cerezo, 2020; Sanchez, Adames, & Mazzula, 2021).

Finally, the recent cultural adaptation research that has foregrounded sociopolitical issues like racism and discrimination (e.g., Robinson et al., 2016; Stewart et al., 2019) suggest that there is considerable potential in the adoption of antioppressive and antiracist perspectives in this work. One challenge, however, is that because these perspectives emphasize dismantling systems of oppression, the systems of care that implement these interventions must also be examined. That is, insofar as medical systems of care perpetuate structural inequities, then the impact of antioppressive interventions will be inherently limited.

Despite these challenges, we remain optimistic about the potential that innovative approaches to mental health service delivery can bring to underserved communities. Integrating cultural adaptation approaches with implementation science perspectives can overcome the limitations within each of these literatures, and increase the reach of effective and culturally sensitive mental health services to all who need them.

Summary points

- Mental health care disparities disproportionately impact individuals from racial and ethnic minority backgrounds, and it is important to attend to these disparities.
- Both cultural adaptation and implementation science researchers have worked toward decreasing disparities in mental health care by adapting the content and or/delivery of CBT interventions to apply them in novel cultural contexts.
- However, both cultural adaptations and implementation science have limitations. Cultural adaptation work tends to focus on changes to manuals without necessarily accounting for the context, while implementation science tends to

look at the context of the delivery of the intervention without necessarily considering changes to the content of the intervention.
- Public health approaches that integrate cultural adaptation and implementation science have significant potential to address disparities, given that the relative strengths and limitations of the two are complementary.
- These integrated approaches should attend to culture and context in (1) the creation and adaptation of the content of the intervention itself, (2) the delivery of the intervention, and (3) the training of the interventionists. We demonstrate how this could happen using the example of the Family Coping Skills Program and the Triangles Course, two CBT-based interventions implemented in novel cultural contexts.
- Future areas of research for the combination of cultural adaptation and implementation science literature include how to evaluate the effectiveness of adaptations, how to incorporate intersectionality into this work, and how to do this work within a broken and oppressive systems of care.

References

Abas, M., Nyamayaro, P., Bere, T., Saruchera, E., Mothobi, N., Simms, V., et al. (2018). Feasibility and acceptability of a task-shifted intervention to enhance adherence to HIV medication and improve depression in people living with HIV in Zimbabwe, a low income country in Sub-Saharan Africa. *AIDS and Behavior, 22*(1), 86–101. https://doi.org/10.1007/s10461-016-1659-4.

Acevedo-Polakovich, I. D., Kassab, V. A., Boress, K. S., Clements, K. V., Stout, S., Alfaro, M., et al. (2017). When context matters: Adaptation for high-risk U.S. Latina/o subgroups. *Journal of Latina/o Psychology, 5*(4), 306–322. https://doi.org/10.1037/lat0000100.

Bartholomew, T. T., Pérez-Rojas, A. E., Bledman, R., Joy, E. E., & Robbins, K. A. (2021). "How could I not bring it up?": A multiple case study of therapists' comfort when black clients discuss anti-black racism in sessions. *Psychotherapy*. https://doi.org/10.1037/pst0000404. Online ahead of print.

Baumann, A., Cabassa, L., & Wiltsey Stirman, S. (2017). Adaptation in dissemination and implementation science. In R. C. Brownson, G. A. Colditz, & E. K. Proctor (Eds.), *Dissemination and implementation research in health: Translating science to practice* Oxford University Press.

Benish, S. G., Quintana, S., & Wampold, B. E. (2011). Culturally adapted psychotherapy and the legitimacy of myth: A direct-comparison meta-analysis. *Journal of Counseling Psychology, 58*(3), 279–289. https://doi.org/10.1037/a0023626.

Blanco, C., Patel, S. R., Liu, L., Jiang, H., Lewis-Fernández, R., Schmidt, A. B., et al. (2007). National trends in ethnic disparities in mental health care. *Medical Care, 45*(11), 1012–1019. https://doi.org/10.1097/mlr.0b013e3180ca95d3.

Cabassa, L. J., & Baumann, A. A. (2013). A two-way street: Bridging implementation science and cultural adaptations of mental health treatments. *Implementation Science, 8*(1), 90. https://doi.org/10.1186/1748-5908-8-90.

Cardemil, E. V. (2010). Cultural adaptations to empirically supported treatments: A research agenda. *The Scientific Review of Mental Health Practice, 7*(2), 8–21.

Cardemil, E. V., Kim, S., Pinedo, T. M., & Miller, I. W. (2005). Developing a culturally appropriate depression prevention program: The family coping skills program. *Cultural Diversity and Ethnic Minority Psychology, 11*(2), 99–112. https://doi.org/10.1037/1099-9809.11.2.99.

Cardemil, E. V., Nelson, T., & Keefe, K. (2015). Racial and ethnic disparities in depression treatment. *Current Opinion in Psychology, 4*, 37–42. https://doi.org/10.1016/j.copsyc.2015.01.021.

Cerezo, A. (2020). Expanding the reach of Latinx psychology: Honoring the lived experiences of sexual and gender diverse Latinxs. *Journal of Latinx Psychology, 8*(1), 1–6. https://doi.org/10.1037/lat0000144.

Chambers, D. (2014). Guiding theory for dissemination and implementation research: A reflection on models used in research and practice. In R. S. Beidas, & P. C. Kendall (Eds.), *Dissemination and implementation of evidence-based practices in child and adolescent mental health* (pp. 9–21). Oxford University Press.

Clauss-Ehlers, C. S., Chiriboga, D. A., Hunter, S. J., Roysircar, G., & Tummala-Narra, P. (2019). APA multicultural guidelines executive summary: Ecological approach to context, identity, and intersectionality. *American Psychologist, 74*(2), 232. https://doi.org/10.1037/amp0000382.

Cook, B. L., Hou, S. S. Y., Lee-Tauler, S. Y., Progovac, A. M., Samson, F., & Sanchez, M. J. (2019). A review of mental health and mental health care disparities research: 2011-2014. *Medical Care Research and Review, 76*(6), 683–710. https://doi.org/10.1177/1077558718780592.

Cook, B. L., McGuire, T., & Miranda, J. (2007). Measuring trends in mental health care disparities, 2000–2004. *Psychiatric Services, 58*(12), 1533–1540. https://doi.org/10.1176/ps.2007.58.12.1533.

Craig, P., Dieppe, P., Macintyre, S., Michie, S., Nazareth, I., & Petticrew, M. (2008). Developing and evaluating complex interventions: The new Medical Research Council guidance. *BMJ, 337*, a1655. https://doi.org/10.1136/bmj.a1655.

Dixon, L. B., & Patel, S. R. (2020). The application of implementation science to community mental health. *World Psychiatry, 19*(2), 173–174. https://doi.org/10.1002/wps.20731.

Domenech Rodríguez, M. M., & Bernal, G. (2012). Frameworks, models, and guidelines for cultural adaptation. In G. Bernal, & M. M. Domenech Rodríguez (Eds.), *Cultural adaptations: Tools for evidence-based practice with diverse populations* (pp. 23–44). American Psychological Association. https://doi.org/10.1037/13752-002.

Edwards, L. M., & Cardemil, E. V. (2015). Clinical approaches to assessing cultural values among Latinos. In K. F. Geisinger (Ed.), *Psychological testing of Hispanics: Clinical, cultural, and intellectual issues* (pp. 215–236). American Psychological Association. https://doi.org/10.1037/14668-012.

Fortuna, L. R., Alegria, M., & Gao, S. (2010). Retention in depression treatment among ethnic and racial minority groups in the United States. *Depression and Anxiety*, *27*(5), 485–494. https://doi.org/10.1002/da.20685.

Galvan, T., & Gudiño, O. G. (2021). Understanding Latinx youth mental health disparities by problem type: The role of caregiver culture. *Psychological Services*, *18*(1), 116–123. https://doi.org/10.1037/ser0000365.

Griner, D., & Smith, T. B. (2006). Culturally adapted mental health intervention: A meta-analytic review. *Psychotherapy: Theory, Research, Practice, Training*, *43*(4), 531–548. https://doi.org/10.1037/0033-3204.43.4.531.

Hammond, W. P. (2010). Psychosocial correlates of medical mistrust among African American men. *American Journal of Community Psychology*, *45*(1–2), 87–106. https://doi.org/10.1007/s10464-009-9280-6.

Hartman, S. J., Miguel-Esponda, G., Watson, A., Cardemil, E., & Rodríguez-Cuevas, F. G. (2018). Facilitators and barriers to the implementation of a cognitive behavioral therapy-informed psychoeducation course led by community health workers in Chiapas, Mexico: A mixed-methods process evaluation. *[Manuscript in preparation]*.

Hays, P. A. (2016). *Addressing cultural complexities in practice: Assessment, diagnosis, and therapy* (3rd ed.). Washington, DC: American Psychological Association.

Hutchison, S. L., Karpov, I., Deegan, P. E., MacDonald-Wilson, K. L., & Schuster, J. M. (2015). Adoption of strategies to improve decision support in community mental health centers. *Implementation Science*, *10*(1), A27. https://doi.org/10.1186/1748-5908-10-S1-A27.

Hwang, W.-C. (2012). Integrating top-down and bottom-up approaches to culturally adapting psychotherapy: Application to Chinese Americans. In G. Bernal, & M. M. Domenech Rodríguez (Eds.), *Cultural adaptations: Tools for evidence-based practice with diverse populations* (pp. 179–198). American Psychological Association. https://doi.org/10.1037/13752-009.

Imel, Z. E., Baldwin, S., Atkins, D. C., Owen, J., Baardseth, T., & Wampold, B. E. (2011). Racial/ethnic disparities in therapist effectiveness: A conceptualization and initial study of cultural competence. *Journal of Counseling Psychology*, *58*(3), 290–298. https://doi.org/10.1037/a0023284.

Jackson, A. P. (2006). The use of psychiatric medications to treat depressive disorders in African American women. *Journal of Clinical Psychology*, *62*(7), 793–800. https://doi.org/10.1002/jclp.20276.

Kearney, L. K., Draper, M., & Barón, A. (2005). Counseling utilization by ethnic minority college students. *Cultural Diversity and Ethnic Minority Psychology*, *11*, 272–285. https://doi.org/10.1037/1099-9809.11.3.272.

Kessler, R. C., Merikangas, K. R., & Wang, P. S. (2010). The epidemiology of mental disorders. In B. L. Levine, K. D. Hennessy, & J. Petrilla (Eds.), *Mental health services: A public health perspective* (3rd ed., pp. 169–200). Oxford University Press.

Kim, J. E., Park, S. S., La, A., Chang, J., & Zane, N. (2016). Counseling services for Asian, Latino/a, and White American students: Initial severity, session attendance, and outcome. *Cultural Diversity and Ethnic Minority Psychology*, *22*(3), 299–310. https://doi.org/10.1037/cdp0000069.

Kohn-Wood, L., & Hooper, L. (2014). Cultural competency, culturally tailored care, and the primary care setting: Possible solutions to reduce racial/ethnic disparities in mental health-care. *Journal of Mental Health Counseling*, *36*(2), 173–188. https://doi.org/10.17744/mehc.36.2.d73h217l81tg6uv3.

Lagomasino, I. T., Stockdale, S. E., & Miranda, J. (2011). Racial-ethnic composition of provider practices and disparities in treatment of depression and anxiety, 2003–2007. *Psychiatric Services*, *62*(9), 1019–1025. https://doi.org/10.1176/ps.62.9.pss6209_1019.

Lewinsohn, P. M., Steinmetz, J. L., Antonuccio, D., & Teri, L. (1984). Group therapy for depression: The coping with depression course. *International Journal of Mental Health*, *13*(3–4), 8–33. https://doi.org/10.1080/00207411.1984.11448974.

McKay, M. M., Sensoy Bahar, O., & Ssewamala, F. M. (2020). Implementation science in global health settings: Collaborating with governmental & community partners in Uganda. *Psychiatry Research*, *283*, 112585. https://doi.org/10.1016/j.psychres.2019.112585.

Miguel-Esponda, G., Bohm-Levine, N., Rodríguez-Cuevas, F. G., Cohen, A., & Kakuma, R. (2020). Implementation process and outcomes of a mental health programme integrated in primary care clinics in rural Mexico: A mixed-methods study. *International Journal of Mental Health Systems*, *14*, 21. https://doi.org/10.1186/s13033-020-00346-x.

Miller, I. W., Ryan, C. E., Keitner, G. I., Bishop, D. S., & Epstein, N. B. (2000). The McMaster approach to families: Theory, assessment, treatment and research. *Journal of Family Therapy*, *22*(2), 168–189. https://doi.org/10.1111/1467-6427.00145.

Miranda, J., Snowden, R. L., & Legha, R. K. (2020). Policy effects on mental health status and mental health care disparities. In H. H. Goldman, R. G. Frank, & J. P. Morrissey (Eds.), *The Palgrave handbook of American mental health policy* (pp. 331–364). Palgrave Macmillan.

Moore, G. F., Audrey, S., Barker, M., Bond, L., Bonell, C., Hardeman, W., et al. (2015). Process evaluation of complex interventions: Medical Research Council guidance. *BMJ*, *350*, h1258. https://doi.org/10.1136/bmj.h1258.

Oakley, L. P., López-Cevallos, D. F., & Harvey, S. M. (2019). The association of cultural and structural factors with perceived medical mistrust among young adult Latinos in rural Oregon. *Behavioral Medicine*, *45*(2), 118–127. https://doi.org/10.1080/08964289.2019.1590799.

Ortega, A. N., Fang, H., Perez, V. H., Rizzo, J. A., Carter-Pokras, O., Wallace, S. P., et al. (2007). Health care access, use of services, and experiences among undocumented Mexicans and other Latinos. *Archives of Internal Medicine*, *167*(21), 2354–2360. https://doi.org/10.1001/archinte.167.21.2354.

Patallo, B. J. (2019). The multicultural guidelines in practice: Cultural humility in clinical training and supervision. *Training and Education in Professional Psychology*, *13*(3), 227–232. https://doi.org/10.1037/tep0000253.

Peters, D., Tran, N., & Adam, T. (2013). *Implementation research in health: A practical guide*. World Health Organization. https://apps.who.int/iris/handle/10665/91758.

Robinson, W. L. V., Case, M. H., Whipple, C. R., Gooden, A. S., Lopez-Tamayo, R., Lambert, S. F., et al. (2016). Culturally grounded stress reduction and suicide prevention for African American adolescents. *Practice Innovations*, *1*(2), 117–128. https://doi.org/10.1037/pri0000020.

Rodríguez-Cuevas, F., Hartman, S. J., Aguerrebere, M., & Palazuelos, D. (2020). Accompanying people with mental illnesses: The role of community health workers in mental health care services in Chiapas, Mexico. In S. O. Okpaku (Ed.), *Innovations in global mental health* Cambridge University Press. https://doi.org/10.1007/978-3-319-70134-9_101-1.

Rodriguez-Seijas, C., Eaton, N. R., & Pachankis, J. E. (2019). Prevalence of psychiatric disorders at the intersection of race and sexual orientation: Results from the National Epidemiologic Survey of Alcohol and Related Conditions-III. *Journal of Consulting and Clinical Psychology, 87*(4), 321–331. https://doi.org/10.1037/ccp0000377.

Ryan, C., Epstein, N. B., Keitner, G., Miller, I. W., & Bishop, D. S. (2005). *Evaluation and treating families: The McMaster approach*. New York: Routledge.

Sanchez, D., Adames, H., & Mazzula, S. (2021). AfroLatinidad: Theory, research, and practice. *Journal of Latinx Psychology, 9*(1). Special issue.

Shelton, R. C., Lee, M., Brotzman, L. E., Wolfenden, L., Nathan, N., & Wainberg, M. L. (2020). What is dissemination and implementation science?: An introduction and opportunities to advance behavioral medicine and public health globally. *International Journal of Behavioral Medicine, 27*(1), 3–20. https://doi.org/10.1007/s12529-020-09848-x.

Simpson, S. M., Krishnan, L. L., Kunik, M. E., & Ruiz, P. (2007). Racial disparities in diagnosis and treatment of depression: A literature review. *Psychiatric Quarterly, 78*(1), 3–14. https://doi.org/10.1007/s11126-006-9022-y.

Smedley, B. D., Stith, A. Y., & Nelson, A. R. (2003). *Unequal treatment: Confronting racial and ethnic disparities in health care*. Washington, DC: Committee on Understanding and Eliminating Racial and Ethnic Disparities in Health Care, Institute of Medicine, National Academy Press. https://doi.org/10.17226/12875.

Soto, A., Smith, T. B., Griner, D., Domenech Rodríguez, M., & Bernal, G. (2018). Cultural adaptations and therapist multicultural competence: Two meta-analytic reviews. *Journal of Clinical Psychology, 74*(11), 1907–1923. https://doi.org/10.1002/jclp.22679.

Stewart, R. W., Orengo-Aguayo, R., Wallace, M., Metzger, I. W., & Rheingold, A. A. (2019). Leveraging technology and cultural adaptations to increase access and engagement among trauma-exposed African American youth: Exploratory study of school-based telehealth delivery of trauma-focused cognitive behavioral therapy. *Journal of Interpersonal Violence, 36*(15–16), 7090–7109. https://doi.org/10.1177/0886260519831380.

Sue, D. W., & Sue, D. (2013). *Counseling the culturally diverse: Theory and practice* (6th). Hoboken, NJ: John Wiley and Sons, Inc.

Tabak, R., Khoong, E., Chambers, D., & Brownson, R. (2013). Models in dissemination and implementation research: Useful tools in public health services and systems research. *Frontiers in Public Health Services and Systems Research, 2*(1). https://doi.org/10.13023/FPHSSR.0201.08.

Vroom, E. B., & Massey, O. T. (2021). Moving from implementation science to implementation practice: The need to solve practical problems to improve behavioral health services. *The Journal of Behavioral Health Services & Research, 49*(1), 106–116. https://doi.org/10.1007/s11414-021-09765-1.

Waltman, S. H., & Sokol, L. (2017). The generic model of cognitive behavioral therapy: A case conceptualization-driven approach. In S. G. Hofmann, & G. J. G. Asmundson (Eds.), *The science of cognitive behavioral therapy* (pp. 3–17). San Diego: Academic Press.

Wittayanukorn, S., Qian, J., & Hansen, R. A. (2014). Prevalence of depressive symptoms and predictors of treatment among U.S. adults from 2005-2010. *General Hospital Psychiatry, 36*, 330–336. https://doi.org/10.1016/j.genhosppsych.2013.12.009.

Yu, K. Y., Pope, S. C., & Perez, M. (2019). Clinical treatment and practice recommendations for disordered eating in Asian Americans. *Professional Psychology: Research and Practice, 50*(5), 279–287. https://doi.org/10.1037/pro0000244.

Chapter 38

CASE STUDY: Cognitive behavior therapy for body dysmorphic disorder in an adult

Marie Drüge and Birgit Watzke
Department of Psychology, University of Zurich, Zurich, Switzerland

Abbreviations

ACT	acceptance and commitment therapy
BABS	Brown Assessment of Beliefs Scale
BDD	body dysmorphic disorder
BMI	body mass index
CBT	cognitive behavioral therapy
DSM	diagnostic and statistical manual of mental disorders
ERP	exposure and response prevention
FKS	Fragebogen Körperdysmorpher Symptome
NICE	National Institute for Health and Care Excellence
PHQ-9	Patient Health Questionnaire 9
SSRI	selective serotonin reuptake inhibitors

Introduction

Body dysmorphic disorder (BDD) is described in the diagnostic and statistical manual of mental disorders, fifth edition (DSM-5) (APA, 2013) as the preoccupation with one or more "defects or flaws" of one's own body part(s) or appearance, which is/are not apparent to others to the same extent. Additionally, BDD involves repetitive and time-consuming behaviors (such as mirror checking, applying special makeup) or mental acts (such as comparing oneself to others or to pictures). These preoccupations may cause severe educational or occupational dysfunction or social isolation (Phillips, Menard, Fay, & Pagano, 2005). Eating disorder must be ruled out as an alternative diagnosis. In the DSM-5, BDD has been reclassified within the obsessive-compulsive spectrum, as BDD has been categorized within the disorder class of somatoform disorders before. The prevalence rates range from 1.7% to 2.9% (Buhlmann et al., 2010) but vary in different samples, such as in student populations (3.3%), psychiatric outpatients (5.8%), or dermatology outpatients (11.3%) (Veale, Gledhill, Christodoulou, & Hodsoll, 2016). The onset is as early as 16 years on average (Gunstad & Phillips, 2003). Although BDD appears to be relatively uncommon as a presenting problem in psychotherapy, patients more often search for help for comorbidities (e.g., accompanying depression). BDD is highly associated with comorbidities such as mood or anxiety disorders, as well as other obsessive and compulsive thoughts or acts, social phobia, and substance use disorders (Gunstad & Phillips, 2003). It is associated with a high burden of disease, such as impaired psychosocial functioning or high suicide risk (Phillips & Menard, 2006). In a prospective study, 57.8% of respondents reported suicidal ideation, and 2.6% attempted suicide within 1 year Phillips & Menard, 2006.

Treatment options for BDD

BDD can be a severe and chronic disorder, especially if left untreated. Without appropriate treatment, it is unlikely to remit (Phillips, Menard, Quinn, Didie, & Stout, 2013). However, recognizing and tactfully diagnosing and communicating the diagnosis of BDD is already an important step of treatment, as BDD is highly associated with a lack of insight or shame, making many willing to hide the disorder, even when there is concern (Veale et al., 1996). The National Institute for Health and Care Excellence (NICE) has developed the only clinical guidelines for BDD diagnostics and treatment (NICE, 2005). These guidelines include specific screening questions, which may help in assessments such as:

1. Do you worry a lot about the way you look and wish you could think about it less?
2. What specific concerns do you have about your appearance?
3. On a typical day, how many hours a day is your appearance on your mind? (More than 1 h a day is considered excessive.)
4. What effect does it have on your life?
5. Does it make it hard to do your work or be with friends?

After diagnosing BDD, two treatment approaches are found to be efficacious, namely, serotonin reuptake inhibitors (SSRIs, Williams, Hadjistavropoulos, & Sharpe, 2006) and cognitive behavioral therapy (CBT) (Harrison, Fernández de la Cruz, Enander, Radua, & Mataix-Cols, 2016). The NICE guidelines recommend that CBT be offered if the patient's degree of functional impairment is mild and/or the patient expresses a preference for this approach (NICE, 2005). In more severe cases (e.g., functional impairment) and/or cases in which the patient expresses a preference for the pharmacological approach, fluoxetine fluvoxamine, paroxetine, sertraline, or citalopram is recommended for SSRI treatment. If effective, medication should be continued for at least 12 months to prevent relapse and to make further improvements (NICE, 2005). After inadequate response to treatment with an SSRI alone (within 12 weeks) or CBT alone (including exposure and response prevention (ERP), and more than 10 therapist hours per patient), a combined treatment should be offered (NICE, 2005).

CBT for BDD

There are two main treatment manuals for applying CBT to BDD (Veale & Neziroglu, 2010; Wilhelm, Phillips, & Steketee, 2013). Both manuals contain some of the core principles of CBT when applied to BDD, such as psychoeducation, cognitive restructuring targeting unrealistic negative thoughts, behavioral in vivo exposures/behavioral experiments, and relapse prevention. Additionally, in both manuals, motivational enhancement is important to improve insight and compliance with treatment. In vivo exposures integrate response prevention as a core element to lower avoidance and safety-seeking behavior, as well as to prevent rituals. Some research indicates that ERP is a central component in BDD treatment (Khemlani-Patel, Neziroglu, & Mancusi, 2011). In comparing the two treatment manuals, the model of Wilhelm et al. (2013) might be slightly less cognitive and more behavioral than the cognitive behavioral model developed by Veale and Neziroglu (2010). A German CBT manual using the same components as those of Wilhelm et al. (2013) and Veale and Neziroglu (2010) was referred to in the case study described below. In addition to standard CBT methods, some other techniques show promising results in enhancing CBT for BDD: mindfulness and perceptual retraining are applied to reduce rumination and appearance comparisons; imagery rescripting is employed to transform aversive mental images into gentler images or to construct new, positive images (Willson, Veale, & Freeston, 2016); and methods from acceptance and commitment therapy (ACT) are harnessed to help the patient accept the appearance, or focus more on valued directions and goals (Linde et al., 2015). The following case study illustrates some of the components of CBT as used for BDD.

Introduction to the case study—The initial phase

Anastacia is a 35-year-old trained saleswoman currently living next to her parents' house with her husband and her son and daughter. She was seeking help after suffering a "breakdown," including going to the emergency room during the first coronavirus lockdown. Her main anxiety was to go bald or entailed other people checking her hair, although there was no reason to believe so. In years prior, she had developed and prolonged a ritual on how to brush her hair and count the number of fallen hairs, which took 2 h in the morning, 1 h after lunch, and another 2 h right before going to bed. This ritual felt "like hell" and held her off from spending quality time with her family (her husband and two kids) or engaging in hobbies (horse riding, being creative), but she explained she "had to do it" to handle intrusive thoughts such as "How many hairs did I lose today?" or "Do others recognize that I am going bald?" Additionally, she stopped some of her hobbies due to an expected impact on her hair loss (e.g., she stopped running as she was worried about losing her hair due to mineral loss). She even recognized that she had the same feelings about losing her hair and losing her children (which frightened her), and motivated herself to undergo a specialized treatment, as the more rational part of her realized that this ratio was no longer healthy. Additionally, she developed depressive symptoms (sleeping problems, a loss of interests and pleasure). She did not want to leave the house anymore to go horseback riding. Anastacia had been in psychotherapy before, but did not feel that she had been taken seriously.

Case history

Anastacia grew up in a Catholic family as the oldest of three children. She felt "safe," though talking about emotions or thoughts was not common, and she was used to dealing with problems herself. Her father became sick with lung cancer when she was 15 and died 5 years later. One year after his death, her mother was also diagnosed with cancer (pancreatic cancer) and was moved into hospice after a while. She died 2 years later. Anastacia explained that in both cases, her parents and relatives did not tell her how bad their conditions truly were, and told her they were getting better (until they died). Still, she had felt all along that her parents would not recover. After the loss of her father, she developed anorexia nervosa and started to control her food until her body mass index (BMI) was 16.5. At the age of 21, she met her husband-to-be, who helped her normalize her eating behavior without professional help. They went on holidays, which she needed after "focusing so much on her father's health." As she later could not cope with the situation at the hospice very well, she refused to go there "as often as I should have gone" and tried to enjoy her youth while partying; thus, she also had feelings of guilt. While her mother was still alive, she had a miscarriage but became pregnant again shortly before her mother died. Her older son was born 8 months after her mother's death, which gave her a new purpose in life. During this time, she started seeing a psychiatrist irregularly to deal with the loss. After moving into a house next door to her parents' house, her second child was born 4 years later, and she mainly focused on being a parent. She enjoyed life, but her thoughts about her hair loss had started.

The diagnostic process and treatment options

As Anastacia came to our outpatient unit because of our specialization in BDD, and as she reported corresponding symptoms, the diagnostic process involved specific measures such as the "Fragebogen Körperdysmorpher Symptome" (questionnaire for BDD symptoms, FKS, Buhlmann, Wilhelm, Glaesmer, Brähler, & Rief, 2009) and the Brown Assessment of Beliefs Scale (BABS; Eisen et al., 1998). Further assessments, as well as quality assurance, entailed use of measures such as the Patient Health Questionnaire-9 (Kroenke, Spitzer, & Williams, 2001) to monitor depressive symptoms each month. Among others, her initial scores were 17 on the FKS, indicating BDD, a score of 7 on the BABS, revealing good insight, and a score of 8 on the PHQ-9, denoting depressive symptoms. Based on the psychometric assessment, during the clinical interview, she was diagnosed with BDD and major depression. After discussing the pros and cons of various treatment options, Anastacia became motivated to start cognitive behavioral treatment for BDD (Veale & Neziroglu, 2010; Wilhelm et al., 2013), including exposure therapy and response prevention. The German CBT manual by Brunhoeber (2019) was used to underpin CBT with many vivid examples. Treating her with selective serotonin reuptake inhibitors (SSRIs) was considered; however, Anastacia was not motivated to take an SSRI and read about hair loss as a side effect. The risk of self-harm and suicide was assessed and considered low, as Anastacia expressed a sense of purpose and meaning (e.g., due to her children).

The beginning of therapy

In the first month of outpatient treatment, the sessions focused on Anastacia's resources while building a reliable and trustful therapeutic alliance. Transparency and honesty were explicit, core principles for communication with her. Anastacia had experienced "dishonesty to spare her" during her parents' illnesses, which she described as "the worst." Further, she wanted to be taken seriously regarding her worries about her condition after her experience with a former psychiatrist telling her to "stop making a fuss about the hair." While focusing on psychoeducation (e.g., the interplay of thoughts, emotions, and behavior), a need to reinitiate activities (e.g., horseback riding) became clear. After getting to know her behavior better through a functional analysis of behavior (SORCK modeling Kanfer & Saslow, 1974), she realized that her ritual of brushing the hair was her only time for herself. Therefore, she was motivated to engage in "me time" and talk to her husband about it. Due to the specific conditions of the coronavirus pandemic, it was difficult for her to find space for herself, and she was unable to utilize some of her earlier resources. The focus was changed to her underlying needs (e.g., socializing and time without housekeeping, time without feeling responsible for the children), and she adapted some of her earlier resources (e.g., inviting her friend over for coffee instead of meeting up for dinner in a restaurant). Her mood started to improve again when she realized that she needed to take time out for herself and she learned to delegate responsibility; thus, the time-consuming ritual and worries about her appearance remained stable. Moreover, a body scan (Kabat-Zinn, 1990) was introduced as a method for mindful observations of body sensations to use four times a week. After 6 weeks of weekly treatment, she understood that her preoccupations about her hair were irrational.

BDD-specific psychoeducation

After acknowledging her anxiety over losing her hair and what hair meant to her, and after validating the distress caused by this, the functionality of obsession-related disorders (e.g., learning the difference between physical appearance and body image, self-checking behavior, the nature of associated behaviors such as checking or seeking reassurance) was discussed. Theory A, "I feel bad because of my hair condition," was juxtaposed with theory B, "I feel bad because of my BDD," acknowledging that she had been following theory A for a while, but in therapy, she needed a fair chance to engage with theory B. Anastacia's own psychological model of her BDD was created.

Since more information about individual symptoms became necessary, a daily self-report diary was implemented, which Anastacia used to document BDD-specific situations, avoidance, and safety-seeking strategies, thoughts, and emotions. Examples of the German treatment manual by Brunhoeber (2019) were used to illustrate some of the symptoms. On the basis of Anastacia's psychological model, her individualized treatment plan was developed.

Exposure and response prevention (ERP)

For Anastacia, it was unthinkable to let go of the ritual completely; she was motivated to try to shorten it. Therefore, the goal was to shorten it to 10 min a week. Her husband was involved in this phase and agreed not to wait for her to have breakfast. Within 3 weeks, the morning ritual was shortened from 120 to 90 min. Anastacia felt motivated throughout the process and reported that sometimes she forgot to do the ritual before going to bed. The daily self-report revealed that some of her avoidance- and safety-seeking behaviors impacted her quality of life. Anastacia categorized her behaviors, rated them in terms of difficulty (0–100), and organized them in a hierarchy. Going to the hairdresser was rated the most difficult (100), going for a run was rated less difficult (70), and using a hairband during the session was rated 50. Apart from the hair focusing behaviors, behaviors concerning her appearance—such as leaving the house without makeup to go to therapy (70) or meeting up with her sister (80)—were also included. One after another, within the next 2 months, Anastacia was able to reduce most of her avoidance- and safety-seeking behaviors using an ERP worksheet to plan exposure in advance and to evaluate the exposure afterward. The first instance of such exposure was to tie up her hair during the therapy session, where the ERP worksheets were introduced to her. Some aversive feelings (insecurity, tension) appeared during exposition and were avoided before. Response prevention was also adapted: Anastacia was able to stop counting her hair after brushing it, which was a huge step forward. The morning ritual remained, but was minimized to 30 min per day.

The reconstruction of thoughts

In the first step, Anastacia was taught how to observe her thoughts. She became aware of unhelpful thoughts, such as that others were "staring at my hair to check if I go bald" and she was encouraged to find objective evidence to support this belief (e.g., asking a person whom she trusted what he/she was thinking and at what he/she was looking). In these experiments, she noticed that her thoughts lacked objective evidence. In her daily self-report, Anastacia also observed reappearing thoughts about not being worthy of love. She expressed that one fallen hair made her immediately believe, "I will lose all my hair and will therefore be less loveable." She was working on this black-and-white thinking by normalizing fallen hair through Internet research. Additionally, Socratic dialogue was applied to question the assumption that "a person with bald head is less loveable." After discussing love, she answered her own question that her children do not love her for her hair, but more for what she is, which reassured her. Additionally, she questioned her own need to be perfect, and while questioning perfection, she realized that the rough edges of something actually make it full of life. She found that her image of losing her hair was connected to her parents losing their hair during cancer treatment. After she became aware of the connection, she became aware of the loss of her parents and recognized her own fear of dying.

Value-focused interventions

For Anastacia, hair loss was connected to the existential issues of losing her parents and her fear of dying, which became the focus of the next 3 months of adapting interventions of ACT (Hayes & Lillis, 2012). Since she had a deep feeling of sadness and fear, therapy was focused on fostering acceptance (e.g., by discouraging experiential avoidance), and stimulating action tendencies toward value-based directions. Therefore, the focus shifted to which values were truly important to her, using the key question, "Which values would you like to pass onto your children?" After realizing that empathy and being an authentic person were most important to her, Anastacia derived changes in terms of how to take action in her daily life. Further, she wanted to make her life easier once again while also not forgetting her past. After thoughtfully explaining the

"At my own funeral" method, Anastacia wanted to think further about which values she wanted to dedicate her life to. Because ACT focuses on committing to actions that are in line with one's values, short-term goals for her everyday life (e.g., engaging in communication with her husband) were identified. The sense of an easier life was addressed through playing board games with her children, going sledding in the winter, or swimming in the summer and adopting a dog.

The end of therapy

After 8 months of therapy, Anastacia no longer fulfilled the criteria for BDD or for major depression. The psychometric assessment at posttreatment indicated remission on the PHQ-9 (4), FKS (6), and BABS (2). Thus, a 30-min long morning ritual remained, and Anastacia felt self-assured and well-equipped to continue some of the methods on her own. Three sessions of relapse prevention focused on consolidating her skills and maintaining her goals. After prolonging the time between sessions, treatment ended 11 months after the initial session.

Applications to other areas

The case presented and the interventions and techniques used can be applied to various fields of application and disorders. Especially for body image disorders (e.g., eating disorders) but also for other obsessive-compulsive spectrum disorders, similar interventions (e.g., specific psychoeducation, cognitive restructuring, and especially ERP) are applicable after deriving an individual disorder model and individual therapy planning. Methods of acceptance and mindfulness as well as value orientation are transdiagnostically applicable.

Key facts

- BDD is a severe mental disorder may cause severe educational or occupational dysfunction or social isolation (Phillips et al., 2005).
- Prevalence rates range from 1.7% to 2.9% (Buhlmann et al., 2010).
- The risk of suicide is high: 57.8% of respondents reported suicidal ideation (Phillips & Menard, 2006).
- CBT has been shown effective for BDD (Harrison, et al. 2016) and is therefore the first line-treatment.
- In more severe cases (e.g., functional impairment) SSRI treatment or a combination is recommended (NICE, 2005).
- CBT for BDD includes psychoeducation, cognitive restructuring targeting unrealistic negative thoughts, behavioral in vivo exposures/behavioral experiments, and relapse prevention. (Veale & Neziroglu, 2010; Wilhelm et al., 2013).
- CBT may be enhanced with further techniques: Mindfulness and perceptual retraining are applied to reduce rumination and appearance comparisons; imagery rescripting is employed to transform aversive mental images into gentler images or to construct new, positive images (Willson et al., 2016); and methods from acceptance and commitment therapy (ACT) are used to help the patient accept the appearance, or focus more on valued directions and goals (Linde et al., 2015).

Summary points

- BDD is a severe mental disorder that is still difficult to identify and diagnose.
- Self-harm and suicidal behavior need to be assessed.
- If untreated, BDD is unlikely to remit.
- BDD can effectively be treated with CBT. In more severe cases, SSRIs or a combined treatment (SSRI and CBT) can be offered.
- Insight, suicidality, comorbidities, and shame might complicate treatment.
- CBT for BDD includes specific interventions, such as ERP and the reconstruction of thoughts.
- Third-wave therapies of CBT may enhance treatment, especially in regard to underlying existential issues.

References

American Psychiatric Association. (2013). *Diagnostic and statistical manual of mental disorders* (5th). Arlington, VA: American Psychiatric Publishing.

Brunhoeber, S. (2019). *Kognitive Verhaltenstherapie bei Körperdysmorpher Störung: Ein Therapiemanual*. Göttingen: Hogrefe.

Buhlmann, U., Glaesmer, H., Mewes, R., Fama, J. M., Wilhelm, S., Brähler, E., & Rief, W. (2010). Updates on the prevalence of body dysmorphic disorder: A population based survey. *Psychiatry Research, 178*, 171–175.

Buhlmann, U., Wilhelm, S., Glaesmer, H., Brähler, E., & Rief, W. (2009). *Fragebogen körperdysmorpher Symptome (FKS): Ein Screening Instrument.* Freiburg: Karger.

Eisen, J. L., Phillips, K. A., Baer, L., Beer, D. A., Atala, K. D., & Rasmussen, S. A. (1998). The Brown Assessment of Beliefs Scale: Reliability and validity. *American Journal of Psychiatry, 155*, 102–108.

Gunstad, J., & Phillips, K. A. (2003). Axis I comorbidity in body dysmorphic disorder. *Comprehensive Psychiatry, 44*, 270–276.

Harrison, A., Fernández de la Cruz, L., Enander, J., Radua, J., & Mataix-Cols, D. (2016). Cognitive-behavioral therapy for body dysmorphic disorder: A systematic review and meta-analysis of randomized controlled trials. *Clinical Psychology Review, 48*, 43–51. https://doi.org/10.1016/j.cpr.2016.05.00.

Hayes, S. C., & Lillis, J. (2012). Acceptance and commitment therapy. *Theories of psychotherapy series.* American Psychological Association.

Kabat-Zinn, J. (1990). *Full catastrophe living: Using the wisdom of your body and mind to face stress, pain and illness.* New York: Bantam Dell.

Kanfer, F. H., & Saslow, G. (1974). Behavioral analysis: An alternative to diagnostic classification. *Archives of General Psychiatry, 12*, 529–538.

Khemlani-Patel, S., Neziroglu, F., & Mancusi, L. (2011). Cognitive behavioral therapy for body dysmorphic disorder: A comparative investigation. *International Journal of Cognitive Therapy, 4*, 363–380. https://doi.org/10.1521/ijct.2011.4.4.363.

Kroenke, K., Spitzer, R. L., & Williams, J. B. (2001). The PHQ-9: Validity of a brief depression severity measure. *Journal of General Internal Medicine, 16*(9), 606–613.

Linde, J., Bjureberg, J., Ivanov, V. Z., Djurfeldt, D. R., Ramnerö, J., & Rück, C. (2015). Acceptance-based exposure therapy for body dysmorphic disorder: A pilot study. *Behavior Therapy, 46*(4), 423–431. https://doi.org/10.1016/j.beth.2015.05.002.

National Institute for Health and Care Excellence [NICE]. (2005). Obsessive-compulsive disorder and body dysmorphic disorder: Treatment. Clinical guideline [CG31]. Verfügbar unter: https://www.nice.org.uk/guidance/cg31.

Phillips, K., Menard, W., Quinn, E., Didie, E., & Stout, R. (2013). A 4-year prospective observational follow-up study of course and predictors of course in body dysmorphic disorder. *Psychological Medicine, 43*(5), 1109–1117. https://doi.org/10.1017/S003329171200173.

Phillips, K. A., & Menard, W. (2006). Suicidality in body dysmorphic disorder: A prospective study. *The American Journal of Psychiatry, 163*, 1280–1282.

Phillips, K. A., Menard, W., Fay, C., & Pagano, M. E. (2005). Psychosocial functioning and quality of life in body dysmorphic disorder. *Comprehensive Psychiatry, 46*, 254–260. https://doi.org/10.1016/j.comppsych.2004.10.004.

Veale, D., Boocock, A., Gournay, K., Dryden, W., Shah, F., Willson, R., & Walburn, J. (1996). Body dysmorphic disorder: A survey of fifty cases. *British Journal of Psychiatry, 169*(2), 196–201. https://doi.org/10.1192/bjp.169.2.196.

Veale, D., Gledhill, L. J., Christodoulou, P., & Hodsoll, J. (2016). Body dysmorphic disorder in different settings: A systematic review and estimated weighted prevalence. *Body Image, 18*, 168–186. https://doi.org/10.1016/j.bodyim.2016.07.003.

Veale, D., & Neziroglu, F. (2010). *Body dysmorphic disorder: A treatment manual.* West Sus-sex: Wiley-Blackwell.

Wilhelm, S., Phillips, K. A., & Steketee, G. (2013). *Cognitive-behavioral therapy for body dysmorphic disorder: A treatment manual.* New York: Guilford.

Williams, J., Hadjistavropoulos, T., & Sharpe, D. (2006). A meta-analysis of psychological and pharmacological treatments for body dysmorphic disorder. *Behaviour Research and Therapy, 44*(1), 99–111. https://doi.org/10.1016/j.brat.2004.12.006.

Willson, R., Veale, D., & Freeston, M. (2016). Imagery rescripting for body dysmorphic disorder: A multiple-baseline single-case experimental design. *Behavior Therapy, 47*(2), 248–261. https://doi.org/10.1016/j.beth.2015.08.006.

Chapter 39

Case study: The role of cognitive behavioral therapy in the treatment of postpartum depression

Elena Mamo[a] and Rachel Buhagiar[b]

[a]Psychology Department, Mater Dei Hospital, Msida, Malta [b]Department of Psychiatry, Mount Carmel Hospital, Attard, Malta

Abbreviations

ANRQ   Antenatal Risk Questionnaire
CBT   cognitive behavioral therapy
EPDS   Edinburgh Postnatal Depression Scale
GAD-7   Generalized Anxiety Disorder-7
NICE   National Institute for Health and Clinical Excellence
PMH   perinatal mental health
PPD   postpartum depression

Introduction

A common, yet significant challenge faced by mothers and fathers in their transition to parenthood is the effective communication of their needs and acceptance of support. Despite it being a time of major adjustment, parents often hold back in asking for help from relatives, friends, and also from professionals. Two well-studied barriers to help-seeking include self and societal expectations about parenthood and stigma surrounding mental health in general (Button, Thorton, Lee, Shakespeare, & Ayers, 2017). Importantly, lack of support may be a risk factor for maternal mental health disorders (Howard & Khalifeh, 2020) which are common complications of pregnancy and of the first postpartum year. Indeed, there has been global commitment to support the mental health and well-being of parents and their babies during the perinatal period (Webb et al., 2021).

Cognitive behavioral therapy (CBT), an evidence-based and effective psychological intervention for the treatment of several psychiatric disorders, can also be applied to the perinatal population and adapted to address the unique demands of parenthood (Branquinho et al., 2021; Wenzel & Kleiman, 2015). This collaborative, time-sensitive and goal-oriented modality provides a containing therapeutic frame for the management of perinatal mood and anxiety disorders.

In this case report, the multidisciplinary support and interventions provided to Lisa,[a] a 34-year-old mother who sought support for her depressive symptoms in her second pregnancy, will be discussed. This will encompass a detailed account of Lisa's journey, starting with a specialized perinatal mental health (PMH) assessment followed by a treatment package consisting of 16 sessions of CBT and perinatal psychiatric intervention. The role of these different treatment modalities on the alleviation of the client's mental health struggles will be explained, as well as its impact on the woman and her family.

The identification of perinatal mental health issues

Lisa was approximately 7 months pregnant with her second child when her obstetrician completed a referral to PMH services with the client's consent. It was during one of the routine appointments that the obstetrician noted Lisa to be

---

a. This case has been formulated through the collation of different perinatal cases encountered in clinical practice over the years. All names and identifiable factors have been changed to protect the clients' identity.

withdrawn and somewhat difficult to engage, unlike her usual outgoing self. She exhibited no emotion during her baby's scan. On further probing by the clinician, Lisa disclosed her mental health struggles and how she was lacking the motivation and interest to engage in previously pleasurable activities. She also described being afraid that the "old" emotional and relational issues previously experienced between her and her husband, Peter, would re-surface because of her unstable mental state.

Perinatal mental health services—Initial intake and screening

Within 2 weeks from referral, Lisa had her initial intake with the specialist PMH midwife. During this first encounter, a one-to-one biopsychosocial assessment was conducted, alongside the completion of mental health questionnaires to screen for any needs and to help establish a care plan.

Screening procedure and outcomes

In the initial assessment, Lisa self-completed The Edinburgh Postnatal Depression Scale (EPDS), the Generalized Anxiety Disorder-7 (GAD-7) and the Antenatal Risk Questionnaire (ANRQ). Notwithstanding their nondiagnostic nature, these widely used instruments can help in the detection of depressive and anxiety symptoms, and psychosocial risk factors for perinatal illnesses, respectively (Levis, Negeri, Sun, Benedetti, & Thombs, 2020; Simpson, Glazer, Michalski, Steiner, & Frey, 2014).

EPDS. Lisa scored a total score of 15 (out of a possible 30) in the EPDS, obtaining relatively high scores in the three anxiety statements (item 3, 4, and 5) of the EPDS. Whilst a cut off value of 11 in the EPDS is considered as the threshold for the detection of major depression among perinatal women (Levis et al., 2020), this scale may also be used to identify anxiety symptoms.

GAD-7. Lisa's score in this 7-item measure for generalized anxiety was of 14 (out of a possible 28), one point higher than the recommended optimal cut-off score for this scale (Simpson et al., 2014).

ANRQ. Whilst Lisa's ANRQ total scoring of 18 (out of a possible 67) was below the recommended value for recommending a mental health assessment (Austin, Colton, Priest, Reilly, & Hadzi-Pavlovic, 2013), her high scoring in the question related to her mental health history (question 1), indicated an increased psychosocial risk, and therefore a need for further exploration. Similarly, the EPDS and GAD-7 scores also meant that further follow-up for possible mood and anxiety symptoms was required.

Previous history

Childhood experiences. Lisa's childhood and upbringing were described as "normal" and "uneventful." She retained a good and supportive relationship with her parents and elder sister. She reported normative life stressors along the years which affected her to different extents, but which never entailed professional support.

Relational information. Lisa and Peter have been together since the age of 21 and married 8 years later. She described their current relationship as stable. However, she also described her first pregnancy as a trying time in their relationship, during which they grew apart and became less supportive towards one another.

Previous perinatal experience. During her first antenatal period, Lisa reported feeling "not quite herself." She started experiencing slowly progressive emotional and mood changes, which persisted throughout pregnancy and worsened in the postpartum period. Although this was a planned and much wanted first pregnancy, Lisa remembered the disappointment and self-stigma for not feeling "as a mother should." Additionally, the lack of communication with her husband resulted in escalating marital tension. At around 8 months postpartum, Lisa's depressive symptoms started to resolve, as she gradually re-introduced her previous work routine with the support of her family and colleagues.

Clinical presentation—From pregnancy to postpartum

Pregnancy. In this second pregnancy, there was evidence of the re-emergence of Lisa's previously experienced difficulties, consisting of tearfulness, loneliness and self-isolation on most days. She started to actively avoid any gatherings with family and friends. Lisa felt a growing sense of lethargy, finding it difficult to concentrate on work and house chores. These depressive symptoms interfered with her personal hygiene as she often did not feel the need to shower or to change unless probed by her family.

Furthermore, the neediness of her son, James frequently left her feeling irritable and angry. This resulted in Peter taking over the parental role, evoking feelings of guilt and inadequacy within her. These emotional struggles also hindered her ability to bond with her unborn. She described mixed emotions with every fetal movement and was unable to build up representations or images of her child and of herself as a mother.

Lisa was determined to get better and remained future-oriented throughout her recovery journey. She expressed positive support from her husband and identified her parents and sister as being always readily available. Another protective factor was her work environment which had greatly helped in her previous postpartum experience.

Postpartum period. The transition from a mother of one to a mother of two was quite daunting for Lisa, even more so with her progressive depressive illness which persisted postpartum.

Lisa reported a lack of euphoria with the arrival of her daughter Kate into this world. She felt that the birth experience did not meet her expectations and described it as "traumatic." Lisa experienced her baby as being "inconsolable," making her unable to positively attach to her new daughter no matter her efforts. She was also struggling to cope with motherhood tasks like feeding the baby and tending to her toddler's needs. This resulted in feelings of guilt and avoidance behaviors that reinforced her clinical presentation of postpartum depression (PPD).

It is important to note that throughout the client's treatment, there was never any evidence of any thoughts of harm to self, her unborn or others. The importance of assessing for any suicide risk, including suicidal thoughts and self-harm behavior, and/or for any risk to the infant during the perinatal period cannot be emphasized enough. Such an assessment ensures the safety of the woman and her baby, and lays the foundation for a safety plan to be developed, and appropriate follow-up to be arranged.

Multidisciplinary support

The client's presenting difficulties and her screening outcomes indicated the need for further professional support. In line with the National Institute for Health and Care Excellence (NICE) guidelines (NICE, 2014), Lisa's formulated care plan consisted of psychotherapy combined with psychiatric consultation to address her persistent symptomatology, as well as to minimize her heightened risk of another episode of PPD in this current perinatal period (Robertson, Grace, Wallington, & Stewart, 2004). This plan was collaboratively discussed and established with Lisa.

Psychotherapeutic support—Lisa's CBT journey

Systematic reviews and metaanalysis have consistently shown the superiority of CBT as a psychotherapeutic modality for the alleviation of symptoms of depression and anxiety in the perinatal period (Branquinho et al., 2021; Cuijpers et al., 2021). Thus, CBT is the treatment of choice for most women, as long as there is goodness of fit between the clients' needs and the modality (Huang, Zhao, Qiang, & Fan, 2018).

Initial phase—Goal setting, case conceptualization and treatment plan

The therapeutic process and principles of CBT were discussed with Lisa, starting antenatally and continuing postpartum. Motivational interviewing techniques (Miller & Rollnick, 2013) helped to address Lisa's concerns about her ability to actively engage in the process of CBT amidst her various other responsibilities, her decreased motivation and feelings of fatigue. Here, an emphasis was placed on the importance of adaptation, the role of supportive factors, and the adoption of other strategies, such as the collaborative inclusion of Peter even in sessions. Another significant support which Lisa retained throughout her treatment was using a therapy journal. This served both as a reminder of the issues covered within sessions, including set homework, but also as a space for her to express herself, practice skills and document her journey.

Case conceptualization. Throughout the initial assessment, the automatic thoughts and core beliefs were identified (Fig. 1). The case conceptualization was constructed and used to support Lisa in understanding further the connection between her thoughts, emotions and behaviors. This then guided the setting of the therapeutic goals according to the changes that Lisa wanted to accomplish from her treatment plan. The core beliefs explored were "I am not good enough" and "I am a failure." Fig. 1 shows the adapted case conceptualization encompassing both antenatal and postnatal periods.

Therapeutic goals. Eight therapeutic goals, divided into three overarching targets, were identified in the initial phase. These were prioritized as follows:

Vulnerability Factors
- Previous marital difficulties
- Self-insecurities (low self-esteem)
- History of untreated PND
- Perfectionism & high expectations

Cultural Factors
- Pressure to cope with work and home
- Expectation to be ok, perfect

Precipitating Factors
- Loss of control, worsening of symptoms
- Difficulty coping with work (antenatal)
- James' temperament
- Lack of sleep & lethargy
- Introduction of baby Kate (postnatal)

Perinatal Depression & Anxiety

Maintained by:

Negative cognitive-behavioral triad

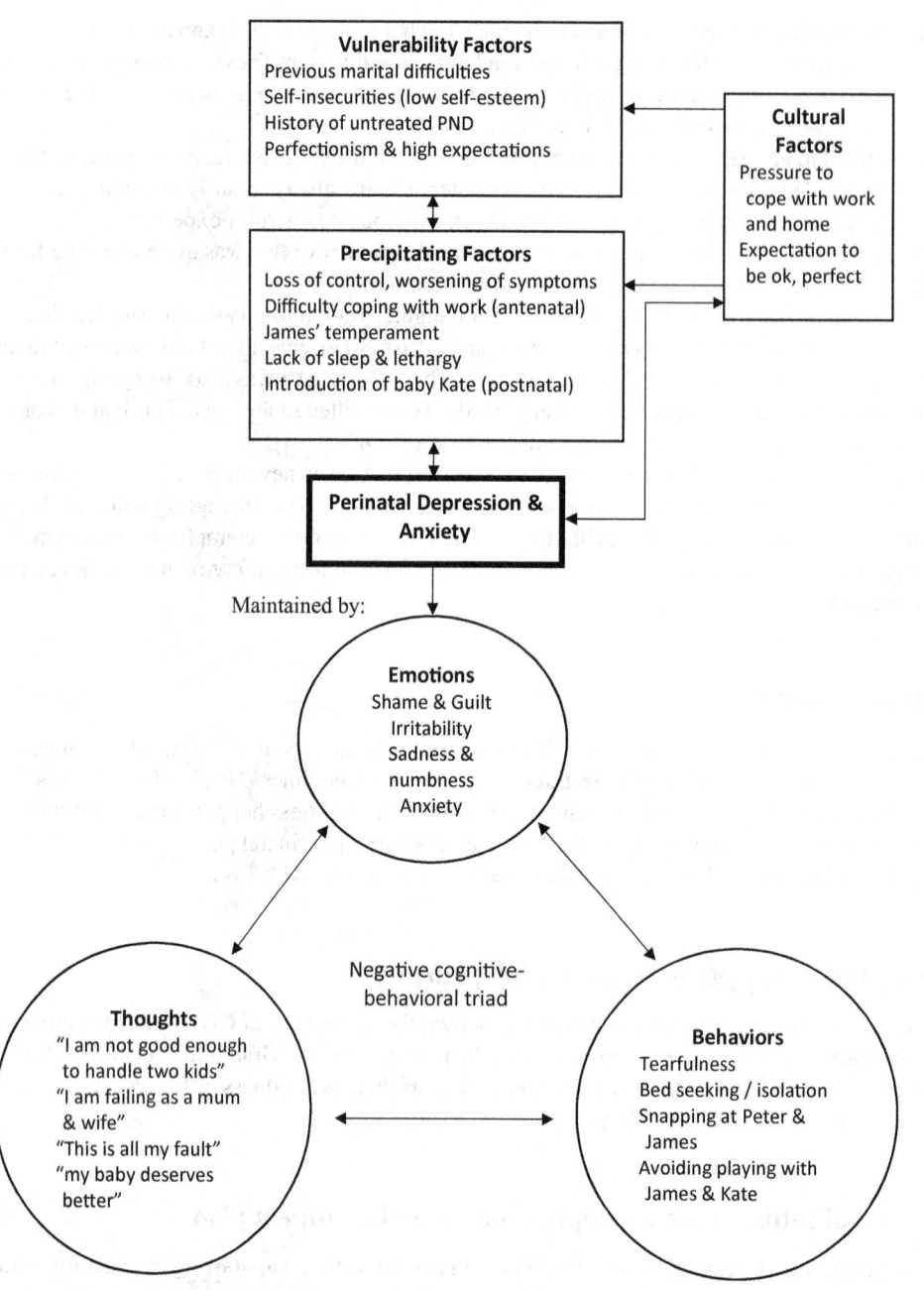

Emotions
- Shame & Guilt
- Irritability
- Sadness & numbness
- Anxiety

Thoughts
- "I am not good enough to handle two kids"
- "I am failing as a mum & wife"
- "This is all my fault"
- "my baby deserves better"

Behaviors
- Tearfulness
- Bed seeking / isolation
- Snapping at Peter & James
- Avoiding playing with James & Kate

FIG. 1 Lisa's case conceptualization. This figure is taken from the model of PPD which indicate the factors influencing Lisa's antenatal and PPD along with the cognitive behavioral triad maintaining and exacerbating the PPD. *(With permission from Milgrom, J., Martin, P. R., & Negri, L. M. (1999). Treating postnatal depression: A psychological approach for health care practitioners. Chichester: Wiley.)*

1. Feeling myself again (decreasing anxiety and bed-seeking behavior; increasing self-confidence and sense of accomplishment; feeling motivated again to tend to personal self-care).
2. To be a good (enough) mother (to bond with baby Kate; to be patient and play with James).
3. To feel better in my relationships (to communicate better with Peter; to go out with friends and family).

Middle phase—Delving into CBT interventions

During these sessions, Lisa was frequently tearful and overwhelmed, needing a space to allay her disappointments and frustrations. This was a delicate but important phase of the therapeutic process, essential in mirroring flexibility and acceptance.

During this difficult transition, the importance of psychotropic intervention as part of her treatment plan was highlighted to the client. Despite Lisa's complete initial resistance to this option, her negative automatic thoughts and beliefs regarding psychotropic medication were addressed and this barrier was overcome. This step was also key to support the effective progress of CBT.

A combination of behavioral techniques, cognitive restructuring and relaxation techniques were used to address the identified goals. With this, the adaptation of homework tasks to make them more manageable and realistic for Lisa was imperative in her progress.

Cognitive interventions. The CBT 5-part model was primarily used to support Lisa's understanding of cognitive processes. Automatic thought record keeping was used to help Lisa learn how to identify and challenge her automatic thoughts (Fig. 2). Two valuable adaptations here included the completion of the thought record sheet during the session and the building of coping cards to use in between the sessions. Such skills of becoming more aware of the triggering thoughts ("I cannot handle this," "I am doing it all wrong"), and the behavioral and physiological responses which followed (escaping when Peter was around, crying and increased agitation), allowed for effective re-structuring of these thoughts (alternative thoughts such as; "I am doing the best I can," "James is communicating with me in the way he knows how"), while also bringing about a reduction in the client's anxiety symptoms and associated bed-seeking behavior.

Behavioral activation. In light of Lisa's avoidance behaviors, focus was placed on increasing her engagement in pleasurable activities. Peter's support was crucial here to help her create a weekly planner and to follow through with it. Role-play was included in sessions to support her confidence and enhance her communication skills with Peter and significant others. In this way, she was able to share her needs with her husband more effectively, whilst also strengthening her marital relationship.

Behavioral experiments were given as homework to support Lisa's goal of increasing her self-confidence. Examples of these tasks included keeping a daily log of pleasant activities (simplified through a checklist created together within the session) and engaging in simple activities with the children. This helped her build evidence of her capabilities as a mother through the observation of her children's reactions. Peter was encouraged to take and share with her photos or videos as evidence of this, alongside his own feedback about the observed interactions. Eventually, Lisa was able to appreciate her efforts. There was a shift in the way she would talk about herself and her roles, which included *liking* how she is as a mother and taking the initiative to meet friends on a few occasions.

Situation	Emotions (0-100)	Automatic thoughts (images)	Evidence that supports the hot thought	Evidence that does not support the hot thought	Alternative/ balanced thought	Emotions (0-100)
What, who, when, where?	What did you feel? Rate each mood (0-100%)	What as going through your mind just before you started to feel this way? Any other thoughts, images? Circle the hot thought			Write an alternative or balanced thought. Rate how much you believe in each alternative or balanced thought (0-100%)	Rerate mood listed in column 2 as well as any new moods (0-100%)
Sitting in the living room at midday trying to breastfeed Kate and James is crying	Overwhelmed (100%) Irritated (80%) Defeated/depressed (90%)	I cannot handle them. I am not a good mum. I should be able to do this. James is crying because he is unhappy. This is too much. I want to get away	My sister has three children and copes better than I do. James is crying and Kate isn't cooperating.	James is not always crying, we have fun sometimes. Peter and my sister both tell me that I am a good mum. Kate settles once she latches on	James is communicating in the way he knows how (80%) I make mistakes, but I am doing the best I can – I am good enough (60%) It is normal that it gets hectic with two children (90%)	Overwhelmed (40%) Irritated (20%) Defeated/depressed (10%) Relieved (60%)

FIG. 2 Thought record sheet. A cognitive restructuring technique to help clients identify, analyze, and challenge dysfunctional automatic thoughts to be supported by alternative, balanced thoughts. *(With permission from Greenberger, D., & Padesky, C. A. (1995).* Mind over mood: A cognitive therapy treatment manual for clients. *Guilford Press.)*

Relaxation techniques. Psychoeducation on self-care, relaxation and mindfulness helped Lisa to understand the importance of these practices on her own mental health and her parenting style. During the sessions, deep breathing, guided visualizations and progressive muscle relaxation were practiced. Although these skills remained part of her relapse prevention toolbox, during the middle phase of therapy, Lisa felt she could mostly engage in deep breathing techniques and informal mindfulness practices due to time constraints or lack of energy to dedicate to longer relaxation practices.

Late phase—Focus on maintenance and relapse prevention

The final four sessions took place over a period of 5 months, tapering the frequency in preparation for termination. Peter was also involved here, attending for the last three sessions. These sessions centered on consolidating the acquired skills, creating a toolbox of such skills, and formulating a relapse prevention plan which was kept in the client's therapeutic journal. The gradual tapering of the sessions enhanced Lisa's confidence to positively tackle setbacks in her mood, and to notice and address any behavioral patterns which she identified as being unhelpful.

Although Lisa still reported the occasional difficult moment, she was able to successfully deal with these instances individually using the learnt tools and by reaching out to Peter. In fact, by the end of therapy, Lisa's scores on the EPDS and GAD-7 reduced significantly and to a nonclinically meaningful level. Furthermore, Lisa reported marked improvements in her relationship with Peter and James, and she was able to build a positive attachment with Kate. These positive outcomes were consistent with Peter's feedback and the therapist's clinical observations.

Perinatal psychiatric treatment

Lisa was also referred for psychiatric evaluation in pregnancy. NICE (2014) recommends high-intensity psychological intervention, such as CBT with/out medications in the treatment of moderate to severe perinatal depression. Medications are particularly favored when the woman declines psychological intervention, or her symptoms have not responded to the latter modality.

When discussing treatment options, the host of negative consequences of untreated depression for the woman, her unborn, and the rest of the family unit were explained by the psychiatrist. The importance of a careful risk-benefit analysis was emphasized (McAllister-Williams et al., 2017). However, the client's ongoing fear and concern about the reproductive safety of psychotropics persisted, amidst her depressive state, and she continued to show preference for psychotherapeutic support alone, a common occurrence among perinatal women (Goodman, 2009). Stigma associated with medication use was one other major factor accounting for the unacceptability of drug therapy in this client.

As therapy progressed, this ambivalence for pharmacotherapeutic measures started to gradually wane. Her therapist helped the client identify, evaluate, and modify her dysfunctional automatic thoughts related to medications. Such thoughts included, "my baby will be harmed" and "I am weak because I require medications." With time, she started demonstrating increased curiosity about the latter, accepting patient information leaflets, and bringing forward any questions. Moreover, she agreed to include the possibility of antidepressant treatment after delivery in case her mental health difficulties persisted.

The strong therapeutic alliance between the therapist and the client was crucial in supporting Lisa in this difficult decision. Through CBT and skillful communication of key information, Lisa became more empowered to understand and accept her illness, and to appreciate how drug therapy can allow her to cope with her illness in a more successful manner, both on an individual and interpersonal level. Therapy also helped with effective problem solving to optimize medication use, such as using coping cards to challenge previously mentioned dysfunctional automatic thoughts (Fig. 3) and timing breast-feeding according to drug administration. Additionally, the close collaboration between Lisa's therapist and the psychiatrist allowed for a joint teamwork approach for optimal care to be attained (Wenzel & Kleiman, 2015).

Conclusion

This case study highlights the positive personal and familial outcomes for a mother with PPD, achieved through an adapted CBT approach in the management of perinatal distress. Central to this approach is the flexibility of the therapist to adapt to the unique needs of the perinatal period while remaining grounded within the core principles of CBT, as shown in the case of Lisa.

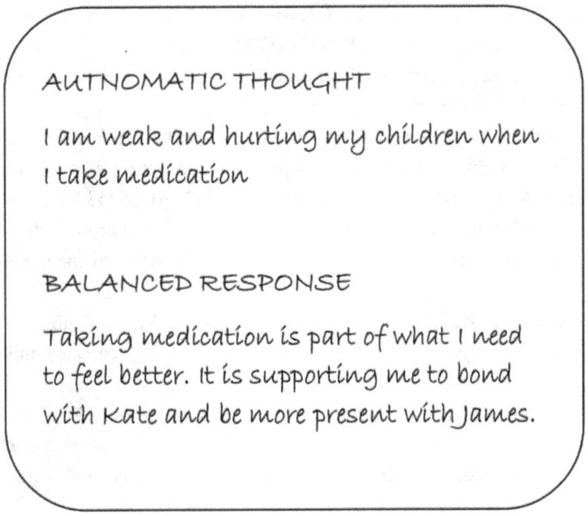

> AUTNOMATIC THOUGHT
>
> I am weak and hurting my children when I take medication
>
> BALANCED RESPONSE
>
> Taking medication is part of what I need to feel better. It is supporting me to bond with Kate and be more present with James.

FIG. 3 Lisa's coping card relating to medication intake. An example of a coping card used as a quick consultation for the work done within the sessions when feelings of guilt and sadness surfaced when taking psychiatric medication. *(Adapted from Wenzel, A., & Kleiman, K. (2015).* Cognitive behavioural therapy for perinatal distress. *New York, NY: Routledge.)*

Although psychological intervention began in pregnancy, the progressive worsening of her depressive symptoms postnatally indicated the need for additional treatment beyond psychotherapy, to ensure further progress. Key factors here were the strong therapeutic alliance between the therapist and Lisa and the application of specific CBT interventions to overcome the client's resistance to psychiatric treatment. This allowed for an evidence-based multidisciplinary approach to the management of PPD and positive outcomes, including Lisa's successful skill acquisition to manage her symptoms, associated distress and family disturbances.

Summary points

1. The perinatal period, from pregnancy to 1 year postpartum, is associated with an increased risk of mental health issues in both men and women, including PPD (O'Hara & Wisner, 2014).
2. The recommended treatment for perinatal depression consists of psychotherapy, namely CBT, combined with pharmacological treatment in case of poor response to or refusal of the former, and/or moderate to severe symptoms (NICE, 2014).
3. This case study outlines the therapeutic journey of Lisa, a pregnant woman, and a mother to a 2-year-old boy, who was identified to be suffering from perinatal depression and anxiety which were successfully treated by means of a combination of psychotherapy and psychiatric interventions.
4. The psychotherapeutic interventions discussed include the application and integration of core CBT components namely cognitive interventions, behavioral activation and relaxation techniques, together with specific adaptations of the standard CBT for depression, to effectively address the client's unique needs within the constraints of the perinatal period.
5. A multidisciplinary, systemic framework approach to the management of PMH disorders which addresses all the family's needs and also includes the partner as an integral support, is recommended.

References

Austin, M. P., Colton, J., Priest, S., Reilly, N., & Hadzi-Pavlovic, D. (2013). The antenatal risk questionnaire (ANRQ): Acceptability and use for psychosocial risk assessment in the maternity setting. *Women and Birth, 26*(1), 17–25.

Branquinho, M., Rodriguez-Muñoz, M. F., Maia, B. R., Marques, M., Matos, M., Osma, J., ... Vousoura, E. (2021). Effectiveness of psychological interventions in the treatment of perinatal depression: A systematic review of systematic reviews and meta-analyses. *Journal of Affective Disorders, 291*, 294–306. https://doi.org/10.1016/j.jad.2021.05.010.

Button, S., Thorton, A., Lee, S., Shakespeare, J., & Ayers, S. (2017). Seeking help for perinatal psychological distress: A meta-analysis of women's experiences. *British Journal of General Practice, 57*(663), E692–E699. https://doi.org/10.3399/bjgp17X692549.

Cuijpers, P., Franco, P., Ciharova, M., Miguel, C., Segre, L., Quero, S., & Karyotaki, E. (2021). Psychological treatment of perinatal depression: A meta-analysis. *Psychological Medicine*, 1–13. https://doi.org/10.1017/S0033291721004529.

Goodman, J. H. (2009). Women's attitudes, preferences, and perceived barriers to treatment for perinatal depression. *Birth Issues in Perinatal Care*, 36(1), 60–69. https://doi.org/10.1111/j.1523-536X.2008.00296.x.

Howard, L. M., & Khalifeh, H. (2020). Perinatal mental health: A review of progress and challenges. *World Psychiatry*, 19(3), 313–327.

Huang, L., Zhao, Y., Qiang, C., & Fan, B. (2018). Is cognitive behavioral therapy a better choice for women with postnatal depression? A systematic review and meta-analysis. *PLoS One*, 13(10). https://doi.org/10.1371/journal.pone.0205243, e0205243.

Levis, B., Negeri, Z., Sun, Y., Benedetti, A., & Thombs, B. D. (2020). Accuracy of the Edinburgh Postnatal Depression Scale (EPDS) for screening to detect major depression among pregnant and postpartum women: Systematic review and meta-analysis of individual participant data. *BMJ*, 371, m4022.

McAllister-Williams, R. M., Baldwin, D. S., Cantwell, R., Easter, A., Gilvarry, E., Glover, V., … Young, A. H. (2017). British Association for Psychopharmacology consensus guidance on the use of psychotropic medical preconception, in pregnancy and postpartum 2017. *Journal of Psychopharmacology*, 1–34.

Miller, W. R., & Rollnick, S. (2013). *Motivational interviewing: Helping people change* (3rd ed.). New York, NY: Guilford Press.

National Institute for Health and Care Excellence (NICE). (2014). *Antenatal and postnatal mental health: clinical management and service guidance.* Retrieved from: https://www.nice.org.uk/guidance/cg192/resources/antenatal-and-postnatal-mental-health-clinical-management-and-service-guidance-pdf-35109869806789.

O'Hara, M. W., & Wisner, K. L. (2014). Perinatal mental illness: Definition, description and aetiology. *Best Practice & Research Clinical Obstetrics & Gynaecology*, 28(1), 3–12. https://doi.org/10.1016/j.bpobgyn.2013.09.002.

Robertson, E., Grace, S., Wallington, T., & Stewart, D. E. (2004). Antenatal risk factors for postpartum depression: A synthesis of recent literature. *General Hospital Psychiatry*, 26(4), 289–295. https://doi.org/10.1016/j.genhosppsych.2004.02.006.

Simpson, W., Glazer, M., Michalski, N., Steiner, M., & Frey, B. N. (2014). Comparative efficacy of the generalised anxiety disorder 7-item scale and the Edinburgh Postnatal Depression Scale as screening tools for generalised anxiety disorder in pregnancy and the postpartum period. *Canadian Journal of Psychiatry*, 59(8), 434–440.

Webb, R., Uddin, N., Ford, E., Easter, A., Shakespeare, J., Roberts, N., … Williams, L. R. (2021). Barriers and facilitators to implementing perinatal mental health care in health and social care settings: A systematic review. *The Lancet Psychiatry*, 8, 521–534.

Wenzel, A., & Kleiman, K. (2015). *Cognitive behavioural therapy for perinatal distress.* New York, NY: Routledge.

Chapter 40

CASE STUDY: Compassion-based cognitive-behavior group therapy for patients with coronary artery disease

Tin-Kwang Lin[a,b], Chin-Lon Lin[a,b], Shu-Shu Wong[c], and Chia-Ying Weng[d]
[a]Department of Internal Medicine, School of Medicine, Tzu Chi University, Hualien, Taiwan, [b]Department of Internal Medicine, Dalin Tzu Chi Hospital, Buddhist Tzu Chi Medical Foundation, Chiayi, Taiwan, [c]Department of Social Work, Chaoyang University of Technology, Taichung, Taiwan, [d]Department of Psychology, National Chung Cheng University, Chiayi, Taiwan

Abbreviations

ANOVA	analysis of variance
ANS	autonomic nervous system
BVP	blood volume pulse
CAD	coronary artery heart disease
CBGT	cognitive-behavior group therapy
CBT	cognitive-behavior therapy
ECG	electrocardiogram
HF	high frequency
HR	heart rate
HRV	heart rate variety
IBI	inter beat interval
LF	low frequency
PSP	psychophysiological stress profile

Introduction

Dispositional hostility has been proposed as a psychosocial risk factor for coronary artery heart disease (CAD) (Kop & Francis, 2007). Meta-analyses indicate that hostility is associated with CAD both in cross-sectional and longitudinal studies (Chida & Steptoe, 2009; Miller, Smith, Turner, Guijarro, & Hallet, 1996). Hostility is associated with increased CAD events in healthy population (Barefoot, Larsen, von der Lieth, & Schroll, 1995; Matthews, Gump, Harris, Haney, & Barefoot, 2004). as well as poor prognosis of higher recurrent rate (Chaput et al., 2002) and higher mortality (Boyle et al., 2004; Chida & Steptoe, 2009) in CAD patients. Therefore, the association between hostility and the development of CAD is well established.

Although the links between hostility and an elevated risk of CAD development are still unclear, one of the physiological mechanism candidates may be an autonomic nervous system (ANS) imbalance. Autonomic imbalance, characterized by an excessively high sympathetic nervous system (SNS) (Krantz & Manuck, 1984) and/or a low parasympathetic nervous system (PNS) level (Thayer & Lane, 2007) may be a pathway to increased morbidity and mortality in CAD (Krantz & Manuck, 1984; Thayer & Lane, 2007; Thayer, Yamamoto, & Brosschot, 2010). Empirical studies demonstrate that both high SNS (Hasking et al., 1986; Kleiger, Miller, Bigger, & Moss, 1987) and low PNS activation (Liao et al., 1997; Tsuji et al., 1996) are associated with the occurrence of CAD.

Evidence also shows links between hostility and autonomic imbalance. A meta-analysis indicates that high-hostile individuals have high SNS and cardiovascular reactivity under stress. Sloan et al. (1994) demonstrated that high-hostile individuals had less vagal modulation during daytime. Experimental studies have also found that, in response to a stressful challenge, high-hostile individuals had greater PNS reduction (Demaree & Everhart, 2004; Sloan et al., 2001). Thus, empirical evidence supports not only the association between hostility and CAD but also the association between hostility and ANS imbalance.

The efficacy of cognitive-behavior therapy (CBT) to reduce hostility is well established (Del Vecchio & O'Leary, 2004; DiGiuseppe & Tafrate, 2003). Furthermore, behavioral interventions that target hostility as part of more comprehensive stress management protocols have been shown to reduce risk factors and even disease end points (Rozanski, Blumenthal, Davidson, Saab, & Kubzansky, 2005). Nonetheless, whether the reduction of hostility can enhance cardiac autonomic regulation is still unknown. Sloan et al. (2010), using a 12-week CBT, reported that reduction of hostility and anger was not accompanied by increases in cardiac autonomic regulation in healthy adults.

In that study, the cardiac autonomic regulation of the young participants, even with high-hostile personality, was still within normal ranges. The mean heart rate was 75.90 ± 8.15; the seated systolic blood pressure was 114.79 ± 10.48; and the seated diastolic blood pressure was 73.36 ± 8.86 for the treatment group. It may be inferred that the cardiac autonomic regulation of those young participants was too healthy to be improved significantly.

To better assess the association between hostility and cardiac autonomic regulation change, the present study was designed to address the above problem. We invited CAD patients to participate in a treatment program to re-examine whether the reduction of hostility was accompanied by increased cardiac autonomic regulation. It was expected that CAD patients, vis-à-vis the healthy young people used in Sloan's study, would have more room to show improvement in autonomic regulation of the heart.

Methods

Study participants

All participants were recruited from cardiology outpatients. A study assistant at the clinic informed them about the compassion-based CBGT and invited them to participate. All participants provided written informed consent. The Institutional Review Board of The Buddhist Dalin Tzu Chi General Hospital approved the study protocols. Data collection took place between July 2009 and June 2013.

Experimental protocol

The hostility, anxiety, depression, and perceived stress and ANS activities of participants were examined before and after the CBGT treatment. The anger recall task (Anderson & Lawler, 1995) was adopted to examine the ANS activities as well as psychophysiological reactions. After a 5-min seated rest period for adaptation, the participants went through the following four 5-min stages:

(1) Baseline: sitting comfortably
(2) Anger recall: recalling an anger event that happened in the past 6 months and still makes them feel angry
(3) Anger description: reporting the anger event
(4) Recovery: sitting comfortably and remaining silent

Participants were instructed to refrain from caffeinated beverages, alcohol, smoking, and excessive exercise during the 24 h prior to the experimental protocol. They completed the psychological questionnaires after the task.

Compassion-based cognitive-behavior group therapy

The CBGT protocol involved eight 2-h sessions that included the following: (1) psycho-education about the associations among anger, perceived stress and CAD risks; (2) diaphragmatic breathing relaxation training with portable biofeedback equipment; (3) cognitive therapy to increase cognitive flexibility and reality testing; (4) behavior therapy to reinforce compassion behaviors, "kind-to-self behavior," and to increase behavioral flexibility by "do something different." Group leaders delivered the individual psychophysiological stress profile (PSP), the psychophysiological reactions during the anger recall task, to each CAD patient in the first session to increase their motivation.

Measurement of outcome variables

Psychological effects of treatment: Multidimensional hostility and psychological distress

Hostility was measured using the 20-item Chinese Hostility Inventory Short-Form (Weng et al., 2008), which contains four dimensions of hostility: (1) hostile cognition ($\alpha = 0.78$); (2) hostile affection ($\alpha = 0.78$); (3) expressive hostility (anger-out) behavior ($\alpha = 0.76$); and (4) suppressive hostility(anger-in) behavior ($\alpha = 0.73$). Cronbach's α was 0.89 for the whole inventory. Anxiety was measured using the 20-item Chinese version of the Trait Anxiety subscale of State and Trait Anxiety Inventory with $\alpha = 0.93$ (Chung & Long, 1984). Depression was measured using the 21-item Chinese Version of the Beck Depression Inventory-II with $\alpha = 0.94$ (Chen, 2000). Perceived stress was measured using the 14-item Perceived Stress Scale-Chinese version with α ranging from 0.84 to 0.86. (Chu & Kao, 2005).

Psychophysiological reactions

Psychophysiological reactions including heart rate, blood volume pulse (BVP), and respiration rate of the participants were collected using the BioGraph Infiniti version 5.0.3 (Thought Technology Ltd., Montreal, Quebec, Canada).

Measurement of autonomic nervous system activities

MP150 Data Acquisition System (GLP-WIN-MP150-WSW-G, Biopac Systems, Inc.) was used to measure ANS activities. Electrocardiograph (ECG) was obtained by 1000 Hz sampling rate. The fast Fourier transformation was then used to transfer the inter-beat-interval (IBI) data to the frequency domain of heart rate variety (HRV) indices: (1) low-frequency (LF) power (0.04–0.15 Hz) as index of both SNS and PNS activities; (2) high-frequency (HF) power (0.15–0.4 Hz) as index of PNS activity (Task Force of the European Society of Cardiology and the North American Society of Pacing and Electrophysiology, 1996; Thayer, Hansen, & Johnsen, 2008). HRV indices were excluded from analysis when ECG artifacts were greater than 20% of IBI. Data from 15 out of 58 patients were excluded due to arrhythmia or ECG artifacts.

Data analysis

Pearson correlation analysis was performed to explore the relationships between hostility scores and psychophysiological reactions. A repeated t-test was used to examine the psychological effects of treatment. A two-way repeated-measures analysis of variance (ANOVA) with Least Significant Difference post hoc comparisons were used to examine the differences among the psychophysiological reactions under four experimental stages (baseline, anger recall, anger description, and recovery) for the experimental manipulation check, and between two time-points (pre and posttreatment) for treatment effects.

Results

We examined the relationships between hostility and autonomic nervous system activities during the four stages of the anger recall task prior to the CBGT program. The Pearson correlation coefficients showed that expressive hostility (anger-out) behavior was negatively correlated with lnHF during baseline ($r = -.28$, $P = .08$) as well as anger recall ($r = -.32$, $P < .05$) stages.

The validation of the experimental manipulation of anger on autonomic nervous system activities

There was a significant main effect of Stage for heart rate ($F_{(3, 33)} = 41.27$, $P < .001$, $\eta2 = .51$) in a two way ANOVA repeated measure. Post hoc test showed that, among the four stages of the anger recall task, the heart rate was elevated during both the anger recall and anger description stages compared to that of baseline (see Table 1). These results indicated that the anger recall task successfully activated the ANS and thus the effect of the experimental manipulation was validated.

Therapeutic effects of CBGT program on hostility levels and psychophysiological reactions

There were significant difference between pre and post of the CBGT program on multidimensional hostility (Cognitive: $t_{(41)} = 3.41$, ES $= 0.22$, $P < .001$; Affective: $t_{(41)} = 5.50$, ES $= 0.43$, $P < .001$; Expressive: $t_{(41)} = 3.77$, ES $= 0.26$, $P < .001$; Suppressive: $t_{(41)} = 5.51$, ES $= 0.43$, $P < .001$), depression ($t_{(41)} = 3.39$, ES $= .26$, $P < .001$), anxiety ($t_{(39)} = 6.03$, ES $= 0.48$, $P < .001$), and perceived stress ($t_{(38)} = 3.39$, ES $= 0.23$. $P < .01$). The results indicated that the CBGT program

TABLE 1 The autonomic nervous system activities during the four stages of anger recall task.

Stage	B	AR	AD	RE		df	F	$\eta 2$	Post hoc
HR (n = 41)									
Time	Mean (SD)				Time	1	0.57	.01	
Pre	65.85 (8.60)	66.91 (8.91)	74.97 (11.93)	65.70 (8.59)	Stage	3	41.27***	.51	AD > AR > B, RE
Post	65.76 (9.16)	66.89 (9.60)	71.03 (10.80)	66.68 (10.20)	Time × Stage	3	6.83**	.15	Post < Pre [in AD]
Ln LF (n = 39)									
Time	Mean (SD)				Time	1	3.54	0.09	
Pre	3.28 (1.01)	3.33 (1.04)	3.93 (0.84)	3.52 (0.87)	Stage	3	19.40***	0.34	AD > RE > B, AR
Post	3.53 (0.94)	3.39 (0.83)	4.04 (0.71)	3.76 (0.87)	Time × Stage	3	0.79	0.02	
Ln HF (n = 39)									
Time	Mean (SD)				Time	1	1.97	0.05	
Pre	3.43 (1.10)	3.30 (1.15)	3.14 (1.06)	3.50 (1.02)	Stage	3	5.13**	0.12	B, AR, RE > AD
Post	3.19 (1.00)	3.26 (0.93)	3.01 (0.84)	3.24 (1.00)	Time × Stage	3	1.25	0.03	
RR (n = 43)									
Time	Mean (SD)				Time	1	17.56***	0.30	Pre > Post
Pre	15.22 (3.02)	14.86 (2.69)	#	14.50 (2.43)	Stage	2	2.96	0.07	
Post	13.17 (3.80)	13.76 (3.35)	#	12.71 (3.14)	Time × Stage	2	1.69*	0.04	
BVP (n = 43)									
Time	Mean (SD)				Time	1	4.43*	0.10	Post > Pre
Pre	34.02 (0.05)	34.00 (0.06)	33.99 (0.08)	34.00 (0.06)	Stage	3	2.25	0.05	
Post	34.04 (0.05)	34.01 (0.06)	34.02 (0.15)	34.02 (0.06)	Time × Stage	3	0.47	0.01	

*$P < .05$; **$P < .01$; ***$P < .001$ ("blod" for the significant value).
AD, anger description; AR, anger recall; B, baseline; BVP, blood volume pulse; RE, recovery; RR, respiratory rate.
#, the respiration rate was not counted due to speaking interference.

was efficacious in reducing hostility and psychological distress. Furthermore, Table 1 gives the respiration rate, heart rate and BVP during the anger recall task pre and post the CBGT program. The significant main effect of Time illustrates that the program reduced the respiration rate ($F_{(1, 37)} = 17.56$, $P < .001$, $\eta 2 = 0.30$) and increased the BVP ($F_{(1, 35)} = 4.43$, $P < 0.05$, $\eta 2 = 0.10$) level. Moreover, the significant interaction effect of HR between Time and Stage ($F_{(3, 34)} = 6.83$, $P < .01$, $\eta 2 = 0.15$) showed the diminished HR increases in the anger description stage after the CBGT program. These results indicated that the treatment alleviated the respiration rate and vasoconstriction responses during the whole anger recall task, and lessened the HR reaction during the anger description stage in CAD patients.

Effects of CBGT program on autonomic nervous system activities and hostility

We examined the correlations between the four dimensions of hostility and ANS activity changes between pre and post the CBGT treatment. The Pearson correlation showed that reduced (pre-post) expressive hostility (anger-out) behavior was

TABLE 2 The correlations between multiple hostilities and Ln HF change between pre and post CBGT treatment during the four stages of anger recall task (N = 39).

	B	AR	AD	RE
Hostility	−0.20	**−0.34***	−0.27	−0.23
Hostile cognition	−0.14	−0.20	−0.17	−0.28
Hostile affect	−0.16	−0.14	−0.19	−0.02
Expressive hostility behavior	**−0.32***	**−0.46****	**−0.38***	**−0.36***
Suppressive hostility behavior	0.08	−0.11	0.01	0.09

*$P < .05$; **$P < .01$ ("blod" for the significant value).
AD, anger description; AR, anger recall; B, baseline; RE, recovery.

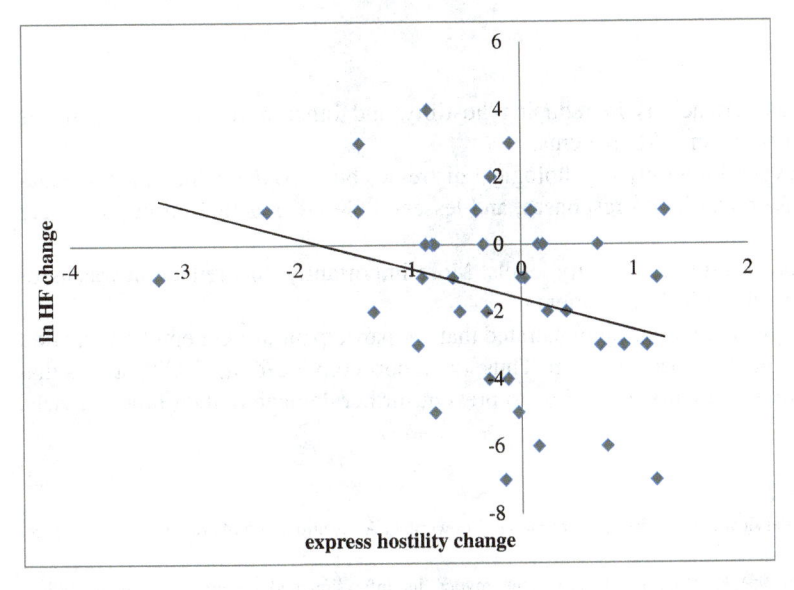

FIG. 1 The correlation between expressive hostility and LnHF change between pre and post CBGT treatment during baseline stage.

correlated with increased (pre-post) lnHF during the four stages ($r = -.32$ to $-.46$, $ps < .05$) of the anger recall task (see Table 2). The results indicated that the increased LnHF concurred with the decreased expressive hostility after the CBGT treatment (Fig. 1).

Conclusion

We used CAD patients to study the therapeutic effect of CBGT on the relationships between hostility and ANS regulation change. We found that psychotherapy reduced multiple hostility levels as shown in previous studies (Del Vecchio & O'Leary, 2004; DiGiuseppe & Tafrate, 2003). More importantly, this reduction was associated with increased vagal modulation of the heart.

Previous empirical evidence indicates not only the association between hostility and CAD but also the association between hostility and ANS imbalance. Nonetheless, whether the reduction of hostility can alter the ANS pathophysiological mechanisms that were thought to link hostility to heart disease still remains unclear. Sloan et al. (2010) were the first to address this issue. They reported that individual psychotherapy successfully reduced hostility level but did not affect ANS regulation.

In order to increase the sensitivity of hostility detection, we adopted the Chinese Hostility Inventory-Short Form (Weng et al., 2008) which includes four dimensions: hostility cognition, hostility affect, expressive hostility (anger-out) behavior, and suppressive hostility (anger-in) behavior. This enabled us to examine the relationships between multidimensional hostility and ANS regulation in detail. We found that, before psychotherapy, expressive hostility behavior was the dimension negatively correlated with lnHF, an index of cardiac parasympathetic modulation, during baseline and anger recall stages.

Furthermore, after psychotherapy, the expressive hostility behavior was the dimension associated with increased cardiac parasympathetic regulation during all the four stages of the anger recall task.

There were some differences between the participants of this study and those of Sloan et al.'s (2010). First, the age range (mean age was 58.42 ± 7.54 ranging from 37 to 70) is higher (mean age $= 30.46 \pm 6.67$ ranging from 20 to 45). Second, CAD patients were invited to participate in this study while healthy adults were invited to participate in Sloan et al.'s. The cardiac autonomic regulation of the young sample of Sloan et al.'s study may have been too healthy to be further improved. That might be one of the reasons explaining the lack of a psychotherapy effect on ANS regulation in Sloan et al.'s study. Furthermore, we delivered the individual report of PSP during the anger recall task to each CAD patient in the first session. That individualized report demonstrated that the participant's anger emotion reaction induced a higher heart rate as well as more severe vasoconstriction indexed by BVP. Those psychophysiological reactions might further increase the recurrent possibility of the life-threatening disease for patients who had previously encountered it. Thus, the report convinced the CAD patients that they could and should practice healthy emotion regulation behavior to prevent further damage to their blood vessels. Their motivation may be stronger than that of the healthy young adults in Sloan et al.'s study.

In conclusion, this study demonstrated that the reduction of expressive hostility through compassion-based CBGT was associated with the increase in cardiac autonomic regulation. This effect might be helpful to the disease prognosis.

Summary points

- The compassion based CBGT program was efficacious in reducing hostility and other psychological distresses including anxiety, depression and perceived stress in CAD patients.
- The compassion based CBGT treatment alleviated not only psychological distresses but also the pathological mechanisms of CAD, including respiration rate, vasoconstriction responses, and lessened the HR reaction during the anger stage in CAD patients.
- The compassion based CBGT program reduced multiple hostility levels. More importantly, this reduction was associated with increased vagal modulation of the heart in CAD patients.
- That individualized report of psychophysiological reactions demonstrated that the participant's anger emotion reaction induced a higher heart rate as well as more severe vasoconstriction. Thus, the report convinced the CAD patients that they could and should practice healthy emotion regulation behaviors to prevent further damage to their blood vessels.

References

Anderson, S. F., & Lawler, K. A. (1995). The anger recall interview and cardiovascular reactivity in women: An examination of context and experience. *Journal of Psychosomatic Research, 39*(3), 335–343.

Barefoot, J. C., Larsen, S., von der Lieth, L., & Schroll, M. (1995). Hostility, incidence of acute myocardial infarction, and mortality in a sample of older Danish men and women. *American Journal of Epidemiology, 142*(5), 477–484.

Boyle, S. H., Williams, R. B., Mark, D. B., Brummett, B. H., Siegler, I. C., Helms, M. J., & Barefoot, J. C. (2004). Hostility as a predictor of survival in patients with coronary artery disease. *Psychosomatic Medicine, 66*(5), 629–632.

Chaput, L. A., Adams, S. H., Simon, J. A., Blumenthal, R. S., Vittinghoff, E., Lin, F., ... Matthews, K. A. (2002). Hostility predicts recurrent events among postmenopausal women with coronary heart disease. *American Journal of Epidemiology, 156*(12), 1092–1099.

Chen, H. Y. (2000). *Guidebook of Beck depression inventory-II-Chinese version*. Chinese Behavioral Science Corporation.

Chida, Y., & Steptoe, A. (2009). The association of anger and hostility with future coronary heart disease: A meta-analytic review of prospective evidence. *Journal of the American College of Cardiology, 53*(11), 936–946.

Chu, L. C., & Kao, S. R. (2005). The moderation of meditation experience and emotional intelligence on the relationship between perceived stress and negative mental health. *Chinese Journal of Psychology, 47*(2), 157–179.

Chung, S. K., & Long, C. F. (1984). A study of the revised state-trait anxiety inventory. *Psychological Testing, 31*, 27–36.

Del Vecchio, T., & O'Leary, K. D. (2004). Effectiveness of anger treatments for specific anger problems: A meta-analytic review. *Clinical Psychology Review, 24*(1), 15–34.

Demaree, H. A., & Everhart, D. E. (2004). Healthy high-hostiles: Reduced parasympathetic activity and decreased sympathovagal flexibility during negative emotional processing. *Personality and Individual Differences, 36*(2), 457–469.

DiGiuseppe, R., & Tafrate, R. C. (2003). Anger treatment for adults: A meta-analytic review. *Clinical Psychology: Science and Practice, 10*(1), 70–84.

Hasking, G. J., Esler, M. D., Jennings, G. L., Burton, D., Johns, J. A., & Korner, P. I. (1986). Norepinephrine spillover to plasma in patients with congestive heart failure: Evidence of increased overall and cardiorenal sympathetic nervous activity. *Circulation, 73*(4), 615–621.

Kleiger, R. E., Miller, J. P., Bigger, J. T., Jr., & Moss, A. J. (1987). Decreased heart rate variability and its association with increased mortality after acute myocardial infarction. *American Journal of Cardiology, 59*(4), 256–262.

Kop, W. J., & Francis, J. L. (2007). Psychological risk factors and pathophysiological pathways involved in coronary artery disease: Relevance to complementary medicine interventions. In *Integrative cardiology complementary and alternative medicine for the heart* (pp. 359–379). NY: McGraw Hill.

Krantz, D. S., & Manuck, S. B. (1984). Acute psychophysiologic reactivity and risk of cardiovascular disease: A review and methodologic critique. *Psychological Bulletin, 96*(3), 435–464.

Liao, D., Cai, J., Rosamond, W. D., Barnes, R. W., Hutchinson, R. G., Whitsel, E. A., … Heiss, G. (1997). Cardiac autonomic function and incident coronary heart disease: A population-based case-cohort study. The ARIC Study. Atherosclerosis Risk in Communities Study. *American Journal of Epidemiology, 145*(8), 696–706.

Matthews, K. A., Gump, B. B., Harris, K. F., Haney, T. L., & Barefoot, J. C. (2004). Hostile behaviors predict cardiovascular mortality among men enrolled in the Multiple Risk Factor Intervention Trial. *Circulation, 109*(1), 66–70.

Miller, T. Q., Smith, T. W., Turner, C. W., Guijarro, M. L., & Hallet, A. J. (1996). A meta-analytic review of research on hostility and physical health. *Psychological Bulletin, 119*(2), 322–348.

Rozanski, A., Blumenthal, J. A., Davidson, K. W., Saab, P. G., & Kubzansky, L. (2005). The epidemiology, pathophysiology, and management of psychosocial risk factors in cardiac practice: The emerging field of behavioral cardiology. *Journal of the American College of Cardiology, 45*(5), 637–651.

Sloan, R. P., Bagiella, E., Shapiro, P. A., Kuhl, J. P., Chernikhova, D., Berg, J., & Myers, M. M. (2001). Hostility, gender, and cardiac autonomic control. *Psychosomatic Medicine, 63*(3), 434–440.

Sloan, R. P., Shapiro, P. A., Bigger, J. T., Jr., Bagiella, E., Steinman, R. C., & Gorman, J. M. (1994). Cardiac autonomic control and hostility in healthy subjects. *American Journal of Cardiology, 74*(3), 298–300.

Sloan, R. P., Shapiro, P. A., Gorenstein, E. E., Tager, F. A., Monk, C. E., McKinley, P. S., … Bigger, J. T., Jr. (2010). Cardiac autonomic control and treatment of hostility: A randomized controlled trial. *Psychosomatic Medicine, 72*(1), 1–8.

Task Force of the European Society of Cardiology and the North American Society of Pacing and Electrophysiology. (1996). Heart rate variability: Standards of measurement, physiological interpretation and clinical use. *Circulation, 93*(5), 1043–1065.

Thayer, J. F., Hansen, A. L., & Johnsen, B. H. (2008). Noninvasive assessment of autonomic influences on the heart: Impedance cardiography and heart rate variability. In L. J. Luecken, & L. C. Gallo (Eds.), *Handbook of physiological research methods in health psychology* (pp. 183–209). Sage Publications, Inc.

Thayer, J. F., & Lane, R. D. (2007). The role of vagal function in the risk for cardiovascular disease and mortality. *Biological Psychology, 74*(2), 224–242.

Thayer, J. F., Yamamoto, S. S., & Brosschot, J. F. (2010). The relationship of autonomic imbalance, heart rate variability and cardiovascular disease risk factors. *International Journal of Cardiology, 141*(2), 122–131.

Tsuji, H., Larson, M. G., Venditti, F. J., Jr., Manders, E. S., Evans, J. C., Feldman, C. L., & Levy, D. (1996). Impact of reduced heart rate variability on risk for cardiac events. The Framingham Heart Study. *Circulation, 94*(11), 2850–2855.

Weng, C. Y., Lin, I. M., Lue, B. H., Chen, H. J., Wu, Y. C., & Cheng, Y. R. (2008). Development and psychometric properties of the Chinese hostility inventory: Short form. *Psychological Testing, 55*(3), 463–487.

Chapter 41

Application of mindfulness-based cognitive therapy and health qigong–based cognitive therapy among Chinese people with mood disorders: A case study

Sunny Ho-Wan Chan[a] and Charlie Lau[b]
[a]School of Health and Social Wellbeing, University of the West of England, Bristol, United Kingdom [b]Department of Rehabilitation Sciences, The Hong Kong Polytechnic University, Hung Hom, Hong Kong

Abbreviations

CBT	cognitive behavior therapy
HQ	health qigong
HQCT	health qigong–based cognitive therapy
MBCT	mindfulness-based cognitive therapy
MBI	mind-body interventions
MM	mindfulness meditation

Introduction

Prevalence of depression and anxiety

Mood disorders such as depression and anxiety are prevalent across the globe. Studies on epidemiology from nationally representative surveys have consistently demonstrated that approximately one-third of the population are affected by mental disorders over their lifetime (Kessler, Berglund, Demler, Jin, & Walters, 2005; Phillips et al., 2009). Within the Chinese population, the first territory-wide epidemiological survey in Hong Kong estimated that the prevalence of common mental disorders in adults was 13.3%, with mixed anxiety and depressive disorder being the most frequent diagnoses (Lam et al., 2015). This finding commensurates with estimation of prevalence of mood disorders in both China and western countries.

Limitation of traditional cognitive behavioral therapy

Owing to the potential dependence on pharmaceutical treatment and its side effects, individuals may employ psychotherapy instead of medication. Although previous studies have revealed that cognitive behavioral therapy (CBT) is an effective psychosocial treatment in depression and anxiety management (Twomey, O'Reilly, & Byrne, 2015; Zhang et al., 2019), a number of reviews have found a steady decline of effect size of CBT in attenuating anxiety and depression since its growing popularity four decades ago (Cuijpers et al., 2011; Johnsen & Friborg, 2015; Lynch, Laws, & McKenna, 2010). It is possible that the CBT approach was over mechanistic and unable to attend to the concerns of the "whole" individual (Gaudiano, 2008), suggesting that CBT by itself might not be sufficient in treating mood disorders.

Alternative forms of therapy: Mindfulness-based cognitive therapy

In light of the limitations of traditional treatments, much attention shifted to alternative forms of therapy. Recently, researchers have shown an increased interest in mind-body interventions (MBIs) in treating depression and anxiety

(Hoch et al., 2012; Kinser, Elswick, & Kornstein, 2014). For example, mindfulness meditation (MM) is considered a contemplative approach to cultivate nonjudgmental moment-to-moment awareness and experiences (Kabat-Zinn, 1990). Studies have documented its effectiveness in alleviating anxiety and depression, especially its potential relevance to psychiatric comorbidity (Khusid & Vythilingam, 2016; Ren, Zhang, & Jiang, 2018; Wielgosz, Goldberg, Kral, Dunne, & Davidson, 2019). Integrating MM with CBT was brought forward to develop awareness of internal experiences and acceptance which is vital to subsequent cognitive restructuring (Radkovsky, McArdle, Bockting, & Berking, 2014). Particularly, the integration of MM and CBT instigates a fabrication of mindfulness-based cognitive therapy (MBCT; Segal, Williams, & Teasdale, 2013). Meta-analyses have demonstrated that MBCT is an effective approach for mental health and well-being (Querstret, Morison, Dickinson, Cropley, & John, 2020) as well as depressive or anxiety symptoms (Goldberg et al., 2019; Thomas, Chur-Hansen, & Turner, 2020).

Alternative forms of therapy: Health qigong cognitive therapy

However, MM only represents the static form of MBIs where mediative body movements are scarcely involved. There is another traditional active form of MBI namely health qigong (HQ; Chan & Tsang, 2019), encompassing both body movement and introspective focus, which is an awareness of breathing and the natural force or energy in the body (Jahnke, Larkey, Rogers, Etnier, & Lin, 2010). Not only have previous studies showed that HQ such as Baduanjin exercises could reduce depression or anxiety in patients with physical or mental illness (Wang et al., 2014; Zou et al., 2018), comparable therapeutic effects of HQ and CBT on depressive and anxiety symptoms have also been reported (Wang et al., 2013). Combining HQ and CBT could advance the application of traditional Chinese meditative movement as one of the major behavioral strategies in combating mood symptoms (Chow & Tsang, 2007). Consistent with a study integrating HQ and CBT revealed beneficial effects on self-perceived personal wellbeing of community-dwelling elderly (Liu & Tsui, 2014), a randomized controlled trial in Hong Kong found that MBCT and health qigong–based cognitive therapy (HQCT) both led to reduction in depressive and anxiety symptoms and improvement in physical and mental health status, perceived stress, sleep quality, and self-efficacy (Chan, Chan, Chao, & Chan, 2020).

Case study: Personal background and history

Twinnie is a 35-year-old married woman living with her husband and two sons aged 8 and 12. She has been working in a design company as a project manager for 5 years. She reported that persistent worry and high level of general anxiety had been her main problems for the past 8 years. It was triggered by watching a piece of tragic TV news which was about the sudden death of a baby, due to suffocation after choking on milk during sleep. Since then, she avoided watching all kinds of news. Her excessive worry extended to her work and family such that she reported having great difficulties in controlling her worry and repeatedly washing her hands, checking electric switches, and door locks at work and at home. She spent hours on washing and checking rituals. She found it difficult to make decisions for daily care of her two sons. Similarly, she was unable to complete daily household tasks. Her husband would like to help her, but she refused his help and insisted on completing tasks on her own. She described herself as a perfectionist and being too much of a "people pleaser."

Case study: Description of recent circumstances

Several months ago, she experienced a number of stressors, including meeting a close friend with terminal illness, multiple deadlines of her projects at work, examination of her children. All of these raised her level of anxiety. In addition to high levels of anxiety, she had been experiencing long-standing symptoms of muscle tension, palpitation, shortness of breath, chest tension, sleep disturbance, and concentration difficulties. Recently, she has reported some physiological responses such as lacking energy, irritable bowel syndrome, and poor appetite. Her worries and anxiety led to significant interference with her daily life. She put off her daily household tasks and childcare activity schedule. She was unable to make decisions because she feared that her final decisions were not good enough. She could only provide minimal function at home and at work in her roles as mother and project manager.

Case study: Description of thought, feelings, and behaviors

Twinnie felt that her life was surrounded by the fear of failure and desire to be perfect. She was striving her best to be the perfect employee, the perfect mother, and the perfect source of comfort and support for her indisposed friend. As a result of her perfectionistic and "people pleasing" tendencies, she struggled to say "No" to others' requests. She felt frustrated and guilty about the limited time she could spend with her children and husband. Moreover, she has been avoiding visiting her close friend with terminal illness because she was anxious that she might make inappropriate conversation, cry, and worsen her friend's condition. This has created a great deal of conflict for her because she was very close to this friend and wanted to visit her friend terribly. However, she was irritable about her failure to cope with her life situation. When she felt worried, she would engage in washing and checking rituals more often. She was worried that her problems would occupy too much of her life and prevent her from excelling in her career and family.

Case study: Initial assessment

Eventually she visited the Generalized Out-Patient Clinic because of her increasing somatic discomfort and was diagnosed with generalized anxiety disorder with obsessive-compulsive features. Mood symptoms, physical and mental health statuses, perceived stress, sleep quality and general self-efficacy were assessed with the Chinese version of the Depression Anxiety and Stress Scale (DASS-21) (Lovibond & Lovibond, 1995; Taouk, Lovibond, & Laube, 2001), Short-form-12 (SF-12) (Lam, Eileen, & Gandek, 2005), Perceived Stress Scale (CPSS) (Ng, 2013), Pittsburgh Sleep Quality Index (PSQI) (Tsai et al., 2005), and General Self-efficacy Scale (CGSS) (Zhang & Schwarzer, 1995), respectively.

Case study: Interventions

Following clinical assessment, she was recommended to attend MBCT for mood symptoms reduction. After attending an 8-week MBCT program, her anxiety symptoms greatly subsided. While she reported increased awareness of her difficulties and struggles, she perceived less stress and was learning to live with them by practicing mindfulness meditation daily. However, she still had various degrees of inexplicable somatic discomfort. After thorough discussion with the therapist, Twinnie showed interest in trying an additional HQCT program. She found her physical health improved significantly after completing an 8-week HQCT program. Interestingly, she also learnt to integrate her bodily sensation awareness into her HQ practice which further helped to alleviate her anxiety symptoms. Moreover, she also reported that she had better sleep after her physical symptoms and discomfort had been alleviated following HQCT.

Summary points

- Common mental disorders are estimated to affect more than 10% of the population in both Hong Kong, China and western countries.
- This chapters describes a case study of a working woman with general anxiety disorder and obsessive-compulsive features.
- The treatment protocol involved mindfulness-based cognitive therapy and health qigong–based cognitive therapy.
- Outcome measures included the Chinese version of the Depression Anxiety and Stress Scale (DASS-21), Short-form-12 (SF-12), Perceived Stress Scale (CPSS), Pittsburgh Sleep Quality Index (PSQI), and General Self-efficacy Scale (CGSS).
- At the end, the patient showed reduced mood symptoms, perceived less stress and had improved physical health, sleep quality and self-efficacy.
- These positive outcomes of mindfulness-based cognitive therapy and health qigong–based cognitive therapy along with an absence of side effects highlight the beneficial effects of these two therapies in combating mood disorders.

References

Chan, S. H. W., Chan, W., Chao, J., & Chan, P. (2020). A randomized controlled trial on the comparative effectiveness of mindfulness-based cognitive therapy and health qigong-based cognitive therapy among Chinese people with depression and anxiety disorders. *BMC Psychiatry*, *20*(1), 590.

Chan, S. H. W., & Tsang, H. W. H. (2019). The beneficial effects of qigong on elderly depression. In S. Y. Yau, & K. F. So (Eds.), Vol. 147. *International review of neurobiology* (pp. 155–188). US: Academic Press.

Chow, Y. W. Y., & Tsang, H. W. H. (2007). Biopsychosocial effects of Qigong as a mindful exercise for people with anxiety disorders: A speculative review. *Journal of Alternative and Complementary Medicine*, *13*, 831–839.

Cuijpers, P., Clignet, F., van Meijel, B., van Straten, A., Li, J., & Andersson, G. (2011). Psychological treatment of depression in inpatients: A systematic review and meta-analysis. *Clinical Psychology Review*, *31*, 353–360.

Gaudiano, B. A. (2008). Cognitive-behavioural therapies: Achievements and challenges. *Evidence-Based Mental Health*, *11*, 5–7.

Goldberg, S. B., Tucker, R. P., Greene, P. A., Davidson, R. J., Kearney, D. J., & Simpson, T. L. (2019). Mindfulness-based cognitive therapy for the treatment of current depressive symptoms: A meta-analysis. *Cognitive Behaviour Therapy*, *48*, 445–462.

Hoch, D. B., Watson, A. J., Linton, D. A., Bello, H. E., Senelly, M., Milik, M. T., ... Kvedar, J. C. (2012). The feasibility and impact of delivering a mind-body intervention in a virtual world. *PLoS One, 7*(3), e33843.

Jahnke, R., Larkey, L., Rogers, C., Etnier, J., & Lin, F. (2010). A comprehensive review of health benefits of qigong and tai chi. *American Journal of Health Promotion, 24*(6), e1–e25.

Johnsen, T. J., & Friborg, O. (2015). The effects of cognitive behavioral therapy as an anti-depressive treatment is falling: A meta-analysis. *Psychological Bulletin, 141*, 747–768.

Kabat-Zinn, J. (1990). *Full catastrophe living: Using the wisdom of your body and mind to face stress, pain, and illness*. New York: Delta Books.

Kessler, R. C., Berglund, P., Demler, O., Jin, R., & Walters, E. E. (2005). Lifetime prevalence and age-of-onset distributions of DSM-IV disorders in the National Comorbidity Survey Replication. *Archives of General Psychiatry, 62*, 593–602.

Khusid, M. A., & Vythilingam, M. (2016). The emerging role of mindfulness meditation as effective self-management strategy, Part 1: Clinical implications for depression, post-traumatic stress disorder, and anxiety. *Military Medicine, 181*, 961–968.

Kinser, P. A., Elswick, R. K., & Kornstein, S. (2014). Potential long-term effects of a mind–body intervention for women with major depressive disorder: Sustained mental health improvements with a pilot yoga intervention. *Archives of Psychiatric Nursing, 28*, 377–383.

Lam, C. L., Eileen, Y. Y., & Gandek, B. (2005). Is the standard SF-12 health survey valid and equivalent for a Chinese population? *Quality of Life Research, 14*(2), 539–547.

Lam, L. C. W., Wong, C. S. M., Wang, M. J., Chan, W. C., Chen, E. Y. H., Ng, R. M. K., ... Bebbington, P. (2015). Prevalence, psychosocial correlates and service utilization of depressive and anxiety disorders in Hong Kong: The Hong Kong Mental Morbidity Survey (HKMMS). *Social Psychiatry and Psychiatric Epidemiology, 50*, 1379–1388.

Liu, Y. W. J., & Tsui, C. M. (2014). A randomized trial comparing Tai Chi with and without cognitive-behavioral intervention (CBI) to reduce fear of falling in community-dwelling elderly people. *Archives of Gerontology and Geriatrics, 59*, 317–325.

Lovibond, S. H., & Lovibond, P. F. (1995). *Manual for the depression anxiety stress scales*. Sydney: Psychology Foundation.

Lynch, D., Laws, K. R., & McKenna, P. J. (2010). Cognitive behavioural therapy for major psychiatric disorder: Does it really work? A meta-analytical review of well-controlled trials. *Psychological Medicine, 40*, 9–24.

Ng, S. M. (2013). Validation of the 10-item Chinese perceived stress scale in elderly service workers: One-factor versus two-factor structure. *BMC Psychology, 1*(1), 1–8.

Phillips, M. R., Zhang, J., Shi, Q., Song, Z., Ding, Z., Pang, S., ... Wang, Z. (2009). Prevalence, treatment, and associated disability of mental disorders in four provinces in China during 2001–05: An epidemiological survey. *Lancet, 373*, 2041–2053.

Querstret, D., Morison, L., Dickinson, S., Cropley, M., & John, M. (2020). Mindfulness-based stress reduction and mindfulness-based cognitive therapy for psychological health and well-being in nonclinical samples: A systematic review and meta-analysis. *International Journal of Stress Management, 27*, 394–411. https://doi.org/10.1037/str0000165.

Radkovsky, A., McArdle, J. J., Bockting, C. L. H., & Berking, M. (2014). Successful emotion regulation skills application predicts subsequent reduction of symptom severity during treatment of major depressive disorder. *Journal of Consulting and Clinical Psychology, 82*, 248–262.

Ren, Z., Zhang, Y., & Jiang, G. (2018). Effectiveness of mindfulness meditation in intervention for anxiety: A meta-analysis. *Acta Psychologica Sinica, 50*, 283–305.

Segal, Z. V., Williams, J. M. G., & Teasdale, J. D. (2013). *Mindfulness-based cognitive therapy for depression* (2nd ed.). New York, NY: Guilford.

Taouk, M. M., Lovibond, P. F., & Laube, R. (2001). *Psychometric properties of a Chinese version of the short depression anxiety stress scales (DASS21)*. Sydney: Report for New South Wales Transcultural Mental Health Centre, Cumberland Hospital.

Thomas, R., Chur-Hansen, A., & Turner, M. (2020). A systematic review of studies on the use of mindfulness-based cognitive therapy for the treatment of anxiety and depression in older people. *Mindfulness*. https://doi.org/10.1007/s12671-020-01336-3.

Tsai, P. S., Wang, S. Y., Wang, M. Y., Su, C. T., Yang, T. T., Huang, C. J., & Fang, S. C. (2005). Psychometric evaluation of the Chinese version of the Pittsburgh Sleep Quality Index (CPSQI) in primary insomnia and control subjects. *Quality of Life Research, 14*(8), 1943–1952.

Twomey, C., O'Reilly, G., & Byrne, M. (2015). Effectiveness of cognitive behavioural therapy for anxiety and depression in primary care: A meta-analysis. *Family Practice, 32*, 3–15.

Wang, C. W., Chan, C. H. Y., Ho, R. T. H., Chan, J. S. M., Ng, S. M., & Chan, C. L. W. (2014). Managing stress and anxiety through qigong exercise in healthy adults: A systematic review and meta-analysis of randomized controlled trials. *BMC Complementary and Alternative Medicine, 14*, 8.

Wang, C. W., Chan, C. L. W., Ho, R. T. H., Tsang, H. W. H., Chan, C. H., & Ng, S. M. (2013). The effect of qigong on depressive and anxiety symptoms: A systematic review and meta-analysis of randomized controlled trials. *Evidence-Based Complementary and Alternative Medicine, 2013*, 716094.

Wielgosz, J., Goldberg, S. B., Kral, T. R. A., Dunne, J. D., & Davidson, R. J. (2019). Mindfulness meditation and psychopathology. *Annual Review of Clinical Psychology, 15*, 285–316.

Zhang, A., Bornheimer, L. A., Weaver, A., Franklin, C., Hai, A. H., Guz, S., & Shen, L. (2019). Cognitive behavioral therapy for primary care depression and anxiety: A secondary meta-analytic review using robust variance estimation in meta-regression. *Journal of Behavioral Medicine, 42*, 1117–1141.

Zhang, J. X., & Schwarzer, R. (1995). Measuring optimistic self-beliefs: A Chinese adaptation of the general self-efficacy scale. *Psychologia: An International Journal of Psychology in the Orient, 38*, 174–181.

Zou, L., Yeung, A., Quan, X., Hui, S. S. C., Hu, X., Chan, J. S. M., ... Wang, H. (2018). Mindfulness-based Baduanjin exercise for depression and anxiety in people with physical or mental illnesses: A systematic review and meta-analysis. *International Journal of Environmental Research and Public Health, 15*(2), 321.

Chapter 42

Case study: Mechanisms of change in cognitive-behavioral therapy for weight loss

Loana Comșa[a] and Oana David[a,b]

[a]Department of Clinical Psychology and Psychotherapy, Babeș-Bolyai University, Cluj-Napoca, Romania, [b]Data Lab: Digital Affective Technologies in Therapy and Assessment, Babes-Bolyai University, Cluj-Napoca, Romania

Abbreviations

BMI	body mass index
CBTWL	cognitive-behavioral therapy for weight loss
SBT	standard behavioral therapy
SCT	social cognitive theory
SDT	self-determination theory
WHO	World Health Organization

Introduction

According to World Health Organization, obesity is "an abnormal or excessive accumulation of fat that can be harmful to health" (WHO, 2021). This disease is one of the most critical health problems that affect people worldwide (Castelnuovo et al., 2017).

Weight loss is an important concern for the people and also for health specialists and thus, reducing the prevalence of obesity has clinical and economic implications (Swift et al., 2018). Weight loss is recommended for multiple health benefits, that already appear at a weight loss of only 2%–3% of initial weight (Swift et al., 2018). A higher decrease of 10% is recommended to improve obesity's comorbid conditions such as fertility, menstrual irregularities, mobility, depression, and quality of life (Ryan & Yockey, 2017).

Behavioral interventions are the most commonly used psychological treatments for weight loss and maintaining lost weight (Teixeira et al., 2010). The efficacy of behavioral treatments is about 10% of weight loss, but this loss is almost always regained (Cooper & Fairburn, 2001).

Cognitive-behavioral therapy is considered the first-line psychological intervention for obesity (Castelnuovo et al., 2017). Its efficacy lies in addition to behavioral treatments, of cognitive techniques such as cognitive restructuring, body image acceptance, weight expectations, problem-solving, and self-efficacy (Cooper & Fairburn, 2001). Recent meta-analyses have revealed that CBT interventions produce a statistically significant effect on weight loss (Comșa, David, & David, 2020; Jacob et al., 2018).

One of the most used cognitive-behavioral protocol interventions was developed by Cooper and Fairburn (2001). This intervention was designed to minimize weight regain by addressing psychological obstacles identified as reasons for failure in effective weight maintenance: weight loss goals and expectations, the abandonment of weight control, and the neglect of the need for active weight maintenance. The treatment is administered one-to-one basis and lasts for 44 weeks.

Intervention and treatment overview

Babes-Bolyai University Psychology Clinic provides psychological counseling and psychotherapy services to its students and employees. One of the most requested interventions in 2019 was weight loss. In order to support patients in need for weight loss, the clinical psychologists of this clinic aimed to adapt an effective CBT protocol to the needs of its clients, creating a shorter group intervention. Although reduced to 8 weeks, the new protocol retains most of the elements of the original protocol (Table 1).

This case study is taken from a program that aimed to develop a shorter, group version of CBTWL (Cooper et al., 2010) conducted by authors between May to July 2019 at the Babes-Bolyai University's Psychology Clinic. Participants were

TABLE 1 Sessions content.

Session	Session topic	Description
0	Pretreatment assessment	Treatment description Anthropological and psychological assessments Start of monitoring: food and physical activities diaries
1	Education	Monitoring observations—discussions about participants' diaries. Identify problematic eating behaviors and address the problems which may interfere with weight loss and maintenance Nutrition information: proteins, carbohydrates, lipids. Information about metabolism. The importance of dietary rules Physical activities: what can we do at home, at the gym, and in the park to facilitate weight loss? Weight loss goals and treatment expectancies Identify barriers. What can interfere with the application of the elements discussed today?
2	Helping participants to implement dietary rules and healthy eating behaviors; introducing cognitive factors: motivation, body image, self-acceptance	Monitoring observations—discussions about participants' diaries. Identify problematic eating behaviors and address the problems which may interfere with weight loss and maintenance Identify barriers. What can interfere with the application of the elements discussed in session 1? Motivation: What are the reasons I would lose weight? Body image—and what if I am not perfect? What does body shape mean to me? What does weight mean to me? Self-acceptance
3	Helping participants to implement dietary rules and healthy eating behaviors; introducing cognitive and behavioral factors: assertiveness, emotional eating, planning, mindful eating Address barriers to weight loss behaviors	Monitoring observations—discussions about participants' diaries. Identify problematic eating behaviors and address the problems which may interfere with weight loss and maintenance • Social situations—assertiveness (it is my body, no one dictates to me) what else can we choose for healthy eating? • Emotional eating—Now I have a problem. Should I make them two? • Planning—how many and how they plan what to eat for the next day? Identify barriers. What can interfere with the application of the elements discussed in session 1? Planning—limiting activating situations Mindfulness eating Motivation—opening FB group
4	Helping participants implement dietary rules and healthy eating behaviors	Monitoring observations—discussions about participants' diaries. Identify problematic eating behaviors and address the problems which may interfere with weight loss and maintenance • Social situations—assertiveness (it is my body, no one dictates to me); what other food can we choose? • Emotional eating—these are two problems; list of adaptive behaviors in case of emotional problems • Planning—how many and how they planned what to eat for the next day • Why do I have food around me that I do not want to eat? Identify barriers. What can interfere with the application of the elements discussed in session 1? Planning—limiting activating situations Mindfulness eating Motivation—feeling good when we made the right decisions Self-acceptance Physical activity Self-efficacy

TABLE 1 Sessions content—cont'd

Session	Session topic	Description
5	Helping participants implement dietary rules and healthy eating behaviors; introducing problem-solving abilities and cognitive restriction	Monitoring observations—discussions about participants' diaries. Identify problematic eating behaviors and address the problems which may interfere with weight loss and maintenance • Diversified and sufficient food • What are the negative elements they experienced? What was hard? Identifying barriers. What can interfere with the application of the elements discussed in session 1? Planning—limiting activating situations: I only eat in certain places, I do nothing when I eat, I leave the table after eating, shopping only with a list Mindfulness eating Motivation Self-acceptance Physical activity Self-efficacy Problem solving—food management — if I ate something caloric now, I would recover later — if I have no choice, I work on quantities — Identifying thoughts—What do I think of when I choose to eat this? • Planning—where I am tomorrow and what I eat: restaurant, home, canteen, in social situations
6	Helping participants implement dietary rules and healthy eating behaviors; introducing problem-solving abilities and cognitive restriction	Monitoring observations—discussions about participants' diaries. Identify problematic eating behaviors and address the problems which may interfere with weight loss and maintenance • Planning—how many and how they planned what to eat for the next day • What are the negative elements they experienced? What was hard? Identifying barriers. What can interfere with the application of the elements discussed in session 1? Planning—limiting activating situations Mindfulness eating Motivation Self-acceptance Physical activity Self-efficacy Problem solving—food management — If I ate something caloric now, I will recover later — if I have no choice, I work on the quantities — Identifying thoughts—What do I think of when I choose to eat this? • Planning—where I am tomorrow and what I eat: restaurant, home, canteen, in social situations
7	Help participants implement dietary rules and healthy eating behaviors; introduce cognitive element—low frustration tolerance	Monitoring observations—discussions about participants' diaries. Identify problematic eating behaviors and address the problems which may interfere with weight loss and maintenance • Planning—how many and how they planned what to eat for the next day • What are the negative elements they experienced? What was hard? • Did the fact that you did differently change the quality of the interaction? Were you proud? • When we have emotional problems, we have to do certain things because they are on the list, not because we have the necessary mood or not Identifying barriers. What can interfere with the application of the elements discussed in session 1? Planning—limiting activating situations; I only eat in certain places; when I eat, I do nothing else; I leave the table after eating; I go shopping only with a list Mindfulness eating Motivation Self-acceptance

Continued

TABLE 1 Sessions content—cont'd

Session	Session topic	Description
		Physical activity Self-efficacy Problem solving—food management - If I ate something caloric now, I'll recover later - if I have no choice, I work on the quantities - Identifying thoughts—What do I think of when I choose to eat this? • Planning—where I am tomorrow and what I eat: restaurant, home, canteen, in social situations
8	Helping participants implement dietary rules and healthy eating behaviors; identifying future barriers Recapitulation of received information and implemented behaviors Weight maintenance	Monitoring observations—discussions about participants' diaries. Identify problematic eating behaviors and address the problems which may interfere with weight loss and maintenance • Eating carbs in the evening—in the long run, this does not lead to maintenance • Planning—how many and how did they plan what to eat for the next day? Rules for what they will have to do when they get out of the routine/they do not have the right choices • What are the negative elements they experienced? What was hard? • When we have emotional problems, we have to do certain things because they are on the list, not because we have the necessary mood or not Identify barriers. What can interfere with the application of the elements discussed so far? What can happen so that they do not follow the rules? What changes occur in their lives in the short term? Planning—limiting activating situations Mindfulness eating Motivation Self-acceptance Physical activity Self-efficacy Problem solving—food management - If I ate something caloric now, I will recover later - if I have no choice, I work on quantities - Identifying thoughts—What do I think of when I choose to eat this? • Planning—where I am tomorrow and what I eat: restaurant, home, canteen, in social situations

The description of the weight loss program.

students or employees of the University, and the weight loss program was provided at no cost to them. To be included in the program, participants had to be ≥ 18 years, not enrolled in other forms of weight loss therapies, and reported that they do not have any known health and clinical problems or take medicines that can affect weight loss. This information was taken at the phone screening with one of the co-therapists, a trained psychologist, who was a master's level student in psychology. The participation consent, the psychological assessments questionnaires, and their demographic details: age, height, weight, and contact details were given by the participants online.

The intervention protocol consisted of the shortened and group-adapted version of a well-known CBT protocol (Cooper & Fairburn, 2001). In this group were enrolled, six participants. Before the group sessions, all participants attended an individual session with a clinical psychologist, the first author. The purpose of this session was to identify the participants' expectations related to therapy, provide details about the program, and start creating the therapeutic alliance. Next, for 7 weeks, 120 minutes of group sessions were administered. At the start of each session, the participants were weighed. At the beginning of the program and the end, participants sent their online psychological assessments. Group sessions were led by the first author, a clinical psychologist, and a master's level student in psychology trained in CBT. The content and topic of the sessions are described in Table 1.

Case example

In this case study, we call our client, Susan, and the identifying information has been altered so that the person cannot be identified to protect the client's identity. This client was selected for this case study because of her background in weight loss attempts. In terms of her cognitive and weight loss results, we considered her a successful case. She was a female participant who was employed at the university where the weight loss program was conducted. Susan was 38 years old, married, with a small child. At the individual session, Susan reported she eats many carbs, and everything remained uneaten from her child. The patient reported a long history of weight loss therapies with nutritionists' lack of empathy and understanding of her struggle. At the beginning of the program, Susan had 114.8 kg and 170 cm in height and stated she did not take any medication.

Bodyweight, fat percentage, and total body water percentage were measured using a Tanita Digital Scale (UM-030) at both assessment time points (baseline and posttreatment).

Eating self-efficacy was measured with the Weight Efficacy Lifestyle Questionnaire—short form (WEL-SF) (Ames, Heckman, Grothe, & Clark, 2012). It is a psychometrically valid measure of self-efficacy for controlling eating and has eight items. The response is based on a Likert scale in 0–10 points: 0 for "Not confident at all" and 10 indicates "Very confident." Higher total scores are associated with higher eating self-efficacy and motivation to make positive lifestyle changes. Cronbach's α for this scale is 0.92.

The irrational food beliefs were measured with the Irrational Food Beliefs Scale (IFB), (Osberg, Poland, Aguayo, & MacDougall, 2008), with 57 items, of which 41 measure irrational beliefs about food and 16 rational ones. The scale has good reliability for both irrational (Cronbach's α = 0.89) and rational (Cronbach's α = 0.70) food beliefs. The responses are made using Likert-type ratings from 1—strongly disagree to 4—strongly agree. The highest scores are significantly associated with weight gain and poor weight loss maintenance.

We used the 21-item Three-Factor Eating Questionnaire revised version (TFEQ-R21) (Cappelleri et al., 2009; Tholin, Rasmussen, Tynelius, & Karlsson, 2005), to measure eating behaviors. The scale measures three aspects of eating behavior: uncontrolled eating (UE), cognitive restraint (CR), and emotional eating (EE). The responses are based on a four-point Likert scale ranging from 1 Definitely true to 4 Definitely false. It has three subscales: one for each type of eating behavior. Higher scores indicate greater UE, CR, or EE.

Susan's results

At the first session, Susan's weight was 114.8 kg; her fat percentage was 46.6%, and her total body water percentage was 39.2%. Her initial BMI was 37.5 kg/m^2.

After eight sessions of CBT group therapy, Susan's uncontrolled eating had decreased. She became more vigilant in response to her increased cognitive restraint. She was also eating emotionally at a lower frequency. The cognitive and behavioral assessments are detailed in Table 2. Regarding cognitive outcomes, CBT also had a beneficial effect on Susan since the level of her irrational food beliefs decreased. Table 3 reveals the weight and fat percentage differences from the beginning of the program until the end, on week 8.

TABLE 2 Susan's cognitive and behavioral assessments.

Time point	Eating self-efficacy	Irrational food beliefs	Uncontrolled eating (%)	Cognitive restraint (%)	Emotional eating (%)
Week 0	75	86	66	61	39
Week 8	77	79	59	66	22

The assessments of self-efficacy, irrational food beliefs, and eating behaviors: uncontrolled eating, cognitive restraint, and emotional eating.

TABLE 3 Susan's weight and fat percentage assessments.

Weight week 0	Weight week 8	Fat percentage baseline	Fat percentage week 8
114.8 kg	110.4 kg	46.6%	40.4%

The weight and fat percentage differences from the beginning of the program until the end, on week 8.

Intervention—Sessions 1–8

In session 1, Susan reported her unsuccessful attempts at weight loss, and she constantly expressed her satisfaction with the topic of discussions in this group session. In session one focus was on education about healthy nutrition and physical activity. A personal trainer taught participants what exercises are good for weight loss at the home, gym, or in the park. Also, in session one, the clinical psychologist started to analyze the participant's food and physical activity diaries. Susan's food journal indicates she eats a lot of high-calorie foods in front of the TV, all through the house, and when feeding her child and her physical activity diaries showed she was doing exercise daily. Her first assessment reveals that she frequently eats uncontrolled despite her attempts to restrict herself.

In session 2, Susan reported she does not have any problems with her body image or accepting herself and identifies her barriers to implementing healthy eating behaviors: her child always leaves food, the grandmother only cooks goodies, and she comes tired from work.

During sessions 3–7, Susan actively participated in the Facebook group created. She learned to plan what to eat the next day, limit difficult situations, implement healthy behaviors, and develop problem-solving skills. Susan completed all food and physical activity diaries and was not absent from any sessions throughout the program.

Susan said that changing her beliefs about food and eating helps her the most. She also experienced increased eating self-efficacy. One of her great perceived successes was that she was not eating what was left of her child, thinking differently about food. One of these differences was related to the fact that eating is not always good and that food is not a source of rewards. At the same time, Susan learned the belief that she can go to the canteen with her colleagues without eating high-calorie foods, that she can choose something low-calorie without the meeting with them decreasing in quality. Moreover, she also noticed that when she has healthy eating behaviors, she feels proud and happy.

In the last meeting, Susan discussed the changes that will occur in her short-term life: changing her actual job and her kid going to kindergarten. He identified possible barriers to the long-term implementation of the elements learned in this program.

From sessions 1 to 8, Susan's weight decreased 3.4% from her initial weight, and her body fat percentage decreased 13%. Also, Susan experienced a decrease in the level of her irrational food beliefs and problematic eating behaviors such as uncontrolled eating and emotional eating. She increased her cognitive restraint and became more careful about what, when, and where she eats during this period.

Discussion

In this study case, we presented the case of Susan, a participant in a weight loss program that aimed to adapt a well-known cognitive-behavioral program to a shorter group version. We considered that Susan successfully completed the weight loss program due to her results. She lost enough weight to obtain health benefits since a 3% loss of the initial weight is considered benefits already appear (Swift et al., 2018). She also experienced a decrease in fat percentage, dysfunctional thoughts and attitudes related to food, and decreased frequency of problematic eating behaviors.

Summary points

- The patient lost enough weight to obtain health benefits.
- Cognitive-behavioral therapy was effective on losing weight.
- Problematic eating behaviors have improved following CBT intervention.
- The client's irrational food beliefs decreased as a result of the CBT intervention.
- The client's self-efficacy increased because of the CBT intervention.
- The program was a shorter version adapted to group therapy of an evidence-based CBT protocol.

References

Ames, G. E., Heckman, M. G., Grothe, K. B., & Clark, M. M. (2012). Eating self-efficacy: Development of a short-form WEL. *Eating Behaviors, 13*(4). https://doi.org/10.1016/j.eatbeh.2012.03.013.

Cappelleri, J. C., Bushmakin, A. G., Gerber, R. A., Leidy, N. K., Sexton, C. C., Lowe, M. R., & Karlsson, J. (2009). Psychometric analysis of the Three-Factor Eating Questionnaire-R21: Results from a large diverse sample of obese and non-obese participants. *International Journal of Obesity, 33*(6). https://doi.org/10.1038/ijo.2009.74.

Castelnuovo, G., Pietrabissa, G., Manzoni, G. M., Cattivelli, R., Rossi, A., Novelli, M., … Molinari, E. (2017). Cognitive behavioral therapy to aid weight loss in obese patients: Current perspectives. *Psychology Research and Behavior Management, 10*, 165–173. https://doi.org/10.2147/PRBM.S113278.

Comşa, L., David, O., & David, D. (2020). Outcomes and mechanisms of change in cognitive-behavioral interventions for weight loss: A meta-analysis of randomized clinical trials. *Behaviour Research and Therapy*, *132*. https://doi.org/10.1016/j.brat.2020.103654, 103654.

Cooper, Z., Doll, H. A., Hawker, D. M., Byrne, S., Bonner, G., Eeley, E., ... Fairburn, C. G. (2010). Testing a new cognitive behavioural treatment for obesity: A randomized controlled trial with three-year follow-up. *Behaviour Research and Therapy*, *48*(8), 706–713. https://doi.org/10.1016/j.brat.2010.03.008.

Cooper, Z., & Fairburn, C. G. (2001). A new cognitive behavioural approach to the treatment of obesity. *Behaviour Research and Therapy*, *39*(5), 499–511. https://doi.org/10.1016/S0005-7967(00)00065-6.

Jacob, A., Moullec, G., Lavoie, K. L., Laurin, C., Cowan, T., Tisshaw, C., ... Bacon, S. L. (2018). Impact of cognitive-behavioral interventions on weight loss and psychological outcomes: A meta-analysis. *Health Psychology*, *37*(5), 417–432. https://doi.org/10.1037/hea0000576.

Osberg, T. M., Poland, D., Aguayo, G., & MacDougall, S. (2008). The Irrational Food Beliefs Scale: Development and validation. *Eating Behaviors*, *9*(1), 25–40. https://doi.org/10.1016/j.eatbeh.2007.02.001.

Ryan, D. H., & Yockey, S. R. (2017). Weight loss and improvement in comorbidity: Differences at 5%, 10%, 15%, and over. *Current Obesity Reports*, *6*(2), 187–194. https://doi.org/10.1007/s13679-017-0262-y.

Swift, D. L., McGee, J. E., Earnest, C. P., Carlisle, E., Nygard, M., & Johannsen, N. M. (2018). The effects of exercise and physical activity on weight loss and maintenance. *Progress in Cardiovascular Diseases*, *61*(2), 206–213. https://doi.org/10.1016/j.pcad.2018.07.014.

Teixeira, P. J., Silva, M. N., Coutinho, S. R., Palmeira, A. L., Mata, J., Vieira, P. N., ... Sardinha, L. B. (2010). Mediators of weight loss and weight loss maintenance in middle-aged women. *Obesity*, *18*(4), 725–735. https://doi.org/10.1038/oby.2009.281.

Tholin, S., Rasmussen, F., Tynelius, P., & Karlsson, J. (2005). Genetic and environmental influences on eating behavior: The Swedish Young Male Twins Study. *American Journal of Clinical Nutrition*, *81*(3). https://doi.org/10.1093/ajcn/81.3.564.

WHO. (2021). *Obesity: Prevention and control*. https://www.who.int/health-topics/obesity#tab=tab_3.

Chapter 43

CASE STUDY: Cognitive-behavioral therapy for Japanese Bipolar II disorder patients

Yasuhiro Kimura
Department of Welfare Psychology, Faculty of Welfare, Fukushima College, Fukushima, Japan

Abbreviations

BD   bipolar disorder
BD2 bipolar II disorder
CBT cognitive-behavioral therapy
QOL quality of life

Introduction

This chapter provides examples of the difficulties in social adaptation faced by Japanese patients with bipolar II disorder (BD2) and the practical use of cognitive-behavioral therapy (CBT) in its treatment.

It has been pointed out that patients with BD have low self-esteem and quality of life (QOL) even if their symptoms remit, resulting in difficulties in social life such as creating meaningful work and interpersonal relationships (Blairy et al., 2004; Sanchez-Moreno et al., 2009; Sierra, Livianos, & Rojo, 2005). In a survey of patients with BD in Japan, approximately 30% were unemployed (Watanabe, Harada, Inoue, Tanji, & Kikuchi, 2016). The goals of the treatment desired by Japanese patients included being able to work and study at their own pace (36%), being able to do something in their daily life (19%), and being able to enjoy hobbies (15%) (Watanabe, Harada, Inoue, Tanji, & Kikuchi, 2016). Thus, it is important to make efforts toward helping patients with BD achieve social adaptation through CBT. In particular, this chapter describes example of how to use patient strengths in interventions. The cases shown in this chapter amalgamated from several cases.

Case 1

Patient information and visit history

Hana was a 33-year-old Japanese woman. The chief complaint that she had was that she repeatedly quit her job and wanted to change her way of thinking. Although she had a pharmacist license, she was unemployed at the beginning of therapy and frequently visited a support center for employment for handicapped people. After graduating from university, she had left eight workplaces in about 10 years, and was unable to continue working stably. For this reason, Hana was recommended by the staff at the support center for employment for handicapped people to receive psychotherapy in order to understand her cognitive pattern. At a psychiatric clinic, she was administered lithium carbonate (600 mg), aripiprazole (1 mg), and venlafaxine hydrochloride (37.5 mg) as outpatient pharmacotherapy.

Growth history and clinical history

Hana was born the second child of three siblings. Hana was lived with her grandparents, parents, and siblings. Her parents hardly interfered with her life because she was the second child. However, her grandparents wanted her to be the best at everything. Hana tried to live up to her grandparents' expectations and had excellent grades up till high school. After entering college, Hana began to feel anxious about whether she was inferior to the others. At the age of 22, Hana was so busy with graduation research and qualification tests that she could not eat, so she visited a psychiatrist and was diagnosed with an adjustment disorder. After graduating, Hana found a job at the hospital of her choice, but she left after a week because she felt she was inferior to those around her. Hana soon got a new job, but she quit after about a week for the same reason. Hana visited another psychiatric clinic and was diagnosed with depression. After that, Hana worked part-time at another hospital for 3 months, but as soon as she became a regular employee, she felt heavy weight of responsibility and resigned. When Hana started working the night shift at her next job, her life rhythm became irregular, and her sleeping hours decreased. She started shopping at higher prices after work and looking for a boyfriend on the Internet. She was gradually unable to continue working and she quit. The treatment for depression continued during this period. At the age of 31, Hana did not get along with her attending psychiatrist, and there were no signs of improvement. As a result, Hana changed her psychiatrist and was diagnosed with bipolar II disorder based on previous hypomanic and current depressive episodes. Although her pharmacotherapy was changed and the mood was stabilized, the problem of adaptation to work did not show improvement. She was, therefore, referred to a support center for employment for handicapped people, and the employment support program there was being attended five times a week.

Case formulation

Hana had never felt a sense of accomplishment before and she was constantly pushing herself to aim for the best. Also, Hana was not able to accept her failures and worried about what other people would think of her. Simultaneously, she was often irritated by others. This was thought to be due to the many "should" thoughts which she said were common sense. Table 1 shows an example of Hana's idea of common sense or "should" thinking, because of which she was only able to focus on negative aspects of herself and others. She could not change her mind when there was one thing she did not like about herself or others. As a result, Hana repeatedly chose to quit her job rather than continue working. On the other hand, Hana was not aware of the problems with her own thinking, until it was pointed out by the staff at the support center for employment for handicapped people.

Intervention

The therapist proposed self-monitoring to Hana to understand herself cognitive patterns. Fortunately, Hana's habit of keeping a diary came up in the conversations with her therapist. Hana's diary was full of worry, incompetence, dislike, and anger. Hana tended to look at her diary entries over the weekends and speculate on the causes of her difficulties. Therefore, the therapist suggested that she write a diary with a frame. In other words, her homework was to write both negative and positive things in her diary. The following are the scripts of the conversation between the therapist and Hana during the sessions following the introduction of self-monitoring.

Therapist: How did you like your homework of changing the way you write your diary?
Hana: When I changed the way I write my diary, I focused on writing down five good things for one bad thing. I realized that there is always something better than what I had focused on. Also, I had thought I was not being able to do anything, but I

TABLE 1 Examples of Hana's "should" thoughts.

I should be acknowledged to do my best
I should make the best choice
I should not fail
My "normal" should be understood
I have to be motivated all the time
Everything should go according to plan
If I am not working, I do not deserve to live
I must not be hated by others
This table is part of Hana's "should" thoughts.

think I was able to do more than I had expected. On the other hand, I realized that even if only one bad thing happened to me, I have a tendency to drag it on.

Therapist: I think this is a great discovery. You noticed that there was something better than what you had focused your thoughts on.

Six months later.

Therapist: Has anything changed since the last time?

Hana: There has not been any big change, but the bad feelings do not last. The time that I am obsessed with them has reduced, and if I write them down in the diary, that was usually the end of I think about it.

Therapist: That is a good trend. How are you feeling?

Hana: Yes. So far, it has been very comfortable. It is strange, but I do not think that there has been a big change in my way of thinking or something.

Therapist: I think focusing on the overall balance means that you are no longer obsessed with negative things. What is the reason for the ease that you feel?

Hana: I think it is the direction of consciousness. It is not something that cannot be changed immediately, but it is something that can be changed, so there is hope for future improvement.

By changing the way she wrote in her diary, Hana began to focus on her positive aspects in what she was doing. The result was a shorter period of negative mood. Hana began to say, "I can do it if I do it," and she started taking an active part in the things she could do, such as housework. Hana began to develop confidence and hope for reemployment. She was rehired 6 months after the end of the therapy.

Case 2

Patient information and visit history

Kumi was a 24-year-old Japanese woman. She came to the therapist with a chief complaint that she wanted to change her thinking that she was a woman of no confidence, no use, and no value. She was clerical employee on a short-term contract. She never had any bad experiences at work, but she cried every morning and was often absent from work because of fatigue. She was thinking about quitting her job. She had quit her previous job as well, and thought the same thing would happen again and again if she did not change her mind. She was receiving 400 mg of sodium valproate and 80 mg of atomoxetine hydrochloride as an outpatient medication at a psychiatric clinic. She consulted with her psychiatrist, was introduced to the therapist, and came for the therapy sessions.

Growth history and clinical history

Kumi lived with her parents. In the family, the father was the main decision-maker, and the mother was constantly worried and overprotective. Kumi had never been scolded by her parents. When she was in junior high school, she started to vaguely think that she had to be liked by the people around her. From junior high to high school, Kumi had a few friends that she could call close friends. However, she was unable to express her opinions and often followed her friends' opinions. She was not bullied, and her grades were excellent. Kumi had no particular problems after entering college. After graduating from college, Kumi got a job in car sales. Her office was busy, though friendly. Kumi thought that she needed to be recognized by those around her as soon as possible, and even though she was not instructed to do so, she overworked, taking work home, and even working at home on holidays. Kumi felt she could do anything when she felt good, so her activities increased, she started to move around busily, and her desire to sleep decreased. After that, she began to cry more often at home and at work, and was diagnosed with bipolar II disorder at the psychiatry clinic recommended to her by her family. Kumi left 6 months after joining the company, and she felt anxious about not working, so she immediately started working for another company. Kumi was often absent from that job too, but she worked there for a year, and then moved to her current job.

Case formulation

Kumi cried every day on her workdays and was absent almost once a week. Kumi had low self-esteem and thought that she did not want everyone to hate her. Kumi wanted to work as long as possible, but she was concerned about receiving negative feedback from people in the company. So, even though her boss told her not to push herself too hard, Kumi continued worked without thinking about her own burden, such as taking work home. Kumi also thought it was natural to change her behavior depending on her mood, so she had a problem deciding whether to go to work or not depending on her mood for the day. As a result, when she missed work, she felt depressed because she thought she was a useless person.

Intervention

The therapist shared that Kumi destabilized her mental and condition because of the burden of her job due to short-term anxiety. The contradiction that previous patterns of behavior, even if were evaluated in the short term, were not evaluated in the long term due to absence or early quitting was also discussed. Kumi gradually began to avoid taking her work home. In addition, by scheduling target behavior, she decided to stop changing her life according to her mood and to stabilize her behavior over the long term. Below is the script from the introductory session.

Kumi: I would like to continue working if possible. But I am anxious, so I get impatient.
Therapist: You know you're trying to overdo it. However, you only care about short-term evaluations.
Kumi: Yes. I am very concerned about what people think at that very moment.
Therapist: Do you have a long-term goal?
Kumi: No, I do not. I thought that if I could achieve our short-term goals, I could naturally achieve my long-term goals, but that was not the case.
Therapist: First, you have a long-term goal and then a short-term goal.
Kumi: Do you mean I do not go too far when I am in a good mood?
Therapist: Yes, I do. Behavior during good times and behavior during bad times are both important.
Kumi: Because if I do too much, I will regret it later.
Therapist: That's what long-term means. When you look back at a moment, you feel different.
Kumi: The day before yesterday, I found a job opening at my previous job, so I contacted my acquaintances and people I had stopped contacting after I left my job. After that, I calmed down and regretted that I had done something unnecessary.
Therapist: On the other hand, it can happen to people when feel sick.
Kumi: I feel like quitting my job.
Therapist: Imagine yourself a year later. What would you think of it a year from now?
Kumi: What a stupid thing I did.
Therapist: Yes. At that moment, it may not be strange, but 1 year later, you feel regret and your self-esteem goes down.
Kumi: I see.
Therapist: The goal is long-term stability. Your goal is to continue working after a year or to be able to live stably to some extent. Think about your daily goals based on your goals for the next year.
Kumi: If my goal is to continue working in 1 year, is it not difficult to achieve my daily goals? I am worried if I can act on my goals when I do not feel well, even if I set a goal for the day.
Therapist: That is a very good question. What do you think will happen if you can't achieve your goal for the day?
Kumi: In my case, when I fail to achieve my goals, I blame myself more and more.
Therapist: I guess so. The target behavior should be a plan you can follow when you are not feeling well.
Kumi: Then they might not do anything.
Therapist: Really?
Kumi: I might at least listen to music.
Therapist: That is right.
Kumi: To tell you the truth, I was planning to do something similar recently, like listening to music from work to the station every day, and watching my favorite anime when I got home, but I was too busy that I forgot to do it without realizing it.
Therapist: I think that is a very good approach. Take advantage of your plan and think about your target behavior for the day. But let's start with something you can do even if you are busy.
Kumi: That means only listening to music on the way home. Is that all?
Therapist: That is all. Continue it for a certain period of time. Also, when you set a target behavior, do not include criteria that you often think about, such as being able to concentrate or being able to move well. You cannot do that when you are not feeling well. What is the least you can do even if you feel bad?
Kumi: Well, I can go to work.
Therapist: The target behavior is to go to work. Even if you go to work and do not do anything, you should evaluate it as achieving your goal.
Kumi: Do I not include my performance in my evaluation?
Therapist: No problem. If you plan to go to work, you should evaluate it as a success when you go.
Kumi: All right.

TABLE 2 Kumi's daily target behaviors.
I wake up every morning at a fixed time
I go to work
I relax by listening to music after work
I practice cognitive restructuring
This table is a list of target behaviors for Kumi on work days.

After the conversation, the goals for the day were set. Table 2 lists Kumi's target behavior. The therapist asked Kumi not to include in her daily evaluation anything that was not listed in the target behavior. Kumi checked the list every time she completed a target behavior, but if she could do all of them, she would get to reward herself with expensive chocolates, and after 15 completions of target behavior, she would get something fancy to drink in her favorite coffee shop.

Since the standards for success became clear, Kumi started to work regardless of her mood or physical condition, and her attendance rate became stable. One month after the intervention, she went to work day after day and cried only 1 day in the morning. Kumi gained confidence by continuing to achieve her goals, began to evaluate herself, and began to adapt to her work. Kumi later extended the term of her employment contract twice and was able to work at the same job for 2 years.

Discussion

These two cases illustrate the use of CBT in Japanese patients with BD2 who have difficulty adjusting to society.

Through self-monitoring, Hana was able to look at the positive side and feel confident. Hana kept a daily diary before she came to the therapy, so she was able to deal with self-monitoring smoothly. When she looked back on her psychotherapy, Hana said she was very happy that the therapist had incorporated what she had already been doing into her therapy, rather than forcing something new on her.

Kumi has been able to work stably without being swayed by her mood by scheduling her target behavior. In addition, by setting her goal at the behavioral level, she gained confidence by clarifying whether she achieved her goals or not. According to Kumi, this intervention was the most useful one in her psychotherapy sessions. The target behaviors was determined taking into account the fact that Kumi had already planned the contents.

While it is desirable for therapists to provide interventions in CBT, it is the patient who ultimately implements the intervention. The usability of the intervention for the patient is an important factor. Focusing on a patient's resources and strengths increase the success rate of the intervention. It is easier to intervene using actions the patient is already performing or aspiring for than to teach techniques that are not in the patient's behavioral repertoire.

The dysfunctional thoughts and behaviors of BD patients prevent them from achieving their desired therapeutic goals. These vulnerabilities can affect work and relationships, even when the symptoms were stable. Improvement in self-efficacy is associated with improvement in the QOL of BD patients (Abraham, Miller, Birgenheir, Lai, & Kilbourne, 2014) and is a desirable outcome. For the Japanese BD patient, it is important for the improvement in the QOL to adapt to work and interpersonal relationships, and the acquisition of specific coping skills by CBT may be one way to contribute to the social adaptation of patients with BD.

Summary points

- BD patients face low self-esteem and quality of life that can be improved by using CBT.
- This chapter provides examples of the difficulties faced by Japanese patients with BD2 in social adaptation and the practical use of CBT.
- In the first example, the use of self-monitoring that focuses on the positive aspects of the self gave rise to hope for reemployment.
- The second example is a case in which stable attendance was achieved by scheduling the target behavior.
- Focusing on patient resources and strengths reduces the difficulty of implementing interventions.
- Acquisition of specific coping skills by CBT can contribute to the social adjustment of patients with BD.

References

Abraham, K. M., Miller, C. J., Birgenheir, D. G., Lai, Z., & Kilbourne, A. M. (2014). Self-efficacy and quality of life among people with bipolar disorder. *The Journal of Nervous and Mental Disease, 202*, 583–588.

Blairy, S., Linotte, S., Souery, D., Papadimitriou, G. N., Dikeos, D., Lerer, B., … Mendlewicz, J. (2004). Social adjustment and self-esteem of bipolar patients: A multicentric study. *Journal of Affective Disorder, 79*, 97–103.

Sanchez-Moreno, J., Martinez-Aran, A., Tabares-Seisdedos, R., Torrent, C., Vieta, E., & Ayuso-Mateos, J. (2009). Functioning and disability in bipolar disorder: An extensive review. *Psychotherapy and Psychosomatics, 78*, 285–297.

Sierra, P., Livianos, L., & Rojo, L. (2005). Quality of life for patients with bipolar disorder: Relationship with clinical and demographic variables. *Bipolar Disorders, 7*, 159–165.

Watanabe, K., Harada, E., Inoue, T., Tanji, Y., & Kikuchi, T. (2016). Perceptions and impact of bipolar disorder in Japan: Results of an internet survey. *Neuropsychiatric Disease and Treatment, 12*, 2981–2987.

Chapter 44

Treating social anxiety with the MISA program: A case study

Isabel C. Salazar and Vicente E. Caballo
Department of Personality, Assessment and Psychological Treatment, Faculty of Psychology, University of Granada, Granada, Spain

Abbreviations

APA	American Psychiatric Association
APD	avoidant personality disorder
AUDIT	Alcohol Use Disorders Identification Test
BDI-II	Beck Depression Inventory-II
CBT	cognitive-behavioral therapy
CSISA	Clinical Semi-Structured Interview for Social Anxiety
DSM-5	Diagnostic and Statistical Manual of mental disorders—5th edition
EQ	Experiences Questionnaire
ICD-11	International Classification of Diseases 11th revision
LSAS-SR	Liebowitz Social Anxiety Scale-Self-report
MISA	Multidimensional Intervention for Social Anxiety
PSQ	Personal Sensibility Questionnaire
PSWQ	Penn State Worry Questionnaire
SAD	Social anxiety disorder
SAQ	Social Anxiety Questionnaire for adults
SOSAQ	Social Skills Assessment Questionnaire
SST	social skills training
WHO	World Health Organization
WHOQOL-Bref	World Health Organization Quality of Life-Bref

Case presentation

James, a 40-year-old male, has been living with his partner for 6 years. He comes to the clinic stating that he has been having a very difficult time for many years due to anxiety when interacting with other people and having to expose himself in public.

Anxiety arises when having to interact with strangers and acquaintances (e.g., former college classmates/teachers, neighbors, relatives). He identifies that he has greater difficulty when he is physically tired, e.g., from overwork or not sleeping well the night before, or when the interaction is one-to-one. With respect to women (especially if he finds them attractive), he also experiences anxiety. He claims to have it even worse with his friends' partners, and his thoughts turn to what both women and his friends might think about him. At work, anxiety occurs when he has to deal with colleagues, bosses and customers. James is an engineer, works in a family business, and his older brother is his boss. The anxiety worsened from a year ago when he was appointed on behalf of the company at professional events and dealing directly with clients.

In all situations he experiences shortness of breath, choking, acceleration of the heart, trembling of hands and legs. Sometimes, he blushes and his voice trembles. He comments that the worst thing is that people notice his symptoms

---

☆ This study is part of the I+D+i project with reference RTI2018–093916-B-I00 funded by MCIN/AEI/10.13039/501100011033/ and FEDER "A way of doing Europe." Financial assistance from the Foundation for the Advancement of Behavioral Clinical Psychology (FUNVECA) is also acknowledged.

and his nervousness. He notes that he has difficulty looking in the eyes and takes off his glasses to avoid seeing the other person. When he walks down the street and sees someone he knows, he makes a detour and if he attends meetings, he is anxious to leave as soon as possible. For oral presentations in his work, he takes beta-blockers.

James starts to feel nervous hours (and even days) before some social events, he rushes the situation over and over in his head and tries to think of stock phrases (or on specific topics) to use in his conversations. He considers himself an introvert and very critical of himself, and as time goes by, he is feeling sadder and more ashamed for "getting the way he does" for what he says "is nonsense," for not being able to relate calmly to anyone, and for being "stuck" professionally.

Anamnesis

Anxiety problems began in adolescence, when he was about 14 years old, but he reports that they have increased over time. He began having problems interacting with strangers and speaking in public; he says he felt "weird" and thought others made fun of his appearance; he says he sought to socialize but was unable to do so. In the past he used alcohol and paroxetine to manage anxiety.

Previous treatments

James received pharmacological treatment (paroxetine) for anxiety for 3 years and stopped 10 years ago. During this time his physical symptoms decreased, and he had a number of disinhibited behaviors (e.g., excessive alcohol and drug use, indiscriminate sexual intercourse), but when he stopped treatment, his previous symptoms reappeared, and new ones were added. Subsequently, she attended Gestalt therapy and had the feeling that he could be happy even though his symptoms remained.

Assessment and diagnosis

The goal of the assessment was to understand the nature of James' problems. This included identifying the clinical picture, operationalizing the problematic responses (cognitions, emotions, and behaviors), as well as the personal and environmental factors that are maintaining them.

Information was gathered through the *Clinical Semi-Structured Interview for Social Anxiety* (CSISA; Salazar & Caballo, 2017), the *Social Anxiety Questionnaire for adults* (SAQ; Caballo et al., 2012, Caballo et al., 2015; Caballo, Salazar, Arias, et al., 2010; Caballo, Salazar, Irurtia, et al., 2010), the *Liebowitz Social Anxiety Scale-Self Report* (LSAS-SR; Liebowitz, 1987), the *Social Skills Assessment Questionnaire* (SOSAQ; Caballo et al., 2017), the *Penn State Worry Questionnaire* (PSWQ; Meyer, Miller, Metzger, & Borkovec, 1990), the *Beck Depression Inventory-II* (BDI-II; Beck, Steer, & Brown, 1996), the *Personal Sensitive Questionnaire* (PSQ; Caballo & Salazar, 2019), the *Alcohol Use Disorders Identification Test* (AUDIT; Babor, Higgins-Biddle, Saunders, & Monteiro, 2001), the *Rosenberg Self-esteem Scale* (RSES; Rosenberg, 1965), the *World Health Organization Quality of Life-Bref* (WHOQOL-Bref; WHOQOL Group, 1996), the *Experiences Questionnaire* (EQ; Fresco et al., 2007), and the *Self-monitoring for anxiety, fear, shame, or nervousness in social situations* (Caballo, Salazar, & Garrido, 2018). Tables 1 and 2 summarize the results obtained.

Regarding SAD, the clinical diagnosis was made by means of the CSISA as well as the identification of problematic social situations in the five dimensions of social anxiety and the maintenance factors. This information is congruent with that obtained with the SAQ and the LSAS-SR. With the SAQ, we see how it exceeds the cut-off point defined for males globally and in all dimensions of social anxiety, and in the LSAS-SR we see that the levels of anxiety are higher than those of avoidance, but, jointly, they exceed the cut-off point established for the SAD.

As for other psychological problems, typical characteristics of an avoidant personality disorder (APD items), and severe depressive symptoms (CSISA and BDI-II) (e. g., sadness most of the time, irritability, less interest in some people and activities that were previously enjoyable, feelings of failure, problems in concentration and decision making) were identified. Also were found other anxiety symptoms (CSISA and PSWQ) (e. g., subjective feeling of overwhelm, tension and excessive worries about different aspects of his life), a high level of sensitivity to external and internal stimuli and events, showing behaviors typical of behavioral inhibition (PSWQ). In the past he had obsessive and agoraphobic symptoms as well as problems related to excessive alcohol and drug use, all of them after the onset of SAD symptoms (CSISA). No current self-esteem problems (RSES), nor alcohol abuse (AUDIT) were detected. There were also no current or past medical problems relevant to or related to SAD.

Regarding social skills (SOSAQ), deficits were identified in interacting with strangers, expressing positive feelings, interacting with attractive people, remaining calm in embarrassing situations, speaking in public with authority

TABLE 1 Summary of relevant aspects of social anxiety (pre/posttest and 12-month follow-up).

Name: *James*

Sex: *Male* Age: *40* Partnership status: *Living in couple* Occupation: *Engineer*

Social anxiety dimensions	Persistent presence of the fear[a]		Severity of anxiety/ avoidance[b]		SCISA Severity of interference/ distress[b]		Duration[c]	Other health issues[d]		SAQ Score			SAQ Above cutoff point[e]			Patient's target area[f]	Severity rating[g]	
	Pre	Post	Pre	Post	Pre	Post		Pre	Post	Pre	Post	12m	Pre	Post	12m		Pre	Post
1. Interactions with strangers	Yes	No	8/4	5/2	7/8	5/4	26 years			26	16	15	Yes	No	No	X	9	4
2. Interactions with the opposite sex	Yes	No	8/6	4/4	9/8	4/4	10 years			27	19	16	Yes	No	No	X	9	4
3. Assertive expression of annoyance, disgust or displeasure	Yes	No	8/2	4/2	6/5	3/3	10 years	No	No	26	17	15	Yes	No	No	X	9	2
4. Criticism and embarrassment	Yes	Yes	8/9	6/4	8/7	6/4	16 years			25	22	18	Yes	Yes	No	X	9	5
5. Speaking in public/ Talking with persons in authority	Yes	Yes	10/10 8/5	9/6 6/4	8/8 6/8	6/5 5/4	23 years 28 years			24	21	16	Yes	Yes	No	X	9	5
Total/In general										128	95	80	Yes	Yes	No		9	4

Notes: [a]Write "Yes/No"; [b]score given by patient; [c]specify length of time experiencing anxiety (pretest); [d]write "Yes/No" whether there are other health issues that may explain the symptoms of anxiety; [e]write "Yes/No" based on the cutoff score according to sex; [f]mark with a cross if the patient has identified the dimension as a therapeutic target (pretest); [g]therapist's evaluation of the dimension's level of seriousness, based on the interview and the questionnaire. Use the following scale:

0——1——2——3——4——5——6——7——8——9——10
None Mild Moderate Severe Very severe

Current diagnosis of SAD: Pre-test: Yes ☒ No ☐ Subclinical ☐   Post-test: Yes ☐ No ☐ Subclinical ☒

Observations:

Pre-test: *LSAS-SR-Total= 82 (yes, social anxiety); LSAS-SR-Anxiety= 49; LSAS-SR-Avoidance= 33*

Post-test: *LSAS-SR-Total= 65 (yes, social anxiety); LSAS-SR-Anxiety= 40; LSAS-SR-Avoidance= 25*

12 m follow-up: *LSAS-SR-Total= 46 (no, social anxiety); LSAS-SR-Anxiety= 26; LSAS-SR-Avoidance= 20*

TABLE 2 Summary of relevant aspects of other psychological problems pre/posttest and 12-month follow-up.

Summary of relevant aspects of other psychological problems pre/post-test and 12-month follow-up

Name: *James*   Sex: *Male*   Age: *40*   Partnership status: *Living in couple*   Occupation: *Engineer*

PSWQ	Pre	Post	12 m
Score:	68	55	50
Yes, worries	x	x	
No			x
AUDIT			
Score:	5	5	5
Yes, risk consumption			
No	x	x	x
APD ítems			
Score:	28	26	25
Yes	x		
No		X	x
PSQ			
Score:	132	122	120
Yes, High sensity	x		
No		x	x
RSES*			
Score:	28	26	27
> 35: High			
24-35: Normal	x	x	x
< 24: Low			
BDI-II			
Score:	27	12	9
0 - 9: Non-depressive			x
10 - 15: Light		x	
16 - 23: Moderate			
24 - 63: Severe	x		

WHOQOL-BREF	Pre		Post		12 m	
Domain	Sc.	No	Sc.	No	Sc.	No
Physical health	7	X	13	X	15	
Psychological	10	X	15		16	
Social relationships	9	X	15		16	
Enviroment	12	X	14	X	15	

EQ*	Pre		Post		12 m	
Subscale	Sc.	No	Sc.	No	Sc.	No
Decentering[a]	23	X	35		40	
Rumination[b]	—	—	—	—	—	—

SOSAQ	Pre		Post		12 m	
Skill	Sc.	No	Sc.	No	Sc.	No
1. Interacting with strangers	6	X	9		12	
2. Express positive feelings	12	X	13		13	
3. Coping with criticism	16		16		15	
4. Interact people attractive	8	X	12		13	
5. Keep. calm embarrass. sit.	6	X	7	X	8	X
6. Public speak/Talk auth. fig.	5	X	9	X	11	
7. Coping with embarrass.	11		11		11	
8. Standing up for rights	7	X	12		13	
9. Apologizing	14		15		15	
10. Refusing request	15		15		15	

Notes: PSWQ= Penn State Worry Questionnaire (cutoff point: men= 54); BDI-II= Beck Depression Inventory-II; AUDIT= Alcohol Use Disorders Identification Test (cutoff point: men≥ 8); RSES= Rosenberg Self-esteem Scale (*for men with social anxiety); PSQ= Personal Sensibility Questionnaire; WHOQOL-BREF= World Health Organization Quality of Life-Bref; EQ= Experiences Questionnaire; APD= Avoidant personality disorder; SOSAQ= Social skills Questionnaire; Sc.= Score No = Does not reach the cut-off point (indicating difficulties in that area).

Depressive disorder: Yes ☐ No ☐ Subclinical ☒   **Started before SAD?** Yes ☐ No ☒
Generalized anxiety disorder: Yes ☐ No ☐ Subclinical ☒   **Started before SAD?** Yes ☐ No ☒
Drug abuse disorder: Yes ☐ No ☒ Subclinical ☐   **Started before SAD?** Yes ☐ No ☐
Borderline personality disorder: Yes ☐ No ☒ Probable ☐
Bipolar disorder: Yes ☐ No ☒ Probable ☐
Schizophrenia/other psychotic disorder: Yes ☐ No ☒ Probable ☐
Other psychological issues: *Past: Depressive, obsessive and agoraphobic symptoms; excessive alcohol and drug use. None prior to SAD.*

Other medical issues: *He takes beta-blockers for performance situations. 10 years ago she finished a pharmacological treatment for anxiety (paroxetine) that he had for 3 years.*

figures and standing up for rights. During the interview, some difficulty in maintaining eye contact was observed, and some gestures and smiles did not synchronize to what he was saying.

Finally, James negatively values his quality of life (WHOQOL-BREF) in Physical health, Psychological, Social relationships, and Environment.

Clinical diagnosis

James presents with *social anxiety disorder* (social phobia) (300.23 [6B04]) according to DSM-5 (APA, 2013) and ICD-11 (World Health Organization, 2021). The social anxiety problem is observed both globally and in all five dimensions:

"Interaction with strangers," "Interaction with the opposite sex," "Assertive expression of annoyance, disgust, or displeasure," "Criticism and embarrassment," and "Public speaking/Interaction with people in authority." Anxiety levels are very high and always (or almost always) occur in different social situations. The most anxious situations are those of public speaking and he avoids them completely, preventing him from advancing professionally. In the other dimensions, the degree of anxiety is somewhat lower but still high. When James has no choice but to remain in the feared social situations, he does so with a high degree of distress, performing some safety behaviors, or when he can, he avoids them. The anxiety and avoidance are causing him clinically significant distress and are interfering with his daily functioning and the ability to achieve some of his goals. For example, she finds it more difficult to perform his job if he must interact with women and has stopped attending important professional events, and in the long term he finds himself unable to work in a company other than the family business or to start and own her own business. In terms of affective relationships, he has doubts about what his current relationship situation would be like if he did not fear being rejected by other women, considering that he has missed opportunities to meet both women and other people. In situations in which he feels upset or disagrees, if he does not express such feelings, he feels angry with himself and reproaches himself for not doing so. Generally speaking, frustration, sadness, feelings of failure, and anhedonia (depressive symptoms) are emotional consequences of the way he deals with social situations.

Problems *started* when he was about 14 years old (although he had some difficulties as a child) and have been increasing since then. It started with those situations related to being embarrassed or criticized and interacting with strangers, with people whom he knew little or who were known to him but he did not see them often; then fears related to speaking in public and interacting with people in authority became more prominent, and the most recent (after discontinuing treatment with paroxetine) are fears related to interacting with the opposite sex and with situations in which assertive expression of discomfort, disgust, or displeasure is required.

These features are also congruent with the characteristic symptoms of APD, according to the DSM-5.

Functional analysis

James presents symptoms characteristic of anxiety, from the subjective appraisal described by himself as overwhelm, nervousness, fear, anguish, or anxiety, accompanied by the physical reactions of choking sensation, chest tightness, lump in the throat, tachycardia, facial flushing, hand tremors, voice trembling, sweating, muscle tension (to the point of feeling stiffness in face and body) and at the end of the day physical pain (a "whiplash").

At the *cognitive* level there is excessive concern about pleasing others or what others think of him (e.g., that a woman thinks he is attracted to her, not inspiring confidence or giving a negative image at work), about carrying the weight of the conversation, about being noticed as nervous or having a bad time, about appearing strange, alienated, or unpleasant. He makes negative anticipations about the outcome (failure, rejection) and about his performance (clumsiness, inability). Once the social situations are over, ruminations appear around how they "should" have been, accompanied by negative criticisms toward himself (due to the standards used and expectations).

At the *behavioral* level, it is identified to participate little in conversations or to interrupt them inappropriately, to talk too fast (in a rushed manner), to avoid personal questions, to take off glasses so as not to see clearly, and to use a sense of humor. Avoid initiating conversations regardless of whether it is an acquaintance/unknown or the gender of the other person.

These anxiety reactions are *triggered* by multiple and varied social situations, covering the entire spectrum raised by the five dimensions of social anxiety. Some of the most anxiety-provoking situations are: making new friends, talking to people you do not know at parties and gatherings, feeling watched by a person of the opposite sex, asking a person of the opposite sex to go out with you, telling someone that you have hurt their feelings, telling someone that their behavior is bothering you and stop it, being teased in public, being criticized, having to speak at a meeting, speaking in public and participating in a meeting with people in authority.

Anxious reactions do not occur only during the feared social situation, but some occur before and are maintained after the social event. As *maintenance factors* we can identify the negative reinforcement that is produced by decreasing the intensity of the symptoms by using escape behaviors (e.g., taking off glasses so as not to see the other person clearly, looking away, talking too fast to finish earlier and leave; going to the bathroom), avoidance behaviors (e.g., changing sidewalks or making a detour to avoid meeting a known person in the street; making excuses not to give lectures, serve customers or meet with others) or safety behaviors (e.g., standing near windows/doors or outside to get some air). In some cases, his anxiety decreases if he is with his partner and feels supported by her, or if he is with his friends and lets them carry the conversation and be the center of attention; he also prefers meeting spaces to be outdoors or open spaces (if he is at a small table, where physical proximity to others is greater, then he experiences more anxiety).

TABLE 3 Functional analysis for anxiety in social situations.

Situation	Cognitive response	Emotional and physical response (0 = "not at all" to 10 = "very much")	Behavioral response	Immediate consequences	After thoughts	Secondary emotions
A girl comes into the office interested in taking some lessons from me. I was talking to a colleague	I can't give her any more time, I made a commitment I have to talk to her, otherwise I'll look terrible! I'm going to try to gain time to recover. I hope she leaves	Anxiety, panic, feeling ridiculous. I had a pretty bad time. 10 Very strong tachycardia, shortness of breath, tremors, blushing, pressure in temples, lump in throat, chills	I extended the conversation I was having with my partner (to recover) I talked to the girl, but I was quick and told her an excuse not to give her the lessons	It lowered my anxiety a little to see that the girl reacted with serenity and understanding Relief when she left Relief at not having to teach the classes	Silly me I should have approached the situation better I let the chance to teach slip away again What would the others (the girl, the office mates) have thought?	Discomfort Embarrassment Low self-esteem I feel like a child
My girlfriend and I met at a bar with a couple of friends who recently lost a child	In this situation they need support and naturality, my nervousness and discomfort cannot be noticed. I have to get it right	As soon as my friends walked through the door I noticed I was more nervous than I expected. 8 Rapid heart rate, jaw trembling	I drank two beers pretty quickly Difficulty looking them in the eye We talked about superfluous things	The anxiety went down and I felt more comfortable, but I ended up very tired	It is an exhausting effort I don't understand why it is so difficult if they are my friends	A little pity and anger with me

Table 3 includes a couple of examples of functional analyses with social situations in which James experienced anxiety.

Treatment goals

At the cognitive level

- Decrease concerns about being noticed as nervous, "weird" or "freaked out," about what the other person thinks (e.g., being attracted to you, being stupid, being "odd"), being liked by others, being likable, being interesting or attractive.
- Decrease negative anticipations of possible conflicts with other people or regarding their public appearances.
- Decrease attention to oneself, "forget about me, stop observing me."
- Focusing on the conversation.
- Lowering the level of self-criticism.

At the emotional and physiological level

- Diminish discomfort from physical symptoms.
- Decrease the intensity and some of the physical symptoms and make the recovery faster.
- Decrease stiffness in his face and body.
- Enjoying the conversation in a relaxed way.

- Decrease feelings of guilt toward his partner when he interacts with women.
- Diminish self-esteem lows after things did not go well.
- Losing the fear of not being valued.

At the behavioral level

- Meeting other people.
- Behaving in a more natural way.
- Look into the eyes when conversing with another person.
- Stop talking fast in situations where he is nervous.
- Laughing at oneself (in embarrassing situations).
- Dealing with unexpected situations such as meeting someone on the street and not avoiding them.
- Learn to recognize his mistakes, do not take criticism as a personal attack.
- Expressing disagreement with someone with whom he has had tensions in the past.
- Stop avoiding group exposure (e.g., teaching/lecturing, explaining things).
- Stop taking beta-blockers for public exposures.

Intervention

James participated in the Multidimensional Intervention for Social Anxiety (MISA) program, a group-based cognitive-behavioral treatment with traditional and third-generation strategies. He attended all 15 treatment sessions, each lasting 2:30h. The first five sessions were devoted to learning the core strategies of the program (values, mindfulness, acceptance, defusion, cognitive restructuring, and social skills) and then, in the other 10 sessions, applying them specifically to each dimension of social anxiety. The description of the specific contents of these sessions can be found in Caballo, Salazar, and Garrido (2018); Caballo, Salazar, Garrido, Irurtia, and Hofmann (2018); Salazar, Hofmann, & Caballo (in press).

Initially, all SAD-related issues were addressed through *psychoeducation*. Doubts referring to his anxiety problems and the use of the drugs were resolved. James' experience with the medication was used to explain the negative reinforcement of beta-blockers use by the immediate reduction of physical symptoms when presenting in public, and the difficulty in discontinuing them due to the lack of mastery of psychological strategies to cope with such situations. It is explained that this behavior is a long-term maintenance factor of the social anxiety problem.

The training of each of the strategies was done gradually and at least one exercise was performed in each session. For example, as physical symptoms were a recurring issue of concern for most of the patients (including James), work was done on acceptance of them. Mindfulness of bodily sensations was also used, including specific instructions for them to be aware if such thoughts (or judgments) occurred, to "take note of it," but to refocus their attention on the sensations without getting carried away by their thoughts.

The cognitive component, being a relevant aspect in this disorder, was introduced showing the relationship between it and the emotional and behavioral reactions characteristic of anxiety. The principles of cognitive therapies were used to identify the "dysfunctional thoughts" and to challenge them. They were then taught another alternative for the cognitive aspects, detaching from the dysfunctional thoughts.

Regarding dimensions, Table 4 contains an exercise to work on the interaction with the opposite sex, putting into action several strategies.

An important highlight of the treatment was James' active participation during the group sessions and his commitment to completing homework assignments.

Posttreatment assessment

The SCISA and pretest questionnaires were used for posttreatment evaluation. Tables 1 and 2 show the clinical improvement of James. He no longer meets the criteria (according to DSM-5) for SAD. Persistence of some fear in different social situations remains in only two of the five dimensions of social anxiety. However, the intensity of anxiety, the frequency of avoidance and the degree of interference in their life have decreased significantly in all dimensions, being of moderate severity (Table 1). These results are consistent with those of the SAQ and the LSAS-SR, where the score decreases considerably, but some problems remain. Following James' participation in the MISA program, his perception

TABLE 4 Self-monitoring of exposure to situations of interaction with the opposite sex.

Description of the situation	Level of anxiety before exposure (0–100 SUD)	What I thought before exposure was…	Description of defusion procedure used	Description of mindfulness strategy used	Behavior implemented	Level of satisfaction with the behavior (0 = totally dissatisfied – 100 = highly satisfied)
Talking to a girl at the bus stop	90	You will see what a fiasco My voice is going to shake I won't even be able to start He's going to think I like it If I don't get it right, I'm going to feel awful	I think about my "mind" and I send her away. I tell her that she doesn't know what will happen, that these are just anticipations, that they are not facts, and to leave me alone	As I walk to the bus stop, I focus on my surroundings Just before I address the girl, I focus on my breathing As I talk to the girl, I focus on what she says and how she says it, how she moves	I made sure the girl was not busy I initiated the interaction I started by asking if the bus had passed I commented on some superficial topics (the weather, how long the bus was taking) I maintained eye contact for longer than usual	100, just for having done it. I didn't even try this before. There are things I could improve, but it will come with practice. In this case, I would go down to 70
Talking to a coworker about a work issue	90	She's going to realize that I'm nervous She's going to think I'm getting like this because of her, she's going to think I like her She's going to think I'm stupid	I imagine putting all my thoughts in a box, locking it and throwing it into the sea	As I walk to his desk, I pay attention to the details of the office, who is there and what they are doing As I turn my attention to what we are talking about	I use a phrase to get their attention and comment on the topic (in general) why I am there. I breathe. I wait for her to answer (I don't talk fast) and we talk about the issue	100 for doing so 80 for the behavior itself (susceptible to improvement)

Note. SUD = Subjective Units of Distress Scale.

regarding the presence of typical APD characteristics also changed (Table 2) and the score does not reach the threshold necessary to consider the presence of this disorder.

There was a decrease in constant worries about various aspects of his life (PSWQ) and in depressive symptomatology, being at that time mild (BDI-II). Levels of sensitivity (PSQ) decreased, and self-esteem (RSES) and alcohol and drug consumption (AUDIT) remained at normal levels. As for quality of life (WHOQOL-BREFF), all domains improved and two ("Psychological" and "Social relationships") exceeded the cut-off point.

James significantly improved his ability to defuse (EQ), maintained at adequate levels the social skills that were initially fine (SOSAQ) and improved others, particularly "Expressing positive feelings," "Interacting with attractive people" and "Standing up for rights." He still shows some deficits in the skills "Keeping calm in embarrassing situations," and "Public speaking/Talking with authority figures," which are closely related to the social anxiety problems that remain (Table 2).

Regarding the goals James had set for himself, he commented that many of them had been partially achieved. However, he acknowledged that, although he had said that he did not expect "a cure," his expectations were very high, and he knew that he could achieve many of these goals in the medium to long term if he continued to practice what he had learned.

One-year follow-up

At 12 months after the end of the intervention, the questionnaires were administered again. Regarding social anxiety, the SAQ results show that James has improved in the two dimensions of social anxiety in which he still had difficulties. However, he should continue to work on reducing the levels of anxiety caused by situations related to the "Criticism and embarrassment" dimension. The LSAS-SR data are congruent and also show a significant decrease in the total score, falling below the cut-off point for social anxiety.

Also, an improvement in the social skills that appeared as deficits in the posttreatment was observed, and those social skills that improved in the posttreatment were maintained. Only the social skills of "Keeping calm in embarrassing situations" continue to be deficient and this makes us think about the role they play in the maintenance of anxiety in situations of "Criticism and embarrassment."

Of the other psychological aspects evaluated, the remission of depressive symptoms (BDI-II) is what stands out most. And finally, regarding quality of life, an improvement is observed in the two domains that were affected the previous year "Physical health" and "Environment," with James perceiving a much better quality of life (WHOQOL-BREF).

Discussion

The MISA program is very effective in the treatment of SAD (Caballo et al., 2021). Significant progress can be seen after the 15 group treatment sessions (posttreatment), but progress is certainly much more noticeable within a year of completing the program. It is possible that many of the changes that are proposed for people with SAD require a certain amount of time, especially since SAD affects all areas of life. The MISA program improves not only the symptoms of SAD, but many other aspects of people's lives.

Key facts

- The MISA program is a comprehensive treatment program for SAD and APD (but not only) using traditional CBT and mindful and acceptance-based strategies.
- The MISA program uses a therapist's guide and a patient's workbook all along the treatment.
- The MISA program encourages patient follow-up at 6 months and at 1 year.
- After completion of the intervention, the patient treated with the MISA program continued to improve at 6 months and at 1 year.

Applications to other areas

The wide selection of therapeutic strategies in the MISA program makes it applicable to psychological problems other than SAD and APD. For example, the MISA program has been recently applied to a group of 45 patients with SAD and has been found to be highly effective in the treatment of comorbid conditions such as excessive worry, deficits in social skills, depressive problems, low self-esteem, high personal sensitivity, and even improving quality of life of participant patients (Salazar et al., 2022).

Summary points

- The case of James, a 40-year-old male patient with social anxiety disorder, is described.
- James' anxiety problems span his entire life, with scores above the cutoff point on all five dimensions of social anxiety.
- James participated in the 4-month MISA program, which uses traditional cognitive-behavioral and third-generation strategies.

- James' posttreatment scores were compared to those obtained before treatment, improving in many areas.
- A follow-up was done 1 year after the end of treatment, finding that James continued to improve in many other areas of his life.

References

American Psychiatric Association. (2013). *Diagnostic and statistical manual of mental disorders. DSM-5*. Arlington, VA: Author.

Babor, T. F., Higgins-Biddle, J. C., Saunders, J. B., & Monteiro, M. G. (2001). *AUDIT. The Alcohol Use Disorders Identification Test. Guidelines for use in primary care* (2nd ed.). Geneva, Switzerland: WHO.

Beck, A. T., Steer, R. A., & Brown, G. (1996). *Manual for the beck depression inventory-II*. San Antonio, TX: Psychological Corporation.

Caballo, V. E., Arias, B., Salazar, I. C., Irurtia, M. J., Hofmann, S. G., & CISO-A Research Team. (2015). Psychometric properties of an innovative self-report measure: The social anxiety questionnaire for adults. *Psychological Assessment, 27*(3), 997–1012. https://doi.org/10.1037/a0038828.

Caballo, V. E., & Salazar, I. C. (2019). *Personal sensibility questionnaire (PSQ)*. Unpublished manuscript Granada, Spain: FUNVECA Clinical Psychology Center.

Caballo, V. E., Salazar, I. C., Arias, B., Irurtia, M. J., Calderero, M., & CISO-A Research Team Spain. (2010). Validation of the "social anxiety questionnaire for adults" (SAQ-A30) with Spanish university students: Similarities and differences among degree subjects and regions. *Behavioral Psychology, 18*(1), 5–34.

Caballo, V. E., Salazar, I. C., & CISO-A Research Team Spain. (2017). Development and validation of a new social skills assessment instrument: The social skills questionnaire (CHASO). *Behavioral Psychology, 25*(1), 5–24.

Caballo, V. E., Salazar, I. C., Curtiss, J., Gómez, R. B., Rossitto, A. M., Coello, M. F., Hofmann, S. G., … MISA Research Team. (2021). International application of the "Multidimensional Intervention for Social Anxiety" (MISA) program: I. Treatment effectiveness in patients with social anxiety. *Behavioral Psychology/Psicología Conductual, 29*(3), 517–547. https://doi.org/10.51668/bp.8321301n.

Caballo, V. E., Salazar, I. C., & Garrido, L. (2018). *Programa de Intervención multidimensional para la ansiedad social (IMAS). Libro del paciente [Multidimensional Intervention for Social Anxiety (MISA) program. Patient's workbook]*. Madrid, Spain: Pirámide.

Caballo, V. E., Salazar, I. C., Garrido, L., Irurtia, M. J., & Hofmann, S. G. (2018). *Programa de Intervención multidimensional para la ansiedad social (IMAS). Libro del terapeuta [Multidimensional Intervention for Social Anxiety (MISA) program. Therapist's guide]*. Madrid, Spain: Pirámide.

Caballo, V. E., Salazar, I. C., Irurtia, M. J., Arias, B., Hofmann, S. G., & CISO-A Research Team. (2010). Measuring social anxiety in 11 countries: Development and validation of the social anxiety questionnaire for adults. *European Journal of Psychological Assessment, 26*(2), 95–107. https://doi.org/10.1027/1015-5759/a000014.

Caballo, V. E., Salazar, I. C., Irurtia, M. J., Arias, B., Hofmann, S. G., & CISO-A Research Team. (2012). The multidimensional nature and multicultural validity of a new measure of social anxiety: The Social Anxiety Questionnaire for Adults (SAQ-A30). *Behavior Therapy, 43*, 313–328. https://doi.org/10.1016/j.beth.2011.07.001.

Fresco, D. M., Moore, M. T., van Dulmen, M. H. M., Segal, Z. V., Ma, S. H., Teasdale, J. D., & Williams, J. M. G. (2007). Initial psychometric properties of the experiences questionnaire: Validation of a self-report measure of decentering. *Behavior Therapy, 38*(3), 234–246. https://doi.org/10.1016/j.beth.2006.08.003.

Liebowitz, M. R. (1987). Social phobia. *Modern Problems in Pharmacopsychiatry, 22*, 141–173.

Meyer, T. J., Miller, M. L., Metzger, R. L., & Borkovec, T. D. (1990). Development and validation of the Penn State Worry Questionnaire. *Behaviour Research and Therapy, 28*, 487–495. https://doi.org/10.1016/0005-7967(90)90135-6.

Rosenberg, M. (1965). *Society and the adolescent self-image*. Princeton, NJ: Princeton University Press.

Salazar, I. C., & Caballo, V. E. (2017). *Entrevista clínica semiestructurada para la ansiedad social (ECSAS) [Clinical Semi-Structured Interview for Social Anxiety (CSISA)]*. Unpublished manuscript Granada, Spain: FUNVECA Clinical Psychology Center.

Salazar, I.C., Hofmann, S.G., & Caballo, V.E. (in press). Social anxiety: Linking cognitive behavioral therapy and strategies of third-generation therapies. In C.R. Martin, V.B. Patel, & V. Preedy (Eds.), Handbook of cognitive behavioral therapy by disorder: Case studies and application for adults. Elsevier.

Salazar, I. C., Caballo, V. E., Arias, V., Curtiss, J., Rossitto, A. M., Gómez, R. B., Hofmann, S. G., … MISA Research Team. (2022). International application of the "Multidimensional Intervention for Social Anxiety" (MISA) program: II. Treatment effectiveness for social anxiety-related problems. *Behavioral Psychology/Psicología Conductual, 30*(1), 19–49. https://doi.org/10.51668/bp.8322102n.

WHOQOL Group. (1996). *WHOQOL-Bref. Introduction, administration, scoring and generic version of the assessment*. Geneva, Switzerland: Programme on Mental Health. World Health Organization.

World Health Organization. (2021). *ICD-11. International classification of diseases 11th revision. The global standard for diagnostic health information*. https://icd.who.int/en.

Chapter 45

Application of mindfulness-based cognitive therapy on suicidal behavior: A case study

Debasruti Ghosh[a], Saswati Bhattacharya[b], Saurabh Raj[c], Tushar Singh[d], Sunil K. Verma[e], and Yogesh K. Arya[d]

[a]Department of Psychology, MDDM College (Babasaheb Bhimrao Ambedkar Bihar University), Muzaffarpur, Bihar, India [b]Tara Neuropsychiatry Clinic and Counseling Center, Ghaziabad, Uttar Pradesh, India [c]Department of Psychology, Ramdayalu Singh College (Babasaheb Bhimrao Ambedkar Bihar University), Muzaffarpur, Bihar, India [d]Department of Psychology, Faculty of Social Sciences, Banaras Hindu University, Varanasi, UP, India [e]Department of Applied Psychology, Vivekananda College, University of Delhi, New Delhi, India

Abbreviations

CBT cognitive-behavior therapy
MCBT mindfulness-based cognitive therapy
FFMQ five facet mindfulness questionnaire

Introduction

Suicidal thoughts are one of the most difficult thoughts that one has to deal with. It expresses the overwhelmed state of an individual, wherein they are unable to cope, and conclude that death can only resolve the problems that they are going through. According to The International Association for Suicide Prevention (2006), death by suicides is way more than deaths that have taken place in terrorist acts, wars, and interpersonal violence combined. The literature has mentioned several models that have explained suicidal cognition as a result of cognitive reactivity, ruminative thinking, thought suppression, and other cognitive processes (Williams, 2008). Several therapeutic approaches have been used to address suicidal behavior or cognitions that are associated with certain conditions like depression and borderline personality disorder. Intensive therapies like dialectical behavior therapy have been effective in reducing deliberate self-harm behavior in individuals diagnosed with borderline personality disorder (Linehan, Armstrong, Suarez, Allmon, & Heard, 1991). Cognitive-behavior therapy (CBT) is another effective psychotherapeutic approach that has been useful in addressing suicidal tendencies in depressive relapse (Berk, Henriques, Warman, Brown, & Beck, 2004; Teasdale et al., 2000). However, the complex nature of suicidal causes needs comprehensive techniques that can have greater and long lasting effect.

An example of the "third wave" of cognitive-behavioral psychotherapies is mindfulness-based cognitive therapy (MBCT; developed by Segal, Teasdale, and Williams (2004)), that emphasizes awareness and acceptance of problems in order to develop changes in individual (Hayes, Follette, & Linehan, M. (Eds.)., 2004). MBCT has been used in both clinical and nonclinical settings to address a range of problems (Baer, 2003). MBCT techniques focus on awareness and mindful acceptance of unwanted thoughts and feelings in a nonjudgmental manner that enhances coping skills in an individual. MBCT mechanisms focus on choosing appropriate coping skills rather than changing them which makes it appropriate for prevention of crisis situations like suicide. This chapter discusses the use of MBCT techniques on a 19-year-old girl with suicidal attempt.

Case summary of the client

MK was a 19-year-old female who was doing her graduation when she came to see the psychiatrist. She was residing in a semi urban area, in a lower middle class joint family setting. She was second in order among her four siblings. Her father owned a small business, while her mother was a homemaker. She was referred for psychotherapy because of low mood and suicide attempt which she tried by hanging herself in the previous month.

MK was very ambitious, performed well in studies, but her family was willing to get her married as soon as possible due to their poor economic condition. Two months back her father asked her to see a marriage proposal. She had two sisters after her; therefore the pressure of marriage was high. She was not willing to get married and had plans to complete her studies and get a job and live in a good city. Additionally, she said that "she becomes extremely anxious" with the thoughts of life after marriage. She narrated that one of her cousin's whom she was very close to had a very difficult (horrible in her words) married life, wherein her in-laws abused her both physically and verbally. MK was unable to convince her family members that she was not willing to get married, and wanted to pursue her studies. She felt extremely sad at her helpless condition and off lately preferred not talking to anyone at home. She described that the day she attempted suicide using her scarf, she was extremely overwhelmed with the thoughts of her future and finding no way out she wanted to just escape. She reported difficulty with in falling asleep, and had difficulty in concentrating for her upcoming exams as she finds herself preoccupied with these thoughts. However, her ability for other daily activities, hygiene and appetite was fairly usual.

Interview and assessment

After the interview, MK was assessed for suicidal risk and depressive symptoms. She had a fairly orthodox family with strict parenting since childhood. She has been shy and introvert, quietest among her siblings. In her words *"We had a strict childhood. I and my sisters did not have much liberty to go out or mingle with others. We were taught household skills since childhood while my brother enjoyed everything. My father would say a daughter belongs to her in-laws. So, she should learn how to run a house rather than studying."* She further mentioned that she had difficulty in interacting with elders and strangers. She was dependent on others for making major decisions, and was hesitant to share her disagreements especially in front of her family members because she felt that they would reject her. She revealed that whenever she feels overwhelmed, the thoughts of killing herself often come to her.

The risk assessment was done for suicide using modified scale for suicidal ideation (Miller, Norman, Bishop, & Dow, 1986) and Beck depression inventory (Beck, Steer, & Brown, 1996). According to the scores obtained MK seemed to have minimal depressive features and mild suicidal ideation (see Table 1). In addition to the above assessments she was asked to complete the five facet mindfulness questionnaire (FFMQ) which was used in therapy sessions to assess the process of change (Baer, Smith, Hopkins, Krietemeyer, & Toney, 2006).

Case conceptualization

Wenzel and Beck (2008) in their cognitive model of suicidal behavior mentioned that suicidal attempts are more common when situational triggers are backed by a predisposing vulnerability. The combination of these mechanisms leads to the activation of dysfunctional thinking pattern and an individual may not find way out of it and concludes that suicide is the only way to get rid of their current problems. In this case MK has been a shy, introvert child from the very beginning. Also, her social interactive ability as well as dependability on others provides a view that her problem solving skills, social skills and decision making ability was inadequate (*predisposing vulnerabilities*). She also closely witnessed her cousin's

TABLE 1 The assessment scores on clinical features of the case (MK).

Assessments		Scores
Modified scale for suicidal ideation		7
Beck depression inventory		10
FFMQ	Observing	3
	Describing	5
	Acting with awareness	3
	Nonjudging	3
	Nonreactivity	4
	Total	18

FFMQ = five facet mindfulness questionnaire.

traumatic marriage life. As her father has got a marriage proposal (*situational trigger*), it has led to activation of dysfunctional thinking pattern and cognitive distortions. Some examples of distortions displayed by MK are:

Catastrophization: "*My married life will be also like my cousin.*"
Jumping to conclusions: "*I won't ever be able to pursue my dreams.*"
Labeling: "*I will be a loser as I am in my life.*"
Mind reading: "*People will say the marriage failed due to me.*"

These thoughts have led to the fixation of attention and continuous rumination led to activation of cognitive processes associated with suicidal acts. Her problem solving deficits and poor distress tolerance added on her vulnerability, causing her to see ending her life as the only option (see Fig. 1). Additionally, cognitive reactivity and ruminative thinking have been implicated as the precursors of suicidal thinking (Williams, 2008). All this while MK felt low and her mood predominantly remained sad. The sad mood worked as a catalyst in activating MK's dysfunctional thinking patterns and limited her problem solving abilities.

Formulation

DISPOSITIONAL VULNERABILITIES: introvert, passive, gives in to authority figure, poor decision making, poor social skills, and poor distress tolerance

SITUATIONAL STRESSOR
Seeing marriage proposal by father

LIFE STRESSOR
Traumatic marital experience of her cousin

COGNITIVE PROCESS ASSOCIATED WITH PSYCHIATRIC DISTURBANCES:

Catastrophisation: " My married life will be like my cousin"

Jumping to conclusion: " I will never be able to pursue my dreams"

Labeling: "I will be a loser in marriage as I am in my life"

Problem solving deficits, poor distress tolerance, faulty cognitive styles

COGNITIVE PROCESS ASSOCIATED WITH SUICIDAL ACTS:

FAULTY COGNITIVE STYLES:

"My future is doomed"

"People will blame me for my failed marriage like they did for my cousin"

" I am a burden to my family, I am better off dead"

"Nobody will listen to me, a girl does not have the right to be independent in our culture"

SUICIDAL ACT

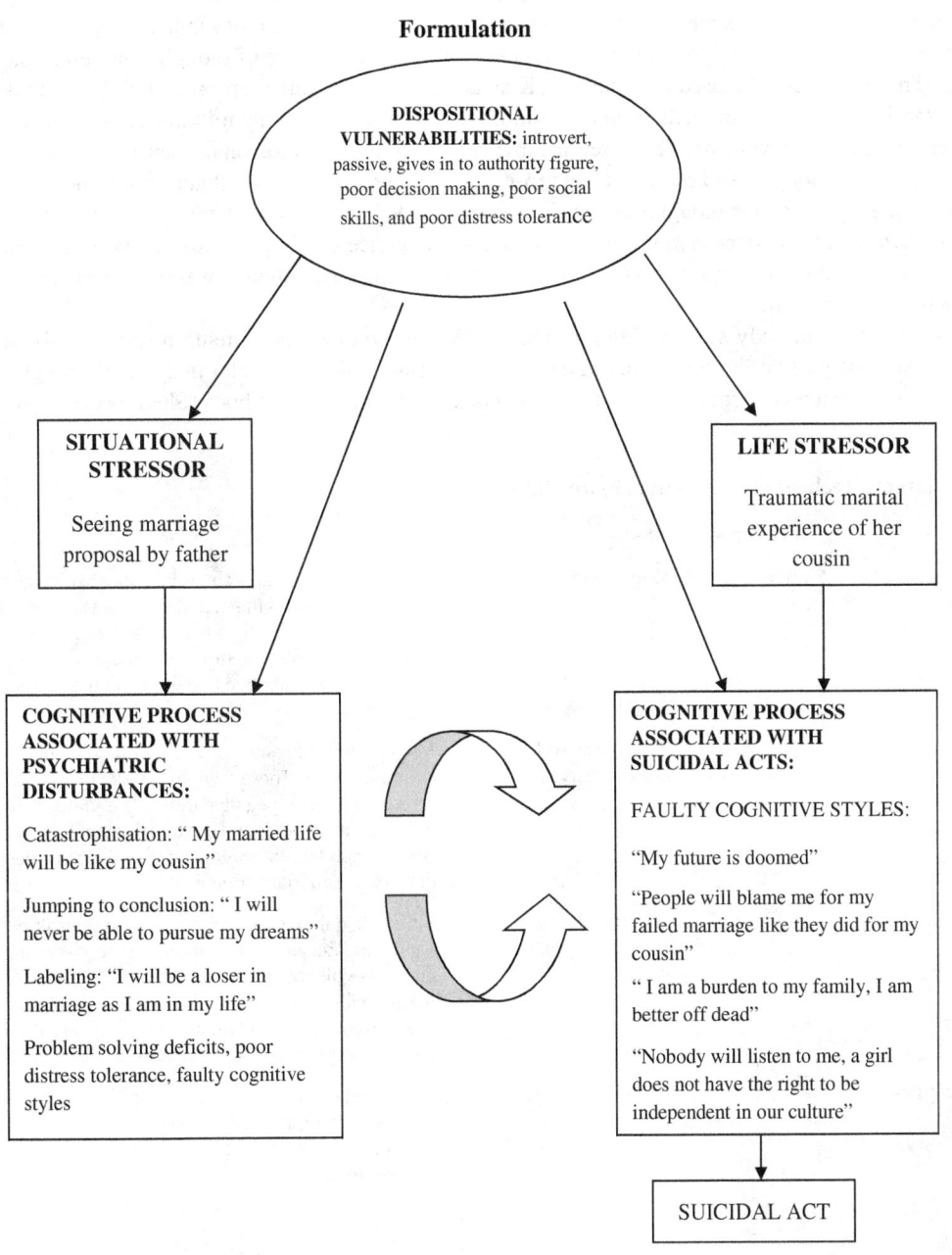

FIG. 1 Conceptualization of MK's suicidal behavior according to Wenzel and Beck's model explaining suicidal behavior (2008).

Planning of techniques

The conceptualization gave a broader view of bases for MK's basis of negative thoughts and suicidal ideation. On the basis of conceptualization and interview, the targets of the therapy were planned for MK. The therapy course followed the MBCT program as described by Segal, Williams, and Teasdale (2018) that includes both formal and informal practices. Table 2 presents the techniques and rationale for choice of techniques according to the targets planned. The therapy sessions spanned over 8 weeks and every week involved one session which lasted for about 45 min to 1 h. Homework assignments were also given that required 1 h per day of mindfulness exercise. The first posttreatment follow-up took place after 2 weeks and the second one after 2 months.

Session summary

The therapy sessions were planned as per the MBCT procedures entailed by Segal, Williams, and Teasdale (2002). The first five sessions included establishing a therapeutic alliance with the client along with psychoeducation about the illness, orientation about the process of therapy to begin with and understanding the nature of thoughts and emotions in a nonjudgmental manner. The first session focused on making MK aware about the automatic pilot mode of the mind and its effects. This session comprised of body scan meditation and mindful eating exercise. In the mindful eating session, MK followed the raisin exercise by eating a few raisins. The exercise involved directing attention on the sensations and movements associated with eating, and on thoughts and emotions that arose while eating. MK revealed that initially she was concentrate, but as the raisin exercise progressed, the thoughts of future were coming to her and she lost her focus. She expressed *"When I bit the raisin, the sweetness reminded me of how my mother makes kheer (rice pudding) for my birthday…but she too wants me to get married…that moment I was again into the thoughts."* The therapist explained how thoughts can be triggered and can take one away from the present.

The session 2 began with body scan meditation where MK tried to direct her attention sequentially on various body parts. She was asked focused on the sensations of each body part in a nonjudgmental manner. If thoughts and emotions arose, she was directed to notice them and gradually bring back her attention to the body. She reported that she felt restless

TABLE 2 The targets, techniques and rationale for the therapy sessions.

Target	Technique	Rationale
1. Awareness of the apprehensive thoughts (*Automatic pilot mode of the mind*)	Body scan and raisin exercise	MK was extremely apprehensive about her life after her marriage. She also recounted the experiences of the past from her cousin. She had difficulty in focusing on her studies and remained preoccupied. Therefore these mindfulness practices will enable her to be more attentive and focus on the present moment experiences
2. Handling negative thoughts and negative emotions	Sitting meditation, pleasant and unpleasant experiences diary, breathing spaces	MK reported of feeing sad constantly and thoughts of helplessness surrounded her. These techniques will enable her to be more aware about how negative thoughts and negative emotions influence each other and are transitory in nature. This awareness is very important to develop an outsider's perspective in order to develop appropriate coping skills
3. Developing a problem solving approach	Ambiguous scenario exercise, Sitting meditation (*thoughts are not facts*)	MK's suicidal attempt or thoughts was a result of her inability to handle the negative thoughts. These cognitive and mindfulness practices will enable her to acknowledge her difficult thoughts. Additionally, it will enable her to take a metacognitive stance, wherein she can reflect on the impact of her internal and external contexts, especially their reactions on the negative experiences
4. Dealing with stressful situations	Mindful walking, mindful stretching exercises, yoga	MK's reactions to aversive situation were either an escape (*suicidal act*) or avoidance. Hence, these practices will help to develop skills of coming back rather than drifting away with stressful thoughts

and uneasy in the first session stating "*I did not feel good initially. My mind drifted to a lot of thoughts those related to the past. Felt like choking and it felt heavy also. I was restless.*" The therapist went on to acknowledge her sensations and gradually equating with the thoughts. This was followed by a thoughts and feelings exercise. MK acknowledged that how she has been giving in to her thoughts because her mood.

Mindful stretching that included yoga poses and other stretching exercises were done in session 3. MK revealed that she found herself more focused in session 3. She stated "*Today I could bring back my focus to my breathing and at one point of time I felt calm and composed as if I am living the moment.*" Furthermore, this session included a 3-min breathing space exercise in which MK practiced mindful awareness of her internal experiences for short periods. This was done to generalize mindfulness practices in daily activities. She reported to have felt difficulty initially but experienced a calming effect at the end. Fourth session included cognitive therapy elements that began with a discussion of how thoughts impact our feelings and adopt a ruminative pattern. MK was encouraged to notice these thoughts and let them pass without labeling or being controlled by them. After the exercise a breathing space practice was done and MK described "*This is what happens with me, I get so much lost in these problems, I have never been able to see my strengths, the other aspects of my life. My focus was just my marital life. I could have asked my father for some more time, maybe he wouldn't have denied.*" In session 5, stretching and yoga practices were given through which MK could connect to her internal experiences and had started understanding the broader approach. In sitting meditation practice (sessions 3–5), a sequential focus approach was encouraged on the sensations and movements of breathing, body, and sounds in the environment.

The practices in sessions 6–8 the practices were focused on cultivating a different relationship with unpleasant experiences, especially those related to suicidal rumination. Beginning in session 6, the sitting meditation exercise was expanded in order to intentionally bring a problem in mind or to identify related cognitive distortions. Once the problem was brought into focus the client was asked to observe the related sensations and emotions carefully. She was also instructed not to change or eliminate these negative cognitions. She explained her experience "*They are really disturbing...but now I know that they make me feel bad...the moment I return back to on feeling my sensations...they don't seem that disturbing.*" She could understand the transitory nature of thoughts and emotions. In these sessions the 3-min breathing space was extended with the perspective to practice it during moments of stress. This practice enables to focus on unpleasant internal experiences and accept then in an open manner. Also, through this exercise mindful awareness is developed, that enables an individual to make constructive choices while handling stressors and crisis situations.

Subsequently, therapist and MK together planned a schedule of her daily activities. The activities that could trigger negative thoughts and moods were reduced and those activities were identified that lead to feelings of pleasure and mastery and she was encouraged to participate in it. Further, the role of these activities in giving rise to adaptive and maladaptive thought patterns were explained in details. In session 8, a prevention plan for suicidal impulse was chalked out, wherein MK was explained to use mindfulness skills to identify her triggers for negative emotions and ruminations. The training that she received on observing the associated sensations of the negative thoughts and feelings and allowing them pass; and generate choices to deal with the stressors was explained.

Homework always included daily practice of one or more mindfulness exercises. She was initially very irregular with the homework and the need for the same was emphasized. She started to do the homework regularly from second session onward and her ability to be mindful improved subsequently.

Treatment outcome and critical evaluation

Overall, the results showed substantial improvement in MK's suicidal thoughts and overall depressive cognitions in terms of frequency, duration and distress (see Table 3). She showed moderate improvement in certain facets of FFMQ like Observing, Nonjudging, and Nonreactivity. During the sessions, MK reported about the frequency of her mindfulness practice reliably, but her practice diary entries were inconsistent and insufficient. The gap of information was complemented by interviewing about her experiences. This revealed MK's unwillingness to sustain the prescribed duration of mindfulness practice over the follow-up period. This could be one of the possible reasons for lower scores in the facets of Describing and Acting with Awareness in FFMQ. Initially, she faced difficulty to focus on her inner experiences but later she improved with every session. Although she continued to experience depressive ruminations but she successfully refrains from reacting and mindfully responds each time when such urges dominate. She also could understand that killing herself does not solve the problem, as much of the things are beyond her control. By the end of the session, she could be aware of her possible triggers, cognitive distortions and how to address them nonjudgmentally. During the follow-up sessions, she was encouraged to practice MBCT skills to communicate assertively. The therapist reviewed her experiences of practices MBCT skills at home and accordingly the plan for future follow-up sessions were planned.

TABLE 3 The assessment pretherapy, posttherapy, and follow-up scores on clinical features of the case (MK).

Assessments		Pretherapy scores	Posttherapy scores	Follow-up scores (after 2 weeks)	Follow-up (after 2 months)
Modified scale for suicidal ideation		7	2	2	1
Beck depression inventory		10	4	3	3
FFMQ	Observing	3	6	6	5
	Describing	5	4	4	5
	Acting with awareness	3	4	4	5
	Nonjudging	3	5	5	4
	Nonreactivity	4	6	5	6
	Total	18	25	24	25

Concluding remarks

MBCT has been implicated in suicidal behavior with depressive individuals (Chesin et al., 2016; Forkmann et al., 2014) and in general population also (Le & Gobert, 2015; Raj et al., 2019). In this case study MBCT has been useful for overall changes and suicidal ideation, however, certain personality characteristics like poor distress tolerance, emotional dysregulation, need for dependency require more intensive intervention. Therefore, MBCT can be combined with other forms of intensive therapy to enhance its efficacy in this area. Also, this case study could not explain the treatment's mechanisms of action. The current case study, despite its limitation, highlights one way in which individuals such as MK can learn to handle difficult situations in a more adaptive manner. However, the usefulness of MBCT is retained when the client is motivated to continue mindfulness practices on an everyday basis. Research in this area is limited; hence, future studies must look into long-term benefits of MBCT.

Application to other areas

MBCT has earned its recognition as the preferred treatment modality across the globe which resulted in its applicability to extend from the field of clinical to a nonclinical populations. Other than the treatment modality for clinical disorders like depression, anxiety, panic disorder (Fumero, Peñate, Oyanadel, & Porter, 2020), MBCT has also been applied to intervene against the fear of disease recurrence among cancer patients (Luberto et al., 2019), to school students and teachers to enhance acceptance and cognitive flexibility (Felver et al., 2017). Furthermore, it has been applied in the prison system to reduce impulse control, self-harm and recidivism and enhance the ability to cope with difficult feelings and sensations among the inmates and prison staff (Maroney et al., 2021).

Key fact

- Target 3.4 of the Sustainable Development Goal of the United Nations focuses on minimizing premature mortality from non-communicable diseases by one third, which is to be done through promotion, prevention, and treatment of mental health and well-being (SDG, 2021). The suicide mortality rate is one of the two indicators for achieving this target.

Summary points

- MBCT techniques help in enhancing awareness and develop appropriate coping skills.
- Suicidal behavior can be triggered in individuals with vulnerabilities when they face situational stressors.
- The techniques used in the case helped in creating an outsider's perspective and see her problems in a nonjudgmental manner.

- MBCT techniques helped identifying the role of rumination and aversions in developing suicidal cognition and enhanced problem solving and decision making skill.
- Future researches must focus on modified MBCT intervention strategies to address crisis in suicidality and also investigate long term impact of MBCT in such problems.

References

Baer, R. A. (2003). Mindfulness training as a clinical intervention: A conceptual and empirical review. *Clinical Psychology: Science and Practice, 10*, 125–143. https://doi.org/10.1093/clipsy.bpg015.

Baer, R. A., Smith, G. T., Hopkins, J., Krietemeyer, J., & Toney, L. (2006). Using self-report assessment methods to explore facets of mindfulness. *Assessment, 13*(1), 27–45.

Beck, A.T., Steer, R.A., & Brown, G.K. (1996). Beck depression inventory (BDI-II) (Vol. 10). Pearson, s15327752jpa6703_13.

Berk, M. S., Henriques, G. R., Warman, D. M., Brown, G. K., & Beck, A. T. (2004). A cognitive therapy intervention for suicide attempters: An overview of the treatment and case examples. *Cognitive and Behavioral Practice, 11*(3), 265–277.

Chesin, M. S., Benjamin-Phillips, C. A., Keilp, J., Fertuck, E. A., Brodsky, B. S., & Stanley, B. (2016). Improvements in executive attention, rumination, cognitive reactivity, and mindfulness among high–suicide risk patients participating in adjunct mindfulness-based cognitive therapy: Preliminary findings. *The Journal of Alternative and Complementary Medicine, 22*(8), 642–649.

Felver, J. C., Doerner, E., Jones, J., Kaye, N. C., & Merrell, K. W. (2017). Mindfulness in school psychology: Applications for intervention and professional practice. In B. A. Gaudiano (Ed.), *Mindfulness: Nonclinical applications of mindfulness: Adaptations for school, work, sports, health, and general well-being* (pp. 111–134). Routledge/Taylor & Francis Group.

Forkmann, T., Wichers, M., Geschwind, N., Peeters, F., van Os, J., Mainz, V., & Collip, D. (2014). Effects of mindfulness-based cognitive therapy on self-reported suicidal ideation: Results from a randomised controlled trial in patients with residual depressive symptoms. *Comprehensive Psychiatry, 55*(8), 1883–1890.

Fumero, A., Peñate, W., Oyanadel, C., & Porter, B. (2020). The effectiveness of mindfulness-based interventions on anxiety disorders. A systematic meta-review. *European Journal of Investigation in Health, Psychology and Education, 10*(3), 704–719. https://doi.org/10.3390/ejihpe10030052.

Hayes, S. C., Follette, V. M., & Linehan, M. (Eds.). (2004). *Mindfulness and acceptance: Expanding the cognitive-behavioral tradition*. Guilford Press.

Le, T. N., & Gobert, J. M. (2015). Translating and implementing a mindfulness-based youth suicide prevention intervention in a Native American community. *Journal of Child and Family Studies, 24*(1), 12–23.

Linehan, M. M., Armstrong, H. E., Suarez, A., Allmon, D., & Heard, H. L. (1991). Cognitive-behavioral treatment of chronically parasuicidal borderline patients. *Archives of General Psychiatry, 48*(12), 1060–1064.

Luberto, C. M., Hall, D. L., Chad-Friedman, E., & Park, E. R. (2019). Theoretical rationale and case illustration of mindfulness-based cognitive therapy for fear of cancer recurrence. *Journal of Clinical Psychology in Medical Settings, 26*(4), 449–460. https://doi.org/10.1007/s10880-019-09610-w.

Maroney, M., Luthi, A., Hanney, J., Mantell, A., Johnson, D., Barclay, N., … Crane, R. (2021). Audit of a mindfulness-based cognitive therapy course within a prison. *Journal of Correctional Health Care: The Official Journal of the National Commission on Correctional Health Care, 27*(3), 196–204. https://doi.org/10.1089/jchc.18.09.0048.

Miller, I. W., Norman, W. H., Bishop, S. B., & Dow, M. G. (1986). The modified scale for suicidal ideation: Reliability and validity. *Journal of Consulting and Clinical Psychology, 54*(5), 724.

Raj, S., Sachdeva, S. A., Jha, R., Sharad, S., Singh, T., Arya, Y. K., & Verma, S. K. (2019). Effectiveness of mindfulness based cognitive behavior therapy on life satisfaction, and life orientation of adolescents with depression and suicidal ideation. *Asian Journal of Psychiatry, 39*, 58–62.

SDG. (2021). *United Nations*. Department of Economic and Social Affairs Sustainable Development. https://sdgs.un.org/goals. (Accessed May 2022).

Segal, Z. V., Teasdale, J. D., & Williams, J. M. G. (2004). Mindfulness-based cognitive therapy: Theoretical rationale and empirical status. In S. C. Hayes, V. M. Follette, & M. M. Linehan (Eds.), *Mindfulness and acceptance: Expanding the cognitive-behavioral tradition* (pp. 45–65). Guilford Press.

Segal, Z. V., Williams, J. M. G., & Teasdale, J. D. (2002). *Mindfulness-based cognitive therapy for depression: A new approach to preventing relapse*. Guilford Press.

Segal, Z. V., Williams, M., & Teasdale, J. (2018). *Mindfulness-based cognitive therapy for depression*. Guilford Publications.

Teasdale, J. D., Segal, Z. V., Williams, J. M. G., Ridgeway, V. A., Soulsby, J. M., & Lau, M. A. (2000). Prevention of relapse/recurrence in major depression by mindfulness-based cognitive therapy. *Journal of Consulting and Clinical Psychology, 68*, 615–623.

The International Association for Suicide Prevention. (2006). *Guidelines for suicide prevention*. Available at: http://www.med.uio.no/iasp/english/guidelines.html. Accessed June 2006.

Wenzel, A., & Beck, A. T. (2008). A cognitive model of suicidal behavior: Theory and treatment. *Applied and Preventive Psychology, 12*(4), 189–201.

Williams, J. M. G. (2008). Mindfulness, depression and modes of mind. *Cognitive Therapy and Research, 32*(6), 721–733.

Chapter 46

Recommended resources for cognitive-behavioral therapy in different disorders

Vinood B. Patel[a], Rajkumar Rajendram[b,c], and Victor R. Preedy[d]

[a]School of Life Sciences, University of Westminster, London, United Kingdom [b]College of Medicine, King Saud bin Abdulaziz University for Health Sciences, Riyadh, Saudi Arabia [c]Department of Medicine, King Abdulaziz Medical City, King Abdullah International Medical Research Center, Ministry of National Guard Health Affairs, Riyadh, Saudi Arabia [d]Faculty of Life Science and Medicine, King's College London, London, United Kingdom

Abbreviations

APA American Psychiatric Association
CBT cognitive behavioural therapy
DSM Diagnostic and Statistical Manual of Mental Disorders
ICD International Classification of Diseases

Introduction

The Diagnostic and Statistical Manual of Mental Disorders (DSM) defines 157 distinct disorders of mental health (American Psychiatric Association, 2013). These conditions can cause significant morbidity and mortality (Rajendram, Patel, & Preedy, 2022). However, consideration must also be given to the fact that there are other classifications such as the International Classification of Diseases (ICD) and there is an ongoing scientific dialogue in which classifications can change. The pharmacological and/or nonpharmacological approaches to the management of a mental disorder are often determined by its classification.

In view of the significant side effects associated with medications for neuropsychiatric diseases; interest in psychotherapies has increased. One of many psychotherapies; cognitive-behavioral therapy (CBT), has well-established roles in the management of many disorders of mental health (Nakao, Shirotsuki, & Sugaya, 2021; Thoma, Pilecki, & McKay, 2015). As a consequence, there are many scientific evaluations of the efficacy of CBT in different conditions and indeed using different delivery platforms (for example: Lau, Yen, Wong, Cheng, & Cheng, 2022; Li et al., 2022; Matsumoto, Hamatani, & Shimizu, 2021; Parrish et al., 2021; Zamboni et al., 2021). Pharmaceuticals and psychotherapy are not mutually exclusive. Indeed, multimodal approaches to mental disease are commonplace.

The knowledge and awareness of psychotherapies such as CBT has developed tremendously over the past few years. Remaining abreast of recent developments can seem like a Sisyphean task even for experienced researchers, scientists and clinicians. We have therefore produced tables containing resources as recommended by researchers and practitioners. The aim of this compilation is to assist those colleagues who are interested in understanding more about CBT. Essentially, the information provided here draws upon the wealth of experience and "know how" accrued over many years. The list below acknowledges all the experts who helped to prepare these valuable resources.

Resources

Tables 1–5 list the most up-to-date information on the regulatory bodies (Table 1), professional societies (Table 2), books (Table 3), journals (Table 4), emerging technologies, platforms and other resources (Table 5), and different research, clinical or other centers (Table 6) that are relevant to an evidence-based approach to CBT and other psychotherapies. Some organizations are listed in more than one table as they occasional fulfill multiple roles.

TABLE 1 Regulatory bodies or organizations dealing with mental health, cognitive behavioral therapy, other psychotherapies or health related matters in different disorders.

Regulatory body or organization	Web address
American Psychiatric Association	https://www.psychiatry.org/
American Psychological Association (APA)	https://www.apa.org
Body Dysmorphic Disorder Foundation	https://bddfoundation.org/
Centre for Perinatal Psychology—Australia	https://www.centreforperinatalpsychology.com.au/
Centre of Perinatal Excellence (COPE)	https://www.cope.org.au
Japanese Ministry of Health, Labour, and Welfare (JMHLW)	https://www.mhlw.go.jp/english/index.html
Mental Health America	https://www.mhanational.org/
National Alliance on Mental Illness	https://www.nami.org/
National Institute for Health and Care Excellence (NICE)	https://www.nice.org.uk
National Institute of Mental Health	https://www.nimh.nih.gov/
National Institute of Mental Health and Neuro Sciences (NIMHANS)	https://nimhans.co.in/
National Institute on Drug Abuse	https://www.drugabuse.gov/
Postpartum Support International	https://www.postpartum.net/
State Mental Health Authority, Delhi	http://smhadelhi.org/
Substance Abuse and Mental Health Service Administration (SAMHSA)	https://www.samhsa.gov/
World Federation for Mental Health	https://wfmh.global/
World Confederation of Cognitive and Behavioral Therapies	https://wccbt.org

This table lists the regulatory bodies and organizations involved with cognitive behavioral therapy, other psychotherapies and associated specialities or interests. The links were accurate at the time of going to press but may move or alter. In these cases, the use of the "Search" tabs should be explored at the parent address or site. In some cases, links direct the reader to pages related to CBT within parent sites. Some societies and organizations have a preference for shortened terms, such as acronyms and abbreviations. See also Table 2.

Other resources

Many important documents pertaining to CBT are held in The Wellcome Collection (https://wellcomecollection.org/collections). The British Library also lists material on topics related to disorders of mental health and psychotherapies. Its holdings are, to some extent, digital and many can be accessed online (https://www.bl.uk/).

Other chapters on resources (recommended by authors and practitioners) may also be relevant to CBT. These include posttraumatic stress disorders (Rajendram, Patel, & Preedy, 2015), depression (Rajendram, Patel, & Preedy, 2021a); the neuroscience of development (Rajendram, Patel, & Preedy, 2021b), the neuroscience of aging (Rajendram, Patel, & Preedy, 2021c) and the neurobiology and behavior of pain (Rajendram, Patel, & Preedy, 2021d).

Summary points

The management of mental disorders (which may include CBT) is often determined by their classification, for example, in the Diagnostic and Statistical Manual of Mental Disorders.

Over 150 conditions are defined in the Diagnostic and Statistical Manual of Mental Disorders.

Pharmacological and nonpharmacological (e.g. psychotherapy) approaches are not mutually exclusive. That is, CBT can be administered alongside medications in the treatment of mental disease.

CBT is a well-established form of psychotherapy that is applicable to many mental disorders.

Without the guidance contained in this resource it would be a Sisyphean task to keep up with the ever expanding of the awareness and knowledge of CBT.

TABLE 2 Professional societies relevant to cognitive behavioral therapy or other psychotherapies in different disorders.

Society name	Web address
Academy of Cognitive Behavioral Treatments	https://www.academyofct.org/
American Psychological Association (Cognitive Behavioral Therapy For PTSD)	https://www.apa.org/ptsd-guideline/patients-and-families/cognitive-behavioral
Aotearoa-New Zealand Association for Cognitive Behavioral Therapy	http://www.cbt.org.nz/
Asian Cognitive Behavioral Therapy Association (ACBTA)	www.asiancbt.weebly.com
Association for Behavioral and Cognitive Therapies (ABCT)	https://www.abct.org/Home/
Association for Contextual Behavioral Science	https://contextualscience.org/
Association of Cognitive and Behavioral Psychotherapy (Russia)	https://www.associationcbt.org/
Australian Association for Cognitive and Behavior Therapy (AACBT)	www.aacbt.org.au
Beck Institute for Cognitive Behavioral Therapy	https://beckinstitute.org/
British Association for Behavioral & Cognitive Psychotherapies (BABCP)	https://www.babcp.com/
Bulgarian Association for Cognitive-Behavioral Psychotherapy	http://www.bacbp.org/en/Home
Canadian Association of Cognitive and Behavioral Therapies (CACBT)	https://cacbt.ca
Chinese Association of Cognitive Behavior Therapy (CACBT)	http://www.cacbt.org/aboutus.htm
Cognitive Behavioral Psychotherapy Ireland[a]	https://cbti.ie/
Egyptian Association of Cognitive Behavioral Therapy	https://eacbt.info/
Estonian Association for Cognitive Behavioral Therapy	https://www.ekka.ee/
European Association for Behavioral and Cognitive Therapies (EABCT)	www.eabct.eu
Indian Association for Cognitive Behavior Therapy	http://www.iacbt.org/
International Association for Cognitive Psychotherapy (IACP)	www.the-iacp.com
International Society of Interpersonal Psychotherapy	https://interpersonalpsychotherapy.org/
Irish Association for Behavioral & Cognitive Psychotherapies	https://babcp.com/IABCP
Japanese Association for Behavior Analysis (JABA)	http://www.j-aba.jp/
Japanese Association for Cognitive Therapy (JACT)	http://jact.umin.jp/
Japanese Association of Behavioral and Cognitive Therapies (JABCT)	http://jabt.umin.ne.jp/
Japanese Association of Behavioral and Cognitive Therapies (JABCT)	http://jabt.umin.ne.jp/
Japanese Association of Certified Public Psychologists (JACPP)	https://www.jacpp.or.jp/
Latin-American Association of Analysis, Behavioral Modification and Cognitive and Behavioral Therapies (ALAMOC)	www.alamoc-web.org
Malaysia Cognitive Behavioral Therapy Association	http://cbtmalaysia.net/
National Association of Cognitive-Behavioral Therapists (NACBT)	https://www.nacbt.org/
New York City Cognitive Behavior Therapy Association	http://www.nyc-cbt.org/
Northern California cognitive-behavioral therapy	https://www.nccbt.net/
Philadelphia Behavior Therapy Association	https://www.philabta.org/
Psychopharmacology and Substance Abuse (APA division 28)	https://www.apadivisions.org/division-28
Society of Clinical Psychology, American Psychological Association Division 12	https://div12.org/
World Confederation of Cognitive and Behavioral Therapies	https://wccbt.org/

This table lists some societies and organizations devoted to cognitive-behavioral therapy and other psychotherapies. Please note, occasionally the location of the websites or web address changes. Some societies and organizations have a preference for shortened terms, such as acronyms and abbreviations. See also Table 1.
[a]This may be a site in transition.

TABLE 3 Books on cognitive behavioral therapy or other psychotherapies in different disorders.

Book title	Authors or editors	Publisher	Year of publication
Acceptance and Mindfulness in Cognitive Behavior Therapy: Understanding and Applying the New Therapies	Herbert JD, Forman EM	John Wiley & Sons, Inc.	2011
ACT Made Simple: An Easy-To-Read Primer on Acceptance and Commitment Therapy	Harris R	New Harbinger Pubns Inc	2019
An Introduction to Cognitive Behavior Therapy: Skills and Applications Third Edition	Kennerley H, Kirk J, Westbrook D	SAGE Publications Ltd	2017
An Introduction to Modern CBT Psychological Solutions to Mental Health Problems	Hofmann SG	John Wiley & Sons	2011
Anxiety and Worry Workbook: The Cognitive Behavioral Solution 1st Edition	Clark DA, Beck AT	The Guilford Press	2011
Behavioral Treatment for Substance Abuse in People with Serious and Persistent Mental Illness: A Handbook for Mental Health Professionals	Bellack AS, Bennet ME, Gearon JS	Routledge	2006
Bipolar Disorder: A Family-Focused Treatment Approach Second Edition	Miklowitz D	Guilford Press	2008
Bipolar Workbook, Second Edition: Tools for Controlling Your Mood Swings	Basco M	Guilford Press	2015
Body Dysmorphic Disorder: A Treatment Manual	Veale D	Wiley-Blackwell	2010
Brief Cognitive Behavior Therapy	Curwen B, Palmer S, Ruddell P	Sage	2018
CBT Practitioner's Guide To ACT: How to Bridge the Gap Between Cognitive Behavioral Therapy and Acceptance and Commitment Therapy	Ciarrochi JV, Bailey A	New Harbinger Publications	2008
Classics In Psychotherapy. Interpersonal Psychotherapy. A Clinician's Guide (Second Edition)	Stuart S, Robertson M	Routledge	2012
Cognitive Behavior Therapy, Third Edition: Basics and Beyond Third Edition	Beck JS	The Guilford Press	2020
Cognitive Behavior Therapy: Applying Empirically Supported Techniques in Your Practice	O'Donohue WT, Fisher JE	John Wiley & Sons	2008
Cognitive Behavioral Therapy for Overcoming Bad Habits: Easy Strategies of CBT for Quitting Three Big Addictions Including Smoking, Alcohol Addiction, and Internet Addiction	Loxely S	–	2020
Cognitive Behavior Therapy for Psychiatric Problems: A Practical Guide	Hawton KE, Salkovskis PM, Kirk JE, Clark DM	Oxford University Press	1989
Cognitive Behavior Therapy: A Guide for The Practising Clinician, Volume 1	Simos G	Routledge	2014
Cognitive Behavioral Therapy for Perinatal Distress	Wenzel A, Kleiman K	Routledge	2014
Cognitive Therapy for Bipolar Disorder: A Therapist's Guide to Concepts, Methods and Practice Second Edition	Lam D, Jones S, Hayward P	Wiley-Blackwell	2010
Cognitive Therapy for Challenging Problems: What to Do When the Basics Don't Work Reprint Edition	Beck JS, Beck AT	The Guilford Press	2011
Cognitive Therapy for Challenging Problems: What to Do When the Basics Don't Work	Beck JS	Guilford Press	2005

TABLE 3 Books on cognitive behavioral therapy or other psychotherapies in different disorders—cont'd

Book title	Authors or editors	Publisher	Year of publication
Cognitive Therapy for Suicidal Patients: Scientific and Clinical Applications	Wenzel A, Brown GK, Beck AT	American Psychological Association	2009
Cognitive Therapy of Substance Abuse Revised Ed. Edition	Beck AT, Wright FD, Newman CF, Liese BS	The Guilford Press	2001
Cognitive Therapy with Chronic Pain Patients 1st Edition	Winterowd C, Beck AT, Gruener D	Springer Publishing Company	2003
Cognitive Therapy: Basics and Beyond	Beck JS, Beck AT	Guilford press.	1995
Cognitive-Behavior Therapy for Severe Mental Illness: An Illustrated Guide	Wright J, Turkington D, Kingdon D, Basco MR	American Psychiatric Publishing	2009
Cognitive-Behavioral Therapy for Body Dysmorphic Disorder: A Treatment Manual	Wilhelm S	Giulford	2013
Cognitive-Behavioral Therapy for Smoking Cessation: A Practical Guidebook to The Most Effective Treatments	Perkins K, Conklin C, Levine M	Routledge: Taylor & Francis Group	2008
Cognitive-Behavioral Therapy	Craske MG	American Psychological Association.	2010
Core Progra A Cognitive Behavioral Guide to Depression	Paterson R, Alden L, Koch W	Changeways Clinic	2008
Enhanced Cognitive-Behavioral Therapy for Couples: A Contextual Approach	Epstein NB, Baucom DH	American Psychological Association.	2002
Evidence-Based Practice of Cognitive Behavioral Therapy	Dobson D, Dobson K	The Guildford Press	2016
General Principles and Empirically Supported Techniques of Cognitive Behavior Therapy	William T. O'Donohue, Jane E. Fisher	John Wiley & Sons	2009
Getting Unstuck in Act: A Clinician's Guide to Overcoming Common Obstacles in Acceptance and Commitment Therapy	Russ Harris	New Harbinger Pubns Inc	2013
Global Mental Health and Psychotherapy: Adapting Psychotherapy for Low- and Middle-Income Countries (1st Edition)	Stein DJ, Bass JK, Hofmann SG (Eds.)	Elsevier	2019
Group Therapy Manual for Cognitive Behavioral Treatment of Depression	Muñoz R, Miranda J	Rand	2000
Handbook of Cognitive-Behavioral Therapies, Fourth Edition	Dobson KS, Dozois DJA	The Guilford Press	2019
Handbook of Perinatal Clinical Psychology. From Theory to Practice.	Quatraro RM, Grussu P	Routledge	2020
Integrated Treatment for Mood and Substance Use Disorders	Westermeyer J, Weiss R, Ziedonis D	The Johns Hopkins University Press	2003
Learning Cognitive-Behavior Therapy: An Illustrated Guide (Core Competencies in Psychotherapy) (Core Competencies in Psychotherapy) 2 Rev. IIIrd Edition	Wright JH, Brown GK, Thase ME, Ramirez Basco (Authors), Gabbard GO (Editor)	Amer Psychiatric Pub	2017

Continued

TABLE 3 Books on cognitive behavioral therapy or other psychotherapies in different disorders—cont'd

Book title	Authors or editors	Publisher	Year of publication
Massachusetts General Hospital Handbook of Cognitive Behavioral Therapy	Petersen TJ, Sprich SE, Wilhelm S	Springer	2016
Mindfulness-Based Cognitive Therapy for Depression	Segal ZV, Williams M, Teasdale J	Guilford Publications	2018
Mindfulness-Based Cognitive Therapy with People at Risk of Suicide	Williams JMG, Fennell M, Crane R, Silverton S	Guilford Publications	2017
Mindfulness-Based Treatment Approaches: Clinician's Guide to Evidence Base and Applications	Baer RA	Elsevier	2015
Online Cognitive Behavioral Therapy: An E-Mental Health Approach to Depression and Anxiety	Alavi N, Omrani M	Springer International Publishing	2019
Postpartum Depression Workbook. Strategies to Overcome Negative Thoughts, Calm Stress, And Improve Your Mood	Burd A	Rockridge Press	2020
Practicing Positive CBT: From Reducing Distress to Building Success	Bannink F	Wiley-Blackwell	2012
Pregnancy & Postpartum Anxiety Workbook. Practical Skills to Help You Overcome Anxiety, Worry, Panic Attacks, Obsessions and Compulsions	Wiegartz PS, Gyoerkoe KL	New Harbinger Publications, Inv.	2009
Process-Based CBT: The Science and Core Clinical Competencies of Cognitive Behavioral Therapy	Hayes SC, Hofmann SG	New Harbinger Publications	2018
Psychoeducation Manual for Bipolar Disorder	Colom F, Vieta E	Cambridge University Press	2006
Recovery-Oriented Cognitive Therapy for Serious Mental Health Conditions 1st Edition	Beck AT, Grant P, Inverso E, Brinen AP, Perivoliotis D	The Guilford Press	2020
Standard and Innovative Strategies in Cognitive Behavior Therapy	de Oliveira IR	InTech	2012
Textbook Of Addiction Treatment	el-Guebaly N, Carrà G, Galanter M, Baldacchino AM	Springer	2021
Treating Bipolar Disorder: A Clinician's Guide to Interpersonal and Social Rhythm Therapy	Frank E	Guilford Press	2005
Treating Postnatal Depression. A Psychological Approach for Health Care Practitioners	Milgrom J, Martin P R, Negri LM	John Wiley & Sons Ltd	1999
Trial-Based Cognitive Therapy: A Manual for Clinicians	De Oliveira IR	Routledge	2014
Wiley Handbook of Cognitive Behavioral Therapy	Hofmann S	Wiley-Blackwell	2013

This table lists books relevant to cognitive-behavioral therapy and other psychotherapies.

Mini-dictionary of terms

Diagnostic and Statistical Manual of Mental Disorders: The American Psychiatric Association (APA) publishes this diagnostic tool. The fifth (current) edition was published in 2013. It contains a taxonomy of mental disease and is the authority on psychiatric diagnosis in many countries worldwide.

Psychotherapy: a generic term for any treatment that involves talking to a healthcare provider specializing in mental disease (e.g., psychologist or psychiatrist). Cognitive-behavior therapy is one form of psychotherapy.

TABLE 4 Relevant journals publishing original research and review articles related to cognitive-behavioral therapy in different disorders.

Frontiers in Psychiatry
Journal of Affective Disorders
BMJ Open
Trials
BMC Psychiatry
Behavior Research and Therapy
Behavior Therapy
Journal of Medical Internet Research
International Journal of Environmental Research and Public Health
Cochrane Database of Systematic Reviews
Internet Interventions
Journal of Consulting and Clinical Psychology
Behavioral and Cognitive Psychotherapy
Psychiatry Research
PLoS One
Cognitive and Behavioral Practice
Frontiers in Psychology
Psychotherapy Research
International Journal of Eating Disorders
Journal of Anxiety Disorders
Cognitive Behavior Therapist
Cognitive Therapy and Research
Lancet Psychiatry
Clinical Psychology and Psychotherapy
Psychological Medicine
JAMA Psychiatry
JMIR Mental Health
World Psychiatry
Clinical Case Studies
Depression and Anxiety

Journals publishing original research and review articles related to cognitive-behavioral therapy. Included in this list are the top 30 journals which have published the most number of articles on cognitive-behavior therapy over the past 5 years. Data derived from Scopus.

Regulatory body: A government agency or public institute/organization that is responsible for legally or nonlegally regulating various aspects of peoples' lives.

Professional body: An organization that aims to further a specific profession, those engaged in and affected by that profession and interests of the public.

Research group: A group consisting of individual researchers working together on a topic or issue.

TABLE 5 Other resources of interest or relevance for health care professionals or patients related to cognitive-behavioral therapy or psychotherapy in different disorders.

Name of resource or organization	Web address
Albert Ellis Institute, New York, USA	http://www.albertellis.org
American Psychological Association Cognitive Behavioral Therapy for PTSD	https://www.apa.org/ptsd-guideline/patients-and-families/cognitive-behavioral
Black Dog Institute, Australia	https://www.blackdoginstitute.org.au/
Caregiver Mental Health Knowledge Sharing Series—Focus on Sub-Saharan Africa	https://caregivermentalhealth.org/
Centers for Disease Control and Prevention	https://www.cdc.gov/tobacco/quit_smoking/index.htm
Centre for Perinatal Psychology	https://www.centreforperinatalpsychology.com.au/
Cognitive Behavioral Therapy (CBT). Royal College of Psychiatrists	https://www.rcpsych.ac.uk/mental-health/treatments-and-wellbeing/cognitive-behavioural-therapy-(cbt)
Cognitive Therapy/Cognitive Behavioral Therapy for Major Depressive Disorder. Ministry of Health, Labour and Welfare	https://www.mhlw.go.jp/kokoro/speciality/manual1.html
David Burn's website—author of Feeling Good	https://feelinggood.com/
Get Self Help	https://www.getselfhelp.co.uk
i4Health Palo Alto University's Institute for International Internet Interventions for Health	https://i4health.paloaltou.edu/
Institute for Cognitive Behavior Therapy & Research	https://www.zoominfo.com/c/institute-for-cognitive-behavior-therapy-research/25789180
Japanese Manuals Of Cognitive Behavioral Therapy for Mood and Anxiety Disorders. Japanese Ministry Of Health, Labour, and Welfare (JMHLW)	https://www.mhlw.go.jp/stf/seisakunitsuite/bunya/hukushi_kaigo/shougaishahukushi/kokoro/index.html
Mental Health Information in Japanese—Anxiety, Panic and Phobias. Royal College of Psychiatrists	https://www.rcpsych.ac.uk/mental-health/translations/japanese/anxiety-phobia
Mental Health Information in Japanese-Bipolar Disorder. Royal College of Psychiatrists	https://www.rcpsych.ac.uk/mental-health/translations/japanese/bipolar-disorder
Mental health Information in Japanese—Cognitive Behavioral Therapy. Royal College of Psychiatrists	https://www.rcpsych.ac.uk/mental-health/translations/japanese/cognitive-behavioural-therapy
Mental health Information in Japanese—Depression. Royal College of Psychiatrists	https://www.rcpsych.ac.uk/mental-health/translations/japanese/depression
Mothers and Babies Program, Northwestern University	https://www.mothersandbabiesprogram.org/
Mothers and Babies: Mood and Health Research Lab, George Washington University	https://mbp.columbian.gwu.edu/
National Center for Cognitive Behavior Therapy and Research (NCCBTR), Tokyo, Japan	https://www.ncnp.go.jp/cbt/english/
Oxford Cognitive Therapy Centre	https://www.octc.co.uk/
Perinatal Mental Health Care Pathways. National Collaborating Centre for Mental Health	https://www.rcpsych.ac.uk/docs/default-source/improving-care/nccmh/perinatal/nccmh-the-perinatal-mental-health-care-pathways-appendices-helpful-resources.pdf?sfvrsn=ea69882c_4
Psychological Treatments for Depression. Beyond Blue	https://www.beyondblue.org.au/the-facts/depression/treatments-for-depression/psychological-treatments-for-depression
Redwood Center for Cognitive Behavior Therapy and Research	https://redwoodcbt.com/
Smokefree	https://smokefree.gov/tools-tips/get-extra-help/free-resources
Society of Clinical Psychology	https://div12.org/psychological-treatments/treatments/cognitive-therapy-for-depression/

TABLE 5 Other resources of interest or relevance for health care professionals or patients related to cognitive-behavioral therapy or psychotherapy in different disorders—cont'd

Name of resource or organization	Web address
Swinburne University of Technology	https://www.swinburne.edu.au/study/courses/units/Cognitive-Behaviour-Therapy-and-Research-PSY80007/local
What is psychology? The British Psychological Society	https://www.bps.org.uk/public/what-is-psychology

This table lists other resources of interest or relevance to cognitive-behavioral therapy and other psychotherapies. Please note, occasionally the location of the websites or web address changes.

TABLE 6 Treatment and research centers, clinics, and other centers.

Treatment and research centers, clinics, and other centers	Web address
Academy of Cognitive & Behavioral Therapies	https://www.academyofct.org/
American Institute for Cognitive Therapy	https://www.cognitivetherapynyc.com/
Association for Behavioral and Cognitive Therapies (ABCT)	https://www.abct.org/Home/
Beck Institute for Cognitive Behavior Therapy	https://beckinstitute.org/
Behavioral Medicine Clinic (Moffit Cancer Center)	https://moffitt.org/patient-family/managing-symptoms/psychological-and-psychiatric-services/
Capital Institute for Cognitive Therapy	http://cognitivetherapydc.com/
Center for Cognitive Behavior Therapy	http://centerforcbt.org/
Center for Mindfulness Studies	https://www.mindfulnessstudies.com/personal/depression-anxiety/mbct/
Center for Substance Abuse Treatment	https://www.samhsa.gov/about-us/who-we-are/offices-centers/csat
Centre for Addiction and Mental Health (CAMH)	http://www.camh.ca/
Changeways Clinic	https://www.changeways.com/index.html
Cleveland Clinic Cognitive behavioral therapy	https://my.clevelandclinic.org/health/treatments/21208-cognitive-behavioral-therapy-cbt
Cochrane Database of Systematic Reviews: smoking cessation interventions for smokers with current or past depression	https://www.cochranelibrary.com/cdsr/doi/10.1002/14651858.CD006102.pub2/full
Cognitive Behavioral Therapy Center of New Orleans	https://cbtnola.com/
Manhattan Center for cognitive-behavioral therapy	https://manhattancbt.com
Mayo Clinic. Cognitive behavioral therapy	https://www.mayoclinic.org/tests-procedures/cognitive-behavioral-therapy/about/pac-20384610
Mindfulness-integrated Cognitive Behavior Therapy	https://www.mindfulness.net.au/
National Center for Cognitive Behavior Therapy and Research	https://www.ncnp.go.jp/cbt/
Ottawa Institute of Cognitive Behavioral Therapy	https://www.ottawacbt.ca
Oxford Mindfulness Centre	https://www.oxfordmindfulness.org/
Riverside Natural Health Centre Trent Bridge, Nottingham	http://www.thecbtclinic.com/
University of Sydney	https://www.sydney.edu.au/medicine-health/our-research/research-centres/pain-management-research-institute/postgraduate-and-short-courses-in-pain-management/putting-cognitive-behavioural-therapy-skills-into-practice.html

This table lists some treatment and research centers, clinics, and other centers relevant to cognitive-behavioral therapy and other psychotherapies as suggested by contributors. Please note, occasionally the location of the websites or web address changes. This list of clinics is included to provide general information only. It is important to point out that it does not constitute any recommendation or endorsement of the activities of these facilities, or other resources listed in this chapter, by the authors or editors of this book.

Acknowledgments

We thank the following authors for their contributions to the development of this resource. We apologies if some of the suggested material was not included in this chapter or has been moved to different sections.

Aonso-Diego, Gema
Arya, Yogesh Kumar
Buhagiar, Rachel
Chalah, Moussa A.
Chan, Sunny Ho-Wan
Drüge, Marie
Felice, Elena
Ghosh, Debasruti
González-Roz, Alba
González-Roz, Alba
Goyal, Deepika
Kimura, Yasuhiro
Kimura, Yasuhiro
Kondo, Masaki
Lau, Charlie
Le, Huynh-Nhu
Mackillop, James
Mamo, Elena
Rabemananjara, Kantoniony
Raj, Saurabh
Secades-Villa, Roberto
Singh, Tushar
Tanoue, Hiroki
Verma, Sunil Kumar
Watzke, Birgit
Weidberg, Sara
Yoshinaga, Naoki

References

American Psychiatric Association. (2013). *Diagnostic and statistical manual of mental disorders* (5th ed.). Arlington, VA: American Psychiatric Association.
Lau, Y., Yen, K. Y., Wong, S. H., Cheng, J. Y., & Cheng, L. J. (2022). Effect of digital cognitive behavioral therapy on psychological symptoms among perinatal women in high income-countries: A systematic review and meta-regression. *Journal of Psychiatric Research, 146*, 234–248.
Li, X., Laplante, D. P., Paquin, V., Lafortune, S., Elgbeili, G., & King, S. (2022). Effectiveness of cognitive behavioral therapy for perinatal maternal depression, anxiety and stress: A systematic review and meta-analysis of randomized controlled trials. *Clinical Psychology Review, 92*.
Matsumoto, K., Hamatani, S., & Shimizu, E. (2021). Effectiveness of videoconference-delivered cognitive behavioral therapy for adults with psychiatric disorders: Systematic and meta-analytic review. *Journal of Medical Internet Research, 23*(12).
Nakao, M., Shirotsuki, K., & Sugaya, N. (2021). Cognitive-behavioral therapy for management of mental health and stress-related disorders: Recent advances in techniques and technologies. *BioPsychoSocial Medicine, 15*(1), 16.
Parrish, J. M., Jenkins, N. W., Parrish, M. S., Cha, E. D. K., Lynch, C. P., Massel, D. H., et al. (2021). The influence of cognitive behavioral therapy on lumbar spine surgery outcomes: A systematic review and meta-analysis. *European Spine Journal, 30*(5), 1365–1379.
Rajendram, R., Patel, V. B., & Preedy, V. R. (2022). Recommended resources on cognitive behavioral therapy. In C. R. Martin, V. R. Preedy, & V. B. Patel (Eds.), *Cognitive behavioral therapy across disorders*. USA: Elsevier. In press.
Rajendram, R., Patel, V. B., & Preedy, V. R. (2015). Recommended resources on post-traumatic stress disorder. In C. R. Martin, V. R. Preedy, & V. B. Patel (Eds.), *Comprehensive guide to post-traumatic stress disorders*. Germany: Springer.
Rajendram, R., Patel, V. B., & Preedy, V. R. (2021a). Recommended resources on the neuroscience of depression: Genetics, cell biology, neurology, behavior, and diet. In C. R. Martin, L.-A. Hunter, V. B. Patel, V. R. Preedy, & R. Rajendram (Eds.), *The neuroscience of depression. Genetics, cell biology, neurology, behaviour and diet*. USA: Academic Press, Elsevier.
Rajendram, R., Patel, V. B., & Preedy, V. R. (2021b). Recommended resources on the neuroscience of development. In C. R. Martin, & V. R. Preedy (Eds.), *Neuroscience of development disease*. UK: Elsevier.
Rajendram, R., Patel, V. B., & Preedy, V. R. (2021c). Recommended resources on the neuroscience of aging. In V. B. Patel, & V. R. Preedy (Eds.), *Neuroscience of aging*. UK: Elsevier.

Rajendram, R., Patel, V. B., & Preedy, V. R. (2021d). The physiology, neurobiology and behaviour of pain: Recommended resources. In R. Rajendram, C. R. Martin, V. B. Patel, & V. R. Preedy (Eds.), *The physiology, neurobiology and behaviour of pain*. UK: Elsevier.

Thoma, N., Pilecki, B., & McKay, D. (2015). Contemporary cognitive behavior therapy: A review of theory, history, and evidence. *Psychodynamic Psychiatry, 43*, 423–461.

Zamboni, L., Centoni, F., Fusina, F., Mantovani, E., Rubino, F., Lugoboni, F., et al. (2021). The effectiveness of cognitive behavioral therapy techniques for the treatment of substance use disorders: A narrative review of evidence. *The Journal of Nervous and Mental Disease, 209*(11), 835–845.

Index

Note: Page numbers followed by *f* indicate figures, *t* indicate tables, and *b* indicate boxes.

A

Acceptance and commitment therapy (ACT)
 body dysmorphic disorder (BDD), 442, 444–445
 diabetes-related distress and HbA1c, 109–113, 115
 dizziness, 121
 persistent postural-perceptual dizziness, 130–131
 social anxiety, 267–268
Acceptance training, 270, 276
Acrophobia, 53
 five trials, 55–57, 59–60
 symptoms, 57–58, 60
 VR-CBT or VRET, 59t
 VR environment, 54, 60f
Adjustment disorder (AjD), 65
 assessment, 67
 classification and diagnostic criteria, 66
 course and impact, 66–67
 future directions, 72
 ICTs for treatment, 68–70
 internet-based treatments, 70–72
 prevalence, 66
 psychological treatment, 67–68
Agencies, Funds, and Programs (AFPs), 246
Alcohol use disorders identification test (AUDIT), 269
Anger management, 299
Antares Foundation's Managing Stress for Humanitarian Workers—Guidelines for Good Practice, 247, 248f
Antenatal Risk Questionnaire (ANRQ), 448
Antidepressants, for persistent postural-perceptual dizziness, 125
Antiepileptic drug (AED), 137
 related sexual disorders, 138
Anxiety, 18
 coronary artery disease (CAD), 308
 prevalence, 321
 treatments, 5–6, 7t
Anxiety disorders
 bipolar disorder, 79, 87–88, 332
 mindfulness-based cognitive therapy, 79, 81–84, 83f
Anxiety disorders (ADs), Italian mental health services, 352–353
 agoraphobia, 343
 application, in public services, 352–353

CBT
 "enriched" CBT, for anxiety management and reasoning bias modification training, 351t
 "first generation" Italian CBT published studies, 345–346t
 Italian "second generation," of CBT intervention, 348–349
 online and digitized CBT intervention, 350
 training, 344–346
 treatments, 344, 352
 COVID-19 pandemic to digitalized CBT for, 350–351
 generalized anxiety disorder (GAD) frequency, 343
 pharmacological treatment, 343
 postdisaster distress transdiagnostic CBT, 349
 prevalence, 343
 Tuscany replication studies, 347–348
Application-based CBT, 229
Attachment security, 221, 230
Automatic pilot mode, 285, 286f, 289
Autonomic nerve system (ANS) dysfunction, 309–310
Avoidant/restrictive food intake disorder (ARFID), 91, 92f
 aversive consequences, 93, 94f
 cognitive-behavioral conceptualization, 92–95, 92f
 cognitive-behavioral therapy, 95–99, 95t
 assessment and treatment, 96
 aversive consequences, 98
 lack of interest in food/eating, 98–99
 overview, 95–96
 sensory sensitivity, 98
 stages, 97–98
 key facts, 101–102
 sensory sensitivity, 93, 93f
Axis I disorder, 417
Axis II disorder, 417

B

Baduanjin, 322, 325–326
Basic Collaboration Onsite, 13
Beck Depression Inventory, 226
Beck Depression Inventory-II (BDI-II), 269, 294, 337, 422, 427
Behavioral activation (BA), 295, 298, 300
Behavioral Activation for Depression Scale (BADS-SF), 294
Behavioral factors, 25

Behavioral techniques, of persistent postural-perceptual dizziness, 130
Behavioral therapy
 obsessive-compulsive disorder (OCD), 205–206
 for smoking cessation, 295
Bergen 4-day treatment (B4DT), 212
Bidirectional models, 294
Bio psycho socio spiritual model, 36, 37f
Bipolar disorder, 479
 fatigue
 case formulation, 477–478
 growth history and clinical history, 477
 intervention, 478–479
 patient information and visit history, 477
 in Japan
 bipolar I disorder, 331
 bipolar II disorder, 331
 CBT effectiveness, 332
 cyclothymic disorder, 331
 epidemiology, 332
 maintenance treatment, 333
 medical care for patients with, 333
 prevalence, 332t
 psychosocial treatment, 334–335
 psychotherapy, 331
 recommended treatment, 333t
 social status, 332–333
 psychotherapy, cognitive pattern
 case formulation, 476
 growth history and clinical history, 475–476
 intervention, 476–477
 patient information and visit history, 475
Blended treatments, for adjustment disorder, 72
Blood volume pulse (BVP), 312
Body dysmorphic disorder (BDD), 214
Body dysmorphic disorder (BDD), CBT, 445
 comorbidities, 441
 depressive symptoms, 442
 diagnostic process and treatment options, 443
 eating disorder, 441
 exposure and response prevention (ERP), 444
 German CBT manual, 443
 initial phase, 442
 prevalence rates, 441
 psychoeducation, 444
 SORCK modeling, 443

Index

Body dysmorphic disorder (BDD), CBT *(Continued)*
  therapy
    beginning of, 443
    end of, 445
  thoughts, reconstruction of, 444
  treatment options for, 441–442
  value-focused interventions, 444–445
Body image, 25
Body mass index (BMI), 21
Body scan, 323, 326
Borderline personality disorder (BPD)
  application of cognitive behavior therapy in, 417–419
  multidimensional intervention for social anxiety (MISA) program, 276–277
  psychotherapy and, 417
  treatment procedure, 419
Brain disease. *See* Epilepsy
Brain injury, 185, 187
  negative stereotypes, 187–189
The British Library, 500
Brooding, 282–284, 289
Brown Assessment of Beliefs Scale (BABS), 443
Burnout, 247, 251, 256, 260

C

Cancer
  patients
    diagnosis and treatment, 43–44
    psychosocial interventions for, 44–45, 44f
  survival, CBT, 45–46, 46t
    key facts, 47
Cannabis, 421–422
  abstinence, 423–425
  treatment, 422
Carbon monoxide (CO) "smokerlyzer", 295–296
Causal models, 293
CBT-informed psychoeducation course, 434, 435t
Cellular level injury response, 187
Chronic dizziness, 127
Chronic subjective dizziness (CSD)
  CBT, 127
  prevalence, 122
  psychological factors, 125
  RCT, 126
  risk for, 125
  SSRI, 125
Cigarette smoking. *See* Smoking
Clinical Research Unit for Anxiety Disorders (CRUFAD) in Sydney, 347
Clinical Semi-Structured Interview for Social Anxiety (CSISA), 269, 482
Close Collaboration with Some System Integration, 13
Cognitive-behavioral conceptualization, of avoidant/restrictive food intake disorder, 92–95, 92f
Cognitive-behavioral interventions
  depression preventive intervention, 434
  women's mental health conditions, in sub-Saharan Africa, 395–396, 403–404
    adjunctive nonpharmacological treatments, 403–404
    developmental lifespan, 403–404
    HIV and comorbidity and CBT, 396–398, 397t
    perinatal depression (PD), 401, 402t
    prevention interventions, 401–403
    psychoeducational and prevention tools, 403–404
    trauma and gender-based violence, 398–401, 399t
Cognitive behavioral therapy (CBT), 4, 45, 53, 68, 70, 79–80, 106, 121, 138, 141, 143, 156–157, 190, 417–419, 421–427, 431
  adjustments to, 38
  aid workers, 256
  cancer survival works, 45–46, 46t
  carbon monoxide levels, 423f
  case formulation, 422
  case studies, 168–169, 170f
  clinical assessment, 422
  comparison with MBCT and, 81t
  coronary artery disease (CAD), 315, 315f, 317–318
  cultural adaption, 34–35
  cultural competence, 36–38
  directed masturbation, 157
  on DRD in DM patients, 109–113, 109f, 110–112t
  duration and setting, 45
  effectiveness and efficacy, 189
  focus on, 43
  insomnia, 410
  internet, 177–178
  intervention programs in DM with DRD, 114–115
  in Japan
    assessment, 337
    behavioral experiments, 338–339
    case formulation, 338
    coping skills development, at prodrome, 339–340, 339t
    daily routine and sleep habits, establishment of, 338
    eclectic therapies in, 336
    implementation rate, for bipolar disorder, 335–336, 335f
    practice and procedures, 336–339
    psychoeducation, 337
    self-monitoring, 338
    therapeutic goal setting, 338
    treatment module, for bipolar disorder, 335
    and universal health insurance, 336
  limitation, 321
  loneliness, 176–177, 177f
  methods, 45
  for mTBI recovery, 188–189
  multiple sclerosis, 198–199
    treatment response, 199
  obsessive-compulsive disorder (OCD)
    components, 207–208
    general psychiatric in-patient services, 209
    in-patient/residential treatment *(see* In-patient/residential treatment, for OCD)
    long-term outcome, 211
    model for, 206, 206f
    out-patient CBT, 208
    treating team, 209
  pelvic-floor exercises, 158
  persistent postural-perceptual dizziness, 126–129, 128–129f
  physical and mental health, 175–176
  postpartum depression (PPD), role in *(see* Postpartum depression (PPD))
  posttraumatic stress disorder (PTSD), 241
    barriers, 239
    improving outcomes, 240
    for insomnia (CBT-I), 237, 240t
    for nightmares, 237–238, 238t
    obstructive sleep apnea (OSA), 238–239, 239f
    trauma-focused cognitive-behavioral therapy (TF-CBT), 236
  for PPCS, 189–190
  problems of scalability and accessibility of virtual reality, 55
  protocol, 69–70
  psychotherapeutic intervention, 170–171
  resources
    books, 499, 502–504t
    The British Library, 500
    journals, 499, 505t
    organizations, 499, 500t, 506–507t
    professional societies, 499, 501t
    regulatory bodies, 499, 500t
    treatment and research centers, 499, 507t
    The Wellcome Collection, 500
  session-by session, 423–427, 424t
  for smoking cessation, 293, 301, 421
    behavioral combination treatments, 295
    core components and strategies in, 295–296, 296t
    empirical evidence on, 294–295
    relapse prevention, 299, 301
  social anxiety disorder (SAD), 267
  for specific phobia, 53–54
  techniques, 67–68
  treatment response, 189
  trends in publications, 4–6f, 5–6
  urine cotinine levels, 423f
Cognitive-behavioral therapy for insomnia (CBT-I), 166–169, 167f
  evidence base, 165
  posttraumatic stress disorder, 237, 240t
  sleep hygiene, 165
  sleep restriction, 165
  stimulus control, 165
Cognitive-behavioral therapy for insomnia (CBT-I), epilepsy, 412–413
  educational content, 412t
  high prevalence and complications, 409
  nonpharmacological methods, 409
  outcome assessment, 411, 413t
  pathophysiological model, 409

prevalence, 409
principles, 410
procedure, 411
program content, 411
psychiatric and psychological services, 410–411
SHSH application, 411
technology use, in treatment of, 411
Cognitive-behavioral therapy for weight loss (CBTWL), 472
  CBT group therapy, 471
  cognitive and behavioral assessments, 471, 471t
  cognitive restraint, 472
  eating self-efficacy, 471–472
  healthy nutrition, 472
  intervention and treatment, 467–470, 468–470t
  physical activity, 472
Cognitive-behavior therapy for ARFID (CBT-AR), 95–99, 95t
  assessment and treatment, 96
  aversive consequences, 98
  case example, 99–101
  key facts, 101–102
  lack of interest in food/eating, 98–99
  overview, 95–96
  sensory sensitivity, 98
  stages, 97–100
Cognitive defusion, 277
Cognitive factors, 25
Cognitive processing therapy (CPT), 400
Cognitive reactivity, 282–284, 283–284t
Cognitive restraint (CR), 471–472
Cognitive restructuring (CR), 271, 276, 349
  obsessive-compulsive disorder (OCD), 206–207, 209–210, 213–214
  smoking and depression, 298, 300
Cognitive techniques, of persistent postural-perceptual dizziness, 129–130, 130t
Cognitive therapy (CT), 80
  obsessive-compulsive disorder (OCD), 206
  strategies, 84
  techniques, 81
Collaborative care model (CCM), 15, 18
Co-located care, 13–14
Combination therapy
  postpartum depression, 225
  posttraumatic stress disorder (PTSD), 241
Common factor models, 294
Community-based interventions, 434
Community health workers (CHWs), 436–437
Comorbid vestibular disorder, 124
Compañeros En Salud (CES), 434
Compassion-based cognitive-behavioral group therapy (CBGT), 315f, 317–318
Compassion-based cognitive-behavior group therapy (CBGT), CAD patients, 460
  autonomic nervous system activities, measurement of, 457
    experimental manipulation of anger, validation of, 457, 458t
    and hostility, CBGT program effects, 458–459

  behavioral interventions, 456
  cardiac autonomic regulation, 456
  data analysis, 457
  meta-analyses, 455–456
  methods
    experimental protocol, 456
    study participants, 456
  multidimensional hostility and psychological distress, 457
  psychophysiological reactions, 457
  therapeutic effects of CBGT program, on hostility levels, 457–458
Compassion theory, 314
Complex insomnia, 239
Composite International Diagnostic Interview, 294
Compulsions, 205–207, 213
Consumer-based automated
  virtual reality cognitive behavior therapy (VR-CBT), 56–58t, 60–62
Contingency management (CM), 295
Continuous positive airway pressure (CPAP), 238–239, 241
Cooccurring condition, 300
Coordinated care, 13
Copenhagen Psychosocial Questionnaire (COPSOQ), 251, 260
Coping skills training, 45
Coronary artery disease (CAD), 317–318
  acknowledging difficulty to change, 312
  altered unhealthy behaviors and lifestyle, 311
  anxiety, 308
  autonomic dysfunctions, 309–310
  compassion-based cognitive-behavioral group therapy (CBGT), 315, 315f, 317–318
  controlling impulsive behavior, 314
  depression, 308
  endothelial dysfunction, 311
  hostility and anger, 308
  hypertension and high blood pressure, 310
  hypothalamic-pituitary-adrenal (HPA) axis, 310
  psychological factors, 307–308
  psychopathological mechanisms, 309, 309f
  psychophysiological stress profile (PSP), 311, 312f
  psychotherapy, 311
  reinforcing compassion behavior, 314
  stress, 308–309
Cotinine urinalysis, 297
Critical incident stress debriefings, 255
Cultural adaptations, 431, 434
  CBT, 34–35
  fundamental areas, 36–37, 36f
  intervention, 435–436
  limitations, 434t
  literature on, 432–433
  psychotherapies, 31–33
  research, 433–434
  types and processes, 32
Culture, 34
Customization, 168

D

Decentering, 284, 288–289
Defusion, 271, 276
Depression, 18, 434
  childhood trauma, 435
  coronary artery disease (CAD), 308
  prevalence of, 321
  symptoms and causes of, 434
  tobacco smoking, 300
    assessment tools for smokers, 300t
    bidirectional models, 294
    causal models, 293
    CBT (see Cognitive behavioral therapy (CBT))
    cognitive restructuring, 298
    common factor models, 294
    diagnostic tools and screening measures for, 294
    transdiagnostic models, 294
  treatments, 5–6, 7t
Diabetes mellitus (DM), 105–106
  CBT intervention programs in, 114–115
  CBT on DRD, 109–113, 109f, 110–112t
  key facts, 115–116
Diabetes online therapy (DOT), 115
Diabetes-related distress (DRD), 106–109, 108t, 114t
  assessment tools, 113–114
  CBT intervention programs in DM with, 114–115
  in DM patients, 109–113, 109f, 110–112t
  factors associated with, 107–109
  Problems Areas in Diabetes Scale (PAID), 107t
Diagnostic and Statistical Manual of Mental Disorders (DSM), 1, 163–164, 186, 221, 235, 499, 504
Diagnostic and Statistical Manual of Mental Disorders, Fifth Edition (DSM-5), 66, 221, 365–366, 441
Diagnostic and Statistical Manual of Mental Disorders-III, 366
Dialectical behavior therapy (DBT), 282
Differential activation hypothesis, 284–285
Directed masturbation
  CBT, 157
Discrepancy based processing, 284
Disorders of mental health
  classification, 1
  Disability-Adjusted Life Years (DALYs)
    England, 1–3, 2f
    Qatar, 1–3, 3f
    USA, 1–3, 3f
  Global burden of disease, 1, 2f
  key facts, 6
  treatment (see Cognitive behavioral therapy (CBT))
Disparities, 431, 432t, 436
  definitions, 432
  mental healthcare system, 431, 433
    importance of attending, 436–437
Documentation, 17
Dysfunctional thoughts, 487

E

Eating disorder
  avoidant/restrictive food intake disorder (ARFID), 93–95, 94f
  cognitive-behavior therapy for ARFID (CBT-AR), 98–99
Edinburgh Postnatal Depression Scale (EPDS), 221, 222f, 226, 401, 448
Effort-reward imbalance (ERI), 251
Electroconvulsive therapy, 210
Emotional dysregulation, 417
Emotional eating (EE), 471
Emotional factors, 25
Emotional pain, 284–285
Engagement, barriers, 432
Engaging media for mental health applications (EMMA), 68–69
Environmental Reward Observation Scale (EROS), 294
Epilepsy
  effect of sexual function, 136–137
  key facts, 143–144
  prevalence of sexual dysfunction, 137
  problems in quality of life, 136
  sexual disorders treatment, 138
  sexual dysfunction mechanism, 137–138
Epilepsy-related sexual dysfunction, 137
Escitalopram, 236
Ethno-cognitive behavioral therapy (Ethano-CBT), 31, 34–35
Ethnopsychotherapy, 31
  need for, 33
Exacerbating factors, of persistent postural-perceptual dizziness, 123–124, 123t
Excerpta Medica dataBASE (EMBASE), 5
Experiences questionnaire (EQ), 269
Exposure and response prevention (ERP), 205–206, 213, 444
Exposure, in multidimensional intervention for social anxiety (MISA) program, 271, 276
Exposure, relaxation, and rescripting therapy (ERRT), 237–238, 238t
Exposure treatment, 53–54, 61
  VRET vs., 54

F

Face-to-face intervention, 198–199
  vs. group intervention, 198
Fagerström Test for Cigarette Dependence (FTCD), 422
Family Coping Skills Program (FCSP), 434–435
  Triangles Course, 436
  women to attend, 436
Fasting plasma glucose (FPG), 105, 115
Fatalism, 33
Fatigue, multiple sclerosis. See Multiple sclerosis (MS)
Female orgasmic disorder (FOD), 151
Female sexual distress scale-revised (FSDS-R), 156
Female sexual dysfunctions (FSDs), 148, 148f
  assessment, 152–156, 153–154t
  sessions, 153–156
  DSM-5 diagnostic criteria, 149–152, 150t
  key facts, 158
  model, 148–149, 149f
  questionnaires, 156
  treatment, 156–158
Female sexual function index (FSFI), 156
Female sexual interest/arousal disorder (FSIAD), 150–151
Five facet mindfulness questionnaire (FFMQ), 492, 495
512 Psychological Intervention Model, 256
"Fragebogen Körperdysmorpher Symptome" (FKS), 443
Freeman's paranoia hierarchy, 348, 349f
Functional analysis of behavior, 294, 296–297, 297f, 300–301
Functional vestibular disorder, 121–122

G

Gender-based violence (GBV) against women, 398–401, 399t
Generalized anxiety disorder (GAD), 84, 84–85b
Generalized anxiety disorder-7 (GAD-7), 448
Genito-pelvic pain/penetration disorder, 151–152
"Getting along" domain, 350
Global Burden of Disease study, 1
Glycemic control, 106, 109, 114–115
Graded exposure, 207, 214
GRID-Hamilton Rating Scale, 356–358
Group CBT (gCBT), 229
Group intervention, face-to-face intervention vs., 198
Group therapy, obsessive-compulsive disorder (OCD), 210

H

Habit reversal therapy, 214
Habituation, 205–206, 208, 213
Hamilton Depression Rating Scale (HAMD), 337
Head-mounted display (HMD), 54, 61
Health qigong (HQ), 322, 326
Health qigong based cognitive therapy (HQCT), 322, 326–327
  mood disorders, among Chinese adults
    applications to other areas, 327
    beneficial effects, 323
    content of 8-week group program, 325, 325–326t
    improved health status, 323
    mood symptoms, reduction in, 322
    physical health, 323
    somatization tendency, 323
    vs. traditional cognitive therapy, 464
Hemoglobin A1c (HbA1c), 107–109, 110–112t, 115
High blood pressure, 310
Humanitarian and disaster relief responders, MHPSS intervention for, 245, 259–261
  guidelines and their limitations, 247
  limitations of research and and considerations for implementation
  cost-effectiveness and funding, 259
  help-seeking and stigma, barriers reduction to, 259
  vulnerable groups and cultural contexts, applications and adaptations to, 259
  psychological wellbeing, comprehensive and systematic framework for, 247
  organizational policy and standards of practice, 251
  perideployment (see Perideployment)
  postdeployment resilience-building and posttraumatic growth, 256–258
  predeployment, 253–254
  prevention and intervention strategies, 247, 249–250t
  reducing workplace psychosocial stressors, 251–253
Human sexual response stages, 136
Hypertension, 310
Hyposexuality, 136–137
Hypothalamic-pituitary-adrenal (HPA) axis, 310

I

IASC Intervention Pyramid, 247, 249f
Imagery rehearsal therapy (IRT), 237–238, 238t
Imagery rescripting and exposure therapy (IRET), 237–238
Immersive VR, 54, 61
Impaired glucose tolerance (IGT), 105
Implementation science, 260, 431, 433–434
  limitations, 434t
  literature, 433–434
Improving Access to Psychological Therapies in Chiba (Chiba-IAPT), 360
Infanticide, 221, 230–231
Information and Communication Technologies (ICTs), 68–70
Informed consent, 17
Inhibitory learning, 53–54
In-patient/residential treatment, for OCD, 208, 211, 214
  advantages/limitations of, 211, 212t
  cognitive behavior therapy (CBT), 209–210
  comorbidities, interventions for, 210
  discharge planning, 210
  electroconvulsive therapy, 210
  evidence for, 211
  family education and support, 210
  goal of, 209
  group therapy, 210
  home-based treatment, 209
  indications for, 212, 212t
  in-patient setting, 208
  intensive CBT, 208–209, 214
  medication management, 210
  occupational therapy, 210
  partial hospitalization, 209
  predictors of response, 211
  repetitive transcranial magnetic stimulation (rTMS), 210
  residential service, 208
  transcranial direct current stimulation (tDCS), 210
  vocational rehabilitation, 210

Insomnia, 18, 164–165, 167f, 237, 241
   cognitive behavior therapy, 168–169, 170f
   key facts, 172
   nonpharmacological management, 164–165
   psychotherapeutic intervention, 168f
Insomnia Severity Index, 237
Integrated care, 15
Integrative adapt therapy (IAT), 256
Intensive CBT, 208–209, 214
International Association for Suicide Prevention, 491
International Classification of Diseases (ICD), 1, 163–164, 499
   ICD, Tenth Edition (ICD-10), 221
International Classification of Diseases 11th (ICD-11), 66, 235, 365–366
International Classification of Mental and Behavioral Disorders, 136
International Committee of the Red Cross (ICRC) societies, 246
Internet-based CBT (iCBT) services, 177–178, 181, 229, 350, 411
   loneliness, 175–176
Internet-based treatments, of adjustment disorder (AjD), 70–72
Internet-delivered mindfulness-based cognitive therapy (iMBCT), 86
Interpersonal psychotherapy (IPT), 256
Interpersonal sensitivity, 348
Interpersonal theory of suicide, 285
Irrational Food Beliefs Scale (IFB), 471
Italian Mental Health Law 180, in 1978, 352
Italian National Health System, 352

J

Japan
   bipolar disorder
      bipolar I disorder, 331
      bipolar II disorder, 331
      CBT effectiveness, 332
      cyclothymic disorder, 331
      epidemiology, 332
      medical care for patients with, 333
      prevalence, 332t
      psychosocial treatment, 334–335
      psychotherapy, 331
      recommended treatment, 333t
      social status, 332–333
   cognitive-behavioral therapy (CBT)
      assessment, 337
      behavioral experiments, 338–339
      case formulation, 338
      coping skills development, at prodrome, 339, 339t
      daily routine and sleep habits, establishment of, 338
      eclectic therapies in, 336
      implementation rate, for bipolar disorder, 335–336, 335f
      practice and procedures, 336–339
      psychoeducation, 337
      self-monitoring, 338
      therapeutic goal setting, 338
      treatment module, for bipolar disorder, 335
      and universal health insurance, 336
   schizophrenia in, CBT, 379
      antipsychotic medication, 377
      characteristics of clinical trials, 381t
      long-term hospitalization, 378–379
      practice and procedure, 382
      prevalence, 377
      psychosocial intervention, 377
      research on CBT, 380–381
      spread of CBT, 379–380
   tinnitus, 393
      CBT, 387–388, 388–392t, 393
      evidence level and recommended strength, 387–388
      practice and procedures, 389–391, 389–392t
      psychological effects, 387
      single nucleotide polymorphism, 387
      treatment, 387, 392
Japanese Association for Behavior Analysis (JABA), 355–356
Japanese Association for Cognitive Therapy (JACT), 355–356
Japanese Association of Behavioral and Cognitive Therapies (JABCT), 355–356
"Jumping to conclusions" bias (JTC), 348

L

Lack of interest, in food/eating
   avoidant/restrictive food intake disorder (ARFID), 93–95, 94f
   cognitive-behavior therapy for ARFID (CBT-AR), 98–99
Landslide effect, 285, 286f, 289
Latinx, 432–433
   approach, 436
   culture, 436
   ethnicity, 437
Loneliness, 175
   cognitive behavior therapy, 176–177, 177f
   internet-based CBT, 175–176
   key facts, 180

M

Major depressive disorder (MDD), 220–221, 220t, 293, 301
Male sexual dysfunctions, 156
Manager-specific training, 254–255
Manic-depressive illness, 332
Maternal deaths, in UK, 221–222, 223f
Maternal postpartum depression, 230–231
   adverse effects, 221
   barriers to care for mothers, 225
   clinical presentation, 220–221
   definition and prevalence, 220
   diagnosis and diagnostic systems, 221
   screening, 221
   severe PPD, 221–223
   treatment, 225
Maternal suicide, 221–222, 223f
MBCT-S program, 142–143
Medicine, 18
Mental health and psychosocial support (MHPSS), for humanitarian and disaster relief personnel, 245, 259–261
   guidelines and their limitations, 247
   limitations of research and and considerations for implementation
      cost-effectiveness and funding, 259
      help-seeking and stigma, barriers reduction to, 259
      vulnerable groups and cultural contexts, applications and adaptations to, 259
   psychological wellbeing, comprehensive and systematic framework for, 247
   organizational policy and standards of practice, 251
   perideployment (see Perideployment)
   postdeployment resilience-building and posttraumatic growth, 256–258
   predeployment, 253–254
   prevention and intervention strategies, 247, 249–250t
   reducing workplace psychosocial stressors, 251–253
Mental health outcomes, 80–81, 86
Mental health services (MHSs). See Anxiety disorders (ADs), Italian mental health services
Metaanalyses, 25, 32–33, 38
Metacognitive awareness, 288–289
Metacognitive training (MCT), 348, 380, 382
MiBCT team, 141
Mild traumatic brain injury (mTBI), 185, 186f
   key facts, 191
   recovery, 185–186
      CBT, 188–189
      models of typical and atypical, 186–187
      psychological approaches, 187–189
Mind-body interventions (MBIs), 321–322, 326, 463–464
Mindful breathing, 323, 326
Mindful eating, 286–287, 289
Mindful movement, 323, 326
Mindfulness, 270, 276–277
Mindfulness-based cognitive therapy (MBCT), 80, 86, 138–140, 195, 198, 200, 256, 321–322, 326–327, 463–464
   anxiety disorders, 81–84, 83f
   application to other areas, 290
   comparison with CBT and, 81t
   eight-session group course, 80–81, 82t
   female sexual dysfunctions, 156–157
   for generalized anxiety disorder (GAD), 84, 84–85b
   key facts, 87
   mood disorders, among Chinese adults
      applications to other areas, 327
      content of 8-week group program, 323–324, 324–325t
      improved health status, 323
      mood symptoms, reduction in, 322
      somatization tendency, 323
   sexual function, 140, 142
      evidence-based experiences, 140–142

Mindfulness-based cognitive therapy (MBCT) *(Continued)*
  steps, 139t, 140
  suicidal behavior, 290
    on associated psychological mechanisms, 288
    awareness and attention, 285
    Buddhist tradition mindfulness, 285
    on cognitive mechanisms, 288
    cognitive reactivity, 282–284
    definition, 282
    elements, 282
    example, 285, 286f
    mindfulness, definition of, 285
    perceived burdensomeness, 285
    process and variations of, 286–288, 287t
    ruminative thinking, 282–285
    suicidal clients, challenges, 289
    suicidal ideation, effect on, 288
    thought suppression, 282–284
    thwarted belongingness, 285
    transdiagnostic process, 282–284, 283–284t
  on suicidal behavior, 496–497
    case conceptualization, 492–493
    clinical and nonclinical settings, 491
    clinical features, 492t, 496t
    cognitive-behavioral psychotherapies, third wave of, 491
    coping skills, 491
    depression and borderline personality disorder, 491
    dialectical behavior therapy, 491
    economic condition, 492
    extreme anxiousness, 492
    interview and assessment, 492, 494–495
    planning of techniques, 494
    treatment outcome and critical evaluation, 495
Mindfulness-based interventions (MBIs), 80
  adverse events and contraindications, 85–86
Mindfulness-based stress reduction (MBSR), 44, 80–83, 195, 198, 200, 267
Mindfulness meditation (MM), 321–322, 326, 463–464
Mindfulness mobile health (mHealth) apps, 86
Mobile Application Rating Scale (MARS), 86
Mobile health interventions, 229
Mobile VR, 55–58
Models of integrated care, 15
Mood and anxiety disorders, in Japan, 361, 363 CBT
  academic societies, development of, 355–356
  cultural adaptations to, 361
  for depression, Japanese version, 362t
  internet-based CBT, 356
  randomized clinical trials, 356–358, 357t
  research on, 356–358
  social anxiety disorder, Japanese version, 362t
  training in, 360
  Western-originated CBT, 361
  health insurance scheme, clinical practice under, 358–360
  Japanese Anxiety Children/Adolescents Cognitive-Behavior Therapy program, 362t
  lifetime prevalence, 355
  pharmacotherapy, 355
  Western-developed protocols/assessment tools, 361
Mood disorders
  among Chinese adults, 465
    anxiety, 464
    behaviors, 464–465
    depression and anxiety, prevalence of, 463
    feelings, 464–465
    health qigong based cognitive therapy (HQCT) *(see* Health qigong based cognitive therapy (HQCT))
    health qigong–based cognitive therapy (HQCT), 464
    initial assessment, 465
    interventions, 465
    interventions, candidates for, 322
    mindfulness-based cognitive therapy (MBCT) 463–464 *(see* Mindfulness-based cognitive therapy (MBCT))
    personal background and history, 464
    study objective, 322
    thoughts, 464–465
    traditional cognitive behavioral therapy, limitation of, 463
  cognitive behavioral therapy (CBT), limitation of, 321
  health qigong based cognitive therapy (HQCT), 322
  mindfulness-based cognitive therapy (MBCT), 321–322
  prevalence, 321
Mother-child relationship, 229
Motivation, 24
Multidimensional Intervention for Social Anxiety (MISA) program, 267–268, 276–277
  assessing program outcome, instruments for, 269
  borderline personality disorder (BPD), 276–277
  contents, 272, 272–274t
  "criticism and embarrassment" dimension, 274–276
  elements, 268
  instruments, for assessing program outcome, 269
  structure, 268, 268f
  therapeutic strategies, 269, 269f, 276
    acceptance training, 270
    cognitive restructuring and defusion, 271
    exposure, 271
    homework assignments, 272
    mindfulness training, 270
    psychoeducation, 270, 270t
    social skills training (SST), 268, 271, 276
    values, education in, 270
  training in dimensions, 272–273

Multidimensional perfectionism scale (MPS), 269
Multiple sclerosis (MS), 195–197
  cognitive behavioral therapy, 198–199
  cortico-striato-thalamocortical loop, 197f
  distinction between primary and secondary, 196f
  mechanisms and management, 197
"MumMoodBooster", 229

N

National Health and Nutrition Examination Survey (NHANES), 431
National Institute for Health and Care Excellence (NICE), 255, 441–442
National Institute of Mental Health and NeuroSciences (NIMHANS), 210
National-level nonstate armed groups (NSAGs), 251
New fathers' depression rate, 219
Nicotine fading procedure, 295, 295f, 297, 301, 422
Nightmares, 237–238, 238t, 241
9-item Patient Health Questionnaire (PHQ-9), 294
Nongovernmental Organizations (NGOs), 246–247
Nonpharmacological management, of insomnia, 164–165

O

Obesity, 21
  key facts, 25–27
Obsessions, 205–207, 210, 213
Obsessive beliefs questionnaire, 206
Obsessive-compulsive disorder (OCD), 214
  behavior therapy, 205–206
  Bergen 4-day treatment (B4DT), 212
  cognitive behavior therapy (CBT) components, 207–208
    general psychiatric in-patient services, 209
    in-patient/residential treatment *(see* In-patient/residential treatment, for OCD)
    long-term outcome, 211
    model for, 206, 206f
    out-patient CBT, 208
    treating team, 209
  cognitive therapy, 206
  definition, 205
  family interventions in, 208
  prevalence of, 205
  structured assessment, tools for, 207, 207t
Obstructive sleep apnea (OSA), 238–239, 239f, 241
Occupational mental health, of humanitarian aid and disaster responders, 245–247
Occupational stress, 247
Occupational therapy, 210
Online psychological treatment, 229
Overgeneral memory, 284–285, 289
Overweight, 21

P

Panic disorder (PD)
  with agoraphobia, 347–348
  without agoraphobia, 347–348
Parental psychiatric disorders and child outcomes, 221, 223f
Parental suicide, 219
Parent-child relationship, 229
Paroxetine, 236
Paternal postpartum depression
  barriers to care for fathers, 225
  cognitive behavioral therapy (CBT), 225
  definition and prevalence, 224
  and family system, 224
  predictors of, 224
Paternal suicide, 219
Patient Health Questionnaire-9 (PHQ-9), 443
Patients with multiple sclerosis (PwMS), 195–198
  fatigue in, 200
PEERS social skills program, 176
Pelvic-floor exercises, 158
Penn state worry questionnaire (PSWQ), 269
People with epilepsy (PWE), 135–137
Perceived burdensomeness, 283–284t, 285
Perideployment
  crisis intervention and psychological treatment
    critical incident stress debriefing, mixed evidence for, 255
    evidence-based psychotherapy interventions, 256, 257f
  mental health monitoring and support
    manager-specific training, 254–255
    social support and team cohesion, enhancement of, 255
Perinatal depression. See Postpartum depression (PPD)
Perinatal mental health (PMH), 219, 230, 453
  clinical presentation
    postpartum period, 449
    pregnancy period, 448
  depressive symptoms, 447
  identification, 447–448
  maintenance and relapse prevention, 452
  multidisciplinary support, 449
  perinatal psychiatric treatment, 452
  previous history
    childhood experiences, 448
    previous perinatal experience, 448
    relational information, 448
  psychotherapeutic modality, CBT as
    behavioral activation, 451
    case conceptualization, 449
    cognitive interventions, 451
    relaxation techniques, 452
    therapeutic goals, 449–450
  screening procedure and outcomes, 448
Perinatal mood disorders, 221
Perinatal period, 219, 230
Perinatal psychiatric disorders, 219
Persistent postconcussion symptoms (PPCS), 185–187
  CBT programs, 188–190
  client considerations, 189–190
  psychological approach, 187–188
  psychological therapy, 187–188
  treatment approaches, 190
Persistent postural-perceptual dizziness (PPPD), 121, 123t
  acceptance and commitment therapy, 130–131
  antidepressant, 125
  behavioral techniques, 130
  cognitive behavioral therapy, 126–127
  cognitive techniques, 129–130, 130t
  diagnostic criteria, 122–123, 122t
  exacerbating factors, 123–124, 123t
  key facts, 131
  pathological hypothesis, 124–125
  precipitant factors, 124
  prevalence, 122
  psychoeducation and cognitive behavioral model, 127–129, 128–129f
  psychological factors, 125
  relaxation techniques, 129
  symptoms of, 123t
  treatment for comorbid vestibular disorder, 124
  vestibular rehabilitation (VR), 125–126
Personality disorder (APD), 268
Personal sensibility questionnaire (PSQ), 269
Pharmacotherapy, 237
Phobic postural vertigo (PPV), 121–122
Pittsburgh Sleep Quality Index, 237
Poor diet control, 107–109
Postconcussion syndrome (PCS), 186–187, 190
Postnatal depression. See Postpartum depression (PPD)
Postpartum blues, 230
Postpartum depression (PPD), 219, 231
  cognitive behavioral therapy (CBT), 230–231
    application-based CBT, 229
    early phase, 226–227, 228f
    effectiveness of, 226
    group CBT (gCBT), 229
    internet-based CBT (iCBT), 229
    late phase and relapse prevention, 229
    middle phase, 227–229
    mobile health interventions, 229
    "problem-focused" and "time-sensitive" characteristics, 226
    session structure in, 226, 227f
    therapeutic relationship, 226
    therapist's approach, 226
  definition, 230
  maternal PPD, 230–231
    adverse effects, 221
    barriers to care for mothers, 225
    clinical presentation, 220–221
    definition and prevalence, 220
    diagnosis and diagnostic systems, 221
    screening, 221
    severe PPD, 221–223
    treatment, 225
  paternal PPD
    barriers to care for fathers, 225
    definition and prevalence, 224
    and family system, 224
    predictors of, 224
    treatment, 225
  psychological and pharmacological treatments, 224–225, 230
Posttraumatic stress disorder (PTSD), 247, 251, 255, 371–372
  cognitive behavioral therapy (CBT), 241
    barriers, 239
    improving outcomes, 240
    for insomnia (CBT-I), 237, 240t
    for nightmares, 237–238, 238t
    obstructive sleep apnea (OSA), 238–239, 239f
    trauma-focused cognitive-behavioral therapy (TF-CBT), 236
  definition, 235, 241
  diagnostic criteria, 235
  in Pakistan
    adaptive CBT protocols, 369–370, 370t
    aspects, 367–368
    challenges, 369
    cognitive distortions, 366
    first line treatment, 369
    healthcare practitioners, 372
    history, 365–366
    major phases, in CBT, 366
    non-systematic narrative review, 371
    prevalence, 368, 368t
    psychotherapeutic interventions, 365–366, 368–369
    social costs, 368
    sources, 367t, 369t
    symptomology, 368, 370t
    trans-diagnostic-based approach, 366
    trauma-focused CBT protocols, 366–367, 370–371
    western CBT protocols, 369–370
  pathophysiology, 235
  pharmacologic treatment, 236
  phenotypes, 235
  and sleep disorders, 237
Prazosin, 236
Precipitant factors, of persistent postural-perceptual dizziness, 124
Predeployment
  psychoeducational training and preparedness, 254
  psychological screening, 253–254
Prevention of Mother-to-Child Transmission (PMTCT), 396–398
Primary care
  behavioral health treatment, 13
  billing, 16
  cognitive behavior therapy (CBT)
    adaptation, 16
    barriers, 15
  common conditions, 12
  mental health
    barriers, 12–13
    concerns, 11–12
Primary care behavioral health model (PCBH), 15, 19

Problem areas in the Diabetes (PAID) scale, 106, 107t, 113, 113t
Problem Management Plus (PM+), 400
Problem-solving therapy (PST), 45, 398
Professional body, 505
Psychiatric daycare programs, 379
Psychoeducation
  borderline personality disorder (BPD), 277
  humanitarian and disaster relief personnel, MHPSS intervention for, 254
  obsessive-compulsive disorder (OCD), 207, 212
  persistent postural-perceptual dizziness, 127–129, 128–129f
  posttraumatic stress disorder (PTSD), 236–238
  social anxiety disorder (SAD), 270, 270t, 275
  suicidal behavior, 289
Psychological distress, 44–46
Psychological first aid (PFA), 256, 260
Psychological treatment, for adjustment disorder, 67–68
Psychophysiological stress profile (PSP), 311, 312f
Psychosocial interventions, for cancer patients, 44–45, 44f
Psychosocial support systems, 43–44
Psychosocial treatment, for bipolar disorder in Japan
  family support, 335
  psychoeducation program, 334
  rework program, 334, 340
Psychotherapeutic intervention, for BPD, 417
Psychotherapy, 109, 504
  coronary artery disease (CAD), 311
  postpartum depression (PPD), 225
  posttraumatic stress disorder (PTSD), 238t
Puerperal depression. *See* Postpartum depression (PPD)

R

Randomized controlled trials (RCTs), 83–86, 125–126, 198, 205–206, 237–238
  anxiety disorders, 358
  chronic subjective dizziness, 126
  mood disorders, 356–358, 357t
  ongoing multicenter, 199
  pilot, 199
Rapid eye movement (REM), 171
Rational emotive behavior therapy (REBT), 271
Regulatory body, 505
Relaxation techniques, for persistent postural-perceptual dizziness, 129
Remission, 417, 419
Repetitive transcranial magnetic stimulation (rTMS), 210
Research group, 505
Rivermead Postconcussion Symptoms Questionnaire (RPQ), 190
Rumination, 80–81, 84
Ruminative thinking, 282–284, 283–284t

S

Schizophrenia, CBT, 379, 382–383
  antipsychotic medication, 377
  characteristics of clinical trials, 381t
  clinical practice, 381–382
  long-term hospitalization, 378–379
  practice and procedure, 382
  prevalence, 377
  psychosocial intervention, 377
  research, 380–381
    on group CBT for, 380–381
    on individual CBTp, for at-risk status, 381–382
  spread of CBT, 379–380
Screening, brief intervention, and referral to treatment (SBIRT), 15, 18
Secondary traumatic stress, 251
Security and risk management, 251
Selective serotonin reuptake inhibitors (SSRIs), 79, 125, 205, 225, 443
Self-efficacy, 24
Self-guided virtual reality therapy, 55, 61
Self-medication hypothesis, 293
Sensate focus, 157
Sensory sensitivity
  avoidant/restrictive food intake disorder (ARFID), 93, 93f
  cognitive-behavior therapy for ARFID (CBT-AR), 98
Sertraline, 236
Sexual dysfunction, 136, 143
  concerns, 140–141
  female, 140
  patients with epilepsy, 137–138
  synergistic effect, 138
  treatment, 138
  in women, 142
Sexual response, 147–151
  female, 148
  model, 158
Sitting meditation, 323, 326
Sleep disordered breathing (SDB), 237
Sleep disorders, 237
Sleep hygiene, 165
Sleeping pills, 164
Sleep restriction, 165
Smartphone CBT app, 356–357
Smoking, 301, 301f
  behavioral activation (BA), 295, 298, 300
  depression, 300
    assessment tools for smokers, 300t
    bidirectional models, 294
    causal models, 293
    cognitive-behavioral therapy (CBT) (*see* Cognitive behavioral therapy (CBT), for smoking cessation)
    cognitive restructuring, 298, 300
    common factor models, 294
    diagnostic tools and screening measures for, 294
    transdiagnostic models, 294
  depressive disorder and depressive symptoms, 293
  feedback of cigarette smoking, 297
  mental health problems, 293
  prevalence rates, 293
  preventable death, cause of, 293
  self-monitoring of cigarette consumption, 297
  smoking cessation treatments, 293
  social skills assertiveness
    anger management, 299
    relapse prevention, 299, 301
    social support, 299
    weight concerns and exercise planning, 299
  stimulus control, 298, 301
Smoking cessation, 421
Social anxiety disorder (SAD), 175–177, 276, 347
  acceptance and commitment therapy (ACT), 267–268
  age of onset, 267
  associated factors, 267
  cognitive-behavioral therapy (CBT), 267
  course, 267
  definition, 265
  group-based CBT, 179–180
  mindfulness-based stress reduction (MBSR), 267
  multidimensional intervention for social anxiety (MISA) program, 267–268, 277
    acceptance training, 270
    assessing program outcome, instruments for, 269
    cognitive restructuring and defusion, 271
    contents, 272, 272–274t
    "criticism and embarrassment" dimension, 274–276
    elements, 268
    exposure, 271
    homework assignments, 272
    instruments, for assessing program outcome, 269
    mindfulness training, 270
    psychoeducation, 270, 270t
    social skills training (SST), 268, 271
    structure, 268, 268f
    training in dimensions, 272–273
    values, education in, 270
  prevalence and comorbidity, 267
  social anxiety dimensions, 266f
    annoyance, disgust/displeasure, assertive expression of, 266
    embarrassmen/criticism, 266
    opposite sex, interaction with, 266
    speaking in public/Interaction with persons of authority, 267
    strangers, interaction with, 266
Social anxiety disorder (SAD), with MISA program, 489–490
  anamnesis, 482
  assessment and diagnosis, 482–485
  clinical diagnosis, 484–485
  criticism and embarrassment dimension, 489
  functional analysis, 485–486, 486t
  interaction, 481
    with opposite sex, 487, 488t
  intervention, 487
  one-year follow-up, 489

posttreatment assessment, 487–489
previous treatments, 482
social events, 482
treatment goals
  at behavioral level, 487
  at cognitive level, 486
  emotional and physiological level, 486–487
Social anxiety questionnaire for adults (SAQ), 269
Social phobia. *See* Social anxiety disorder (SAD)
Social skills questionnaire (SOSAQ), 269
Social skills training (SST), 176, 179–180, 268, 271, 276
SOLUS
  internet-based, 176
  intervention, 178–179, 178*t*
  treatment, 181
SORCK modeling, 443
Southampton adaptation framework, 34–35
Specific phobia, 53
  cognitive behavioral therapy, 53–54
  effects and quality of automated VR, 57–60, 57–58*t*
  symptoms, 61
  virtual reality cognitive behavioral therapy, 54–55, 56*t*
Stand-alone virtual reality therapy, 55, 61
Standard behavioral treatment (SBT), 22
Stepped-care psychological treatment model, 18, 256, 257*f*
Stimulus control, 298, 301
  cognitive behavior therapy for insomnia (CBT I), 165
Subclinical depression, 294–295, 301
Subjective units of distress (SUDS), 207
Sudanese Red Crescent Society, 255
Suicidal behavior, 290
  acceptance and commitment therapy (ACT), 282
  adjunctive treatment, 282
  definition, 281–282
  dialectical behavior therapy (DBT), 282
  differential activation hypothesis, 284–285
  discrepancy based processing, 284
  driven doing mode, 284
  emotional pain, 284–285
  environmental triggers, 281–282
  individual level causes, 281–282
  mental health awareness, 281–282
  mindfulness-based cognitive therapy (MBCT)
    on associated psychological mechanisms, 288
    awareness and attention, 285
    Buddhist tradition mindfulness, 285
    on cognitive mechanisms, 288
    cognitive reactivity, 282–284
    definition, 282
    elements, 282
    example, 285, 286*f*
    mindfulness, definition of, 285
    perceived burdensomeness, 285
    process and variations of, 286–288, 287*t*
    ruminative thinking, 282–285

"solution blindness" situation, 285
suicidal clients, challenges, 289
suicidal ideation, effect on, 288
thought suppression, 282–284
thwarted belongingness, 285
transdiagnostic process, 282–284, 283–284*t*
overgeneral memory, 284
prevalence, 281–282
preventive approach, 282
psychological pain and hopelessness, 284–285
psychopathology, symptom/consequential action of, 281–282
rebound effect, 282–284
recurring depression, clients with, 282–284
sense of urgency, 284–285
social and cultural risk factors, 285
social determinants, 281–282
suicidal mind, 282–284
"Sullatha suttha" concept, 284–285
Suicidal ideation, 281–285, 288–289
Suicidality, 221–222, 231
"Sullatha suttha" concept, 284–285
Supportive-expressive group therapy (SEGT), 44–45
Survival, cancer, 43–44, 44*f*
  CBT, 45–46, 46*t*
Symptoms cluster, 197
Systematic desensitization, 157–158

T

Telephone-delivered education intervention, 198–199
"Telepsychiatric" intervention, 350
Therapeutic relationship, 226–227, 230
Therapist-assisted computerized CBT (TacCBT) intervention for young adults, 350
Thought record sheet, 451*f*
Thought suppression, 282–284, 283–284*t*
3-min breathing space, 323, 326
Thwarted belongingness, 283–284*t*, 285
Tinnitus, Japan, 393
  CBT, 387–388, 388–392*t*, 393
  evidence level and recommended strength, 387–388
  practice and procedures, 389–391, 389–392*t*
  psychological effects, 387
  single nucleotide polymorphism, 387
  treatment, 387, 392
Tinnitus retraining therapy (TRT), 387
Tobacco, 423–425
  CBT protocol, 421–422, 425–426
  concepts, 425
  consumption, 422
Tobacco use disorder, in smokers, 301, 301*f*
  behavioral activation (BA), 295, 298, 300
  depression, 300
    assessment tools for smokers, 300*t*
    bidirectional models, 294
    causal models, 293
    cognitive-behavioral therapy (CBT) (*see* Cognitive behavioral therapy (CBT), for smoking cessation)
    cognitive restructuring, 298, 300

common factor models, 294
diagnostic tools and screening measures for, 294
transdiagnostic models, 294
depressive disorder and depressive symptoms, 293
feedback of cigarette smoking, 297
mental health problems, 293
prevalence rates, 293
preventable death, cause of, 293
self-monitoring of cigarette consumption, 297
smoking cessation treatments, 293
social skills assertiveness
  anger management, 299
  relapse prevention, 299, 301
  social support, 299
  weight concerns and exercise planning, 299
stimulus control, 298, 301
Transcranial direct current stimulation (tDCS), 210, 213
Transdiagnostic models, 294
Transdiagnostic process, 282–284, 283–284*t*, 289
Trastorno Adaptativo Online (TAO), 70–72, 71*f*
  blended application of, 72
  practical case, 72
  treatment, 71, 71*t*
Trauma-focused cognitive-behavioral therapy (TF-CBT)
  components, 236
  examples, 236
  intervention, 236
  limiting factor for, 236
Trauma Risk Management (TRiM), 255
Traumatic brain injury (TBI), 185
Trauma, women's mental health conditions, in sub-Saharan Africa, 398–401, 399*t*
Treatment as usual group (TAU), 350
Triangles Course, 434–435
  case, 436–437
  FCSP and, 436
  participants, 436–437
Trichotillomania, 214
21-item Three-Factor Eating Questionnaire revised version (TFEQ-R21), 471
Type 2 diabetes mellitus (T2DM), 106, 109

U

Uncontrolled eating (UE), 471
UNHCR Peer Support Personnel Network, 255
Universal health insurance system, 333, 336, 340

V

Vestibular rehabilitation (VR), of persistent postural-perceptual dizziness, 125–126
Virtual reality (VR), 53–54, 61
  adjustment disorder (AjD), 68–69
  effects and quality of automated, 57–60
  limitations, 61
Virtual reality cognitive behavior therapy (VR-CBT), 60–61
  for acrophobia, 59*t*

Virtual reality cognitive behavior therapy (VR-CBT) *(Continued)*
consumer-based automated, 56–58*t*
key facts, 62
problem of scalability and accessibility of, 55
specific phobia, 54–55, 56*t*
Virtual reality exposure therapy (VRET), 54, 56–59*t*, 57–58
Vocational rehabilitation, 210

W

Web-based interventions, 198–199
Weight Efficacy Lifestyle Questionnaire—short form (WEL-SF), 471

Weight loss
cognitive behavioral therapy, 22
characteristics, 22
key facts, 27
limitations, 24
measurement instruments, 26–27*t*
mechanisms of change *vs.* outcomes effect sizes, 23–24
outcome measures, 24
protocols, 24, 25*t*
research, 23
theories, 23
treatments, 21–22
Weight, WHO classification of, 21, 22*t*

The Wellcome Collection, 500
Workplace psychosocial hazards, 246, 251
Workplace psychosocial stressors, 251–253, 252–253*t*, 260
World Health Organization (WHO), 21
Worries, 80–81
in anxiety disorders, 81–83
GAD, 84
MBCT, 84

Y

Young Mania Rating Scale (YMRS), 337